Lecture Notes in Computer Science 3741

Commenced Publication in 1973
Founding and Former Series Editors:
Gerhard Goos, Juris Hartmanis, and Jan van Leeuwen

Editorial Board

David Hutchison
  Lancaster University, UK
Takeo Kanade
  Carnegie Mellon University, Pittsburgh, PA, USA
Josef Kittler
  University of Surrey, Guildford, UK
Jon M. Kleinberg
  Cornell University, Ithaca, NY, USA
Friedemann Mattern
  ETH Zurich, Switzerland
John C. Mitchell
  Stanford University, CA, USA
Moni Naor
  Weizmann Institute of Science, Rehovot, Israel
Oscar Nierstrasz
  University of Bern, Switzerland
C. Pandu Rangan
  Indian Institute of Technology, Madras, India
Bernhard Steffen
  University of Dortmund, Germany
Madhu Sudan
  Massachusetts Institute of Technology, MA, USA
Demetri Terzopoulos
  New York University, NY, USA
Doug Tygar
  University of California, Berkeley, CA, USA
Moshe Y. Vardi
  Rice University, Houston, TX, USA
Gerhard Weikum
  Max-Planck Institute of Computer Science, Saarbruecken, Germany

Ajit Pal   Ajay D. Kshemkalyani
Rajeev Kumar   Arobinda Gupta (Eds.)

Distributed Computing – IWDC 2005

7th International Workshop
Kharagpur, India, December 27-30, 2005
Proceedings

 Springer

Volume Editors

Ajit Pal
Indian Institute of Technology Kharagpur
Department of Computer Science and Engineering
Kharagpur, WB 721 302, India
E-mail: apal@cse.iitkgp.ernet.in

Ajay D. Kshemkalyani
University of Illinois at Chicago, Department of Computer Science
851 S. Morgan Street, Chicago, IL 60607-7053, USA
E-mail: ajayk@cs.uic.edu

Rajeev Kumar
Indian Institute of Technology Kanpur
Department of Computer Science and Engineering
Kanpur, UP 208 016, India
E-mail: raj@iitk.ac.in

Arobinda Gupta
Indian Institute of Technology Kharagpur
Department of Computer Science and Engineering
and School of Information Technology
Kharagpur, WB 721 302, India
E-mail: agupta@iitkgp.ac.in

Library of Congress Control Number: 2005937698

CR Subject Classification (1998): C.2, D.1.3, D.2.12, D.4, F.2, F.1, H.4

ISSN   0302-9743
ISBN-10   3-540-30959-4 Springer Berlin Heidelberg New York
ISBN-13   978-3-540-30959-8 Springer Berlin Heidelberg New York

This work is subject to copyright. All rights are reserved, whether the whole or part of the material is concerned, specifically the rights of translation, reprinting, re-use of illustrations, recitation, broadcasting, reproduction on microfilms or in any other way, and storage in data banks. Duplication of this publication or parts thereof is permitted only under the provisions of the German Copyright Law of September 9, 1965, in its current version, and permission for use must always be obtained from Springer. Violations are liable to prosecution under the German Copyright Law.

Springer is a part of Springer Science+Business Media

springer.com

© Springer-Verlag Berlin Heidelberg 2005
Printed in Germany

Typesetting: Camera-ready by author, data conversion by Scientific Publishing Services, Chennai, India
Printed on acid-free paper   SPIN: 11603771   06/3142   5 4 3 2 1 0

General Chairs' Message

It all started as a small sapling, sowed in 1999, as the first workshop. Within a short span of time, it grew considerably in dimension, intensity, participation, and impact. IWDC is presently a recognized name among the symposia and conferences in the area of Distributed Computing and bears the testimony of serious works of the researchers and academics working in the area. One salient feature of IWDC is that it is held in the different academic centers of India and has acted as a great impetus in fostering research in the area of Distributed Computing in India. The Seventh International Workshop on Distributed Computing, IWDC 2005, is being held at the Indian Institute of Technology, Kharagpur, one of the premier technology institutes in India. Kharagpur is very conveniently placed out of the bustle of the big metropolis, yet in close proximity to Kolkata, a major city in India, thus providing an ideal environment for academic brainstorming. As the General Chairs of the conference, we extend to you a very hearty welcome to IWDC 2005.

Over the years, IWDC has retained its character of not being drowned in the magnitude of the conference and losing the scope of close interactions and academic discussions. We are sure that this year will be no exception . The Program Committee has painstakingly carried out the review process and has set up a program that is rich in content and quality. The Organizing Committee has put in all efforts to make the conference smooth sailing and provide you with a comfortable stay. Putting all the pieces together is not an easy task and the efforts would not have been successful without the benevolence of the sponsors.

We are thankful to our sponsors HP India Ltd., Tata Consultancy Ltd., Microsoft Research, Capgemini, and General Motors, for their generous support in making the conference a success. Our sincere thanks are due to the Program Chairs Ajit Pal and Ajay Kshemkalyani for their efforts in putting together an exciting program. We are grateful to the Keynote Chair Sajal Das for arranging five high-quality keynote talks by eminent persons in this field. We are thankful to David Peleg, Walter Brooks, Taieb Znati, and Viktor Prasanna for delivering keynote talks in the conference. We thank L.M. Patnaik for agreeing to deliver the A.K. Choudhury Memorial Lecture at the conference this year. As in previous years, we have been able to set up a number of interesting and relevant tutorials. This has been made possible through the untiring efforts of the Tutorial Chairs Somprakash Bandyopadhyay and Archan Mishra. We are also grateful to Arup Acharya, R. Badrinath, Sajal K. Das, and Pradip K. Srimani for agreeing to present tutorials at the conference. Special thanks are due to our Publication Chair, Rajeev Kumar, who has put a tremendous amount of effort into compiling the final proceedings. We also thank the Finance Chair, Shamik Sural, for organizing financial support for the conference, and our Publicity Chairs, Abhijit Das, and Sandip Sen, for the great work they did in publicizing the event across the world.

We also thank all the members of the Organizing Committee, Arobinda Gupta, Ajit Pal, Shamik Sural, Indranil Sengupta, Abhijit Das, Dipankar Sarkar, Jayanta Mukhopadhyay, Pallab Dasgupta, Debasish Samanta, Soumya Ghosh, Arijit Bishnu, Pabitra Mitra, and Sudeshna Sarkar, for their efforts. We are grateful to the Indian Institute of Technology Kharagpur, and in particular to Prof. S.C. De Sarkar, Head, School of Information Technology and Deputy Director, IIT Kharagpur, for extending the logistic support to the conference. Prof. De Sarkar is also a member of the Steering Committee of IWDC. Last but not the least, we also thank Prof. Sukumar Ghosh, who is heading the IWDC Steering Committee, for his guidance and his continuous support and advice.

No academic meeting can achieve its desired end without the contributions of the authors, reviewers, and the participants. We extend our heartfelt thanks to all of them for making the event a success. We sincerely hope that this event will be a valuable addition to the Distributed Computing research endeavor.

December 2005

Anupam Basu
Michel Raynal

Program Chairs' Message

On behalf of the Program Committee of the 7th International Workshop on Distributed Computing (IWDC) 2005, it is our great pleasure to welcome all of you to Kharagpur, India. Our goal has been to put together a rich technical program, including high-quality technical papers and state-of-the-art tutorials and invited talks.

The conference received 253 submissions in response to the call for papers. About 10% of the submissions were from Europe and the Americas, 41% from India, and 49% from the rest of Asia and Australia. The Program Committee, comprising 53 distinguished experts in the field, with the help of several external reviewers, provided very detailed and rigorous reviews in most cases, and we were able to obtain at least three reviews for almost every paper in a timely manner. Based on three weeks of intense deliberations over the reviews, 30 regular papers (12 pages each) and 35 short papers (6 pages each) were accepted for the program. This represents an acceptance rate of about 25.7%. Finally, for various reasons, a few additional papers were excluded, and 28 regular papers and 33 short papers were finally accepted for inclusion in the technical program, which has been organized in two tracks and is spread across 13 sessions. The program also contains four invited keynote papers and talks by David Peleg (Weizmann), Walter Brooks (NASA), Viktor K. Prasanna (USC), and Taieb Znati (U. Pittsburgh). The traditional A.K. Choudhury Memorial Lecture will be delivered by Lalit M. Patnaik (IISc). Tutorials on state-of-the-art themes will be given by R. Badrinath, Sajal Das, Arup Acharya, and Pradip K. Srimani.

We would like to express our sincere thanks to all whose efforts and participation have made this conference possible. Firstly, we thank all the authors who submitted their work to the conference. We are greatly indebted to the PC members and the external reviewers for submitting detailed reviews. We thank the Keynote Speakers for accepting our invitation. Thanks to Keynote Chair Sajal Das for organizing the invited talks from highly eminent researchers. We also thank the Tutorial Speakers for agreeing to provide a valuable service to the research community. Tutorial Chairs Somprakash Bandyopadhyay and Archan Mishra have organized a great tutorial program. Organizing Chair Arobinda Gupta not only organized all the hospitality and logistics down to the smallest detail but also played an important role in coordinating the paper review process.

Publication Chair Rajeev Kumar, as well as Arobinda Gupta, deserve special thanks for their painstaking efforts in editing the proceedings. Thanks to our student volunteers Plaban Bhowmick and Sushanta Karmakar who have put immense efforts into format checking and correcting the files to bring them to the present form. We also thank Anirban Sarkar for developing the Web pages and the online submission software, and for customizing it dynamically during the submission and review process.

Once again we welcome all the delegates to the exciting technical program of IWDC 2005 and hope you enjoy the pleasant winter of Kharagpur in December.

December 2005

Ajit Pal
Ajay Kshemkalyani

Executive Committee

Steering Committee Chair
Sukumar Ghosh, University of Iowa, Iowa City, USA

General Chairs
Anupam Basu, Indian Institute of Technology Kharagpur, India
Michel Raynal, IRISA, France

Program Chairs
Ajit Pal, Indian Institute of Technology Kharagpur, India
Ajay Kshemkalyani, University of Illinois at Chicago, USA

Organizing Chair
Arobinda Gupta, Indian Institute of Technology Kharagpur, India

Keynote Chair
Sajal Das, University of Texas at Arlington, USA

Tutorial Chairs
Archan Misra, IBM T. J. Watson Research Center, USA
Somprakash Bandyopadhyay, Indian Institute of Management Calcutta, India

Finance Chair
Shamik Sural, Indian Institute of Technology Kharagpur, India

Publicity Chairs
Abhijit Das, Indian Institute of Technology Kharagpur, India
Sandip Sen, University of Tulsa, USA

Publication Chair
Rajeev Kumar, Indian Institute of Technology Kanpur, India

Program Committee

A.L. Ananda	National University of Singapore, Singapore
Ajay Kshemkalyani (*Co-chair*)	University of Illinois at Chicago, USA
Ajit Pal (*Co-chair*)	Indian Institute of Technology Kharagpur, India
Ambuj K. Singh	University of California, Santa Barbara, USA
Amitava Bagchi	Indian Institute of Management Calcutta, India
Arunabha Sen	Arizona State University, USA
Arup Acharya	IBM T. J. Watson Research Center, USA
Asim K. Pal	Indian Institute of Management Calcutta, India
Ayalvadi Ganesh	Microsoft Research, UK
Bhabani P. Sinha	Indian Statistical Institute Kolkata, India
Bhaskaran Raman	Indian Institute of Technology Kanpur, India
Biswanath Mukherjee	University of California, Davis, USA
Boaz Patt-Shamir	Tel Aviv University, Israel
Chandan Mazumdar	Jadavpur University, India
Christof Fetzer	Technical University of Dresden, Germany
C. Pandu Rangan	Indian Institute of Technology Madras, India
C. Siva Ram Murthy	Indian Institute of Technology Madras, India
Cyril Gavoille	Université Bordeaux, France
Debashis Saha	Indian Institute of Management Calcutta, India
G. Sajith	Indian Institute of Technology Guwahati, India
Goutam Chakraborty	Iwate Prefectural University, Japan
Indranil Sengupta	Indian Institute of Technology Kharagpur, India
James H. Anderson	University of North Carolina, Chapel Hill, USA
Jiannong Cao	The Hong Kong Polytechnic University, Hong Kong
Joy Kuri	Indian Institute of Science Bangalore, India
Krithi Ramamritham	Indian Institute of Technology Bombay, India
Kwangjo Kim	Information and Communications University, Korea
Luís Rodrigues	University of Lisbon, Portugal
Mahbub Hassan	University of New South Wales, Australia
Mainak Chatterjee	University of Central Florida, Orlando, USA
Masafumi Yamashita	Kyushu University, Japan
Mukesh Singhal	University of Kentucky, USA
Nabanita Das	Indian Statistical Institute Kolkata, India
Nabendu Chaki	Calcutta University, India

Pascal Felber	Université de Neuchâtel, Switzerland
Peng Ning	North Carolina State University, USA
Pradip K. Das	Jadavpur University, India
Pradip K. Srimani	Clemson University, USA
R. Badrinath	Hewlett-Packard, India
Rahul Banerjee	Birla Institute of Technology and Science Pilani, India
Rajkumar Buyya	University of Melbourne, Australia
Ravi Prakash	University of Texas at Dallas, USA
Ricardo Jimenez-Peris	Technical University of Madrid, Spain
Roberto Baldoni	Università di Roma "La Sapienza", Italy
Roy Friedman	Technion Israel Institute of Technology, Israel
Samir R. Das	Stony Brook University, USA
Santosh Shrivastava	University of Newcastle Upon Tyne, UK
Saswat Chakraborty	Indian Institute of Technology Kharagpur, India
Soma Chaudhuri	Iowa State University, USA
Sridhar Iyer	Indian Institute of Technology Bombay, India
Sriram Pemmaraju	University of Iowa, USA
Subir Bandyopadhyay	University of Windsor, Canada
Tetsuro Ueda	ATR Lab, Japan

Additional Reviewers

A. Prasad Sistla
Aad van Moorsel
Aaron Block
Abhijit Das
Adnan Noor Mian
Ahmet Murat Bagci
Albert Chung
Amitabha Ghosh
Anirban Sengupta
Anshul Sehgal
Arobinda Gupta
Ashfaq Khokhar
B.D. Sahu
Bao Hong Shen
Barbara Di Eugenio
Bartlomiej Sieka
Bhed Bahadur Bista
Bheemarjuna Reddy
Bing Zhang
Bin Wu
Bo Xu
Cheng Hui
C. Ramakrishna
D. Goswami
D. Manivannan
D. Roychoudhury
Debasish Chakraborty
Dingbang Xu
Dipyaman Banerjee
Donggang Liu
Filipe Araújo
Geetha Manjunath
Graham Morgan
Habib Ammari
Hafiz Malik
Hideyuki Uehara
Himadri Sekhar Paul

Huaizhi Li
Huaping Shen
Hugo Miranda
Hyun Jung Choe
Jaewon Kang
James Riely
John Calandrino
Jose Rufino
Jun-Won Ho
Kun Sun
Leonardo Querzoni
M. Baseem Hassan
Mansi Thoppian
Mansoor Mohsin
Marc Schiely
Marta Patiño-Martínez
Mauricio Papa
Mridul Sankar Barik
Nadeem Ahmed
Nandini Mukherjee
Nathan Fisher
Nirmalya Roy
Oliver Yu
P. Mitra
Pan Wang
Parama Bhaumik
Paul Ezhilchelvan
Peter Davis
Piotr Gmytrasiewicz
Pradip De
Preetam Ghosh
Rajeev Kumar
Ranjita Bhagwan
Rashid Ansari
S. Bandyopadhyay
S.V. Rao
Sabyasachi Saha

Saikat Chakrabarti
Sajal K. Das
Samiran Chattopadhyay
Sandip Sen
Sarmistha Neogy
Sasthi C. Ghosh
Shamik Sural
Shao Tao
Shashank Khanvilkar
Shashidhar R. Gandham
Siuli Roy
Soumya Ghosh
Sridhar K.
Srikant Kuppa
Stéphane Airiau
Stuart Wheater
Subhas C. Nandy
Subir Biswas
Suhua Tang
Sukumar Ghosh
Suli Zhao
Takashi Watanabe
Teddy Candale
Uday Chakraborty
Uma Maheswari Devi
Umesh Deshpande
V.N. Venkatakrishnan
Venkata Giruka
Wei Zhang
Wu Xiuchao
Yan Sun
Yaoping Ruan
Yongwei Wang
Zbigniew Jerzak
Zhibin Wu

Table of Contents

Keynote Talk I

Distributed Coordination Algorithms for Mobile Robot Swarms: New Directions and Challenges
David Peleg 1

Session I A: Theory

Labeling Schemes for Tree Representation
Reuven Cohen, Pierre Fraigniaud, David Ilcinkas, Amos Korman, David Peleg 13

Single-Bit Messages are Insufficient in the Presence of Duplication
Kai Engelhardt, Yoram Moses 25

Safe Composition of Distributed Programs Communicating over Order-Preserving Imperfect Channels
Kai Engelhardt, Yoram Moses 32

Efficiently Implementing LL/SC Objects Shared by an Unknown Number of Processes
Prasad Jayanti, Srdjan Petrovic 45

Placing a Given Number of Base Stations to Cover a Convex Region
Gautam K. Das, Sandip Das, Subhas C. Nandy, Bhabani P. Sinha 57

Session I B: Sensor Networks I

A State-Space Search Approach for Optimizing Reliability and Cost of Execution in Distributed Sensor Networks
Archana Sekhar, B.S. Manoj, C. Siva Ram Murthy 63

Protocols for Sensor Networks Using COSMOS Model
Zhenyu Xu, Pradip K. Srimani 75

CLUR-Tree for Supporting Frequent Updates of Data Stream over Sensor Networks
Soon-Young Park, Jung-Hyun Kim, Yong-Il Jang, Jae-Hong Kim, Soon-Jo Lee, Hae-Young Bae 87

Optimizing Lifetime and Routing Cost in Wireless Networks
M. Julius Hossain, Oksam Chae 93

Multipath Source Routing in Sensor Networks Based on Route Ranking
Chun Huang, Mainak Chatterjee, Wei Cui, Ratan Guha 99

Reliable Time Synchronization Protocol in Sensor Networks Considering Topology Changes
Soyoung Hwang, Yunju Baek 105

A.K. Choudhury Memorial Lecture

The Brain, Complex Networks, and Beyond
L.M. Patnaik 111

Session II A: Fault Tolerance

An Asynchronous Recovery Algorithm Based on a Staggered Quasi-Synchronous Checkpointing Algorithm
D. Manivannan, Q. Jiang, J. Yang, K.E. Persson, M. Singhal 117

Self-stabilizing Publish/Subscribe Protocol for P2P Networks
Zhenyu Xu, Pradip K. Srimani 129

Self-stabilizing Checkpointing Algorithm in Ring Topology
Partha Sarathi Mandal, Krishnendu Mukhopadhyaya 141

Performance Comparison of Majority Voting with ROWA Replication Method over PlanetLab
Ranjana Bhadoria, Shukti Das, Manoj Misra, A.K. Sarje 147

Self-refined Fault Tolerance in HPC Using Dynamic Dependent Process Groups
N.P. Gopalan, K. Nagarajan 153

Session II B: Optical Networks

In-Band Crosstalk Performance of WDM Optical Networks Under Different Routing and Wavelength Assignment Algorithms
V. Saminadan, M. Meenakshi 159

Modeling and Evaluation of a Reconfiguration Framework in WDM Optical Networks
Sungwoo Tak, Donggeon Lee, Passakon Prathombutr, E.K. Park 171

On the Implementation of Links in Multi-mesh Networks Using WDM
Optical Networks
   Nahid Afroz, Subir Bandyopadhyay, Rabiul Islam,
   Bhabani P. Sinha 183

Distributed Dynamic Lightpath Allocation in Survivable WDM
Networks
   A. Jaekel, Y. Chen 189

Protecting Multicast Sessions from Link and Node Failures in
Sparse-Splitting WDM Networks
   Niladhuri Sreenath, T. Siva Prasad 195

Session III A: Peer-to-Peer Networks

Oasis: A Hierarchical EMST Based P2P Network
   Pankaj Ghanshani, Tarun Bansal 201

GToS: Examining the Role of Overlay Topology on System Performance
Improvement
   Xinli Huang, Yin Li, Fanyuan Ma 213

Churn Resilience of Peer-to-Peer Group Membership: A Performance
Analysis
   Roberto Baldoni, Adnan Noor Mian, Sirio Scipioni,
   Sara Tucci-Piergiovanni 226

Uinta: A P2P Routing Algorithm Based on the User's Interest and the
Network Topology
   Hai Jin, Jie Xu, Bin Zou, Hao Zhang 238

Session III B: Wireless Networks I

Optimal Time Slot Assignment for Mobile Ad Hoc Networks
   Koushik Sinha 250

Noncooperative Channel Contention in Ad Hoc Wireless LANs with
Anonymous Stations
   Jerzy Konorski 262

A Power Aware Routing Strategy for Ad Hoc Networks with Directional Antenna Optimizing Control Traffic and Power Consumption
   Sanjay Chatterjee, Siuli Roy, Somprakash Bandyopadhyay,
   Tetsuro Ueda, Hisato Iwai, Sadao Obana 275

Power Aware Cluster Efficient Routing in Wireless Ad Hoc Networks
  Sanjay Kumar Dhurandher, G.V. Singh 281

A New Routing Protocol in Ad Hoc Networks with Unidirectional Links
  Deepesh Man Shrestha, Young-Bae Ko 287

Keynote Talk II

Impact of the Columbia Supercomputer on NASA Science and
Engineering Applications
  *Walter Brooks, Michael Aftosmis, Bryan Biegel, Rupak Biswas,
  Robert Ciotti, Kenneth Freeman, Christopher Henze, Thomas Hinke,
  Haoqiang Jin, William Thigpen* 293

Session IV A: Sensor Networks II

Hierarchical Routing in Sensor Networks Using k-Dominating Sets
  Michael Q. Rieck, Subhankar Dhar 306

On Lightweight Node Scheduling Scheme for Wireless Sensor Networks
  Jie Jiang, Zhen Song, Heying Zhang, Wenhua Dou 318

Clique Size in Sensor Networks with Key Pre-distribution Based on
Transversal Design
  Dibyendu Chakrabarti, Subhamoy Maitra, Bimal Roy 329

Session IV B: Wireless Networks II

Stochastic Rate-Control for Real-Time Video Transmission over
Heterogeneous Network
  *Jae-Woong Yun, Hye-Soo Kim, Jae-Won Kim, Youn-Seon Jang,
  Sung-Jea Ko* 338

An Efficient Social Network-Mobility Model for MANETs
  *Rahul Ghosh, Aritra Das, P. Venkateswaran, S.K. Sanyal,
  R. Nandi* 349

Design of an Efficient Error Control Scheme for Time-Sensitive
Application on the Wireless Sensor Network Based on IEEE 802.11
Standard
  *Junghoon Lee, Mikyung Kang, Yongmoon Jin, Gyungleen Park,
  Hanil Kim* 355

Agglomerative Hierarchical Approach for Location Area Planning in a PCSN
  Subrata Nandi, Purna Ch. Mandal, Pranab Halder, Ananya Basu ... 362

Keynote Talk III

A Clustering-Based Selective Probing Framework to Support Internet Quality of Service Routing
  Nattaphol Jariyakul, Taieb Znati 368

Session V A: Network Security

A Fair and Reliable P2P E-Commerce Model Based on Collaboration with Distributed Peers
  Chul Sur, Ji Won Jung, Jong-Phil Yang, Kyung Hyune Rhee 380

An Efficient Access Control Model for Highly Distributed Computing Environment
  Soomi Yang 392

Cryptanalysis and Improvement of a Multisignature Scheme
  Manik Lal Das, Ashutosh Saxena, V.P. Gulati 398

Key Forwarding: A Location-Adaptive Key-Establishment Scheme for Wireless Sensor Networks
  Ashok Kumar Das, Abhijit Das, Surjyakanta Mohapatra, Srihari Vavilapalli 404

New Anonymous User Identification and Key Establishment Protocol in Distributed Networks
  Woo-Hun Kim, Kee-Young Yoo 410

Session V B: Grid and Networks

Semantic Overlay Based Services Routing Between MPLS Domains
  Chongying Cao, Jing Yang, Guoqing Zhang 416

Effective Static Task Scheduling for Realistic Heterogeneous Environment
  Junghwan Kim, Jungkyu Rho, Jeong-Oog Lee, Myeong-Cheol Ko 428

eHSTCP: Enhanced Congestion Control Algorithm of TCP over High-Speed Networks
  Young-Soo Choi, Hee-Dong Park, Sung-Hyup Lee, You-Ze Cho 439

Keynote Talk IV

Programming Paradigms for Networked Sensing: A Distributed Systems' Perspective
    Amol Bakshi, Viktor K. Prasanna 451

Session VI A: Middleware and Data Management

Deadlock-Free Distributed Relaxed Mutual-Exclusion Without Revoke-Messages
    Sukhamay Kundu 463

Fault Tolerant Routing in Star Graphs Using Fault Vector
    Rajib K. Das 475

Optimistic Concurrency Control in Firm Real-Time Databases
    Anand S. Jalal, S. Tanwani, A.K. Ramani 487

Stochastic Modeling and Performance Analysis for Video-On-Demand Systems
    Vrinda Tokekar, A.K. Ramani, Sanjiv Tokekar 493

A Memory Efficient Fast Distributed Real Time Commit Protocol
    Udai Shanker, Manoj Misra, A.K. Sarje 500

A Model for the Distribution Design of Distributed Databases and an Approach to Solve Large Instances
    Héctor Fraire H., Guadalupe Castilla V., Arturo Hernández R., Claudia Gómez S., Graciela Mora O., Arquimedes Godoy V. 506

Session VI B: Mobility Management

Tracking of Mobile Terminals Using Subscriber Mobility Pattern with Time-Bound Self Purging Indicators and Regional Route Maps
    R.K. Ghosh, Saurabh Aggarwala, Hemant Mishra, Ashish Sharma, Hrushikesha Mohanty 512

SEBAG: A New Dynamic End-to-End Connection Management Scheme for Multihomed Mobile Hosts
    B.S. Manoj, Rajesh Mishra, Ramesh R. Rao 524

Efficient Mobility Management for Cache Invalidation in Wireless Mobile Environment
    Narottam Chand, R.C. Joshi, Manoj Misra 536

Analysis of Hierarchical Multicast Protocol in IP Micro Mobility Networks
Seung Jei Yang, Sung Han Park 542

Efficient Passive Clustering and Gateway Selection in MANETs
*T. Shivaprakash, C. Aravinda, A.P. Deepak, S. Kamal,
H.L. Mahantesh, K.R. Venugopal, L.M. Patnaik* 548

Mobile Agent Based Message Communication in Large Ad Hoc Networks Through Co-operative Routing Using Inter-agent Negotiation at Rendezvous Points
Parama Bhaumik, Somprakash Bandyopadhyay 554

Network Mobility Management Using Predictive Binding Update
*Hee-Dong Park, Yong-Ha Kwon, Kang-Won Lee, Young-Soo Choi,
Sung-Hyup Lee, You-Ze Cho* 560

Session VII: Distributed Articial Intelligence

Planning in a Distributed System
Rajdeep Niyogi, Sundar Balasubramaniam 566

Using Inertia and Referrals to Facilitate Satisficing Distributions
Teddy Candale, Ikpeme Erete, Sandip Sen 572

Privacy Preserving Decentralized Method for Computing a Pareto-Optimal Solution
Satish K. Sehgal, Asim K. Pal 578

Author Index 585

Distributed Coordination Algorithms for Mobile Robot Swarms: New Directions and Challenges

David Peleg*

Department of Computer Science,
The Weizmann Institute of Science,
Rehovot, Israel
david.peleg@weizmann.ac.il

Abstract. Recently there have been a number of efforts to study issues related to coordination and control algorithms for systems of multiple autonomous mobile robots (also known as *robot swarms*) from the viewpoint of distributed computing. This paper reviews the literature in the area and discusses some open problems and future research directions.

1 Introduction

Mobile robots have been developed for over half a century, beginning in the 1950's with pioneering projects such as Shannon's electromechanical mouse Theseus, Grey Walter's tortoise and Stanford's Shakey (cf. [4]). Applications for such robots abound, including industrial tasks (e.g., moving materials around), military operations (e.g., surveillance or automated supply lines), search and rescue missions, space exploration (e.g., Sojourner's Mars Pathfinder mission in 1997 or the recent automated transfer vehicle project of the European Space Agency), as well as a variety of home applications, from babysitters and pets to smart appliances such as vacuum cleaners and lawn mowers. Mobile robots come in all shapes, sizes and designs, and vary in their motion type, sensors, handling mechanisms, computational power and communication means.

Systems of multiple autonomous mobile robots (often referred to as *robot swarms*) have been extensively studied throughout the past two decades (cf. [17, 8, 24, 27, 12, 5, 41]). The motivating idea is that for certain applications it may be preferable to abandon the use of a single, strong and costly robot in favor of a group of tiny, functionally simple and relatively cheap robots. For instance, it may be possible to use a multiple robot system in order to perform certain tasks that require spreading over a large area, and thus cannot be performed by a single robot. Also, robot swarms may be the preferred alternative in hazardous environments, such as military operations, chemical handling and toxic spill cleanups, search and rescue missions or fire fighting. In such situations, one may also be willing to accept the possibility of losing a fraction of the units in the swarm. Multiple robot systems may also be used for simple repetitive tasks that humans find extremely boring, tiresome of repelling.

---
* Supported by the Israel Science Foundation (grant No. 693/04).

Autonomous mobile robot systems have been studied in a number of different disciplines in engineering and artificial intelligence. Some notable examples for directions taken include the Cellular Robotic System [23], swarm intelligence [7], the self-assembly machine [26], social interaction and intelligent behavior [25], behavior based robot systems [27, 28, 6], multi robot learning [29, 30], and ant robotics [42]. See [8] for a survey of the area.

Robot swarms typically consist of robots that are very small, very simple and very limited in their capabilities. More specifically, they have weak energy resources, limited means of communication and limited processing power. In fact, a common and recurring metaphor is that of insect swarms, and a number of algorithms and methodologies developed for robot swarms draw their inspiration from this metaphor.

While most of the research efforts invested in mobile robots to date were dedicated to engineering aspects (focusing on mobility and function), it is clear that the transition from a single robot to a swarm of robots necessitates some changes also in the approach taken towards the control and coordination mechanisms governing the behavior of the robots. In particular, dealing with the movements of robots in a swarm raises some algorithmic problems that do not exist when considering a single mobile robot. The individual robots must coordinate their movements at least partially, in order to avoid colliding or constricting each other, and to optimize the performance of the entire swarm.

Typical coordination tasks studied in the literature include the following. *Gathering* is the task where starting from any initial configuration, the robots should gather at a single point (within a finite number of steps). A closely related problem is *convergence*, requiring the robots to converge to a single point, rather than gather at it (namely, for every $\epsilon > 0$ there must be a time t_ϵ by which all robots are within distance of at most ϵ of each other). *Pattern formation* requires the robots to arrange themselves in a simple geometric form such as a circle, a simple polygon or a line segment. *Flocking* is the task of following a designated leader. Additional coordination tasks include partitioning, spreading, exploration and mapping, patrolling and searching, and avoiding collisions or bottlenecks.

Most of the experimental studies of multiple robot systems dealt with a fairly small group of robots, typically less than a dozen. A system of that size can usually be controlled centrally, relying on ad-hoc heuristic protocols. Indeed, algorithmic aspects were usually handled in such systems in an implicit manner, mostly ignoring issues such as correctness proof or complexity analysis. However, multi-robot systems envisioned for the future will consist of tens of thousands of small individual units, and such systems can no longer be controlled by a central entity in an efficient way. While hierarchical approaches may be developed, it seems that certain tasks may need to be managed in a fully decentralized manner.

Subsequently, over the last decade there have been a number of efforts to study issues related to the coordination and control of robot swarms from the point of view of distributed computing (cf. [31, 39, 40, 37, 3]), and in particular, to model an environment consisting of mobile autonomous robots and study the capabilities the robots must have in order to achieve their common goals.

This development is fascinating in that it provides the "distributed computing" community with a distributed model that is fundamentally different in some central ways from most of the traditional distributed models, including the model assumptions, the research problems one is required to solve, and the typical concerns one is faced with in trying to solve those problems.

The current paper reviews this exciting area of research and its main developments over the last decade, discusses some of the central obstacles and difficulties, and outlines two main directions for future research.

2 Review of the Literature

2.1 Common Models for Distributed Coordination Algorithms

A number of computational models for robot swarms were proposed in the literature, and several studies dealt with characterizing the influence of the chosen model on the ability of a robot swarm to perform certain basic tasks under different constraints. The general setting consists of a group of mobile robots which all execute the same algorithm in order to perform a given coordination task.

Robot operation cycle: Each robot operates individually in cycles consisting of the following three steps.

- **Look**: identify the locations of the other robots and form a map of the current configuration on your private coordinate system (the model may assume either a perfect vision or a limited visibility range),
- **Compute**: execute the given algorithm, obtaining a *goal point* p_G,
- **Move**: move towards the point p_G. (It is sometimes assumed that the robot might stop before reaching its goal point p_G, but is guaranteed to traverse at least some minimal distance, unless reaching the goal first.)

The "look" and "move" steps are carried out identically in every cycle, independently of the algorithm used; algorithms differ only in their "compute" step.

In most papers in this area (cf. [38, 39, 21, 12]), the robots are assumed to be *oblivious* (or memoryless), namely, they cannot remember their previous states, their previous actions or the previous positions of the robots. Hence the algorithm employed by the robots for the "compute" step cannot rely on information from previous cycles, and its only input is the current configuration.

The robots are also assumed to be indistinguishable, so when looking at the current configuration, each robot knows its own location but does not know the identity of the robots at each of the other points. Furthermore, the robots are assumed to have no means of directly communicating with each other.

The synchronization model: With respect to time, three main models have been considered. The first [37, 40], hereafter referred to as the *semi-synchronous* model, is partially synchronous: all robots operate according to the same clock cycles, but not all robots are necessarily active in all cycles. Robots that are awake at a given cycle may measure the positions of all other robots and then make a local computation and move instantaneously accordingly. The activation of the different robots can be thought of as managed by a hypothetical "scheduler", whose only "fairness" obligation is that each robot must be activated and given a chance to operate infinitely often in any infinite execution. The second, closely related model of [31, 32, 34], hereafter referred to as the *asynchronous* model, differs from the semi-synchronous model in that each robot acts independently in a cycle composed of *four* steps: Wait, Look, Compute, Move. The length of this cycle is finite but not bounded. Consequently, there is no bound on the length of the walk in a single cycle, and different cycles of the same robot may vary in length. The third model is the *synchronous* model [40], in which robots operate by the same clock and all robots are active on all cycles.

2.2 Known Results on Distributed Coordination Algorithms

Much of the theoretical research on distributed algorithms for mobile robots was focused on attempting to answer the question: "how restricted can the robots be and still be able to accomplish certain cooperative tasks?" In other words, the primary motivation of the studies presented, e.g., in [37, 40, 31, 32, 39] was to identify the *minimal* capabilities a collection of distributed robots must have in order to accomplish certain basic tasks and produce interesting interaction.

Various aspects of coordination in autonomous mobile robot systems have been studied in the literature. A basic task that has received considerable attention is the gathering problem. This problem was discussed in [39, 40] in the semi-synchronous model, where it was shown that gathering *two* oblivious autonomous mobile robots without common orientation is impossible. In contrast, an algorithm for gathering $N \geq 3$ robots was presented in [40]. In the fully asynchronous model, a gathering algorithm for $N = 3, 4$ robots is given in [33, 12], and for arbitrary $N \geq 5$ the problem is solved in [11]. Gathering was studied also (in both the semi-synchronous and asynchronous models) in an environment of limited visibility. Visibility conditions are modeled via a (symmetric) *visibility graph* representing the visibility relation between the robots. The problem was proven to be unsolvable when the visibility graph is not connected [21]. A convergence algorithm for any N in limited visibility environments is presented in [2]. A gathering algorithm in the asynchronous model is described in [21], under the assumption that all robots share a compass (i.e., agree on a direction in the plane). The natural gravitational algorithm based on going to the center of gravity, and its convergence properties, were studied in [15, 14] in the semi-synchronous and asynchronous models respectively. Gathering without the ability to detect multiplicity but with unlimited memory is studied in [10], and gathering without both capabilities is shown to be impossible in the asynchronous model in [35].

Formation of geometric patterns was studied in [3, 37, 39, 40, 16, 19, 9, 22]. The algorithms presented therein enable a group of robots to self-arrange and spread itself nearly evenly along the form shaped. The task of flocking, requiring the robots to follow a predefined leader, was studied in [33].

Searching a (static or moving) target in a specified region by a group of robots in a distributed fashion is a natural application for mobile robot systems. Two important related tasks, studied in [37], are even distribution, namely, requiring the robots to spread out uniformly over a specified region, and partitioning, where the robots must split themselves into a number of groups. Finally, the wake-up task requires a single initially awake robot to wake up all the others. A variant of this problem is the Freeze-Tag problem studied in [5, 41].

3 Future Directions

3.1 Modifications in the Robot Model

The existing body of literature on distributed algorithms for autonomous mobile robot systems represents a significant theoretical base containing a rich collection of tools and techniques. The main goals of initial research in this area were to obtain basic understanding and develop a pool of common techniques and methodologies, but equally importantly, to explore and chart the border between the attainable and the unattainable under the most extreme model, representing the weakest possible type of robots in the harshest possible external environment.

Consequently, the models adopted in these studies assume the robots to be very weak and simple. In particular, these robots are generally (although not always) assumed to be oblivious. They are also assumed to have no common coordinate system, orientation, scale or compass, and no means of explicit communication (not even of a limited type, such as receiving broadcasts from a global beacon). It is also assumed that these robots are anonymous, namely, have no identifying characteristics. Also, the robots are usually taken to be dimensionless, namely, treated as points. This implies that robots do not obstruct each other's visibility or movement, i.e., two robots whose timed trajectories intersect will simply pass "through" each other. (This is not necessarily a "weak" property, but it is an unrealistic assumption nontheless.)

These assumptions lead to challenging "distributed coordination" problems since the only means of communication is through using "positional" or "geometric" information, yielding a novel variant of the classical distributed model (which is based on direct communication). The resulting questions are interesting from a theoretical point of view, as they allow us to explore the theoretical limits of robot swarms. Moreover, it is often advantageous to develop algorithms for the weakest robot types possible, as an algorithm that works correctly for weak robots will clearly work correctly in a system of stronger robot types.

On the other hand, the extremely weak model often leads to cumbersome, artificial and sometimes impractical algorithmic solutions. Moreover, towards the practical application of such algorithmic techniques, it is necessary to develop a methodology supporting modularity and allowing multi-phase processes. This becomes difficult if the robots are assumed to be completely memoryless. In fact, it seems that tasks even slightly more involved than the basic ones studied in the literature might pose insurmountable barriers under such weak assumptions. Consider a two-stage project requiring the robots to gather and then perform some follow-up task. The feasibility of such a project is unclear: as the robots are deaf, mute, and forgetful, it seems doubtful that they can accomplish much once they do meet each other. Furthermore, even if they do try to embark on the follow-up task after gathering, their obliviousness will repeatedly force them to immediately resume their attempts to gather.

It is thus clear that the focus on extremely weak robots limits the practicality of many of the distributed algorithms presented in the literature for autonomous mobile robot systems, despite their importance as a base of algorithmic ideas, paradigms and techniques for multi-robot coordination. Subsequently, future research in this area should focus on modifying the model in order to allow a more accurate representation, taking into account the fact that actual robots are usually not so helpless. It is expected that a rigorous algorithmic theory based on accurate assumptions and realistic models may lead to simpler and more practical algorithms which can be readily used within experimental and real systems.

Understandably, it does not make sense to expect the emergence of a single unifying model covering the entire spectrum of possible applications. Nevertheless, let us outline some of the main characteristics a realistic model should have, with a number of possible variations in certain aspects.

A central modification in the model that has to be examined involves the effects of equipping each robot with a small amount (say, $O(1)$ bits) of stable memory. The most immediate benefit is that this will allow the (possibly significant) simplification of most existing algorithms for robot coordination. The reason for this is that many of the complications present in those algorithms were necessary to overcome this lack of memory, and once robots can save state, those complications can be dispensed with. The effect of this change should be systematically investigated across all coordination tasks studied in the literature. A second advantage of introducing memory is that allowing the robots some stable memory may facilitate the modular composition of a number of sub-procedures into a single algorithm, since this stable memory may allow the robots to recognize the computational phase they're in at any given moment. It may be interesting to consider also partial changes along this line, such as allowing the robot to maintain partial history (say, remember the last k cycles).

A second modification concerns the assumption that the robots in a swarm lack common orientation. In many natural settings, the robots may enjoy at least a partial agreement on their orientation. For instance, they may agree on the North, or use a common unit of distance or a common point of reference. It could be interesting to examine the effects of such partial orientation agreements on the solvability and computational complexity of simple coordination tasks. Our initial studies in this direction indicate that with respect to the gathering problem, each of these assumptions may suffice to improve the situation, either by making the problem solvable in settings where impossibility holds otherwise, or by facilitating a simpler solution.

Another interesting question concerns examining which problems can be solved more efficiently or in a simpler manner when the robots are allowed a partial means of explicit communication. This relaxation is also expected to cause a dramatic change in the efficient solvability of various coordination problems. Since the robots are expected to operate in difficult environments and on rugged terrains, it makes sense to focus on restricted communication forms. For example, in certain scenarios a robot may be allowed to communicate only with robots within a limited range (say, radius r from its location), or only with robots to which its line of sight is unobstructed.

Even in settings where explicit communication is infeasible or prohibitively expensive, it may be possible (and desirable) to incorporate in the model some simple means of identification and signalling, such as marking (at least some of) the robots with colors, flags or visible indicator lights. Such modifications may be simple to implement and yet may positively affect the ease of solving some coordination problems, hence this direction deserves thorough examination.

Another assumption that may need to be discarded is that robots are dimensionless, and can pass each other without colliding. A more realistic assumption is that two (or more) robots moving towards each other will stop once meeting (say, by colliding) or shortly before (say, through some "soft halt" mechanism allowing robots to detect a near-collision and halt).

3.2 Introducing Fault Tolerance

While the classical model is rather restrictive on the one hand, it is perhaps somewhat "too optimistic" on the other, in that it assumes perfectly functioning robots. As future robot swarms are expected to comprise of cheap, simple and relatively weak robots and operate under harsh conditions, the issue of resilience to failure becomes crucial, since in such systems one cannot possibly rely on assuming fail-proof hardware or software.

When considering the issue of coping with faults, we may classify the problems that need to be dealt with into two types: problems that occur regularly during the normal operation of every robot as a result of its inherent imperfections, and problems resulting from the *malfunction* of some robots. Next we discuss these two fault types and possible ways to overcome them.

Overcoming Robot Imperfections. The common robot model makes the assumption that the configuration map obtained by a robot observing its surroundings is perfect. In fact, certain algorithmic solutions proposed in the literature rely critically on this assumption. In practice, however, the robot measurements suffer from nonnegligible inaccuracies in both distance and angle estimations. (For instance, the accuracy of range estimation in sonar sensors is about $\pm 1\%$ and the angular separation is about $3°$, cf. [36].) The same applies to the precision of robot movements, as a variety of mechanical factors, including unstable power supply, friction and force control, make it hard to control the exact distance a robot traverses in a single cycle, or to predict it with high accuracy.

Another unrealistic assumption is that robots are capable of carrying out infinite precision calculations over the reals. For instance, this assumption underlies the distinction between the gathering and convergence problems. In fact, it is sometimes assumed that the robots have unlimited computational power. The fact that in reality robots cannot perform perfect precision calculations may seem insignificant, since floating point arithmetic can be carried to very high accuracy with modern computers. However, this may prove to be a serious problem. For instance, the point that minimizes the sum of distances to the robots' locations (also known as the Weber point) may be used to achieve gathering. However, this point is not computable, due to its infinite sensitivity to location errors. More generally, the correctness of many of the distributed coordination algorithms presented in the literature is proven by relying on basic properties from Euclidean geometry. Unfortunately, these properties are often no longer valid when measurement or calculation errors occur. To illustrate this point, consider Algorithm 3-Gather presented in [1], which gathers three robots using several simple rules. One of these rules states that if the robots form an obtuse triangle, then they move towards the vertex with the obtuse angle. Thus, as shown in [13], this algorithm might fail to achieve even convergence in the presence of angle measurement errors of at least 15°. Similar problems arise with other algorithms described in the literature.

Subsequently, for the "next-generation" model of robot swarms, it is desirable to discard these unrealistic assumptions and examine whether efficient algorithmic solutions can still be obtained for coordination problems of interest.

An initial study [13] examines a model in which the robot's location estimation and movements are imprecise, with imprecision bounded by some accuracy parameter ϵ known at the robot's design stage. The measurement imprecisions can affect both distance and angle estimations. Formally, the robot's distance estimation is ϵ-*precise* if, whenever the real distance to an observed point in the robot's private coordinate system is D, the measurement d taken by the robot for that distance satisfies $(1-\epsilon)d < D < (1+\epsilon)d$. A similar imprecision is allowed for angle estimations.

Several impossibility results are established in [13], limiting the maximum inaccuracy that still allows convergence. Specifically, it is shown that gathering is impossible for any number of robots assuming inaccuracies in both distance and angle measurements, even in a fully synchronous model and when the robots have unlimited memory and are allowed to use randomness. (If angle measurements are always exact, then impossibility of gathering is known only for $N = 2$ robots, and is conjectured for any N.) Hence at best, only the weaker requirement of convergence can be expected. Actually, it seems reasonable to conjecture that even convergence is impossible for robots with large measurement errors. The exact limits are not completely clear. Some rather weak limits on the possibility of convergence are given in [13], where it is shown that for a configuration of $N = 3$ robots having an error of $\pi/3$ or more in angle measurement, there is no deterministic algorithm for convergence even assuming exact distance estimation, fully synchronous model and unlimited memory. On the other hand, an algorithm is presented in [13] for convergence under bounded imprecision (specifically, $\epsilon < 0.2$ or so) in the synchronous and semi-synchronous models.

Some natural questions to be explored further include the following. First, the precision required of the robots for the algorithm of [13] to work correctly is still significant, and improved techniques are necessary for overcoming this. Second, it would be interesting to obtain similar results in the asynchronous setting. Third, similar techniques should be developed for other coordination tasks, such as pattern formation, search, etc.

It may also be interesting to examine distributed coordination algorithms with an eye towards complexity, trying to develop variants that are both simple and resource efficient in terms of internal computation costs at each robot. One specific aspect of this is discarding the assumption of infinite precision in real computations, and settling for approximations. This may necessitate some relaxations in the definitions of certain common tasks (such as gathering at a single point or forming perfect geometric objects) to fit these weaker assumptions.

Overcoming Robot Malfunctions. Robot swarms are intended to operate in tough and hazardous environments, so it is to be expected that certain robots may malfunction. Indeed, one of the main attractive features of robot swarms is their potential for enhanced fault tolerance through inherent redundancy. For example, a fault tolerant algorithm for gathering should be required to ensure that even if some fraction of the robots fails in any execution, all the *nonfaulty* robots still manage to gather at a single point within a finite time, regardless of the actions taken by the faulty ones.

Perhaps surprisingly, however, this aspect of multiple robot systems has been explored to very little extent so far. In fact, almost all the results reported in the literature rely on the assumption that all robots function properly and follow their protocol without any deviation.

One exception concerns *transient* failures. As observed in [40, 37, 20], any algorithm that works correctly on oblivious robots is necessarily *self-stabilizing*, i.e., it guarantees that after any transient failure the system will return to a correct state and the goal will be achieved. Another fault model studied in [37] considers restricted sensor and control failures, and assumes that whenever failures occur in the system, the identities of the faulty robots become known to all robots. Unfortunately, this assumption might not hold in many typical settings, and in case unidentified faults do occur in the system, it is no longer guaranteed that the algorithms of [37, 40] remain correct.

Following traditional approaches in the field of distributed computing, it is interesting to study robot algorithms under the *crash* and *Byzantine* fault models. In order to pinpoint the effect of faults, all other aspects of the model can be left unchanged, following the basic models of [37, 31]. In the Byzantine fault model it is assumed that a faulty robot might behave in arbitrary and unforeseeable ways. It is sometimes convenient to model the behavior of the system by means of an *adversary* which has the ability to control the behavior of the faulty robots, as well as the "undetermined" features in the behavior of the nonfaulty processors (e.g., the distance to which they move). In the crash fault model, it is assumed that the only faulty behavior allowed for a faulty robot is to crash, i.e., stop functioning. This may happen at any point in time during the cycle, including any time during the movement towards the goal point.

In [43], an algorithm is given for the *Active Robot Selection Problem (ARSP)* in the presence of *initial* crash faults. The ARSP creates a subgroup of nonfaulty robots from a set that includes also initially crashed robots and enables the robots in that subgroup to recognize one another.

A systematic study of the gathering problem in failure-prone robot systems is presented in [1]. Under the crash fault model, it is shown in [1] that the gathering problem with at most one crash failure is solvable in the semi-synchronous model. Considering the Byzantine fault model, it is shown that it is impossible to perform a successful gathering in the semi-synchronous or asynchronous model even in the presence of a single fault. For the synchronous model, an algorithm is presented for solving the gathering problem in N-robot systems whenever the maximum number of faults f satisfies $3f + 1 \leq N$.

In general, the design of fault-tolerant distributed control algorithms for multiple robot systems is still a largely unexplored direction left for future study. Particularly, a number of questions are left open in [1]. In the synchronous model, while the algorithm of [1] does solve the problem even with Byzantine faults, its complexity is prohibitively high, rendering it impractical except maybe for very small systems. Hence it is desirable to look for a simpler and faster algorithm. In the asynchronous and semi-synchronous models, the techniques of [1] are inadequate for handling more than a single fault, again limiting their applicability rather drastically, and it is interesting to investigate approaches for extending these techniques to multiple failures. More generally, as the asynchronous model captures a more faithful representation of typical actual settings, we view the derivation of suitable algorithms for performing various coordination tasks in this model in the presence of multiple crash faults as one of the central directions of research in this area. Turning to Byzantine faults in the asynchronous and semi-synchronous models, as such faults make gathering impossible, a plausible alternative is to try to solve the slightly weaker problem of convergence.

Moreover, as the initial study of [1] was limited to the gathering problem, it would be interesting to investigate also the fault-tolerance properties of currently available algorithms for other tasks described above (e.g., formation of geometric patterns). Specifically, a central theme of both theoretical and practical significance concerns identifying the maximum number of faults under which a solution for a particular coordination problem is still feasible. It would be attractive to develop a general theory answering this question, similar to the theory developed for the analogous question in classical distributed systems.

References

1. N. Agmon and D. Peleg. Fault-tolerant gathering algorithms for autonomous mobile robots. In *Proc. 15th ACM-SIAM Symp. on Discrete Algo.*, 1063–1071, 2004.
2. H. Ando, Y. Oasa, I. Suzuki, and M. Yamashita. A distributed memoryless point convergence algorithm for mobile robots with limited visibility. *IEEE Trans. Robotics and Automation*, 15:818–828, 1999.

3. H. Ando, I. Suzuki, and M. Yamashita. Formation and agreement problems for synchronous mobile robots with limited visibility. In *Proc. IEEE Symp. of Intelligent Control*, 453–460, 1995.
4. R.C. Arkin. *Behavior-Based Robotics*. MIT Press, 1998.
5. E. Arkin, M. Bender, S. Fekete, J. Mitchell, and M. Skutella. The freeze-tag problem: How to wake up a swarm of robots. In *Proc. 13th ACM-SIAM Symp. on Discrete Algorithms*, 2002.
6. T. Balch and R. Arkin. Behavior-based formation control for multi-robot teams. *IEEE Trans. on Robotics and Automation*, 14, 1998.
7. G. Beni and S. Hackwood. Coherent swarm motion under distributed control. In *Proc. DARS'92*, 39–52, 1992.
8. Y.U. Cao, A.S. Fukunaga, and A.B. Kahng. Cooperative mobile robotics: Antecedents and directions. *Autonomous Robots*, 4(1):7–23, 1997.
9. I. Chatzigiannakis, M. Markou, and S.E. Nikoletseas. Distributed circle formation for anonymous oblivious robots. In *Proc. 3rd Workshop on Experimental and Efficient Algorithms*, LNCS 3059, 159–174, 2004.
10. M. Cieliebak. On the feasibility of gathering by autonomous mobile robots. In *Proc. 6th Latin American Symp. on Theoret. Inform.*, LNCS 2976, 577–588, 2004.
11. M. Cieliebak, P. Flocchini, G. Prencipe, and N. Santoro. Solving the robots gathering problem. In *Proc. 30th ICALP*, 1181–1196, 2003.
12. M. Cieliebak and G. Prencipe. Gathering autonomous mobile robots. In *Proc. 9th Int. Colloq. on Struct. Info. and Commun. Complex.*, 57–72, 2002.
13. R. Cohen and D. Peleg. Convergence of autonomous mobile robots with inaccurate sensors and movements. Tech. Rep. MSC 04-8, Weizmann Inst. of Science, 2004.
14. R. Cohen and D. Peleg. Convergence properties of the gravitational algorithm in asynchronous robot systems. In *Proc. 12th ESA*, LNCS 3221, 228–239, 2004.
15. R. Cohen and D. Peleg. Robot convergence via center-of-gravity algorithms. In *Proc. 11th Colloq. on Struct. Inf. and Comm. Complex.*, LNCS 3104, 79–88, 2004.
16. X. Defago and A. Konagaya. Circle formation for oblivious anonymous mobile robots with no common sense of orientation. In *Proc. 2nd ACM Workshop on Principles of Mobile Computing*, 97–104. ACM Press, 2002.
17. M. Erdmann and T. Lozano-Pdrez. On multiple moving objects. *Algorithmica*, 2:477–521, 1987.
18. M. Erdmann and T. Lozano-Perez. On multiple moving objects. In *Proc. IEEE Conf. on Robotics and Automation*, 1419–1424, 1986.
19. P. Flocchini, G. Prencipe, N. Santoro, and P. Widmayer. Hard tasks for weak robots: The role of common knowledge in pattern formation by autonomous mobile robots. In *Proc. 10th Int. Symp. on Algorithms and Computation*, 93–102, 1999.
20. P. Flocchini, G. Prencipe, N. Santoro, and P. Widmayer. Distributed coordination of a set of autonomous mobile robots. In *Proc. IEEE Intelligent Vehicles Symp.*, 480–485, 2000.
21. P. Flocchini, G. Prencipe, N. Santoro, and P. Widmayer. Gathering of autonomous mobile robots with limited visibility. In *Proc. 18th STACS*, 247–258, 2001.
22. B. Katreniak. Biangular circle formation by asynchronous mobile robots. In *Proc. 12th Colloq. on Struct. Info. and Commun. Complex.*, LNCS 3499, 185–199, 2005.
23. Y. Kawauchi, M. Inaba, and T. Fukuda. A principle of decision making of cellular robotic system (CEBOT). In *Proc. IEEE Conf. on Robotics and Automation*, 833–838, 1993.
24. Y. Kuniyoshi, S. Rougeaux, M. Ishii, N. Kita, S. Sakane, and M. Kakikura. Cooperation by observation - the framework and basic task patterns. In *Proc. Int. Conf. on Robotics and Automation*, 767–774, 1994.

25. M.J. Mataric. *Interaction and Intelligent Behavior.* PhD thesis, MIT, 1994.
26. S. Murata, H. Kurokawa, and S. Kokaji. Self-assembling machine. In *Proc. IEEE Conf. on Robotics and Automation*, 441–448, 1994.
27. L.E. Parker. Designing control laws for cooperative agent teams. In *Proc. IEEE Conf. on Robotics and Automation*, 582–587, 1993.
28. L.E. Parker. On the design of behavior-based multi-robot teams. *J. of Advanced Robotics*, 10, 1996.
29. L.E. Parker and C. Touzet. Multi-robot learning in a cooperative observation task. In *Distributed Autonomous Robotic Systems 4*, 391–401, 2000.
30. L.E. Parker, C. Touzet, and F. Fernandez. Techniques for learning in multi-robot teams. In T. Balch and L.E. Parker, editors, *Robot Teams: From Diversity to Polymorphism*. A. K. Peters, 2001.
31. G. Prencipe. CORDA: Distributed coordination of a set of autonomous mobile robots. In *Proc. 4th Eur. Res. Seminar on Adv. in Distr. Syst.*, 185–190, 2001.
32. G. Prencipe. Instantaneous actions vs. full asynchronicity: Controlling and coordinating a set of autonomous mobile robots. In *Proc. 7th Italian Conf. on Theoretical Computer Science*, 185–190, 2001.
33. G. Prencipe. *Distributed Coordination of a Set of Autonomous Mobile Robots.* PhD thesis, Universita Degli Studi Di Pisa, 2002.
34. G. Prencipe. The effect of synchronisity on the behavior of autonomous mobile robots. *Theory of Computing Systems*, 2004.
35. G. Prencipe. On the feasibility of gathering by autonomous mobile robots. In *Proc. 12th Colloq. on Struct. Info. and Commun. Complex.*, LNCS 3499, 246–261, 2005.
36. SensComp Inc. *Spec. of 6500 series ranging modules.* http://www.senscomp.com.
37. K. Sugihara and I. Suzuki. Distributed algorithms for formation of geometric patterns with many mobile robots. *J. of Robotic Systems*, 13(3):127–139, 1996.
38. I. Suzuki and M. Yamashita. Agreement on a common x-y coordinate system by a group of mobile robots. In *Proc. Dagstuhl Seminar on Modeling and Planning for Sensor-Based Intelligent Robots*, 1996.
39. I. Suzuki and M. Yamashita. Distributed anonymous mobile robots - formation and agreement problems. In *Proc. 3rd Colloq. on Struct. Info. and Commun. Complex.*, 313–330, 1996.
40. I. Suzuki and M. Yamashita. Distributed anonymous mobile robots: Formation of geometric patterns. *SIAM J. on Computing*, 28:1347–1363, 1999.
41. M. Sztainberg, E. Arkin, M. Bender, and J. Mitchell. Analysis of heuristics for the freeze-tag problem. In *Proc. 8th Scand. Workshop on Alg. Theory*, 270–279, 2002.
42. I.A. Wagner and A.M. Bruckstein. From ants to a(ge)nts. *Annals of Mathematics and Artificial Intelligence*, 31, special issue on ant-robotics:1–5, 1996.
43. D. Yoshida, T. Masuzawa, and H. Fujiwara. Fault-tolerant distributed algorithms for autonomous mobile robots with crash faults. *Systems and Computers in Japan*, 28:33–43, 1997.

Labeling Schemes for Tree Representation

Reuven Cohen[1], Pierre Fraigniaud[2,*], David Ilcinkas[2,*],
Amos Korman[1], and David Peleg[1]

[1] Dept. of Computer Science, Weizmann Institute, Israel
{r.cohen, amos.korman, david.peleg}@weizmann.ac.il
[2] CNRS, LRI, Université Paris-Sud, France
{pierre, ilcinkas}@lri.fr

Abstract. This paper deals with compact label-based representations for trees. Consider an n-node undirected connected graph G with a predefined numbering on the ports of each node. The *all-ports* tree labeling $\mathcal{L}_{all}$ gives each node v of G a label containing the port numbers of all the *tree* edges incident to v. The *upward* tree labeling $\mathcal{L}_{up}$ labels each node v by the number of the port leading from v to its parent in the tree. Our measure of interest is the worst case and total length of the labels used by the scheme, denoted $M_{up}(T)$ and $S_{up}(T)$ for $\mathcal{L}_{up}$ and $M_{all}(T)$ and $S_{all}(T)$ for $\mathcal{L}_{all}$. The problem studied in this paper is the following: Given a graph G and a predefined port labeling for it, with the ports of each node v numbered by $0, \ldots, \deg(v) - 1$, select a rooted spanning tree for G minimizing (one of) these measures. We show that the problem is polynomial for $M_{up}(T)$, $S_{up}(T)$ and $S_{all}(T)$ but NP-hard for $M_{all}(T)$ (even for 3-regular planar graphs). We show that for every graph G and port numbering there exists a spanning tree T for which $S_{up}(T) = O(n \log \log n)$. We give a tight bound of $O(n)$ in the cases of complete graphs with arbitrary labeling and arbitrary graphs with symmetric port assignments. We conclude by discussing some applications for our tree representation schemes.

1 Introduction

This paper deals with compact label-based representations for trees. Consider an n-node undirected connected graph G. Assume that we are given also a predefined numbering on the ports of each node, i.e., every edge e incident to a node u is given an integer label $l_u(e)$ in $\{0, \ldots, \deg(u) - 1\}$ so that $l_u(e) \neq l_u(e')$ for any two distinct edges e and e' incident to u. In general, one may consider two types of schemes for representing a spanning tree in a given graph. An *all-ports* tree representation has to ensure that each node in the graph knows the port numbers of all its incident tree edges. An *upward* tree representation has to ensure that each node in the graph knows the port number of the tree edge connecting it to its parent. Such representations find applications in the areas of data structures, distributed computing, communication networks and others.

---

* Supported by project "PairAPair" of the ACI Masses de Données, project "Fragile" of the ACI Sécurité et Informatique, and project "Grand Large" of INRIA.

Corresponding to the two general representation types discussed above, we consider two label-based schemes. The *all-ports* tree labeling $\mathcal{L}_{all}$ labels each node v of G by a label containing the port numbers of all the *tree* edges incident to v. The *upward* tree labeling $\mathcal{L}_{up}$ labels each node v of G by the number of the port connected to the edge e of T leading from v toward the root. We use the standard binary representation of positive integers to store the port numbers.

Our measure of interest is the worst case or average length of the labels used by tree labeling schemes. Let us formalize these notions. Given a graph G (including a port numbering) and a spanning tree T for G,

- the sum of the label sizes in the labeling $\mathcal{L}_{up}$ (respectively, $\mathcal{L}_{all}$) on T is denoted by $S_{up}(T)$ (resp., $S_{all}(T)$);
- the maximum label size in the labeling $\mathcal{L}_{up}$ (respectively, $\mathcal{L}_{all}$) on T is denoted by $M_{up}(T)$ (resp., $M_{all}(T)$).

This paper studies the following problem. Given a graph G and a predefined port labeling for it, with the ports of each node v numbered $0, \ldots, \deg(v) - 1$, select a rooted spanning tree T for G minimizing (one of) these measures.

We show that there are polynomial time algorithms that given a graph G and a port numbering, construct a spanning tree T for G minimizing $M_{up}(T)$ or $S_{up}(T)$. Moreover, we conjecture that for every graph G, and any port numbering for G, there exists a tree T spanning G, for which $S_{up}(T) = O(n)$. In other words, we conjecture that there is a tree for which the upward labeling requires a constant number of bits per node on average. We establish the correctness of this conjecture in the cases of complete graphs with arbitrary labeling and arbitrary graphs with symmetric port assignments. For arbitrary graph, we show a weaker algorithm, constructing for a given graph G (with its port numbering) a spanning tree T with $S_{up}(T) = O(n \log \log n)$.

Turning to all-port labeling schemes, for any spanning tree T the labeling $\mathcal{L}_{all}$ has average label size $O(\log \Delta)$ in graphs of maximum degree Δ, which is optimal on some n-node graphs of maximum degree Δ. It turns out that here there is a difference between the measures $S_{all}(T)$ and $M_{all}(T)$. We show that there is a polynomial time algorithm that given a graph G and a port numbering, constructs a tree T minimizing $S_{all}(T)$. In contrast, the problem of deciding, for a given graph G with a port numbering and an integer k, whether there exists a spanning tree T of G satisfying $M_{all}(T) \leq k$ is NP-hard. This holds even restricted to 3-regular planar graphs, and even for fixed $k = 3$. Nevertheless, denoting the smallest maximum degree of any spanning tree for the graph G by δ_{min}, there is a polynomial time approximation of the tree of minimum $M_{all}(T)$, up to a multiplicative factor of $O(\log \Delta / \log \delta_{min})$.

We conclude by discussing some applications for our tree representation schemes, including basic distributed operations such as broadcast, convergecast and graph exploration. A number of well-known solutions to these problems (cf. [11], [1], [12]) are based on maintaining a spanning tree for the network and using it for efficient communication. All standard spanning tree constructions that we are aware of do not take into account the memory required to store the spanning tree, and subsequently, the resulting tree may in general require a total of up to $O(n \log \Delta)$ memory bits over an n-node network of maximum degree Δ. Using the tree representations developed herein may improve the memory requirements of storing the tree representation. For instance, for applications that require only an upward tree representation, our construction yields a total memory requirement of $O(n \log \log n)$ bits, which is lower in high degree graphs. These applications are discussed in more detail in Section 4.

The all-port labeling scheme is particularly convenient for broadcast applications because it minimizes the number of messages. For less demanding tasks such as graph exploration, more compact labeling schemes can be defined. In particular, [3] describes a labeling scheme which uses only three different labels and allows a finite automaton to perform exploration in time at most $O(m)$ on m-edge graphs.

2 Upward Tree Labeling Schemes with Short Labels

2.1 Basic Properties

Let us first establish a naive upper bound on $S_{up}(T)$ and $M_{up}(T)$. In the basic upwards tree labeling scheme, the label kept at each node v is the port number of the tree edge leading from v toward the root. Hence no matter which tree is selected, the label assigned to each node v by the upwards tree labeling scheme uses at most $\lceil \log \deg(v) \rceil$ bits. This implies the following bounds. (Throughout, some proofs are omitted.)

Lemma 1. *For every n-vertex graph G of maximum degree Δ, and for every spanning tree T of G, we have (1) $M_{up}(T) \leq \lceil \log \Delta \rceil$, and (2) $S_{up}(T) \leq \sum_v \lceil \log \deg(v) \rceil$.*

Note that the second part of the lemma implies that in graph families with a linear number of edges, such as planar graphs, the average label size for any spanning tree is at most $O(1)$.

Given $G = (V, E)$, let $\boldsymbol{G} = (V, X)$ be the directed graph in which every edge $\{u, v\}$ in E corresponds to two arcs (u, v) and (v, u) in X. The arcs of $\boldsymbol{G}$ are weighted according to the port numbering of the edges in G, i.e., the arc (u, v) of $\boldsymbol{G}$ has weight

$$\omega(u, v) = \begin{cases} 1, & p = 0, \\ \lfloor \log p \rfloor + 1, & p \geq 1, \end{cases}$$

where p is the port number at u of the edge $\{u, v\}$ in G. That is $\omega(u, v)$ is the number of bits in the standard binary representation of positive integers required to encode[1] port number p.

Finding a spanning tree T minimizing $M_{up}(T)$ is easy by identifying the smallest k such that the digraph $\boldsymbol{G}_k$ obtained from $\boldsymbol{G}$ by removing all arcs of

---

[1] Note that this encoding is not a prefix coding and therefore might not be decodable. However, efficient encoding methods exist which are asymptotically optimal (cf. [8]) and therefore the overall results are also valid for such encoding.

weight greater than $\lfloor \log k \rfloor + 1$, contains a spanning tree directed toward the root. Thus we have the following.

Proposition 1. *There is a polynomial time algorithm that, given a graph G and a port numbering, constructs a spanning tree T for G minimizing $M_{up}(T)$.*

Similarly, applying any Minimum-weight Spanning Tree (MST) algorithm for digraphs (cf. [2], [7]) on $\boldsymbol{G}$ with weight function ω, we get the following.

Proposition 2. *There is a polynomial time algorithm that, given a graph G and a port numbering, constructs a spanning tree T for G minimizing $S_{up}(T)$.*

There are graphs for which the bound on M_{up} specified in Lemma 1 is reached for any spanning tree T (e.g., a graph composed of two Δ-regular graphs linked by a unique edge labeled Δ at both of its extremities). However, this is not the case for S_{up}, and we will show that, for any graph, there is a spanning tree T for which $S_{up}(T)$ is much smaller than the bound in Lemma 1.

2.2 Complete and Symmetric Graphs

First, consider the case of a complete graph with arbitrary labeling. We show that there exists a spanning tree T of it, for which $S_{up}(T) = O(n)$. We establish the claim by presenting an algorithm that yields a labeling of this cost. The algorithm is a variant of Kruskal's minimum-weight spanning tree (MST) algorithm (cf. [4]). The algorithm maintains a collection of rooted directed tree with the edges of each tree directed towards its root. Initially, each vertex forms a tree on its own. The algorithm merges these trees into larger trees until it remains with a single tree giving the solution.

The algorithm operates in phases. Let size(T) denote the size (number of nodes) of the tree T. A tree T is *small* for phase $k \geq 1$ if size(T) < 2^k.

Each phase k of the algorithm consists of four steps. At the beginning of the phase, we identify the collection of small trees for the phase: $\mathcal{T}_{\text{small}}(k) = \{T \mid \text{size}(T) < 2^k\}$. Second, for each tree $T \in \mathcal{T}_{\text{small}}(k)$ with root $r(T)$, we look at the set $S(T)$ of outgoing edges that connect $r(T)$ to nodes in other trees $T' \neq T$, and select the edge $e(T)$ of minimum weight in $S(T)$. (Note that $S(T) \neq \emptyset$ since the graph is complete.) Third, we add these edges to the collection of trees, thus merging the trees into 1-factors. Formally, a *1-factor* is a weakly-connected directed graph of out-degree 1. Intuitively, a 1-factor is a directed subgraph consisting of a directed cycle and a collection of directed trees rooted at the nodes of the cycle. Figure 1 illustrates two 1-factors. Finally, for the last of the four steps, in each 1-factor we arbitrarily select one of the edges on the cycle and erase it, effectively transforming the 1-factor back into a rooted directed tree. This process is continued until a single tree remains, which is the desired tree.

Claim. Denote the collection of trees at the beginning of the kth phase, $k \geq 1$, by $T_1^k, \ldots, T_{m_k}^k$.

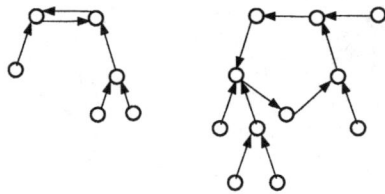

Fig. 1. Two 1-factors

1. $\sum_{j=1}^{m_k} \text{size}(T_j^k) = n$ for every $k \geq 1$;
2. $\text{size}(T_j^1) = 1$ for every $1 \leq j \leq n$ (observe that $m_1 = n$);
3. $\text{size}(T_j^k) \geq 2^{k-1}$ for every $k \geq 1$ and $1 \leq j \leq m_k$;
4. $m_k \leq n/2^{k-1}$ for every $k \geq 1$.
5. The number of phases is at most $\lceil \log n \rceil$.

Observe that when selecting the outgoing edge $e(T_j^k)$ for the root $r(T_j^k)$ on the kth phase, the only outgoing edges of $r(T_j^k)$ excluded from consideration are the $\text{size}(T_j^k) - 1$ edges leading to the other nodes in T_j^k. Hence even if all of these edges are "lighter" than the edges leading outside the tree, the port number used for $e(T_j^k)$ is at most $\text{size}(T_j^k) - 1$, hence:

$$\begin{cases} \omega(e(T_j^k)) = 1 & \text{if } k = 1 \\ \omega(e(T_j^k)) \leq \lfloor \log(\text{size}(T_j^k) - 1) \rfloor + 1 & \text{if } k > 1 \end{cases}$$

Moreover, we have $\log \text{size}(T_j^k) < k$ because outgoing edges are selected only for small trees, and thus we have $\omega(e(T_j^k)) \leq k$. Hence the total weight C_k of the edges added to the structure throughout the kth phase satisfies

$$C_k \leq \sum_{T_j^k \in \mathcal{T}_{\text{small}}(k)} k = k \cdot |\mathcal{T}_{\text{small}}(k)| \leq k \cdot m_k.$$

By Part 4 of Claim 2.2, $C_k \leq kn/2^{k-1}$, and the total weight C of the resulting tree satisfies $C = \sum_{k \geq 1} C_k \leq \sum_{k \geq 1} kn/2^{k-1} \leq 4n$. We have the following.

Proposition 3. *On the complete graph (with an arbitrary port numbering), there exists a spanning tree T for which $S_{up}(T) = O(n)$.*

Next, we consider another interesting and potentially applicable special case, namely, arbitrary graphs with symmetric port assignment.

Proposition 4. *On graphs with symmetric port assignments (i.e., where for every edge $e = \{u, v\}$, the port numbers of e at u and v are identical), there exists a spanning tree T for which $S_{up}(T) = O(n)$.*

Proof. For graphs with symmetric port assignments, we again present an algorithm that yields a labeling of cost $O(n)$. The algorithm is a variant of the one

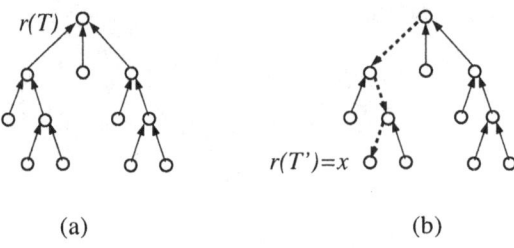

Fig. 2. (a) The tree T. (b) The tree T'.

used for proving Property 3. The general structure of the algorithm is the same, i.e., it is based on maintaining a collection of rooted directed tree and merging them until remaining with a single tree. The main difference has to do with the fact that since the graph is not complete, it may be that for the small tree T under consideration, the set $S(T)$ is empty, i.e., all the outgoing edges of the root $r(T)$ go to nodes *inside* T. Therefore, an additional step is needed, transforming T into a tree T' on the same set of vertices, with the property that the new root, $r(T')$, has an outgoing edge to a node outside T'.

This is done as follows. We look for the lightest (least port number) outgoing edge from some node x in T to some node outside T. Note that such an edge must exist so long as T does not span the entire graph G, as G is connected. Let $p(T) = (v_1, v_2, \ldots, v_j)$ be the path from $r(T)$ to x in T, where $r(T) = v_1$ and $v_j = x$. Transform the tree T into a tree T' rooted at x by reversing the directions of the edges along this path. (See Figure 2 where dashed edges represent the path from the original root to T.) Observe that by symmetry, the cost of T' is the same as that of T, so the proof can proceed as for Property 3. □

2.3 Arbitrary Graphs

For the general setting, we show the universal bound of $O(n \log \log n)$ on S_{up}. Again, the algorithm yielding this cost is a variant of the one used for proving Property 3. As in the proof of Property 4, since the graph is not complete, it may be that for the small tree T under consideration, all the outgoing edges of the root $r(T)$ go to nodes *inside* T. It is thus necessary to transform T into a tree T' on the same set of vertices so that the new root $r(T')$ has an outgoing edge to a node outside T'. However, it is not enough to pick an arbitrary outgoing edge and make its internal endpoint the new root because, in the absence of symmetry, the reversed route may be much more expensive than the original path, thus causing the transformed tree to be too costly.

Instead, the transformation is performed as follows (cf. Fig. 3). We look for the shortest path (in hops) from the current root $r(T)$ to the node in T that is the closest to the root, and that has an outgoing edge to a node outside T. Moreover, all the nodes of the path must be in T. (Such a path must exist so long as T does not span the entire graph G, as G is connected.) Let this path be $p(T) = (v_1, v_2, \ldots, v_j)$, where (1) $r(T) = v_1$, (2) $v_1, \ldots, v_j \in T$, and (3) v_j has a

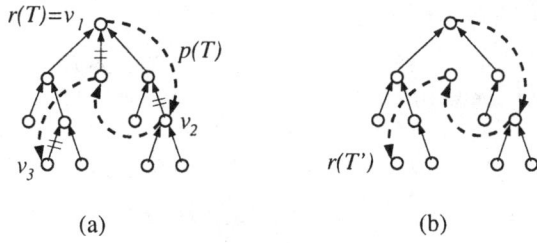

Fig. 3. (a) The tree T and the escape path $p(T)$ (dashed). (b) The tree T'.

neighbor $z \notin T$. For every $1 \leq i \leq j-1$, we add the edge (v_i, v_{i+1}) of $p(T)$ to T. In turn, for $2 \leq i \leq j$, we remove from T the (unique) outgoing edge of v_i in T, (v_i, w_i). The resulting subgraph is a directed tree T' rooted at $r(T') = v_j$. (Note that in case the original root $r(T)$ has an outgoing edge to some node z outside T, this transformation uses $p(T) = (r(T))$ and leaves T unchanged.)

Clearly, applying these transformations on the small trees in each phase incurs additional costs. To estimate them, we bound from above the additional cost incurred by adding the paths $p(T)$ for every tree $T \in \mathcal{T}_{\text{small}}(k)$ in every phase k. For such a tree T with $p(T) = (v_1, v_2, \ldots, v_j)$, denote the set of nodes whose outgoing edge was replaced (hence whose labels may increase) by $A(T) = \{v_1, v_2, \ldots, v_{j-1}\}$, and let $A_k = \bigcup_{T \in \mathcal{T}_{\text{small}}(k)} A(T)$.

Partition the nodes of the graph G into classes by their degrees, setting

$$D_\ell = \{v \mid 2^{\ell-1} < \deg(v) \leq 2^\ell\}$$

for $\ell \geq 0$. Define $A^\ell(T) = A(T) \cap D_\ell$ and $A_k^\ell = A_k \cap D_\ell = \bigcup_{T \in \mathcal{T}_{\text{small}}(k)} A^\ell(T)$.

Claim. $\sum_{v \in A(T)} \deg(v) \leq 3 \cdot \text{size}(T)$ for every phase $k \geq 1$ and tree $T \in \mathcal{T}_{\text{small}}(k)$.

Proof. Note that the nodes of $A(T)$ have all their neighbors inside T, hence their degrees in the (undirected) subgraph $G(T)$ induced by the nodes of T are the same as their degrees in G. Since $p(T)$ is a shortest path from v_1 to v_j in $G(T)$, we have that every node w in $G(T)$ has at most 3 neighbors in $p(T)$ (otherwise it would provide a shortcut yielding a shorter path between v_1 and v_j in $G(T)$, contradicting the assumption). Thus the number of edge ports in the nodes of $p(T)$ is at most $3 \cdot \text{size}(T)$. □

Claim. $|A_k^\ell| \leq 3n/2^{\ell-1}$ for every phase k.

To effectively bound the cost increases, we rely on the following observation. A node v may participate in several paths $p(T)$ throughout the construction. Each time, it may replace its outgoing edge with a new one. Nevertheless, the cost it incurs in the final tree is just the cost of its *final* outgoing edge, since all the other outgoing edges added for it in earlier phases were subsequently replaced. Denote this cost by $X(v)$. (For nodes that did not incur such costs at all throughout the execution, let $X(v) = 0$.) For a set of nodes W, let $X(W) = \sum_{v \in W} X(v)$.

Claim. For every ℓ, (1) $X(D_\ell) \leq \ell \cdot |D_\ell|$, and (2) $X(D_\ell) \leq 3\ell n \lceil \log n \rceil / 2^{\ell-1}$.

Proof. By the definition of D_ℓ, we have $X(v) \leq \lceil \log \deg(v) \rceil \leq \ell$ for every node $v \in D_\ell$. Part (1) of the claim follows. For part (2), we first note that Claim 2.2 holds for the setting of the T_j^k in arbitrary graphs. Hence, the number of phases is at most $\lceil \log n \rceil$ by Claim 2.2, item 5. Therefore $X(D_\ell) \leq \sum_{k=1}^{\lceil \log n \rceil} X(A_k^\ell) \leq \sum_{k=1}^{\lceil \log n \rceil} \ell \cdot |A_k^\ell|$. Using Claim 2.3 we get $X(D_\ell) \leq \lceil \log n \rceil \cdot \ell \cdot 3n/2^{\ell-1}$. □

Claim. The total additional cost incurred by the nodes is $O(n \log \log n)$.

Proof. Partition the total additional cost X into $X = X_L + X_H$ where $X_L = \sum_{\ell \leq \log \log n} X(D_\ell)$ and $X_H = \sum_{\ell > \log \log n} X(D_\ell)$. Note that by item (1) of Claim 2.3, $X_L \leq \sum_{\ell \leq \log \log n} \ell \cdot |D_\ell| \leq n \log \log n$. Also, by item (2) of Claim 2.3, $X_H \leq \sum_{\ell > \log \log n} 3\ell n \lceil \log n \rceil / 2^{\ell-1} \leq 3n \lceil \log n \rceil \cdot \frac{2 \log \log n}{\log n} = O(n \log \log n)$. □

Consequently, we have the following.

Theorem 1. *There is a polynomial time algorithm that given a graph G and a port numbering constructs a tree T spanning G in which $S_{up}(T) = O(n \log \log n)$.*

3 All-Ports Tree Labeling Schemes with Short Labels

Let us now turn our attention to S_{all} and M_{all}. Any spanning tree T enables the construction of a labeling $\mathcal{L}_{all}$ with average label size $O(\log \Delta)$ in graphs of maximum degree Δ. This is optimal in the sense that there are n-node graphs of maximum degree Δ and port numberings for which $S_{all}(T) = \Omega(n \log \Delta)$ for any spanning tree T. For instance, take a bipartite graph $G = (V_1, V_2, E)$ where $V_i = \{(i,x), x = 0, \ldots, n-1\}$, $i = 1,2$, and $\{(1,x),(2,y)\} \in E$ if and only if $(y-x) \bmod n \leq \Delta - 1$. Then, label any $\{(1,x),(2,y)\} \in E$ by $l = (y-x) \bmod n$ at $(1,x)$, and by $\Delta - l$ at $(2,y)$. For any tree T spanning G, at least one of the two labels at the extremity of every edge of T is larger than $\lfloor \Delta/2 \rfloor$, and therefore $S_{all}(T) \geq \Omega(n \log \Delta)$.

However, for many graphs, one can do better by selecting an appropriate spanning tree T. Assign a weight $\omega(l) + \omega(l')$, where $\omega(x) = 1$ for $x = 0$ and $\lfloor \log x \rfloor + 1$ for $x \geq 1$, to every edge e where l and l' are the port numbers of e at its two endpoints. It is easy to check that running any MST algorithm returns a tree T minimizing $S_{all}(T)$. Thus, we have the following.

Proposition 5. *There is a polynomial time algorithm that given a graph G and a port numbering constructs a tree T minimizing $S_{all}(T)$.*

On the other hand, by a reduction from the Hamiltonian path problem in 3-regular planar graphs, we have the following negative result.

Proposition 6. *The following decision problem is NP-hard.*
  Input: A graph G with a port numbering, and an integer k;
  Question: Is there a spanning tree T of G satisfying $M_{all}(T) \leq k$.
This result holds even restricted to cubic planar graphs, and even for fixed $k = 3$.

Obviously, one way to obtain a tree T with small $M_{all}(T)$ is to construct a spanning tree with small maximum degree. Finding a spanning tree with the smallest maximum degree δ_{min} in an arbitrary graph G is NP-hard. However, it is known (cf. [9]) that a spanning tree with maximum degree at most $\delta_{min} + 1$ can be computed in polynomial time. Hence we have the following.

Theorem 2. *There is a polynomial time algorithm that given a graph G and a port numbering constructs a spanning tree T for G satisfying $M_{all}(T) = O(\delta_{min} \log \Delta)$.*

On the other hand, any tree T^* minimizing M_{all} in a graph G has a degree $\Delta_{T^*} \geq \delta_{min}$. Thus $M_{all}(T^*) \geq \sum_{i=1}^{\Delta_{T^*}} \log i \geq \sum_{i=1}^{\delta_{min}} \log i \geq \Omega(\delta_{min} \log \delta_{min})$. Hence we obtain a polynomial time approximation of the optimal tree for $\mathcal{L}_{all}$, up to a multiplicative factor of $O(\log \Delta / \log \delta_{min})$.

4 Applications of Tree Labeling Schemes

Let us now discuss the applicability of our tree representation schemes in various application domains, mainly in the context of distributed network algorithms. Hereafter we consider an n-vertex m-edge graph G of maximum degree Δ, such that the smallest maximum degree of any spanning tree for G is δ_{min}.

4.1 Information Dissemination on Spanning Trees

A number of fundamental distributed processes involve collecting information upwards or disseminating it downwards over a spanning tree of the network. Let us start with applications of our tree representation schemes for these operations.

Broadcast. The broadcast operation requires disseminating an information item initially available at the root to all the vertices in the network. Given a spanning tree of the graph, this operation can be performed more efficiently than by the standard flooding mechanism (cf. [11], [1], [12]). Specifically, whereas flooding requires $O(m)$ messages, broadcasting on a spanning tree can be achieved using only $O(n)$ messages.

Broadcast over a spanning tree can be easily performed given an all-ports tree representation scheme, with no additional communication overheads. Consider the overall memory requirements of storing such a representation. Using an arbitrary spanning tree may require a total of $O(n \log \Delta)$ memory bits throughout the entire network and a maximum of $O(\Delta \log \Delta)$ memory bits per node. In contrast, using the constructions of Property 5 or Theorem 2, respectively, yields the following bounds.

Corollary 1. *For any graph G, it is possible to construct an all-port spanning tree representation using either optimal total memory over the entire graph or maximum memory $O(\delta_{min} \log \Delta)$ per node, in a way that will allow performing a broadcast operation on the graph using $O(n)$ messages.*

Upcast and Convergecast. The basic *upcast* process involves collecting information upwards to the root over a spanning tree. This task is rather general, and refers to a setting where each vertex v in the tree has an input item x_v and it is required to communicate all the different items to the root. Analysis and applications of this operation can be found, e.g., in [12]. Any representation for supporting such operation must allow each vertex to know its parent in the tree.

Again, using an arbitrary spanning tree may require a total of $O(n \log \Delta)$ memory bits throughout the network. Observe, however, that the upcast process does not require knowing the children so it can be based on an upwards tree representation scheme. Given such a representation, the upcast process can be implemented with no additional overheads in communication. Hence using the construction of Theorem 1 we get the following.

Corollary 2. *For any graph G, it is possible to construct an upwards tree representation using $O(n \log \log n)$ memory bits over the graph in a way that will allow performing an upcast on the graph using $O(n)$ messages.*

A more specialized process, known as the *convergecast* process, involves collecting information of the same type upwards over a spanning tree. This process may include the computation of various types of global functions. Suppose that each vertex v in the graph holds an input x_v and we would like to compute some global function $f(x_{v_1}, \ldots, x_{v_n})$ of these inputs. Suppose further that f is a *semigroup function*, namely, it enjoys the following two properties: (1) $f(Y)$ is well-defined for any subset $Y \subseteq \{x_{v_1}, \ldots, x_{v_n}\}$ of the inputs, (2) f is associative and commutative.

A semigroup function f can be computed efficiently on a tree T by a convergecast process, in which each vertex v in the tree sends upwards the value of the function on the inputs of the vertices in its subtree T_v, namely, $f_v = f(X_v)$ where $X_v = \{x_w \mid w \in T_v\}$. An intermediate vertex v with k children $w_1, \ldots, w_k$ computes this value by receiving the values $f_{w_i} = f(X_{w_i})$, $1 \leq i \leq k$, from its children, and applying $f_v \leftarrow f(x_v, f_{w_1}, \ldots, f_{w_k})$, relying on the associativity and commutativity of f. The message and time complexities of the convergecast algorithm on a tree T are $O(n)$ and $O(Depth(T))$, respectively, matching the obvious lower bounds. For a more detailed exposition of the convergecast operation and its applications see [12].

Observe that the convergecast process requires each vertex to receive messages from all its children before it can send a message upwards to its parent. This implies, in particular, that a vertex needs to know the number of children it has in the tree. This means that when using the spanning tree T, the label size at each node v has another component of $\log(\deg_T(v))$. Hence the maximum label size increases by $\log \delta_{min}$, and the average label size increases by $\frac{1}{n} \sum_v \log(\deg_T(v)) = O(1)$.

Here, too, using an arbitrary spanning tree would require a total of $O(n \log \Delta)$ memory bits throughout the network. In contrast, using the construction of Theorem 1 we get the following.

Corollary 3. *For any graph G, it is possible to construct an upwards tree representation using a total of $O(n \log \log n)$ memory bits over the entire graph in a way that will allow performing a convergecast operation on the graph in time at most $Diam(G)$ using $O(n)$ messages.*

4.2 Fast Graph Exploration

Graph exploration is an operation carried out by a finite automaton, simply referred to in this context as a *robot*, moving in an unknown graph $G = (V, E)$. The robot has no a priori information about the topology of G and its size. The robot can distinguish between the edges of the currently visited node by their port numbers. The robot has a transition function f, and a finite number of states. If the robot enters a node of degree d through port i in state s, then it switches to state s' and exits the node through port i', where $(s', i') = f(s, i, d)$. The objective of the robot is to *explore* the graph, i.e., to visit all its nodes.

The tree labeling schemes allow fast exploration. Specifically, the all-ports labeling scheme $\mathcal{L}_{all}$ allows exploration to be performed in time at most $2n$ in n-node graphs. The upward labeling scheme $\mathcal{L}_{up}$ allows exploration to be performed in time at most $4m$ in m-edge graphs.

More compact labeling schemes can be defined for graph exploration. In particular, [3] describes a labeling scheme using only 2 bits per node. However, this latter scheme yields slower exploration protocols, i.e., ones requiring $20m$ steps in m-edge graphs.

Suppose our graph G has a spanning tree T. As a consequence of [6], if the labels allow the robot to infer at each node v, for each edge e incident to v in G, whether e belongs to T, then it is possible to traverse G perpetually, and traversal is ensured after time at most $2n$. Indeed, the exploration procedure in [6], which applies to trees only, specifies that when the robot enters node v by port i, it leaves the node by port $(i+1) \bmod d$ where $d = \deg(v)$. In the case of general graphs, exploration is performed as follows. When the robot enters node by port i, it looks for the first j in the sequence $i+1, i+2, \ldots$ such that port $j \bmod d$ is incident to a tree-edge and leaves the node by port $j \bmod d$.

Clearly, this exploration procedure performs a DFS traversal of T. Hence, as a corollary of [6], using the all-ports labeling scheme $\mathcal{L}_{all}$, we get the following.

Corollary 4. *It is possible to label the nodes of every graph G in polynomial time, with labels of maximum size $O(\delta_{min} \log \Delta)$ and average size $O(\log \Delta)$, in a way that will allow traversal of the graph in time $2n$ by a robot with no memory.*

The following result shows that exploration can be performed with smaller labels, using the upward labeling scheme on a spanning tree of the graph.

Lemma 2. *Consider a node-labeled m-edge graph G, with a rooted spanning tree T. It is possible to perform traversal of G within time at most $4m$, terminating at the root of T.*

By Lemma 2, using a labeling $\mathcal{L}_{up}$ on an arbitrary spanning tree and relying on Lemma 1 and Theorem 1, we get the following.

Corollary 5. *It is possible to label the nodes of every graph G with labels of maximum size $O(\log \Delta)$ and average size $O(\log \log n)$ in a way that will allow traversal of the graph in time at most $4m$.*

By Lemma 1, the scheme uses labels of *total* size at most $\sum_v \lceil \log \deg(v) \rceil$. This means, in particular, that in graph families with a linear number of edges, such as planar graphs, the average label size for any spanning tree is $O(1)$.

References

1. H. Attiya and J. Welch. *Distributed Computing: Fundamentals, Simulations and Advanced Topics*. McGraw-Hill, 1998.
2. Y. Chu and T. Liu. On the shortest arborescence of a directed graph. *Science Sinica* 14, pages 1396–1400, 1965.
3. R. Cohen, P. Fraigniaud, D. Ilcinkas, A. Korman and D. Peleg. Label-Guided Graph Exploration by a Finite Automaton. In *Proc. 32nd Int. Colloq. on Automata, Languages & Prog. (ICALP)*, LNCS 3580, pages 335–346, 2005.
4. T.H. Cormen, C.E. Leiserson, and R.L. Rivest. *Introduction to Algorithms*. MIT Press/McGraw-Hill, 1990.
5. A. Czumaj and W.-B. Strothmann. Bounded-degree spanning tree. In *Proc. 5th European Symp. on Algorithms (ESA)*, LNCS 1284, pages 104–117, 1997.
6. K. Diks, P. Fraigniaud, E. Kranakis, and A. Pelc. Tree Exploration with Little Memory. In *Proc. 13th Ann. ACM-SIAM Symp. on Discrete Algorithms (SODA)*, pages 588–597, 2002.
7. J. Edmonds. Optimum branchings. *J. Research of the National Bureau of Standards* 71B, pages 233–240, 1967.
8. P. Elias. Universal Codeword Sets and Representations of the Integers. *IEEE Trans. Inform. Theory* 21(2):194–203, 1975.
9. M. Fürer and B. Raghavachari. Approximating the minimum degree spanning tree within one from the optimal degree. In *Proc. 3rd Ann. ACM-SIAM Symp. on Discrete Algorithms (SODA)*, pages 317–324, 1992.
10. M. Garey, D. Johnson, and R. Tarjan. The planar Hamiltonian circuit is NP-complete. *SIAM Journal on Computing* 5(4):704–714, 1976.
11. N. Lynch. *Distributed Algorithms*. Morgan Kaufmann, 1995.
12. D. Peleg. *Distributed Computing: A Locality-Sensitive Approach*. SIAM, 2000.

Single-Bit Messages are Insufficient in the Presence of Duplication[*]

Kai Engelhardt[1] and Yoram Moses[2],[**]

[1] School of Computer Science and Engineering
The University of New South Wales, and NICTA[***]
Sydney, NSW 2052, Australia
kaie@cse.unsw.edu.au

[2] Department of Electrical Engineering
Technion, Haifa, 32000 Israel
moses@ee.technion.ac.il

Abstract. Ideal communication channels in asynchronous systems are reliable, deliver messages in FIFO order, and do not deliver spurious or duplicate messages. A message vocabulary of size two (i.e., single-bit messages) suffices to encode and transmit messages of arbitrary finite length over such channels. This note proves that single-bit messages are insufficient once channels potentially deliver duplicate messages. In particular, it is shown that no protocol allows the sender to notify the receiver which of three values it holds, over a bidirectional, reliable, FIFO channel that may duplicate messages. This implies that messages must encode some additional control information, e.g., in the form of headers or tags.

1 Introduction

Ideal communication channels in asynchronous systems are reliable, deliver messages in FIFO order, and do not deliver spurious or duplicate messages. Single-bit messages suffice to encode and transmit messages of arbitrary finite length over unidirectional channels of this type. When only the FIFO requirement is relaxed (so that messages may be reordered), the same can be achieved over a bidirectional channel. Fekete and Lynch proved that reliable end-to-end communication (data link) is impossible for (fair) lossy FIFO channels without messages containing header information [5]. The results of Wang and Zuck show that, in non-FIFO models with duplication or loss, reliable end-to-end communication is impossible unless there are more different packet types than there are different potential messages sequences to transmit [8]. We consider the impact of duplication, and prove a result closely related to Fekete and Lynch for a seemingly better-behaved model we call RELDFI. Namely, we show that no protocol allows

---

[*] Work was partially supported by ARC Discovery Grant RM02036.
[**] Work on this paper happened during a sabbatical visit to the School of Computer Science and Engineering, The University of New South Wales, Sydney, NSW 2052, Australia.
[***] National ICT Australia is funded through the Australian Government's *Backing Australia's Ability* initiative, in part through the Australian Research Council.

the sender to notify the receiver which of three values it holds, over an asynchronous, bidirectional, reliable, FIFO channel that may duplicate messages. While single-bit protocols exist for transmitting a binary value over a duplicating channel, our result implies that these cannot be composed to implement a data-link layer, without using a larger set of message types. Intuitively, to transmit more complex messages or to implement a data-link layer, messages must encode some additional control information, e.g., in the form of headers or tags. A general theory of composition for this model, in which messages are assumed to have headers, is presented in [3].

This note is devoted to proving the following theorem.

Theorem. *Let P be a protocol for two processes that uses only single-bit messages over a single bi-directional, finitely-duplicating FIFO channel between the sender S and the receiver R. Then P cannot transmit more than two distinct values from S to R.*

Since data-link layers enable the transmission of all finite sequences of bits, our theorem yields

Corollary 1. *No data-link protocol exists in the model of the previous theorem.*

2 Preliminary Definitions

Processes and local runs. We consider systems consisting of two processes, a *sender*, S, and a *receiver*, R. We let X range over $\{S, R\}$ and denote by $\overline{X}$ the other process. Each process X has a set Σ_X of initial states, a message set M_X, and a set A_X of *internal actions*. The *moves of* X consist of its internal actions A_X as well as *send actions* $\mathsf{snd}(m)$ for $m \in M_X$. An *event (of X)* is a move of X or a *delivery* $\mathsf{dlv}(m)$ of a message $m \in M_{\overline{X}}$ sent by $\overline{X}$.

A *local run of X* is an infinite sequence $x = \langle v, e_0, \ldots \rangle$ where $v \in \Sigma_X$ and the e_i events of X, infinitely many of which are moves of X (and the remaining ones are deliveries to X). This assumption prevents crash behavior or denial-of-service scenarios. A *run* $r = (s, l, \delta)$ consists of local runs s and l of S and R, respectively, and a *matching function* δ mapping delivery events to send events. More formally, δ is a function from pairs (X, j) to indices k where the j'th event in x is a delivery and the k'th event in the other local state $\overline{x}$ is a send of the same message. Moreover, δ satisfies:

Interleaving. There exists a total ordering of all events in s and l extending the orders of events in s and l such that $\delta(e)$ precedes e, for all e in the domain of δ.
FIFO. δ is monotone, i.e., for $j < k \in \mathbb{N}$ if both e_j and e_k are delivery events to X then $\delta(X, j) \leq \delta(X, k)$. This prevents re-ordering of messages.
Reliability. δ is surjective, in other words, every send is related to at least one delivery. This prevents message loss.
Finite Duplication. Every send event is related by δ to only finitely many deliveries. This prevents infinite duplication of messages.

Observe that our assumption that δ is a total function prevents spurious message from being delivered.

Local states. A *local state* of X is a non-empty finite prefix $x(k) = \langle v, e_0, \ldots, e_{k-1} \rangle$ of a local run $x = \langle v, e_0, \ldots \rangle$ of X. Observe that no information is discarded from the local state of a process over time. Hence, processes have *perfect recall* and thus, in a precise sense, accumulate knowledge as efficiently as possible.[1]

Protocols. A *protocol P* associates with each process a function from that process's local states to its actions. In particular, the behavior of processes is deterministic.[2] A *run of P* is a run $r = (s, l, \delta)$ where, for each process X and $k \in \mathbb{N}$, the $k + 1$'st event in X's local state $x \in \{s, l\}$ is either a delivery or an occurrence of the action $P(X)(x(k))$ prescribed by the protocol for the preceding local state $x(k)$. These definitions imply that processes cannot prevent messages from being delivered to them. Thus they are *input-enabled* in the sense of Lynch and Tuttle [7].

Executions. The crux of the proof of our impossibility result will consist of the construction of runs as limits of chains of finite approximations of runs, which we call finite runs. A *finite run of P* is a triple (a, b, β) where a and b are local states of S and R, respectively, and β is a matching function restricted to these local states, that is, it maps delivery events in a and b to send events in b and a, respectively. Moreover, β satisfies the conditions called **Interleaving**, **FIFO**, and **Finite Duplication**, but not necessarily **Reliability** from above, with a, b, and β substituted for s, l, and δ, respectively. One finite run (a', b', β') is a prefix of another (a, b, β) if a' and b' are prefixes of a and b, respectively, and $\beta' \subseteq \beta$. A *chain* is a sequence $(c_i)_{i \in \mathbb{N}}$ of finite runs where c_i is a prefix of c_{i+1} for all $i \in \mathbb{N}$.

Knowledge. For a given protocol P, we can talk about what processes *know*[3] w.r.t. P by considering the set of all runs of P. Specifically, we say that the receiver *knows the sender's initial value*, denoted by $K_R v$, at a local state b (w.r.t. P) if there exists a value $v \in \Sigma_S$ such that in every run of P in which the state b appears, the sender's initial state is v. Thus, the fact that R is in state b implies that the sender's value is necessarily v. We say that a protocol P *transmits n values* if $|\Sigma_S| = n$ and in every run of P the receiver eventually knows the sender's initial value. Formally, this is expressed as: for all runs $r = (s, l, \delta)$ of P there exists $k \in \mathbb{N}$ such that for all runs $r' = (s', l', \delta')$ of P satisfying $l(k) = l'(k)$ we have that $s(0) = s'(0)$.

Our main result can now be rephrased as: *If $|M_S| = 2$ then no protocol can transmit 3 values in* RELDF1. The remainder of the paper is devoted to the proof of this theorem.

3 Proof of the Theorem

Let $|\Sigma_S| = 3$, let $M_S = \{0, 1\}$, and, w.l.o.g., assume that Σ_R is a singleton set. Fix a protocol P and assume, by way of contradiction, that P transmits three values. All finite runs and runs mentioned will be ones of P.

---

[1] For the purpose of proving an impossibility result, perfect recall is preferred over a more explicit notion of local state based on variables. Any modifications to a more general form of local state can be simulated based on the protocol, initial state, and messages received [2].

[2] The restriction to deterministic protocols is again motivated by the kind of result we are after. Should a non-deterministic protocol P solve a transmission problem reliably then so does any deterministic protocol compatible with P.

[3] Our notion of knowledge here coincides with the formal notion of knowledge in the sense of [6, 4].

Lemma 2. *Every finite run can be extended to a run.*

Proof. Let $c = (s, l, \delta)$ be a finite run of P. For $X \in \{S, R\}$ let $\langle m_0^X, \ldots, m_{i_X}^X \rangle$ be the sequence of messages sent by $\overline{X}$ in c outside the range of δ (i.e., not yet delivered in c). Define[4] $c' = (s \frown \tau_S, l \frown \tau_R, \delta \cup \delta_S \cup \delta_R)$, where τ_X is $\langle \mathsf{dlv}(m_0^X), \ldots, \mathsf{dlv}(m_{i_X}^X) \rangle$ and δ_X matches the k'th of these deliveries to the k'th unmatched send of $\overline{X}$ in c. Construct the run r as the limit of the sequence of finite runs $(c_i)_{i \in \mathbb{N}}$ defined as follows. Let $c_0 = c'$ and obtain c_{k+1} inductively from c_k by having each process make the move prescribed by P, and if that move is a send event then a delivery of this message appears immediately after the current move of the other process. The limit r of the c_i is indeed a run of P. □

Lemma 3. *Let $r = (s, l, \delta)$ be a run of P. If $\mathsf{K_R} v$ holds at $l(k)$ then $l(k)$ contains a delivery.*

Proof. Let $r = (s, l, \delta)$ be a run and let $k \in \mathbb{N}$ such that $l(k)$ does not contain a delivery. Notice that $l(k)$ is uniquely determined by k. For each $v \in \Sigma_S$, construct the finite run, $c^{(v)} = (s^{(v)}, l^{(v)}, \delta^{(v)})$ by performing k moves for the sender and the receiver but without delivering a single message should any be sent. Each $c^{(v)}$ can be extended to a run by Lemma 2. Observe that each receiver state $l^{(v)}$ equals $l(k)$. It follows that $\mathsf{K_R} v$ does not hold at $l(k)$. □

A delivery event e to R in a run $r = (s, l, \delta)$ of P is called an *alternation* either if it is the first delivery to R or if its content is distinct from that of the preceding delivery to R. We also call a send event by S an alternation if the earliest delivery matched to it is an alternation. In particular, the first send by S and the first delivery to R are alternations.

Proof (of the theorem). We construct a pair of chains $(c_i)_{i \in \mathbb{N}}$ and $(d_i)_{i \in \mathbb{N}}$ of finite runs of P with different initial sender states but identical local states for R in each pair (c_i, d_i). Let $i \in \mathbb{N}$ and let l_i be R's local state in both c_i and d_i. Since c_i and d_i are finite runs, each of them can be extended to a run by Lemma 2. Since the sender has different initial states in these runs, $\mathsf{K_R} v$ does not hold at l_i. As we shall show, the limit of at least one of these chains is a run. In that run the sender's value is never transmitted, contradicting the assumption that P transmits three values.

Outline of the proof: Our first step is to find two values for which the first message sent by the sender is the same. Then, we generate the two chains $(c_i)_{i \in \mathbb{N}}$ and $(d_i)_{i \in \mathbb{N}}$ of finite runs starting from these two sender values, respectively. The intuition underlying the second step is as follows. We maintain an invariant that in c_i and d_i the receiver has the same local state and is scheduled to move at the same local states (which will occur at odd steps of our construction). Since the protocol P is deterministic, R performs the same actions in both chains. Moreover, every message sent by R is delivered immediately. More delicate is the handling of the sender S, whose moves occur at even steps of the construction. If P prescribes the same move for S in both finite runs, then this move is taken, and, if the move is a send, the message is delivered to R. If S is prescribed a send in one finite run that repeats the most recent value delivered to R, then this message is delivered to R and is regarded by δ as a duplicate delivery in the finite run in

---

[4] Given two sequences σ and τ, we use $\sigma \frown \tau$ to denote the result of appending τ at the end of σ.

which the message was not sent. Finally, if S should send an alternation in one of the finite runs (say c_i) but not in the other, then this message is delayed and the sender is suspended in the corresponding (say c) chain. From this point on, in even steps of the construction, S moves only in the finite runs in which it is not suspended (d), until an alternation is sent by S there. In case this never happens, the limit of the chain in which S continues to move is a legal run in which the value is never transmitted. Indeed, S is guaranteed to move infinitely often in at least one of the chains (possibly both), and such a chain will yield the desired contradiction. To make the above intuition precise, we shall use a simple automaton to help determine in which of the chains S should move at even steps of the construction.

Step 1: Fix $\lambda \in \Sigma_R$. This will R's initial state in all finite runs and runs considered from now on. For each of S's three initial states, we start a finite run of P and stop it as soon as S sends its first message. Until then, both S and R move in lock step. Every message sent by R in, say, step k is delivered to S right after its k'th move. We claim that the sender eventually sends a message in each of these finite runs. Assume by way of contradiction that in one such finite run e the sender does not send any messages. Observe that e contains infinitely many moves by both processes and every message sent is delivered. Thus e is a run. By Lemma 3, however, $K_R v$ never holds in e and hence the value is not transmitted. Since the messages sent by S are single bits, in the finite runs starting from at least two of the three values, say v and w, the first message sent by S is the same.

Step 2: Next we construct two chains of finite runs c_i and d_i with initial sender values v and w, respectively. In each step i of the construction, we define two finite runs, $c_i = (s_i, l_i, \delta_i)$ and $d_i = (s'_i, l_i, \delta'_i)$ in which R's state is the same. Initially, $s_0 = \langle v \rangle$, $s'_0 = \langle w \rangle$, $l_0 = \epsilon$, and $\delta_0 = \delta'_0 = \emptyset$. The whole construction is symmetric. We focus on constructing c_i. We distinguish odd-numbered steps from even-numbered ones.

Odd-numbered steps: A step $i = 2k+1$ of the construction contains a move by R. If that move is a send then the step also contains a delivery of that message to S. More formally, let $e = P(R)(l_{i-1})$. Define $l_i = l_{i-1} \frown \langle e \rangle$. If e is not a send then $s_i = s_{i-1}$ and $\delta_i = \delta_{i-1}$. Otherwise, if e is $\mathsf{snd}(b)$ then $s_i = s_{i-1} \frown \langle \mathsf{dlv}(b) \rangle$ and $\delta_i = \delta_{i-1} \cup \{(S, |s_i|) \mapsto |l_i|\}$.

Even-numbered steps: A step $i = 2k+2$ of the construction handles a move by S. In this case, however, S might perform a move in just one of the finite runs, or in both.

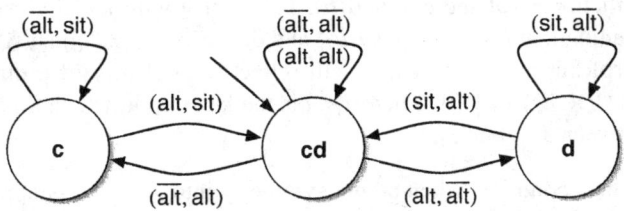

Fig. 1. The construction automaton A

Who moves and how is determined by an auxiliary 3-state automaton and by P. The state σ_i of the automaton in step i is one of **c**, **d**, and **cd**, where the occurrence of a letter in a state's name indicates that the sender moves in the corresponding finite run. (See Fig. 1.) For instance, if $\sigma_{i-1} = $ **c** then the sender only moves between c_{i-1} and c_i but not between d_{i-1} and d_i. The initial state of the automaton is **cd**. Odd moves do not affect the automaton state, i.e., $\sigma_{2k+1} = \sigma_{2k}$ for all k.

It is convenient to consider the sender's behavior m in the step from c_{i-1} to c_i, depending on $e = P(S)(s_{i-1})$ and σ_{i-1}, to be one of {alt, skip, rpt, sit}. Intuitively, alt stands for the receipt of an alternation; skip stands for an internal action not involving communication; rpt indicates the receipt of a message that is not an alternation; sit means that this sender does not participate in the current step. If $\sigma_{i-1} = $ **d** then $m = $ sit. Otherwise we define m as follows. If $e = $ skip then $m = $ skip. If $e = $ snd(b) and this send is an alternation then $m = $ alt. Otherwise this send repeats the preceding message, whence we define $m = $ rpt. We define m' based on $e' = P(S)(s'_{i-1})$ and σ_{i-1} analogously.

The transition function of the automaton is described in Fig. 1. Its transitions are labeled with pairs, the first component of which describes m and the second describes m', where alt stands for skip or rpt.

We can now specify the i'th step of the construction based on m, m' and σ_{i-1} as follows.

- If $m = $ sit then $s_i = s_{i-1}$ and otherwise $s_i = s_{i-1} \frown \langle e \rangle$.
- If $\sigma_{i-1} = $ **cd**, $m = m' = $ alt, and $e = $ snd(b), then the alternation is delivered immediately in both chains, that is, $l_i = l_{i-1} \frown \langle \text{dlv}(b) \rangle$, $\delta_i = \delta_{i-1} \cup \{(R, |l_i|) \mapsto |s_i|\}$, and δ'_i is obtained analogously.
- If $\sigma_{i-1} = $ **cd** and $m = $ alt but $m' \neq $ alt then the alternation is not delivered immediately but the sender is suspended from making moves in the following s_j by the automaton entering state **d**. As long as no alternation is encountered in the following s'_j, the automaton state **d** is preserved. When a matching alternation occurs, the pending message is finally delivered, as is the matching alternation, and the automaton returns to **cd**. Formally, if $\sigma_{i-1} = $ **cd**, $m = $ alt, and $m' = $ skip, then $l_i = l_{i-1}$ and $\delta_i = \delta_{i-1}$.
- If $\sigma_{i-1} = $ **d**, $m' = $ alt, and $e' = $ snd(b) then $l_i = l_{i-1} \frown \langle \text{dlv}(b) \rangle$ and δ_i will reflect the delivery of the pending alternation, that is, $\delta_i = \delta_{i-1} \cup \{(R, |l_i|) \mapsto |s_j|\}$, where $j < i$ is the last step in which S moves between s_{j-1} and s_j. Moreover, $\delta'_i = \delta'_{i-1} \cup \{(R, |l'_i|) \mapsto |s'_i|\}$.
- If $m = $ skip and $m' \in $ {sit, skip} then $l_i = l_{i-1}$ and $\delta_i = \delta_{i-1}$.
- Suppose that $m = $ rpt and $e = $ snd(b). This repeat will be delivered to R immediately, meaning that $l_i = l_{i-1} \frown \langle \text{dlv}(b) \rangle$ and $\delta_i = \delta_{i-1} \cup \{(R, |l_i|) \mapsto |s_i|\}$.
- If $m' = $ rpt but $m \neq $ rpt then δ_i will reflect the delivery of a duplicate, that is, $\delta_i = \delta_{i-1} \cup \{(R, |l_i|) \mapsto |s_j|\}$, where $j < i$ is the last step in which S performs a snd() action between s_{j-1} and s_j.

The description is complete when taking symmetry into account, swapping the roles of s_i, δ_i, e, **d**, and m with s'_i, δ'_i, e', **c**, and m', respectively.

Let $c = \lim_i c_i$ and $d = \lim_i d_i$. Observe that the construction has established the following properties.

1. Both c_i and d_i are finite runs of P, for every i.
2. The receiver moves infinitely often in both c and d.
3. The sender moves infinitely often in at least one of c and d, because every even-numbered step of the construction contributes such a move to at least one of them.
4. All messages sent by the receiver are delivered.
5. In both, c and d, if the sender moves in step i then all messages it sent earlier have been delivered. It follows that once the sender performs infinitely many moves, all of its messages are delivered.

It follows that at least one of c and d is a run of P and that run the sender's value is never transmitted. □

The theorem is as strong as can be expected, since two values can trivially be transmitted in RELDFI using a one-bit message. Moreover, it is straightforward to show that a message set of size 3 suffices to transmit arbitrary values, as well as infinite sequences of values. (One message can serve as a delimiter.) The Alternating-Bit Protocol transmits arbitrary sequences of bits using 4 different messages in all FIFO models we consider [1].

Acknowledgment

We thank Alan Fekete for his helpful comments on an earlier draft.

References

1. K. A. Bartlett, R. A. Scantlebury, and P. T. Wilkinson. A note on reliable full-duplex transmission over half-duplex links. *Communications of the ACM*, 12(5):260–261, 1969.
2. B. Coan. A communication-efficient canonical form for fault-tolerant distributed protocols. In *Proc. 5th ACM Symp. on Principles of Distributed Computing*, pages 63–72. ACM Press, 1986.
3. K. Engelhardt and Y. Moses. Safe composition of distributed programs communicating over order-preserving imperfect channels. In *this volume*.
4. R. Fagin, J. Y. Halpern, Y. Moses, and M. Y. Vardi. *Reasoning About Knowledge*. MIT-Press, 1995.
5. A. Fekete and N. Lynch. The need for headers: An impossibility result for communication over unreliable channels. In J. C. M. Baeten and J. W. Klop, editors, *CONCUR '90: Theories of Concurrency: Unification and Extension*, volume 458 of *LNCS*, pages 199–215. Springer-Verlag, 1990.
6. J. Y. Halpern and Y. Moses. Knowledge and common knowledge in a distributed environment. *Journal of the ACM*, 37(3):549–587, July 1990.
7. N.A. Lynch and M. Tuttle. An introduction to input/output automata. *CWI Quarterly*, 2(3):219–246, Sept. 1989.
8. D.-W. Wang and L. D. Zuck. Tight bounds for the sequence transmission problem. In *PODC '89: Proceedings of the eighth annual ACM Symposium on Principles of Distributed Computing*, pages 73–83. ACM Press, 1989.

Safe Composition of Distributed Programs Communicating over Order-Preserving Imperfect Channels*

Kai Engelhardt[1] and Yoram Moses[2],**

[1] CSE, UNSW, and NICTA***, Sydney, NSW 2052, Australia
[2] Department of Electrical Engineering, Technion, Haifa 32000, Israel

Abstract. The fundamental question considered in this paper is when program Q, if executed immediately after program P, is guaranteed not to interfere with P and be safe from interference by P. If a message sent by one of these programs is received by the other, it may affect and modify the other's execution. The notion of *communication-closed layers (CCLs)* introduced by Elrad and Francez in 1982 is a useful tool for studying such interference. CCLs have been considered mainly in the context of reliable FIFO channels (without duplication), where one can design programs layers that do not interfere with any other layer. When channels are less than perfect such programs are no longer feasible. The absence of interference between layers becomes context-dependent. In this paper we study the impact of message duplication and loss on the safety on the safety of layer composition. Using a communication phase operator, the *fits after* relation among programs is defined. If program Q fits after P then P and Q will not interfere with each other in executions of $P*Q$. For programs P and Q in a natural class of programs we outline efficient algorithms for the following: (1) deciding whether Q fits after P; (2) deciding whether Q *seals* P, meaning that Q fits after P and no following program can communicate with P; and (3) constructing a *separator S* that both fits after P and satisfies that Q fits after $P*S$.

1 Introduction

Much of the distributed algorithms literature is devoted to solutions for individual tasks. Often, the issues involved in solving an individual task in models of interest deserve considerable attention and may involve a great deal of work. In large distributed applications, subtasks often need to be implemented so that one task can make use of the output of another, in a sequential manner. For example, a leader election algorithm may precede a vote coordinated by the chosen leader, and the result of this vote may determine updates to local copies of a distributed database. While these tasks may be partially concurrent, each uses the outcome of preceding tasks, and the overall logical

---

* Work was partially supported by ARC Discovery Grant RM02036.
** Work on this paper happened during a sabbatical visit to the School of Computer Science and Engineering, The University of New South Wales, Sydney, NSW 2052, Australia.
*** National ICT Australia is funded through the Australian Government's *Backing Australia's Ability* initiative, in part through the Australian Research Council.

structure is clearly sequential. Nevertheless, efficiency in a distributed setting is obtained via concurrency, and this precludes strictly sequential execution of the tasks, in which a task does not start executing at any node before all previous tasks have completed at all nodes. A high degree of concurrency imposes the need to guarantee non-interference among concurrent tasks. In current technologies this is often obtained by way of a data-link layer that provides an abstraction of reliable FIFO communication. In effect, such an abstraction layer implements a fresh set of virtual channels for every subtask, and incurs high translation costs for every message. A basic question is whether and when the overhead paid for such a layer is justified. Under what assumptions can we achieve safe composition of programs at a significantly reduced cost? In a previous work [4] we studied safe composition in the REL model of communication, when channels are reliable and non-duplicating but may reorder messages. It was shown that safe composition can be achieved in REL at a much smaller price than required by a data-link layer implementation, or the use of global barriers. The current paper is devoted to a study of program composition in models with order-preserving (FIFO) channels that are imperfect due to possible message duplication and/or message loss.

Sequential ordering (or *layering*) of programs for asynchronous message passing systems has received a modest amount of attention in the literature, and this has focused almost exclusively on the RELFI model in which channels are reliable, non-duplicating, and FIFO [3], [1], [18]. Indeed, Gerth and Shrira showed that in RELFI it is possible to design off-the-shelf components that are guaranteed to safely compose with each other [7]. Once any of the assumptions about the channels are relaxed, such components are no longer viable. Recall that the definition of CCL refers to the full program context of the layer in question.

This paper considers composition in models with asynchronous order-preserving (aka FIFO) channels. These consist of FFI (in which the FIFO channels are lossy but fair and non-duplicating), RELDFI (reliable channels with finite duplication), and FDFI (lossy but fair channels with finite duplication). As we show, in these models it is possible to safely compose layers by ensuring compatibility at the interfaces between adjacent layers, without requiring an analysis of the full program context. We shall analyze composition in the RELDFI model first, and then show that a similar analysis applies to the other two models.

Communication-Closed Layers. We consider a *layering* operator $*$ for composing distributed programs in the spirit of Janssen, Poel, and Zwiers [10]. Pratt defines a similar operator on pomsets which he calls *local concatenation* [17]. Our layering operator $*$ acts as sequential composition between programs of the same process and as parallel composition between programs of different processes. Thus, if P and P' are both programs of the same process i, then the actions of P are executed before those of P' in the program $P * P'$. If Q is a program of a process $j \neq i$, then $P * Q$ is a program in which P and Q execute concurrently. Layering does not impose any precedence constraints between actions in P and those of Q. Hence the program $P * Q$ is equivalent to $Q * P$. Thus, for general programs S and T, the composition $S * T$ results, for every process i, in all actions of i induced by S preceding those actions of i induced by T.

Because of the potential concurrency of layers in programs constructed using layered composition, it is essential to ensure that messages from one layer are not mistakenly considered as being sent by a different layer. The notion of a communication-closed layer, introduced by Elrad and Francez in [3], is a useful tool for studying such interaction and interference between components of a distributed program. Let $L = (L_1 * \ldots * L_n)$ where each L_i is a program of process i. Let $P = Q * L * Q'$. Then L is a *communication-closed layer (CCL) in P* if no command in any L_i ever communicates with a command in Q or Q' in executions of P. If a program P can be decomposed into a sequence of CCLs then every execution of P consists of a sequence of executions of P's layers in order. Hence, reasoning about P can be reduced to reasoning about each of its layers in isolation. This approach has been investigated further and applied to a variety of problems [1], [10], [16], [8], [18], [9], [13], [14]. An introductory exposition to the proof technique based on communication-closed layers can be found in [2].

Safe layering. We defined in [4] that a program Q *seals* a program P if the two do not communicate and furthermore no succeeding program R can communicate with P in any isolated execution of $P * Q * R$. Given a sequence $L^{(1)}, \ldots, L^{(m)}$ of programs, if $L^{(k+1)}$ seals $L^{(k)}$ for $1 \leq k < m$, then every layer $L^{(k)}$ will be a CCL in an isolated execution of $P = L^{(1)} * \ldots * L^{(m)}$. In this paper we introduce and characterize two additional properties of interest in this context. One is the notion of a program Q *fitting after* P, which means simply that Q does not communicate with P in executions of $P * Q$. Fits after can be used to construct CCLs in the following fashion. If $L^{(k+1)}$ fits after $(L^{(1)} * \ldots * L^{(k)})$ for $1 \leq k < m$, then every layer $L^{(k)}$ will be a CCL in an isolated execution of $P = L^{(1)} * \ldots * L^{(m)}$. While sealing is a more powerful notion of compatibility, fitting after is more appropriate when considering smaller program segments (as layers), since it does not require Q to cover all channels on which P can communicate. There are cases in which we need to compose programs P and Q that are incompatible and we do not have the freedom to modify them. In this case we can search for a *separator* between P and Q, which is a program S satisfying (1) S fits after P and (2) Q fits after $P * S$. Thus, in isolated executions of $P * S * Q$, all three will be CCLs.

Example 1 (Value transmit). Consider the problem of transmitting a value v by process i to process j in an asynchronous message-passing system. If communication is reliable, a simple solution is given by a program $\text{VT}_{i \to j}$ (for *value transmit* from i to j) consisting of a single send statement for i on the channel from i to j, and a single receive statement for j. How do things change if we wish to transmit two values? Intuitively, we can perform $\text{VT}_{i \to j}$ twice. This will indeed work if and only if the channel is reliable, communication from i to j is FIFO, and non-duplicating (RELFI).

Observe that $\text{VT}_{i \to j}$ will produce the desired behavior in RELFI if it is executed when the channel from i to j is empty. Indeed, when this occurs the channel will again be empty once $\text{VT}_{i \to j}$ has completed its execution. In a precise sense, the $\text{VT}_{i \to j}$ program can serve as an off-the-shelf component for being a CCL in a larger program. Once we go beyond RELFI, the situation becomes more complex. Executing $\text{VT}_{i \to j}$ in RELDFI when the channel is empty will successfully transmit the value from i to j. It does not, however, guarantee to leave the channel empty when it terminates. Indeed, in this model, once a message is sent over a channel, we can never again be sure that the channel is empty. Hence $\text{VT}_{i \to j}$ is not an off-the-shelf component in RELDFI. As we shall formalize in Th. 2 there are only non-communicating or diverging off-the-shelf components in this model.

The need for headers. Fekete and Lynch proved that reliable end-to-end communication is impossible in FFI (and thus in FDFI) without messages containing header information [6]. In [5] we show that it is impossible in RELDFI to reliably transmit more than a single bit by only exchanging one-bit messages between two processes. This implies that a message set consisting of at least three different packets is necessary to transmit sequences of bits reliably in RELDFI. In light of these results we shall restrict attention to tagged messages in the three FIFO models we consider here.

Contributions. The main contributions are: 1. We introduce the *fits-after* relation among programs. It suggests a new method for safely composing distributed applications from a sequence of distributed programs (or layers). 2. Based on the fits-after relation, we define *separators* between distributed programs. Since universal barriers do not exist in most models other than RELFI, such tailor-made separators are necessary for composing layers P and Q where Q does not fit after P. 3. We study safe composition of programs in the FIFO models RELDFI, FFI and FDFI. In each of them we define the signature of a program to capture the composition behavior (for a natural class of programs). We show: (a) that only trivial programs are tail communication closed, (b) how to characterize fits after and sealing, (c) that each pair of programs can be separated, and (d) that each program can be sealed properly.

2 Distributed Programs with Layering

In this section we define a syntax for distributed programs with layering, a run-based model of asynchronous message passing, a notion of a program occurring over an interval within a run, and a notion of refinement between programs. These technical preliminaries will be used in the next section to conduct an analysis of safe composition.

Let $n \in \mathbb{N}$ and $\mathbb{P} = \{1, \ldots, n\}$ be a set of processes. Throughout the paper, n will be reserved for denoting the number of processes. Let $(Var_i)_{i \in \mathbb{P}}$ be mutually disjoint sets of *program variables (of process i)*. Let $\mathcal{L}$ be propositional logic over the set of arithmetic expressions over Var_i and *tag expressions* $\text{PEEK}_{j \leftarrow i}$. We define a syntactic category Prg of *programs*, where $x \in Var_i$, $e \in Expr_i$, $i, j \in \mathbb{P}$, t is a tag value, and $\phi \in \mathcal{L}$, by:

$$Prg \ni P ::= \varepsilon \mid x := e \mid \text{SND}^t_{i \to j}(e) \mid \text{RECV}^t_{j \leftarrow i}(x) \mid [\phi] \mid \tau P \mid P * P \mid P + P \mid P^\omega$$

The formal semantics of Prg is given in the Section 2.3. The intuitive meaning of these constructs is as follows. The symbol ε denotes the *empty program*, which takes no time to execute. *Assignment statement* $x := e$ evaluates expression e and assigns its value to variable x. The $\text{SND}^t_{i \to j}(e)$ statement sends a message containing the value of e and tagged with t on the channel from i to j. This enables the delivery of that message to process j. Communication is asynchronous, and sending is non-blocking. Finitely many enabled deliveries can happen spontaneously. Each delivery of a message sent from i to j appends that message to a message buffer $\text{buf}_{j \leftarrow i}$ for that channel. The *tagged receive* $\text{RECV}^t_{j \leftarrow i}(x)$ behaves as follows. It busily waits until there is a message tagged t in $\text{buf}_{j \leftarrow i}$. During the busy waiting messages can be delivered to j. Once $\text{buf}_{j \leftarrow i}$ contains a message tagged t, the $\text{RECV}^t_{j \leftarrow i}(x)$ stores the content of the first such message in x and removes that message together with any preceding messages from $\text{buf}_{j \leftarrow i}$. (Note that the preceding messages will necessarily have tags different from t.) If a message tagged t never appears in $\text{buf}_{j \leftarrow i}$, then j will perform infinitely many no-ops in the course of busy waiting and thus diverge.

The *guard* $[\phi]$ takes no time to execute and expresses a constraint on the execution of the program: in a run of the program, ϕ must hold at this location. We call τ the *phase quantifier*. It is a specification construct that needs to be eliminated before a program can be considered an implementation. The program τP behaves the same as P, with the additional restriction that it does not communicate with statements outside P. For the purposes of τ, a tagged receive $\text{RECV}^t_{j \leftarrow i}(x)$ is considered to communicate with the send statement $\text{SND}^t_{i \rightarrow j}$ generating the message it stored in x. It is not considered communicating with the statements generating preceding messages removed in the process of receiving. The symbol "+" denotes (demonic) nondeterministic choice. By P^ω we denote zero or more (possibly infinitely many) layered repetitions of P. More conventional programming constructs such as conditionals and loops are defined in the usual manner, e.g., **if** ϕ **then** P **else** Q **fi** abbreviates $[\phi]P + [\neg\phi]Q$ and **while** ϕ **do** P **od** stands for $([\phi]P)^\omega[\neg\phi]$.

2.1 A Model of Asynchronous Message Passing

States and events. A *send record (for i)* is a tuple $(i \rightarrow j, t, v)$, which records sending a message with contents v and tag t from i to j. Similarly, $(i \leftarrow j, t, v)$ is a *deliver record*, $(i \leftarrow j, v)$ is a *receive record*, and $(i \leftarrow j)$ is a *drop record (for i)*. We call all these records *communication records (for i)*. Note that receive and drop records ignore tags and that drop records ignore message contents. A *local state (for process i)* is a mapping from Var_i to values and from the reserved variable h_i to a sequence of communication records for i.

Local runs. A *local run (for process i)* is an infinite sequence of local states. We identify an *event (of i)* with the transition from one local state in a local run of i to the next. An event is either a *send*, *deliver*, *receive*, *drop*, or an *internal* event. An *idle* event is an internal event that preserves the local state.

Buffers. The *message buffer* $\text{buf}_{j \leftarrow i}$ for channel $\text{chan}_{i \rightarrow j}$ at *local point* (r_j, k) is a queue comprising of all but the first ℓ messages delivered to j from i by that local point, where ℓ is the total number of receive and drop events in (r_j, k). Thus a delivery event appends a message to the end of the queue whereas receives and drops can only take place when the message buffer is non-empty. They result in the message at the head of the queue being removed, and, in the case of receives, the message being removed is received.

Global runs. A *(global) run* is a tuple $r = ((r_i)_{i \in \mathbb{P}}, \delta_r)$ of local runs — one for each process $i \in \mathbb{P}$ — plus a *matching function* δ_r associating a send event with each deliver event in r. The k'th event of process i in r is referred to by $\text{E}^r_i(k)$. It starts at the local point $(r_i, k-1)$ and ends at (r_i, k). The mapping δ_r is restricted such that:

1. If $\delta_r(e) = e'$ and e is a deliver event of process j resulting in the appending of $(j \leftarrow i, t, v)$ to j's message history then e' is a send event of process i appending the send record $(i \rightarrow j, t, v)$ to i's message history.
2. Messages are delivered in order. More formally, for each pair (i, j) of distinct processes, if $\delta_r(\text{E}^r_j(k_1)) = \text{E}^r_i(k'_1)$ and $\delta_r(\text{E}^r_j(k_2)) = \text{E}^r_i(k'_2)$ and $k_1 < k_2$ then $k'_1 \leq k'_2$.

3. Messages are duplicated at most a finite number of times, that is, for each send event e' in r, the set $\{\, e \mid \delta_r(e) = e' \,\}$ of deliver events mapped to the same send event is finite.
4. Lamport's causality relation[1] $\xrightarrow{L}$ induced by δ_r on the events of r is an irreflexive partial order, hence acyclic.

The first condition captures the property that messages are not corrupted in transit. The second condition prevents reordering of messages. That distinct deliver events can be matched with the same send allows for duplication of messages, however, the third condition prevents infinite message duplication. The fact that the function δ_r is total precludes the reception of spurious messages. Further restrictions on δ_r can be made to capture additional properties of the communication medium such as reliability, fairness, non-duplication, etc. We use $\hat{\delta}_r$ to refer to the function from receive and drop events to send events induced by δ_r. The fourth condition on δ_r implies that the partial order on all events in a run induced by the sequential nature of the local runs can be extended to a total order (i.e., an interleaving of the local runs) such that each deliver event is preceded by the matching send event.

Cuts and channels. Write $\mathbb{N}_+$ for $\mathbb{N} \cup \{\infty\}$. A *cut* is a pair (r,c) consisting of a run r and a $\mathbb{P}$-indexed family $c = (c_i)_{i \in \mathbb{P}}$ of $\mathbb{N}_+$-elements. We write "$\leq$" for the component-wise extension of the natural ordering on $\mathbb{N}_+$ to cuts within the same run. A cut is *finite* if all its components are.

Say that an event $E_i^r(k)$ is *in* a cut (r,c) if $k \leq c_i$. A cut (r,c) corresponds to the, possibly implausible, situation in which the events in the cut have occurred for each process $i \in \mathbb{P}$.

The *(message) balance between i and j*, denoted $\mathrm{bal}_{i \to j}$, at a cut (r,c) is a signed sequence of tagged messages, namely those messages from i to j that have been delivered in (r,c) but not been sent yet or vice versa. A positive sign indicates that the sequence consists of messages sent but not yet delivered whereas a negative sign indicates that the sequence lists delivered messages that were not yet sent in (r,c). We define the *channel* $\mathrm{chan}_{i \to j}$ at a cut (r,c) to be the pair consisting of the message buffer $\mathrm{buf}_{j \leftarrow i}$ and the message balance $\mathrm{bal}_{i \to j}$ at (r,c). A channel is *empty* whenever both of its components are.

2.2 Semantics of Formulas

Evaluating the tag expression $\mathrm{PEEK}_{j \leftarrow i}$ at a cut (r,c) results in the tag value of the head of $\mathrm{buf}_{j \leftarrow i}$ if that buffer is empty and $\bot$ otherwise. Finally, a formula $\phi \in \mathcal{L}$ holds at (r,c), and we write $(r,c) \models \phi$, if ϕ holds in standard propositional logic when, for each $i \in \mathbb{P}$, program variables in Var_i are evaluated in the local states $r_i(c_i)$ if c_i is finite. The value of a variable in Var_i in (r,c) is considered unspecified if $c_i = \infty$.

---

[1] In [12] Lamport defined a "happened before" relation $\xrightarrow{L}$ on the set of events occurring in a run r of a distributed system. The relation $\xrightarrow{L}$ is defined as the smallest transitive relation subsuming (1) the total orders on the events of process i given by the local run r_i, and (2) the relation between send and deliver events induced by the matching function, that is, $\{\, (e_1, e_2) \mid \delta_r(e_2) = e_1 \,\}$.

2.3 Semantics of Programs

We define the meaning of programs by stating when a program occurs over an interval. We do so for a more general class of programs, called *basic programs*, which contains (untagged) drop $\text{DRP}_{j \leftarrow i}$ and receive $\text{RCV}_{j \leftarrow i}(x)$ statements in place of the tagged receives, which are defined as follows.

$$\text{RECV}^t_{j \leftarrow i}(x) = \text{FLUSH}^t_{j \leftarrow i} * \text{RCV}_{j \leftarrow i}(x) \text{ , where}$$
$$\text{FLUSH}^t_{j \leftarrow i} = \text{AWAIT}(\text{PEEK}_{j \leftarrow i} \neq \bot) *$$
$$\textbf{while } \text{PEEK}_{j \leftarrow i} \neq t \textbf{ do } \text{DRP}_{j \leftarrow i} * \text{AWAIT}(\text{PEEK}_{j \leftarrow i} \neq \bot) \textbf{ od}$$
$$\text{AWAIT}(\phi) = \textbf{while } \neg \phi \textbf{ do } \text{SKIP } \textbf{od} \qquad \text{SKIP} = x := x$$

An *interval* consists of two cuts (r, c) and (r, d) over the same run with $c \leq d$, which we denote for simplicity by $r[c, d]$. An event is in $r[c, d]$ if it is in (r, d) but not in (r, c). We define the *occurrence relation* ⊩ between intervals and programs such that deliver events can occur in addition to the events prescribed by the program text. Formally, we define that $r[c, d] \Vdash P$ iff there exist c', d' such that

- $c \leq c' \leq d' \leq d$,
- $r[c', d'] \Vvdash P$ as defined below,
- there are finitely many events in $r[c, c']$ and $r[d', d]$, all of which are deliver events.

Intuitively, this assumption makes processes "input enabled" at all times. Program $P \in Prg$ is *embedded* into interval $r[c, d]$, denoted $r[c, d] \Vvdash P$, iff:[2]

$r[c, d] \Vvdash \varepsilon$ if $c = d$.
$r[c, d] \Vvdash x := \mathsf{e}$ if c_i is finite, $d = c[i \mapsto c_i + 1]$, and $r_i(d_i) = r_i(c_i)[x \mapsto v]$, where v is the value of e in $r_i(c_i)$.
$r[c, d] \Vvdash \text{SND}^t_{i \rightarrow j}(\mathsf{e})$ if c_i is finite, $d = c[i \mapsto c_i + 1]$, and $r_i(d_i) = r_i(c_i)[h_i \mapsto r_i(c_i)(h_i) \cdot \langle (i \rightarrow j, t, v) \rangle]$, where v is the value of e in $r_i(c_i)$.
$r[c, d] \Vvdash \text{DRP}_{i \leftarrow j}$ if c_i is finite, $d = c[i \mapsto c_i + 1]$, the message buffer $\text{buf}_{i \leftarrow j}$ is non-empty at (r_i, c_i), and $r_i(d_i) = r_i(c_i)[h_i \mapsto r_i(c_i)(h_i) \cdot \langle (i \leftarrow j) \rangle]$
$r[c, d] \Vvdash \text{RCV}_{i \leftarrow j}(x)$ if c_i is finite, $d = c[i \mapsto c_i + 1]$, the message buffer $\text{buf}_{i \leftarrow j}$ is non-empty at (r_i, c_i), and $r_i(d_i) = r_i(c_i)[h_i \mapsto r_i(c_i)(h_i) \cdot \langle (i \leftarrow j, v) \rangle, x \mapsto v]$.
$r[c, d] \Vvdash [\phi]$ if $c = d$ and $(r, c) \models \phi$.
$r[c, d] \Vvdash \tau P$ if $r[c, d] \Vdash P$ and no send or receive event in $r[c, d]$ is matched by $\hat{\delta}_r$ with a receive or send event outside $r[c, d]$.
$r[c, d] \Vvdash P * Q$ if there exists c' satisfying $c \leq c' \leq d$ such that $r[c, c'] \Vdash P$ and $r[c', d] \Vdash Q$.
$r[c, d] \Vvdash P + Q$ if $r[c, d] \Vdash P$ or $r[c, d] \Vdash Q$.
$r[c, d] \Vvdash P^\omega$ if, intuitively, an infinite or finite number (possibly zero) of iterations of P occur over $r[c, d]$. More formally, $r[c, d] \Vvdash P^\omega$ if there exists a finite or infinite sequence $(c^{(k)})_{k \in I}$ such that I is a non-void prefix of $\mathbb{N}_+$, $c^{(0)} = c$, $c^{(k)} \leq c^{(k')}$ for all $k < k' \in I$, $\bigsqcup_{k \in I} c^{(k)} = d$, and $r[c^{(k)}, c^{(k+1)}] \Vdash P$ for all $k, k + 1 \in I$.

---
[2] We denote by $f[a \mapsto b]$ the function that agrees with f on everything but a, and maps a to b.

2.4 Refinement

A *model* is a set of runs. For instance, RELDFI denotes the the set of all runs $(r_1, \ldots, r_n, \delta_r)$ such that δ_r is a monotone surjective function. The fact that it is monotone prohibits reordering of messages whereas surjectivity guarantees reliability. Given a model Γ, we say that *P refines Q in* Γ, denoted $P \sqsubseteq_\Gamma Q$, iff $r[c,d] \Vdash P$ implies $r[c,d] \Vdash Q$, for all $r \in \Gamma$ and $c,d \in (\mathbb{N}_+)^\mathbb{P}$. In other words, every execution of P (in a Γ run) is also one of Q, regardless of what happens before and after. Therefore, we may replace Q by P in any larger program context. This definition of refinement is in particular appropriate for stepwise top-down development of programs from specifications. The refinement relation on programs is transitive (in fact a pre-order) and all programming constructs are monotone w.r.t. the refinement order.

3 Composing Layers

As we explained in detail in [4], the phase operator τ allows us to delineate the interactions that a layer can have with other parts of the program. When combined with refinement it is useful for defining CCL and related notions. For example, we can express that the program L is a CCL in the program $Q * L * Q'$ w.r.t. Γ by:

$$\tau(Q * L * Q') \sqsubseteq_\Gamma Q * \tau L * Q' .$$

Say that S is a *barrier* in Γ if,

$$\tau(Q * S * Q') \sqsubseteq_\Gamma \tau Q * \tau S * \tau Q' \text{, for all } P \text{ and } Q.$$

Inspired by [7] we defined in [4] that *P is tail communication closed (TCC)* in Γ if

$$\tau(P * Q) \sqsubseteq_\Gamma \tau P * \tau Q \text{, for all programs } Q.$$

Hence, inductively we obtain that each layer P_i is a CCL in $P_1 * \ldots * P_m$ whenever all P_i are TCC in Γ. We can show that only trivial programs are TCC in RELDFI [3].

Theorem 2. *Let $P \in Prg$. If there is a communication event (i.e., send or receive) in a finite interval $r[c,d]$, where $r \in$ RELDFI, $r[c,d] \Vdash P$, and all channels are empty in (r,c), then P is not TCC in RELDFI.*

Corollary 3. *Barriers do not exist in RELDFI.*

The above theorem captures an essential difference between RELDFI and RELFI w.r.t. composability of programs. Since nontrivial programs are not TCC in RELDFI, an alternative methodology is required for determining when composing programs in RELDFI yields desirable behavior.

---
[3] Proofs had to be omitted from this version. The full version is available at ftp://ftp.cse.unsw.edu.au/pub/users/kaie/EM2005b.pdf.

3.1 Fitting After

Say that *Q fits after P (in* Γ*)* if $\tau(P*Q) \sqsubseteq_\Gamma \tau P * \tau Q$. Intuitively, this is the case when P and Q do not communicate. For example, any program of the form $\text{VT}^t_{i\to j} = \text{SND}^t_{i\to j} * \text{RECV}^t_{j\leftarrow i}$ fits after any program of the form $\text{VT}^{\bar{t}}_{i\to j}$ in RELDFI, where $\bar{t} \neq t$. Consequently, a program of the form $\text{VT}^t_{i\to j} * \text{VT}^{\bar{t}}_{i\to j}$ fits after itself in RELDFI. A rather pathological example is that ε fits after any program and vice versa.

Before discussing a more general class of programs, we consider fitting after and related notions for straight-line balanced programs. Formally, a *straight-line* program is one that is free of guards, choice, and loops. A *balanced* program is a program, where, on each channel, the sequences of tags on send and receive statements are equal. The shortest communicating balanced programs are those of the form $\text{VT}^t_{i\to j}$. By BSL we refer to balanced straight-line programs.

Characterizing fits-after. Next we turn to the topic of characterizing the fits-after behavior of BSL programs P. For each channel $\text{chan}_{i\to j}$ there are two important aspects to characterize. We denote by $\overleftarrow{P}_{i\to j}$ the tag t that the first tagged receive $\text{RECV}^t_{j\leftarrow i}$ on this channel in P expects, and by $\overrightarrow{P}_{i\to j}$ the tag t' of the last send $\text{SND}^{t'}_{i\to j}$ in P on $\text{chan}_{i\to j}$, if they exist. (The notation $\bot$ is used when such a tag does not exist.) We define the *signature* of P as a function associating with every channel $\text{chan}_{i\to j}$ the pair $(\overleftarrow{P}_{i\to j}, \overrightarrow{P}_{i\to j})$. Consequently, the size of P's signature is in $O(n^2)$ and it can be computed in time $O(|P|)$. The following theorem yields an $O(n^2)$ algorithm for deciding the fits-after relation.

Theorem 4. *Let P and Q be BSL programs such that* τP *occurs over a finite* RELDFI *interval. Then Q fits after P in* RELDFI *iff, for all channels* $\text{chan}_{i\to j}$*, we have that either* $\overleftarrow{Q}_{i\to j} \neq \overrightarrow{P}_{i\to j}$ *or that both are* $\bot$.

To see that the requirement in Th. 4 for P to have a terminating execution is not redundant consider $P = \text{RECV}^t_{j\leftarrow i}(x) * \text{RECV}^t_{i\leftarrow j}(y) * \text{SND}^t_{i\to j}(3) * \text{SND}^t_{j\to i}(4)$. Any Q fits after this P in RELDFI because P necessarily diverges for i and j in any execution of $\tau(P*Q)$.

Separators. Program S is a *separator* between programs P and Q if S fits after P and Q fits after $P * S$. Intuitively, a separator is a tailor-made barrier. Given two BSL programs P and Q, we can construct a separator S between P and Q as follows. For each channel $\text{chan}_{i\to j}$ on which Q does not fit after P, that is, satisfying $\overleftarrow{Q}_{i\to j} = \overrightarrow{P}_{i\to j} \neq \bot$, the separator contains a layer of the form $\text{SND}^t_{i\to j} * \text{RECV}^t_{j\leftarrow i}$ for some tag $t \neq \overrightarrow{P}_{i\to j}$. Consequently, the separator is BSL and can be constructed in $O(n^2)$ time and space.

Theorem 5. *Each pair of BSL programs can be separated by a terminating BSL program in* RELDFI.

Sealing. In [4] we defined a notion of *sealing* that formalizes the concept of a program S serving as an impermeable layer between P and later layers such that no later communication will interact with P. This notion is stronger than fitting after. We defined that *S seals P in* Γ if, for all programs Q, $\tau(P*S*Q) \sqsubseteq_\Gamma \tau P * \tau(S*Q)$. A seal S for P is *proper* if S never diverges after P, that is, for all $r \in \Gamma$ and c,d,d', whenever $r[c,d] \Vdash \tau P$, and $r[d,d'] \Vdash S$ and d is finite then so is d'.

It follows that sealing implies fitting after. A program Q that merely fits after P may leave untouched channels on which P communicates whereas a seal must re-use the channel with a different tag. If, however, a BSL program fits after itself in Γ, then it also seals itself in Γ. Observe that the program $P = \text{VT}^t_{i \to j} * \text{VT}^{\bar{t}}_{i \to j}$ of our example properly seals itself in RELDFI. Lemma 3 of [4] also states that if P properly seals itself in RELDFI then P properly seals P^ω in RELDFI. It follows that the infinite repetition of P solves the sequence transmission problem in RELDFI. To decide sealing it once again suffices to inspect the signatures of the two programs involved. The following theorem claims for sealing what Th. 4 does for fitting after. Its proof is analogous.

Theorem 6. *Let $P, Q \in Prg$ be BSL such that $\tau(P * Q)$ occurs over a finite RELDFI interval. Then Q properly seals P in RELDFI iff $\overrightarrow{P}_{i \to j} \neq \bot$ implies that both $\overleftarrow{Q}_{i \to j} \neq \overrightarrow{P}_{i \to j}$ and $\overleftarrow{Q}_{i \to j} \neq \bot$, for all channels $\text{chan}_{i \to j}$.*

The next theorem shows that BSL programs can always be properly sealed.

Theorem 7. *All BSL programs can be properly sealed in RELDFI.*

3.2 Beyond Straight-Line Programs

The restriction to straight-line programs can be easily overcome, however, requiring the sequences of tags on send and receive statements to be equal seems justified in any account of safe composition.[4] For example, a program consisting of a single $\text{RECV}_{j \leftarrow i}$ will either diverge for j or preclude future layers that send on $\text{chan}_{i \to j}$ from being a CCL by receiving one message sent on that channel. In the full paper we define a notion of balance for general programs P, which, roughly speaking, amounts to requiring that, for every channel, the sequence of tags of messages sent on the channel are equal to the corresponding sequence of tags received, in every execution of the program. For non-straight-line programs, different executions can give rise to different tag sequences.

The *joint state* at a cut (r, c) is the tuple $r(c) = (r_i(c_i))_{i \in \mathbb{P}}$ of the local states of the processes at (r, c), where $r_i(\infty) = \omega$. Let $P \in Prg$ (i.e., not necessarily BSL).

Let $\overleftarrow{P}_{i \to j}(\sigma)$ be the set of tags of the first tagged receives in P on $\text{chan}_{i \to j}$ that appear in intervals $r[c, d]$ such that $r[c, d] \Vdash \tau P$ and $\sigma = r(c)$. Similarly, let $\overrightarrow{P}_{i \to j}(\sigma)$ be the set of tags of the last sends in P on $\text{chan}_{i \to j}$ that appear in intervals $r[c, d]$ such that $r[c, d] \Vdash \tau P$ and $\sigma = r(d)$. Note that, $\overleftarrow{P}_{i \to j}(r(c)) = \emptyset$ whenever $c_j = \infty$. The characterization of fits after in Th. 4 can now be generalized to the following.

Theorem 8. *Let $P, Q \in Prg$ be balanced. Then Q fits after P in RELDFI iff, for all channels $\text{chan}_{i \to j}$ and joint states σ we have that $\overleftarrow{Q}_{i \to j}(\sigma) \cap \overrightarrow{P}_{i \to j}(\sigma) = \emptyset$.*

Algorithmically, this theorem is of limited use because the condition characterizing the fits-after property is undecidable for general balanced programs.

Separators do not exist for all pairs of balanced programs. For example, two instances of the the non-deterministic choice between $\text{VT}_{i \to j}$ for *all* possible tags t are not separable. Many pairs of programs can be separated, by relatively simple means.

---

[4] A minor generalization is possible: extra send statements can be compensated for.

For instance, if, on each channel, no $\overleftarrow{Q}_{i \to j}(\sigma)$ contains $t_{(i,j)}$ and no $\overrightarrow{P}_{i \to j}(\sigma)$ contains $t'_{(i,j)}$ then S containing a layer $\mathrm{VT}_{i \to j}^{t'_{(i,j)}} * \mathrm{VT}_{i \to j}^{t_{(i,j)}}$ for each $\mathrm{chan}_{i \to j}$ is a separator between balanced programs P and Q.

3.3 Models with Lossy Channels

Having characterized fitting after and separation in RELDFI, we now turn to FFI and FDFI. The potential of losing messages in these models is typically compensated for by retransmitting messages. Channels therefore contain duplicates of messages much in the same way as in RELDFI. As a result, program composition in these models has many of the same characteristics as in RELDFI.

Let us focus on the basic problem of transmitting a message e from sender i to receiver j in FFI. No balanced program as defined above is suitable for solving this problem since no bounded number of send statements guarantees a delivery in FFI (and thus FDFI). A standard solution to this problem is to retransmit the message until i *knows* that the value of e is (guaranteed to be) delivered to j [15]. Retransmitting in turn introduces multiple deliveries of the same message much in the spirit of the duplication possible in RELDFI. In FFI, receiving an acknowledgment message from j informs i of the reception (and thus preceding delivery) of e. (Recall that there are no spurious messages in our models.) In a rather ad hoc manner, we could extend Prg to Prg' by replacing the send and tagged receive actions by the *acknowledgment-awaiting send* $\mathrm{ACKSND}_{i \to j}^{t}(\mathrm{e})$ and the *acknowledging receive* $\mathrm{ACKRECV}_{j \leftarrow i}^{t}(x)$. When executing $\mathrm{ACKSND}_{i \to j}^{t}(\mathrm{e})$ process i sends t-tagged messages containing the value of e to j repeatedly until i receives a t-tagged acknowledgment from j. When executing $\mathrm{ACKRECV}_{j \leftarrow i}^{t}(x)$ process j first performs $\mathrm{RECV}_{j \leftarrow i}^{t}(x)$ followed by repeatedly sending a t-tagged acknowledgment to i until detecting a non-t-tagged message in the buffer.

Koo and Toueg proved that in FFI no necessarily communicating program can terminate [11]. This also applies to Prg'. Observe that in order to terminate any acknowledging receive requires the delivery of a differently tagged message in addition to the one it receives. Consequently no last layer of a communicating program can terminate.

For programs in Prg' composition in FFI (and FDFI) is very similar to that in RELDFI. In particular, only trivial and necessarily diverging programs are TCC. An essentially identical notion of a signature of a balanced program in Prg' can be defined, leading to results analogous to Th. 4 and 5. A variant of Th. 7 holds: all balanced programs in Prg' can be sealed in FFI by a not necessarily diverging program. This is as much as could be expected, given that proper seals only exist for trivial and necessarily diverging programs. The complexities of computing all composition-related issues carry over from the RELDFI case to FFI. Finally, the behavior of Prg' programs as far composition is concerned is exactly the same in FDFI as it is in FFI.

The nature of seals. It is instructive to compare the nature of seals in RELDFI, FFI, and FDFI to their nature in the model REL of reliable non-duplicating channels with potential reordering [4]. The latter crucially depends on establishing a causal dependence between receives in one layer and sends on the same channel in later layers. The standard way to achieve this is by sending an acknowledgment in the reverse direction. In the former models, however, sealing is achieved by switching the message tag. It therefore requires only half-duplex channels. Acknowledgments play a different role in the lossy FIFO models. They inform the sender of the arrival of a message and thus allow him to progress.

4 Conclusion and Future Work

Much of the distributed algorithms literature is devoted to solutions for individual tasks. Often, the issues involved in solving an individual task in models of interest deserve considerable attention and may involve a great deal of work. At the end of the day, however, such solutions must be combined to form larger applications. Answers to the questions of when, whether, and how distributed programs can be composed are therefore crucial to methods for correctly designing distributed applications.

In [4] we introduced a powerful notion of *sealing* which captures the property of one program serving as an impermeable layer between a given predecessor program and any following program. The correct behavior of an application composed from a number of layers, each of which seals its predecessor, follows from the correct behavior of each of the layers in isolation. In this paper we introduce *fits after*, a weaker notion than sealing, which applies more broadly, in particular to smaller programs. A seal will often consist of a sequence of smaller programs each of which fits after the prefix of the layer accumulated so far.

No interesting program is TCC in any of RELDFI, FFI, and FDFI. This implies that universal barriers do not exist in these models. We show in this paper that, for given pairs of BSL programs, tailor-made barriers, which we call *separators*, always exist in all of the above models. We also outline efficient algorithms for deciding the fits-after and seals relations as well as constructing separators. For general balanced programs, we defined a suitable generalization of signatures and characterized the fits-after relation in terms of signatures.

References

1. C. T. Chou and E. Gafni. Understanding and verifying distributed algorithms using stratified decomposition. In D. Dolev, editor, *PODC '88*, pp. 44–65. ACM Press, 1988.
2. W.-P. de Roever, F. de Boer, U. Hannemann, J. Hooman, Y. Lakhnech, M. Poel, and J. Zwiers. *Concurrency Verification*. Cambridge University Press, 2001.
3. T. Elrad and N. Francez. Decomposition of distributed programs into communication-closed layers. *Science of Computer Programming*, 2(3):155–173, Dec. 1982.
4. K. Engelhardt and Y. Moses. Causing communication closure: Safe program composition with non-FIFO channels. In *DISC 2005*, *LNCS 3724*, pp. 229–243. Springer-Verlag, 2005.
5. K. Engelhardt and Y. Moses. Single-bit messages are insufficient in the presence of duplication. In *this volume*.
6. A. Fekete and N. Lynch. The need for headers: An impossibility result for communication over unreliable channels. In *CONCUR '90*, *LNCS 458*, pp. 199–215. Springer-Verlag, 1990.
7. R. Gerth and L. Shrira. On proving communication closedness of distributed layers. In *FSTTCS 1986*, *LNCS 241*, pp. 330–343, Springer-Verlag, 1986.
8. W. Janssen. *Layered Design of Parallel Systems*. PhD thesis, University of Twente, 1994.
9. W. Janssen. Layers as knowledge transitions in the design of distributed systems. In *TACAS 1995*, NS-95-2 in Notes Series, pp. 304–318, Dept. of Comp. Sci., U. of Aarhus, 1995.
10. W. Janssen, M. Poel, and J. Zwiers. Action systems and action refinement in the development of parallel systems. In *CONCUR '91*, *LNCS 527*, pp. 298–316, 1991.

11. R. Koo and S. Toueg. Effects of message loss on the termination of distributed protocols. *Information Processing Letters*, 27(4):181–188, 1988.
12. L. Lamport. Time, clocks, and the ordering of events in a distributed system. *Communications of the ACM*, 7:558–565, 1978.
13. B. Meenakshi and R. Ramanujam. Reasoning about message passing in finite state environments. In *ICALP, LNCS 1853*, pp. 487–498. Springer-Verlag, 2000.
14. B. Meenakshi and R. Ramanujam. Reasoning about layered message passing systems. *Computer Languages, Systems & Structures*, 30(3-4):171–206, 2004.
15. Y. Moses and O. Kislev. Knowledge-oriented programming. In *PODC 93*, pp. 261–270. ACM Press, 1993.
16. M. Poel and J. Zwiers. Layering techniques for development of parallel systems. In *CAV '92, LNCS 663*, pp. 16–29. Springer-Verlag, 1992.
17. V. R. Pratt. Modelling concurrency with partial orders. *International Journal of Parallel Programming*, 15(1):33–71, 1986.
18. F. A. Stomp and W.-P. de Roever. A principle for sequential reasoning about distributed algorithms. *Formal Aspects of Computing*, 6(6):716–737, 1994.

Efficiently Implementing LL/SC Objects Shared by an Unknown Number of Processes*

Prasad Jayanti and Srdjan Petrovic

Department of Computer Science,
Dartmouth College, Hanover, New Hampshire, USA
{prasad, spetrovic}@cs.dartmouth.edu

Abstract. Over the past decade, a pair of instructions called load-linked (LL) and store-conditional (SC) have emerged as the most suitable synchronization instructions for the design of lock-free algorithms. However, current architectures do not support these instructions; instead, they support either CAS (e.g., UltraSPARC, Itanium) or restricted versions of LL/SC (e.g., POWER4, MIPS, Alpha). To bridge this gap, a flurry of algorithms that implement LL/SC from CAS have appeared in the literature. Some of these algorithms assume that N, the maximum number of participating processes, is fixed and known in advance. Others make no such assumption, but are either non-blocking (not wait-free), implement small LL/SC objects, or require that a process performs $O(N)$ work to join the algorithm. Specifically, no constant-time, word-sized, wait-free LL/SC algorithm that does not require the knowledge of N exists. In this paper, we present such an algorithm.

1 Introduction

1.1 Background

In shared-memory multiprocessors, multiple processes running concurrently on different processors cooperate with each other via shared data structures (e.g., queues, stacks, counters, heaps, trees). Atomicity of these shared data structures has traditionally been ensured through the use of locks. To perform an operation, a process obtains the lock, updates the data structure, and then releases the lock. Locking, however, has several drawbacks, including deadlocks (each of two processes waits for a lock currently held by the other), priority inversion (a low priority process holds a lock needed by a high priority process, and the low priority process is preempted by a medium priority process), and convoying (a descheduled process that holds a lock causes other processes to wait). Locking also limits parallelism: even when operations update disjoint parts of the data structure, they are applied sequentially, one after the other. Finally, lock-based implementations are not fault-tolerant: if a process crashes while holding a lock, other processes can end up waiting forever for the lock.

---

* This work is partially supported by the NSF Award EIA-9802068.

Wait-free implementations were conceived to overcome the above drawbacks of locking [1], [2]. A wait-free implementation guarantees that every process completes its operation on the data structure in a bounded number of its steps, regardless of whether other processes are slow, fast, or have crashed. This bound (on the number of steps that a process executes to complete an operation on the data structure) is the *time complexity* (of that operation). A weaker form of implementation, known as *non-blocking* implementation [2], guarantees that if a process p repeatedly takes steps, then the operation of *some* process (not necessarily p) will eventually complete. Thus, non-blocking implementations guarantee that the system as a whole makes progress, but admit starvation of individual processes.

It is a well understood fact that whether lock-free algorithms (i.e., wait-free or non-blocking) can be efficiently designed depends crucially on what synchronization instructions are available for the task. As we describe in the next section, the synchronization instructions supported by modern machines are not well suited for the task. The goal of this paper (and a lot of recent research) has been to remedy this situation by implementing more useful synchronization instructions.

1.2 Weaknesses of Hardware Synchronization Instructions

Most modern machines support either a *compare&swap* (CAS) instruction (e.g., UltraSPARC [3], Itanium [4]), or a pair of instructions *RLL/RSC* (e.g., POWER4 [5], Alpha [6] processors). Neither of these instructions are well suited for the design of shared data structures. To understand why, we must first look at their semantics.

The instruction $CAS(X, u, v)$ checks if location X has value u; if so, it changes the value to v and returns *true*, else it returns *false* and leaves the value unchanged. In practice, CAS is most commonly used as follows. First, we would read some value A from a location X; then, we would perform some computation (which may involve reading other locations) and compute the new value to be stored into X; finally, we would use CAS to attempt to change location X from A to the new value. Most often, our intent is for CAS to succeed only if between the read and the CAS the location X hasn't been changed. However, it is quite possible that the location X changes from A to some value B, and then back to A again between the read and the CAS; in that case, CAS will succeed, even though our intent was for it to fail. This undesirable behavior is known in the literature as the ABA-problem [7], and has greatly complicated the design of shared data structures.

Next, we turn to the instructions RLL/RSC, which are also supported on many modern machines. The RLL and RSC instructions act like read and conditional-write, respectively. More specifically, the $RLL(X)$ instruction by process p returns the value of the location X, while the $RSC(X, v)$ instruction by p checks whether some process updated the location X since p's latest RLL, and if that isn't the case it writes v into X and returns *true*; otherwise, it returns *false* and leaves X unchanged.

Due to their semantics, the RLL/RSC instructions do not suffer from the ABA-problem. However, they impose two severe restrictions on their use [8]: (1)

there is a chance of *RSC* failing spuriously: *RSC* might fail even when it should have succeeded, and (2) a process is not allowed to access any shared variable between its *RLL* and the subsequent *RSC*. Due to these restrictions, it is hard to design algorithms based on these instructions.

1.3 Solution: LL/SC Instructions

The instructions LL/SC have the same semantics as RLL/RSC, except that they do not impose any restrictions on their use. For this reason, they are very well suited for the design of shared data structures. Some examples of recent LL/SC-based lock-free algorithms are [9], [10], [11], [12], [13], [14], [15], [16].

However, despite the desirability of LL/SC, no processor supports these instructions in hardware because it is impractical to maintain (in hardware) the state information needed to determine the success or failure of each process' SC operation on each word of memory. Thus, there is a gap between what algorithm designers want (namely, LL/SC) and what multiprocessors actually support (namely, CAS or RLL/RSC). To bridge this gap, we must efficiently emulate LL/SC instructions in software, which gives rise to the following research problem:

Research Problem: Design a wait-free algorithm that implements LL/SC memory words from memory words supporting either CAS or RLL/RSC operations.

The above problem has been extensively studied in the literature [17], [18], [19], [20], [21], [22], [23], [24], [25], [8]. The most efficient algorithm for implementing LL/SC from CAS is due to Moir [8]. His algorithm runs in constant time and has no space overhead. However, it can only implement small (e.g., 24 to 32 bit) LL/SC objects, which are inadequate for storing pointers, large integers and doubles. This size limitation is due to the fact that Moir's algorithm stores a sequence number along with the object's value in the same memory word. Since sequence number could take up to 32 to 40 bits, only 24 to 32 bits are left for the value field.

Elsewhere, we presented an algorithm that implements a *word-sized* LL/SC object from a word-sized CAS object and registers (e.g., 64-bit LL/SC on a 64-bit machine) [23]. This algorithm stores a value and a sequence number in separate memory words, thus enabling values to be as big as 64 bits. The algorithm implements both LL and SC in $O(1)$ time and uses $O(N)$ space, where N is the maximum number of processes that the algorithm is designed to handle. Although these space requirements are modest when a single LL/SC object is implemented, the algorithm does not scale well when the number of LL/SC objects to be supported is large. In particular, in order to implement M LL/SC objects, the algorithm requires $O(NM)$ space. Furthermore, the algorithm requires that N is known in advance. Removing these two drawbacks has been the focus of some recent research [19], [22], [25] which we describe below.

Doherty, Herlihy, Luchangco, and Moir [19] present an algorithm that uses only $O(N + M)$ space and does not require knowledge of N, but is only nonblocking and not wait-free. Michael's [25] algorithm, on the other hand, is wait-free and does not require knowledge of N, but uses $O(N^2 + M)$ space. The main drawback of this algorithm is the time complexity of the SC operation: although the *expected amortized* running time of SC is only $O(1)$, the *worst-case* running time of SC is $O(N^2)$. In [22], we designed a wait-free algorithm that has a space complexity of $O(N^2 + M)$, while still maintaining the $O(1)$ worst-case running time for LL and SC. This algorithm too does not require knowledge of N. In the following, we refer to this algorithm and the algorithm by Michael [25] by the names JP and MIC, respectively.

The drawback of algorithms JP and MIC is that a process needs to perform $O(N)$ work in order to join the algorithm. Therefore, JP and MIC are not constant-time algorithms. In this paper, we present an algorithm that removes this drawback: our algorithm allows a process to join the algorithm in $O(1)$ time, while still maintaining the $O(1)$ running time for LL and SC. To the best of our knowledge, the algorithm presented in this paper is the only constant-time wait-free algorithm that implements a word-sized LL/SC object *without requiring the knowledge of N*.

We note that in certain other dimensions the algorithms JP and MIC do better than our algorithm: (1) when implementing M variables, the space complexity of algorithms JP and MIC is $O(N^2 + M)$, whereas the space complexity of our algorithm is $O(N^2 + NM)$; (2) JP and MIC allow processes to join and leave the algorithm, while our algorithm allows only joins; and (3) JP and MIC are good for implementing multiword LL/SC objects, while our algorithm implements only word-sized objects.

In terms of techniques, we achieve our result by employing two ideas: (1) we use our earlier LL/SC algorithm [23] as a base for the new algorithm, and (2) we use a novel notion of *dynamic arrays* that we introduced earlier in [22]. For completeness, we explain both ideas again in this paper.

Organization for the Rest of the Paper. We present our main result in two steps. First, we restate our earlier algorithm [23] that implements a 64-bit LL/SC object for a known N. Building on this algorithm, we present a more general algorithm that works without the knowledge of N. These two algorithms are described in Sections 2 and 3.

2 LL/SC for a Known N

Figure 1 presents our earlier algorithm that implements a 64-bit LL/SC objects shared by a fixed number of processes N [23]. We begin by providing an intuitive description of how this algorithm works.

2.1 How the Algorithm Works

The algorithm implements a 64-bit LL/SC object $\mathcal{O}$. Central to the implementation is the variable X that supports CAS and $read$ operations. In addition, there are four atomic registers at each process p—$\text{val}_p[0]$, $\text{val}_p[1]$, oldval_p and oldseq_p—that are written to only by p but may be read by any process. The meanings of these variables are described as follows.

The algorithm associates a tag with every successful SC operation on $\mathcal{O}$. A tag consists of a process id and a sequence number. Specifically, the tag associated with

Types
  valuetype = 64-bit number
  seqnumtype = $(64 - \log N)$-bit number
  xtype = **record** *pid*: $0 .. N - 1$; *seqnum*: seqnumtype **end**
Shared variables
  X: xtype (X supports *read* and *CAS* operations)
  For each $p \in \{0, \ldots, N - 1\}$, we have four single-writer, multi-reader registers:
    $\text{val}_p[0]$, $\text{val}_p[1]$, oldval_p: valuetype; oldseq_p: seqnumtype
Local persistent variables at each $p \in \{0, \ldots, N - 1\}$
  x_p: xtype; seq_p: seqnumtype
Initialization
  $X = (0, 1)$; $\text{val}_0[1] = v_{init}$, the desired initial value of $\mathcal{O}$; $\text{oldseq}_0 = 0$; $seq_0 = 2$
  For each $p \in \{1, \ldots, N - 1\}$ $seq_p = 1$

procedure LL($p, \mathcal{O}$) **returns** valuetype
1: $x_p = X$
   Let $(q, k) = (x_p.pid, x_p.seqnum)$
2: $v = \text{val}_q[k \bmod 2]$
3: $k' = \text{oldseq}_q$
4: **if** $(k' = k - 2) \vee (k' = k - 1)$ **return** v
5: $v' = \text{oldval}_q$
6: **return** v'

procedure SC($p, \mathcal{O}, v$) **returns** boolean
7: $\text{val}_p[seq_p \bmod 2] = v$
8: **if** CAS(X, x_p, (p, seq_p))
9:   $\text{oldval}_p = \text{val}_p[(seq_p - 1) \bmod 2]$
10: $\text{oldseq}_p = seq_p - 1$
11: $seq_p = seq_p + 1$
12: **return** *true*
13: **else return** *false*

procedure VL($p, \mathcal{O}$) **returns** boolean
14: **return** $X = x_p$

Fig. 1. An unbounded implementation of the 64-bit LL/SC object $\mathcal{O}$ using a 64-bit CAS object and 64-bit registers, taken directly from our earlier paper [23]

a successful SC operation is (p, k) if it is the kth successful SC operation by process p. The variable X always contains the tag corresponding to the latest successful SC.
  Suppose that the current value of X is (p, k) (which means that the last successful SC was performed by p and p performed k successful SC operations so far). The algorithm ensures that the value written by the kth successful SC by p is in $\text{val}_p[0]$ if k is even, or in $\text{val}_p[1]$ if k is odd; i.e., the value is made available in $\text{val}_p[k \bmod 2]$. The registers oldval_p and oldseq_p hold an older value and its sequence number, respectively. Specifically, if p has so far performed k successful SC operations, oldseq_p and oldval_p contain, respectively, the number $k - 1$ and the value written by the $(k - 1)$th successful SC by p.
  In addition to the shared variables just described, each process p has two persistent local variables, seq_p and x_p, described as follows. The value of seq_p is the sequence number of p's next SC operation: if p has performed k successful SC operations so far, seq_p has the value $k + 1$. (Thus, sequence numbers in our algorithm are local: p's sequence number is based on the number of successful SC's performed by p, not by the system as a whole.) The value of x_p is the value of X read by p in its latest LL operation.
  Given this representation, the variables are initialized as follows. Let v_{init} denote the desired initial value of the implemented object $\mathcal{O}$. We pretend that

process 0 performed an "initializing SC" to write the value v_{init}. Accordingly, X is initialized to $(0,1)$, $\text{val}_0[1]$ to v_{init}, oldseq_0 to 0, and seq_0 to 2. For each process $p \neq 0$, seq_p is initialized to 1. All other variables are arbitrarily initialized.

We now explain the procedure $\text{SC}(p, \mathcal{O}, v)$ that describes how process p performs an SC operation on $\mathcal{O}$ to attempt to change $\mathcal{O}$'s value to v. First, p makes available the value v in $\text{val}_p[0]$ if the sequence number is even, or in $\text{val}_p[1]$ if the sequence number is odd (Line 7). Next, p tries to make its SC operation take effect by changing the value in X from the tag that p had witnessed in its latest LL operation to the tag corresponding to its current SC operation (Line 8). If the CAS operation fails, it follows that some other process performed a successful SC after p's latest LL. In this case, p's SC must fail. Therefore, p terminates its SC procedure, returning *false* (Line 13). On the other hand, if CAS succeeds, then p's current SC operation has taken effect. To remain faithful to the previously described meanings of the variables oldval_p and oldseq_p, p writes in oldval_p the value written by p's earlier successful SC (Line 9) and writes in oldseq_p the sequence number of that SC (Line 10). (Since the sequence number for p's current successful SC is seq_p, it follows that the sequence number for p's earlier successful SC is $seq_p - 1$, and the value written by that SC is in $\text{val}_p[(seq_p - 1) \bmod 2]$; this justifies the code on Lines 9 and 10.) Next, p increments its sequence number (Line 11) and signals successful completion of the SC by returning *true* (Line 12).

We now turn to the procedure $\text{LL}(p, \mathcal{O})$ that describes how process p performs an LL operation on $\mathcal{O}$. In the following, let $\text{SC}_{q,i}$ denote the ith successful SC by process q and $v_{q,i}$ denote the value written in $\mathcal{O}$ by $\text{SC}_{q,i}$. First, p reads X to obtain the tag (q, k) corresponding to the latest successful SC operation, $\text{SC}_{q,k}$ (Line 1). Since $\text{SC}_{q,k}$ wrote $v_{q,k}$ into $\text{val}_q[k \bmod 2]$, and since $\text{val}_q[k \bmod 2]$ is not modified until q initiates an SC operation with $seq_q = k+2$, it follows that at the instant when p performs Line 1, the variable $\text{val}_q[k \bmod 2]$ holds the value $v_{q,k}$. Furthermore, the value of $\text{val}_q[k \bmod 2]$ is guaranteed to be $v_{q,k}$ until q completes $\text{SC}_{q,k+1}$.

So, in an attempt to learn $v_{q,k}$, p reads $\text{val}_q[k \bmod 2]$ (Line 2). By the observation in the previous paragraph, if p is not too slow and executes Line 2 before q completes $\text{SC}_{q,k+1}$, the value v read on Line 2 will indeed be $v_{q,k}$. Otherwise the value v cannot be trusted. To resolve this ambiguity, p must determine if q has completed $\text{SC}_{q,k+1}$ yet. To make this determination, p reads the sequence number k' in oldseq_q (Line 3). If $k' = k - 2$ or $k' = k - 1$, it follows that $\text{SC}_{q,k+1}$ has not yet completed even if it had been already initiated (because, by Line 10, $\text{SC}_{q,k+1}$ writes k into oldseq_q). It follows that the value v obtained on Line 2 *is* $v_{q,k}$. So, p terminates the LL operation, returning v (Line 4).

If $k' \geq k$, q must have completed $\text{SC}_{q,k+1}$, its $(k+1)$th successful SC. It follows that the value in oldval_q is $v_{q,k}$ or a later value (more precisely, the value in oldval_q is $v_{q,i}$ for some $i \geq k$). Therefore, the value in oldval_q is a legitimate value for p's LL to return. Accordingly, p reads the value v' of oldval_q (Line 5) and returns it (Line 6). Although v' is a recent enough value of $\mathcal{O}$ for p's LL to legitimately return, it is important to note that v' is not the current value of $\mathcal{O}$. This is because the algorithm moves a value into oldval_q only after it is no longer the current value. Since the value v' that p's LL returns on Line 6 is not the current value, p's subsequent SC must fail (by the specification of LL/SC). Our algorithm satisfies this requirement because, when p's subsequent SC performs Line 8, the CAS operation fails since x_p is (q,k) and the value of X is not (q,k) anymore (the value of X is not (q,k) because, by the first sentence of this paragraph, q has completed its $(k+1)$th successful SC). This completes the description of how LL is implemented.

The VL operation by p is simple to implement: p returns *true* if and only if the tag in X has not changed since p's latest LL operation (Line 14). The following theorem summarizes the above discussion.

Theorem 1 ([23]). *The wait-free algorithm in Figure 1 implements a linearizable 64-bit LL/SC object from a single 64-bit CAS object and an additional six registers per process. The time complexity of LL, SC, and VL is $O(1)$.*

3 LL/SC for an Unknown N

In this section, we present a modified version of the algorithm in Figure 1 that does not require N to be known in advance. In particular, the algorithm supports a new operation, $Join(p)$, that allows a process p to join the algorithm at any given time. If K is the maximum number of processes that have joined the algorithm so far, then the space complexity of the algorithm is $O(K^2 + KM)$. The time complexity of procedures Join, LL, SC, and VL is $O(1)$.

The algorithm is given in two steps. First, we introduce an important building block of the algorithm, namely, an implementation of a dynamic array that supports constant-time read and write operations (with some restrictions). Then, we present our main result, namely, an algorithm that implements the LL/SC object shared by an unknown number of processes. These two steps are described in Sections 3.1 and 3.2.

3.1 Dynamic Arrays

A dynamic array is just like a regular array except that it places no bounds on the highest location that can be written. In particular, a process can write into the ith location of the dynamic array, for any natural number i. At all times, the size of the array must stay proportional to the highest location written so far. Furthermore, all reads and writes in the array must complete in $O(1)$ time. In this paper, we consider only a weaker version of dynamic array that has the following restrictions: (1) all writes into the same location write the same value, (2) a write into a location i must precede a read on that location, and (3) a write into a location i must precede a write into location $i+1$. We capture the above restrictions in an object that we call a *DynamicArray* object. This object is formally defined as follows.

A DynamicArray object supports two operations: $write(i,v)$ and $read(i)$. The $write(i,v)$ operation writes value v into the ith location of the array, while the $read(i)$ operation returns the value stored in the ith location of the array. The following restrictions are placed on the usage of *read* and *write*:

- Before $write(k+1,*)$ is invoked, at least one $write(k,*)$ must complete.
- Before $read(k)$ is invoked, at least one $write(k,*)$ must complete.
- If $write(k,v)$ and $write(k,v')$ are invoked, then $v = v'$.

As it turns out, a DynamicArray object can be implemented efficiently (i.e., in $O(1)$ time) from a CAS object and registers [22]. Because of space constraints, we omit the details of the implementation from this paper and only state the main theorem.

Theorem 2 ([22]). *There exists a wait-free implementation of a DynamicArray object $\mathcal{D}$ from a word-sized CAS object and registers. The time complexity of read and write operations on $\mathcal{D}$ is $O(1)$. The space used by the algorithm at any time t is $O(nK)$, where n is the number of processes executing the algorithm at time t and K is the highest location written in $\mathcal{D}$ prior to time t.*

3.2 The Unknown-N LL/SC Algorithm

We now present our main result, namely, the algorithm that implements an LL/SC object shared by an unknown number of processes. The algorithm is presented in Figure 2. Below, we describe how the algorithm works.

Recall that the algorithm in Figure 1 stores the process id and a sequence number together in a central variable X. The assumption is that, once a process p reads a process id q from X, it can immediately locate all the shared variables owned by q, namely, $\texttt{oldseq}_q$, $\texttt{oldval}_q$, $\texttt{val}_q[0]$, and $\texttt{val}_q[1]$. Although the exact mechanism as to how p learns the locations of q's shared variables is not described in the algorithm, it is easy to see that the following approach will do: maintain an array A of length N, with one entry for each process, and store in each entry $\texttt{A}[r]$ the address of the block containing r's shared variables. To lean the location of q's shared variables, p simply reads the address stored in $\texttt{A}[q]$.

To simulate the above approach in the new algorithm (where N is not known in advance), we keep a dynamic array $\mathcal{D}$ in place of the static array A. This array will grow in size as more processes keep joining the algorithm. When a process p joins the algorithm, it first obtains a name – say, i – which will be its unique index into the array $\mathcal{D}$. Next, p inserts the address of the block that contains p's shared variables into location i of array $\mathcal{D}$. From this point onwards, p uses its name i in place of its process id, i.e., it writes i into X instead of p. It is easy to see that if some process q reads i from X, it can learn the location of p's shared variables by simply consulting the ith location in array $\mathcal{D}$.

The main challenge in implementing the above scheme is to allow concurrent processes to obtain unique names in $O(1)$ time. Below, we explain how the algorithm addresses this issue. We start by describing the variables used by the algorithm.

Each process p maintains a block of memory where it keeps its shared variables, namely, $\texttt{oldseq}$, $\texttt{oldval}$, $\texttt{val}[0]$, and $\texttt{val}[1]$ (which have the same meaning as in the algorithm in Figure 1), as well as a new variable $\texttt{name}$ which holds p's name. The address of this memory block is kept in p's local variable loc_p. In addition to loc_p, each process p maintains the following three (local) variables:

Types
　　valuetype = 64-bit value
　　xtype = **record** *type*: $\{0,1\}$; **if** (*type* == 1) (*ptr*: *blocktype)
　　　　　　　　　　else (*name*: 20-bit number; *seqnum*: 43-bit number) **end**
　　blocktype = **record** name: 20-bit number; val[0], val[1], oldval: valuetype;
　　　　　　　　　　oldseq: 43-bit number **end**
Shared variables
　　X: xtype; $\mathcal{D}$: **dynamic array of** *blocktype; N: 20-bit number
Local persistent variables at each p
　　x_p: xtype; seq_p: 43-bit number; loc_p: blocktype; $first_p$: boolean
Initialization
　　loc = malloc(sizeof *blocktype); $loc{\to}$name = 0; $loc{\to}$oldseq = 0;
　　$loc{\to}$val[1] = v_{init}, the desired initial value of $\mathcal{O}$; X = $(1, loc)$; N = 0

procedure LL$(p, \mathcal{O})$ **returns** valuetype	**procedure** SC$(p, \mathcal{O}, v)$ **returns** boolean
1: x_p = X	17: loc_p.val[seq_p mod 2] = v
2: **if** ($x_p.type$ == 1)	18: **if** ($first_p$)
3: $\quad l = x_p.ptr$	19: $\quad loc_p$.name = N
4: $\quad k = 1$	20: $\quad$ **if** ($succ$ = CAS(X, x_p, $(1, \&loc_p)$)))
5: $\quad$ da_write($\mathcal{D}, l{\to}name, l$)	21: $\quad\quad first_p$ = $false$
6: $\quad$ CAS(N, $l{\to}$name, $l{\to}$name + 1)	22: **else** $succ$ = CAS(X, x_p,
7: **else** l = da_read($\mathcal{D}, x_p.name$)	$\quad\quad\quad\quad\quad (0, loc_p.name, seq_p))$
8: $k = x_p.seqnum$	23: **if** ($succ$)
9: $v = l{\to}$val[k mod 2]	24: $\quad loc_p$.oldval =
10: $k' = l{\to}$oldseq	$\quad\quad\quad loc_p$.val[$(seq_p - 1)$ mod 2]
11: **if** $(k' = k - 2) \vee (k' = k - 1)$ **return** v	25: $\quad loc_p$.oldseq = $seq_p - 1$
12: $v' = l{\to}$oldval	26: $\quad seq_p = seq_p + 1$
13: **return** v'	27: $\quad$ **return** $true$
	28: **return** $false$
procedure Join(p)	
14: $seq_p = 1$	**procedure** VL$(p, \mathcal{O})$ **returns** boolean
15: loc_p.oldseq = 0	29: **return** (X == x_p)
16: $first_p$ = $true$	

Fig. 2. An unbounded implementation of the 64-bit LL/SC object $\mathcal{O}$ shared by an unknown number of processes

(1) seq_p, which stores p's sequence number, (2) x_p, which stores the value of X that p had read in its latest LL operation, and (3) $first_p$, which holds value *true* if p hasn't yet performed a successful SC operation, and *false* otherwise.

In addition to the variables stored at each process, there are two global shared variables, namely, X and N. Variable N stores an unbounded integer, and is used by processes to acquire names. Variable X stores the following information: if q is the latest process to perform a successful SC, then X holds either (1) a pair $(1, b)$, where b is a pointer to q's block of memory, or (2) a tuple $(0, k, s)$, where k is q's name and s is a sequence number.

We now explain the procedure SC$(p, \mathcal{O}, v)$ that describes how a process p performs an SC operation on $\mathcal{O}$ to attempt to change $\mathcal{O}$'s value to v. First, p

makes available the value v in $loc_p.\texttt{val}[0]$ if the sequence number is even, or in $loc_p.\texttt{val}[1]$ if the sequence number is odd (Line 17). Next, p checks whether it had previously performed at least one successful SC (Line 18). If it hasn't, then p reads variable N to obtain a name, and saves that name in $loc_p.\texttt{name}$ (Line 19). (Multiple processes reading N at the same time may get the same name; however, only one process will actually keep that name, as we explain below.) Next, p tries to make its SC operation take effect by changing the value in X from the value that p had witnessed in its latest LL operation to a value $(1, \&loc_p)$ (Line 20). If the CAS operation succeeds, then p's SC is successful. Since it is p's *first* successful SC, p updates variable $first_p$ to *false* (Line 21). Furthermore, p keeps the name it had read from N. If, on the other hand, p's CAS fails, then p's SC has failed and so p terminates its SC procedure by returning *false* (Line 28). Furthermore, p discards the name it had read from N. Therefore, of all processes that had read N at the same time, only one process (namely, the process that performed a successful CAS on X) actually keeps that name; all other processes abandon it, and attempt to capture a name again during their subsequent SC operations.

If p had previously performed a successful SC (Line 18), p attempts to change the value in X to a value $(0, i, s)$, where i is p's name and s is a sequence number (Line 22). Again, if the CAS operation succeeds, then p's SC is successful. Otherwise, p's SC has failed, and p terminates its SC procedure by returning *false* (Line 28).

If p's SC is successful (for the first time or not), p performs the same steps as in the algorithm in Figure 1. Namely, it (1) writes into $loc_p.\texttt{oldval}$ the value written by p's earlier successful SC (Line 24), (2) writes into $loc_p.\texttt{oldval}$ the sequence number of that SC (Line 25), and (3) increments its sequence number (Line 26). Finally, p signals successful completion of the SC by returning *true* (Line 27).

We now turn to the procedure LL$(p, \mathcal{O})$ that describes how a process p performs an LL operation on $\mathcal{O}$. In the following, let $SC_{q,i}$ denote the ith successful SC by process q, and $v_{q,i}$ denote the value written in $\mathcal{O}$ by $SC_{q,i}$. First, p reads the current value x of variable X (Line 1). Suppose that $SC_{q,k}$ is the latest successful SC operation to write into X before p reads X. Then, if $x = (1, l)$ (Line 2), we have $k = 1$ (Line 4). Furthermore, l is the address of the memory block containing q's shared variables. Since it is possible that l has not yet been inserted into the DynamicArray $\mathcal{D}$, p inserts l into $\mathcal{D}$ (Line 5) and increments N to a value that is by one greater than the value of q's name (Line 6). By doing so, q ensures that before another process obtains a new name, the following holds: (1) the address of p's memory block has been inserted into $\mathcal{D}$, and (2) variable N is by one greater than p's name. As a result, each process obtains a name that is unique, and all entries in array $\mathcal{D}$ are written (for the first time) in sequential order: entry j is written before entry $j + 1$, for all $j \geq 0$.

If $x = (0, i, k)$ (Line 2), then we have $k > 1$. Furthermore, by the above argument, the address l of q's memory block has already been written into location i of array $\mathcal{D}$. So, p simply reads that location to obtain l (Line 7).

Notice that, in both of the above cases, p is able to obtain address l of q's memory block: either directly from X (Line 3), or indirectly from $\mathcal{D}$ (Line 7). From this point onwards, p proceeds in the same way as in the algorithm in Figure 1. In particular, p first reads $l\rightarrow$val$[k \bmod 2]$ to try to learn $v_{q,k}$ (Line 9). Next, p reads the sequence number k' in $l\rightarrow$oldseq (Line 10). If $k' = k - 2$ or $k' = k - 1$, then SC$_{q,k+1}$ has not yet completed, and the value v obtained on Line 9 *is* $v_{q,k}$. So, p terminates the LL operation, returning v (Line 11). If $k' \geq k$, q must have completed SC$_{q,k+1}$. Hence, the value in $l\rightarrow$oldval is $v_{q,k}$ or a later value (more precisely, the value in $l\rightarrow$oldval is $v_{q,i}$ for some $i \geq k$). Therefore, the value in $l\rightarrow$oldval is not too old for p's LL to return. Accordingly, p reads the value v' of $l\rightarrow$oldval (Line 12) and returns it (Line 13).

The VL procedure is self-explanatory. Based on the above, we have the following theorem.

Theorem 3. *The wait-free algorithm in Figure 2 is linearizable. The time complexity of Join, LL, SC, and VL is* $O(1)$. *The space complexity of the algorithm is* $O(K^2 + KM)$, *where* K *is the total number of processes that have joined the algorithm.*

References

1. Herlihy, M.: Wait-free synchronization. ACM TOPLAS **13** (1991) 124–149
2. Lamport, L.: Concurrent reading and writing. Communications of the ACM **20** (1977) 806–811
3. International, S.: (The SPARC Architecture Manual) Version 9.
4. Corporation, I.: Intel Itanium Architecture Software Developer's Manual Volume 1: Application Architecture. (2002) Revision 2.1.
5. Group, I.S.: IBM e server POWER4 System Microarchitecture. (2001)
6. Site, R.: Alpha Architecture Reference Manual. Digital Equipment Corporation. (1992)
7. Center, I.T.W.R.: System/370 Principles of operation. (1983) Order Number GA22-7000.
8. Moir, M.: Practical implementations of non-blocking synchronization primitives. In: Proceedings of the 16th Annual ACM Symposium on Principles of Distributed Computing. (1997) 219–228
9. Afek, Y., Dauber, D., Touitou, D.: Wait-free made fast. In: Proceedings of the 27th Annual ACM Symposium on Theory of Computing. (1995) 538–547
10. Barnes, G.: A method for implementing lock-free shared data structures. In: Proceedings of the 5th Annual ACM Symposium on Parallel Algorithms and Architectures. (1993) 261–270
11. Herlihy, M.: A methodology for implementing highly concurrent data structures. ACM Transactions on Programming Languages and Systems **15** (1993) 745–770
12. Jayanti, P.: f-arrays: implementation and applications. In: Proceedings of the 21st Annual Symposium on Principles of Distributed Computing. (2002) 270 – 279
13. Jayanti, P.: An optimal multi-writer snapshot algorithm. In: Proceedings of the 37th annual ACM symposium on Theory of computing. (2005) 723–732
14. Moir, M.: Transparent support for wait-free transactions. In: Proceedings of the 11th International Workshop on Distributed Algorithms. (1997) 305–319

15. Moir, M.: Laziness pays! Using lazy synchronization mechanisms to improve non-blocking constructions. Distributed Computing **14** (2001) 193–204
16. Shavit, N., Touitou, D.: Software transactional memory. In: Proceedings of the 14th Annual ACM Symposium on Principles of Distributed Computing. (1995) 204–213
17. Anderson, J., Moir, M.: Universal constructions for large objects. In: Proceedings of the 9th International Workshop on Distributed Algorithms. (1995) 168–182
18. Anderson, J., Moir, M.: Universal constructions for multi-object operations. In: Proceedings of the 14th Annual ACM Symposium on Principles of Distributed Computing. (1995) 184–194
19. Doherty, S., Herlihy, M., Luchangco, V., Moir, M.: Bringing practical lock-free synchronization to 64-bit applications. In: Proceedings of the 23rd Annual ACM Symposium on Principles of Distributed Computing. (2004) 31–39
20. Israeli, A., Rappoport, L.: Disjoint-Access-Parallel implementations of strong shared-memory primitives. In: Proceedings of the 13th Annual ACM Symposium on Principles of Distributed Computing. (1994) 151–160
21. Jayanti, P., Petrovic, S.: Efficient wait-free implementation of multiword LL/SC variables. (To appear in 25th International Conference on Distributed Computing Systems (ICDCS 2005))
22. Jayanti, P., Petrovic, S.: Efficiently implementing a large number of LL/SC variables. Technical Report TR2005-446, Dartmouth College Computer Science Department (2005)
23. Jayanti, P., Petrovic, S.: Efficient and practical constructions of LL/SC variables. In: Proceedings of the 22nd ACM Symposium on Principles of Distributed Computing. (2003)
24. Luchangco, V., Moir, M., Shavit, N.: Nonblocking k-compare-single-swap. In: Proceedings of the fifteenth annual ACM symposium on Parallel algorithms and architectures. (2003) 314–323
25. Michael, M.: Practical lock-free and wait-free LL/SC/VL implementations using 64-bit CAS. In: Proceedings of the 18th Annual Conference on Distributed Computing. (2004) 144–158

Placing a Given Number of Base Stations to Cover a Convex Region

Gautam K. Das, Sandip Das, Subhas C. Nandy, and Bhabani P. Sinha

Indian Statistical Institute, Kolkata 700 108, India

Abstract. An important problem of mobile communication is placing a given number of base-stations in a given convex region, and to assign range to each of them such that every point in the region is covered by at least one base-station, and the maximum range assigned is minimized. The algorithm proposed in this paper uses Voronoi diagram, and it works for covering a convex region of arbitrary shape. Experimental results justify the efficiency of our algorithm and the quality of the solution produced.

1 Introduction

In a mobile radio network, a set of base-stations are appropriately positioned in a desired area, and their transmission ranges are assigned. The mobile terminals communicate with its nearest base-station, and the base-stations communicate with each other over scarce wireless channels in a multi-hop fashion. Each base-station emits signals periodically, and all the mobile terminals within its range can identify it as its nearest base-station after receiving such signals. We study the problem of positioning the base-stations and the assignment of transmission ranges such that the entire area under consideration is covered, and the total power consumed by all the base-stations is minimum.

We assume that, the region to be covered is a convex polygon in 2D, the number of base-stations is given *a priori*, and the range assigned to each of them is same. If the range of a base-station is ρ, it can communicate with all the mobile terminals present in the circular region of radius ρ and centered at the position where the base-station is located. Our problem is to minimize ρ by identifying the positions of the base-stations appropriately.

It is slightly different from the well-known k-center problem in 2D, where we need to place a set S of k supply points on the plane such that the maximum Euclidean distance of a demand point from its nearest supply point is minimized. For a given set D of n demand points, the k-center problem can be solved using parametric search technique when k is small. For a fixed value of k, the best known algorithm for this problem runs in $O(n^{O(\sqrt{k})})$ time [4]. But, if k is a part of the input, then the problem becomes NP-complete [2]. In our case, the set of demand points D is the entire convex region under consideration, and the problem is referred to as a *covering problem* in the literature. Two variations of this problem are studied:

(i) finding the minimum number of unit-radius circles that are necessary to cover a given square, and

(ii) finding the arrangement (positioning) of the members in S and determining a real number ρ such that the circles of radius ρ centered at positions in S can cover the unit square, but for any real number $\rho' < \rho$, there exists no arrangement of S which can cover the entire unit square.

In [12], a lower bound was given for problem (i); it says that if m is the minimum number of unit circles required for covering a square with each side of length σ, then $\frac{3\sqrt{3}}{2}m > \sigma^2 + c\sigma$, where $c > \frac{1}{2}$. Substantial studies have been done on problem (ii) [3], [6], [7], [8], [9], [11]. The objective was to cover a unit square region with a given number (say k) of equal radius circles with minimum radius. In [9], simulated annealing approach was used to obtain near-optimal solutions for the unit square covering problem for $k \leq 30$. As it is very difficult to get a good stopping criteria for a stochastic global optimization problem, they used heuristic approach to stop their program. It is mentioned that, for $k = 27$ their algorithm runs for about 2 weeks to achieve the stipulated stopping criteria. For $k > 28$, the time requirement is very high. So, they have changed their stopping criteria, and presented the results. In [8], the same approach is adopted for covering a equilateral triangle of unit edge length with circles of equal radius, and results are presented for different values of k.

We have adopted a geometric approach using Voronoi diagram for solving the same problem in a more general situation, where the region to be covered may be a convex polygon of arbitrary shape. Experimental results say that our algorithm terminates in a fraction of a second for reasonably large values of k. We could compare our results when the region to be covered is a square or an equilateral triangle and when k is small (≤ 30). The solutions produced by our algorithm are favorably comparable with that of [8], [9]. Thus, our algorithm will be very useful in practical applications.

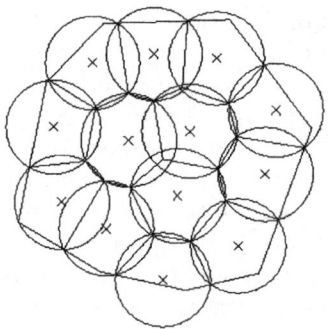

Fig. 1. Illustration of our problem

2 Algorithm

Consider a set of points $P = \{p_1, p_2, \ldots, p_k\}$ inside a convex polygon Π where the i-th base-station is located at point p_i. We use $VOR(P)$ to refer the Voronoi diagram [1] of the set of points P, and $vor(p_i)$ to denote the Voronoi polygon of a point $p_i \in P$. Since we need to establish communication inside Π, if a part of the region $vor(p_i)$ goes outside Π for some i, then the region $vor(p_i) \cap \Pi$ is used as $vor(p_i)$. Note that, all the points inside $vor(p_i)$ are closer to p_i than any other point $p_j \in P$, $j \neq i$. Thus, all these points will communicate with p_i. As the base stations are of equal range, our objective is to arrange the points in P inside Π such that the maximum range required (ρ) among the points in P is minimized. Our algorithm is an iterative one. At each step, it perturbs the point set P as described below, and finally, it attains a local minimum.

In each iteration, we compute $VOR(P)$ [1], and then compute the circumscribing circle C_i of each $vor(p_i)$ using the algorithm proposed in [10]. Let r_i denote the radius of C_i. In order to cover a convex polygon by a base-station with minimum range, we need to place the base-station at the center of the circumscribing polygon of that convex region with range equal to the radius of that circle. Thus, for each $i = 1, 2, \ldots, k$, we move p_i to the center of C_i and assign range r_i to it. Next, we compute $\rho = max\{r_i, i = 1, 2, \ldots, k\}$.

Lemma 1. *At each iteration, (i) the newly assigned position of each point p_i lies inside the corresponding $vor(p_i)$, and (ii) the value of ρ decreases.*

Remark 1. The iteration terminates when the value of ρ reaches to a local minima, or in other words, $\rho_{new} = \rho_{old}$ is attained.

We also apply a refinement step to improve the solution. Note that, if a point (base-station) p_i is on the boundary of Π, then at least 50% of the area of C_i lies outside Π, and hence this region need not be covered. This indicates, the scope of further reduction in the area of C_i. Thus, if a point goes very close to the boundary of Π, we move it to the centroid of Π, whose coordinate is computed as $(\frac{1}{m}\sum_{j=1}^{m} x_\alpha, \frac{1}{m}\sum_{j=1}^{m} y_j)$, where m is the number of vertices of Π. It can be shown that, the centroid of a convex region is always inside that region.

It is observed that, such a major perturbation moves the solution from a local minima, and it leads to a scope of further reduction in ρ. We again continue iteration with this initial placement until it again reaches another local minima.

Theorem 1. *The worst case time complexity of an iteration is $O(k \log k)$.*

Proof: The factors involved in this analysis are (i) computing $VOR(P)$, which can be done in $O(k \log k)$ time [1], and (ii) computing C_i for all $i = 1, 2, \ldots, k$, which needs $O(k)$ time due to the fact that each edge appears in at most two Voronoi cells, and computing the circular hull of a convex polygon needs time linear in its number of edges [10]. □

It is observed that the number of iterations needed to reach to a local optima from an initial configuration is reasonably small. The overall time complexity depends on the number of times we apply the refinement step.

3 Experimental Results

An exhaustive experiment is performed with several convex shapes of the given region and with different values of k. It is easy to show that, for a given initial

Table 1. Covering a unit square

k	ρ_{opt} using method in [9]	ρ^*_{opt} using our method	k	ρ_{opt} using method in [9]	ρ^*_{opt} using our method
4	0.35355339059327376220	0.353553	18	0.16063966359715453523	0.160682
5	0.32616054400398728086	0.326165	19	0.15784198174667375675	0.158345
6	0.29872706223691915876	0.298730	20	0.15224681123338031005	0.152524
7	0.27429188517743176508	0.274295	21	0.14895378955109932188	0.149080
8	0.26030010588652494367	0.260317	22	0.14369317712168800049	0.143711
9	0.23063692781954790734	0.230672	23	0.14124482238793135951	0.141278
10	0.21823351279308384300	0.218239	24	0.13830288328269767697	0.138715
11	0.21251601649318384587	0.212533	25	0.13354870656077049693	0.134397
12	0.20227588920818008037	0.202395	26	0.13176487561482596463	0.132050
13	0.19431237143171902878	0.194339	27	0.12863353450309966807	0.128660
14	0.18551054726041864107	0.185527	28	0.12731755346561372147	0.127426
15	0.17966175993333219846	0.180208	29	0.12555350796411353317	0.126526
16	0.16942705159811602395	0.169611	30	0.12203686881944873607	0.123214
17	0.16568092957077472538	0.165754			

Table 2. Covering a equilateral triangle

k	ρ_{opt} using method in [8]	ρ^*_{opt} using our method	k	ρ_{opt} using method in [8]	ρ^*_{opt} using our method
4	0.2679491924311227065	0.267972	21	0.0962250448649376274	0.099165
5	0.2500000000000000000	0.250006	22	0.0951772351261450917	0.095877
6	0.1924500897298752548	0.192493	23	0.0937742911094478264	0.094625
7	0.1852510855786008545	0.185345	24	0.0923541375945022204	0.093982
8	0.1769926664029649641	0.177045	25	0.0906182448311340175	0.091688
9	0.1666666666666666667	0.166701	26	0.0887829248953373781	0.090231
10	0.1443375672974064411	0.144681	27	0.0868913397937031505	0.088238
11	0.1410544578570137366	0.141252	28	0.0824786098842322521	0.086795
12	0.1373236156889236662	0.137633	29	0.0818048133956910115	0.084545
13	0.1326643857765088351	0.133379	30	0.0808828500258641436	0.082246
14	0.1275163863998600644	0.127829	31	0.0798972448089536737	0.081665
15	0.1154700538379251529	0.115811	32	0.0788506226168764215	0.080457
16	0.1137125784440782042	0.114574	33	0.0776371221483728244	0.079604
17	0.1113943099632405880	0.112141	34	0.0763874538343494465	0.078827
18	0.1091089451179961906	0.109890	35	0.0751604548962267707	0.076918
19	0.1061737927289732618	0.107288	36	0.0721687836487032206	0.075950
20	0.1032272183417310354	0.104049			

placement of P, at each iteration the value of ρ is decreased. As the process reaches a local minima, the quality of the result completely depends on the initial choice of the positions of P. We have studied the problem with random distribution of P. It shows that in an ideal solution, the distribution of points is very regular. So, while working with unit square region, we choose the initial

Table 3. Performance evaluation of the algorithm

k	ρ^*_{opt}	$\rho_{average}$	std. devn.	Time (in sec.)	k	ρ^*_{opt}	$\rho_{average}$	std. devn.	Time (in sec.)
4	0.353553	0.395284	0.040423	0.052	18	0.160682	0.164347	0.001092	0.351
5	0.326165	0.326247	0.000201	0.073	19	0.158345	0.160797	0.000885	0.377
6	0.298730	0.309837	0.008433	0.090	20	0.152524	0.156772	0.000877	0.405
7	0.274295	0.27603	0.001668	0.107	21	0.149080	0.153131	0.001253	0.436
8	0.260317	0.26131	0.003079	0.124	22	0.143711	0.148640	0.000582	0.465
9	0.230672	0.231119	0.000540	0.143	23	0.141278	0.145498	0.001738	0.499
10	0.218239	0.218244	0.000004	0.164	24	0.138715	0.142105	0.001507	0.531
11	0.212533	0.213855	0.000894	0.184	25	0.134397	0.139549	0.001572	0.557
12	0.202395	0.205567	0.000908	0.206	26	0.132050	0.136489	0.001618	0.587
13	0.194339	0.194960	0.000645	0.228	27	0.128660	0.133725	0.001298	0.623
14	0.185527	0.189217	0.001722	0.258	28	0.127426	0.131589	0.001357	0.655
15	0.180208	0.182782	0.001883	0.279	29	0.126526	0.129241	0.000964	0.688
16	0.169611	0.174669	0.003178	0.303	30	0.123214	0.127069	0.000881	0.719
17	0.165754	0.168231	0.002336	0.327					

placement of the points in P as follows: compute $m = \lfloor \sqrt{k} \rfloor$. If $m^2 = k$, we split the region into $m \times m$ cells, and in each cell place a point of P randomly. If $k - m^2 < m$, then split the region into m rows of equal width. Then, arbitrarily choose $(k - m^2)$ rows and split each of these rows into $(m + 1)$ cells; the other rows are split into m cells. Now place one point in each cell. If $k - m^2 > m$, then split the square into $m + 1$ rows, and each row is split into m or $m + 1$ rows to accommodate all the points in P.

For each k, we have chosen 1000 initial instances. For each of these instances, we have run our algorithm, and have computed ρ_{min} which is the minimum value of ρ observed during the experiment. Finally, we report ρ^*_{opt} = minimum value of ρ_{min} over all the 1000 instances. Thus, ρ^*_{opt} indicates the minimum value of ρ that is achieved by our experiment. In Table I, we have compared ρ^*_{opt} with the value of ρ_{opt} obtained by the algorithm in [9] for different values of k.

We have also compared our method with that of [8] when the region is an equilateral triangle. The experimental results for different values of k appear in Table II. Figure 1 demonstrates the output of our algorithm for covering a given convex polygon with 13 circles.

In order to present the performance of our heuristic, we report the minimum, average and standard deviation of the value of ρ_{min} over all the 1000 instances for different values of k with unit square region (see Table III). We have performed the entire experiment in SUN BLADE 1000 machine with 750 MHz CPU speed, and have used LEDA [5] for computing the Voronoi diagram. The average time for processing each instance is also given. Similar results are observed with equilateral triangular area; so it is not specifically mentioned.

Experimental results indicate that the solutions produced by our algorithm are very close to those of the existing results on this problem where the region is a square [9] and an equilateral triangle [8]. This is highly acceptable in the context of our application. It is mentioned in [8,9] that for a reasonably large value of k (≥ 27), it need to run several weeks to get the solution, whereas our method needs a fraction of a second. This is very important in this particular application.

References

1. M. de Berg, M. Van Kreveld, M. Overmars and O. Schwarzkopf, Computational Geometry Algorithms and Applications, Springer-Verlag, 1997.
2. R. J. Fowler, M. S. Paterson and S. L. Tanimoto, Optimal packing and covering in the plane are NP-complete, Information Processing Letters 12 (1981) 133 - 137
3. A. Heppes and J. B. M. Melissen, Covering a rectangle with equal circles ,Periodica Mathematica Hungarica 34 (1997) 65 - 81
4. R. Z. Hwang, R. C. T. Lee and R. C. Chang, The slab dividing approach to solve the Euclidean p-center problem, Algorithmica 9 (1993)1 - 22
5. K. Mehlhorn and S. Nher, The LEDA Platform of Combinatorial and Geometric Computing, Cambridge University Press, 1999.
6. J. B. M. Melissen and P. C. Schuur, Covering a rectangle with six and seven circles, Discrete Applied Mathematics 99 (2000) 149 - 156
7. J. B. M. Melissen and P. C. Schuur, Improved covering a rectangle with six and seven circles, Electronic J. on Combinatorics 3 (1996) R32
8. K. J. Nurmela, Conjecturally optimal coverings of an equilateral triangle with up to 36 equal circles, Experimental Mathematics 9 (2000)
9. K. J. Nurmela and P. R. J. Ostergard, Covering a square with up to 30 Equal Circles, Research Report HUT-TCS-A62, Laboratory for Theoretical Computer Science, Helsinky University of Technology, 2000.
10. N. Megiddo, Linear-time algorithms for linear programming in R^3 and related problems, SIAM Journal on Computing 12 (1983) 759 - 776
11. T. Tarnai and Z. Gasper, Covering a square by equal circles, Elementary Mathematics 50 (1995)167 - 170
12. S. Verblunsky, On the least number of unit circles which can cover a square, Journal of the London Mathematical Society 24 (1949) 164 - 170

A State-Space Search Approach for Optimizing Reliability and Cost of Execution in Distributed Sensor Networks

Archana Sekhar[1], B.S. Manoj[2], and C. Siva Ram Murthy[3]

[1] McKinsey & Company, Mumbai 400 021, India
Archana_Sekhar@mckinsey.com
[2] Department of Electrical and Computer Engineering,
University of California at San Diego, San Diego, CA 92093, USA
bsmanoj@ucsd.edu
[3] Department of Computer Science and Engineering,
Indian Institute of Technology Madras, Chennai 600 036, India
murthy@iitm.ac.in

Abstract. Sensor networks are increasingly being used for applications which require fast processing of data, such as multimedia processing. Distributed computing can be used on a sensor network to reduce the completion time of a task and distribute the energy consumption equitably across all sensors. The distribution of task modules to sensors should consider not only the time and energy savings, but must also improve reliability of the entire task execution. We formulate the above as an optimization problem, and use the A^* algorithm with improvements to determine an optimal static allocation of modules among a set of sensors. We also suggest a faster but suboptimal algorithm, called the greedy A^* algorithm. Both algorithms have been simulated, and the results have been compared in terms of energy savings, decrease in completion time of the task, and the deviation of the sub-optimal solution from the optimal one. The sub-optimal solution required 8-35% less computation, at the cost of 2.5-15% deviation from the optimal solution in terms of average energy spent per sensor node. Both the A^* and greedy A^* algorithms have been shown to distribute energy consumption more uniformly across sensors than centralized execution. The greedy A^* algorithm is found to be scalable, as the number of evaluations in determining the allocation increases linearly with the number of sensors.

1 Introduction

Sensor networks consist of a large number of small, lightweight, highly resource-constrained wireless devices called sensors. Typical scenarios of application of sensors include habitat monitoring, intrusion detection, chemical and meteorological sensing, and military use [1]. Sensors have limited processing capability and battery power, and are typically not equipped with rechargeable or replaceable power sources. With an increase in data- and computation-intensive applications such as multimedia processing and data collaboration, which may be subject to time constraints, distributed applications play a vital role in sensor networks. A given task can be split into modules and allocated to a group of sensor nodes considering (a) minimum completion time, (b) minimum energy consumed per node, (c) increased reliability. The process of splitting a task into modules introduces computation and communication overheads. In a centralized execution, one sensor spends a large amount of energy to complete a complex task. By distributing it, each sensor spends some energy towards the task completion. This distribution of energy consumption ensures that the whole network is equally involved in computation and collaboration, which is preferred in sensor networks, since it avoids the premature death of some sensors due to battery drain.

Distributed sensor networks have generated research interest in recent times. In [2], an architecture called SensorWare has been proposed which utilizes the computation, communication, and sensing resources available in sensor nodes using lightweight and mobile scripts. EnviroTrack [3] is an environmental computing paradigm proposed for sensor networks, which develops embedded systems of massively distributed, disposable sensors for habitat monitoring and intrusion tracking. Distributed computing systems have been explored in the wired domain, as a set of interconnected processors which together perform a specific task. Algorithms have been studied to allocate modules efficiently to different processors [4]. The lifetimes of sensors and communication links are assumed to follow exponential distribution, the simplest life distribution model [6]. Running applications which require high computational resources often proves to be difficult on single sensor nodes, due to their limited processing capabilities. On the other hand, such computation-intensive applications are increasing in number and importance, which makes it imperative to explore the possibilities of distributing the computation across nearby sensors. Distributed real-time applications on sensors are being studied in the context of military applications. Multi-spectral image analysis is used to derive surveillance information from wavelengths outside the visible range of the spectrum. This requires algorithms such as auto-correlation and the fast Fourier transform to be executed in a distributed manner. Such applications make distributed computation on sensor networks extremely essential. The organization of the rest of this paper is as follows: We present our work in Section 2. Our simulation results are presented in Section 3, and we summarize our findings in Section 4.

2 Our Work

We have presented an optimization problem formulation for the distribution of tasks on a sensor network. The major costs involved in the execution of any task are those of computation and communication. A task is split into modules, and is then allocated to the sensors in the network. We have considered a static allocation of tasks, where the split-up into modules, and the expected communication between modules, is known *a priori*. Also, the distribution of modules is performed among a known set of sensors. A central entity such as a base station (BS) could program sensors to perform certain modules of the task, and communicate the results to it. Alternatively, a sensor could itself distribute a task among its neighbors. While computation costs depend on the capability of the processor and the total processing required for a task, communication costs depend on the bandwidth available between two nodes and the inter-module communication between the modules running on them. We have considered a heterogeneous network in which nodes have different processing speeds and communication and computation costs. Node failure has been assumed to be mainly due to battery drain, since the network is highly power-constrained. Link reliability has also been considered. Consider a task such as intruder-tracking or a multimedia application, which has to be split into modules and distributed among the nodes of a sensor network. Let there be n nodes available, labeled $N_0, N_1, ..., N_{n-1}$. The task T is split into m modules i.e., $T = M_0, M_1, ..., M_{m-1}$. Several methods for splitting an application in to modules (or tasks) can be found in [7]. We consider a heterogeneous network, where the nodes have different processing capabilities. Let the processing speed of each node be recorded in a matrix $PROC$ of order $1 \times n$, where $PROC[i]$ represents the processing capability of node N_i, for $i = 0, 1, ..., n-1$. The communication links between different nodes also have different speeds, represented by the $n \times n$ matrix $LINK$, where $LINK[i][j]$ is the speed of the link between node N_i and N_j, for i, j such that $0 \leq i, j < n$. The diagonal entries of the $LINK$ matrix are set to infinity, since the speed of communication within a node is much faster than that across nodes (the network links are not required for communication within a node). Each module M_i has a certain computation requirement $COMP[i]$, where $0 \leq i < m$. The inter-module communication requirement is represented by an $m \times m$ matrix IMC, where $IMC[i][j]$ is the communication requirement between modules M_i and M_j, for i, j such that $0 \leq i, j < m$. The maximum computational load that a node can handle is given by $LOAD[i]$ where $0 \leq i < n$ and the maximum available energy of a node is given by $ENERGY[i]$ where $0 \leq i < n$. Let the energy required for unit computation on node N_i be $E_{COMP}[i]$ and that for unit communication be $E_{COMM}[i]$. Typically, faster nodes have a higher value of E_{COMP}.

Since the nodes of the network have different processing capabilities, execution of a module on different nodes will entail different costs. To model this, an $m \times n$ matrix $exec$ is used, where $exec[i][j]$ represents the execution cost of module M_i on node N_j, $0 \leq i < m, 0 \leq j < n$. The entries of the matrix $exec$ are filled up as $exec[i][j] = COMP[i]/PROC[j]$ where $0 \leq i < m, 0 \leq j < n$. Similarly, non-identical communication links result in different communication costs when modules are executed on different nodes. A 4-dimensional matrix $comm$ is used to model the communication costs. $comm[i][j][k][l]$ is the communication cost incurred due to inter-module communication between modules M_i and M_j when they are executed on nodes N_k and N_l, respectively, for $0 \leq i, j < m$ and $0 \leq k, l < n$. It is assumed that the communication cost between modules executing on the same node is 0. The matrix $comm$ is filled up using the equation $comm[i][j][k][l] = IMC[i][j]/LINK[k][l]$ for $0 \leq i, j < m, 0 \leq k, l < n$. Since the denominator term $LINK[k][l]$ is set to ∞ for $k = l$, the communication cost within the same node goes to 0. All the m modules are to be assigned to the n nodes, and the assignment is represented by an $m \times n$ binary matrix X.

$$X[i][j] = 1 \; if \; M_i \text{ is assigned to } N_j$$
$$= 0 \;\; \text{otherwise} \tag{1}$$

Since a given module is assigned to one and only one node, the row-sum of any row of the assignment matrix must be 1. Hence

$$\sum_{j=1}^{n} X[i][j] = 1 \tag{2}$$

The computation cost of the task is given by

$$\sum_{i=1}^{m}\sum_{j=1}^{n} X[i][j]exec[i][j] \tag{3}$$

The total communication cost of the task is

$$\sum_{i=1}^{m}\sum_{j=1}^{m}\sum_{p=1}^{n-1}\sum_{q>p} X[i][p]X[j][q]comm[i][j][p][q] \tag{4}$$

An important feature of our modeling is the inclusion of reliability as a criterion for the assignment of modules to nodes. The reliability of a node N_k, $0 \le k < n$ in a time interval t is $e^{-\lambda_k t}$ where λ_k is the failure rate of node N_k [6]. The failure rate is inversely proportional to the available energy of the node. Hence, it has been modeled as the reciprocal of the available energy. The time for which a module M_i runs on a node N_k under a given assignment X is $exec[i][k]$. Hence, the total running time of the modules on a node under X is given by $\sum_{i=1}^{m} X[i][k]exec[i][k]$. The reliability of the node N_k is thus given by

$$R_k(T,X) = exp(-\lambda_k \sum_{i=1}^{m} X[i][k]exec[i][k]) \tag{5}$$

Similarly, link reliability is also modeled to account for the vagaries of the wireless medium. A matrix μ is used to model the failure rate of paths between any two nodes. $\mu[p][q]$ denotes the failure rate of the path between nodes N_p and N_q. Then, the reliability of the path is given by

$$R_{pq}(T,X) = exp(-\mu[p][q] \sum_{i=1}^{m}\sum_{j=1}^{m} X[i][p]X[j][q]comm[i][j][p][q]) \tag{6}$$

Then the reliability of the entire task is given by the product of all the individual node reliabilities and link reliabilities. Hence

$$R(T,X) = [\prod_{k=1}^{n} R_k(T,X)][\prod_{p=1}^{n-1}\prod_{q>p} R_{pq}(T,X)] \tag{7}$$

Using Equations (5) and (6), this can be rewritten as

$$R(T,X) = exp(-RelCost(X)) \tag{8}$$

The term *RelCost* must be minimized to ensure that the reliability of the entire task is maximized. Hence, using the expressions for the computation, communication, and reliability costs (from Equations (3), (4), and (8)) the objective function of the task assignment is to minimize the cost

$$\sum_{i=1}^{m}\sum_{j=1}^{n} X[i][j]exec[i][j] +$$

$$\sum_{i=1}^{m}\sum_{j=1}^{m}\sum_{p=1}^{n-1}\sum_{q>p} X[i][p]X[j][q]comm[i][j][p][q] + RelCost(X) \qquad (9)$$

Equivalently, substituting for $RelCost(X)$, the objective is to minimize

$$\sum_{i=1}^{m}\sum_{j=1}^{n}(1+\lambda_j)X[i][j]exec[i][j] + \sum_{i=1}^{m}\sum_{j=1}^{m}\sum_{p=1}^{n-1}\sum_{q>p}$$
$$(1+\mu[p][q])X[i][p]X[j][q]comm[i][j][p][q]) \qquad (10)$$

Besides the row-sum constraint on the assignment matrix (Equation (2)), the modules executed on a node must satisfy two other resource constraints – the total energy required must be less than the available energy at the node and the total computational load offered must be within the capacity of the node. These are represented by the following inequality constraints.

$$\sum_{i=1}^{m} X[i][k]exec[i][k]E_{COMP}[k] +$$

$$\sum_{i=1}^{m-1}\sum_{p=1}^{n}\sum_{j=1,j>i}^{m} X[i][k]comm[i][j][k][p]E_{COMM}[k] \leq ENERGY[k] \qquad (11)$$

$$\sum_{i=1}^{m} X[i][k]exec[i] \leq LOAD[k] \qquad (12)$$

The optimization problem is now formulated, with the objective as in Equation (10), and constraints of Equations (2), (11), and (12). This is a generic problem formulation, which reduces to simpler special cases depending on the values given to parameters μ and λ. If the sensors are assumed to have ample energy, and hence are very reliable, then the values of $\lambda[i]$ go to 0. Similarly, if the communication links are also assumed to be reliable, the μ matrix is set to 0.

2.1 Computation of Optimal Module Allocation

We use the A^* algorithm [8] to find an optimal allocation of modules among a set of sensors. Each vertex x in the search tree represents a partial allocation of modules to sensors. A goal vertex represents a complete allocation of all modules. Every vertex x has an associated cost function $f(x)$, which is a lower bound on the minimum cost of a complete allocation which includes the partial allocation A_x at vertex x. Any goal vertex in the sub-tree rooted at x will have a cost greater than $f(x)$. $f(x) = g(x) + h(x)$, where $g(x)$ is the cost of the partial allocation A_x and $h(x)$ is a lower bound on the minimum cost of a path from vertex x to a goal vertex. $h(x)$ is calculated by making a temporary allocation of all the unallocated modules, and summing up their computation costs and their communication costs *only with* the modules already allocated in the partial allocation $A(x)$. The search begins with the null allocation, where no module has been assigned a sensor. At each stage in the search, the vertex with minimum $f(x)$ is expanded, until a goal vertex is reached. The order in which modules are allocated to sensor nodes greatly affects the required computation for the solution search. Suppose there are k independent modules (which do not have inter-module communication among themselves). Then the tentative allocation represented by vertex x at level $m - k$ itself is a goal vertex, since the only costs induced further in the subtree rooted at x are the computation costs, which are already included in the calculation of $f(x)$. This restricts the search to only $m - k$ levels of the search tree. In order to ensure feasibility of the temporary allocation, the energy and load constraints must also be checked in the computation of $h(x)$. Finding the maximal set of independent modules is an NP-complete problem [9]. We use the algorithm independent-module-set heuristic to find a set of independent modules, as presented by Sinclair [5] and on the ordered set of modules produced by this algorithm, the A^* algorithm is applied.

Algorithm Independent-module-set

1. $M = $ all modules, $I = \phi$
2. Compute the degree of each module
3. While(M contains more than 1 module)
    a. Find a module x in M of minimum degree, Remove x from M and add to I
    b. $\forall\ y \in M$ such that x and y communicate
        i. Remove y from M
        ii. $\forall\ z$ in M such that y and z communicate, reduce degree of z by 1
end while
4. Insert last remaining module in I

Algorithm Optimal-module-allocation

1. Set terminating_level $= m - k$, order the modules in M using Independent-module-set
2. Insert root vertex(ϕ, ϕ, ..., ϕ) in a list OPEN. Set $f(r) = 0$ and vertex_level $= 0$
3. While (vertex_level != terminating_level)
    a. Move the vertex x with least $f(x)$ to a list CLOSED
    b. if(vertex_level(x) < terminating_level)
        i. Expand x by assigning next unassigned module to all sensor nodes
        ii. Insert all feasible new vertices into OPEN
        iii. vertex_level of each new vertex $=$ vertex_level(x) $+1$
end while
4. Return the assignment of vertex x

2.2 Greedy A^* Algorithm

The A^* algorithm guarantees optimal allocation of modules, but at the expense of evaluations of many solution points (vertices) in the search tree. Since the execution of the algorithm itself could drain the resources of a sensor node, in this case, a simpler sub-optimal solution, given by the greedy A^* algorithm, can be preferred. We, in Step $3b.ii.$ of Optimal-module-allocation, instead of inserting all new feasible vertices into the $OPEN$ list, only the least cost vertex is inserted. This greedy approach, of exploring only the least-cost path, is called the greedy A^* algorithm. Consider the following example in which a task T involving 80 units of computation. This is now to be distributed among 4 nodes (sensors) $N_0, N_1, N_2,$ and N_3, with processing speeds 2,4,1, and 3, respectively. The cost of computation on each processor is proportional to the speed. Hence, the matrix E_{COMP} is $[2, 4, 1, 3]$. Suppose the task can be split into 5 modules, with computational loads [20,25,20,15,20]. The inter-module communication cost between modules and the interlink bandwidth between nodes are given by the matrix IMC and LINK, respectively.

$$IMC = \begin{bmatrix} 0 & 4 & 2 & 0 & 0 \\ 4 & 0 & 0 & 3 & 1 \\ 2 & 0 & 0 & 0 & 2 \\ 0 & 3 & 0 & 0 & 0 \\ 0 & 1 & 2 & 0 & 0 \end{bmatrix} \qquad LINK = \begin{bmatrix} \infty & 4 & 1 & 3 \\ 4 & \infty & 3 & 2 \\ 1 & 3 & \infty & 2 \\ 3 & 2 & 2 & \infty \end{bmatrix}$$

All nodes are assumed to have a starting energy of 500 units, and can take a maximum computational load of 100 units. The energy for communication E_{COMM} is assumed to be 4 units from all nodes on all links to other nodes. If the task is executed in a centralized fashion, assuming it is run on the fastest node (of speed 4), the completion time will be $80/4 = 20$ time units. The only energy spent will be for computation on node N_1. The energy per unit computation is 4 units ($E_{COMP}[1] = 4$), hence total energy spent is $(80 \times 4) = 320$ units of energy. Applying the ordering algorithm on the modules, the order obtained is [2,1,4,0,3]. Applying the A^* algorithm, after evaluation of 24 solution points in the solution tree, the optimal solution is determined as shown in columns 1 and 2 of Table 1. This allocation entails an execution time of 16.25 time units, and the energy spent at nodes N_1 and N_3 are 296 and 141 units, respectively. The completion time in the distributed allocation is less than the centralized execution, and the energy spent by the fastest node (node 1) is also reduced.

On the other hand, using the greedy A^* algorithm to explore only the least-cost path down the search tree, a solution is obtained after evaluation of 12 solution points. The allocation is shown in columns 1 and 3 of Table 1. The solution is sub-optimal, with completion time 21.67 time units. The energy consumption at nodes N_1 and N_3 is 176 and 231 units, respectively. The completion time of the greedy A^* allocation is close to that of the centralized execution, and the energy spent by node 1 is decreased. While the greedy A^* algorithm reduces the number of solution points evaluated in determining the module distribution, it may not provide the least completion time of tasks. Comparing the energy consumed in the centralized and distributed execution scenarios, node 1 spends 320 units in the centralized case, but only 296 units using A^* and 176 units using greedy A^* algorithm. This illustrates that the energy spent by a single sensor is reduced, and the load is partially shared by other sensors, *e.g.* node 3 spends 141 units of energy in the A^* allocation and 231 units in the greedy A^* allocation, respectively.

3 Results

The working of the A^* and greedy A^* algorithms was studied using simulations in C++. A task of 100 units of computation was split into 2, 3, 4, or 5 modules. The division of the task into modules introduces both computation and communication overheads. The added computation on each module was generated by a uniform random distribution of 1 to 5 units. The IMC cost matrix was generated as a uniform random distribution between 1 and 10 units of communication. To account for heterogeneity of nodes in the network, the speed of each node was a random integer between 1 and 5, and the cost of computation on a node was proportional to its speed. In our simulations, we have assumed the cost of computation equal to the speed. The subset of nodes among which the modules are to be distributed varies in size from 2 to 5. The cost of communication between any two nodes was specified by the $P:C$ ratio. The $P:C$ values of 1:5, 1:3, and 1:1 were used, indicating that communication is 5, 3, or 1 time(s) as expensive as computation. The bandwidth of links connecting any two nodes was uniformly distributed between 1 and 5 units. The initial state of the network, in terms of capacity of the nodes and available energy, was also modeled using a random distribution. Nodes had an initial computation capacity distributed in the range (800, 1200) and energy in the range (500, 800). The reliability of links, represented by the μ matrix, was uniformly distributed in (0, 1). The failure rate of the nodes was inversely proportional to the available energy of the nodes. The orthogonal factors which defined the input configurations were the number of modules (2, 3, 4, or 5), number of nodes (2, 3, 4, 5, or 6), and the $P:C$ ratio (1:5, 1:3, or 1:1). Each configuration was run on 10 random seeds. Hence, both the optimal A^* and the sub-optimal greedy A^* algorithms were run for $4 \times 5 \times 3 \times 10 = 600$ times. Distributing a task among a set of nodes results in faster completion of the task compared to executing it in a centralized form on 1 node. This was demonstrated by the difference in completion time of

Table 1. Module Allocation

Module	A^* Node	Greedy A^* Node
0	3	1
1	1	3
2	1	3
3	3	1
4	1	3

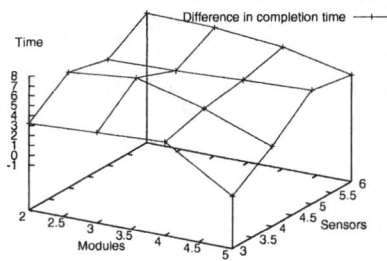 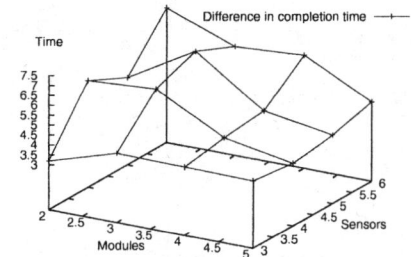

Fig. 1. Completion time of task with $P:C = 1:5$

Fig. 2. Completion time of task with $P:C = 1:1$

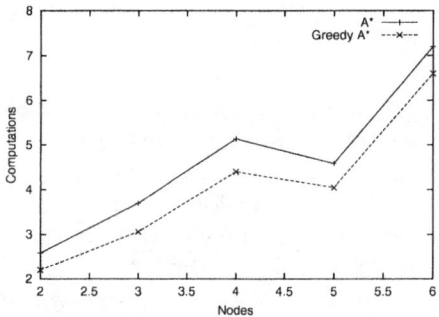 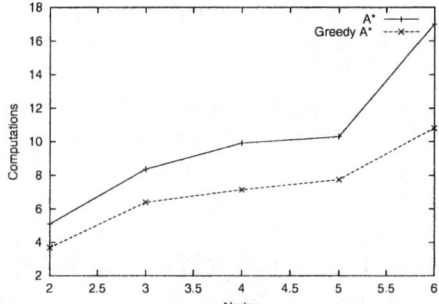

Fig. 3. Solution point evaluations for allocation of 2 modules

Fig. 4. Solution point evaluations for allocation of 5 modules

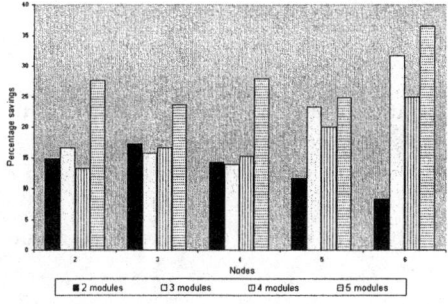 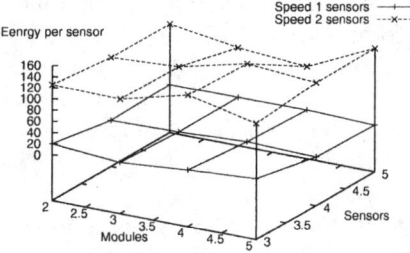

Fig. 5. Percentage savings in computation of sub-optimal solution by greedy A^* algorithm

Fig. 6. Energy spent with $P : C = 1:1$

the task under the centralized and distributed scenarios, as shown in Figures 1 and 2. The $P : C$ values in the two sets of results are 1:5 and 1:1, respectively. In these graphs, for most configurations of number of modules and nodes, the distributed execution of the task results in an earlier completion time, in spite of an increased computation overhead. The metric for comparison between the

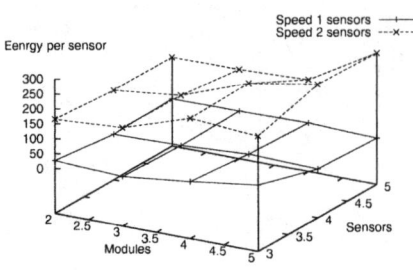
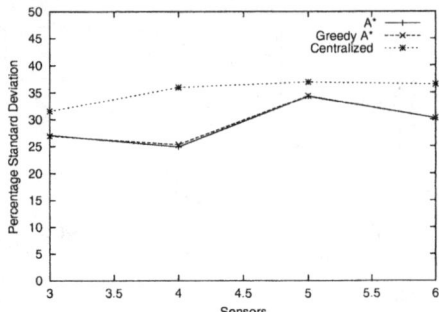

Fig. 7. Energy spent with $P : C = 1{:}5$

Fig. 8. Percentage standard deviation of energy consumed with fastest node executing centralized task

optimal A^* algorithm and the greedy A^* algorithm was the number of solution points evaluated till the goal node is reached. The results for 2 and 5 module allocations are shown in Figures 3 and 4. In these cases, the number of computations required for the optimal solution is more than that for the sub-optimal solution. Also, there is a trend of increase seen in the number of evaluations required for greater number of nodes and modules. The percentage savings in computations given by the sub-optimal solution over the optimal is shown in Figure 5. The deviation of the sub-optimal solution from optimality, in terms of total energy spent, is shown in Table 2. For the distribution of modules among 2, 3, 4, or 5 nodes, the average energy consumed is computed for the A^* and the greedy A^* algorithms. Averages have been computed over all $P : C$ values, and over different number of modules to be allocated. The entries in bold are the total energy consumed by all nodes and the percentage difference between the optimal and sub-optimal solution.

The distribution of a task across nodes results in a more uniform energy consumption across sensors compared to executing the entire task on a single sensor in a centralized fashion. In Figure 8, the fastest sensor is assumed to be chosen for

Table 2. Energy spent by nodes using the optimal allocation and the sub-optimal allocation

Node	5-nodes		4-nodes		3-nodes		2-nodes	
	A^*	Greedy A^*	A^*	Greedy A^*	A^*	Greedy A^*	A^*	Greedy A^*
Node 0	170.66	171.90	134.59	132.28	180.32	184.72	157.89	190.78
Node 1	192.78	195.97	170.09	154.54	202.57	205.48	271.70	304.08
Node 2	178.82	197.45	218.04	298.20	173.00	181.96	**429.59**	**494.86**
Node 3	105.93	107.66	198.33	219.29	**555.97**	**572.16**	**Increase - 15.2%**	
Node 4	110.28	113.12	**721.05**	**804.31**	**Increase - 2.9 %**			
Total	**758.40**	**786.10**	**Increase - 11.5 %**					
	Increase - 3.6 %							

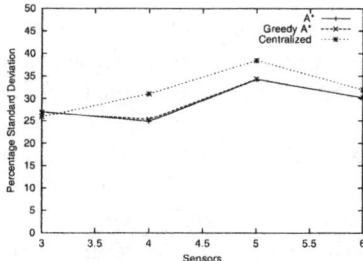 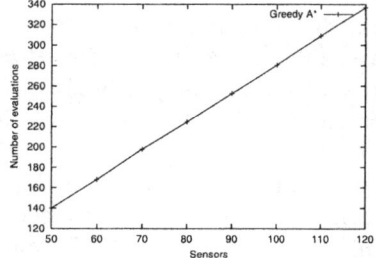

Fig. 9. Percentage standard deviation of energy consumed with node with maximum available energy executing centralized task

Fig. 10. Solution point evaluations of greedy A^* algorithm

execution of the centralized task, while in Figure 9, the sensor with the maximum available energy is chosen for centralized execution. Both graphs show that the A^* and greedy A^* algorithms distribute the energy consumption more equitably across all sensors. In the case of 3 sensors in Figure 9, the centralized algorithm has a lower percentage standard deviation, but the completion time is adversely affected, as seen earlier in Figure 1. Such a situation may occur when the node with maximum available energy is slower than the other nodes. Since the A^* and greedy A^* algorithms evaluate a combined objective of completion time and reliability, equitable energy consumption is traded off against faster completion time. In order to compare the energy spent by nodes of different capabilities, a simplified scenario was considered, where nodes are of only two different speeds, one set of nodes twice as fast as the other. In a given group of nodes, both kinds were assumed to be equally likely. As expected, the nodes of higher speed, which consume higher energy for computation and communication, spent more energy, since they contribute more to reducing the completion time. The results are shown in Figures 6 and 7, for $P : C$ ratios of 1:1 and 1:5. While it was possible to run both the A^* and the greedy A^* algorithms for a small number of modules and sensors, the A^* algorithm took an inordinately long time for a larger number of modules and sensors. Hence, only the greedy A^* algorithm was run for larger number of sensors and modules. The greedy algorithm was employed on 50 to 120 sensors, and the task was split into 20 to 40 modules. Figure 10 shows the number of solution point evaluations and the average energy spent per node using the greedy A^* algorithm. The increase is almost linear in the number of sensors, which shows the high scalability of the greedy A^* algorithm.

4 Summary

In this paper, we have analyzed and formulated the problem of distributing the modules of a task among a group of sensors. We proposed an algorithm to optimally and reliably allocate the modules. Simulations have demonstrated that the completion time of tasks is reduced by distributing them across sensors and that the energy spent is equitably distributed across sensors. We have also proposed the greedy A^* algorithm to reduce the computation involved in finding an allocation. The greedy A^* algorithm explores only the least-cost path of the search tree in the solution space. The solution produced is sub-optimal, but simulations show that the deviation from optimality is low (about 15%). Both the A^* and greedy A^* algorithms distribute the modules such that the energy consumption is shared across sensors more uniformly than centralized execution. This leads to uniform depletion of resources in the network, and reduces the possibility of faster nodes dying out earlier. The greedy A^* algorithm was found to be highly scalable, showing only a linear increase in the number of solution point evaluations with increase in the number of sensors.

References

1. Akyildz I.F., Su W., Sankarasubramaniam Y., and Cayirci E.: A Survey on Sensor Networks. IEEE Communications Magazine, vol. 40, no. 8, pp. 102-114, August (2002).
2. Boulis A. and Srivastava M.B.: Enabling Mobile and Distributed Computing in Sensor Networks. Technical Report, EE Department, University of California at Los Angeles, (2001).
3. Abdelzaher T., Blum B., Cao Q., Chen Y., Evans D., George J., George S., Gu L., He T., Krishnamurthy S., Luo L., Son S., Stankovic J., Stoleru R., Wood A.: EnviroTrack : Towards an Environmental Computing Paradigm for Distributed Sensor Networks. Department of Computer Science, University of Virginia, (2003).
4. Kartik S. and C. Siva Ram Murthy: Task Allocation Algorithms for Maximizing Reliability of Distributed Computing Systems. IEEE Transactions on Computers, vol. 46, no. 6, pp. 719-724, June (1997).
5. Sinclair J.B.: Efficient Computation of Optimal Assignments for Distributed Tasks. Journal of Parallel and Distributed Computing, vol. 4, pp. 342-361, (1987).
6. Papoulis A.: Probability, Random Variables, and Stochastic Processes. McGraw-Hill, Inc., New York (1984).
7. Sarkar V.: Partitioning and Scheduling Parallel Programs for Multiprocessors. MIT Press, MA, USA, (1989).
8. Nilsson N.J.: Problem Solving Methods in Artificial Intelligence. McGraw-Hill, New York, (1977).
9. Garey M.R. and Johnson D.S.: Computers and Intractability: A Guide to the Theory of NP-Completeness, Freeman, San Francisco, (1979).

Protocols for Sensor Networks Using COSMOS Model*

Zhenyu Xu and Pradip K. Srimani

Department of Computer Science,
Clemson University, Clemson, SC 29634–0974

Abstract. Authors in [1] have recently introduced an interesting model, COSMOS (Cluster-based heterOgeneouS MOdel for Sensor networks) for sensor networks; COSMOS is a hierarchical network architecture that consists of a large number of low cost sensors with very limited computation capability and a smaller number of more powerful "clusterheads". The clusterheads can communicate between each other in an asynchronous fashion while the low capability sensors under each clusterhead operate in a synchronous way with their respective clusterheads. Our purpose in the present paper is to design several protocols for benchmark programs like broadcast, matrix multiplication and matrix chain multiplication using this model and provide detailed complexity analysis of these protocols. Our results further illustrates the usefulness of the model for use in sensor networks.

1 Introduction

Wireless sensor networks [2], [3], [4], [5] consist of large number of tiny low-cost sensors that are used to sense natural phenomenon. These sensors have limited computation power as well as limited communication capability. We need specialized computing and communication protocols that can effectively adapt to these limitations of the sensor nodes.

Authors in [1] have recently introduced an interesting model, COSMOS (Cluster-based heterOgeneouS MOdel for Sensor networks) for sensor networks; COSMOS is a hierarchical network architecture that consists of a large number of low cost sensors with very limited computation capability and a smaller number of more powerful "clusterheads". The clusterheads can communicate between each other in an asynchronous fashion while the low capability sensors under each clusterhead operate in a synchronous way with their respective clusterheads. Our purpose in the present paper is to design several protocols for benchmark programs like broadcast, and matrix multiplication using this model and provide detailed complexity analysis of these protocols.

2 The COSMOS Model

COSMOS model has been introduced in details in [1]. COSMOS assumes that the sensors are uniformly distributed in a two dimensional plane. The total area

---
* The work was supported by an NSF Award # ANI-0219485.

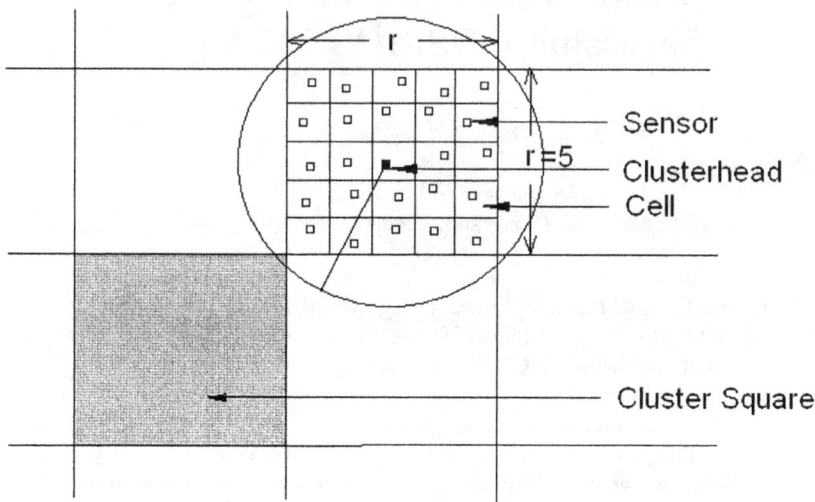

Fig. 1. Clusters in COSMOS

is arranged as a grid of cells where each sensor occupies a cell. The sensors are organized into clusters, each cluster with a clusterhead which has a broader transmission range and more computational power than individual sensors. Within a cluster, the communication is single hop and its size is determined by the transmission range of the sensor. We assume the size of each cluster is $r \times r$, where the transmission range of the sensor is at least $r/\sqrt{2}$, as shown in Figure 1. The concept of clustering the sensors can also be applied in arbitrary networks [6]. However, the properties of particular topology such as mesh is utilized to simplify the computation and communication, as was done previously in [7] (the model was a strict arrangement of sensors in a mesh).

We assume the clusterhead knows the size of the sensor network and its own position (column and row index) in the mesh network. Each sensor has unit memory, unit processing power, and unit bandwidth. Each clusterhead has $m \geq r^2$ memory, $c \geq r^2$ processing power and $b \geq r^2$ bandwidth. This enables the clusterhead to transmit or receive b data elements in one time step, either from other clusterheads or from the sensors within its cluster. All sensors in a cluster are time synchronized with their clusterhead. The communication between clusterheads is asynchronous using message passing.

2.1 Performance Metrics

To evaluate the proposed algorithms using the COSMOS computational model, we use three metrics of performance: Time complexity, Energy dissipation, and Message complexity. These metrics were introduced in [1]; we briefly describe them in the following:

Definition 1. *Time complexity of an algorithm in the COSMOS model is defined to be the total execution time of the longest weighted execution chain on the clusterheads and sensors in the network.*

Time complexity includes the time taken to transmit, receive, or locally calculate data on clusterheads and sensors. The unit of data is the smallest data item on which computation or communication is performed. Since a clusterhead is more powerful in terms of computing power and bandwidth than a sensor, computation and communication at clusterheads are assigned higher weights. Each computation and communication of one unit of data at a sensor node is normalized to unity. The computation of one unit of data at a clusterhead is assigned a weight of $1/c$ (a clusterhead is c times computationally more powerful than a sensor). Similarly, communication of one unit of data at a clusterhead is assigned a weight of $1/b$ (a clusterhead has b times more bandwidth than that of a sensor).

Definition 2. *Total energy dissipation of an algorithm is defined to be the sum of energy consumed at sensors and clusterheads.*

We define the energy used to transmit, receive, or locally compute on one unit of data to be one unit of energy. [This assumes that the size of the sensor network small; the transmission energy is dominated by a range independent constant.

Definition 3. *Message complexity of an algorithm is defined to be the total number of messages transmitted in the execution of algorithm.*

A sensor always transmits and receives one unit of data in one message, since it has only one unit of memory. The message transmitted between clusterheads may contain multiple units of data.

2.2 System Primitives

We assume a underlying protocol provides reliable message passing between the sensors and clusterheads. Following system primitives are provided by the underlying protocol.

- **send (i, j, x)**. The send primitive transmits the data x from the current clusterhead to another clusterhead labeled $S_{i,j}$ within the transmission range. Both clusterheads maintains a local variable x. By calling this system primitive, the current clusterhead sends a message that contains the data in its local variable x. This message is received by clusterhead $S_{i,j}$, and $S_{i,j}$ stores the data in its own local variable x.

  It is apparent that the execution time of $\text{send}(i, j, x)$ is $|x|/b$, where $1/b$ is the weight of transmitting one unit of data between clusterheads, and the $|x|$ is the size (number of units) of the data to be sent; and, the energy consumed in this process is $|x|$.

- call (i, j, proc(args_list)). This is a system primitive of RPC (remote procedure call). By calling this system primitive, the current clusterhead sends a message to a neighboring clusterhead $S_{i,j}$, indicating that $S_{i,j}$ will invoke the local procedure $proc$ with parameters $args\_list$.

We assume the RPC message is short enough to be treated as one unit of data. Thus the execution time of this system primitive is $1/b$, and the energy consumed in this process is 1.

It is possible to use different frequency to transfer data messages and RPC messages. In this case, there will be no collision between the two types of messages. In this paper, we assume only one frequency is used two transfer both types of messages so that only one clusterhead can be sending at the same time in the neighborhood of a particular clusterhead, no matter what type of the message to be sent.
- wait (t). This system primitive simply let the clusterhead wait t units of time, without doing anything.

Throughout the paper, we use the notations shown in Table 1.

Table 1. Notations

Symbol	Description
n	number of sensors in network
S	clusterhead
r	number of rows and columns of sensors in each cluster
b, c	weight of computation and communication cost on clusterhead
m_1, m_2	number of rows and columns of clusters in network
a, b	the row and column index of some particular cluster
s, t	the row and column index of some particular cluster
i, j, k	iteration index of row and column of clusters

3 One to All Data Broadcasting

Consider a two dimensional $m_1 \times m_2$ mesh of clusterheads, where $S_{a,b}$ denotes the specific clusterhead, $1 \leq a \leq m_1, 1 \leq b \leq m_2$. A clusterhead $S_{a,b}$ has some local data x. This data item x can be of any type; typically, it may be an array of integers or it may have a size of r^2 where the clusterhead collects data from all the r^2 sensor nodes that are attached to this clusterhead.

Without lost of generality, let the type of x be an array of integers. If only a single unit of data is to be broadcasted, the size of x is 1. Otherwise if more than one unit of data are to be broadcasted, the size of x is the number of data units. For example, when broadcasting the information collected from all the sensors attached to $S_{a,b}$, the size of x is r^2.

The COSMOS model does not include multicast as a feature, which can be used to flood the data from one clusterhead to all neighboring clusterheads in one step. Because multicast is not available in the network, we have to deploy strategies to minimize the time and energy needed. The most important issue in data broadcasting is the message collision. To prevent the message collision, there can be only one clusterhead sending the data at the same time, within the neighborhood of any clusterhead.

Consider the data broadcasting in a row of clusterheads. Assume the clusterheads are labeled $S_{0,0}, S_{0,1}, S_{0,2}, \ldots, S_{0,m}$, and $S_{0,0}$ contains the original data. In the first round, $S_{0,0}$ sends data and RPC to $S_{0,1}$, and in the second round, $S_{0,1}$ send the data and RPC to $S_{0,2}$. In $m-1$ rounds, all the clusterheads will get the data. Now consider these clusterheads will further send the data to the other nodes in the same column. Since $R = r$, $S_{0,i}$ and $S_{0,i+1}$ can send messages to $S_{1,i}$ and $S_{1,i+1}$ relatively in the same time, without incurring collision. So the strategy is, first send the data to all the clusterheads in the same row, then these clusterheads send to all other clusterheads in the same column.

3.1 Algorithm

The pseudo code for the data broadcasting algorithm, **Broadcast(a, b, x)**, is shown in Figure 2. This algorithm broadcasts data x from the clusterhead $S_{a,b}$ to all the clusterheads in the network, where $1 \leq a \leq m_1, 1 \leq b \leq m_2$ and $m_1 \times m_2$ are the size of the mesh of clusterheads. Before the algorithm executes, only $S_{a,b}$ has the data x. When the algorithm ends, all the clusterheads have a local copy of x.

The algorithm **Broadcast(a, b, x)** has three parameters. Parameters a and b are the coordinates of the clusterhead that contains the data to be broadcast. Parameter x is the data.

We use the first parameter a to denote the row coordinate and the second parameter b to denote the column coordinate of the clusterhead. The coordinates are integers that are known to all the clusterheads. Thus when we say "all clusterheads on row a", we refer to all clusterheads of the form $S_{a,j}$, where $1 \leq j \leq m_2$. We use this naming convention in the remainder of this paper.

Broadcast(a, b, x) uses two subroutines: ColBroadcast(a, b, x), which sends the data x to a column, and RowBroadcast(a, b, x), which sends x to a row. Initially, Broadcast(a, b, x) is called on $S_{a,b}$, which contains the data x.

3.2 Time Complexity

The algorithm can be divided into two phases. In the first phase, the data is sent to all clusterheads on row a. The clusterheads that get the data wait until data reaches all clusterheads on row a. After that, phase 2 starts and the data is sent along the columns.

Theorem 1. *In a $m_1 \times m_2$ mesh of clusterheads that contains n sensors, the time complexity of **Broadcast(a, b, x)** is $O(\sqrt{n})$.*

Proof. In **RowBroadcast**, each clusterhead takes $3/c$ units of time to do the comparison, $|x|/b$ units of time to transmit the data and $1/b$ unit of time to perform the RPC, then it starts waiting. So the execution time of RowBroadcast is $max(m_2 - b, b) \times (|x|/b + 1/b + 3/c)$. The upper bound is $m_2(|x|/b + 1/b + 3/c)$.

```
Following code is executed on cluster S_{i,j}:

RowBroadcast(a, b, x)
Begin
        if  j ≥ b ∧ j < m_2 ∧ i = a  then
                send(i, j+1, x)
                call(i, j+1, RowBroadcast(a,b,x))
        if  j ≤ b ∧ j > 0 ∧ i = a  then
                send(i, j-1, x)
                call(i, j-1, RowBroadcast(a,b,x))
        if  j > b ∧ j ≤ m_2 ∧ i = a  then
                wait(max(m_2 − j, 2b − j + 1))
        else if  i = a  then
                wait(max(m_2 − (2b − j + 1), j))
        else
                wait(max(m_2 − b, b − 1))
End

ColBroadcast(a, b, x)
Begin
        if  i ≥ a ∧ i < m_i  then
                send(i+1, j, x)
                call(i+1, j, ColBroadcast(a,b,x))
        if  i ≤ a ∧ i > 0  then
                send(i-1, j, x)
                call(i-1, j, ColBroadcast(a,b,x))
End

Following code is executed on cluster S_{a,b},
which contains the original data to be broadcast:

Broadcast(x)
Begin
        RowBroadcast(a, b, x)
        ColBroadCast(a, b, x)
End
```

Fig. 2. Algorithm 2: One to All Broadcast Algorithm

Similarly, the execution time of ColBroadcast is $m_1(|x|/b + 1/b + 2/c)$. So the total execution time is $m_2(|x|/b + 1/b + 3/c) + m_1(|x|/b + 1/b + 2/c)$, which is $O(m_1 + m_2)$. For a $\sqrt{n}/r \times \sqrt{n}/r$ mesh, $m_1 = m_2 = \sqrt{n}/r$, time complexity is $O(2\sqrt{n}/r) = O(\sqrt{n})$.

3.3 Energy Dissipation

Theorem 2. *In a $m_1 \times m_2$ mesh of clusterheads that contains n sensors, the energy dissipation of **Broadcast(a, b, x)** is $O(n)$.*

Proof. Each clusterhead receives one data message and one RPC message, except for clusterhead $S_{a,b}$. So total number of data or RPC messages sent is $\frac{n}{r^2}-1$. Each data message contains $|x|$ units of data, and each RPC message contains 1 unit of data, so the energy dissipation for transmitting messages is $(\frac{n}{r^2}-1) \times |x| = O(n)$.

In RowBroadcast, the number of comparisons performed on each clusterhead is 3. In ColBroadcast, the number of comparisons performed on each clusterhead is 2. So the total number of computations is $3 \times m_2 + 2 \times m_1 \times m_2$. Each computation on clusterhead takes 1 unit of energy. So the energy dissipation for computation is $O(m_1 \times m_2) = O(n)$.

3.4 Message Complexity

Theorem 3. *In a $m_1 \times m_2$ mesh of clusterheads that contains n sensors, the message complexity of **Broadcast(a, b, x)** is $O(n)$.*

Proof. Except for the initial clusterhead $S_{a,b}$, each clusterhead receives one data message and one RPC message. So total number of messages transmitted is $\frac{2n}{r^2} - 2$. Thus the message complexity is $O(n)$.

4 All to All Data Broadcasting

The All to All data broadcasting in COSMOS model is defined as all the clusterheads transmits data to every other clusterheads. It is possible to implement the all-to all data broadcasting by repeating the One to All data broadcasting $m_1 \times m_2$ times. However, this approach is not time efficient. Two non-interfering clusterheads can be scheduled to transmit different data at the time to save execution time.

4.1 Data Structures and Algorithm

As in One to All data broadcasting, we assume each clusterhead maintains an integer variable x that contains the data to be broadcast to all other clusterheads.

Furthermore, to store the data comes from other clusterheads, each clusterhead also maintains an integer array $Y[1..m_1][1..m_2]$ of size $m_1 \times m_2$. For each clusterhead $S_{i,j}$, we define a procedure **sync**(α, β), where the parameters α and β can take values as shown in Table 2.

Consider all the clusterheads on row i. To prevent the collision, when $S_{i,j}$ is executing sync(0, 1), $S_{i,j+1}$ and $S_{i,j+2}$ cannot execute sync(0, 1). However, $S_{i,j+3}$ can execute sync(0, 1), as well as other clusterheads in the same column. This is shown in figure 3.

In the first and second round, all clusterheads on column $j, j + 3, j + 6, \ldots$ execute sync(0, 1) and sync(0, -1). This sends the data on those columns to adjacent columns. In the third and fourth round, all clusterheads on column $j + 1, j + 4, j + 7, \ldots$ execute sync(0, 1) and sync(0, -1). In the fifth and sixth round, all clusterheads on column $j + 2, j + 5, j + 8, \ldots$ execute sync(0, 1) and

Table 2. The sync(α, β) procedure

α	β	Definition	Description
1	0	send($i+1, j, Y[1..i][1..m_2]$)	sends the upper part (up to row i) of y to the lower neighbor of clusterhead $S_{i,j}$.
-1	0	send($i-1, j, Y[i..m_1][1..m_2]$)	sends the lower part (up to row i) of y to the upper neighbor of clusterhead $S_{i,j}$.
0	1	send($i, j+1, Y[1..m_1][1..j]$)	sends the left part (up to column j) of y to the right neighbor of clusterhead $S_{i,j}$.
0	-1	send($i, j-1, Y[1..m_1][j..m_2]$)	sends the right part (up to column j) of y to the left neighbor of clusterhead $S_{i,j}$.

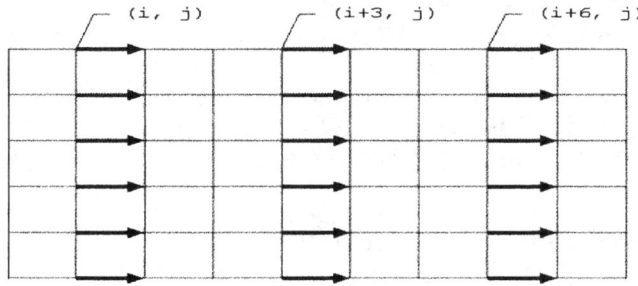

Fig. 3. Executing sync(0, 1) every three columns. Arrows denote the data transmission.

sync(0, -1). After the six rounds, each clusterhead contains the correct data value from its left and right neighbors.

This process is repeated $m_2/3+1$ times. After these rounds, each clusterhead contains all the data from the clusterheads on the same row. Then all clusterheads execute sync(1, 0) and (-1, 0), in the same way of every three rows, to transfer the data to the entire mesh. The formal algorithm is presented in figure 4.

4.2 Time Complexity

Theorem 4. *In a $m_1 \times m_2$ mesh of clusterheads that contains n sensors, the time complexity of **All2AllBroadcast(x)** is $O(n^{3/2})$.*

Proof. In the first loop of **All2AllBroadcast**, each clusterhead executes $m_2/3+1$ times of sync(0, 1) and $m_2/3+1$ times of sync(0, -1). In sync(0, 1), $|x| \times m_1 \times (j+1)$ units of data are transmitted. In sync(0, -1), $|x| \times m_1 \times (m_2-j)$ units of data are transmitted. So the execution time in the first loop is $|x| \times m_1 \times (m_2+1) \times (m_2/3+1) \times 1/b = O(m_1 \times m_2^2)$.

Similarly, in the second loop, the execution time is $|x| \times (m_1+1) \times m_2 \times (m_1/3+1) \times 1/b = O(m_1^2 \times m_2)$.

For a $\sqrt{n}/r \times \sqrt{n}/r$ mesh, $m_1 = m_2 = \sqrt{n}/r$, time complexity is $O(m_1 \times m_2^2 + m_1^2 \times m_2) = O(n^{3/2})$.

```
Following code is executed on cluster S_{i,j}:

All2AllBroadcast(x)
Begin
        Y[i][j] = x
        for  k = 0 to m_2/3  do
                ⎧ wait(j mod 3)
                ⎨ sync(0, 1)
                ⎩ sync(0, -1)
                  wait(2 - (j mod 3))
            for  k = 0 to m_1/3  do
                ⎧ wait(i mod 3)
                ⎨ sync(1, 0)
                ⎩ sync(-1, 0)
                  wait(2 - (i mod 3))
End
```

Fig. 4. All to All Broadcast Algorithm

Recall that the time complexity of One to All broadcast is $O(n^{1/2})$. If simply apply $m_1 \times m_2$ times One to All broadcast on each clusterhead, the time complexity will be $O(n^{1/2} \times n^2) = O(n^{5/2})$. So algorithm **All2AllBroadcast** is more time efficient.

4.3 Energy Dissipation

Theorem 5. *In a $m_1 \times m_2$ mesh of clusterheads that contains n sensors, the energy dissipation of **All2AllBroadcast(x)** is $O(n^{3/2})$.*

Proof. In the first loop of **All2AllBroadcast**, each clusterhead executes $m_2/3 + 1$ times of sync(0, 1) and $m_2/3 + 1$ times of sync(0, -1). In sync(0, 1), $|x| \times m_1 \times (j+1)$ units of data are transmitted. In sync(0, -1), $|x| \times m_1 \times (m_2 - j)$ units of data are transmitted. So the energy dissipation in the first loop is $|x| \times m_1 \times (m_2 + 1) \times (m_2/3 + 1) = O(m_1 \times m_2^2)$.

Similarly, in the second loop, the energy dissipation is $|x| \times (m_1 + 1) \times m_2 \times (m_1/3 + 1) = O(m_1^2 \times m_2)$.

For a $\sqrt{n}/r \times \sqrt{n}/r$ mesh, $m_1 = m_2 = \sqrt{n}/r$, total energy dissipation is $O(m_1 \times m_2^2 + m_1^2 \times m_2) = O(n^{3/2})$.

4.4 Message Complexity

Theorem 6. *In a $m_1 \times m_2$ mesh of clusterheads that contains n sensors, the message complexity of **All2AllBroadcast(x)** is $O(\sqrt{n})$.*

Proof. In the first loop of **All2AllBroadcast**, each clusterhead executes $m_2/3 + 1$ times of sync(0, 1) and $m_2/3 + 1$ times of sync(0, -1). So the number of messages transmitted is $m_2 \times 2/3 + 2$. Similarly, in the second loop, the number of messages transmitted is $m_1 \times 2/3 + 2$.

For a $\sqrt{n}/r \times \sqrt{n}/r$ mesh, $m_1 = m_2 = \sqrt{n}/r$, So total number of messages transmitted is $\sqrt{n}/r \times 4/3 + 4$. Thus the message complexity is $O(\sqrt{n})$.

5 Matrix Multiplication

Given two matrices $A_{m \times m}$ and $B_{m \times m}$, the matrix multiplication $C = A \times B$ can be calculated as $C_{ij} = \sum_{1 \leq k \leq m} A_{ik} B_{kj}$, where $1 \leq i, j \leq m$. In the COSMOS model, the matrix multiplication does the following: if for all the sensors, A_{ij} and B_{ij} is stored on cluster row s column t, and inside the cluster the sensor on row p column q, where $i = (s-1)r + p, j = (t-1)r + q$. Then after the matrix multiplication is done, the result C_{ij} is stored in the same way.

5.1 Data Structure

Each sensor keeps three integer variables a, b, and c. Before the algorithm is started, a and b contain the corresponding element of matrix A and B. After the algorithm is finished, c contains the element of matrix $C = AB$. Each clusterhead keeps following variables: Integer arrays $X[1..m][1..m]$ and $Y[1..m][1..m]$ of size $m \times m$, which store elements of A and B that get from the sensors within the cluster and from other clusterheads. An integer array $z[1..m][1..m]$ of size $m \times m$, which stores computed elements of the result matrix C. Two integer variables s and t that denote the index of the clusterhead in the mesh.

5.2 Algorithm

The first step is aggregating data a and b from the sensors to array x and y of its clusterhead. For the clusterhead at row s and column t, it stores all the a elements to $X[(s-1)r+1..s \times r][(t-1)r+1..t \times r]$, and stores all the b elements to $Y[(s-1)r+1..s \times r][(t-1)r+1..t \times r]$. The next step uses the All to All data broadcasting to send the block of A and B matrices to all the clusterheads. The clusterhead $S_{s,t}$ can then calculate the block $z[(s-1)r+1..s \times r][(t-1)r+1..t \times r]$ as follows: $z[i][j] = \sum_{0 \leq k < m} X[i][k] \times Y[k][j]$. Finally, $z[i][j]$ is distributed to the sensor on row i column j in the cluster.

5.3 Time Complexity

Lemma 1. *DataAggregation takes $8r^2 + 6$ time steps.*

Proof. [1] gives the aggregation schedule: It takes $2r^2$ time for one cluster to aggregate data: In the first r^2 time, each sensor is assigned a rank order. In the next r^2 time, sensors send data x to clusterhead in the rank order.

Following code is executed on cluster $S_{s,t}$:

MM(a, b)
Begin
        do DataAggregation(a, $X[(s-1)r+1..s \times r][(t-1)r+1..t \times r]$)
        do DataAggregation(b, $Y[(s-1)r+1..s \times r][(t-1)r+1..t \times r]$)
        do All2AllBroadcast(x)
        do All2AllBroadcast(y)
        calculate $z[(s-1)r+1..s \times r][(t-1)r+1..t \times r]$ using data in $X[1..m][1..m]$
and $Y[1..m][1..m]$.
        send result in $z[(s-1)r+1..s \times r][(t-1)r+1..t \times r]$ back to sensors.
End

Fig. 5. Matrix Multiplication Algorithm

To avoid collision, adjacent clusters must not aggregate data in same period. The aggregation sequence is scheduled in this way: In time $0 \leq t < 2r^2$, all clusters on even row and even column do data aggregation. In time $2r^2 + 2 \leq t < 4r^2 + 2$, all clusters on even row and odd column do data aggregation. In time $4r^2 + 4 \leq t < 6r^2 + 4$, all clusters on odd row and odd column do data aggregation. In time $6r^2 + 6 \leq t < 8r^2 + 6$, all clusters on odd and even column do data aggregation. The 6 time steps are required to notify neighboring clusterheads to start the data aggregation. Thus the total time is $8r^2 + 6$.

Theorem 7. *In a $m \times m$ mesh that contains n sensors, the time complexity of **Matrix Multiplication** is $O(n^{5/2})$.*

Proof. By lemma 1, data aggregation takes $8r^2 + 6$ time steps. The All to All data broadcasting takes $O(n^{5/2})$ time steps. The matrix multiplication is then concurrently performed on all the clusterheads. It takes $2m/c$ time steps to compute one element in z, so in all it takes $2r^2 m/c$ time steps to compute all $r \times r$ elements. The last step is the reverse of data aggregation, so it also takes $8r^2 + 6$ time steps. Adding all together, the time complexity is $O(n^{5/2})$.

5.4 Energy Dissipation

Theorem 8. *In a $m \times m$ mesh that contains n sensors, the energy dissipation of **Matrix Multiplication** is $O(n^{3/2})$.*

Proof. The total number of data transmissions in data aggregation step is $2n$. The total number of data transmissions in last step (reverse data aggregation) is n. So the energy dissipation is $O(n)$ for these steps. The energy dissipation in **All2AllBroadcast** is $O(r^2) \times O(n^{3/2}) = O(n^{3/2})$. Calculating $z[1..r][1..r]$ in each cluster takes $O(m \times r^2)$ calculations in all. Since r is fixed size of the cluster, it is a constant. So the energy dissipation on calculation is $n \times O(m) = O(n^{3/2})$. Therefore Total energy dissipation is $O(n) + 2n^{3/2}/b + 2n = O(n)$.

5.5 Message Complexity

Theorem 9. *In a $m_1 \times m_2$ mesh that contains n sensors, the message complexity of **Matrix Multiplication** is $O(n)$.*

Proof. The total number of transmissions in data aggregation and reverse data aggregation step is $3n$. Each of the textbfAll2AllBroadcast step sends $O(n^{1/2})$ messages. So the total number of transmissions is $O(n) + O(n^{1/2}) = O(n)$.

References

1. M. Singh and V. K. Prasanna. A hierarchical model for distributed collaborative computation in wireless sensor networks. In *the International Parallel and Distributed Processing Symposium (IPDPS'03)*, 2003.
2. M. Tubaishat and S. Madria. Sensor networks: an overview. *IEEE Potentials*, 22(2):20–23, April 2003.
3. I.F. Akyildiz, W. Su, Y. Sankarasubramaniam, and E. Cayirci. A survey on sensor networks. *IEEE Communication Magazine*, August 2002.
4. K. Sohrabi, J. Gao, V. Ailawadhi, and G.J. Pottie. Protocols for self-organization of a wireless sensor network. *IEEE Personal Communications*, pages 16–27, October 2000.
5. C. Intanagonwiwat, R. Govindan, and D. Estrin. Directed diffusion: a scalable and robust communication paradigm for sensor networks. In *Proceedings of the ACM MobiCom '00*, pages 56–67, Boston, MA, 2000.
6. W.R. Heinzelman, A. Chandrakasan, and H. Balakrishnan. Energy-efficient communication protocol for wireless microsensor networks. In *IEEE Proceedings of the Hawaii International Conference on System Sciences*, pages 1–10, January 2000.
7. L. Schwiebert, S.K.S. Gupta, and J. Weinmann. Research challenges in wireless networks of biomedical sensors. In *MobiCom '01: Proceedings of the 7th annual international conference on Mobile computing and networking*, pages 151–165, New York, NY, USA, 2001. ACM Press.

CLUR-Tree for Supporting Frequent Updates of Data Stream over Sensor Networks[*]

Soon-Young Park[1], Jung-Hyun Kim[1], Yong-Il Jang[1], Jae-Hong Kim[2], Soon-Jo Lee[3], and Hae-Young Bae[1]

[1] Dept. of Computer Science and Information Engineering, Inha University,
253 Yonghyun-dong, Nam-gu, Incheon, 402-751, Korea
{sunny, jungkim, himalia}@dblab.inha.ac.kr, hybae@inha.ac.kr
[2] Dept. of Computer Engineering, Youngdong University,
12-1 Seolgue-ri, Youngdong-eup, Youngdong-gun, Chungbuk, 370-701, Korea
jhkim@youngdong.ac.kr
[3] Dept. of Computer Science and Information Engineering, Seowon University,
231 Mochung-dong, Heungduk-gu, Cheongju-si, Chungbuk, 361-742, Korea
sjlee@seowon.ac.kr

Abstract. Data streams from sensors are usually characterized as continuous, with very frequent updates. Queries over those data streams need to be processed in near real-time. So it is needed to design the index structure for supporting the frequent updates and fast retrieval of data efficiently. In this paper, CLUR-Tree (Cache-conscious Lazy Update R-Tree) is proposed, which is a spatial index for efficient processing of frequent updates of data streams in locality preserving monitoring applications. CLUR-Tree has two characteristics. First, it excludes index reconstruction overhead by permitting modification of only the index node of the sensor which moves out of the corresponding MBR (Minimum Bound Rectangle). Second, it reduces the key spaces by applying new compression method for MBR used as key in R-Tree and by considering cache to prevent bottleneck due to speed difference between main memory and CPU. The experimental results indicate that the proposed CLUR-Tree enhances update performance and gives a good retrieval performance simultaneously.

1 Introduction

On new database management environments such as for sensor networks, data streams from sensors are fed into a database management system (DBMS) [9], [11]. Data stream from sensors are usually characterized as continuous, with very fast updates. A main memory based DBMS can be used for supporting this environment [3], [7], [8]. But generally, cost of updating operation in the traditional DBMS is more than search operation. And there are still no functions and structures for efficient processing of dynamic updating of the data stream [2], [7].

---

[*] This research was supported by the MIC (Ministry of Information and Communication), Korea, under the ITRC (Information Technology Research Center) support program supervised by the IITA (Institute of Information Technology Assessment).

In particular, R-Tree [5] is not considered suitable for modifying only the sensing data from the sensor of a specific location contrary to data insertion and deletion. Therefore the design of the spatial index structure in main memory is needed for efficient processing of the dynamic updating of the data stream like sensing data.

In this paper, CLUR-Tree (Cache-conscious Lazy Update R-Tree) is proposed, which is a new index structure for efficient processing of frequent updates of data stream over sensor networks. The proposed CLUR-Tree index is a kind of modified R-Tree and has the following two characteristics. First, it excludes index reconstruction overhead, which happens because of the splitting and merging of index node. Second, it optimizes the key spaces by considering cache block size. A new compression method is used to translate MBR into relative MBR of various lengths integer. The proposed CLUR-Tree enhances update performance of index compared with existing index and gives good retrieval performance simultaneously. Therefore the proposed CLUR-Tree can be used to efficiently process frequent updates of data stream over sensor networks.

The remainder of this paper is organized as follows. Section 2 briefly reviews related work. Section 3 presents the proposed CLUR-Tree, indexing technique to reduce update cost for efficient processing of stream data. Section 4 presents the experimental results to compare proposed CLUR-Tree with existing approaches. Finally, conclusion and future work are discussed in Section 5.

2 Related Work

2.1 R-Tree Based Indexes

An R-Tree is an approximately height-balanced search tree [1], which is widely used for handling spatial data in traditional database systems.

Assuming that we consider spatial objects embedded in 2-dimensional space, the spatial extent of each data object is represented by a MBR (Minimum Bounding Rectangle). Leaf nodes in the R-Tree contain entries of the form (*oid*, *mbr*), where *oid* is a pointer to the object in the databases and *mbr* is the MBR of the object. Non-leaf nodes contain entries of the form (*mbr*, *ptr*), where *mbr* is the MBR that bounds all the MBRs in the child node and *ptr* is a pointer to a child node in the tree.

2.2 Cache-Conscious Indexes

To overcome the speed gap between the CPU and DRAM, cache is used [6]. Several researches have shown that cache behavior is important for main memory index structure, and they are classified into two types - using the pointer elimination technique and using compression of node keys.

In CSB^+-Tree (Cache Sensitive B^+-Tree), nodes are increased by using the pointer elimination technique. The CSB^+-Tree puts all the child pointers for a given node contiguously in an array and stores only the pointer to the first child node [10]. But the pointer elimination technique cannot be directly applied to multidimensional index structures such as the R-Tree, because MBR is much larger than pointer size.

The CR-Tree (Cache-Conscious R-Tree) is a modified R-Tree by compression of MBR. Compressed MBR of the CR-Tree is a quantized relative representation of the object MBR [4, 6]. But because quantized compression technique translates absolute coordinates into quantized relative coordinates of fixed size without consideration of the area, more errors happen as it goes to the root node.

3 CLUR-Tree: Cache-Conscious Lazy Update R-Tree

In this section, firstly TRMBR (Transformed Relative MBR) technique is designed as a compression method of MBR. And then CLUR-Tree applying the TRMBR has been proposed.

3.1 TRMBR Technique: Compression Method of MBR

In general, R-Tree has a characteristic that the internal nodes consist of (*rectangle, pointer*) pairs, the *pointer* points to a node one level below in the tree, and the rectangle is a MBR of all the objects in the subtree pointed by the *pointer*. The MBR of the parent node includes the MBR of its child node. Therefore the MBR of the child node in R-Tree can be represented relatively as MBR of its parent node. In this paper, the TRMBR is a transformed relative MBR and works by applying the conversion function to relative MBR.

The conversion function for TRMBR must be selected taking the distribution characteristics of the object and the size of the area which the index is constructed, and it must reduce a cache miss by increasing fan-out of node as much as possible. When MBR of parent node R0 is as (*R0.xl, R0.yl, R0.xh, R0.xh*) and MBR of the first child node R1 of the R0 is as (*R1.xl, R1.yl, R1.xh, R1.xh*), the conversion function in this paper is as follows.

$$getTRMBR(MBR) = (<\sqrt{R0.xl-R1.xl}>,<\sqrt{R0.yl-R1.yl}>,<\sqrt{R0.xh-R0.xh}>,<\sqrt{R0.yh-R1.yh}>) \quad (1)$$

Fig. 1 (a) shows the absolute coordinates in real map, and R0 is parent node of R1 and R2. Fig. 1 (b) shows the relative MBR (RMBR) of the child node R1 and R2 to the parent node R0. And Fig. 1 (c) shows the TRMBR which is transformed MBR from the RMBR using conversion function of formula (1). This form is applied to every node in the index.

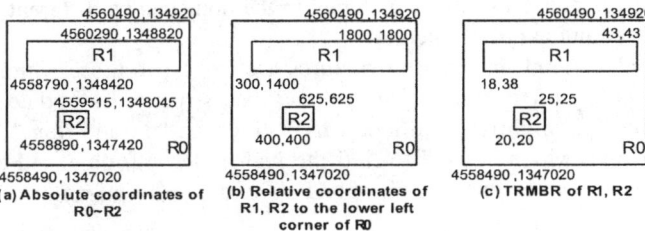

Fig. 1. Absolute coordinates, TRMBR of node R0 and its child node R1, R2

TRMBR technique in this paper compresses MBR of the object to 2, 4, 8 bytes and compressed MBR is stored in entry instead of original MBR. In case TRMBR is 16 and less, MBR is represented as 2 bytes, and in case TRMBR is 256 and less, MBR is represented as 4 bytes, and in case TRMBR is 65536 and less, MBR is represented as 8 bytes.

As entries are assigned according to compressed size of MBR dynamically, a node's fan-out is increased. Using TRMBR technique, accuracy of retrieval can be increased because it reduces the error in compression by considering the area.

3.2 Structure of the CLUR-Tree

Based on TRMBR technique, node structure of the CLUR-Tree is illustrated in Fig. 2. The node of the CLUR-Tree is fixed to make the efficiency of the cache increase as a multiple of the cache size. TRMBR is used for increasing the fan-out of the node which is of fixed size. MBR is stored at the node as shown in Fig. 2 (a), which is used to recalculate the MBR of the entry at the retrieval, deletion, and insertion operations. In a leaf or non-leaf node, TRMBR obtained by the conversion function is stored as shown in Fig. 2 (b) and 2 (c).

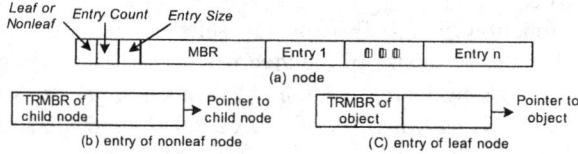

Fig. 2. The node structure of the CLUR-Tree

TRMBR can be calculated differently according to the area of the entry in the node and the entry of the node of the CLUR-Tree is dynamically allocated according to the calculated TRMBR.

The structure of the CLUR-Tree is as follows. The node is fixed to the multiple of the cache size. For example, n entries of 8 bytes lengths are included in a root node of fixed size, $2n$ entries of 4 byte lengths are included in a node of lower level, and $4n$ entries of 2 byte length are included in a leaf node because contained area becomes small as it goes to a leaf node. In case the contained area is different among nodes in the same level, the number of entry included in the node can be different. Then 'Entry Size' field is used in the node structure.

Almost all algorithms and index structures are similar to those which are used in other R-Tree variants. However algorithm and data structure in update operation of the CLUR-Tree are slightly modified. CLUR-Tree has an additional access path, called Hash Table, which is used to find the leaf node with object id and position. Each entry has a pointer to the corresponding entry in a leaf node of the CLUR-Tree.

Entries of the node in CLUR-Tree can be changed according to TRMBR. Therefore when an object is inserted and deleted, it is very important that the value of 'Entry Size' is not changed for minimizing cost of the node reconstruction. That is, in case of standard R-Tree, when we decide the position of the new object to be inserted, we first consider that the node including the object has a minimal extension of the node. But in case of CLUR-Tree, we first consider that the value of 'Entry Size' is not changed though node has been extended more.

When an object is updated, firstly we find a leaf node through the Hash Table, and check whether the MBR of the leaf node contains a new position or not. If the new position is in the MBR, we modify only the position of the object in the entry. Otherwise, we delete an old position and insert a new one. The CLUR-Tree can reduce the update cost for large number of objects, since it prevents unnecessary traversal and modifications from the root node of the R-tree.

4 Performance Evaluation

In this section, we will present the result of some experiments to analyze the performance of the CLUR-Tree with respect to the search performance and the update performance. For comparing and verifying the effectiveness of the proposed CLUR-Tree, we implemented the ordinary R-Tree and CR-tree respectively in C++ on Windows XP PC with Pentium 4 2.4G Hz CPU, 1 GB memory and 80 GB HDD. In this evaluation, query selectivity was fixed within 0.01 % of the data space. MBR size is 16 bytes and TRMBR size is about one-fourth of the MBR size even if it is variable. And, data set was made by moving object generator developed by ourselves.

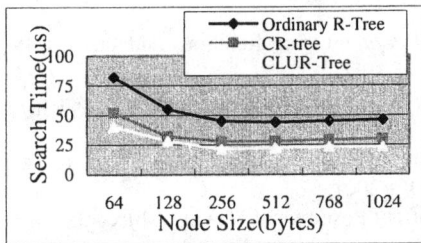
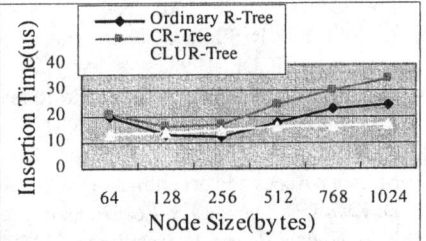

Fig. 3. Search Performance   **Fig. 4.** Update Performance

In the first experiment, the search performance is compared in terms of the time spent processing. Fig. 3 shows that the search time quickly approaches the minimum, and then increases slowly. It results from increasing the number of accessed nodes as node size is small. For all node sizes, CLUR-Tree displayed the search performance more than two times than that of R-Tree. In the second experiment, we inserted 10,000 objects into the index bulk-loaded with uniform data set to measure the update performance. Fig. 4 shows the measured insertion time. For a given node size, the CLUR performed similar to or better than the R-Tree.

The large part of the time required in insertion is used to find the proper leaf node of the tree. In case of the R-Tree and the CR-tree, the update performance is dropped because number of node is increased. But, the proposed technique made performance enhancement by keeping the Hash Table structure.

5 Conclusion

In this paper CLUR-Tree has been proposed, which is a new index structure for efficient processing of frequent updates of data stream over sensor networks. The proposed CLUR-Tree index is a modified R-Tree to manage stream data efficiently and has following two characteristics. First, it excludes index reconstruction overhead by permitting to modify only the index node of sensor which moves out of the corresponding MBR. Second, it adjusts the key spaces by considering cache to prevent bottleneck, and by applying new compression method. It translates MBR into transformed relative MBR of various lengths integer using conversion function.

The proposed CLUR-Tree enhances update performance of index compared to existing index techniques and gives good retrieval performance simultaneously. Therefore the proposed CLUR-Tree can used to efficiently process frequent updates of data stream over sensor networks.

References

1. R. Bayer and E. McCreight: Organization and Maintenance of Large Ordered Indices, Proceedings of ACM SIGFIDET, 1970.
2. D. Carney, U. Cetintemel, M. Cherniack, and C. Convey: Monitoring Stream – A New Class of Data Management Applications, Proceedings of VLDB, 2002.
3. M. Demirbas and H. Ferhatosmanoglu: Peer-to-Peer Spatial Queries in Sensor Networks, Proceedings of P2P, 2003.
4. J. Goldstein, R. Ramakrishnan, and U. Sharft: Compressing Relations and Indexes, Proceedings of ICDE, 1998.
5. A. Guttaman, R-trees: A Dynamic Index Structure for Spatial Searching, Proceedings of SIGMOD 1984.
6. K. Kim, S. K. Cha, and K. Kwon: Optimizing Multidimensional Index Trees for Main Memory Access, Proceedings of ACM SIGMOD, 2001.
7. D. Kwon, S. Lee, and S. Lee: Indexing the Current Positions of Moving Objects Using the Lazy Update R-tree, Proceedings of MDM, 2002.
8. M. L. Lee, W. Hsu, C. S. Jensen, B. Cui, and K. L. Teo: Supporting Frequent Updates in R-Trees: A Bottom-Up Approach, Proceedings of VLDB, 2003.
9. S. Madden and M. J. Franklin: Fjording the Stream: An Architecture for Queries over Streaming Sensor Data, Proceedings of ICDE, 2002.
10. J. Rao and K. Ross: Making B+-trees Cache Conscious in Main Memory, Proceedings of ACM SIGMOD, 2000.
11. Y. Yao and J. Gehrke: Query Processing for Sensor Networks, Proceedings of CIDR, 2003.

Optimizing Lifetime and Routing Cost in Wireless Networks

M. Julius Hossain and Oksam Chae*

Department of Computer Engineering, Kyung Hee University,
1 Seochun-ri, Kiheung-eup, Yongin-si, Kyonggi-do, South Korea, 449-701
mdjulius@yahoo.com, oschae@khu.ac.kr

Abstract. This paper presents a new routing approach for wireless networks based on the combination of both lifetime and routing cost. As the nodes in the wireless sensor and ad hoc networks are limited in power, a power failure occurs if a node has insufficient remaining energy to send a message. So, it is important to minimize the energy expenditure as well as to balance the remaining battery power among the nodes. In ad hoc networks, movement of nodes also causes frequent disconnections of routes and thus effects on network stability. Cost effective routing algorithms attempt to minimize the total power needed while lifetime prediction routing algorithms try to balance the remaining energies among the nodes in the networks. However, because of ignoring other parameter, each method fails to achieve the objective of other. The proposed routing protocol suggests a tradeoff between these two parameters, and ensures a balanced utilization to achieve maximum overall performance.

1 Introduction

Wireless sensor and ad hoc networks are likely to be widely deployed in various applications including remote monitoring, online information processing, and communication among the soldiers on the battle field and disaster relief personnel. The nodes in these networks are equipped with limited battery power, which makes energy a crucial consideration to prolong its lifetime. The lifetime of the node is limited by its residual energy and in order to increase the lifetime, minimum battery power should be used. Cost-effective routing protocols ensure that a packet from a source to a destination gets routed along the most energy efficient path possible. These approaches frequently select efficient path having nodes with very short remaining energy and result an early death of some nodes as well as network disconnection. In mobile ad hoc networks, mobility of nodes also results frequent disconnection of routing path. In both cases a significant topological change is taken place in the network and would require reorganizing the network and re-routing of packets [1], [2].

In case of cost effective routing protocols, the probability of a node within the transmission range to be selected as a forwarding node is proportional to the degree of that node, where degree of a node is the number of neighboring nodes with in its transmission range. So, nodes with higher degree might die soon since they are likely to be used in most cases [3], [4]. Lifetime prediction routing protocols mainly consider

---

* Corresponding author.

the residual energy of the individual nodes in a path and are aimed at maximizing the network lifetime by finding routing solution that minimizes the variance of the remaining energies of the nodes in the network [2], [4]. However, these approaches often select path having much higher cost than cost effective algorithms do. To achieve a tradeoff between the routing cost and network stability, we propose a new routing technique that combines the best features of these two routing approaches.

2 Overview of the Proposed Method

The proposed Effective and Energy Balance Routing (BEER) protocol is a reactive routing protocol like DSR [5] and it attempts to minimize the total transmission power needed and to avoid nodes with a short battery's remaining lifetime. It finds a tradeoff between the cost and the lifetime of each of the possible paths. Incase of wireless sensor networks where nodes are static, lifetime parameter is calculated only from the residual energy or battery. In ad hoc networks, the nodes in the network may move; hence lifetime of a path is calculated from both residual battery energy and predicted time before disconnection due to the mobility of nodes. Transmission cost can be determined from energy, hop count, delay, link quality as well as other factors. Hop count is mostly used parameter to measure energy requirement of a routing task. However, if nodes can adjust their transmission power based on the distance of their neighbors, different energy levels can be used depending on distance between nodes [6]. The distance between neighboring nodes can be estimated on the basis of incoming signal strengths or directly communicating with a satellite, using global positioning system [7]. We used the later approach to determine the transmission cost.

3 Effective and Energy Balanced Routing

3.1 The Network Model

We model a wireless network by a triplet, $N = (V, E, C)$, where $V = \{v_1,....,v_n\}$, represents nodes, $E \subseteq V \times V$, represents set of edges $\{(v_i, v_j), 1 \leq i, j \leq n\}$, that connect all the nodes, and $C: E \rightarrow R$ (Rational number) is a weight function for each edge (v_i, v_j) that indicates the transmission cost of a data packet between node v_i and v_j. Each node in the network has a unique identification number. Data are broadcast to all nodes inside its transmission range. In case of sensor networks, nodes as well as edge cost are static. In ad hoc networks and the edge cost between any two nodes may change over time. The lifetime of node may also change over time. However, for the ease of presentation, we assume a static network during the route discovery phase.

3.2 Selection of Path with Static Nodes

Let us assume that the maximum possible lifetime of any node is L and the maximum possible transmission cost between any two nodes is C. We define a scaling factor:

$$\sigma = \frac{L}{C}.\tag{1}$$

σ contributes to generate meaningful path selection parameter and also helps to add other parameters like mobility with it. Let there be n paths ($\pi_1, \pi_2, \ldots \pi_n$) from source to destination. The lifetime of a path is bounded by the lifetime of all the nodes along the path. So, the lifetime of a path π_i is defined as:

$$\tau_i = Min(T_j(t))\ldots\ldots\{j \in \pi_i\}.\tag{2}$$

$T_j(t)$ is the predicted lifetime of node j in path π_i at time t. The cost of a path is the sum of all the costs calculated between two consecutive nodes along the path from source to the destination. Cost of a path π_i is defined as:

$$\chi_i = \sum_{j=1}^{m(\pi_i)-1} c_{\pi_{ij,j+1}}(t).\tag{3}$$

where, $m(\pi_i)$ is the number of nodes in path π_i and $c_{\pi_{ij,j+1}}$ is the cost between node j and $j+1$ of the path π_i at time t. Our path selection parameter β is represented by

$$\beta_i = \frac{\tau_i}{\sigma \chi_i}.\tag{4}$$

BEER selects a path, which has the largest β i.e. max (β_i). If more than one path having the highest β are found, any one can be selected. Thus, the proposed method is inclined to select a path having higher lifetime τ and lower cost χ. Figure 1 displays an instance of wireless network represented by a graph. Nodes are marked with their lifetime values and edges are labeled with transmission cost.

In this instance there are six paths from source, S to destination, D. They are SABD, SABCFGD, SEFCBD, SEFGD, SCFGD and SCBD, where the total cost and

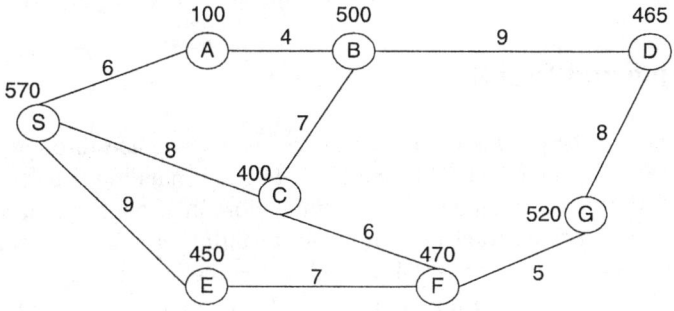

Fig. 1. An instance of Wireless Network

lifetime pairs of the paths are (19,100), (36, 100), (38, 400), (29, 450), (27, 400) and (24, 400) respectively. Power Aware Routing (PAR) [8] selects the path SABD, having cost 19 and lifetime 100 while the route, SEFGD, is chosen by Lifetime Prediction Routing (LPR) [4], having lifetime 450 and cost 29.

Let us assume maximum possible cost (C) between any two nodes is 15 and maximum possible lifetime (L) of any node is 600. So the scaling factor σ becomes 40. Hence, using BEER algorithm, the selection parameter β for the paths SABD, SABCFGD, SEFCBD, SEFGD, SCFGD and SCBD are 0.1316, 0.0694, 0.2632, 0.3879, 0.3704 and 0.4167 respectively. The path SCBD possesses the highest value of β. So, BEER protocol will select the path, SCBD, having cost 24 and lifetime 400.

3.3 Selection of Path with Moving Nodes

In mobile ad hoc networks, each host may change its position and thus routes are subject to frequent disconnections. However, nodes in the network exhibit some degree of regularity in the mobility pattern. By exploiting the non-random traveling pattern of mobile nodes, future state of network topology can be predicted. Various approaches are taken to enhance the stability of routing protocols using mobility prediction. In [9], the amount of time two mobile hosts p and q will stay connected, $\lambda_{p,q}$ is predicted from their initial positions, speeds and moving directions. So we define the predicted connection time of a path π_i at time t as:

$$\lambda_i = Min(\lambda_{p, p+1}(t))\ldots\ldots\{p \in \pi_i \wedge 1 \leq p \leq m(\pi_i) - 1\} . \tag{5}$$

If λ_i is greater than L we use equation 4 to calculate β_i, else we use:

$$\beta_i = \frac{\alpha * \tau_i + (1-\alpha)\lambda_i}{\sigma \chi_i} . \tag{6}$$

where, α is an adjusting parameter determines the relative importance between residual energy and mobility value. Usually α varies from 0.8 to 1. A network having most of the connections are of long duration may use higher value of α.

4 Simulation and Results

The performance of the proposed BEER protocol is investigated through simulation and is compared with that of the LPR and PAR. In our simulation, we considered up to 25 nodes distributed randomly over the simulation area; confined in a 400X400 m^2. Every node has a fixed transmission power resulting in a 50 m of transmission range. Random connections were established between nodes within the transmission range. In case of simulating ad hoc networks, we use "random waypoint" model to generate node movement, where the motion is characterized by two factors: maximum speed and pause time. The lifetime of a node is varied between 1 and 600 while the transmission cost between two neighboring nodes is varied between 2 and 11.

Each packet received or transmitted has a cost factor. If the cost factor is n then $n-1$ is considered as the cost at the transmitter node and remaining unit cost goes to the receiving node. So transmission band may vary from 1 to 10 where receiving band is unit cost for all the nodes in the network. We run simulations for 150 times for networks considering both static and moving nodes; and average the resultant data to obtain the final data.

Results of simulation in cost perspective are depicted in Figure 2. It can be noticed that PAR performs the best in this perspective. Transmission cost of BEER lies between PAR and LPR and it is a bit closer to PAR. Figure 3 shows the time for individual node to run out of power. In PAR, first power failure occurs shortly, as some nodes are frequently selected by neighboring nodes. LPR maintains the longest lifetime for individual node among the three protocols as nodes are selected based on the remaining energy of that node. Figure 4 and Figure 5 show the average network lifetime in low and high node density, respectively. In low density three curves are closer, as there are less routing options to choose.

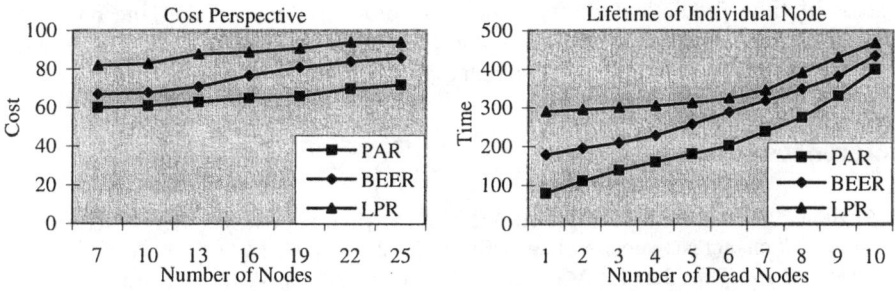

Fig. 2. Comparison of cost among three related protocols

Fig. 3. Lifetime of individual node among three related protocols

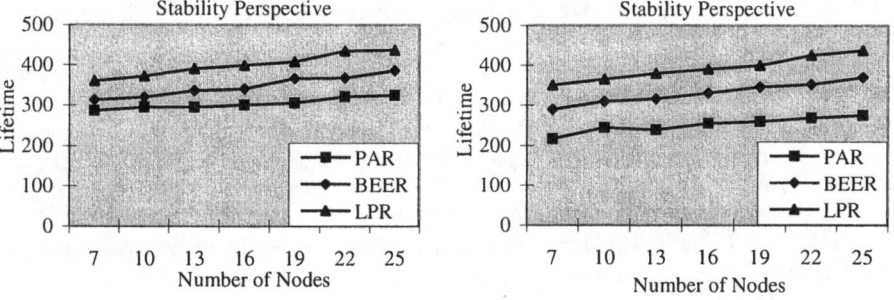

Fig. 4. Comparison of lifetime/stability among three related protocols (low node density)

Fig. 5. Comparison of lifetime/stability among three related protocols (high node density)

We consider network lifetime until 40% of total nodes die. Some of the nodes, alive at this point are also rendered unreachable due to the lack of forwarding nodes. From the figures shown above, we can conclude that PAR offers minimum cost but its network stability is poor. On the other hand the LPR has maximum network lifetime or stability but it suffers from high routing cost. BEER does not suffer extremely from either of the routing cost or network stability, as it is not biased by a single parameter. Thus, it maintains a balance between the two and offers cost-effective routing maintaining maximum possible network stability.

5 Conclusions

In this paper, we elaborate a cost effective and energy balanced routing protocol where routing problem is formulated as maximizing the network lifetime while minimizing the routing cost. We notice that the proposed BEER protocol may select a path with cost little higher than a path with least cost and a path having little less lifetime than a path having the highest lifetime. But it increases the network lifetime up to about 22% than that of power aware routing and cut routing cost up to 30% from the cost of lifetime prediction routing. Thus, the proposed method cuts the cost short while tries to maintain maximum possible lifetime of the network and thus emphasizes the advantage of combined approach over power only or lifetime only methods.

References

1. Akyildiz, I.F., Su, W., Sankarasubramaniam, Y., Cayirci, E.: Wireless Sensor Networks: A Survey. Computer Networks, Vol. 38. Issue 4, (2002) 393-422
2. Chang, J. H., Tassiulas, L.: Maximum Lifetime Routing in Wireless Sensor Networks. IEEE/ACM Transactions on Networking, Vol. 12. No. 4 (2004) 609 - 619
3. Huda, M.N., Hossain, M.J., Yamada, S., Kamioka, E., Chae, O.S.: Cost-Effective Lifetime Prediction Based Routing Protocol for MANET. Lecture Notes in Computer Science, Vol. 3391 (2005) 170-177
4. Maleki, M.; Dantu, K.; Pedram, M.: Lifetime Prediction Routing in Mobile Ad-Hoc Networks. Proceedings of IEEE WCNC, Vol. 2 (2003) 1185 - 1190
5. Johnson, D. B., Maltz, D. A.: Dynamic Source Routing in Ad Hoc Wireless Networks. Mobile Computing, Kluwer Academic Publishers (1996) 153–181
6. Rodoplu, V., Meng, T.H.: Minimum Energy Mobile Wireless Networks. IEEE Journal on Selected Areas in Communications, Vol. 17, No. 8 (1999) 1333-1344
7. Capkun, S., Hamdi, M., Hubaux, J.P.: GPS-Free Positioning in Mobile Ad-hoc Networks. Hawaii International Conference on System Sciences (2001) 1-10
8. Singh, S., Woo, M., Raghavendra, C. S.: Power-Aware Routing in Mobile Ad-hoc Networks. Proceedings of ACM/IEEE MOBICOM (1998) 181-190
9. Su, W., Lee, S.J., Gerla, M.: Mobility Prediction and Routing in Ad hoc Wireless Networks. International Journal of Network Management, Vol. 11 (2001) 3-30

Multipath Source Routing in Sensor Networks Based on Route Ranking

Chun Huang, Mainak Chatterjee, Wei Cui, and Ratan Guha

School of Computer Science,
University of Central Florida,
Orlando, FL 32816-2450
{chuang, mainak, weicui, guha}@cs.ucf.edu

Abstract. Multipath source routing is an effective way to exploit the redundant routes that are usually common in dense sensor networks. In this paper, we present a multipath source routing algorithm that uses a ranking technique to distinguish between the quality of different routes for the same source-destination pair. A ranking coefficient is calculated for each route based on three different metrics- energy, delay and reliability. The number of parallel routes that is considered is governed by the minimum reliability requirements. Simulation experiments are conducted that show that multipath routing can increase the reliability, and dissipate energy more evenly among the nodes.

1 Introduction

The advancement of wireless communication technologies coupled with the techniques for miniaturization of electronic devices have enabled the development of low-cost, low-power, multi-functional sensor networks [1]. The sensor nodes sense the environment and pass the information to a destination node (usually called a sink) through a single route as obtained by the underlying routing algorithm. For reduced complexity and overhead, such *single-path* routing algorithms are usually used in ad hoc and sensor networks. To avoid the dependency on the single route, we propose to select and use multiple routes to transfer packets from the source to the sink. Multipath routing is not a new concept and has been proposed as an alternative to single shortest path routing to distribute load and alleviate congestion in the network [2]. In multi-path routing, traffic bound to a destination is split across multiple paths to that destination. In other words, multipath routing uses multiple good paths instead of the best path. This mechanism also ensures that traffic load is distributed over the network to achieve load balancing and improve end-to-end delay.

In this paper, we first try to characterize the *rank coefficient* of a route. Rank is just a relative measure of how good or bad a route is with respect to some performance metrics like remaining (battery) energy, packet loss, and end-to-end delay. We discover multiple paths from the source to the destination (sink). We define the ranking metric as a linear combination of the three metrics and rank each route. Since, the possible number of routes can potentially be very large, we restrict ourselves to a subset consisting of at most K routes which are chosen based on their ranks. When a source node has packets for a destination node, it distributes them among the K routes. The fraction of packets going through each route is inversely proportional to the routing coefficient. To increase the robustness of the packet losses, we use forward error correction (FEC) codes. Simulation experiments are conducted that show the benefit of using multiple routes. The most important observation is the saving in energy of the nodes which extends the lifetime of the network.

2 Multipath Routing

Since this research does not deal with any particular routing algorithm, the proposed technique is generic enough and can be applied to any routing algorithm. We will assume that routes between source-destination pairs are obtained through algorithms such as DSR [4]. We also assume that there are multiple routes available for the same source-destination pair. This assumption can be justified by the fact that the density of sensor deployment is usually high and each node potentially has many neighbors. The existence of multiple routes can be obtained by using techniques such as [2] or extending DSR to incorporate multiple routes.

Cache Design: We suppose that each sensor node has a route cache consisting of two parts- one for routing and the other for local recovery. The route part stores the routing table. It contains request-id, parent-id and route number. Request-id is a unique id generated by the sink. As the name suggests, the local recovery part is used for local recovery of routes. Since we extend DSR to incorporate multiple routes, the route cache at each node must be re-designed.

Route Discovery: We assume that the sink issues route discovery only when it needs to send a query message. The sink first initiates a route discovery by broadcasting its query packet with a unique request-id and its node-id. When an intermediate sensor node receives this request, it checks the routing table to see if it has this request-id. If not, it will consider the node that forwarded the request as its parent and record it in its table. If a node receiving a query request is a destination node, it records the information and the route number. The destination node will save the information of all the preceding nodes and number the route number.

Route Reply: Destination sends route reply message back to its parents; the reply message will also contain information about the remaining energy, hop number, and delay. Each parent node that gets the message will record the route number, add its energy value, add its queuing delay time and increase the hop number to the message and forward the message to their parents. The route reply travels from the sink to the source node. Our algorithm guarantees node disjoint routes since each node (except source node) only sends reply message once.

3 Ranking of Routes

During transmission, the sink records the transmission status of each route. Based on the performance of each of the K routes, the sink ranks each route and periodically sends the ranking information to the source. The rank parameters considered here are nothing but the QoS parameters that we are interested in. They are energy consumption, end-to-end delay, and reliability (throughput). Our final goal is to improve the network lifetime for which considering the routes with more average residual energy is important. Let us now consider the factors that we use for ranking the routes. Let K be number of routes that are used. The determination of K will be discussed later.

Residual Energy: We define residual energy of a node as the amount of energy remaining at that node. We assume that at the time of network activation, all the nodes have equal amount of energy. With the lapse of time, energy will be depleted by the nodes- the amount of which will depend on the activity of the node. Let the average residual energy for the ith route be given by E_i. We normalize the average residual energy for the K routes and define the normalized residual energy for route i as $\hat{E}_i = \frac{E_i}{\sum_{i=1}^{i=K} E_i}$.

End-to-end Delay: The end-to-end delay is mainly governed by the queueing delay at every intermediate node. The queueing delay at a node is usually hard to calculate because it not only depends on the packet generation rate of that node but also the activities of the neighboring nodes. For estimating the queueing delay, we use the congestion related information at each node. This information is nothing but the MAC buffer state occupancy which is conveyed by the nodes in their beacon signals. So, for every route, the sink can calculate the the total delay that is expected at all the intermediate nodes. If T_i is the total delay for route i, then the normalized delay for route i is defined as $\hat{T}_i = \frac{T_i}{\sum_{i=1}^{i=K} T_i}$.

Reliability: For reliability, we consider the packet loss probability of each route. As the sink receives the packets along multiple routes, it calculates the ratio of packets lost for every route. If the packet loss probability for route i as P_i, then the normalized packet loss probability for route i is defined as $\hat{P}_i = \frac{P_i}{\sum_{i=1}^{i=K} P_i}$.

Overall Ranking Metric: So far, we have defined three metrics which are somewhat independent of each other. It can be noted that a high value is E_i is desirable, where as $\hat{T}_i$ and $\hat{P}_i$ should be low. We propose a linear combination of the three for the overall ranking of the routes. We used the normalized values for each factor such that all the factors have the same bounds, i.e., between 0 and 1. Thus the ranking coefficient for route i is defined as

$$R_i = \alpha(1 - \hat{E}_i) + \beta \hat{T}_i + \gamma \hat{P}_i \qquad (1)$$

where α, β, and γ are the weighing or tuning parameters for the three metrics respectively. Also, $\alpha + \beta + \gamma = 1$. The rank coefficients, R_i's, when sorted in the ascending order gives the ranks.

Packet Distribution: With the relative ordering of all the K routes being known, it is important that this route diversity be exploited for distributing the load over the network. We do so by making use of all the routes in a proportional manner. According to the rank of each route, the source distributes a fraction of packets along different routes. Since, lower the R_i better the route, we use the inverse ratio of the rank coefficients to calculate the fraction of the packets that would be routed through route i. Thus, if $R_1, R_2, \cdots, R_K$ are the rank coefficients for routes $1, 2, \cdots, K$, then the ratio in which packets are distributed are in the ratio $R_1^{-1}, R_2^{-1}, \cdots, R_K^{-1}$. Therefore, the fraction of packets through route i is given by $f_i = \frac{R_i^{-1}}{S}$, where $S = \sum_{i=1}^{i=K} R_i^{-1}$. Obviously, $\sum_{i=1}^{i=K} f_i = 1$.

Determination of K: Thus far, we dealt with K routes, but we never discussed how to determine K. It is intuitive that the number of routes, K, has a close relationship with the reliability that the network must operate. We impose the reliability requirement must be such that all packets are expected to arrive at the destination. Since we propose to use FEC coding, we can still achieve the desired level of reliability even if there are packet losses. If we use h redundancy packets for n original packets, then we can afford to loose h packets out of the total $N = n + h$ packets. These N packets are distributed among the K routes such that $N_i = Nf_i$ packets are routed through route i. We choose K routes such that the expected number of packets arriving at the destination is greater than n. Thus, $\sum_{i=1}^{K} \sum_{l=0}^{N_i} \binom{N_i}{l} P_i^l (1 - P_i)^{N_i - l} \geq n$. Recall, P_i is the packet loss probability for route i. The inner sum calculates the expected number of packet received through route i, and the outer sum finds the total number of packets received over all the K routes. K must be such chosen that the expected number of packets over all the K routes must be at least n.

4 Improving Reliability Through FEC

Forward error correction (FEC) is a method which is usually used to recover packets that get corrupted during transmission. The correction capability of these codes will depend on the kind of codes and the length of the code used. Since this paper does not deal with FEC codes, the simplest simplest of codes- *block codes* will be used. In block codes, M redundancy bits are added to the information bearing N bits. (Note that the extra M bits are generated using a generator matrix operating on the N bits.) In this paper, we use FEC on the packet level and not bit level. If we consider a packet of $N + M$ bits, then the resulting bit loss probability is given by [3] $b = \sum_{i=M+1}^{M+N} \binom{M+N}{i} b_p^i (1 - b_p)^{M+N-i}$, where, b_p is the bit loss probability before decoding and b is the decoded bit error probability.

5 Simulation Model and Results

To evaluate the performance of routing efficiency when multiple routes are used, we conducted simulation experiments where every sensor node was initialized with the same amount of energy. The bit error rates of each route was varied from 0 to 0.2. To calculate the power consumed for transmitting and sensing, a simple first order radio model was used [5], where the radio dissipates $E_{elec} = 50$ nJ/bit to power the transmitter/receiver circuitry and $E_{amp} = 100$ pJ/bit/m^2 for the transmit amplifier to achieve an acceptable E_b/N_0. Therefore, to transmit a k-bit message over a distance of d meters, the energy expended is

$$E_{Tx}(k, d) = kE_{Elec} + kd^2 E_{amp} \qquad (2)$$

To receive a k-bit message, the energy expended is $E_{Rx}(k) = kE_{Elec}$. We used different values of K. The ranking coefficients are calculated for the required number of routes and packets are distributed accordingly. The density of the nodes and the transmission range are so set that the number of hops range from 5 to 15. To investigate the rate at which the energy is consumed, we assume that every node is initialed with just enough power to transmit 1000 packets. We discuss the results with respect to network lifetime and reliability.

Lifetime: We compared the lifetime for 2 different cases. The first one is the lifetime measured using single route, and the second case is using single route with one backup route in case of the route failure. We compared the results with the proposed multipath routing scheme. We assumed that the sink recalculates the rank after every 250 packets. We show how the lifetime is affected in terms of both average remaining energy and the worst route remaining energy for $K = 4$ and 8 in figures 1 and 2 respectively.

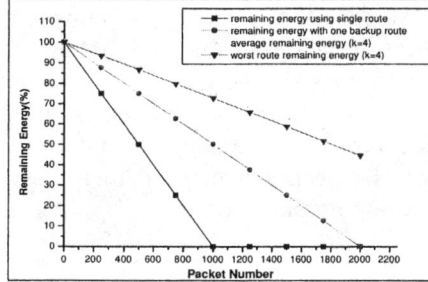 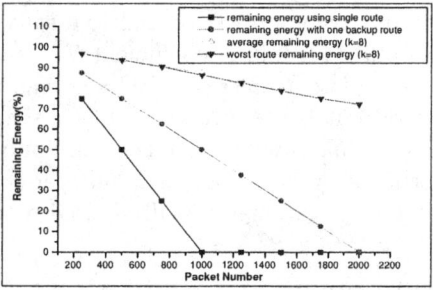

Fig. 1. Remaining energy for K = 4   **Fig. 2.** Remaining energy for K = 8

We observe that for a single route, the energy of the route is used up very fast, i.e., after 1000 packets are transmitted, there is no energy available in that route signifying a dead route. With one extra route as backup, the energy usage is better, but the route dies after 2000 packets were transmitted. Results improved on using multiple routes. Our multipath routing can distribute the packets to different routes according to their residual energy and also dynamically adjusts the distribution as and when their rank changes.

Reliability: We use the packet loss probability as a measure of reliability. We set the packet loss probability as 1% and check the block loss rate when block size change from 4 to 16 and redundant packet changes from 0 (i.e., no FEC) to 3. Figure 3 shows the loss probability with and without FEC. From the plot, we can see that without using FEC technique ($h = 0$), the probability to loose a packet is much higher than applying FEC. Figure 4 shows the loss probability when the number of redundant packets is changed from 1 to 3. This provides guideline on how to select the number of redundant packets.

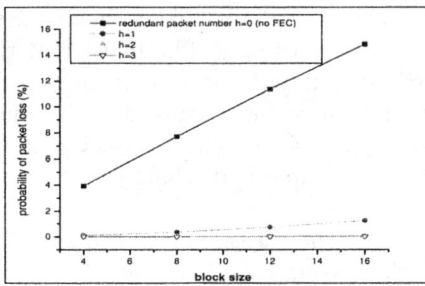
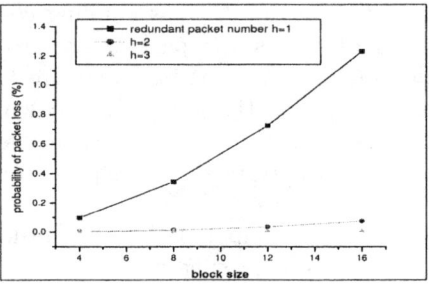

Fig. 3. Loss with and without FEC **Fig. 4.** Loss with different redundancy

6 Conclusions

In this paper, we presented a multipath source routing algorithm that exploits the relative goodness of multiple routes. We devised a ranking mechanism that computes a ranking coefficient for each route based on a linear combination of three different metrics. FEC was also used to increase the reliability. The number of routes used was such chosen that the expected number of packets arriving at the sink, would meet the minimum reliability requirements. Simulation experiments were conducted that show that the proposed method increases the reliability and energy is dissipated more evenly among the nodes.

References

1. Akyildiz I.F., Su W., Sankarasubramaniam Y., and Cayirci E.: A Survey on Sensor Networks. IEEE Comm. Magazine, Vol. 40, No. 8, August (2002), pp. 102-114.
2. De S., Qiao C., and Wu H.: Meshed multipath routing with selective forwarding: An efficient strategy in wireless sensor networks. Elsevier Computer Networks, Special Issue on Wireless Sensor Networks, vol. 43, no. 4, Nov. (2003), pp. 481-497.
3. Sklar B.: Digital Communications. 2nd ed. Prentice Hall.
4. Johnson D.B., and Maltz D.A.: Dynamic Source Routing in Ad Hoc Networks. Mobile Computing, Eds: T. Imielinski and H. Korth, Kulwer, (1996), pp. 152-81.
5. Salhieh A., and Schwiebert L.: Power aware metrices for wireless Sensor networks. International Journal of Computers and Applications, Vol. 26, No. 4, (2004).

Reliable Time Synchronization Protocol in Sensor Networks Considering Topology Changes*

Soyoung Hwang and Yunju Baek

Department of Computer Science and Engineering,
Pusan National University, Busan 609-735, South Korea
{youngox, yunju}@pnu.edu

Abstract. In this paper, we propose a reliable time synchronization protocol (RTSP) in sensor networks considering topology changes. Due to movement of sensor nodes, running out of energy or crashes in the network, the topology of sensor networks changes very frequently. In the proposed method, synchronization error is decreased by creating a hierarchical tree with lower depth and reliability is improved by maintaining and updating the information of candidate parent nodes. The RTSP reduces recovery time and cost compared to the TPSN (Timing-sync Protocol for Sensor Networks) when there are changes in topology. Simulation results show that RTSP has about 10% better performance than TPSN in synchronization accuracy. The number of messages in RTSP is 10%~30% lower than that in TPSN when there are topology changes.

1 Introduction

As in any distributed computer system, time synchronization is a critical issue in sensor networks. Time synchronization is a prerequisite for sensor network applications such as object tracking, consistent state updates, duplicate detection, and temporal order delivery. In addition to these domain-specific requirements, sensor network applications often rely on synchronization as typical distributed system do: for secure cryptographic schemes, coordination of future action, ordering logged events during system debugging, and so forth [1]. Traditional time synchronization methods in distributed systems can not be applied to the sensor networks directly because of the characteristic of sensor networks with limited computation and energy.

In the first stage of research on time synchronization in sensor networks, most approaches are based on the synchronization model such as event ordering or relative clock. These methods do not synchronize the sensor node clocks but generate a right chronology of events or maintain relative clock of nodes. From a viewpoint of the network topology, synchronization coverage is limited in a

---

* This work was supported by the Regional Research Centers Program (Research Center for Logistics Information Technology), granted by the Korean Ministry of Education and Human Resources Development.

single broadcast domain; however, typical wireless sensor networks operate in areas larger than the broadcast range of a single node, so network-wide time synchronization is needed essentially. Besides, adjusting the local clock has better efficiency than maintaining relative clock since it requires more memory capacity and communication overheads [2]. The FTSP [3] and the TPSN [4] are the representative ones which meet these requirements.

FTSP achieves robustness against node and link failures by utilizing periodic flooding of synchronization message and implicit dynamic topology update. On the other hand, TPSN does not handle dynamic topology changes; however, FTSP can not be applied generally since the synchronization accuracy in FTSP is seriously affected by the analyzed source of delays and uncertainties which are varied according to changes of the systems. The synchronization accuracy of network-wide multi-hop synchronization is a function of the construction and depth of the tree. The synchronization error is propagated hop by hop. Therefore new approaches are required to reduce the synchronization error and to manage dynamic topology changes.

This paper proposes a reliable time synchronization protocol in sensor networks considering topology changes. The topology of sensor networks changes frequently due to moving of sensor nodes, running out of energy or physical crashes in the network. In the proposed method, synchronization error is decreased by creating hierarchical tree with lower depth and reliability is improved by maintaining and updating information of candidate parent nodes. The RTSP reduces recovery time and costs - communication overheads - comparing to TPSN [4] when there are changes of topology.

2 Reliable Time Synchronization Protocol

In the following we present our scheme called *Reliable Time Synchronization Protocol* (RTSP) in sensor networks. It is assumed that nodes in the network have unique ID, but it does not need that each node is aware of the neighbor set as in the TPSN. The management of neighbor nodes is included in the operations of the protocol.

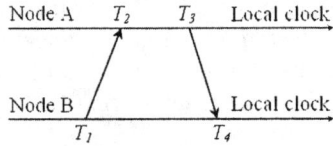

Fig. 1. Measuring delay and offset

As in the NTP, the roundtrip delay and clock offset between two nodes A and B are determined by a procedure in which timestamps are exchanged via wireless communication links between them. The procedure involves the four most recent timestamps numbered as shown in Figure 1. The measured roundtrip delay δ and clock offset θ of B relative to A are given by [5]

$$\delta = (T_4 - T_1) - (T_3 - T_2), \; \theta = \frac{(T_2 - T_1) + (T_3 - T_4)}{2}.$$

2.1 The First Phase: Hierarchical Topology Setup

In the first phase, a hierarchical topology is created in the network. This phase works to create a tree structure with lower depth and candidate parent list is generated to manage failure of nodes in the network.

Step 1: The root node initiates topology setup phase. Level *0* is assigned to the root node. It broadcasts topology setup message with its ID and its level.

Step 2: A node receives topology setup message during pre-defined time interval. (Root node discards this message.) It selects a parent with the lowest level number from received messages and stores other information to the candidate parent list according to the level number. Then it broadcasts topology setup message with its ID and its level.

Step 3: Each node in the network performs step 2 and eventually every node is assigned level.

Step 4: When a node does not receive topology setup message or a new node joins the network, it waits for some time to be assigned a level. If it is not assigned a level within that period, it broadcasts topology setup request message and then performs step 2 with reply of its neighbors.

2.2 The Second Phase: Synchronization and Handling Topology Changes

In the second phase, a node belonging to level i synchronizes with its parent node which is belonging to level $i-1$ by exchanging time-stamp messages. When a node can not communicate with its parent, it selects another parent in the candidate list and performs synchronization.

Step 1: The root node initiates synchronization phase by broadcasting synchronization message.

Step 2: On receiving synchronization message, nodes belonging to level *1* exchange time-stamp message with the root node and adjust the local clock and then broadcast synchronization message.

Step 3: On receiving synchronization message, each node belonging to level i exchanges time-stamp message with its parent and performs step 2. Eventually every node is synchronized. Once it receives a synchronization message, it discards additional messages from other upper level nodes.

Step 4: When a node can not communicate with its parent, it selects another parent in the candidate list, updates its own level - if it is needed - and performs step 3. The level of its child nodes will be updated when they execute synchronization. If the candidate list is empty, it performs step 4 of the topology setup phase ahead. Candidate list can be updated periodically by listening to communications of neighbors.

When the root node fails, a node which has the lowest ID in the next level takes over its role. The synchronization accuracy may be improved by utilizing the concepts of MAC layer time-stamping as in the TPSN, and the random back-off mechanism can be adapted to avoid the collision of wireless links.

3 Performance Evaluation

In order to evaluate the performance of the proposed method, we established a simulation model in the NESLsim based on the PARSEC platform [6, 7]. N nodes are deployed in a uniformly random fashion over a sensor terrain of size 100x100. Each node has a transmission range of 28. The number of nodes, N, is varied from 100 to 300 with each increase of 50. All other parameters are arranged with the same value in the TPSN simulation. The setup includes a CSMA MAC. The radio speed is 19.2kb/s, similar to the UC Berkeley MICA Motes, and every packet has a fixed size of 128bits. A node is chosen randomly to act as the root node. The granularity of the node clocks, which is the minimum accuracy attainable, is $10\mu s$. The clock model used in simulations has been derived from the characteristics of the oscillators used in sensor nodes. The frequency drift is varied randomly with time, within the specified range, to model the temporal variations in temperature. All sensor node clocks drift independently of each other. There is an initial random offset uniformly distributed over 2 seconds among the sensor node clocks to capture the initial spatial temperature variations and the difference in the boot up times [8]. All results are averaged over hundred simulation runs. The performance is compared to the TPSN. The synchronization error is defined as the difference between the clocks of the sensor nodes and the root node.

In Fig. 2, the number of messages processed during the simulation and the synchronization accuracy are presented when there is no failure of nodes. In almost the same number of messages, the RTSP has better performance in synchronization accuracy. This is the effect of the tree depth. Usually RTSP has 1~2 lower depth than TPSN.

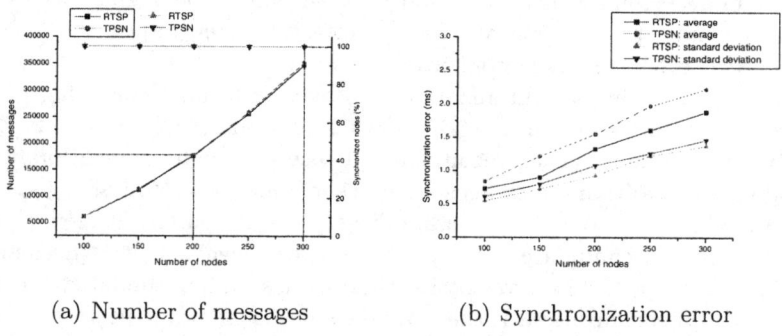

(a) Number of messages   (b) Synchronization error

Fig. 2. Without failure of nodes

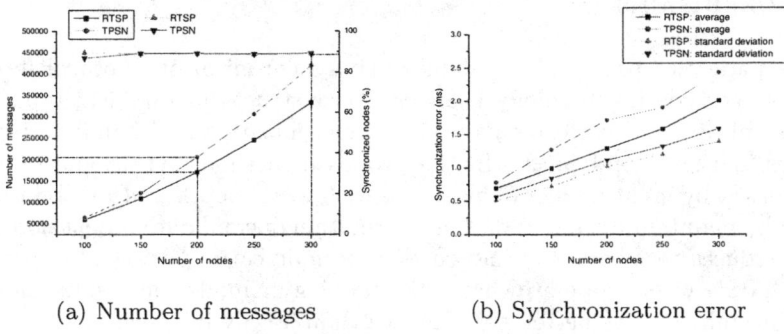

(a) Number of messages   (b) Synchronization error

Fig. 3. 10% failure of nodes

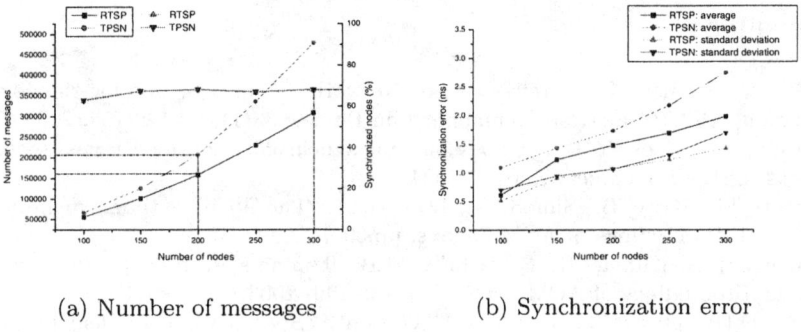

(a) Number of messages   (b) Synchronization error

Fig. 4. 30% failure of nodes

Fig. 3 and Fig. 4 show the number of messages processed during the simulation, synchronized proportion of nodes and synchronization accuracy when there are 10% and 30% failure of nodes respectively. In sensor networks, sensor nodes can fail easily such as nodes may move, may run out of energy and may be destroyed physically. This failure of nodes leads to topology changes. In the simulation, node failure means that there are topology changes. In a similar proportion of synchronized nodes to the entire nodes, RTSP reduces the number of messages and shows better performance in synchronization accuracy. In sensor networks, communication is one of the dominant factors in energy efficiency; therefore, communication overheads must be reduced to save energy. The RTSP reduces the number of messages and improves the synchronization accuracy by handling dynamic topology changes through the candidate parent list.

As can be seen in the results, the performance of RTSP gets better than TPSN as the failure rate (topology change) is increased. At 10% failure out of 300 nodes, the number of messages in the RTSP is 20% lower than that in the TPSN. At 30% failure out of 300 nodes, the number of messages in the RTSP is decreased by 35% compared to that in the TPSN.

4 Conclusions

In this paper we proposed a reliable time synchronization protocol in sensor networks considering topology changes. It constructs hierarchical topology in the first phase, and performs pair-wise synchronization and handling topology changes in the second phase. In the proposed method, synchronization error is decreased by creating hierarchical tree with lower depth and reliability is improved by maintaining and updating information of candidate parent nodes. The RTSP reduces recovery time and costs - communication overhead - comparing to the TPSN when there are changes of topology. Simulation results show that RTSP has about 10% better performance than TPSN in synchronization accuracy. The number of message in the RTSP is 10%~35% lower than that in the TPSN when there are topology changes.

References

1. Elson, J., Romer, K.: Wireless Sensor Networks: A new regime for time synchronization, ACM Computer Communication Review 33(1), pp.149-154, 2003.
2. Hwang, S.Y, Baek, Y.J.: A survey on time synchronization for wireless sensor networks, ESLAB Technical Report, 2004.
3. Maroti, M., Kusy, B., Simon, G., Ledeczi, A.: The flooding time synchronization protocol, Proceedings of ACM SenSys, pp.39-49, 2004.
4. Ganeriwal, S. Kumar, R., Srivastava, M.B.: Timing-sync protocol for sensor networks, Proceedings of ACM SenSys, pp.138-149, 2003.
5. Mills, D.L: Network Time Protocol (Version 3) Specification, Implementation and Analysis, RFC1305, 1992.
6. PARSEC User Manual, http://pcl.cs.ucla.edu/projects/parsec, 1999.
7. Ganeriwal, S., Tsiatsis, V., Schurgers, C., Srivastava, M.B.: NESLsim: A parsec based simulation platform for sensor networks, NESL, 2002.
8. Ganeriwal, S., Kumar, R., Adlakha, S., Srivastava, M.B.: Network-wide time synchronization in sensor networks, NESL Technical Report, 2003.

The Brain, Complex Networks, and Beyond

L.M. Patnaik

Indian Institute of Science,
Bangalore 560012
lalit@micro.iisc.ernet.in

(Prof. A K Choudhury Memorial Lecture)

Abstract. This presentation covers a synthesizing overview of the structural organisation of the brain, viewed as a complex network. Such an organisation is encountered in social, information, technological, and biological networks. The underlying conclusions may, in future, lead to interesting studies in the areas of cognition, and distribution computing. It is also hoped that the brain network structure studied through scale-free, small world, and clustering concepts may facilitate better understanding and design of brain-computer interface (BCI) systems.

1 Introduction

I deem it an honour to deliver the Prof. A K Choudhury Memorial Lecture at the Seventh International Workshop on Distributed Computing. Prof. Choudhury was an outstanding researcher who has made pioneering contributions in diverse areas in the broad discipline of Electrical Sciences. Notable among the areas where he made some of his excellent contributions are, control and system theory, fault diagnosis, computer hardware and logic design, network and circuit theory. As a befitting tribute to this great scholar, I have chosen a topic of interdisciplinary nature covering some of the above areas.

Networks in the human brain possibly work similar to those in the internet. Networks often have very many nodes with very few links, and very few nodes with very many links. The brain is one of the most challenging complex systems. The neurons are massively interconnected to each other. To understand the complexity of the nervous system, we need to characterize its network structure. Networks are described by simply defining a set of nodes and connections (edges) between them. A wide variety of such systems are scale-free, where the connectivity distribution takes a power-law form. What makes such networks complex is not only their size but also the interaction of architecture or the interconnection topology and dynamics. In many networks, cluster of nodes group into tightly coupled neighborhoods, but maintain short distances among nodes in the entire network. Such a situation leads to what is known as 'small world' within the network [1]. For many networks, the degree of individual nodes forms a distribution that decays as a power law, producing a 'scale-free architecture' characterized by highly connected nodes (hubs).

Nervous systems generate and integrate information from several external and internal sources in real-time. The aim of this presentation is to review some insights into connectivity of the brain. Though many studies have been reported on single neuron networks, what is of more interest from a computer network paradigm is the large-scale networks of the cerebral cortex, enabling us to study links between neural organization and cognition.

2 Cerebral Cortex: A Network Structure

Large-scale connections of the cerebral cortex of mammalians have been studied. This area is neither completely connected with each other nor randomly linked. On the contrary, a specific and intricate organisation is revealed in this region. The functional roles of brain regions are specified by their inputs and outputs. At the next level of organization, i.e. neural circuits linking small sets of connected brain areas; we need to look for patterns of local interconnections occurring with a higher frequency in real networks than in the randomized networks.

2.1 Connection Patterns

Mammalian connection patterns studied through graph theoretical techniques have revealed interesting features. Cortical connection patterns exhibit small world features, which have short path lengths and high clustering coefficients [3-4]. The average shortest path is given by the mean of the entries of the distance (adjacency) matrix. The clustering coefficient C_i of a node is calculated as the number of existing connections between the node's neighbors divided by all their possible connections. High clustering and short path lengths can be found across cortical organisation. Graph theory tools provide insights into the functioning of neural architectures. In-degree and out-degree specify the amount of functional convergence and divergence of a given region. On the other hand, the clustering coefficient indicates the degree to which the area is part of functionally related regions. The path length metric between two regions of the brain represents the potential 'functional proximity'. Inter-cluster connections link areas with one another in all shortest paths and are important for structural stability of cortical networks [5].

2.2 Functional Networks

Deterministic clustering method has been used to combine cross-correlations between fMRI signals, and graph theory formalism. Image voxels are represented by nodes of a graph, and the corresponding temporal correlation matrix represents the weight matrix of the edges between the nodes. Based on fMRI data, a network can be implemented, where those voxels that are functionally linked are 'connected'. Their degree distribution and the probability of finding a link versus the distance decay as a power law. The corresponding characteristic length is short, although the clustering coefficient is much larger.

A possible link between network organisation and cognition is likely to exist. Our future understanding of human cognition will benefit from the studies on complex brain networks. One may still be interested in answering questions such as, are all cognitive processes carried out in distributed networks? Are some cognitive processes carried out in more restricted processes?

Predominantly brain activity is spatio-temporal. It is hard to analyse such systems using numerical techniques. Thus attempts have been made to study such systems using the concepts of complex networks consisting of nodes and links with specific topological properties. During any given task, the networks are constructed using magnetic resonance brain activity. This activity is measured, at each time step, from several brain sites.

2.3 Scale-Free Networks

Networks with power-law degree distributions have been the focus of a great deal of attention. They are sometimes referred to as scale-free networks, although it is only their degree distributions are scale-free. In any real network, some nodes are more highly connected than others are. To quantity this effect, let p_k denote the fraction of nodes that have k links. Here k is called the degree and p_k is the degree distribution. For many real networks, p_k decays much more slowly than a Poisson distribution and is given by a power law $p_k \sim 1/k^\gamma$. These networks are 'scale-free' by analogy with fractals and other situations where power laws arise and no single characteristic scale can be defined.

The brain creates and reshapes continuously complex functional networks, during behavior or at rest. These networks have been studied, using functional magnetic resonance imaging in humans. The degree distribution of the network for a subject in finger tapping task is found to demonstrate the scale-free property of the network. This property implies that there are always a small but finite number of brain sites having broad access to most other brain regions. The scale-free character is unaltered for tasks engaging different brain states corresponding to tasks such as listening to music.

2.4 The Small-World Effect

A direct demonstration of the small-world effect is the fact that most pairs of vertices in most networks seem to be connected by a short path through the network. If one considers the spread of information across a network, the small-world effect implies that the spread will be fast on most networks. If it takes six steps for a rumour to spread from any person to any other person, then the rumour will spread much faster than if it takes a hundred steps. From a calculation of path length and clustering coefficient, the small-world structure of the brain network can be demonstrated.

2.5 Wiring in Networks

Graphs that result from selection for complex dynamics can be placed in a physical space such that the wiring cost is low. However, in real brains possibly the positioning of vertices precedes the edge formation between them. But complex dynamics should consider low wiring costs too. Evolution exerts pressure on connectivity to reduce the overall wiring length, or to maximize connectivity while minimizing volume or to place brain components in order to minimize wiring length. It is unlikely that evolutionary pressure on wiring alone is responsible for the specific patterns of connectivity we notice today. Possibly the anatomical structures have evolved to accommodate specific kinds of functional and dynamic interactions supporting adaptive behavior. Relationship between wiring and functional connectivity could be investigated by embedding graphs in two-or three-dimensional space, by incorporating explicit development rules in the wiring process or by including conduction delay type temporal features. As diverse sources of environmental information need to be integrated, and varied output patterns are required for adaptive behavior, there is a need for the selection for neural architecture capable of matching signals as well as for degenerate pathways to increase robustness against failure. Consequently the complexity of the neural circuits will increase.

3 Neuronal Characterization by Fractal Dimension

Fractal dimension has been used to characterize the neurons. Such a computation makes extensive use of automated image analysis system and the approach is extremely useful in studying multiple neurons connected through a network. The fractal dimension D may not always be an adequate descriptor of a neuron. For example, two neurons may appear visually very different from one another, yet having the same fractal dimension. Moreover, a complex structure such as a neuron can be a mixture of different fractals, each one with a different fractal dimension. Attempts have been made to use multifractals as a more comprehensive methodology to provide information about the distribution of fractal dimension in biological systems [6].

4 Brain-Computer Interface (BCI)

Electroencephalographic activity or other electrophysiological measures of brain function might provide a new channel for sending messages to the external world - a brain-computer interface (BCI) [7]. Such systems provide a supportive communication and control technology for those with severe neuromuscular disorders, such as brainstem stroke, and spinal cord injury. These systems can provide users, who may be completely paralyzed, with basic communication capabilities so that they can express their wishes to people attending to them; they can even operate keyboard and mouse. These signals typically include cortical potentials and cortical neuronal activity recorded by implanted electrodes. The user encodes the commands in these signals and the BCI system derives the commands from the signals.

The signal features used in present-day BCIs reflect identifiable brain events like firing of the synchronized and rhythmic synaptic activation in sensorimotor cortex that produces a mu rhythm. Knowledge of these events can help guide BCI development. The location and function of the cortical area generating a rhythm or an evoked potential can indicate how it should be recorded and how to eliminate the effects of non-CNS (Central Nervous System) artifacts.

Most BCIs use electrophysiological signal features representing brain events that are well-defined both anatomically and physiologically. These include rhythms reflecting oscillations in particular neuronal circuits (mu and beta rhythms from sensorimotor cortex), potentials evoked from particular brain regions by specific stimuli, or action potentials generated by particular cortical neurons.

User training may be the most important and least understood factor affecting the BCI capabilities of different signal features. BCI signal features are not normal or natural brain output channels. They are artificial output channels created by BCI systems. Thus it is not clear to what extent these artificial outputs will observe known principles. For example, mu rhythms and other features generated in sensorimotor cortex, may be more useful than alpha rhythms generated in visual cortex. Initial efforts have focused on neurons in motor cortex. Other cortical areas need exploration.

Some of the above issues can be addressed if the complex brain network is studied extensively, both mathematically and by simulation studies, using the network principles discussed earlier.

5 Conclusions

The structure of brain networks is a result of the combined forces of natural selection and natural activity during evolution and development from computational and information theoretical concepts. Brain has to solve the problem of information extraction from inputs and the generation of coherent states that allow coordinated action. This imposes severe constraints on the set of possible cortical connection patterns. More empirical and computational work is needed to develop the functional principles underlying the structural connection patterns in the cortex. There may be more ways in which structural properties of brain networks influence the dynamical and informational patterns neurons can generate. Dynamical patterns generated by brain networks underlie cognition and perception operators. Some aspects of vision seem to be embedded in structural connectivity of the thalamocortical network. Network analysis may enable us to understand the computational power of the brain. Also if the studies of brain image classification [8] are suitably integrated into brain network analysis, there may be a scope to identify regions of the brain responsible for malfunctions such as epilepsy, Schizophrenia, Parkinson's, Huntington's, Alzheimer's etc.

It is envisaged that such studies may be of mutual interest to the neuroscience and distributed computing research communities, to learn more about the performance of such complex systems. Brain-computer interface has been attracting significant attention in recent years and network-centric studies of the brain may, in future, throw open several challenging issues.

References

1. Watts, D.J., Strogatz, S.H.: Collective Dynamics of 'Small World' Networks. Nature. 393 (1998) 440-442
2. Barabasi, A.L, Albert, R.: Emergence of Scaling in Random Networks, Science. 286 (1999) 509-512

3. Hilgetag, C.C. et .al.: Anatomical Connectivity Defines the Macaque Monkey and the Cat..Philosophical Transactions of the Royal Society of London B Biological Sciences. 335 (2000) 91-110
4. Sporns, O. and Tononi, G.: Classes of Network Connectivity and Dynamics, Complexity. 7 (2002) 28-38
5. Kaiser, M., and Hilgetag, C.C.: Edge Vulnerability in Neural and Metabolic Networks. Biological Cybernetics. 90 (2004) 311-31
6. Smith, T. G., Lange G.D., Mark W. B.: Fractal Methods in Cellular Morphology n-Dimensions, Lacunarity, and Multifractals. Journal of Neuroscience Methods. 69 (1996) 123-36
7. Mc Farland, D .J., Sarnacki, W.A., Wolpaw, J.R.: Brain-Computer Interface (BCI) Operation: Optimizing Information Transfer Rates. Biological Psychology. 36 (2003) 237-251
8. Patnaik, L .M.: Daubechis-4 Wavelet with SVM as an Efficient Method for Classification of Brain Images. Journal of Electronic Imaging. 14 (2005) 1-7

An Asynchronous Recovery Algorithm Based on a Staggered Quasi-Synchronous Checkpointing Algorithm*

D. Manivannan, Q. Jiang, J. Yang, K.E. Persson, and M. Singhal

Computer Science Department, University of Kentucky, Lexington, KY 40506
{manivann, richardj, jyang2, karl, singhal}@cs.uky.edu

Abstract. Checkpointing and rollback recovery are established techniques for handling failures in distributed systems. Under synchronous checkpointing, each process involved in the distributed computation takes checkpoint almost simultaneously. This causes contention for network stable storage and hence degrades performance. To overcome this problem, checkpoint staggering under which checkpoints by various processes are taken in a staggered manner, has been proposed. In this paper, we propose a staggered quasi-synchronous checkpointing algorithm which reduces contention for network stable storage without any synchronization overhead. We also present an asynchronous recovery algorithm based on the checkpointing algorithm.

1 Introduction

In distributed computing systems, checkpointing and rollback recovery are well-established techniques for handling failures [1], [2], [3], [4], [5], [6], [7], [8]. Existing checkpointing algorithms can be classified into three main categories – *asynchronous*, *synchronous* and *quasi-synchronous* [9]. In *asynchronous* checkpointing, processes take checkpoints periodically without any coordination. However, when a failure occurs, recovery may suffer from *domino effect*, in which processes roll back recursively in order to roll back the system to a consistent global state. Moreover, multiple checkpoints need to be kept in stable storage and some or all the checkpoints taken may not be part of any consistent global checkpoint and hence are useless.

In *synchronous* checkpointing schemes, domino-free recovery is achieved by sacrificing process autonomy and incurring extra synchronization overhead during checkpointing. In this approach, processes synchronize their checkpointing activity so that a globally consistent set of checkpoints is always maintained in the system [1], [10], [5]. Under *quasi-synchronous* (or *communication-induced*) checkpointing [11], [12], [13], [6] processes are allowed to take checkpoints (called

---

* This material is based in part upon work supported by the US National science Foundation under Grant No. IIS-0414791. Any opinions, findings, and conclusions or recommendations expressed in this material are those of the authors and do not necessarily reflect the views of the National Science Foundation.

basic checkpoints) asynchronously, as well as reduce the number of useless checkpoints by forcing processes to take additional checkpoints (called *forced checkpoints*) at appropriate times. Hence, they have the advantages of both synchronous and asynchronous checkpointing algorithms.

Quasi-synchronous checkpointing mitigates the problems with synchronous and asynchronous algorithms. However, contention for stable storage is still a problem when several processes take checkpoints simultaneously. This can significantly impact the checkpointing *overhead* and extend the total execution time of the distributed computation [14], [15]. Contention for stable storage can be mitigated by *staggering* the checkpoints [16]. *Staggered* checkpointing attempts to prevent two or more processes take checkpoints at the same time and reduce contention for stable storage. To the best of our knowledge, checkpoint *staggering* has previously been proposed for only *synchronous*, or coordinated, checkpointing algorithms [16], [15].

Objectives

In this paper, we present a staggered quasi-synchronous checkpointing algorithm that takes basic checkpoints in a staggered manner to reduce contention for stable storage. We also present a basic recovery algorithm based on the checkpointing algorithm.

Organization

The rest of the paper is organized as follows. In Section 2 we present the system model, background and related work. Section 3 describes our staggered quasi-synchronous checkpointing algorithm. In Section 4 we present our basic recovery algorithm. Section 5 concludes the paper.

2 System Model, Background and Related Work

In this section, we present the system model and background required. We also briefly review the related work.

2.1 System Model

A distributed computation consists of N sequential processes denoted by P_0, P_1, P_2, $\cdots$, P_{N-1} running concurrently on a set of computers in the network. Processes do not share global memory or a global physical clock. Message passing is the only way for processes to communicate with one another. The computation is asynchronous: each process evolves at its own speed and messages are exchanged through communication channels, whose transmission delays are finite but arbitrary. We assume that messages are not lost, altered or spuriously introduced. Processes are fail-stop. All failures are detected immediately and result in halting failed processes and initiating recovery action [8].

Execution of a process is modeled by three types of events – the send event of a message, the receive event of a message, and an internal event. The states of processes depend on one another due to interprocess communication. Lamport's *happened before* relation [17] on events, $\xrightarrow{hb}$, is defined as the transitive closure of the union of two other relations: $\xrightarrow{hb} = (\xrightarrow{xo} \cup \xrightarrow{m})^+$. The $\xrightarrow{xo}$ relation captures the order in which local events of a process are executed. The ith event of any process P_p (denoted $e_{p,i}$) always executes before the $(i+1)^{st}$ event: $e_{p,i} \xrightarrow{xo} e_{p,i+1}$. The $\xrightarrow{m}$ relation shows the relation between the send and receive events of the same message: if a is the send event of a message and b is the corresponding receive event of the same message, then $a \xrightarrow{m} b$ [7].

2.2 Background

A local checkpoint of a process is a recorded state of the process in stable storage. A checkpoint of a process is considered as a local event of the process for the purpose of determining the existence of happened before relation among states of processes. Each checkpoint of a process is assigned a unique sequence number. The checkpoint of process P_p with sequence number i is denoted by $C_{p,i}$.

The send and the receive events of a message M are denoted respectively by $send(M)$ and $receive(M)$. So, $send(M) \xrightarrow{hb} C_{p,i}$ if message M was sent by process P_p before taking the checkpoint $C_{p,i}$. Also, $receive(M) \xrightarrow{hb} C_{p,i}$ if message M was received and processed by P_p before taking the checkpoint $C_{p,i}$. $send(M) \xrightarrow{hb} receive(M)$ for any message M. The set of events in a process that lie between two consecutive checkpoints is called a checkpoint interval. Next, we present the definition of a consistent global checkpoint.

Definition 1. *A set $S = \{C_{0,m_0}, C_{1,m_1}, \cdots, C_{N-1,m_{N-1}}\}$ of N checkpoints, one from each process, is said to be a **consistent global checkpoint**[1] if $C_{p,m_p} \not\xrightarrow{hb} C_{q,m_q}$ for all p,q, $0 \leq p,q \leq N-1$.*

Z-paths and their Properties

In [7], Netzer and Xu give a necessary and sufficient condition for a given set of checkpoints to be part of a consistent global checkpoint. They introduce the notion of *zigzag paths*, which is a generalization of *causal paths*[2] induced by the Lamport's happened before relation. A zigzag path (or a Z-path for short) between two checkpoints is like a causal path, but a Z-path allows a message in the sequence to be sent before the previous one in the path is received.

We use the notation $A \xrightarrow{zp} B$ to indicate the existence of a Z-path from checkpoint A to B. Note that the existence of Z-paths is a transitive relation. In other

---
[1] Also called a **a consistent cut**.
[2] A causal path from a checkpoint A to checkpoint B exists if and only if there exists a sequence of messages $m_1, m_2, \cdots, m_n$ such that m_1 is sent after A, m_n is received before B, and m_i is received by some process before the same process sends m_{i+1} ($1 \leq i < n$).

words, if $A \stackrel{zp}{\leadsto} B$ and $B \stackrel{zp}{\leadsto} C$, then $A \stackrel{zp}{\leadsto} C$. A checkpoint C is said to be in a *Z-cycle* if $C \stackrel{zp}{\leadsto} C$. An important property of Z-paths is that it captures the precise requirement for a set of checkpoints to be a part of a consistent global checkpoint as stated in the following theorem due to Netzer and Xu [7]. We state the theorem using the notations introduced above.

Theorem 1. *A set of checkpoints S can be extended to a consistent global checkpoint if and only if for any two checkpoints $A, B \in S$ (not necessarily distinct) neither $A \stackrel{zp}{\leadsto} B$ nor $B \stackrel{zp}{\leadsto} A$ holds.*

Proof. Proof can be found in [7]. □

In particular, if we take S to be a set containing a single checkpoint in Theorem 1, it follows that a checkpoint can be part of a consistent global checkpoint if and only if it does not lie on a Z-cycle. So, we have the following Corollary.

Corollary 1. *A checkpoint of a process is part of a consistent global checkpoint if and only if it does not lie on a Z-cycle.*

So, checkpoints that lie on a Z-cycle are *useless*. An efficient quasi-synchronous checkpointing algorithm tries to minimize the useless checkpoints while minimizing the number of forced checkpoints.

2.3 Related Work

In this section we briefly review previous work related to staggered quasi-synchronous checkpointing.

Chandy and Lamport [1] propose a synchronous checkpointing algorithm. Their algorithm assumes the channels to be FIFO. The checkpointing process is initiated by a coordinator. The coordinator first records its own state (takes a checkpoint) and then sends a marker message along all outgoing channels before sending any other messages. If a process that receives the marker has not already recorded its state, it immediately records the state of the incoming channel as empty and then records its state. It then resends the marker along all its outgoing channels. If a process that receives the marker has already taken a checkpoint, it merely records the messages received (along the channel on which the marker was received) since its last checkpoint as the state of that channel. The algorithm guarantees that the checkpoints taken form a consistent global checkpoint. However, contention for stable storage can occur as a result of multiple processes taking checkpoints simultaneously.

Plank [16] observes that, to a certain degree, the Chandy-Lamport (C-L) algorithm [1] staggers checkpoints when marker messages (initially sent by the coordinator) only reach neighboring processes, which in turn resend the marker to their neighbors. In contrast, the staggered behavior is eliminated if all processes simultaneously receive a marker message from the coordinator directly. Plank proposes a variation of the C-L algorithm that staggers a *limited* number of checkpoints, depending on the network topology. Plank assumes a connected, but not necessarily complete, underlying interconnection network. Clearly, in this approach a completely connected topology would subvert staggering. A network sweeping algorithm is also used to route messages through neighboring nodes, and to ensure a consistent global state. Once all processes have finished sweeping, and notified the coordinator, the local checkpoints are committed and a consistent global state is obtained from the set of local checkpoints. The algorithm successfully maintains a consistent global state in a coordinated manner similar to Chandy-Lamport [1]. Moreover, contention for stable storage is proportional to the degree of connectivity in the underlying network topology.

Based on Plank's observation, Vaidya [15] proposes another synchronous checkpointing algorithm that staggers *all* checkpoints. Like Plank [16] and Chandy-Lamport [1], Vaidya uses a coordinator to initiate the checkpointing process. The algorithm has two phases. In the first phase, the coordinator P_0 takes a physical checkpoint and sends a *take_checkpoint* message to the next process P_1. Upon receipt of the *take_checkpoint* message, process P_i takes a physical checkpoint and resends it to process P_j, where $i>0$ and $j = (i+1)$ mod n. The phase is terminated when the coordinator P_0 receives the *take_checkpoint* message from the last process P_{n-1}. In the second phase, the channel states, called by author as logical checkpoints, are recorded. The set of logical checkpoints, together with the physical checkpoints, form a consistent global state. The algorithm successfully staggers all physical checkpoints. However, contention for stable storage exists for taking the logical checkpoints. In the next section, we present our staggered quasi-synchronous checkpointing algorithm which reduces contention for stable storage without any synchronization overhead.

3 Our Staggered Quasi-Synchronous Checkpointing Algorithm

In this section, we present our staggered quasi-synchronous checkpointing algorithm which not only makes all checkpoints useful but also reduces contention for stable storage by taking basic checkpoints in a staggered manner. Since all checkpoints taken are useful, the algorithm ensures the existence of a *recovery line*[3] containing any checkpoint of any process. This property of the algorithm helps bound rollback during recovery due to a failure.

3.1 The Algorithm

Informal Description of the Algorithm

Under our algorithm, each process takes basic checkpoints asynchronously. In addition, to prevent useless checkpoints, processes take forced checkpoints upon the reception of some messages. Each checkpoint is assigned a unique sequence number. The sequence number assigned to a basic checkpoint is the current value of a local counter (an integer variable). Since the sequence numbers assigned to

---
[3] A consistent global checkpoint.

basic checkpoints are picked from the local counters which are incremented periodically, the sequence numbers of the latest checkpoints of all the processes will differ by at most one as long as the local clocks do not drift more than half the checkpoint time interval. This property helps in advancing the recovery line. When a process P_p sends a message, it appends the sequence number of its current checkpoint to the message. When a process P_q receives a message, if the sequence number appended to the message is greater than the sequence number of the latest checkpoint of P_q, then, before processing the message, P_q takes a checkpoint and assigns the sequence number received in the message as the sequence number of the checkpoint taken. When it is time for a process to take a basic checkpoint, it skips taking a basic checkpoint if its latest checkpoint has a sequence number greater than or equal to the current value of its counter (this situation could arise as a result of the forced checkpoints or drift in local clocks). This strategy helps to reduce the checkpointing overhead, i.e., the number of checkpoints taken. An alternative approach to reduce the number of checkpoints would be to allow a process to delay processing a received message until the sequence number of its latest checkpoint is greater than or equal to the sequence number received in the message.

If several processes take checkpoints simultaneously, they will contend for access to the stable network storage. The network contention can be reduced by taking checkpoints in a staggered manner. Next, we illustrate our approach for taking basic checkpoints in a staggered manner. We assume that there are a total of N processes $P_0, P_1, \ldots, P_{N-1}$, involved in the distributed computations we consider. Each process has a unique process id. For example, process P_p (where $0 \leq p < N$) has process id p. We also assume that it takes at most t (maximum checkpoint latency) time units to take a checkpoint and send it to the stable network storage in the absence of contention for stable storage. Each process takes one checkpoint (either basic or forced) within each checkpoint interval X. A local variable $next_p$ keeps track of the current number of checkpoint intervals by incrementing by 1 at the end of each checkpoint interval. $next_p$ is initialized to 1. We denote the local clock at the site in which process P_p is running as C_p. The current time at clock C_p is denoted by $V(C_p)$. For simplicity, we assume that $V(C_p)$ is initialized to 0.

Within each checkpoint interval of length X time units, a process takes a basic checkpoint some time during the second half of the interval if it has not taken a forced checkpoint yet. The second half of the interval is divided into several time slots. The size of each slot T is at least t (maximum checkpoint latency) plus δ (maximum local clock drift) time units. So, T is defined as follows:

$$T = t + \delta \tag{1}$$

The number of slots within a checkpoint interval, denoted by γ, is given by Equation 2.

$$\gamma = \lfloor X/(2T) \rfloor \tag{2}$$

We assume that X is chosen such that $T << X$. For example, if $N \leq 15$ and $T = 10$ seconds, it may be ideal to choose $X = 5$ minutes so that there would be

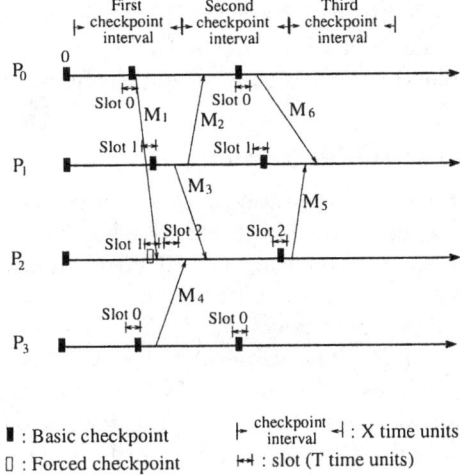

Fig. 1. An example illustrating basic checkpoints taken in a staggered way

15 available slots during the later half of each 5 minute interval and each process can have its own slot to take a checkpoint. However, if there are more processes than time slots, then we use a round-robin method to reduce the contention for stable storage. This can be achieved as follows:

> A process P_p takes a basic checkpoint when its local time $V(C_p) = (next_p - 1) * X + X/2 + (p \bmod \gamma) * T$ if there is no forced checkpoint already taken in the period of time from $(next_p - 1) * X$ to $V(C_p)$.

So far we discussed how our algorithm takes basic checkpoints in a staggered manner to reduce contention. It is also possible to reduce contention between basic and forced checkpoints. In Section 3.2, we discuss one such optimization.

Next, we illustrate the basic idea behind how basic checkpoints are taken in a staggered manner using an example.

An Example. In Figure 1, each checkpoint interval is X time units long. Each slot is of length T and there are a total of 3 slots in the second half of each checkpoint interval. There are a total of 4 processes involved in the distributed computation. To take a basic checkpoints in a staggered manner, we require that process P_0 takes its basic checkpoint in time slot 0 in each checkpoint interval, process P_1 in slot 1, P_2 in slot 2, and P_3 in slot 0[4]. If no forced checkpoints are taken in any checkpoint interval, only process P_0's basic checkpoint slot (i.e., slot 0) collides with process P_3's basic checkpoint slot (slot 0) in each checkpoint interval, assuming synchronized clocks. This is illustrated in the second checkpoint interval in Figure 1. In the presence of forced checkpoints, illustrated in the first checkpoint interval of Figure 1, process P_0's

---

[4] P_3 and P_0 take checkpoint in the same time slot. This is because, 0 mod 3 = 3 mod 3.

basic checkpoint slot (slot 0) collides with P_3's basic checkpoint slot, and P_1's basic checkpoint slot (slot 1) also collides with P_2's forced checkpoint slot. The forced checkpoint, taken by P_2 in slot 1, is due to the reception of message M_1 from process P_0.

Formal Description of the Algorithm

The example given above is very simple, but it illustrates the main idea behind taking basic checkpoints in a staggered manner. Next, we present the staggered quasi-synchronous checkpointing algorithm formally. The variable $next_p$ of process P_p represents its local counter. It keeps track of the current number of checkpoint intervals at process P_p. $V(C_p)$ denotes the current value of local clock at the site of P_p. The value of the variable sn_p represents the sequence number of the latest checkpoint of P_p at any time. So, whenever a new checkpoint is taken, the checkpoint is assigned a sequence number and sn_p is updated accordingly. Also, $C.sn$ denotes the sequence number assigned to the checkpoint C and $M.sn$ denotes the sequence number piggybacked with message M.

The Staggered Quasi-Synchronous Checkpointing Algorithm

Data Structures at Process P_p

$V(C_p) := 0;$ {Current value of local clock, initialized to 0.}
$sn_p := 0;$ {Sequence number of the current checkpoint, initialized to 0. This is updated every time a new checkpoint is taken.}
$next_p := 1;$ {Sequence number to be assigned to the next basic checkpoint, initialized to 1}

When it is time for process P_p to increment $next_p$
$next_p := next_p + 1;$ {$next_p$ is incremented at periodic time intervals of X time units}

When process P_p sends a message M
$\quad M.sn := sn_p;$ {sequence number of the current checkpoint appended to M}
$\quad$ send $(M);$

Process P_q, upon receiving a message M from process P_p
$\quad$ if $sn_q < M.sn$ then {if sequence number of the current checkpoint is less than
$\quad\quad$ Take checkpoint C; checkpoint number received in the message, then
$\quad\quad$ $C.sn := M.sn;$ take a new checkpoint before processing the message}
$\quad\quad$ $sn_q := M.sn;$
$\quad$ Process the message.

When it is time for process P_p to take a basic checkpoint (i.e., When $V(C_p) = (next_p - 1) * X + X/2 + (p \bmod \gamma) * T)$
$\quad$ if $next_p > sn_p$ then {skips taking a basic checkpoint if $next_p \leq sn_p$ (i.e., if it already
$\quad\quad\quad\quad sn_p := next_p;$ took a *forced* checkpoint with sequence number $\geq next_p$)}
$\quad\quad\quad\quad$ Take checkpoint C;
$\quad\quad\quad\quad C.sn := sn_p;$

Theorem 2. *The staggered quasi-synchronous checkpointing algorithm presented above makes every checkpoint useful.*

Proof: We only need to prove that none of the checkpoints lies on a Z-Cycle by Corollary 1. Let C be any checkpoint. Suppose C lies on a Z-cycle, then there exists a sequence of messages $M_1, M_2, \cdots, M_n$ that forms a Z-path from C to itself. In particular, M_1 is sent after the checkpoint C is taken and M_n is received before the checkpoint C is taken. Thus $M_1.sn \geq C.sn$. Since a message M is received and processed by a process only after it had taken a checkpoint with sequence number $\geq M.sn$, it follows from the definition of Z-paths that $M_i.sn \geq C.sn \, \forall i, 1 \leq i \leq n$. In particular, $M_n.sn \geq C.sn$. This is impossible since M_n is received before the checkpoint C is taken and there is no checkpoint with sequence number $\geq C.sn$ that precedes C, since $C.sn \leq M_n.sn$ and all checkpoints that precede C have sequence numbers $< C.sn$. Hence, our assumption that C is on a Z-cycle is incorrect and hence every checkpoint is useful. □

When processes take basic checkpoints in a staggered manner, contention for stable storage is reduced. If there are no forced checkpoints taken, then the degree of contention can be easily computed. In the absence of forced checkpoints, the degree of contention, for stable network storage, DC_{nw}, can be defined as follows:

$$DC_{nw} = \begin{cases} 0 & \text{if } N \leq \gamma \\ N/\gamma & \text{otherwise} \end{cases} \quad (3)$$

When forced checkpoints are present, more than one process may take checkpoint in some time slots even if $N \leq \gamma$, while fewer checkpoints (or none at all) are taken in other slots. The degree of network contention can not be easily computed, and depends on the communication pattern as well as the values of N, X, and T. So, we analyze performance of our algorithm under various scenarios using simulation.

3.2 An Optimization

In the staggered quasi-synchronous checkpointing algorithm presented above, effort is made to stagger basic checkpoints. However, nothing is done to reduce the contention that arises when forced and basic checkpoints are taken simultaneously by two different processes. We propose an optimization to handle this situation when the number of time slots is *at least twice* as many as the number of processes. In this case, we can reduce the probability of a basic checkpoint and a forced checkpoint being taken in the same slot significantly. This is achieved by allowing processes to take basic checkpoints in the even (or odd) numbered slots within each checkpoint interval, while the forced checkpoints are taken in the odd (or even) numbered slots within the same checkpoint interval. Contention for stable storage is reduced because basic checkpoints are taken in different time slots.

Next, we describe this optimization formally, where the number of slots within each checkpoint interval is at least twice as many as the number of processes (i.e.,

$\gamma \geq 2N$). For simplicity, we only provide rules which differ from the algorithm in Section 3.1:

(1): Each process P_p takes a basic checkpoint at a specified even-numbered slot within each checkpoint interval, if no forced checkpoint has been taken within the same checkpoint interval yet.

(2): When process P_p receives a message with sequence number greater than its current checkpoint sequence number, it checks whether the value of its local clock is within an odd-numbered slot in (the later half of) the current checkpoint interval. It takes a forced checkpoint if it is currently in an odd-numbered slot. Otherwise, it delays to process the message and takes a forced checkpoint at the next odd-numbered slot and then processes the message.

4 Recovery Algorithm

In this section, we first present a basic recovery algorithm that rolls back the processes to checkpoints that form a consistent global checkpoint. Due to space restriction, we do not present a comprehensive recovery algorithm which handles the different types of messages (such as lost messages, delayed messages, etc) appropriately. We also do not present the performance evaluation of the algorithm for the same reason.

4.1 The Basic Recovery Algorithm

The basic recovery algorithm presented below only rolls back processes to a consistent global checkpoint when a process fails. It does not necessarily restore the system to a consistent global state.

The Basic Recovery Algorithm

When process P_p fails
    Roll back to the latest checkpoint C;
    send $roll\_back\_to(C.sn)$ to all the other processes;

Process P_q on receiving $roll\_back\_to(n)$ message
    If $sn_q \geq n$ **then**
        Find the checkpoint C of P_q such that $C.sn = n$;
        Roll back to C;
        $sn_q := C.sn$;
        Discard all the checkpoints taken after C;
    Else   {In this case the process does not roll back at all}
        Take a checkpoint C;   {It takes a checkpoint and proceeds normally}
        $C.sn := n$;
        $sn_q := C.sn$;   {update sn_q}

Note that under the basic recovery algorithm, a failed process rolls back to its latest checkpoint, say with sequence number n, and all other processes roll back to their checkpoint with sequence number n as well. The set of checkpoints with the same sequence number to which the processes roll back form a consistent global checkpoint because a message sent by a process after taking a checkpoint with sequence number n is never received by a process before taking a checkpoint with sequence number n. Even though, the basic recovery algorithm rolls back the processes to a consistent global checkpoint, it may not restore the system to a consistent state. For example, due to rollback, a process may have undone the event $receive(M)$ of some message M while the sender of M might not have undone $send(M)$.

5 Conclusion

In this paper we presented a staggered quasi-synchronous checkpointing algorithm that makes every checkpoint useful, and reduces contention for stable storage significantly. In contrast to previous staggered checkpointing algorithms, our approach does not require explicit coordination. We also studied the performance of our algorithm with varied approaches for selecting time-slots for basic and forced checkpoints. Our simulation results indicate that the adaptive optimization of our algorithm performs the best. We also presented a comprehensive recovery algorithm based on the checkpointing algorithm. For handling the lost messages due to rollback, messages are logged selectively and optimistically at both sender and receiver. Thus, our approach does not have the disadvantages of simple optimistic or pessimistic message logging but has the advantages of both of them; and this advantage comes with very low overhead as our performance evaluation indicates.

References

1. Chandy, K.M., Lamport, L.: Distributed Snapshots : Determining Global States of Distributed Systems. ACM Transactions on Computer Systems **3** (1985) 63–75
2. Elnozahy, E.N., Zwaenepoel, W.: Manetho: Transparent Rollback-recovery with Low Overhead, Limited Roll-back and Fast Output Commit. IEEE Transactions on Computers **41** (1992) 526–531
3. Helary, J.M.: Observing Global States of Asynchronous Distributed Applications. In: Proceedings of 3rd International Workshop on Distributed Algorithms, LNCS 392, Berlin: Springer (1989) 124–134
4. Johnson, D.B., Zwaenepoel, W.: Recovery in Distributed Systems Using Optimistic Message Logging and Checkpointing. Journal of Algorithms **11** (1990) 462–491
5. Koo, R., Toueg, S.: Checkpointing and Roll-back Recovery for Distributed Systems. IEEE Transactions on Software Engineering **SE-13** (1987) 23–31
6. Manivannan, D., Singhal, M.: Asynchronous Recovery Without Using Vector Timestamps. Journal of Parallel and Distributed Computing **62** (2002) 1695–1728
7. Netzer, R.H.B., Xu, J.: Necessary and Sufficient Conditions for Consistent Global Snapshots. IEEE Transactions on Parallel and Distributed Systems **6** (1995) 165–169

8. Strom, R.E., Yemini, S.: Optimistic Recovery in Distributed Systems. ACM Transactions on Computer Systems **3** (1985) 204–226
9. Manivannan, D., Singhal, M.: Quasi-Synchronous Checkpointing: Models, Characterization, and Classification . IEEE Transactions on Parallel and Distributed Systems **10** (1999) 703–713
10. e Silva, L.M., Silva, J.G.: Global Checkpointing for Distributed Programs. In: Proceedings of Symposium on Reliable Distributed Systems. (1992) 155–162
11. Baldoni, R., Helary, J.M., Mostefaoui, A., Raynal, M.: A Communication Induced Algorithm that Ensures the Rollback Dependency Trackability. In: Proceedings of the 27th International Symposium on Fault-Tolerant Computing, Seattle. (1997)
12. Kim, K.H.: A Scheme for Coordinated Execution of Independently Designed Recoverable Distributed Processes. In: Proceedings of 16th IEEE Symposium on Fault-Tolerant Computing. (1986) 130–135
13. Manivannan, D., Singhal, M.: A Low-overhead Recovery Technique using Quasi-synchronous Checkpointing. In: Proceedings of the 16th IEEE International Conference on Distributed Computing Systems, Hong Kong (1996) 100–107
14. Vaidya, N.: On Checkpoint Latency. In: Proceedings of the Pacific Rim International Symposium on Fault-Tolerant Systems. (1995)
15. Vaidya, N.: Staggered Consistent Checkpointing. IEEE IEEE Transactions on Parallel and Distributed Systems **10** (1999) 694–702
16. Plank, J.: Efficient Checkpointing on MIMD Architectures. PhD thesis, Priceton University (1993)
17. Lamport, L.: Time, Clocks and Ordering of Events in Distributed Systems. Communications of the ACM. **21** (1978) 558–565

Self-stabilizing Publish/Subscribe Protocol for P2P Networks*

Zhenyu Xu and Pradip K. Srimani

Department of Computer Science,
Clemson University, Clemson, SC 29634–0974

Abstract. In this paper, we develop a new self-stabilizing (fault tolerant) protocol for publish/subscribe scheme in a P2P network. We provide a complexity analysis of the recovery (stabilization) time of the protocol after arbitrary failures in the network. The protocol converges in at most $n^2(\Delta + 1)m + n^3 - n$ time in the worst case where n, m, and Δ denote respectively the number of nodes, edges, and the maximum degree of a node in the system graph (network). We also propose a a space efficient way to utilize this self-stabilizing publish/subscribe scheme, which allows flexibility in implementations.

1 Introduction

Publish/Subscribe has become a popular method of distributing information in the P2P networks. In a P2P system, the number of information sources is usually large and hence the problem of how to obtain the desired information in the system, is of great importance to the peers.

The publish/subscribe system involves two different kinds of processes: information producer and information consumer. The producer is responsible of announcing to the network what information the producer introduces into the system. The consumer, on the other hand, announces what information the consumer is interested in, and retrieves this information accordingly.

When implementing the publish/subscribe scheme in a P2P network, brokers play an important role. The brokers gather the announcements from the information producers and the subscriptions from the information consumers. With this knowledge, brokers match the information publisher and subscriber. There are two types of brokers: centralized broker and distributed broker. In the centralized approach, every node in the P2P network talks to the unique broker in the system. In the distributed approach, there are multiple brokers where each of them is responsible for a part of the subscriptions. Distributed brokers are more desirable in real life as they are capable of adapting to the network scaling and topology changes. But, this distributed approach needs to handle the additional problem of sharing data between the brokers. Traditional solutions include multicast tree and dynamic routing [1], [2].

---

* The work was supported by an NSF Award # ANI-0219485.

Various publish/subscribe protocols have been proposed in the literature. Castro et al. presented a group based multicast protocol, Scribe [3], on top of Pastry [4], which route the message based on numeric node ID of the peers. In [5], Fox and Pallickara presented the Narada brokering system, where the brokers route the message through the shortest path in a hierarchical server/peer topology. Bayeux [2] is a multicast protocol presented by Zhuang et al. It organizes the information consumers into a multicast tree rooted at the information provider, and route the message according to the suffix of the node ID.

In a recent paper [6], the authors introduced an interesting approach of implementing publish/subscribe system. Their scheme is anonymous (nodes do not have unique IDs), decentralized, modular, and self-organizing. Most importantly, only local information is needed at each peer node to construct the organization. The approach starts with building a logical directed acyclic graph (DAG), which determine the priority of the peers. Only the privileged peers are allowed to disseminate information. The algorithm has a built-in mechanism to assign privileges to the peer nodes and is designed in such a way that the privileged peer, once activated, will relinquish its privilege, by changing the logical DAG of peers; thus, every node in the DAG will eventually get activated infinite times. This liveness property associated with and starvation-freeness s the unique feature of this publish/subscribe scheme [6]. However, this approach requires the system to be initialized to start with a logical DAG of the peers (by adjusting local state variables at peer nodes). If the initial state of the system is not legitimate (the peer nodes do not form a DAG), or there is temporary corruption of the local state variables, then the algorithm is not guaranteed to satisfy the properties of liveness and lack of starvation. In other words, the approach is not self-stabilizing and not tolerant to error.

The publish/subscribe algorithm proposed in [6], is a localized (actions at nodes are based on local knowledge [7]) distributed algorithm but it is not fault tolerant. Self-stabilization is a relatively new paradigm for designing fault tolerant localized distributed algorithms for networks; it is an *optimistic* way of looking at system fault tolerance and scalable coordination, because it provides a built-in safeguard against transient failures that might corrupt the data in a distributed system. The concept was introduced by Dijkstra in 1974 [8], and Lamport [9] showed its relevance to fault tolerance in distributed systems in 1983; a good survey of early self-stabilizing algorithms can be found in [10] and Herman's bibliography [11] also provides a fairly comprehensive listing of most papers in this field. Our purpose in this paper is to design a self-stabilizing publish/subscribe protocol for P2P networks; the network can start from any arbitrary state (no initialization or global reset is necessary for starting the protocol), the protocol can recover from an arbitrary data corruption at any number of nodes and the protocol is a localized distributed algorithm (each node needs to have knowledge only of the states of its immediate neighbors). We achieve this objective by modifying the algorithm of [6] with the concept of unison from [12]. An unison system is one where each node has a clock variable that is assigned value $i+1$ iff the clock variables on all neighboring nodes has the value

i or $i+1$. By applying the unison concept, we show that the publish/subscribe scheme becomes now self-stabilizing. We provide a detailed worst case time complexity analysis of the bounded unison system (using a bounded clock variable). It is to be noted that the unison system was proved to converge in finite time ([12]), but no complexity analysis was done in previous works. We also show that it is possible to carry numerous topics with fixed resources while using the publish/subscribe schemes in P2P networks. By providing a trade-off between transfer time and system resources we can attain flexibility on publish/subscribe scheme; we show that the resulting protocols are more efficient. This property will prove useful when designing restricted systems. Comparing to Pastry [4] and Scribe [3], our protocol further ensures every peer in the multicast group gets the privilege to publish or subscribe data.

2 Logical DAG and the Edge Reversal Algorithm [6]

The original method proposed in [6] was based on the assumption that a logical DAG is imposed on the system graph by the node variables. Each node has two variables: an integer identifier lid and an integer variable val, where $val \in \{0, 1, 2\}$. The logical orientation of the edges in DAG is defined as follows:

Definition 1. *The relation $\prec$ is defined as:*

$$x \prec y \overset{\text{def}}{=} y = (x+1) \bmod 3$$

Definition 2. *The logical orientation of the edges $\rightarrow$ is defined as:*

$$q \rightarrow p \text{ iff } (val_p \prec val_q) \vee (val_p = val_q \wedge lid_p < lid_q)$$

Definition 3. *A sink is defined as a node such that:*

$$sink(p) \text{ iff } \forall q \in N(p), q \rightarrow p$$

For any value of the pair (val, lid) at node p and q, it is guaranteed that either $p \rightarrow q$ or $q \rightarrow p$ is true. Only a sink node is privileged to move (take actions). Once a sink node moves, it resigns the sink status by reversing the logical orientation on all the edges incident at this node. This is done by the algorithm shown in Figure 1.

R1: if $sink(i) \wedge (\forall j \in N(i), val_i = val_j)$
　　then $val_i := (val_i + 1) \bmod 3$
R2: if $sink(i) \wedge (\exists j \in N(i), val_i \prec val_j)$
　　then $val_i := max_{j \in N(i)}(val_j)$ and $lid_i := min\{k \in 0..n|$
　　　$\forall j \in N(i), val_i \prec val_j \Rightarrow k > lid_j\}$

Fig. 1. Re-orientation Algorithm

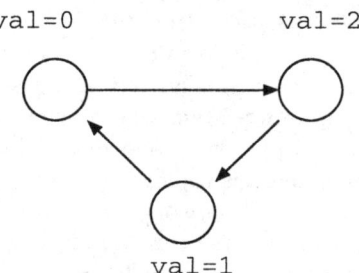

Fig. 2. Example of cycle in system

2.1 Initial State Dependency

It is shown in [6] that if the initial global system state denoted by the local variables at each node such that the induced directed subgraph is acyclic, then the algorithm 1 will maintain this acyclicity, i.e. every time a sink node moves and the system enters into a new global system state, there will always be a sink node in the new state, each node will become a sink node infinitely often and the publish/subscribe protocol will properly function.

However, when error occurs in transactions, or the system starts from illegitimate state, it is possible that the induced directed system graph by the local variables at each node is not acyclic and hence there may not be any sink node; this results in a situation that no node can move. An example is given in Figure 2

By the definition of $\prec$, we know that $0 \prec 1 \prec 2 \prec 0$. In the above example, the three nodes in the network have val values set at $0, 1, 2$ respectively, thus forming a logical directed cycle; no node is privileged to move, so the system stops functioning. The problem is inherent in the $\prec$ relation. Since every edge must have logical orientation explicitly defined by the local variables, the relationship $\prec$ need be a total order. To maintain a acyclic digraph, there will always exist a local minimum. On the other hand, the protocol needs to re-orient an edge by changing the variables on one node only, i.e., for every possible $x \prec y$, we should be able to find a z such that $y \prec z$. this will require infinite elements in the domain of the variable.

In order to use of the re-orientation, authors in [6] uses the lexicographical ordering of two variables (val, lid) for comparison, where the variable val assumes values of positive integers modulo 3. Such wrapping restricted the values to a finite domain, but introduced the necessity of starting system from a fixed initial state. (since the values are from a finite domain, to satisfy the re-orientation requirement, there must be a subset of values $V = \{v_1, v_2, \ldots v_s\}$, such that $v_1 \prec v_2 \prec \ldots \prec v_s \prec v_1$; thus, if this is the configuration of the initial values on the nodes in a cycle, there will be no sink in system).

3 New Method to Determine Priority

We propose a new method to determine the priority of the nodes. Such a method should have following desired properties:

- Liveness: At any given time, there is at least one node in the system privileged to move.
- Starvation freeness: Each node in the system must be privileged infinitely often.
- Self-stabilization: For any arbitrary initial state of the system, it converges to a legitimate state in finite time.

3.1 Bounded Unison System

The bounded unison system is defined in [12] as follows:

Definition 4. v *is a clock variable which can take the value* $0 \ldots Z-1$. v *is maintained on every node* p. *We denote it as* v_p.

Definition 5. *A relation* $\prec$ *is defined as*

$$x \prec y \text{ iff } (y-x) \text{ mod } Z \leq n$$

here n *is the number of nodes in system.*

Definition 6. *A relation* $\asymp$ *is defined as*

$$x \asymp y \text{ iff } not(y \prec x) \wedge not(x \prec y)$$

Definition 7. *A system is a bounded unison system iff for every node* p *in the system,* p *can only change its* v_p *when privileged:*
$\forall q \in N(p), v_q = v_p \vee v_q = v_p + 1 \text{ mod } Z$. *Here* Z *is a predetermined constant greater than* n^2.

Definition 8. *A legitimate state of a bounded unison system is defined as:*

$$\forall p, \forall q \in N(p), |v_p - v_q| \leq 1 \text{ mod } Z$$

A node gets the priority when it is privileged. We present the algorithm for a node to resign its priority:

This algorithm will converge to legitimate states, and then moving from one legitimate state to another legitimate state for infinite times. This can be proved by showing the number of execution of R1 on each node is finite. The proof is given in [12]. We will show the proof in the next section, and give a bound of convergence. By the definition of legitimate state, every node gets privileged infinitely often.

Algorithm 3 is a replacement of algorithm 1. We construct the publish/subscribe algorithm based on algorithm 3 in exactly the same way that algorithm 1 is applied.

R1: if $\exists j \in N(i), v_j \asymp v_i \wedge v_i > v_j$
    then $v_i := 0$
R2: if $\forall j \in N(i), v_i \prec v_j$
    then $v_i := v_i + 1 \text{ mod } Z$

Fig. 3. Algorithm 3: Unison Re-orientation

4 Analysis

4.1 Correctness

In this section, we present the proof of the algorithm's satisfaction of publish/subscribe requirement.

Lemma 1. *Algorithm 3 meets the* liveness *requirement.*

Proof. The proof is given by [12]. For any 2 given nodes p and q on an edge (p, q), the relation between v_p and v_q is one of the following three:

$$v_p \prec v_q \quad v_q \prec v_p \quad v_p \asymp v_q$$

If there exists an edge (p, q), such that $v_p \asymp v_q$, by the definition of $\asymp$ we know that $|v_p - v_q| > n$. Thus $v_p \neq v_q$, either p or q will be privileged by R1.

If there does not exist such an edge (p, q), $v_p \asymp v_q$, assume there is no node in the system gets privileged, i.e. $\forall p, \exists q \in N(p) : v_q \prec v_p$. By the definition of $\prec$, we know that $(v_p - v_q) \bmod Z \leq n$. Since we choose Z such that $Z > n^2$, it requires at least $n + 1$ nodes in the system. Contradiction. So there always be at least one node that is privileged.

Lemma 2. *Algorithm 3 meets the* starvation-free *requirement.*

Proof. Authors in [12] proved that algorithm 3 will converge to legitimate states in finite time. After that, the system will evolve within the legitimate states for infinite long time. This implies that at least one of the nodes, say i, will get privileged infinite times.

When system is in legitimate state, only R2 is executed on every node. So each time R2 is executed, v_i is increased by 1. Suppose from time t to t', v_i is increased three times to $v_i' = v_i + 3$. For any node $j \in N(i)$, $v_j \leq v_i + 1$ at time t, and $v_j' \geq v_i' - 1$ at time t'. Therefore $v_j' \geq v_j + 1$, node v is privileged at least one time between t and t'. Since i is privileged infinite times, j is also privileged infinite times.

Thus for a node that get privileged infinite times, every adjacent node is privileged infinite times. Because the system is a connected graph, eventually every node gets privileged infinite times.

Lemma 3. *Rule R1 is executed at most* $(\Delta + 1)m$ *times, where* Δ *is the max degree and m is the number of edges.*

Proof. Assume R1 is executed on node i. Let $v_i(t)$ be the value of v_i before the move, and $v_i(t+1)$ be the value of v_i after the move. There must be an adjacent node j, such that $v_j(t) \asymp v_i(t)$, and $v_i(t) > v_j(t)$.

Define following two invariants:

$$\psi_1 \stackrel{\text{def}}{=} |\{(i, j) \in E | v_j \asymp v_i \wedge v_i > v_j \wedge v_j > 0\}|$$

ψ_1 is the number of edges (i, j) such that j's existence makes i to execute R1, and $v_j > 0$.

$$\psi_2 \stackrel{\text{def}}{=} |\{(i,j) \in E | v_j \asymp v_i \wedge v_i > v_j \wedge v_j = 0\}|$$

ψ_2 is the number of edges (i,j) such that j's existence makes i to execute R1, and $v_j = 0$.

For any such $v_j(t) > n$, the execution of R1 on node i decreases ψ_1 by 1, and increases ψ_2 by at most $deg(i)$.

For any such j, $n \geq v_j(t) > 0$, the execution of R1 on node i decreases ψ_1 by at least 1, and does not change ψ_2.

For any such j, $v_j(t) = 0$, the execution of R1 on node i does not change ψ_1, by at least 1, and decreases ψ_2 by 1.

And the execution of R2 won't change both ψ_1 and ψ_2. Therefore ψ_1 is non-increasing, and ψ_2 is increased only when ψ_1 is decreased.

The upper bound of ψ_1 is the number of edges m. The upper bound of ψ_2 is the same. For ψ_1 to decrease to 0, ψ_2 is increased at most Δm. When R1 is executed, either or both ψ_1 and ψ_2 is decreased. So the total number of executions of R1 is at most $(\Delta + 1)m$.

Theorem 1. *Algorithm 3 meets the self-stabilization requirement. The system will converge to legitimate state within $n^2(\Delta+1)m + n^3 - n$ moves. After that, all moves will lead system to another legitimated state.*

Proof. By Lemma 3, R1 is executed at most $(\Delta + 1)m$ times.

Consider two executions of R1 on the same node i. There may be R2's executed on i between these two executions of R1. Let $v_i(t)$ be the value of v_i after the first R1, $v_i(t+1)$ be the value after next R2, and so on. Let $v_i(t+T)$ be the value of v_i after the second R1,

If $v_j(t) > n$, the next move on j will be R2. And i will not move until R2 has been executed on j. If $v_j(t) \leq n$, j won't move until after several R2's, $v_i = v_j$ or $v_i = v_j + 1$. Since $v_i = v_j + 1$ comes after R2 on i, before that R2 we still have $v_i = v_j$. So, after R1 is executed, only one node among i and j can move, until $v_i = v_j$. And the number of moves before $v_i = v_j$ is less than $n+1$.

After $v_i = v_j$, i and j will keep $|v_i - v_j| \leq 1$ by executing R2, until one of them executes R1.

Therefore if R2 is continuously executed on i more than $n+1$ times, $\forall j \in N(i), |v_i(t+n+1) - v_j(t+n+1)| \leq 1$.

Repeat this step. If R2 is continuously executed on i more than $s(n+1)$ times, then for any node k within distance s from i, $(d(k,i) < s)$, $\forall l \in N(k), |v_k(t+n+1) - v_l(t+n+1)| \leq 1$.

Because the maximum distance between i and any other node is at most $n-1$, the maximum number of continuous executions of R2 on i is $(n-1)(n+1) = n^2 - 1$. If more than this number of R2's are continuously executed, then system will be in a legitimated state.

So there can be at most $n^2 - 1$ R2's between any two consequent R1's on any node i. Since R1 is executed totally at most $(\Delta+1)m$ times, the converge time will be $(n^2-1)(\Delta+1)m + (\Delta+1)m + (n^2-1)n = n^2(\Delta+1)m + n^3 - n$. After this number of moves, system is guaranteed to be in legitimated state.

5 Publish/Subscribe

5.1 Layered Publish/Subscribe

There are multiple topics or contents existing in the system. Node publishes or subscribes the topic or content that it is interested in. Given a priority algorithm as showed in previous sections, a nature way to organize the topics and contents is to assign a virtual layer for each topics or contents [6].

They also showed topic based publish/subscribe and content based publish/subscribe can be established on this same priority adjustment method. So on the next we only show the topic based publish/subscribe scheme. Content based scheme is quite the same.

A virtual layer L^s is defined as a set of variables v_i^s on all nodes i, and the algorithms that adjusts v_i^s. Two variables v_i^s and v_i^t are accessed and modified separately on node i, therefore layers L^s and L^t is independent. The algorithm 4 now works on every layer:

> R1-s: if $\exists j \in N(i), v_j^s \asymp v_i^s \wedge v_i^s > v_j^s$
>   then $v_i^s := 0$
> R2-s: if $\forall j \in N(i), v_i^s \prec v_j^s$
>   then $v_i^s := v_i^s + 1 \ mod \ Z$

Fig. 4. Algorithm 4: Unison Re-orientation on layer L^s

A node i gets priority on layer L^s if the legitimate invariant of bounded unison is hold on v_i^s, and v_i^s is privileged to change.

For each topic, a new virtual layer is created on the graph. The node is allowed to take action to publish or forward information s only when it gets priority on layer s.

5.2 Actions Performed on Privileged Nodes

When sending or forwarding information s, node i send information data to all $j \in N(i)$. The data is then stored in the local buffer buf_j^s of node j.

Definition 9. *A buffer buf_i^s is a local storage on node i. When information data related to layer s is received at node i, it will be put into buf_i^s*

When node i gets priority on layer s, it reads buf_i^s, discards redundant messages, then forward the received messages, and send the new message created by node i itself, if any. The whole process is described in algorithm 5.

A control layer (layer 0) is used to coordinate between the nodes. Layer 0 transfers the information that what topics are running on other layers. A node has to get priority in layer 0 to initial a new layer. When node i gets priority and wants to initial a new layer, it is guaranteed that all previous layer initialization started on other nodes are already traversed to node i. Thus no conflicts will occur.

```
if priority(i, s)
    then read local buffer $buf_i^s$;
            send received information and new
information that are on topic s;
            execute algorithm 4;
```

Fig. 5. Algorithm 5: publish/subscribe on topic s

```
$s_t(t = 1 \ldots k)$ are the topics on layer s
if priority(i, s)
    then read local buffer $buf_i^s$;
            send received information and new
information that are on topic $s_t$;
            execute algorithm 4;
```

Fig. 6. Algorithm 6: publish/subscribe

5.3 Time Space Trade-Off

It can be easily showed that the total number of layers is $L + 1$ if there are L topics in the system. And the memory storage for the variables on each node is L. The network traffic consists of information messages and the value of all v_i^s that are used to maintain a bounded unison. For L layers, each layer will have one set of v_i^s to sent between nodes.

When the number of topic goes up, the number of layers increases in lineal scale, so do the storage and the network traffic to maintain legitimate states. Consider the nodes with limited resources (e.g. in sensor network), sometimes a fixed storage is required. This means to keep the number of layers about the same, while number of topics increases.

In order to handle this, we present the multi-access of the layer. Each layer is assigned several topics, and the node can only publish or forward information of those topics when is gets priority on the related layer.

In extreme condition, only 1 layer is needed. This will reduce the variable storage, but it also has drawbacks. The most apparent drawback is the transfer time. Layers work in parallel. Since there can be several nodes get priority on different layers, the more the number of layers, the more nodes execute publish/subscribe scheme at the same time. Therefore, when number of layer decreases, the time that useful information traverse in the network increases. In the extreme condition, it takes L times to the one-topic-per-layer scheme if all L topics run in single layer.

As a result, we have two optimization metrics, optimal space and optimal time. If t topics use 1 layer, then for L topics, comparing to the one-topic-per-layer scheme, the space needed on each node is $1/t$, and the time is t times.

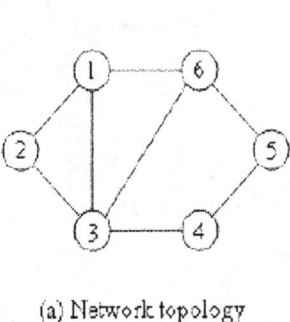
(a) Network topology

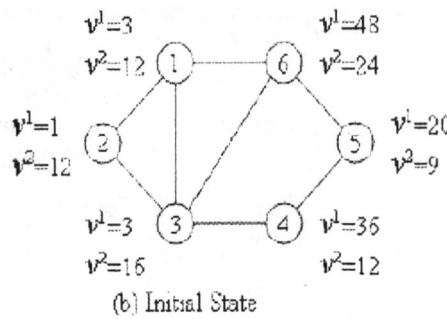
(b) Initial State

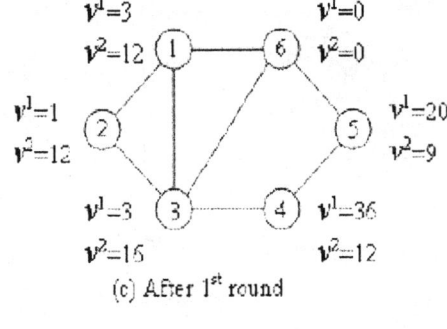
(c) After 1st round

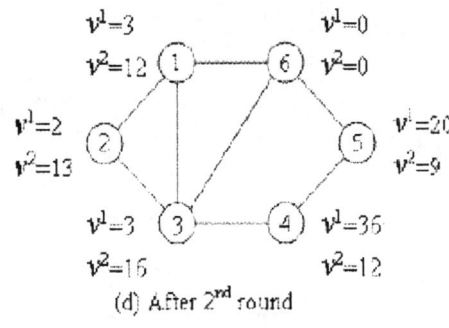
(d) After 2nd round

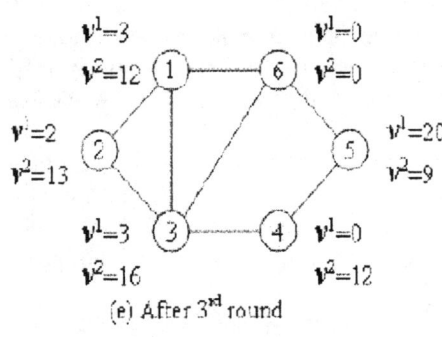
(e) After 3rd round

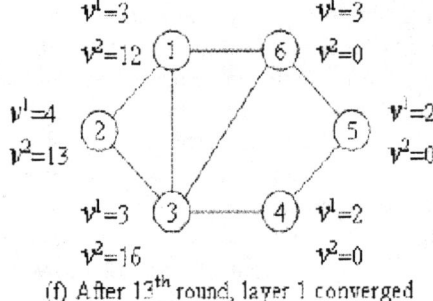
(f) After 13th round, layer 1 converged

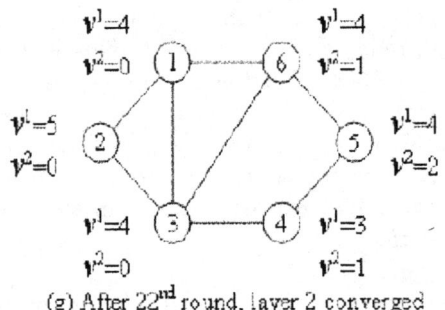
(g) After 22nd round, layer 2 converged

Fig. 7. Example Execution Sequence

6 An Illustrative Example

Figure 7 shows the execution sequence of the publish/subscribe algorithm with two virtual layers on a network of 6 nodes. For each layer $L^s, (s = 1, 2)$, v^s is the variable used in algorithm. We omit the subscript when the context is clear. e.g. The variable v^1 labeled by node 2 will be v_2^1. The Unison Re-orientation algorithm (as shown in Figure 4) executed on layer L^1 consists of rules R1-1 and R2-1, and the same algorithm executed on layer L^2 consists of rules R1-2 and R2-2.

The network topology is shown in (a), and the initial value of v^1 and v^2 of each node are shown in (b). Number of nodes is $n = |V| = 6$. We pick the constant $Z = 50 > 6^2$. So $x \prec y$ iff $(y - x)$ mod $50 \leq 6$.

In the initial state: *node 2 is privileged by R2-1 and R2-2, node 4 is privileged by R1-1, node 6 is privileged by R1-1 and R1-2*. Assume the daemon picks node 6 to move. R1-1 and R1-2 are executed on node 6 and the v values are set to 0. This is shown in (c): *node 1 is privileged by R1-2, node 2 is privileged by R2-1 and R2-2, node 3 is privileged by R1-2, node 4 is privileged by R1-1, node 5 is privileged by R1-1 and R1-2*. Next, assume the daemon picks node 2 to move. After the move: *node 1 is privileged by R1-2, node 2 is privileged by R2-1, node 3 is privileged by R1-2, node 4 is privileged by R1-1, node 5 is privileged by R1-1 and R1-2*, as shown in (d). Next, assume the daemon picks node 4 to move. After the move: *node 1 is privileged by R1-2, node 2 is privileged by R2-1, node 3 is privileged by R1-2, node 5 is privileged by R1-1 and R1-2*, as shown in (e). Next, assume the daemon picks following nodes in sequence: node 5, 6, 4, 6, 2, 6, 2, 5, 4, 5. After these ten moves, layer 1 is in a global legitimate state, but layer 2 is still not converged. The state after the moves is illustrated in (f): *node 1 is privileged by R1-2 and R2-1, node 3 is privileged by R1-2, node 4 is privileged by R2-1, node 5 is privileged by R2-1 and R2-2*. In this state, node 1, 4, 5 get the priority to publish/subscribe on layer 1. i.e. $priority(1, 1) = priority(4, 1) = priority(5, 1) = true$. Assume the daemon then picks node 1, 5, 3, 6, 1, 4, 3, 5, 2. After these moves, both layers are converged, as shown in (g). In this state, node 1, 4, 6 get the priority to publish/subscribe on layer 1, and node 1, 2, 3 get the priority to publish/subscribe on layer 1.

References

1. G. Banavar, T. Chandra, B. Mukherjee, and J. Nagarajarao. An efficient multicast protocol for content based publish subscribe systems. In *Proceedings of the 19th International Conference on Distributed Computing Systems (ICDCS'99)*, 1999.
2. Y. Huang and H. Garcia-Molina. Publish/subscribe in a mobile environment. In *Proceedings of the 2nd ACM International Workshop on Data Engineering for Wireless and Mobile Access*, pages 27–34, 2001.
3. M. Castro, P. Druschel, A. Kermarrec, and A. Rowstron. Scribe: A large-scale and decentralized application-level multicast infrastructure. *IEEE Journal on Selected Areas in Communications*, 20(8):100–110, October 2002.

4. P. Druschel A. Rowstron. Pastry: Scalable, decentralized object location and routing for large-scale peer-to-peer systems. In *In Proceedings of the 18th IFIP/ACM International Conference on Distributed System Platforms (Middleware 2001)*, 2001.
5. G. Fox and S. Pallickara. The narada event brokering system: Overview and extensions. In *PDPTA '02: Proceedings of the International Conference on Parallel and Distributed Processing Techniques and Applications*, pages 353–359, CSREA Press, 2002.
6. A. K. Datta, M. Gradinariu, M. Raynal, and G. Simon. Anonymous publish/subscribe in p2p networks. In *the International Parallel and Distributed Processing Symposium (IPDPS'03)*, 2003.
7. D. Estrin, R. Govindan, J. S. Heidemann, and S. Kumar. Next century challenges: Scalable coordination in sensor networks. In *Mobile Computing and Networking*, pages 263–270, 1999.
8. E. W. Dijkstra. Self-stabilizing systems in spite of distributed control. *Communications of the ACM*, 17:643–644, 1974.
9. L. Lamport. Solved problems, unsolved problems, and non-problems in concurrency. In *Proceedings of the 3rd Annual ACM Symposium on Principles of Distributed Computing*, pages 1–11, 1984.
10. M. Schneider. Self-stabilization. *ACM Computing Surveys*, 25(1):45–67, March 1993.
11. T. Herman. A comprehensive bibliograph on self-stabilization, a working paper. *Chicago J. Theoretical Comput. Sci.*, http://www.cs.uiowa.edu/ftp/selfstab/bibliography.
12. J. Couvreur, N. Francez, and M. Gouda. Asynchronous unison. In *ICDCS*, pages 486–493, 1992.

Self-stabilizing Checkpointing Algorithm in Ring Topology

Partha Sarathi Mandal and Krishnendu Mukhopadhyaya

Advanced Computing and Microelectronics Unit,
Indian Statistical Institute,
203, B T Road, Kolkata 700108, India
{partha_r, krishnendu}@isical.ac.in

Abstract. If the variables used for a checkpointing algorithm have data faults, the algorithm may fail. In this paper, a self-stabilizing checkpointing algorithm is proposed for handling data faults in a ring network. The proposed algorithm can deal with concurrent initiations of checkpointing and at most one data fault per process. However, several processes may be faulty.

1 Introduction

A self-stabilizing distributed system [1],[4] ensures recovery from an *illegitimate* configuration in a finite number of steps. A system may reach an *illegitimate* configuration due to failure or a perturbation in the system.

In this paper, a self-stabilizing checkpointing and *data fault* correction protocols for an unreliable distributed system on a ring network is proposed. Two types of faults, *data fault* and *process fault* are considered. *Data fault* means that the data of a variable is changed or corrupted due to some unreliability of the system. *Process fault*, means that a process in the volatile storage is corrupted and the process can be recovered only using its saved state in the non-volatile storage. If some variables, used by the checkpointing algorithm, are corrupted, then some of the existing checkpointing algorithms will not give a *Consistent Global checkpointing State (CGS)* [8] after rollback. This paper describes self-correction of data-faults in checkpointing algorithms. At most one data fault per process is assumed. That fault may occur any time during the computation. In the worst case, all processes can have data faults concurrently. The system is in a *legitimate* configuration if there is no data fault and there exists a CGS for the system. In this proposed work, in a finite number of steps, system reaches a *legitimate* configuration from an *illegitimate* configuration.

In [2], a scalable, time-independent method to stabilize from k-fault configuration on a tree topology is proposed. Ghosh et al. [3] proposed several ways of measuring the performances of fault-containing self-stabilizing algorithms. Only one [5],[7] or several [10],[11],[8] snapshot collection processes may be active at any point of time. In [7] Vidya used a concept of *logical checkpoint*. In the recovery algorithm of [12], all processes recover from their last existing checkpoints.

2 The Underlying Model

The underlying network topology used in this paper is the same as in [8]. We consider a distributed system consisting of n processes on a ring network. Processes are numbered P_0, P_1, P_2, $\cdots$, P_{n-1} sequentially, in the clockwise direction. In case of checkpointing, process sends checkpointing request ($ckpt\_req$) along the anti-clockwise direction. There is no common clock, shared memory or central coordinator. Message passing is the only mode of communication between any pair of processes. Any process can initiate checkpointing.

We assume that the checkpointing state ($ckpt\_state$) and checkpointing version number (v_no) might be corrupted or changed because of the unreliable system. If a process fails when a data fault is present, the algorithm proposed in [8] will not give a CGS after rollback. Each process maintains a counter, called v_no. Whenever a process takes a logical checkpoint [8], it increments its v_no by one. Each process may store at most two checkpoints (one permanent and one temporary) when checkpointing algorithm is running. Each process maintains a list of unacknowledged messages in a *Message Logging Table (MLT)*.

3 Predicates for Self-stabilization

Process, P_i maintains four variables $prev_i$, $curr_i$, $state\_prev_i$, and $state\_curr_i$ in the stable storage $\forall\ i \in \{1, 2, \cdots, n\}$. The v_no of the previous checkpoint and the v_no of the current checkpoint are stored in $prev_i$ and $curr_i$ respectively. The state variables $state\_prev_i$ and $state\_curr_i$ denote the states of the previous and the latest checkpoints for the process respectively.

Each process maintains two predicates. $pred_1$ is associated with $prev_i$ and $curr_i$ and $pred_2$ is associated with $state\_prev_i$ and $state\_curr_i$.

$pred_1$:   **if** $(curr_i = prev_i + 1)$ **then** $pred_1 = True$ **else** $pred_1 = False$ **end if**
$pred_2$:   **if** $(state\_prev_i = T)$ **then** $pred_2 = False$ **else** $pred_2 = True$ **end if**

If process is in a legitimate state, both $pred_1$ and $pred_2$ should return values $True$. It may be noted here that both the predicates returning values $True$ does not guarantee that there is no error. But such errors are handled later. If one of the predicates return value $False$, the process is in an illegitimate state. We do not consider the case where a single process may have more than one error.

In case where a data fault is detected, if possible, the process corrects itself; otherwise it takes help from the other processes. A process will check its predicates whenever it sends an application message, control message or an application message is passing through the process with an *undecided* information.

4 Data Fault Detection and Correction

Process, P_i checks its predicates before sending an application message and logs the message in the MLT along with its $curr_i$. If $pred_2$ returns $False$, P_i corrects the fault by putting $state\_prev_i = P$. Since at most one data fault in a process is assumed,

$pred_1$ returning $False$ implies that the fault is either in $curr_i$ or in $prev_i$. If $prev_i$ is faulty, then the correct value for $prev_i$ would be $curr_i - 1$. If $curr_i$ is faulty then the correct value for $curr_i$ would be $prev_i + 1$. In this situation P_i can not decide which one would be correct. P_i sends an *undecided* (U) tag with the application message. When a process sends an application message with U tag, it sets the value of the flag to T. If $pred_1$ returns $True$, P_i sends the application message with tag D. P_i sends an application message to the next process with $prev_i$, $curr_i$, $state\_curr_i$, k (receiver id), i (sender id).

When P_j receives a message with U, if $pred_1$ is $True$, then P_j corrects the fault of sender (P_i) of this message. If $pred_1$ is $False$, and if one of the following condition is $True$, then P_j would not be able to correct the fault of P_i and its own. Now, P_j also become undecided.

Condition 1: $((prev_i = prev_j) \wedge (curr_i = curr_j) \wedge (state\_curr_i = state\_curr_j))$
Condition 2: $((prev_i = prev_j + 1) \wedge (curr_i = curr_j + 1) \wedge (state\_curr_i = T) \wedge (state\_curr_j = P))$
Condition 3: $((prev_j = prev_i + 1) \wedge (curr_j = curr_i + 1) \wedge (state\_curr_j = T) \wedge (state\_curr_i = P))$

If none of the above three conditions is $True$, P_j corrects the fault. Given that $pred_1$ is $False$ for P_i, $curr_i \neq prev_i + 1$. Let $S_i^1 = (curr_i - 1, curr_i)$, and $S_i^2 = (prev_i, prev_i + 1)$. The correct value for the ordered pair $(prev_i, curr_i)$ is either S_i^1 or S_i^2. Similarly, the correct value for the ordered pair $(prev_j, curr_j)$, for process P_j, is either $R_j^1 = (curr_j - 1, curr_j)$ or $R_j^2 = (prev_j, prev_j + 1)$.

Let $(prev_i', curr_i') \in S_i^u$ and $(prev_j', curr_j') \in R_j^v$, where $u, v \in \{1, 2\}$. (S_i^u, R_j^v) is correct for some $u, v \in \{1, 2\}$ if and only if one of *Conditions 4, 5 or 6* is $True$.

Condition 4: $((prev_i' = prev_j' + 1) \wedge (curr_i' = curr_j' + 1) \wedge \neg(state\_curr_i = state\_curr_j) \wedge (state\_curr_i = T))$
Condition 5: $((prev_j' = prev_i' + 1) \wedge (curr_j' = curr_i' + 1) \wedge \neg(state\_curr_i = state\_curr_j) \wedge (state\_curr_j = T))$
Condition 6: $((prev_j' = prev_i') \wedge (curr_j' = curr_i') \wedge (state\_curr_i = state\_curr_j))$

If P_j is undecided, it forwards the message to the next process, without changing anything. If P_j is able to correct the fault it overwrites the corrected value in the appropriate variable and changes the message tag from U to D and then forwards the message to the next process.

When P_k receives a message with tag D, if it finds that $pred_1 = False$ then it can correct the fault as follows:

Procedure 1

if $((curr_k = curr_i + 1(-1)) \wedge (state\_curr_k = T(P)))$ **then** $prev_k \leftarrow curr_k - 1$
    if $(state\_curr_i \neq P(T))$ **then** $state\_curr_i \leftarrow P(T)$ **end if end if**
if $((curr_k = curr_i) \wedge (state\_curr_k = P(T)))$ **then** $prev_k \leftarrow curr_k - 1$
    if $(state\_curr_i \neq P(T))$ **then** $state\_curr_i \leftarrow P(T)$ **end if end if**
if $((prev_k = prev_i + 1(-1)) \wedge (state\_curr_k = T(P)))$ **then** $curr_k \leftarrow prev_k + 1$
    if $(state\_curr_i \neq P(T))$ **then** $state\_curr_i \leftarrow P(T)$ **end if end if**

if $((prev_k = prev_i) \wedge (state\_curr_k = P(T)))$ **then** $curr_k \leftarrow prev_k + 1$
  if $(state\_curr_i \neq P(T))$ **then** $state\_curr_i \leftarrow P(T)$ **end if end if**

After correcting the data fault, if $curr_k < curr_i$, P_k takes a temporary checkpoint with $v\_no = curr_i$ and then processes the message. If $curr_k \geq curr_i$, P_k processes the message without taking a checkpoint. After processing a message P_k sends an acknowledgement message (ack_msg) with $state\_curr_i$, $curr_i$, $curr_k$ to P_i.

If P_j receives a message with tag D and finds that $pred_1$ is $False$, then it corrects the data fault (using Procedure 1 with k replaced by j) and forwards the message to the next process without changing the body of the message.

On receiving an ack_msg from P_k, process P_i first makes its correction if $pred_1 = False$. Then it compares $curr_k$ with $curr_i$ of the message logged in the MLT. If $curr_k$ is greater than or equal to the $curr_i$, then the $curr_i$ is replaced by $curr_k$ in the MLT. The message will be deleted when the $curr_i$ of the process becomes greater than the $curr_i$ of the message logged in MLT.

When P_k receives a message with tag U from P_i, if $pred_1 = False$ and one of conditions *1, 2* or *3* is *True* then P_k also becomes undecided. P_k keeps the message for future processing. It passes the message without message data to the next process with i as the changed receiver id of the message.

In the worst case, a message with tag U returns back to P_i, its originator. If there exists at least one i such that $state\_curr_i = T$, P_i will wait for $ckpt\_req$. After receiving $ckpt\_req$, P_i corrects the data fault. Otherwise, all processes have data faults and they are unable to rectify these faults. Several processes may receive such messages with tag U returned to them. Another round of message passing is required to elect one process among them (may be the one with minimum id). This can be done by passing a message round the system by all the processes. So in total there will be $O(n)$ messages and $O(n)$ time. Let P_m be the elected process. As it is impossible to decide which one of $prev_m$ and $curr_m$ is correct, P_m assumes that $prev_m$ is correct. $curr_m$ is replaced by $prev_m + 1$. P_m sends a correction message $(correction\_msg)$ with $curr_m$ and $state\_curr_m$ to other processes.

On receiving $correction\_msg$, P_j takes the following actions:

Procedure 2

if $(state\_curr_j = state\_curr_m)$ **then** $curr_j \leftarrow curr_m$ **and** $prev_j \leftarrow curr_j - 1$
else if $((state\_curr_m = T) \wedge (state\_curr_j = P))$ **then** $curr_j \leftarrow curr_m - 1$ **and**
      $prev_j \leftarrow curr_j - 1$
  else $curr_j \leftarrow curr_m + 1, prev_j \leftarrow curr_j - 1$ **end if**
end if

The $correction\_msg$ is forwarded until it passes through all the processes and it returns back to P_i. The message which was held up due to U tag be processed after recovery.

5 Checkpointing Algorithm

A process without a temporary checkpoint or any data fault may initiate checkpointing. All control messages for the checkpointing are routed in the *anti-clockwise* direction. The following checks are carried out during the initiation.

if $((pred_1 = True) \land (pred_2 = True) \land (state\_curr_i = P))$ **then** take checkpoint
set $initiator\_flag_i \leftarrow T$, $state\_curr_i \leftarrow T$, $prev_i \leftarrow curr_i$, $curr_i \leftarrow curr_i + 1$, $v\_no \leftarrow curr_i$, $send(ckpt\_req, curr_i, i)$ **end if**
On receiving a $ckpt\_req$, if P_j finds $pred_1 = False$, it corrects the fault and takes a checkpoint as per the following procedure.
if $(state\_curr_j = T)$ **then set** $curr_j \leftarrow curr_i$ **and** $prev_j \leftarrow curr_j - 1$ **end if**
if $(state\_curr_j = P)$ **then** take checkpoint **set** $curr_j \leftarrow curr_i$, $prev_j \leftarrow curr_j - 1$, $v\_no \leftarrow curr_j$, $initiator\_flag_j \leftarrow F$, $state\_curr_i \leftarrow T$ **end if**
If both $pred_1$ and $pred_2$ are $True$ then $curr_j$ is compared with the $curr_i$ of the message. A new checkpoint is taken as follows.
if $(curr_j \neq curr_i)$ **then** take a checkpoint **set** $curr_j \leftarrow curr_i$, $prev_j \leftarrow curr_j - 1$, $v\_no \leftarrow curr_j$, $initiator\_flag_j \leftarrow F$, $state\_curr_i \leftarrow T$ **end if**
if $(curr_j = curr_i)$ **then** do not take a checkpoint **end if**
As concurrent initiations of checkpointing are allowed, several $ckpt\_req$ may be received a by a process. The decision to forward, discard or generate a commit message ($commit\_msg$) is taken by the following logic.
if $((initiator\_flag_j = T) \land (j < initiator\_id))$ **then** discard the message **end if**
if $((initiator\_flag_j = T) \land (j = initiator\_id))$ **then** discard the message and send a $commit\_msg$ to the next process. **end if**
if $(j > initiator\_id)$ **then** forward the $ckpt\_req$ to the next process. **end if**
On receiving a $commit\_msg$, P_j takes the following actions:

if $(j \neq i)$ **then** delete the checkpoint with $v\_no = prev_j$, keeping $prev_j$ unchange
   set $state\_curr_j \leftarrow P$, forward the $commit\_msg$ to the next process.
end if

When the $commit\_msg$ returns back to its creator, it stops the message propagation. The checkpointing process is terminated and a CGS, one checkpoint per process with same v_no is established.

6 Correctness and Complexity Analysis

In case of a single data fault in the system, if self-correction is not possible then the next process can correct the data fault. Only two message exchanges are required to correct the fault. This takes $O(1)$ time. Maximum number of messages are exchanged when all processes have data faults, and no process can correct its fault. If messages with tag U are returned to multiple processes then the election procedure takes $O(n)$ messages and hence $O(n)$ time. But the probability of occurrence for such a case is very low. Checkpointing algorithm requires two rounds of message exchanges in case of single and multiple checkpointing initiations. For both single and concurrent checkpointing initiations $O(n)$ message exchanges are required. Proofs of the following results may be found in [9].

Theorem 1. *The system reaches a legitimate configuration from an illegitimate configuration in $O(n)$ steps.*

Lemma 1. *There will not be any missing message and any orphan message in the system.*

Lemma 2. *If $curr_i$ is corrected after a data fault, the set of checkpoints, which would be obtained in case of a process fault, are consistent.*

Theorem 2. *In case of a process fault, the system can roll back to a consistent global state.*

Theorem 3. *The set of checkpoints generated by the proposed algorithm is consistent. The time complexity is $O(n)$. The message complexity is $O(n)$.*

7 Conclusion

In this paper, a self-stabilizing checkpointing scheme in an unreliable distributed system on a ring topology has been proposed. The worst case time and message complexities are both $O(n)$. Earlier concurrent checkpointing algorithms [10], [11] were designed for general topologies. Their worst case message complexities are $O(n^3)$ and this worst case occurs for the ring. Data fault assumed is in the variables used for checkpointing and is due to unreliable system. Single data fault per process is considered; but, multiple processes may have faults. An interesting extension is to consider multiple data faults per process and/or a general topology.

References

1. Dijkstra E.W.: Self stabilizing systems in spite of distributed control. Communications of the ACM, Vol. 17, pp. 643-644, (1974).
2. Ghosh S., He X.: Scalable Self-Stabilization. Journal of Parallel and Distributed Computing, Vol. 62, Issue 5, pp. 945-960, May, (2002).
3. Ghosh S., Gupta A., Herman T., Pemmaraju S.V.: Fault-containing self-stabilizing algorithms. Proc. 15th ACM Symp. Princ. of Distrib. Comput., pp 45-54, (1996).
4. Schneider M.: Self-Stabilization. ACM Computing Surveys, 25(1), pp. 45-67, (1993).
5. Chandy K.M., Lamport L.: Distributed snapshots: Determining global states of distributed systems. ACM Trans. Comput. Syst., 3(1), pp. 63-75, Feb. (1985).
6. Manivannan D., Singhal M.: Quasi-synchronous checkpointing: Models, characterization, and classification. IEEE Trans. on Parallel and Distributed Systems, Vol. 10, No. 7, pp. 703-713, July, (1999).
7. Vidya N.H.: Staggered consistent checkpointing. IEEE Trans. on Parallel and Distributed Systems, Vol. 10, No. 7, pp. 694-702, July, (1999).
8. Mandal P.S., Mukhopadhyaya K.: Concurrent checkpoint initiation and recovery algorithms on asynchronous ring networks Journal of Parallel and Distributed Computing, Vol. 64, Issue 5, pp. 649-661, May, (2004).
9. Mandal P.S., Mukhopadhyaya K.: Self-Stabilizing checkpointing algorithm in ring topology, TR: ACMU/2005/01. Indian Statistical Institute, Kolkata, (2005).
10. Spezialetti M., Kearns P.: Efficient distributed snapshots. Proc. 6th International Conference on Distributed Computing Systems, pp. 382-388, (1986).
11. Prakash R., Singhal M.: Maximal global snapshot with concurrent initiators. Proc. 6th IEEE Symp. Parallel and Distrib. Processing, pp. 334-351, Oct. (1994).
12. Manivannan D., Singhal M.: Asynchronous recovery without using vector timestamps. J. Parallel Distrib. Comput. Vol. 62, Issue 12, pp. 1695-1728, (2002).

Performance Comparison of Majority Voting with ROWA Replication Method over PlanetLab[*]

Ranjana Bhadoria, Shukti Das, Manoj Misra, and A.K. Sarje

Department of Electronics and Computer Engineering,
Indian Institute of Technology Roorkee, India
sarjefec@iitr.ernet.in

Abstract. Since the Web started in 1990, it has shown an exponential growth. It is essential that the Web's scalability and performance keep up with increased demand and expectations. The key to achieving these goals of scalability, robustness and responsiveness lies in the practices of caching and replication. Quorum Consensus is a popular protocol used for data replication. This paper describes an implementation of two special cases of Quorum Consensus protocol, namely Majority Voting and Read-One-Write-All (ROWA) and compares their performance. The performance evaluation was done using a number of systems located at PlanetLab member institutions at different locations over the world. This enabled simulation of real world Internet conditions. The study shows that the ROWA protocol performs better than the Majority Voting under no-site-failure conditions in terms of response time, communication overhead and growing number of users.

1 Introduction

Replication involves creating and maintaining duplicates of a database or file system on different computers, typically servers, to enhance services. Motivations for using replication are [7]: performance enhancement, increased availability and fault Tolerance. A common requirement for replicating data is *replication transparency*. The clients should not be aware of multiple physical copies but feel that operations are being performed on a single database. *Mutual consistency* as well as *internal consistency* [11] must be preserved. Replication of changing data requires protocols toensure that clients receive up-to-date data at all times. Network partitions and disconnected operations reduce data availability. To overcome this problem, users can maintain local copies of heavily used data. Replica failure and recovery also have to be taken into consideration. Many replication control methods have been proposed in the literature [1]. In this paper we focus on two special cases of Quorum Consensus protocol [3], Majority Voting and Read-One-Write-All (ROWA) and compare their performance.

In Quorum Consensus protocol each site is given a nonnegative weight. It assigns two integers to read and write operations on an item X, namely a read quorum (r), and a write quorum (w) that must satisfy the following conditions:

$$r + w > S, \qquad w > S/2,$$

---

[*] This work is partly supported by Intel Technologies India, and the support provided by Planet Lab.

where S is the total weight of all sites at which X resides. To execute a read operation, enough replicas must be read such that their total weight is more than or equal to r. To execute a write operation, enough replicas must be written to so that their total weight is greater than or equal to w. The two conditions for Quorum Consensus mentioned above ensure that there is a non-null intersection between every read quorum and every write quorum. There is always a subset of the servers, with total votes w, that consists of current replicas. Thus, any read quorum gathered is guaranteed to have a current copy of the object.

The benefit of the quorum consensus approach is that it can permit the cost of either reads or writes to be selectively reduced by appropriately defining the quorums. [1], [3]. In read-one-write-all (ROWA) protocol generally all replicas have equal weight. A read requires locking only one replica whereas a write needs all replicas. In majority protocol [2] both operations require a quorum, which constitutes a majority.

2 Implementation

The Client in Fig. 1 requests the front end to process a transaction. The front end provides replication transparency to the clients. It creates a new front end request handler (FERH) for each client transaction. The FERH implements the Quorum Protocol and is the transaction coordinator. It forms read/write quorums and sends these transaction requests to the replicated servers in the corresponding quorums over the Internet on behalf of the client. The responses from the servers are accepted by the FERH and forwarded to the client. Version numbers are used to know whether the server contains the current data or stale data.

The server creates a new request handler for each request received from the FERH (transaction coordinator). The request handler coordinates the processing of the request coordinating with the other modules and sends the response to the FERH. It interacts with the lock manager to handle lock/release requests and deadlock manager to prevent the request from creating a deadlock. In case the transaction needs to be aborted, the request handler initiates a cleanup. The database module is contacted for reading and writing to replicated objects assuming appropriate locks have already been acquired. The lock manager maintains a lock table for handling locks and release requests from the request handler. It responds when the request is granted otherwise stalls it. It also initiates deadlock detection at the deadlock manager whenever a request is not granted. The deadlock manager maintains wait-for-graphs to detect deadlock. The deadlock detection algorithm is run periodically and whenever a lock request is not granted. The protocols have been tested using a number of systems located at PlanetLab member institutions at different locations over the world. PlanetLab [10] is an open, global network test-bed for developing, deploying and accessing planetary-scale services.

Performance Evaluation Parameters: The performance of Majority Voting protocol has been compared with that of the ROWA on the basis of Message Traffic Overhead, Response Time, Scalability and Availability.

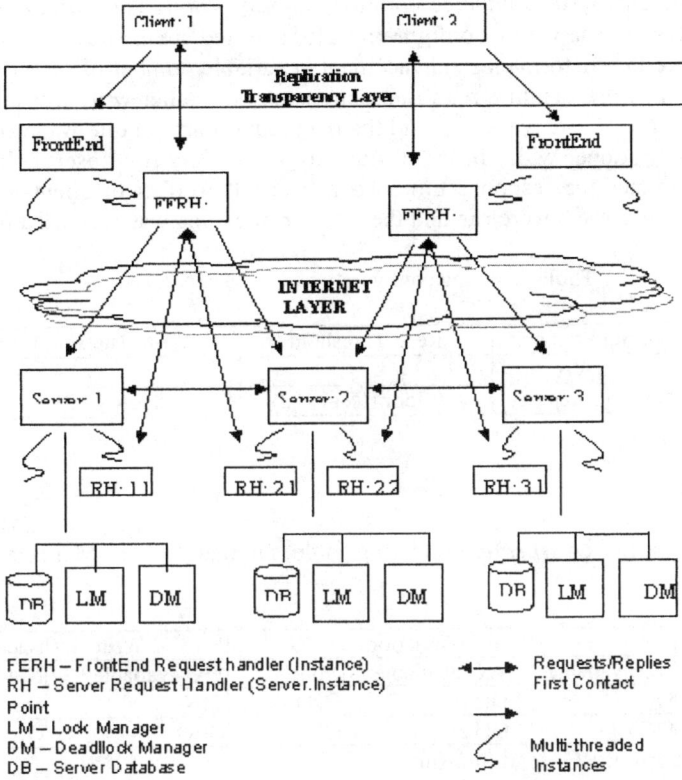

Fig. 1. Overall design for the Quorum Consensus Protocol

3 Experimental Setup

The experiments were carried out on PlanetLab nodes located mostly in the United States of America and few others in India and Netherlands. The machines are connected through the Internet and run Linux. A new instance of the database (files) was used at the replicated servers for each experiment and these were carried out during the day-time (in India) to maintain similar testing conditions. After completion of each experiment, log files containing the response times were copied to the home terminal using SCP [10] and then cleared for the next experiment.

Each read/write client ran for four minutes generating approximately forty read transactions or ten write transactions depending on its type. The transactions from various clients were generated simultaneously. This random transaction generation was simulated using Poisson's distribution. Average of these response times have been used for better confidence in the results. The read and write operations were not symmetric. Writes took more time than reads. Also write operations were given higher priority than reads so that clients always receive up-to-date data. The ratio of clients performing reads to writes was almost 1:3. All servers had equal weights of unity.

All experiments were conducted both for Majority Voting Quorum Consensus protocol and ROWA. The voting configurations [9] selected are as shown in the Table 1.

To measure the performance enhancement, a variable number of servers were used with two clients, one sending read transactions and the other write transactions. The write request arrival rate was fixed and the read request arrival rate was varied. Measurements were done with different number of servers to observe how it enhances/deteriorates the response time. To measure the effect of client-scalability, a fixed number of servers were run and the number of clients was increased linearly.

Table 1. Voting configurations for the experiments

Voting Protocol	Read Threshold (r)	Write Threshold (w)
ROWA	1	N
Quorum (Majority)	floor ((N+1) / 2)	floor(N / 2) + 1

4 Results

Message Exchange Overhead: Following table summarizes the Message Exchange Overhead

Voting Protocol	Read Transaction			Write Transaction	
	RQ available	WQ available	None	WQ available	Not available
ROWA	O(1)	O(1)	O(1)	O(N)	O(N)
Majority	O(1)	O(1)	O(r)	O(w)	O(r + w)

WQ: write quorum; RQ: read quorum

Response Time: The first experiment was performed with fixed write request arrival rate ($\lambda = 0.01$) and by varying the read arrival rate ($\lambda = 6, 8, 10$). The read/write transactions ran for four minutes each simultaneously. The voting configuration was ROWA.

As seen in Fig. 4, the response time for read requests increases with the increase in the number of servers. This is because reads have to wait for the simultaneously running write transactions. As the number of servers increases, the response time for write requests also increases (Fig. 5) due to the overhead involved in write transactions. Writes are not compatible with other writes and read requests whereas reads are compatible with other read requests.

Scalability: The second experiment tested the scalability of ROWA and Quorum protocols with four servers. The number of clients was varied from one to nine. The arrival rates for write request ($\lambda = 0.01$) and read request ($\lambda = 1$) were fixed.

The read transaction response time increases slightly with the increase in the total number of clients as more clients compete for the same resources (Fig. 6). The response for Majority Quorum Consensus Protocol (QC), on the other hand, increases by a huge margin as the read quorum size is $\lfloor(N+1)/2\rfloor$ instead of one (ROWA). Thus reads are much more expensive in the case of Majority Quorum Protocol.

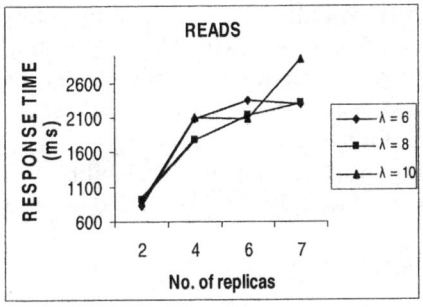
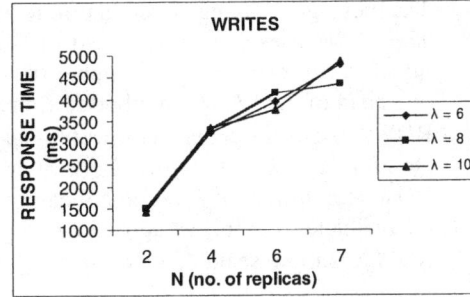

Fig. 4. Read response time versus servers

Fig. 5. Write response time versus servers

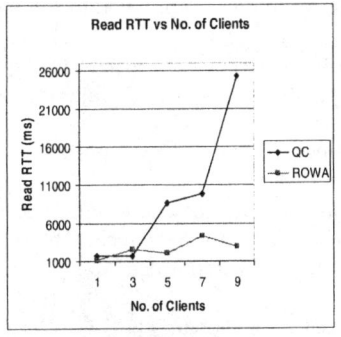
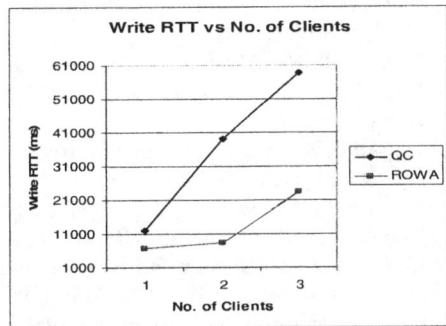

Fig. 6. Read response times for MIT client with different number of requesting clients

Fig. 7. Write response times for the Washington client with different number of requesting clients

Though the write quorum size for the Majority Quorum Consensus protocol is smaller than that of ROWA, still because of the read transactions, Majority writes suffer as shown in Fig. 7.

Availability: The Majority Voting technique can tolerate at most floor $((N + 1) / 2)$ failures, where N is the number of replicated servers, when the optimal voting configuration is used. ROWA, on the other hand, does not tolerate any site failures or network partitions [8].

5 Conclusion

An actual implementation of the quorum consensus method and experimental evaluation of its performance was carried out on the Plant Lab [10] test-bed and included globally distributed nodes, which are members of Planet Lab. These machines are connected through the Internet. This setup provided a realistic network substrate that experiences congestion, failures, and diverse link behaviors and also models realistic client workload. A performance comparison with ROWA protocol was done. The conclusions that can be drawn from the experiments carried out are:

- The message exchange overhead in ROWA is lesser than that in majority Voting. The message exchange overhead of Quorum Consensus (with Majority Voting) increases linearly with the number of replicas of a replicated object whereas the overhead of the ROWA method is almost invariant to the number of replicas.
- ROWA performs better in terms of response time than Majority Voting.
- Quorum Consensus with Majority Voting provides higher availability than ROWA, but the ROWA protocol can be adapted to ROWAA [8] (read-one-write-all available) to improve upon this.
- ROWA is more scalable than Majority Voting.

References

1. Silberschartz A., Korth H. F., Sudarshan S.: Database System Concepts. Fourth Edition, McGraw-Hill, (2002).
2. Thomas R. H.: A Majority Consensus Approach to Concurrency Control for Multiple Copy Databases. ACM Transactions on Database Systems, Vol. 4, No. 2, June (1979), Pages 180 - 209
3. Gifford D. K., Weighted Voting for Replicated Data. Proceedings of 7th Symposium on Operating Systems Principles, December (1979)
4. Kemme B., Alonso G.: A New Approach to Developing and Implementing Eager Database Replication Protocols. ACM Transactions on Database Systems, September (2000)
5. Herlihy M.: Concurrency and Availability as Dual Properties of Replicated Atomic Data. Journal of ACM, Vol. 31. No. 2, April (1990).
6. Gray J.: Notes on Database Operating Systems. IBM Research Laboratory (1977)
7. Coulouris G., Dollimore J., Kindberg T.: Distributed Systems: Concepts and Design. Third Edition; Addison Wesley, (2000).
8. Jimenez-Peris R., Patino-Martinez M., Alonso G., Kemme B.: Are Quorums an Alternative for Data Replication? ACM Transactions on Database Systems, Vol. 28, No. 3, September (2003).
9. Helal A., Bhargava B.: Performance Evaluation of the Quorum Consensus Replication Method. Proceedings of the International Computer Performance and Dependability Symposium (IPDS'95), IEEE Computer Society, (1995).
10. PlanetLab User's guide http://www.planet-lab.org/docs/UsersGuide.php
11. Son S. H.: Replicated Data Management in Distributed Database Systems. SIGMOD RECORD, Vol. 17, No. 4, December (1988).

Self-refined Fault Tolerance in HPC Using Dynamic Dependent Process Groups

N.P. Gopalan and K. Nagarajan

Department of Computer Science and Engineering,
National Institute of Technology,
Tiruchirappalli, TN 620015, India
{gopalan, csk0303}@nitt.edu

Abstract. This paper proposes a novel method for achieving a distributed self-refined fault tolerance by dynamically partitioning the processes into smaller groups, which are mutually disjoint and collectively exhaustive of the whole system. The present model provides tolerance for frequent faults, makes the roll back recovery simple and less time consuming. An optimal checkpoint interval is found using a mathematical approximation and a spare process is made to capture all the in-transit messages when a process fails at its ends. Piggybacking the events of dependent processes on the outgoing messages is used for process grouping. A process with maximum information can scatter chunk values to the other dependent processes in its group. Each process constructs a checkpoint when the received chunk matches with its log.

1 Introduction

The recent trend in high performance computing (HPC) involves the use of clusters and Grids containing a huge number of processors where node and network failures are common. Processes may migrate to other nodes to increase the system performance and facilitate administration. Hence, studies concerning the fault tolerance and process migration at run time assumed significance in the recent past.

In the synchronization of processors using messages, the system tends to be asynchronous with unpredictable message delays and receiver overrun. The coordinated checkpointing protocol requires synchronization of all processors before constructing a recovery line [5]. When one or more processors fail, all others rollback to the most recent checkpoint (without message logging) to arrive at a consistent state. This is economical for communication intensive parallel programs running in small and medium sized environments [2], [5]. The algorithms used for dedicated parallel computing systems [1], [7] cannot be applied to large-scale systems with varied dynamic behaviors and non-FIFO properties as they complicate the system synchronization.

In the uncoordinated checkpointing with message logging (UC-ML) only the failed processes participated in rollback under complicated recovery procedures, garbage collection and domino effect [3], [4] degrading the performance. The Uncoordinated Checkpointing with Event Logging (UC-EL) uses message envelopes containing

Sender/Receiver sequence numbers (*SSN/RSN*) and the list of dependent processes ids (*PIDs*) and are advantages under certain environments [4]. The idea of built-in checkpoints and condor library are introduced in Starfish [1] and Co-check MPI [7] respectively. Systems like MPICH-V [3] and MPICH-V2 [4] used fault tolerant MPI without re-computations and re-transmission of messages. But it was shown that they take more communication time for a full recovery in case of crashes. Elaborate descriptions of protocols related to checkpointing, message logging, and rollback recovery can be found in [6] and are used in fault tolerant MPI.

In this paper, a novel method of dependent process grouping with event logging (DPG-EL) is presented for a large-scale fault tolerant system without synchronization overhead. Using MPI library functions, processes in the cluster are partitioned into smaller groups based on their dependency. The fault tolerance is implemented at the application level and is transparent to the user. An optimal fail-free checkpoint interval is computed using a mathematical approximation for each process and a process with the maximum information (about others in the group) initiates a checkpoint at its end. In case of failures, a stand by spare process replaces the failed process to receive the in-transit messages.

2 Optimal Checkpoint Interval

Let the time interval between two successive checkpoints be T_{Int} and the time required to save the process state information (*PSI*) in the stable storage before the occurrence of a failure be T_{Store}. If T_{Delay} is the delay incurred while transferring a checkpoint to a system with stable storage (SSS), T_{Sys} is the time taken by a system message from a process to reach SSS and T_{Record} is the time taken to save a checkpoint on a SSS; then $T_{Store} = T_{Delay} + T_{Sys} + T_{Record}$. Similar notations can be defined for $T_{Retrive}$ which is the time taken to retrieve the information from SSS. The occurrences of failures are assumed to follow a Poisson process with a failure rate η and mean time between failures (T_{MTBF}) $1/\eta$. The probability density function $P(t)$ for the time interval t between failures is given by $P(t) = \eta e^{-\eta t}$. It is assumed that the initial checkpoint was constructed before the process execution starts. A spare process is kept ready to receive the in-transit messages and deliver them in the same order to the failed process when it restarts. Hence, the total time lost (T_{Los}) due to the occurrence of a failure, checkpointing and information logging is,

$$T_{Los} = \eta \int_0^\infty t e^{-\eta t} dt - \eta T_{Int} \sum_{n=0}^{\infty} n \int_{n(T_{Int}+T_{Store})}^{(n+1)(T_{Int}+T_{Store})} e^{-\eta t} dt + \eta (T_{Retrive} + T_{Rproduce}) \sum_{n=0}^{\infty} \int_{n(T_{Int}+T_{Store})}^{(n+1)(T_{Int}+T_{Store})} e^{-\eta t} dt \quad (1)$$

The equation (1) on integration and simplification becomes

$$T_{Los} = \frac{1}{\eta} + \frac{T_{Int}}{1 - e^{\eta(T_{Int}+T_{Store})}} + \frac{T_{Re\,trive} + T_{Rproduce}}{1 - e^{\eta(T_{Int}+T_{Store})}} \quad (2)$$

The best value of the checkpoint interval (T_{Int}) is one that minimizes the value of T_{Los}. So, differentiating (2) with respective to T_{Int}, and equating to zero,

$$e^{\eta T_{Int}}(1 - \eta T_{Int} - \eta(T_{Retrive} + T_{Rproduce})) = 1 - e^{-\eta T_{Store}} \qquad (3)$$

Retaining up to the 2nd degree terms in $e^{\eta T_{Int}}$ and $e^{-\eta T_{Store}}$ (3) can be written as,

$$\eta^2 T_{Int}^2 + 2\eta^2(T_{Retrive} + T_{Rproduce})T_{Int} + 2(\eta(T_{Retrive} + T_{Rproduce}) - 1) = \eta^2 T_{Store}^2 - 2\eta T_{Store} \qquad (4)$$

Substituting $T_{MTBF} = 1/\eta$, $T_{log} = T_{Retrive} + T_{Rproduce}$ and using $T_{Store} \ll T_{Fmean}$ in (4) solving it, the optimal fail free checkpoint interval is obtained as,

$$T_{Int} = \sqrt{T_{Log}^2 + 2T_{MTBF}(T_{Log} + T_{Store} - T_{MTBF})} - T_{Log} \qquad (5)$$

It is clear from (5) that the optimal checkpoint interval is decided based on the current availability of the resources and fault frequency.

3 Algorithms

When a process *P* sends or receives a message, the sender (with Sender Process id: *Spid*), the receiver (with id: *Rpid*) and their PIDS are dependent on *P* during a communication event. This dependency is accumulated until a checkpoint is constructed. In the proposed model, there are *N* processes with α active and (*N*-α) spare processes. Each process *P* has a *PSI*, containing the set of *PIDs* with its current *SSN* and *RSN* and is used for the application-level process coordination. It captures all causal, non-causal dependent messages with the help of process grouping. Each *PSI* is updated with the occurrence of a communication event. Further, checkpointing and recovery procedures are used to overcome the process failures.

3.1 Process Grouping with Event Logging

1. When the sender process wants to send or receive any message, it increments *SSN* or *RSN*.
2. Before a message is sent, the sender process will identify the receiver's group using *"Process Grouping algorithm"* given below:
    i. **While both *Spid* and *Rpid* are new for an existing group**: The sender(s) will create a new group(s) with *Spid* and *Rpid*. When multiple concurrent senders to a single receiver is present, the multiple groups formed are to be merged.
    ii. **While *Spid* and *Rpid* are in a group**: The sender knows that the group already exists, and *PSI* is not required for piggybacking.
    iii. **While *Spid* is new and *Rpid* already in a group**: The sender is new to the group in existence and the group has to extend its communicator with *Spid*.
    iv. **While *Spid* is old and *Rpid* is new for a group**: The sender is from an already existing group but the receiver may or may not be in an existing group. Accordingly, the causal dependency may be extended or the non-active receiver process may be included in an existing group. When there are multiple receivers, the different groups are to be merged and a common communicator is to be formed.

3. The sender process logs and piggybacks the *PSI* with the outgoing messages.
4. With the receipt of a message, the receiver will update its dependent processes information.

3.2 Checkpoint and Recovery Procedure

The receiver process with the maximum *PSI* may initiate the checkpoint construction procedure anticipating a possible failure and the recovery procedure is self-activated with the failure of a process. This is done as follows:

When $\forall PID \in$ same *Gpid*, where $(1 \leq PID \leq S)$ do {

1. if T_{Int} is reached and $\sum_{PID=1}^{S} SSN = \sum_{PID=1}^{S} RSN$, then using *MPI_Scatter* broadcast the chunk values to all the processes within its group;
2. upon the receipt of chunk value, each process matches the received chunk data with that of its own chunk. When they match, a checkpoint is constructed at the end of its log.
3. if a *PID* with a *Log Record* exists after a checkpoint, compare the check pointed data with the existing log record and if they match, create a temporary log for the set of sent and received messages with their *PIDs*. Delete the log record entries after checkpointing;
4. if a temporary log exists, send it to the stable memory; }
5. When a process fails, assign a standby process to receive the messages meant for it. When it restarts, reload the log information pertaining to its execution from its most recent checkpoint with the Spare process supplying the messages received in the order of arrival.
6. After the restart of a failed process, delete information pertaining to the spare process.

4 Results

The experimental results presented in this section are obtained using a cluster of PCs under Linux 2.4.18. The cluster test bed consists of sixteen 2.8 GHz Pentium IV processor based workstations connected to a 100 Mbps Ethernet. Each workstation has dual processors with 512 MB of main memory and 40 GB of stable storage. The experiment uses a MPI program for the *Gauss-Jordan method* of solving system of linear equations. The linear system is evenly distributed by rows among (N-1) processes from where the results are collected by the *MPI_Allreduce* function call to the rank 0 process. The MPI implementation uses the LAM/MPI version 7.0.4. Test programs were compiled using the GNU GCC version 2.96. Three different linear systems with 4,000, 8,000 and 16,000 processes are considered and the experimental results are shown in Fig 1.

The test programs are executed in three modes:

1. Non-checkpointing execution without failures (source code alone is executed)
2. Checkpointing execution with failures (varying from 1 to 7) and
3. Recovery and restart after a failure.

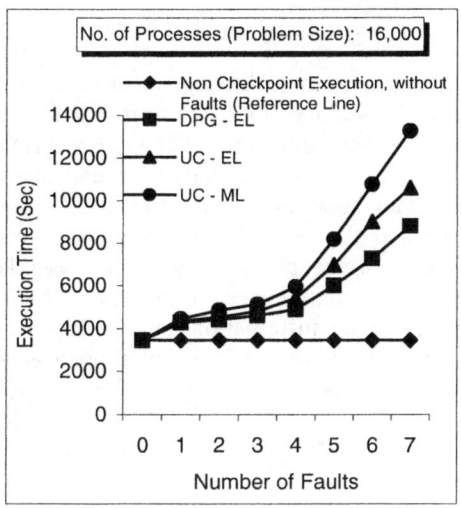
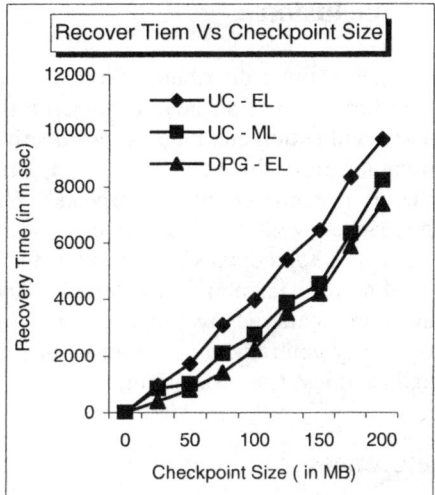

Fig. 1   Fig. 2

The execution time in mode 1 is used for comparing the results of mode-2 with processes grouping, logging and checkpoint procedures as in sections 3.1 and 3.2. The compression is made in two parts: First, the proposed DPG-EL is executed with the test program under Mode-1 with different problem sizes. The results are then compared with those of UC-EL and UC-ML methods of execution [3, 4]. The execution times remain nearly the same for all problem sizes when the number faults do not exceed 2, but vary drastically when the faults are 4 or more. (The study can go up to 7 faults as T_{Store} and $T_{Int} >> T_{Store}$) (Cf. Fig 1). The variations are qualitatively similar for various problem sizes and hence they are not shown separately. Under UC-EL and UC-ML, the execution times are found to be higher by about 16%, 16.8% and 20.5% and 31.17%, 36.7% and 50.7% than those observed in the present DPG- EL model using 4000, 8000 and 16000 processes in action. This may be due to the following reasons:

1. UC-ML suffers from total logging and garbage collection and this may degrade the performance with the increase in failures.
2. In UC-EL, the dependent processes overheads are proportional to the number of failures due to the possible occurrences of domino effect.

In DPG-EL, once the dependent groups are formed, processes do not incur synchronization overheads. The recovery of the failed process is very simple and less time consuming as compared to UC-ML and UC-EL because the groups formed are smaller and the log information are confined to these sub-groups. (Shown in Fig. 2). In addition, recovery times in UC-ML and UC-EL are higher by 52 % and 79.4% than that required for DPG-EL for a checkpoint size of 200 MB.

5 Conclusion

For a self-refined distributed fault tolerance checkpointing, the dependent processes group formation is an advantageous design and it reduces the issues on scalability, garbage collection and huge restart overhead. It is well suited for FIFO and non-FIFO communication channels. It also captures all causal and non-causal dependencies without synchronization overhead. When messages are sent, the *PSI*'s are piggybacked with the computation-messages and the processes (with in a group) construct a forced checkpoint after the receipt of the chunk values. This avoids cascading rollback and reproduction of messages during recovery. Further, the chunk values are scattered by processes with the maximum information and so does not require any centralized co-coordinator process. The recovery of the failed process is simple and less time consuming.

References

1. A. Agbaria and R. Friedman., Starfish: Fault tolerant dynamic MPI programs on clusters of workstations, Proceedings of the 8th IEEE Symposium on High Performance Distributed Computing, pp 31- 42, IEEE CS Press, 1999.
2. L. Alvisi and K. Marzullo., Message Logging: Pessimistic, optimistic, causal and optimal, IEEE Transactions on Software Engineering, 24(2): 149–159, FEB 1998.
3. G. Bosilca et. al. MPICH-V: Toward a scalable fault tolerant MPI for volatile nodes, Proceedings of Super Computing Conference, PP 23-41, ACM/IEEE CS Press, 2002.
4. Bouteiller et. al. MPICH-V2: a fault tolerant MPI for volatile nodes based on pessimistic sender based message logging, Super Computing, 2003.
5. K.M. Chandy and L. Lamport., Distributed snapshots: Determining global states of distributed systems, ACM Transactions on Computing Systems, 3(1): 63-75, Aug. 1985.
6. E.N. Elnozahy, L. Alvisi, Y.M. Wang, and D.B. Johnson., A survey of rollback-recovery protocols in message-passing systems, ACM Computing Surveys, 34(3): 375–408, 2002.
7. G. Stellner., Cocheck: Checkpointing and process migration for MPI. IPPS, pages 526–531, 1996.

In-Band Crosstalk Performance of WDM Optical Networks Under Different Routing and Wavelength Assignment Algorithms

V. Saminadan and M. Meenakshi

Department of Electronics and Communication Engineering,
College of Engineering, Anna University, Chennai 600 025, India
sivan_saminadan@yahoo.co.in

Abstract. The impact of different routing and wavelength assignment algorithms on the in-band crosstalk performance of a 4 x 4 mesh-torus and a 15-node network has been studied. This paper considers both switch-induced crosstalk and the crosstalk induced by the multiplexers and demultiplexers. Fixed routing and fixed-alternate routing of connection requests have been considered. First-fit and random wavelength assignment algorithms have been employed. A crosstalk-aware wavelength assignment has also been considered. In-band crosstalk leads to poor received signal quality at the destination node. This results in increased receiver bit error rate (BER). This implies that some of the routes will deliver a signal quality which is unsatisfactory. To ensure that no resources are wasted on those connections which cannot deliver an unacceptable signal quality, this paper uses an event-driven simulation which incorporates on-line BER calculations. A call request is accepted only if the BER at the destination node is less than 10^{-12}; otherwise it is rejected.

1 Introduction

Establishing a connection in all-optical networks involves selecting a wavelength and a route for that connection with the constraint that the same wavelength is available on all fiber links of the route. This problem of routing a set of connections is referred to as routing and wavelength assignment (RWA) [1]. A connection established in the above manner is called a lightpath (LP). Two lightpaths cannot be assigned the same wavelength on any given link. In this work, lightpaths are established for dynamically arriving call requests. In this paper, wavelength conversion is not assumed at the network nodes.

Various algorithms have been proposed for route selection and wavelength selection. Fixed routing (FR), fixed-alternate routing (FAR) and adaptive routing (AR) are the approaches used for routing the connection requests [1]. In fixed routing, the Dijikstra's algorithm is used to find the shortest path between a given source-destination pair. In fixed-alternate routing, a set of routes to be used between each source-destination pair is statically computed [1]. The routes in this set may be edge-disjoint to ensure fault tolerance [2]. In this paper, the number of routes between each source-destination pair is restricted to two. The routes are edge-disjoint.

Random wavelength assignment (RN), first-fit wavelength assignment (FF), least-used wavelength assignment (LU) and most-used wavelength assignment (MU) algorithms are used to select a free wavelength [1]. In this paper, FF, RN and a crosstalk-aware wavelength assignment scheme (C-RN) are tested for crosstalk performance [3]. The RWA algorithms used in this paper are mentioned below

- Fixed routing and first-fit wavelength assignment(FR/FF)
- Fixed routing and random wavelength assignment(FR/RN)
- Fixed-alternate routing and first-fit wavelength assignment(FAR/FF)
- Fixed-alternate routing and random wavelength assignment(FAR/RN)
- Fixed routing and crosstalk-aware wavelength assignment(FR/C-RN)

A wavelength-routed all-optical network consists of wavelength-routing nodes (WRNs) interconnected by optical fibers. Wavelength-routing nodes (or optical cross-connect nodes) employ erbium-doped fiber amplifiers (EDFAs) to compensate for the signal power loss introduced by the optical fibers. The wavelength-routing nodes and EDFAs may cause significant transmission impairments such as crosstalk generation in the optical space switches of the nodes, generation of amplified spontaneous emission (ASE) noise by EDFA while providing signal amplification, saturation and wavelength dependence of EDFA gains and crosstalk generation due to the Demux/Mux employed in the nodes arising due to the non-ideal separation of wavelengths by the demultiplexer [4], [5], [6]. This paper considers the in-band crosstalk introduced by wavelength-routing nodes and the ASE noise introduced by the EDFAs. In [4], the in-band crosstalk induced by the demux/mux was not considered. In [7], both switch-induced crosstalk and demux/mux intraband crosstalk were considered. In [4], [7] only FR/FF and FR/RN RWA algorithms were considered. This work studies the impact of the different RWA algorithms mentioned above on the crosstalk performance of WDM networks. In [3], only fixed routing of connection requests was considered but MU and LU wavelength assignment were considered.

For each dynamically arriving call request, BER is calculated on candidate routes at an available free wavelength before setting up a call. If the BER is less than 10^{-12}, a call is set up on a lightpath; otherwise it is blocked. An event-driven simulation with on-line BER computation is used to accomplish the above task. The rest of the paper is organized as follows. Section 2 presents the network architecture and also discusses the origination of in-band crosstalk in optical networks, Section 3 discusses the BER calculations, Section 4 presents the results and Section 5 concludes the paper.

2 Network Architecture and Origination of In-Band Crosstalk

A lightpath in the optical network consists of intermediate wavelength-routing nodes (WRNs) between the source and destination nodes, interconnected by fiber segments. Fig. 1 presents a block diagram for a possible realization of a WRN [4]. The constituent optical components in a given wavelength routing node include, in general, a crossconnect switch (XCS), a pair of optical power taps on either side of XCS at each port. The EDFA on the input side compensates (with small signal gain, G_{in}) for the signal attenuation along the input fiber and tap loss. The EDFA on the output side (with small signal gain, G_{out}) compensates exactly for the losses of the XCS. The XCS is realized using an array of demultiplexers, optical wavelength-routing switches (WRS) and multiplexers. Further, multiplexers are realized using power combiners whereas demultiplexers are realized using a combination of power splitters and filters [5], [6].

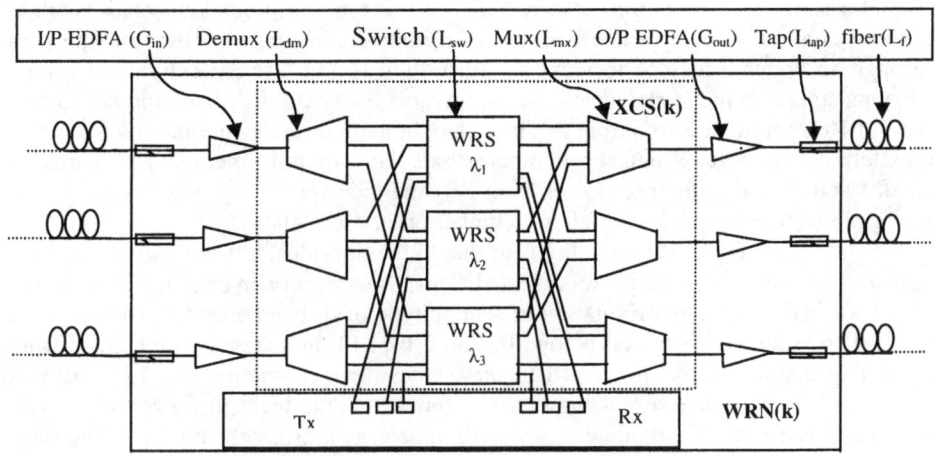

Fig. 1. Realization of a wavelength-routing node

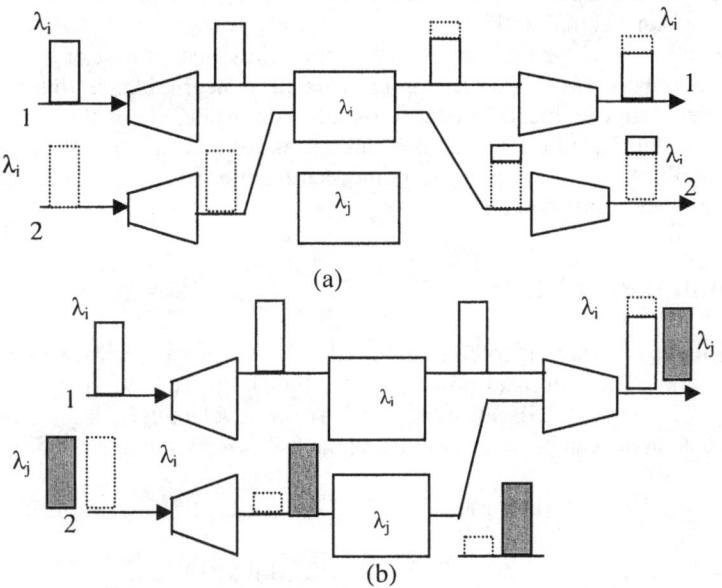

Fig. 2. Types of in-band crosstalk

The term crosstalk represents the effect of other signals on the given signal. Two forms of crosstalk can arise in WDM networks: in-band crosstalk and out-of-band crosstalk [3]. In-band crosstalk effect can be much more severe than out-of-band crosstalk [4]. In this paper, in-band crosstalk effect is considered while establishing the lightpath. In-band crosstalk is widely regarded as a major transmission impairment which limits the BER performance of all-optical networks. Three types of in-band crosstalk can arise in the network [3]. The first type of in-band crosstalk (switch-induced crosstalk) occurs when two or more lightpaths of the same wavelength pass through an optical crossconnect. As an illustration in Fig. 2(a), two lightpaths, both carrying signal on the same wavelength λ_i traverse the OXC: LP_1 from input 1 to output 1, LP_2 from input 2 to output 2. Since they both enter the switching module of λ_i, crosstalk occurs here. When the lightpaths exit the switching module, LP_1 carries a small fraction of interference power from LP_2 and vice versa. The interference power may generate a first order crosstalk or a higher order crosstalk [4].

The other two types of crosstalk occur due to the non-ideal channel isolation of the optical filters in the demultiplexers [5], [6]. This effect occurs on channels that are adjacent to each other. The origination of second type of in-band crosstalk (demux/mux in-band crosstalk) is discussed below. In Fig. 2(b), LP_1 on λ_i traverses the OXC from input 1 to output 1. LP_2 on λ_j and LP_3 on λ_i enter input 2 together and LP_2 will exit output 1. LP_2 will have a leakage power from LP_3. This leakage power will travel with LP_2 via the switch module-λ_j and will appear as a crosstalk for LP_1. The third type of in-band crosalk which also arises due to non-ideal channel isolation of the optical filters has negligible effect and is not considered here. It is to be noted that the first and second types of in-band crosstalk effect arise from another signal which is of the same wavelength as the desired signal. The third type of in-band crosstalk originates from the same signal itself.

In-band crosstalk of type 1 can also be further classified as first order and higher order crosstalk. The effect of higher order crosstalk is negligible. In this paper, only the first order switch induced in-band crosstalk is considered. In this work, optical space switches fabricated on Ti: $LiNbO_3$ substrates have been considered [4], [8]. In this paper, multiple substrate point-to-point architecture, which is a nonblocking architecture, has been considered [4].

3 Computation of BER

Consider a lightpath which is to be established on wavelength λ_i between nodes 1 and N in a network. The outbound powers of the signal ($p_{sig}(k, \lambda_i)$), switch induced crosstalk ($p_{xt}(k, \lambda_i)$) and ASE noise ($p_{ase}(k, \lambda_i)$) on wavelength λ_i, at the output of the kth intermediate node, can be expressed using the following recursive relations [4]:

$$p_{sig}(k, \lambda_i) = p_{sig}(k-1, \lambda_i) L_f(k-1, k) G_{in}(k, \lambda_i) L_{dm}(k) L_{sw}(k) L_{mx}(k) G_{out}(k, \lambda_i) L_{tap}^2, \quad (1)$$

$$p_{xt}(k, \lambda_i) = p_{xt}(k-1, \lambda_i) L_f(k-1, k) G_{in}(k, \lambda_i) L_{dm}(k) L_{sw}(k) L_{mx}(k) G_{out}(k, \lambda_i) L_{tap}^2 + \sum_{j=1}^{J_k} X_{sw} p_{in}(j, k, \lambda_i) L_{sw}(k) L_{mx}(k) G_{out}(k, \lambda_i) L_{tap}, \quad (2)$$

$$P_{ase}(k, \lambda_i) = p_{ase}(k-1, \lambda_i)L_f(k-1, k)G_{in}(k, \lambda_i)L_{dm}(k)L_{sw}(k)L_{mx}(k)G_{out}(k, \lambda_i)L_{tap}^2 + 2n_{sp}[G_{in}(k, \lambda_i)-1]hv_iB_0 L_{dm}(k)L_{sw}(k)L_{mx}(k)G_{out}(k, \lambda_i)L_{tap} + 2n_{sp}[G_{out}(k, \lambda_i)-1]hv_iB_0 L_{tap}. \quad (3)$$

The outbound power of the demux/mux in-band crosstalk ($p_{mt}(k, \lambda_i)$) on wavelength λ_i, at the output of the kth intermediate node, can be expressed using the following recursive relation:

$$P_{mt}(k, \lambda_i) = p_{mt}(k-1, \lambda_i)L_f(k-1, k)G_{in}(k, \lambda_i)L_{dm}(k)L_{sw}(k)L_{mx}(k)G_{out}(k, \lambda_i)L_{tap}^2 + \sum_{q=1}^{Q_k} Mp(q, k, \lambda_i) L_{dm}(k) L_{sw}(k)L_{mx}(k)G_{out}(k, \lambda_i)L_{tap}. \quad (4)$$

The loss and gain variables for various components used above are indicated in Fig. 1. Generally $L_x(k)$ refers to the losses, $G_x(k, \lambda_i)$ refers to EDFA gain at wavelength λ_i. $L_f(k-1, k)$ refers to the loss of the fiber segment connecting the nodes k-1 and k. Further $p_{in}(j, k, \lambda_i)$ is the power of the jth propagating signal at the switch shared by the desired signal (i.e., the switch, WRS-λ_i, for wavelength λ_i) at the kth node contributing to a first-order switch induced in-band crosstalk with J_k being the total number of such crosstalk sources at the kth node. The terms X_{sw} refers to the switch crosstalk ratio and M (filter adjacent channel isolation) represents the fraction of power leaking from a wavelength to the adjacent wavelength due to non-ideal channel isolation of the optical filters in the demultiplexers. Further, $p(q, k, \lambda_i)$ is the power of the qth signal at λ_i which contributes to demux/mux in-band crosstalk. Note that this power is referred at the input of the demultiplexer in the kth node. A fraction of $p(q, k, \lambda_i)$, namely, $M.p(q, k, \lambda_i)$, leaks into an adjacent channel and will travel along with the adjacent channel and will appear as demux/mux in-band crosstalk when this adjacent channel is multiplexed with the desired signal as shown in Fig.2(b). The number of such crosstalk sources is Q_k. B_o is the optical bandwidth, h is Planck's constant, n_{sp} represents the spontaneous emission factor and v_i is the optical frequency at λ_i. The receiver BER at the destination node can then be calculated as given below

$$P_b = 0.25\left[\text{erfc}\left(\frac{I_{s1}-I_{TH}}{\sqrt{2}\sigma_1}\right) + \text{erfc}\left(\frac{I_{TH}}{\sqrt{2}\sigma_0}\right)\right]. \quad (5)$$

The noise variances are given below

$$\sigma_{sxi}^2 = \xi_{pol}R_\lambda^2 b_i p_{sig}^1(N, \lambda_i)p_{xt}^1(N, \lambda_i), \quad (6)$$

$$\sigma_{shi}^2 = 2qR_\lambda(b_i p_{sig}^1(N, \lambda_i) + p_{xt}^1(N, \lambda_i) + p_{mt}^1(N, \lambda_i))B_e, \quad (7)$$

$$\sigma_{smi}^2 = \xi_{pol}R_\lambda^2 b_i p_{sig}^1(N, \lambda_i)p_{mt}^1(N, \lambda_i), \quad (8)$$

$$\sigma_{sspi}^2 = 4R_\lambda^2 b_i p_{sig}^1(N, \lambda_i) p_{ase}^1(N, \lambda_i) B_e/B_o, \qquad (9)$$

$$\sigma_{th}^2 = \eta_{th} B_e. \qquad (10)$$

The signal component of the photocurrent is given by

$$I_{si} = b_i R_\lambda p_{sig}^1(N, \lambda_i). \qquad (11)$$

In the above equations, $p_{sig}^1(N, \lambda_i)$, $p_{xt}^1(N, \lambda_i)$, $p_{mt}^1(N, \lambda_i)$ and $p_{ase}^1(N, \lambda_i)$ are the power referred at the receiver of the destination node. In equations (5) through (11), i in the subscripts represent the data bit (0 or 1) being received. Further $b_i = 0$ or 1 for i = 0 or 1, respectively (assuming perfect laser extinction). B_o and B_e denote the optical and electrical bandwidth respectively. ξ_{pol} is the polarization mismatch factor and is taken as ½ [4]. R_λ is the responsivity of the photodetector (1 A/W). The spectral density of the thermal noise current in the optical receiver is represented by η_{th}. The threshold current is $I_{s1}/2$ assuming perfect laser extinction (i.e., $b_0 = 0$ and $I_{s0} = 0$).

In this work, a 50% mark density of the crosstalk channels is assumed while calculating the beat noise components between signal and crosstalk [9]. The noise variance σ_{sxi}^2 accounts for the beating between the signal and switch-induced crosstalk. The noise variance σ_{smi}^2 arises due to the beating between the signal and demux/mux in-band crosstalk, σ_{sspi}^2 accounts for the beat between signal and ASE noise, σ_{shi}^2 accounts for the shot noise of the digital receiver and σ_{th}^2 accounts for the thermal noise of the digital receiver. Beat noise components between ASE and itself and crosstalk and itself are not dominant and can be neglected.

4 Results and Discussions

The impact of the various RWA algorithms on the crosstalk performance of a 15-node mesh network and on a 4 x 4 mesh-torus is presented. In obtaining these results, EDFA gain saturation is assumed to be absent. This implies that the EDFAs always deliver the desired small signal gain irrespective of the input signal powers and signal wavelengths. This is possible by providing an excess small signal gain at each amplifier in the network which ensures that enough gain is supplied to a signal even though the amplifier may be saturated [4]. It is to be noted that the ASE noise is always present and has been incorporated during BER calculation. Fig. 3 and Fig. 4 show the 15-node mesh network and the 4 x 4 mesh-torus respectively.

The internode distance is 100km in both of the networks. Each edge actually consists of two standard single mode fibers carrying bi-directional traffic. Table 1 presents the values of the system parameters used in the event-driven simulation [3], [4], [7]. The number of wavelengths on each link is 8 and they are: [1546.99, 1547.80, 1548.60, 1549.40, 1550.20, 1551.00, 1551.80 and 1552.60] nm. The signal power per channel is assumed to be 1mW at the transmitter. External modulation is supposed at the transmitters. The bit rate per channel is 2.5 Gbps. In this condition, the chirping of the transmitted signal and chromatic dispersion can be neglected.

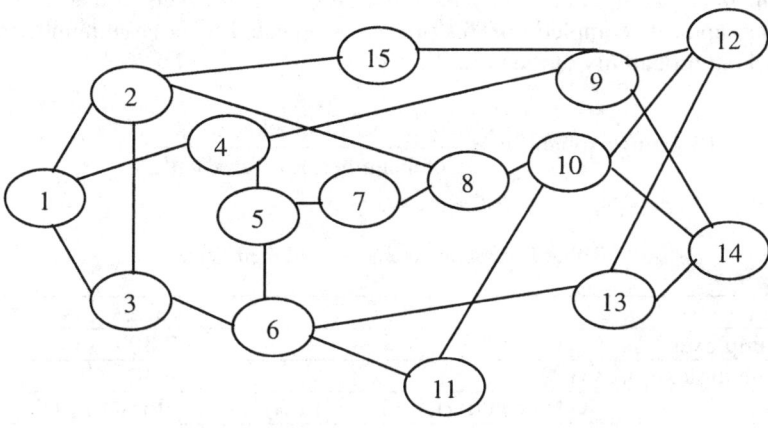

Fig. 3. 15-node mesh network

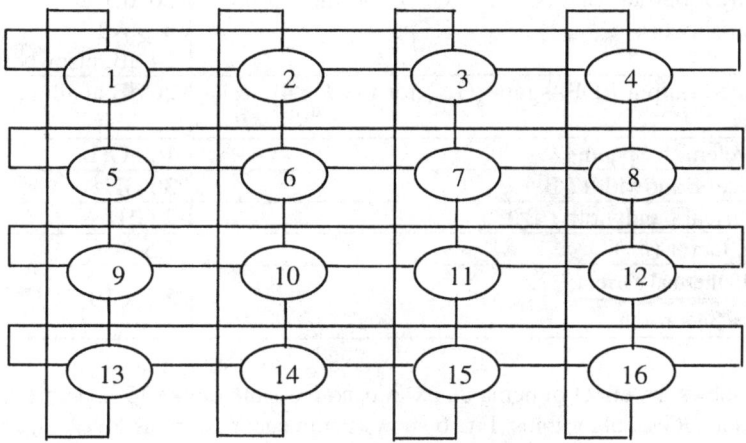

Fig. 4. 4x4 mesh-torus

The event-driven simulation module and the on-line BER-evaluation module used in this paper are similar to [4]. Calls arrive to the network following a Poisson process. The source and destination of the incoming call is determined using a uniform distribution. The call durations are exponentially distributed with a mean of 1. For each dynamically arriving call request, the event-driven simulation module determines a route and a free wavelength using one of the five RWA algorithms discussed in Section 1. If no free wavelength is available, the call is blocked. If a free wavelength is available, simulation is switched over to the on-line BER–evaluation module. Before establishing a lightpath for this call, BER at the destination node of this connection is estimated. If the receiver BER associated with this connection request is less than 10^{-12}, a lightpath is established; otherwise, it is blocked. An admitted call is terminated upon its completion. This process is repeated for a large number of calls. The blocking probability of the network is given by

$$\text{Blocking probability} = \frac{\text{Number of blocked calls}}{\text{Total number of offered calls}}. \qquad (12)$$

Table 1. System parameters and their values

Parameters	Values
Multiplexer loss (L_{mx})	-7 dB
Demultiplexer loss (L_{dm})	-9 dB
Switch loss (L_{sw}) (NxN switch) (Ls = Lw = 1 dB)	$(2\log_2 N)L_s + 4L_w$ dB
Tap loss (L_{tap})	-1 dB
Fiber loss (L_f)	-0.2 dB/km
Desired input EDFA gain for 15 node mesh network and 4 x 4 mesh-torus (G_{in})	22 dB
Desired output EDFA gain (G_{out}) for the 15-node mesh network	26 dB at nodes 2, 6, 9 &10. 24 dB, elsewhere
Desired output EDFA gain (G_{out}) for the 4 x 4 mesh-torus	26 dB at all nodes
Wavelength Spacing	100 GHz
Optical Bandwidth (B_0)	36 GHz
Electrical bandwidth (B_e)	2 GHz
ASE factor (n_{sp})	1.5
$\frac{\text{RMS thermal current}}{\sqrt{\text{Bandwidth}}}, \sqrt{\eta_{th}}$	$5.3 \times 10^{-24} \frac{A}{\sqrt{Hz}}$

Fig. 5 shows the effect of demux/mux in-band crosstalk on the 15-node mesh network under various RWA algorithms. Fig. 6 shows the impact of various RWA algorithms on the demux/mux in-band crosstalk performance of a 4 x 4 mesh-Torus. Each data point on the graph is obtained by simulating one million calls. In obtaining these results, switch-induced crosstalk is assumed to be eliminated (i.e., $X_{sw} = 0$). Filter adjacent channel isolation (M) is assumed to be -25 dB. In these figures, I-FR/FF, I-FR/RN, I-FAR/FF and I-FAR/RN refer to the FR/FF, FR/RN, FAR/FF and FAR/RN RWA algorithms in the absence of any crosstalk. Further, MCT-FR/FF, MCT-FR/RN, MCT-FAR/FF and MCT-FAR/RN refer to the FR/FF, FR/RN, FAR/FF and FAR/RN RWA algorithms in the presence of only demux/mux in-band crosstalk. In the absence of any crosstalk I-FAR/FF shows the best performance, followed by I-FAR/RN, I-FR/FF and I-FR/RN. This implies that I-FAR/FF RWA algorithm blocks the least number of calls due to non-availability of free wavelengths. However in the presence of demux/mux in-band crosstalk, MCT-FAR/RN shows the best performance, followed by MCT-FAR/FF, MCT-FR/RN and MCT-FR/FF. It may be noted that calls may be blocked due to non-availability unavailability of free wavelengths as well as due to the BER exceeding 10^{-12}. At higher loads, the performances of MCT-FAR/RN and MCT-FAR/FF do not differ significantly. Similarly MCT-FR/RN and MCT-FR/FF perform almost alike at higher loads.

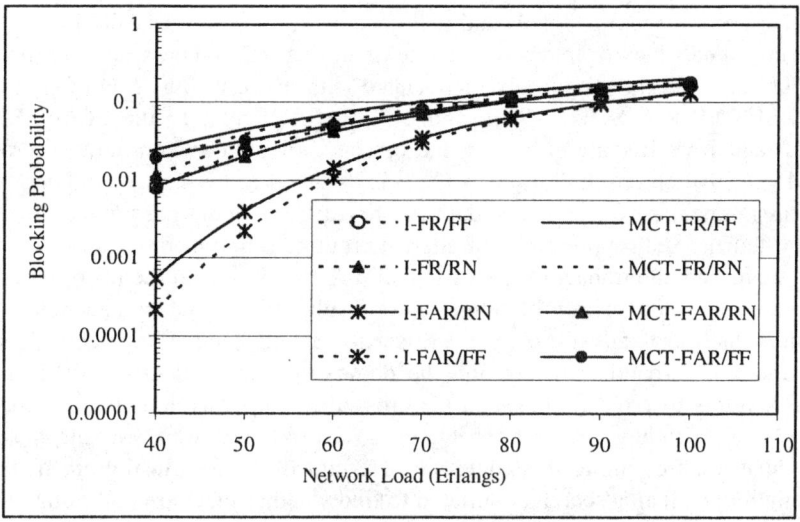

Fig. 5. Impact of the various routing and wavelength assignment algorithms on the demux/mux in-band crosstalk performance of the 15-node mesh network ($X_{sw} = 0$ and $M = -25$ dB)

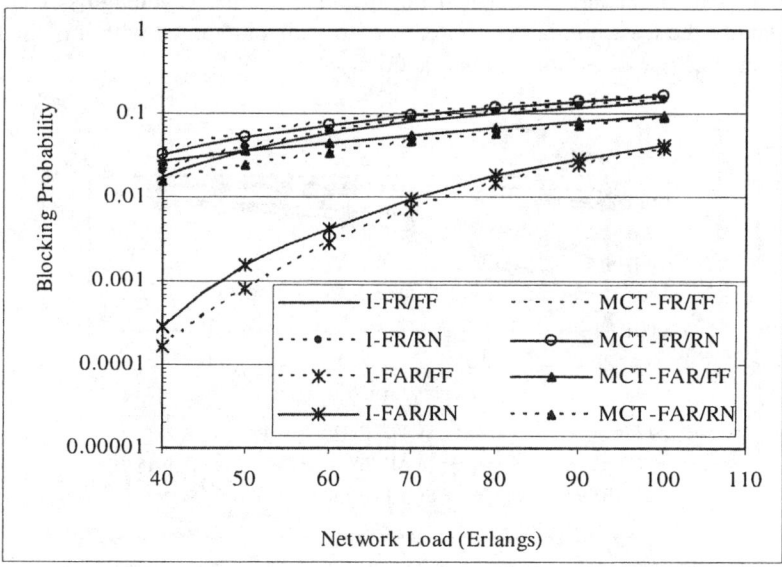

Fig. 6. Impact of the various routing and wavelength assignment (RWA) algorithms on the demux/mux in-band crosstalk performance of the 4x4 mesh-torus ($X_{sw} = 0$ and $M = -25$ dB)

Fig. 7 and Fig. 8 show the worst case effect of switch-induced in-band crosstalk in the 15-node mesh network and in the 4 x 4 mesh-torus respectively. These results are called as worst case effect for the reason discussed below. In obtaining these results, it was assumed that a signal propagating through a switch module will always interfere with other co-propagating signals and will generate first order crosstalk. In reality, the interfering signals may or may not generate first order in-band crosstalk depending on the input ports and the output ports associated with them. In Fig. 7 and Fig. 8, SCT-FR/FF, SCT-FR/RN, SCT-FAR/FF and SCT-FAR/RN refer to the FR/FF, FR/RN, FAR/FF and FAR/RN algorithms in the presence of only switch-induced crosstalk (i.e., M =0). Switch crosstalk ratio is (X_{sw}) is assumed to be -25 dB. Random wavelength assignment performs better than the first-fit wavelength assignment irrespective of whether fixed routing or fixed-alternate routing is assumed.

Fig. 9 presents the impact of the various RWA algorithms on the in-band crosstalk in 15-node mesh network. In obtaining these results, filter adjacent channel isolation (M) and switch crosstalk ratio (X_{sw}) are set to -30 dB each. Fig. 9 considers both switch-induced in-band crosstalk and the demux/mux in-band crosstalk. In Fig. 9, FR/C-RN refers to fixed routing and the crosstalk-aware wavelength assignment. In FR/C-RN, after finding a route connecting a given source-destination pair, the wavelengths that are free along this route are determined. As an illustration, if the free wavelength is λ_k, then a search is initiated to find whether there are other ongoing signals at wavelength λ_k through the crossconnects of the concerned route. The number of such signals is counted. This gives the number of sources contributing switch-induced crosstalk. Similarly, the number of sources contributing to demux/mux in-band crosstalk at wavelength λ_k is also found. The sum of both sources of crosstalk is then found. This procedure is repeated for all the available wavelengths. The wavelength that has the least number of sources of crosstalk associated with it is finally selected. In case of ties, selection is done randomly.

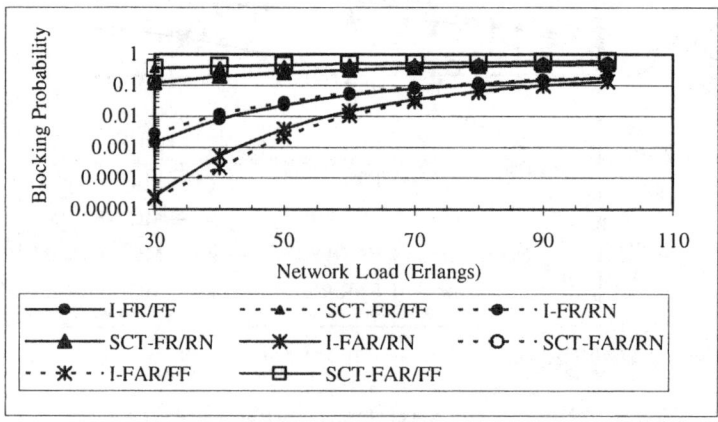

Fig. 7. Impact of the various routing and wavelength assignment algorithms on the switch-induced in-band crosstalk performance of the 15-node mesh network (X_{sw} = -25 dB and M = 0)

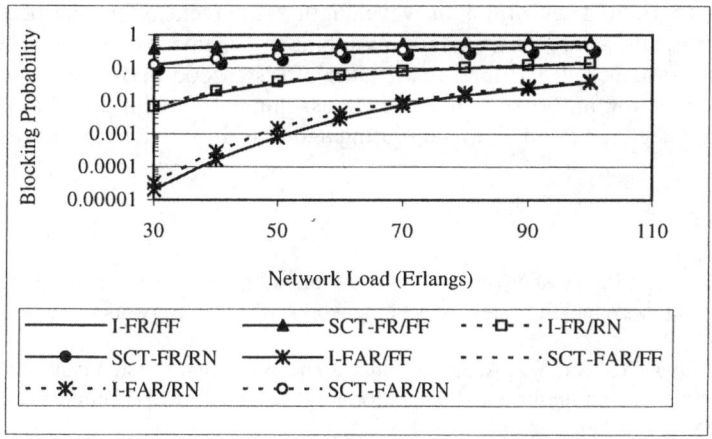

Fig. 8. Impact of the various routing and wavelength assignment (RWA) algorithms on the switch-induced in-band crosstalk performance of the 4x4 mesh-torus (X_{sw} = -25 dB and M = 0)

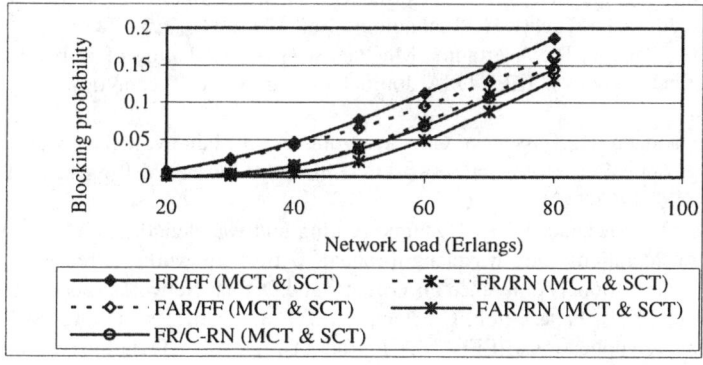

Fig. 9. Impact of the various routing and wavelength assignment (RWA) algorithms on the in-band crosstalk performance of the 15-node mesh network (X_{sw} = -30 dB and M = -3 0 dB)

As can be seen from Fig. 9, FAR/RN exhibits the best performance. This can be explained as follows. Fixed-alternate routing admits more calls into the network than the fixed routing. Random wavelength assignment tends to geographically spread wavelengths across the network such that crosstalk effects are not likely to be severe. Thus the combination of fixed-alternate routing and random wavelength assignment improves the blocking performance in the network.

5 Conclusions

In this paper, the impact of various RWA algorithms on the in-band crosstalk performance of wavelength-routed optical networks has been studied. It is observed that fixed-alternate routing with random wavelength assignment offers the best performance. A crosstalk-aware wavelength assignment scheme is also considered for crosstalk performance. It is found that it also offers a good performance when compared with fixed routing/ first fit wavelength assignment, fixed routing/random wavelength assignment and fixed alternate routing and first fit wavelength assignment.

References

1. Zang, H., Jue, J.P., Mukherjee, B.: A review of routing and wavelength assignment approaches for wavelength-routed optical networks. Optical Networks Magazine, Vol. 1 (2000) 47-60
2. Lee, P., Gong, Y., Gu, W.: Adaptive routing and wavelength assignment algorithms for WDM networks with uniform and nonuniform traffic model. IEEE Communication Letters, Vol.8 (2004) 397-399
3. Deng, T., Subramaniam, S., Xu, J.: Crosstalk-aware wavelength assignment in dynamic wavelength-routed optical networks. Proceedings of the first International conference on Broadband Networks (2004) 140-149
4. Ramamurthy, B., Datta, D., Feng. H., Heritage. J.P., Mukherjee, B.: Impact of transmission impairments on the teletraffic performance of wavelength-routed optical networks. IEEE/OSA Journal of Lightwave Technology, Vol. 17 (1999) 1713-1723
5. Iannone, G., Sabella, R., Avertanne, M., Paolis, G.D.: Modelling of in-band crosstalk in WDM optical networks. IEEE/OSA Journal of Lightwave Technology, Vol. 17 (1999) 1135-1141
6. Zhou, J., Caddadu. R., Cassaccia, G., Cavazzoni, C., O'Mahony, M.J.: Crosstalk in multi-wavelength optical crossconnect networks. IEEE/OSA Journal of Lightwave Technology, Vol. 14 (1996) 1423-1435
7. Saminadan, V., Meenakshi, M.: Dynamic routing and wavelength assignment with signal quality considerations for wavelength-routed optical networks. Proceedings of first IEEE/IFIP International conference on Wireless and Optical Networks (2004) 106-109
8. Papadimitriou, G.I., Papazoglou, C., Pomportsis, A.S.: Optical switching: Switch fabrics, techniques and applications. IEEE/OSA Journal of Lightwave Technology, Vol. 21 (2003) 384-405
9. Takahashi, H., Oda, K., Toba, H.: Impact of crosstalk in arrayed waveguide multiplexer on NxN optical interconnection. IEEE/OSA Journal of Lightwave Technology, Vol. 14 (1996)1097-1105

Modeling and Evaluation of a Reconfiguration Framework in WDM Optical Networks

Sungwoo Tak[1], Donggeon Lee[1], Passakon Prathombutr[2], and E.K. Park[3]

[1] Department of Computer Science and Engineering, Pusan National University,
30, Jangjeon-dong, Geumjeong-gu, Busan, 609-735, Republic of Korea
swtak@pnu.edu
[2] National Electronics and Computer Technology Center, Thailand
[3] School of Computing and Engineering, University of Missouri – Kansas City

Abstract. This paper studies a series of reconfiguration processes corresponding to a series of traffic demand changes in a WDM (Wavelength Division Multiplexing) optical network. The proposed reconfiguration framework consists of two objective functions, a reconfiguration process, and a reconfiguration policy. The two objective functions are AHT (objective function of minimizing Average Hop distance of Traffic) and NLC (objective function of minimizing Number of Lightpath routing Changes). The reconfiguration process finds a set of non-dominated solutions using the PEAP (Pareto Evolutionary Algorithm adapting the Penalty method) that optimizes two objective functions by using the concept of Pareto optimality. The reconfiguration policy picks a solution from the set of non-dominated solutions using the MDA (Markov Decision Action). Experimental results show that our reconfiguration framework incorporating the PEAP and the MDA yields efficient performance in the entire series of reconfiguration processes.

1 Introduction

There are two topologies in a WDM (Wavelength Division Multiplexing) optical network, a physical topology and a virtual topology. The physical topology consists of optical fiber links and photonic nodes. The virtual topology consists of a set of lightpaths that carry optical signals from source nodes to destination nodes for given traffic demands. A process of rearranging the virtual topology to meet new traffic demands is called a reconfiguration process [1]. The reconfiguration process of a virtual topology is a major task when new traffic demands are given in a WDM optical network. When the previous traffic demands are changed over a period of time, the optimal reconfiguration of a virtual topology is required to minimize network cost and maximize network performance. Since the reconfiguration is not a one-time operation, it will be activated whenever the current traffic demands are changed. The consequent reconfiguration problem is how and when to perform a reconfiguration process. A reconfiguration policy should be considered to control the reconfiguration process to generate an optimal virtual topology in the long term. The reconfiguration process and the reconfiguration policy are challenging problems in WDM optical networks. We found three major limitations of the previous reconfiguration methods available in literature. First, reconfiguration process methods proposed in [1] considers only a one-time reconfiguration and not a series of future reconfigurations. We know that once a reconfiguration generates a new virtual topology, it will serve the traffic until the demand changes. Then the next reconfiguration is started over again. The virtual topology must serve well not only for the current traffic demand but also for the traffic changes in the future. The reconfiguration problem in WDM optical networks becomes a series of reconfigurations in the long term and not a one-time reconfiguration. Second, two methodologies are widely used in WDM optical networks: ILP (Integer Linear Programming) methodology used in [2-3] and heuristic methodology used in [4]. The ILP methodology considers only one objective at a time. Additionally, it is not possible for the ILP methodology to find an optimal solution in large-size problem domain. As the complexity and size of problem domain becomes higher and larger, a heuristic methodology has been employed to find a near optimal solution. However, the heuristic methodology can be stuck in a local optimal solution because a rule of thumb or incomplete knowledge based on experience is used to reduce the amount of search. Usually, the heuristic methodology will be accepted if it is able to find a good solution, although the solution is not the best. Third, reconfiguration techniques available in literature have showed good performance for a single objective goal. Reconfiguration techniques available in [1-4] have not addressed their performance in terms of both network performance and network cost. Therefore, a reconfiguration framework that considers both network performance and network cost simultaneously needs to be proposed and evaluated extensively to design an optimal, reconfigurable WDM optical network.

2 Reconfiguration Framework

The reconfiguration framework consists of two objective functions, AHT and NLC, which are described in Section 2.1, the PEAP in Section 2.2, and the MDA in Section 2.3. The reconfiguration process based on the PEAP first finds a set of non-dominated solutions (i.e., virtual topologies) using the concept of evolutionary algorithms. Then the reconfiguration policy based on the MDA picks an optimal solution in the set of non-dominated solutions on the Pareto front.

2.1 Problem Formulation

In this section we formulate the reconfiguration process and policy problems mathematically. The formulation of reconfiguration process and policy problems in this paper is different from a general virtual topology design because it requires not only an objective goal that maximizes network performance but also an objective goal that minimizes the number of changes in a virtual topology. Therefore, the reconfiguration problem considered is a multi-objective problem that considers two objectives, AHT (objective function of minimizing Average Hop distance of Traffic) for network performance and NLC (objective function of minimizing Number of Lightpath routing Changes) for network cost. We assume that the reconfiguration of a virtual topology is only triggered by the change of given traffic demands. Additionally, all nodes are capable of grooming a bunch of low-speed traffic to the available capacity of a lightpath as much as possible. All transceivers are freely tuned to any wavelengths. We do not allow the de-multiplexing of OC-x traffic streams lower than its capacity when the traffic channel is routed through a network. Two or more OC-x traffic streams with the same source and destination nodes may pick a different route. In this section, two objective functions, AHT and NLC, are proposed along with the following parameters, variables, and fundamental constraints. We formulate the reconfiguration policy through a MDA model. The MDA model consists of five elements: 1) a set of decision epochs which are a period of time that triggers the action, 2) a set of states which indicates the status of the network, e.g., a performance parameter and a current traffic demand, 3) a set of actions, 4) a set of states and actions dependent on immediate rewards and costs, and 5) a set of state transition probabilities which relies on the action and the arrival traffic. The reward is the benefit gaining from doing a particular action while the cost is incurred from the action. Let $R_i(H)$ be the reward function of H, where H is the performance variable in the ith reconfiguration round. Let $C_i(\eta)$ be the cost function of η, where η is the number of lightpath routing changes in the ith reconfiguration round. For each state transition with a performed action, we want to maximize the expected outcome O in every reconfiguration round where

$$O = \lim_{y \to \infty} \frac{1}{y} E\left\{ \sum_{i=1}^{y} (R_i(H) - C_i(\eta)) \right\} \quad (1)$$

The reconfiguration policy tells us what action we should select in each state to maximize the expected outcome O. The average hop distance of traffic reflects the performance of grooming OC-x traffic streams. Low OC-x traffic streams are groomed at each edge node in the electrical domain before they are converted to a wavelength, which is carried through a lightpath. The higher the value of average hop distance of traffic streams, the more the network operation cost and propagation delay of traffic streams because of O-E-O (Optical-Electrical-Optical) conversion at intermediate nodes. The AHT is formulated as follows:

$$Min \frac{1}{\sum_{sd,x} \Lambda_{sd}^{x}} \sum_{ij} \sum_{sd,x} \left(x \times \lambda_{sd,ij}^{x} \right) \quad (2)$$

Λ_{sd}^{x} represents demand of OC-x traffic streams between node s and node d. $\lambda_{sd,ij}^{x}$ represents number of OC-x traffic streams from node s to node d being routed on the lightpath ij, where $x \in \{1, 3, 12\}$. The objective goal of equation (2) minimizes the ratio of $\lambda_{sd,ij}^{x}$ to Λ_{sd}^{x}. Therefore, the AHT can minimize the average hop distance of traffic required for the transmission of total OC-x traffic streams between node s and node d. Lightpath routing changes require the additional network operation cost to meet new traffic demands. Lightpath routing changes are costly because of wavelength retuning. The disruption and overhead costs of lightpath routing changes occur during the operation of wavelength retuning. The NLC is formulated as follows:

$$Min \sum_{i,j} \sum_{m,n} \sum_{k} \sum_{w} \left| \sigma_{ij,mn,w}^{k'} - \sigma_{ij,mn,w}^{k} \right| \qquad (3)$$

$\sigma_{ij,mn,w}^{k}$ denotes 1 if there exists a lightpath from node i to node j being routed through fiber link mn and the lightpath uses the kth path and wavelength w, 0 otherwise, where $k \in K$ and $w \in W$. Note that K denotes the number of alternative routing paths and W denotes the number of wavelengths that can be multiplexed on an optical fiber link. The objective goal of equation (3) minimizes the difference between the current lightpath routing $\sigma_{ij,mn,w}^{k}$ and the lightpath routing $\sigma_{ij,mn,w}^{k'}$ produced by a new traffic demand. Two objective functions, the AHT and the NLC, are in conflict. The average hop distance of traffic tends to increase when the number of lightpath routing changes is minimized. Thus, optimizing two competitive objective goals of the AHT and the NLC simultaneously belongs to the multi-objective optimization problem. In the multi-objective optimization problem, there is a set of optimal solutions that non-dominate each other within the set of solutions but dominate other solutions outside of the set of solutions for given multi-objective goals. The set of optimal solutions is known as the Pareto optimal set or the Pareto front.

2.2 PEAP (Pareto Evolutionary Algorithm Adapting the Penalty Method) for Reconfiguration Process

In this section we present the PEAP procedure that optimizes two competitive objective functions, AHT and NLC. The PEAP procedure exploits the concept of chromosomes and generates a set of non-dominated solutions known as a Pareto front. The PEAP simulates a process of natural evolution based on the concept of stochastic optimization. The PEAP is able to capture a Pareto optimal set in a single run. It is also less susceptible to the shape of the Pareto front, so it can search on a problem with the non-convex Pareto front. In the reconfiguration problem, a sequence of lightpath routing changes effects the disruption of traffic and network availability. The PEAP searches all possible sequences of lightpath routing changes because a different sequence of lightpath routing changes affects network performance and cost. In the PEAP, a virtual topology is represented by a chromosome. The chromosome is encoded by the string of $N \times (N-1)$ elements, where N is the total number of nodes in a WDM optical network. The chromosome represents an intermediate virtual topology for given traffic demands. Each cell represents a transmitter unit of lightpath routing from source node i to destination node j where $i \neq j$. The value of each cell represents a path index used for the lightpath routing. If the kth path index is equal to 0, there is no lightpath on the transmitter. Otherwise, the lightpath traverses over the kth path. A set of K-shortest paths are exploited for a set of path indices.

The PEAP optimizes multiple objective goals considered in a reconfiguration process. The PEAP consists of five procedures: (1) an initialization procedure of generating a set of initial chromosomes, (2) a procedure of evaluation, (3) a procedure of a fitness assignment, (4) a procedure of selection, and (5) a procedure of crossover and mutation. The initialization procedure generates a set of chromosomes. The chromosome is an encoded solution to the problem which is presented in a binary format. Each chromosome consists of genes which take on certain values. If the size of chromosome population is too big, it will waste the time to evaluate the chromosomes. If it is too small, an optimal solution may not be found. The evaluation procedure measures how well the chromosome is survived in the generation of next population. The AHT and the NLC are used for an evaluation function and a fitness function in a reconfiguration process. The selection procedure allows good solutions to be kept and bad solutions to be eliminated while maintaining the same population size for the next generation. A tournament selection is used for the selection scheme. The tournament selection divides solutions into two sets and matches up each pair randomly. The winner, which has a better fitness value, is placed in the mating pool whose size is the same as that of initial population. A good fitness solution has a chance to win tournaments. The next procedure is the crossover and the mutation procedures. The crossover procedure yields a recombination of solutions by exchanging segments between pairs of chromosomes. Two chromosomes are randomly picked to change the segments. The value of m is randomly selected and m random crossover points are used. The mutation operation flips binary bits in the chromosome to keep a diversity of chromosomes in the population. To improve the performance of the PEAP, we exploit the Pareto-based fitness assignment strategy and the penalty method. In the next two subsections, the Pareto-based fitness assignment strategy and the penalty method are described in detail.

Two objective functions (AHT and NLC) used for a reconfiguration process are incorporated in the fitness assignment phase to generate the Pareto front. We optimize the goals of the two objective functions using the concept of Pareto optimality. A Pareto optimal outcome cannot be improved without hurting at least one solution. Thus, some of non-dominated solutions need to be utilized to generate an optimal solution. A solution x in the PEAP is said to dominate a solution y if conditions I and II are true; (I) the solution x has the equal or less average hop distance of traffic than the solution y, and the solution x has the equal or lower number of lightpath routing changes than the solution y and (II) there exists one objective that the solution x is better than that of y. The term "better" means the less average hop distance of traffic or the lower number of lightpath routing changes. The PEAP exploits the Pareto-based fitness assignment strategy to determine the reproduction probability of each chromosome. Additionally, it performs clustering to reduce the number of non-dominated solutions while maintaining its characteristics might be necessary or even mandatory. A chromosome is referred to as a solution in this section. The flow of the PEAP procedure is as follows.

1. Generate P_t for given f_{obj} (AHT or NLC) where $|P_t| \leq D$ and $D \geq 1$;
2. $P_t' \leftarrow \emptyset; P_t'' \leftarrow \emptyset$;
3. Find non-dominated solutions ω, where $\omega \in P_t$;
4. $P_t' \leftarrow P_t' \cup \omega$;
5. Find dominated solutions, $\omega' \in P_t'; P_t' \leftarrow P_t' - \omega'$;
6. *if* $|P_t'| > D'$ *then* $P_t'' \leftarrow$ clustering$(P_t', N'); P_t' \leftarrow P_t''; fi$
7. $P_t'' \leftarrow \emptyset; P_t'' \leftarrow P_t'$;
8. *while* $(P_t' \neq \emptyset)$ *do*
9. Select solution $i \in P_t'; P_t' \leftarrow P_t' - i$;
10. $S_i = $ (# of solutions dominated by i) $/ (D + 1); F_i = S_i$;
11. *if* (traffic rerouting occurs) *then* $F_i = F_i + \tau \cdot \sum \Phi(S_i)$, where $\Phi(S_i) = (1+S_i)^2$;

12. **od**
13. $P_t' \leftarrow P_t''$;
14. $P_t'' \leftarrow \emptyset; P_t'' \leftarrow P_t$;
15. **while** $(P_t \neq \emptyset)$ **do**
16.   Select solution $j \in P_t; P_t \leftarrow P_t - j$;
17. $F_j = \rho + \sum S_i$, where $\{i \in P_t' \land [(f_{obj}(i)$ is better than $f_{obj}(j))]\}$ and $\rho \geq 1$;
18.   **if** (traffic rerouting occurs) **then** $F_j = F_j + \tau \cdot \sum \Phi(S_i)$, where $\Phi(S_i) = (1+S_i)^2$;
19. **od**
20. $P_t \leftarrow P_t''$;
21. $P_t'' \leftarrow \emptyset; P_t'' = $ tournament selection procedure (P_t, P_t');
22. **if** $|P_t''| \geq D'$ **then** stop; **else** execute crossover and mutation operations; go to Step 3; *fi*

Steps 1 through 2 generate an initial dominated population P_t with size D and create an empty non-dominated population P_t' with size D'. t denotes the tth population generation. After the dominated population P_t is generated for a given objective function f_{obj} (AHT or NLC), non-dominated solutions ω are found from P_t. In Step 4, ω is copied into P_t'. Step 5 finds dominated solutions ω' within P_t' and deletes ω', which are covered by any other members of P_t'. Hence, the PEAP maintains elites among non-dominated populations. This ensures that only non-dominated solutions are kept in P_t' and carried through the next generation by the elitist property. These allow some of the non-dominated solutions to be continually improved and to be an optimal solution. If the number of stored non-dominated solutions exceeds a given maximum D', Step 6 prunes P_t' by means of clustering. If the number of solutions in P_t' is greater than or equal to D', a clustering process based on the Euclidean distance is executed to reduce $|P_t'|$ into D'. At the beginning of the clustering process, each solution itself is a cluster. Then two clusters with the minimum distance of cluster-center gravity are merged into a bigger cluster. The process of merging clusters is repeated until the number of clusters is reduced to D'. In the final phase of the clustering process, the number of elements in each cluster is reduced to one by keeping a solution which has the minimum average distance from other solutions in the cluster. Other solutions are deleted in the cluster.

The fitness assignment procedure is a two-stage process. First, the fitness values of individuals in the non-dominated set P_t' are evaluated in Steps 8 through 12. Second, the fitness values of individuals in the population P_t are evaluated in Steps 15 through 19. Step 9 selects a solution i, which is a chromosome of non-dominated population P_t'. In Step 10, S_i is a real value, which is proportional to the value of D plus 1, where $S_i \in [0, 1)$. S_i is defined as the average value of solutions dominated by element i. S_i becomes the fitness value for the solution i in Step 10. Step 11 calculates the total traffic flow on the virtual topology, which is a solution of P_t'. If traffic is blocked, a penalty method is applied. The same rule as in Step 11 is also applied to Step 18. More details of the penalty method are described later. Step 16 selects a solution j, which is a chromosome of non-dominated population P_t. For each solution j in P_t, its fitness value F_j is calculated by the summation of average value of S_i and a gain weight factor ρ in Step 17. The gain weight factor ρ is at least one in order to guarantee that solutions of P_t' may have better fitness than solutions of P_t. Since two objective functions (AHT and NLC) need to be minimized, the value of fitness should be minimized. It implies that small fitness values correspond to high reproduction probabilities. Therefore, the probability of selecting solutions of P_t' is greater than that of selecting solutions of P_t. Step 21 executes the tournament selection procedure addressed in this section to eliminate bad solutions. In Step 22, if the maximum number of generations is reached, then the PEAP stops. Otherwise, crossover and mutation operators are applied and then the PEAP goes to Step 3.

After generating virtual topology solutions (each individual of P_t and P_t' represented by a chromosome), it is possible that the number of lightpaths required by a virtual topology is greater than the number of transmitters available in the physical topology. We take a heuristic process. The heuristic process eliminates a lightpath which occupies the lowest traffic. We repeat the heuristic process until the number of lightpaths required by the virtual topology is not greater than the number of available transmitters. The traffic in this process is the sum of OC-x traffic streams required between source and destination nodes. The traffic is routed by the following policies. The traffic is routed over the virtual topology using the K-shortest paths algorithm. The traffic routing starts from the highest streams (e.g., route OC-12 streams first, followed by OC-3 streams and OC-1 streams). Routing bifurcations are allowed in the same OC-x stream level - i.e. an OC-12 stream cannot be broken into four OC-3 streams and routed separately but two OC-12 streams with the same source and destination nodes may use different routes. As many traffic streams as possible are first routed over single-hop lightpaths. The remaining traffic is routed over multiple-hop lightpaths. If all OC-x traffic streams are routed over a single-hop, the average hop distance of traffic is equal to 1. It is the lowest bound of the average hop distance of traffic. Afterwards, we calculate the total traffic flows on the virtual topology. If the flows are blocked in the virtual topology by the distinct wavelength assignment constraint, a penalty is imposed on the virtual topology. As a result, we want to get some information out of infeasible solutions, by degrading their fitness rankings in relation to the degree of constraint violation. We set a penalty function Φ and its penalty coefficient τ to the chromosome if traffic streams are blocked (see steps 11 and 17 in the PEAP procedure). A number of alternatives exist for the penalty function Φ. Note that we consider a multi-objective minimization problem, so a smaller fitness value represents a better solution. Hence, $\Phi(S_i) = (1+S_i)^2$ for violated constraint S_i, which is exploited in Steps 11 and 17 of the PEAP. The penalty function will downgrade the fitness value of a chromosome and cause it to be eliminated in the next generation. Under certain conditions, the unconstrained solution converges to the constrained solution as the penalty coefficient τ approaches infinity. As a practical matter, τ values may be often sized separately for each type of constraint so that moderate violations of the constraints yield a penalty that is some significant percentage of a nominal operating cost. Finally, the PEAP generates a set of non-dominated solutions that is a non-blocking virtual topology. The non-dominated solutions belong to the Pareto front that optimizes multiple objective goals. A reconfiguration policy addressed in Section2.3 picks one of solutions in the Pareto front.

2.3 MDA (Markov Decision Action) for Reconfiguration Policy

We model a reconfiguration policy by the MDA (Markov Decision Action) to pick up one of the solutions in the Pareto front generated by the PEAP. The reconfiguration policy is activated with the MDA. The goal of MDA is to find a reconfiguration policy which produces an optimal decision and an optimal action to be taken in each state. The MDA consists of a set of decision epochs, a set of states, a set of actions, a set of states and actions dependent on immediate rewards and costs, and a set of state transition probabilities. For decision epochs, the time between reconfiguration transitions is assumed discrete. We define a state as the tuple (AHT$_{outcome}$, Ψ) for the MDA. Ψ denotes the virtual topology utilization for given traffic demands. AHT$_{outcome}$ denotes the outcome generated by the AHT. It implies that the MDA considers the virtual topology utilization and the outcome of the AHT. Definition 1 is used in the state description.

Definition 1. Virtual topology utilization Ψ is defined by a ratio of the total amount of traffic routed over the network to the upper bound of virtual topology capacity.

Remark: Let N be the number of optical nodes, T be the maximum number of transceivers, and C be the capacity of lightpaths. Ψ is $\left\{\sum_{sd,x}(x \times \Lambda_{sd}^{x})\right\} / (N \times T \times C)$.

In Definition 1, N, T, and C are constant or rarely changed unless the total network capacity is full. Ψ relies on traffic demand Λ_{sd}^{x} in Definition 1. Additionally, Ψ reflects the Pareto front curve because the reconfiguration process requires more number of light path routing changes to achieve the high virtual topology utilization. Λ_{sd}^{x} is a parameter of the AHT described in equation (2). Therefore, the tuple (AHT$_{outcome}$, Ψ) is defined as a state for the MDA. An action states how to perform the reconfiguration process by picking a solution x on the Pareto front. The Pareto front is the combination of AHT and NLC. We define the set of actions as the different positions of the Pareto front's curve. For each position indicating an action, we select the solution x closest to the pseudo-weight factor calculated by equation (4).

$$w_i = \left(\frac{f_i^{max} - f_i(x)}{f_i^{max} - f_i^{min}}\right) / \sum_{j}^{Obj}\left(\frac{f_j^{max} - f_j(x)}{f_j^{max} - f_j^{min}}\right) \quad (4)$$

The pseudo-weight factor in equation (4) is calculated for each solution on the Pareto front's curve. f_i^{max} and f_i^{min} are the maximum outcome and the minimum outcome of objective function f_i respectively. Obj is the number of objective functions. The outcome o_{ij}^{k} generated in moving from state i to state j for action k is defined as $o_{ij}^{k} = r_{ij}^{k} - c_{ij}^{k}$. The reconfiguration policy determines what action should be selected in each state to maximize o_{ij}^{k}. r_{ij}^{k} and c_{ij}^{k} are the immediate gaining reward and incurring cost respectively when state i is changed to state j using action k. The immediate gaining reward r_{ij}^{k} is defined as $r_{ij}^{k} = \beta \cdot H_{ij}^{k} + c$. H_{ij}^{k} is the average hop distance of traffic when state i is changed to state j using action k. β is a weight assigned to the reward and c is a control factor. The cost c_{ij}^{k} is defined as

$c_{ij}^k = \alpha \cdot \eta_{ij}^k + \gamma$. η_{ij}^k denotes the average number of lightpath routing changes required in the reconfiguration process, where state i is changed to state j using action k. α is a weight assigned to the cost. γ is a one-time cost required for activating the reconfiguration operations. Note that reward and cost functions can be any functions that reflect reconfiguration performance and cost factors such as delay, throughput, packet loss, load balance, management cost, and resource costs. q_i^k shown in equation (5) denotes the expected immediate outcome out of state i for action k. p_{ij}^k denotes the state transition probability from state i to state j for action k. Each outcome o_{ij}^k and transition probability p_{ij}^k has its specific value according to an action k.

$$q_i^k = \sum_{j=1}^{N} p_{ij}^k o_{ij}^k \text{ for } \forall i = 1, 2, 3, \ldots, N \qquad (5)$$

As shown in equation (6), the next state $v_i(n+1)$ from the current state $v_i(n)$ is selected by utilizing three information, p_{ij}^k, o_{ij}^k, and $v_i(n)$. $v_i(n)$ represents the expected total outcome in the nth transition starting from state i.

$$v_i(n+1) = \max_k \left[\sum_{j=1}^{N} p_{ij}^k \{o_{ij}^k + v_j(n)\} \right] \text{ for } \forall i = 1, 2, 3, \ldots, N \qquad (6)$$

$$v_i(n+1) = \max_k \left[q_i^k + \sum_{j=1}^{N} p_{ij}^k v_j(n) \right] \text{ for } \forall i = 1, 2, 3, \ldots, N \qquad (7)$$

Equation (7) is generated by combining equations (5) and (6). We apply the iterative cycle of Howard [5] to find the optimal decision for the MDA. It consists of two operations; the value-determination operation and the policy-improvement operation. These two operations take turn to produce the optimal gain g that represents the optimal reconfiguration policy. The value-determination operation shown in equation (8) exploits p_{ij}^k and o_{ij}^k to produce the value of g, which is the expected optimal outcome.

$$g + v_i(n) = q_i^k + \sum_{j=1}^{N} p_{ij}^k + v_j(n) \text{ for } \forall i = 1, 2, 3, \ldots, N \qquad (8)$$

The policy-improvement operation shown in equation (9) finds the optimal action k. These two operations are executed iteratively until the new g' is not better than the current g under the condition such that $g' - g > \varepsilon$ and ε is a threshold value.

$$\max_k \left[q_i^k + \sum_{j=1}^{N} p_{ij}^k v_j(n) \right] \text{ for } \forall i = 1, 2, 3, \ldots, N \qquad (9)$$

3 Experiments

The 14-node NSFNET network topology is used for the performance measurement of the proposed reconfiguration framework [6]. We assume that each node is working as both an access node and a routing node. The lightpath capacity is OC-192. The total number of wavelengths available over each link is 8. The number of transmitters and receivers per each node is assumed to be 6, thus, there are at most six lightpaths initiated or terminated at each node. Transmitters and receivers are tunable to any wavelengths. We simulate the changes of traffic by swapping the data randomly within each traffic matrix to preserve the values of Ψ. We randomly swap all pairs of data, i.e. $N(N-1)/2$ pairs. The results are new traffic demand matrices used in the next round of the reconfiguration process. Thirty sets of traffic demand matrices are generated for reconfiguration processes. The parameters used in the PEAP are set as follows. The probability of crossover is 0.6. The probability of mutation is 0.01. The dominated population size is 50. The non-dominated population size is 50. In the reconfiguration policy, we reduce state spaces by considering the only traffic demands with the same value of Ψ. Therefore, we can ignore Ψ in the state tuple (AHT$_{outcome}$, Ψ). Now the state is defined by the AHT$_{outcome}$. Since the AHT$_{outcome}$ is a continuous value, we define a discrete state based on a range of the AHT$_{outcome}$ and use the median of an AHT$_{outcome}$ range to represent a state.

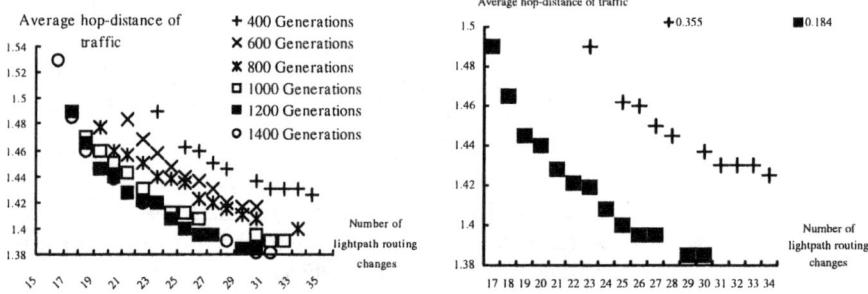

Fig. 1. Pareto front of the reconfiguration at $K = 2$

Fig. 2. Pareto front with $\Psi = 0.355$ and $\Psi = 0.184$

We first find the right number of generations for our experiments. We run the PEAP and plot the Pareto fronts generated in the PEAP as illustrated in Fig. 1. The horizontal axis is the number of lightpath routing changes in the virtual topology and the vertical axis is the average hop distance of traffic. We found that running the PEAP at 1200 generations is enough to generate the optimal Pareto front in our experiments. Fig. 1 shows that the more the number of generations, the better the results. However, the performance of results is saturated when the number of population generations is greater than 1200 generations. Additionally, we find the experimental K value in K-shortest paths. We run the PEAP at 1200 generations with the different values of K. Through the extensive experimentation, $K = 2$ is the right choice. It generates better results than those of $K = 3$ and $K = 4$ because they have a large search space and better Pareto front is not found at $K \geq 3$. As described earlier in this section, the states in the MDA rely on the value of Ψ in experiments. So, we need to compare the Pareto front in terms of the value of Ψ. In Fig. 1, the value of Ψ is 0.355 when the number of generations is 400. When the number of generations is 1200, the value of Ψ is 0.184. In Fig. 2, the value of the average hop-distance of traffic seems worse when the value of Ψ is 0.355. The high value of Ψ implies that the high utilization of virtual topology is accomplished by maximizing the average hop-distance of traffic. The optimal policy is derived from the results generated through the following experimentation. The value of Ψ is set to 0.184, which is near optimal as shown in Fig. 2. Finally, the MDA process is applied to find the optimal decision. The efficiency of the MDA is compared with that of the IHO (Immediate Highest Outcome reconfiguration policy) over thirty sets of traffic demand matrices. The IHO selects the solution on the Pareto front that produces the immediate best outcome in the current state of virtual topology reconfiguration. We run 30 rounds of reconfigurations. The IHO selects the solution in the Pareto front that produces the immediate best outcome in the current state of virtual topology reconfiguration.

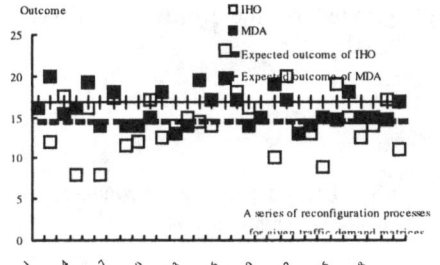
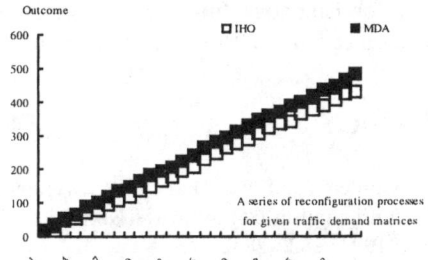

Fig. 3. Individual reconfiguration outcomes   **Fig. 4.** Accumulated reconfiguration outcomes

Fig. 3 shows a series of individual outcomes. Fig. 4 shows a series of accumulated outcomes in round 1 through round 30. Even if the IHO selects the best immediate outcome in the current state of virtual topology reconfiguration, it does not generate better overall outcomes than those of the MDA as shown in Fig. 4. In the long term, the MDA produces better outcomes than the IHO as shown in Fig. 4. In Fig. 4, the total accumulated outcome of the IHO is 425.787 and that of the MDA is 442.947. All of the experiments performed in this paper were carried out using 2.4 GHz Intel based processor. The worst case in the experiments took less than 30 minutes, which is acceptable for the reconfiguration process where traffic demands are changed in at least daily basis. The computational complexity of the PEAP is $O(P^2)$ where P is the population size. The routing computational complexity is $O(N^2)$ where N is the number of nodes in the network. Thus the overall complexity needed in each generation of the PEAP is $O((PN)^2)$.

4 Conclusion

We propose a reconfiguration framework adapting multi-objective optimization in WDM optical networks. The reconfiguration problem in WDM optical networks requires a process of multi-objective optimization because the objective of reconfiguration considers the network performance and the network cost simultaneously. In this paper, the AHT is exploited for the measurement of network performance and the NLC is exploited for the measurement of network cost. The proposed reconfiguration framework includes a reconfiguration process and a reconfiguration policy. The reconfiguration process finds a set of non-dominated solutions using the PEAP that optimizes two objective functions by using the concept of Pareto optimal. The reconfiguration policy picks a solution from the set of non-dominated solutions using the MDA. A case study based on experiments shows that the performance of the PEAP incorporating the MDA is better than that of the IHO in the entire series of reconfiguration processes.

Acknowledgements

This work was supported by the Regional Research Centers Program (Research Center for Logistics Information Technology), granted by the Korean Ministry of Education & Human Resources Development.

References

1. Labourdette, J.F.P., Hart G.W., Acampora, A.S.: Branch-exchange Sequences for Reconfiguration of Lightwave Networks. IEEE Trans. On Communications, Vol. 42, No. 10, (1994) 2822-2832
2. Banerjee, D., Murkherjee, B.: Wavelength-routed Optical Networks. Linear Formulation, Resource Budget Tradeoffs, and a Reconfiguration Study. IEEE/ACM Trans. on Networking, Vol. 8, No. 5, (2000) 598-607
3. Ramamurthy, B., Ramakrishnan, A.: Virtual Topology Reconfiguration of Wavelength-Routed Optical WDM Networks. Proc. of IEEE GLOBECOM, San Francisco, (2000) 1269-1275
4. Zheng, J., Zhou, B., Mouftah, H. T.: Design and Reconfiguration of Virtual Private Networks (VPNs) over All-Optical WDM Networks. Proc. of 11th International Conference on Computer Communications and Networks, Miami, Florida, (2002) 599-602
5. Howard, R.A.: Dynamic Programming and Markov Process, M.I.T. Press, Cambridge, 1960
6. Claffy, K.C., Polyzos, G.C., Braun, H.W.: Traffic characteristics of the T1 NSFNET backbone. Proc. of IEEE INFOCOM, San Francisco, CA, (1993) 885 – 892

On the Implementation of Links in Multi-mesh Networks Using WDM Optical Networks

Nahid Afroz [1], Subir Bandyopadhyay [1], Rabiul Islam [1], and Bhabani P. Sinha [2]

[1] University of Windsor, Windsor, Ontario, Canada
{afroz, subir, rabiul}@uwindsor.ca
[2] Indian Statistical Institute, Kolkata, India

Abstract. In this paper, we have suggested a novel way to implement the interconnections in the Multi-Mesh (MM) network using optical devices. In a traditional, copper-based approach, the long connections between processors in an interconnection network create major limitations with respect to the speed of communication. Our approach for inter-block communication in a MM uses optical communication using Wavelength Division Multiplexing (WDM). Rather than passive stars or free-space optics, used to implement some recent optoelectronic communication schemes for interconnection networks, this design uses wavelength routed fiber-based networks.

1 Introduction

Das, De and Sinha [2] proposed the Multi-Mesh (MM) interconnection network topology recently. In this paper we have outlined a scheme for implementing such a network using optical technology. Traditionally, metal-based electrical connections have been used to realize links in interconnection networks. Copper-based connections to realize complex interconnections is problematic since long copper wires are needed for such topologies which accentuates problems like skin effect, crosstalk, interference, wave reflections and electrical noise due to current changes, and dielectric imperfections [4]. Metal interconnects can cause severe pulse distortions and attenuation, clock skew and random propagation delays and suffer from the technological limitations of communication bandwidth constraints, low interconnect density, long network latencies, and high power requirements [4]. A major advantage of optical communication over electronic communication is that, for relatively shorter distances needed in multi-processor systems, the delay in optical communication is negligible, essentially independent of communication distance. Other advantages of optical interconnections over metal include inherent parallelism, higher bandwidth, ability to propagate in parallel channels without interference, low crosstalk, immunity from electromagnetic interference, lower signal and clock skew, lower power dissipation, potential for reconfigurable interconnects [1], [6].

The Optical Multi-Mesh Hypercube (OMMH) proposed by Louri and Sung [5] and the Optical Transpose Interconnect System (OTIS) proposed by Marsden et al. [7] are two notable interconnection networks based on optical interconnects. Free-space optical interconnects exploiting air space for optical signal propagation [6] and passive stars have been used in such networks [5].

Optical technology has become dominant in large capacity backbone networks. It is technologically impossible to exploit the huge bandwidth of optical fiber using a single high-capacity channel. W*avelength-division multiplexing* (WDM) can be used to define multiple communication channels on optical networks to avoid this problem [8].

In our approach WDM wavelength-routed networks have been used to realize the links between blocks. This is the first known approach to avoid the use of complex alignments needed in free-space optics or the high power [8] of passive star couplers. Due to lack of space, the issue of single faults in the fiber links has not been discussed. In our approach, faults may be handled easily with a small increase in the number of wavelengths needed. Our optical implementation for inter-block connections uses wavelength division multiplexing (WDM). The intra-block links can always be realized using VLSI technology since they require short links of constant length. For effective use in parallel processing, it is essential that the delay along each link is small and uniform (O (1)). Since the inter-block links used in the 3D MM are relatively long, optical links for such inter-block connections may be used to ensure a small uniform delay link.

2 The Physical Topology for Communication in a Multi-mesh

In our scheme we propose to use n^2 routers - one for each of the n^2 blocks. Figure 1 shows part of the physical topology where a square represents a block (which is a $n \times n$ mesh of processors) and an oval represents an optical router. All the routers are arranged in the form of a two-dimensional grid. To simplify the diagram we have not shown the connections from the boundary processors to the routers. As shown in Figure 1, the connection between the routers follows the architecture of a torus. For clarity, we have shown the wrap-around links only for the first and the last rows and columns. Each row and column has similar connections. In figure 1, we have used bi-directional links. To realize a bi-directional link $x \leftrightarrow y$, there will be two unidirectional fibers - one allowing communication from x to y and one for communication from y to x. We will now discuss the topology corresponding to the connections from the boundary processors on the top and the bottom edge of block B_{ij}. The physical topology corresponding to the connections from the boundary processors on the right and the left edge of block B_{ij} are similar. Router R_{ij} will be connected to the corresponding block B_{ij} carrying incoming and outgoing optical signals as follows:

1) the router R_{ij} will be connected to block B_{ij} with one fiber carrying signals from processors $P(i, j, 1, k)$ of block B_{ij} for communication to processor $P(k, j, n, i)$ of block B_{kj}, for all k, $1 \leq k \leq n, k \neq j$. This may be easily achieved by using a multiplexer M_{ij}^{U}, shown in Figure 2, with inputs from processors $P(i, j, 1, k)$, for all k, $1 \leq k \leq n$. The fiber carrying the output of multiplexer M_{ij}^{U} is connected as an input to router R_{ij} as shown in Figure 2.

2) the router R_{ij} will be connected to block B_{ij} with one fiber carrying signals from processors $P(i, j, n, k)$ of block B_{ij} to processor $P(k, j, 1, i)$ of block B_{kj}, for all k, $1 \leq k \leq n$, $k \neq j$. This may be easily achieved by using a multiplexer M_{ij}^D shown in figure 2 with inputs from processors $P(i, j, n, k)$, for all k, $1 \leq k \leq n$. The fiber carrying the output of multiplexer M_{ij}^D is connected an input to router R_{ij} as shown in Figure 2.

3) the router R_{ij} will be connected to block B_{ij} with one fiber carrying signals from processors $P(k, j, n, i)$ of block B_{kj} to processor $P(i, j, 1, k)$ of block B_{ij}, for all k, $1 \leq k \leq n$, $k \neq j$. This may be easily achieved by using a de-multiplexer D_{ij}^U, shown in figure 2 with inputs from processors $P(k, j, n, i)$ for all k, $1 \leq k \leq n$. The fiber carrying the input to de-multiplexer D_{ij}^U is an output from the router R_{ij} as shown in Figure 2.

4) the router R_{ij} will be connected to block B_{ij} with one fiber carrying signals from processors $P(k, j, 1, i)$ of block B_{kj} to processor $P(i, j, n, k)$ of block B_{ij}, for all k, $1 \leq k \leq n$, $k \neq j$. This may be easily achieved by using a de-multiplexer D_{ij}^D, shown in figure 2 with inputs from processors $P(k, j, n, i)$ for all k, $1 \leq k \leq n$. The fiber carrying the input to the de-multiplexer D_{ij}^D is an output from router R_{ij} as shown in Figure 2.

Figure 3 shows the ith column of a Multi-Mesh and the four fiber links between the router R_{i1} and block B_{i1}. Here the links are shown only in one direction (top to bottom). There is also a link in the opposite direction that was omitted for clarity. All the routers have similar connections to the corresponding blocks.

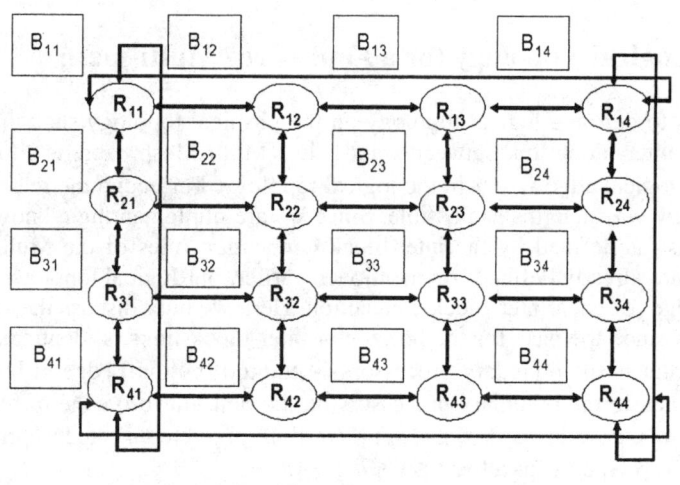

Fig. 1. Connections between Routers in a Multi-Mesh network of order 4

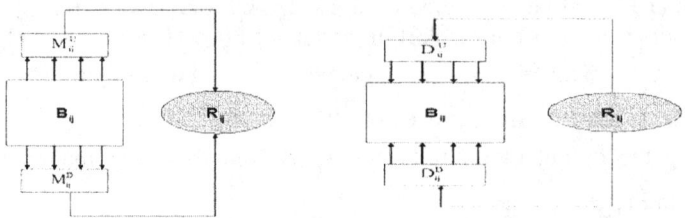

Fig. 2. Connections of multiplexers and demultiplexers to block B_{ij}

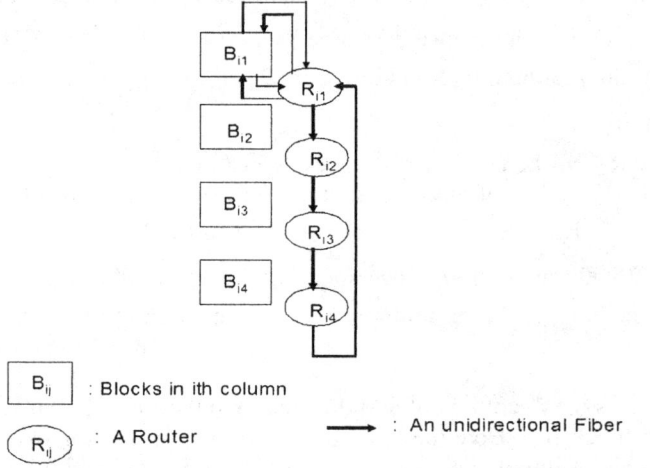

Fig. 3. Connection between router R_{i1} and block B_{i1}

3 The Logical Topology for a Fault-Free Multi-mesh

Our task is to define a logical topology on the physical topology such that, for every undirected inter-block link between x and y in a Multi-Mesh, there is a logical edge $x \to y$ and a logical edge $y \to x$ in the logical topology. For economic reasons, we wish to use as few wavelengths as possible. Since we are implementing a known pattern of connections (as defined by the inter-block connection rules of the Multi-Mesh), the lightpaths are already defined. As mentioned earlier, our logical topology must have a directed edge for each inter-block connection. Here we only discuss the vertical inter-block links since the case for the horizontal inter-block links is identical. In a Multi-Mesh of order n, the boundary processors on the top (bottom) edge of block $B(\alpha, \beta)$, are connected to the boundary processors on the bottom (top) edge of block $B(*, \beta)$. In other words, processors $P(\alpha, \beta, 1, y)$ ($P(\alpha, \beta, n, y)$) are connected to processors $P(y, \beta, n, \alpha)$ ($P(y, \beta, 1, \alpha)$), for all y, $1 \le y \le n$, $y \ne \alpha$.

In our problem, we need two lightpaths from each block $B_{\alpha, \beta}$ to block $B_{y, \beta}$ - one for the connection from processors $P(\alpha, \beta, 1, y)$ to $P(y, \beta, n, \alpha)$ and one for the connection from processors $P(\alpha, \beta, n, y)$ to $P(\alpha, \beta, 1, y)$, for all α, y, $1 \le \alpha, y \le n$. We now look at the ring consisting only of the routers in column number β and the fibers connecting them. We may view the $B_{\alpha,\beta}$, as the end-node connected by the multiplexer collecting lightpaths from all processors on the top edge of the block to router $R_{\alpha,\beta}$. The set of lightpaths from the processors on the top edge define a completely connected ring. Similarly the set of lightpaths from the processors on the bottom edge define another completely connected ring. In summary our problem is to define complete connectivity for a bidirectional ring using a set of wavelengths say $\{\lambda_1, \lambda_2, ... \lambda_K\}$. This constitutes the set of connections from all the processors on the top edge of block in column β. Then we define an independent second set of complete connections by using another set of wavelengths $\{\lambda_{K+1}, \lambda_{K+2}, ... \lambda_{2K}\}$. This second set constitutes the set of connections from all the processors on the bottom edge of block in column β.

Due to the symmetric nature of our network, we have chosen a straight forward route for our lightpaths - we will use only the fibers connecting routers in column β when defining lightpaths from any block in column β to any other block in the same column. The algorithm for assigning routes and wavelengths to each lightpath to define complete connectivity for a bidirectional ring [8] may be used directly here. They also chose a shortest path routing and have described a recursive algorithm to determine the wavelengths needed for complete connectivity[8]. We will use their algorithm which requires $(n^2 - 1)/8$ wavelengths for complete connectivity, giving one lightpath between every pair of end-nodes. Since we need to define two lightpaths from each end-node to every other end-node, we will need $K = (n^2 - 1)/4$ wavelengths.

4 Logical Topology for a Fault-Tolerant Multi-mesh

Due to lack of space, only an outline of our scheme for handling single faults can be described. We have used shared path protection schemes [7] and have shown how define the primary paths as well the backup paths for each of the inter-block connections and have calculated the cost of such a scheme. The primary paths are the same as those used in section 3. The backup paths are routed in such a way that any single faulty link is bypassed. It has been shown that an additional $\lceil n/2 \rceil$ wavelengths are sufficient to achieve shared path protection.

5 Conclusions

In this paper we have described a scheme for realizing the long inter-block connections of a Multi-Mesh using optical technology. The physical topology of a torus network is convenient for realizing these connections. Since wavelength-routed WDM technology has been used, the large power requirements of passive start or the careful alignment needed in free space optics have been avoided. The scheme may be easily extended to handle single link faults in the optical part without changing the routing algorithm.

References

1. R. D. Chamberlain and R. R Krchnavek: Architectures for optically interconnected multicomputers. IEEE Global Telecommunication Conference, GLOBECOM '93, Vol. 2, pp. 1181-1186, 1993.
2. D. Das, M. De and B. P. Sinha: A new network topology with multiple meshes. IEEE Transactions on Computers, Vol. 48, No. 5, pp. 536-551, 1999.
3. K. Hwang and F. A. Briggs: Computer Architecture and Parallel Processing. New York: McGraw-Hill, 1983.
4. A. Louri and H. Sung: An Optical Multi-Mesh Hypercube: a scalable optical interconnection network for massively parallel computing. Journal of Lightwave Technology, Vol.12, Iss. 4, pp. 704 -716, 1994.
5. A. Louri and H. Sung: 3D Optical Interconnects for high-speed interchip and interboard communications. Computer, Vol. 27, pp. 27–37, 1994.
6. G. C. Marsden, P. J. Marchand, P. Harvey and S. C. Esener: Optical transpose interconnection system architectures. Optical Letters, Vol. 18, No. 13, pp. 1083-1085, 1993.
7. S. Ramamurthy and B. Mukherjee: Survivable WDM Mesh Network, Part I-Protection. Proc. of IEEE INFOCOM '99, pp. 744-751, 1999.
8. T. E. Stern, K. Bala: Multi-wavelength Optical Network. Addison Wesley Longman, Inc. 1999.
9. H. S. Stone and J. Cocke. Computer Architecture in the1990s. Computer Vol. 24, No. 9, pp. 30-38, 1991.

Distributed Dynamic Lightpath Allocation in Survivable WDM Networks

A. Jaekel and Y. Chen

University of Windsor,
Windsor, Ont N9B 3P4, Canada
arunita@uwindsor.ca

Abstract. There has been considerable research interest in the use of path protection techniques for the design of survivable WDM networks. In this paper, we present a distributed algorithm for dynamic lightpath allocation, using both dedicated and shared path protection. The objective is to minimize the amount of resources (wavelength-links) needed to accommodate the new connection. We have tested our algorithms on a number of well-known networks and compared their performance to "optimal" solutions generated by ILPs. Experimental results show that our algorithm generates solutions that are comparable to the optimal, but are significantly faster and more scalable than corresponding ILP formulations.

1 Introduction

Optical networks are attractive candidates for wide-area backbone networks, due to their large bandwidth, low attenuation and low error rates 1. A *lightpath* in an optical network is an end-to-end all-optical communication path from a source node to a destination node through a number of intermediate router nodes. Each lightpath must be assigned a route over the physical network, and a specific channel on each fibre it traverses. This is the standard *routing and wavelength assignment* (RWA) problem 2.

A wavelength routed optical network may use either a *static* or a *dynamic* lightpath allocation strategy 3, 4. A number of ILP formulations for solving the RWA in survivable WDM networks have been presented in 5, 6. A centralized heuristic to solve this problem is given in 7. The main problem with such centralized algorithms (both ILPs and heuristics) is that the central agent can quickly become a bottleneck. In this paper, we present a distributed algorithm for dynamic lightpath allocation that allocates resources based only on local knowledge available at each node. We assume that there are no wavelength converters available. Therefore, a lightpath must be assigned the same channel on each fibre it traverses. We use path protection techniques 8-9, so that for each new connection a primary path and an edge-disjoint backup path are established during call setup. We demonstrate through simulations that our algorithms generate solutions comparable to the optimal solutions (generated by the ILP formulations) but are much faster and more scalable than exact ILP formulations.

2 Distributed Algorithm

The centralized algorithms require a single control node that stores information about the state of the entire network and about every connection that is currently active in the network. In the distributed scheme, a node does not have global knowledge of the state of the entire network, but operates based only on "local" information. In this scheme a single node need not know the routes between all source destination pairs, only the routes from itself to other nodes. Similarly, it is not aware of all connections established over the network, or the state of all channels on each edge of the network. It only knows about those connections that are routed through it and those edges that are directly connected to it. Each node stores two main types of information:

i) Network Information: This includes information about the (partial) network topology as well as information about the state of each channel on the outgoing links from the node. The link-state information includes two parameters CurrentState(λ) and NumLP(λ) for each channel λ, on each outgoing edge. The network information stored at a given node x_i consists of the following five fields:

- its own node_id
- node_ids of its adjacent nodes and the outgoing link to be used to connect to these nodes.
- a set of R edge-disjoint routes from itself to all other nodes in the network.
- the set of available channels Λ_e, on each outgoing edge e.
- CurrentState (λ) for each channel on each outgoing link
- NumLP(λ) for each channel on each outgoing link (required for shared protection only).

CurrentState (λ) for a channel on a particular link refers to one of four possible states:

a) CurrentState (λ) = 0 indicates that the channel is "free" and is available for allocation to a new lightpath on the link.

b) CurrentState (λ) = 1 indicates that the channel is "busy" and has already been allocated for a primary or backup lightpath, on that particular link.

c) CurrentState (λ) = 2 indicates that the channel is being considered as a *potential* candidate for allocation to a new lightpath. There is a "lock" on the channel, as it is temporarily reserved for the new lightpath.

d) CurrentState (λ) = 3 indicates that the channel was already assigned to one or more backup lightpaths and is now being considered as a *potential* candidate for allocation to another backup lightpath (needed only for shared path protection).

The value of NumLP(λ), on a particular link, specifies the number of lightpaths that have been assigned to channel λ, on that link. This information is only needed in shared path protection, where more than one lightpath may be assigned to the same channel on a given link. For dedicated protection, the value of NumLP(λ) is always 0 (1) if CurrentState(λ) is 0(1).

ii) <u>Lightpath Information:</u> Each node in the network stores certain information for each lightpath that is routed through it, whether it is already established or currently being setup. For a lightpath from a source node s to a destination node d, each node x_i in its physical route stores a record called LP-record corresponding to the lightpath. An LP-record consists of seven fields, containing the following information about the lightpath:

- Source
- Destination
- ConnectionNumber
- PhysicalRoute
- LightpathType (primary or backup)
- SelectedWavelength (-1 indicates a channel has not yet been assigned)
- LockedChannels (set of channels L_i that have been temporarily reserved on edge $x_i \rightarrow x_{i+1}$ for this lightpath).

2.1 Control Messages

In the distributed approach, each node works independently of the other nodes. Inter-node communication and co-ordination takes place by passing control messages between nodes. Each control message is associated with a specific lightpath and always contains the corresponding LP-record. The messages are processed at each node in the physical route of the lightpath and appropriate actions are taken at each step. There are four types of control messages, as explained below.

InitiateConnectionSetup. In our scheme a request for a new connection is generated randomly at a given time, based on a predetermined probability p, ($0 < p < 1$). When a connection request is generated, a source node s and a destination node d are also selected randomly and an *InitiateConnectionSetup* message is added to the message queue of the selected source node. This type of message is processed at the source node s.

The first step is to assign a unique identifier (C_{new}) to the new request. The combination (s, C_{new}) can be used to uniquely identify a connection in the entire network. Next we select the primary and backup routes. This is done by selecting the two routes which have the maximum number of free channels on their first edge, with the expectation that this will increase the chances of success. Of course, this is not necessarily the best choice, because other edges in the route may be congested and may not have available channels. But, since we are operating based on local information only, it is a "reasonable" choice. If two such routes can be found, the connection setup phase is started by putting locks on the available channels on the appropriate outgoing edge and creating a LP-record for each lightpath (primary and backup) at the source node.

ForwardRequest. A *ForwardRequest* control message is used in the setup phase of a lightpath. It is responsible for forwarding the LP-record form a node x_i to the next node x_{i+1}, along the selected route. At each intermediate node x_i, the message is processed and the usable channels (if any) are reserved on the appropriate outgoing link, before forwarding the LP-record to the next node.

If x_i is the destination node, it means that there is at least one free channel along the entire route. In this case, x_i, selects a wavelength $\lambda \in L_{i-1}$ and sends a *ResponseSignal* message back to x_{i-1}, indicating that the request was successful. If x_i is an intermediate node, then L_{i-1} gives the set of usable channels for the lightpath up to node x_i and $L_i = L_{i-1} \cap \Lambda_e$ is the set of usable channels on link e, where $e = x_i \rightarrow x_{i+1}$. If L_i is empty, then there are currently no usable channels available on edge e and a response signal, indicating failure, is sent back to node x_{i-1}.

If $L_i \neq empty$, then all channels $\lambda \in L_i$ are "locked" on edge e, for the current lightpath, and the LP-record is updated so that LockedChannels = L_i. The updated LP-record is then sent to node x_{i+1} in a *ForwardRequest* control message.

ResponseSignal. This type of message is sent from a destination node or an intermediate node, back towards the source node, along the physical route of a lightpath. It indicates the lightpath setup request failed (SelectedWavelength = -1) or a suitable channel λ_s was found and has been assigned to the lightpath (SelectedWavelength $\neq$ -1).

When a node x_i receives a *ResponseSignal* message indicating a suitable channel λ_s has been found, it releases the locks on all wavelengths $\lambda \in L_i, \lambda \neq \lambda_s$, and updates the local LP-record and status information of the relevant channels on edge $e = x_i \rightarrow x_{i+1}$. Then the *ResponseSignal* message is sent to node x_{i-1}.

There is some additional processing that must be done, when a *ResponseSignal* message is received at the source node s. We know that there are two lightpaths (primary and backup) for each connection request. When the source node receives a *ResponseSignal* for one lightpath, it checks if it has already received a response for the other lightpath. If not, it simply waits until both responses are available. Once responses for both lightpaths have been received, we need to consider three possibilities.

Case 1 (Both responses indicate success): In this case, the connection request is successful and communication can begin along primary path.

Case 2 (Both responses indicate failure): In this case, the connection is blocked and the corresponding entry is deleted from the node.

Case 3 (One indicates success, the other failure): In this case, the connection is also blocked, but the resources allocated to the successful lightpath must be reclaimed. A *FreeResources* control message, containing the appropriate LP-record, is sent to the next node (x_1) along the physical route ρ_{sd}^f of the successful lightpath. Finally, the local copy of the LP-record (for the successful lightpath) and the entry corresponding to the new connection request are both deleted from the source node.

FreeResources. This type of message is used to reclaim resources allocated to a lightpath, when they are no longer needed. A *FreeResources* message is generated at the source node of a connection for one of two reasons: a) a successfully established connection needs to be terminated and the corresponding resources reclaimed b) one of the lightpaths for a new connection request was successfully established, but the other failed (Case 3 above). A node x_i processes a *FreeResources* message by releasing the channel λ_s allocated to the lightpath on its outgoing link. If there are no other lightpaths assigned the same channel on edge $e = x_i \rightarrow x_{i+1}$ (i.e. NumLP(λ_s)=0), then the current state of λ_s is reset to 0. This will always be the case for dedicated path protection, or if the lightpath being considered is a primary lightpath. Finally, the *FreeResources* message is sent to the next node x_{i+1} on the physical route for the lightpath and the local copy of the LP-record is deleted from the node.

3 Experimental Results

In this section, we compare the performance of our algorithm with "optimal" solutions generated from exact ILP formulations as well as a centralized heuristic, in terms of the number of successful connections. Table 1 shows the total number of connections that can be accommodated by the network for dedicated protection and shared protection.

Table 1. Number of successful connections for dedicated and shared path protection

Number of wavelengths per fiber	No. of successful connections established					
	Dedicated			Shared		
	optimal	centralized	distributed	optimal	centralized	distributed
4	13	11	10	17	16	12
8	29	28	23	43	38	36
16	65	62	54	101	85	73
32	142	125	112	222	199	146
64	289	263	251	454	419	293

We see that the performance of the centralized heuristic is typically within 10-15% of the optimal. However, the drop in the number of connections with the distributed algorithm is more noticeable. The lower performance of the distributed algorithm is expected and can be attributed to the following reasons:

i) In the centralized approach (this includes optimal ILP formulations as well as our centralized heuristic), connection requests are presented to the control node sequentially. But in the distributed approach several connections may be in the setup phase simultaneously. This means many channels could be "reserved" and cannot be considered, even if they are ultimately released.

ii) In the centralized approach, each of the R pre-computed physical routes are considered for a lightpath from s to d, based on global knowledge of network conditions. If one route fails, the next one is considered. In the distributed approach, we pre-select a single route for a lightpath (based on incomplete local information only). This can reduce the chances of success.

4 Conclusions

In this paper we have presented a distributed algorithm for dynamic lightpath allocation in survivable WDM networks. In this scheme the network nodes can operate independently, based only on local information, and communicate by passing control messages. We have compared our algorithm with "optimal" solutions, generated from ILP formulations. The simulation results demonstrate that this is a viable and attractive option for practical networks.

Acknowledgement. The work of A. Jaekel was supported by research grants from the Natural Science and Engineering Research Council (NSERC), Canada.

References

1. Stern, T., Bala, K.: Multiwavelength Optical Networks-a Layered Approach. Addison-Wesley (1999).
2. Zang, H., Jue, J.P., Mukherjee, B.: A Review of Routing and Wavelength Assignment Approaches for Wavelength-Routed Optical WDM Networks. Optical Networks Magazine (2000) 47-60.
3. Chlamtac, I. et al..: Lightnets: Topologies for High-Speed Optical Networks. IEEE/OSA J. of Lightwave Tech., Vol. 11. (1993) 951-961.
4. Gerstel, O. et al.: Dynamic Channel Assignment for WDM Optical Networks with Little or No Wavelength Conversion. Proc. 34th Annual Allerton Conf. (1996) 32-43.
5. Zhong, S., Jaekel, A.: Optimal Priority Based Lightpath Allocation for Survivable WDM Networks. Int. Conf. on Computers, Communications and Networks (2004) 17-22.
6. Sahasrabudhe, L., Ramamurthy, S., Mukherjee, B.: Fault Management in IP-over-WDM Networks: WDM Protection Versus IP Restoration. IEEE JSAC., Vol. 20, No. 1. (2002) 21 - 33.
7. Ou, C., Zhang, J., Zang, H., Sahasrabuddhe, L., Mukherjee, B.: New and Improved Approaches for Shared-Path Protection in WDM Mesh Networks. IEEE/OSA J. of Lightwave Tech., Vol. 22, No. (2004) 1223-1232.
8. Ramamurthy, S., Sahasrabudhe L., Mukherjee, B.: Survivable WDM Mesh Networks", IEEE/OSA J. of Lightwave Tech., Vol. 21, No. 4 (2003) 870-883.
9. Sridharan, M., Salapaka, M.V., Somani, A.: A Practical Approach to Operating Survivable WDM Networks. IEEE JSAC, Vol. 20, No. 1 (2002) 34-46.
10. Zang, H., Ou, C., Mukherjee, B.: Path-Protection RWA in WDM Mesh Networks Under Duct-Layer Constraints. IEEE/ACM Trans. on Networking, Vol. 11, No. 2 (2003) 248-258.

Protecting Multicast Sessions from Link and Node Failures in Sparse-Splitting WDM Networks

Niladhuri Sreenath and T. Siva Prasad

Department of Computer Science and Engineering,
Pondicherry Engineering College, Pondicherry 605 014, India
nsreen@yahoo.com, siva542@yahoo.co.in

Abstract. Optical splitting capability at some nodes is necessary to get efficient multicast routing in the wavelength routed wavelength division multiplexing (WDM) networks. There is a growing interest in efficiently protecting multicast sessions against the failure of network components. We propose algorithms for protecting multicast sessions against failure of network components such as links and nodes in a network with sparse splitting and sparse wavelength conversion. The effectiveness of the proposed algorithms is verified through extensive simulation experiments.

1 Introduction

A WDM network employing wavelength routing consists of wavelength routing nodes interconnected by point-to-point fiber links in an arbitrary topology [1]. A lightpath is an optical path established between two nodes in a network, created by the allocation of the same wavelength throughout the path. The requirement that the same wavelength must be used on all the links along a selected path is known as wavelength continuity constraint. A wavelength converter is an optical device, which can convert an optical signal on one wavelength to another wavelength. This type of node is called as a wavelength conversion node or simply a WC-node. A wavelength-routed node may have the capability to tap small amount of optical power from the wavelength channel, which is forwarded by that node. This type of node is called as a Drop and Continue node or simply a DaC-node.

To support multicasting in a WDM network, nodes in the network need to have light (optical) splitting capability. A node with splitting capability can forward an incoming message to more than one outgoing link. If a network has splitting capability at all nodes, then it is referred to as a network with full splitting capability. A network with a few split-capable nodes is called a network with sparse splitting capability. The multicast capability at the routing nodes can also be achieved by converting the optical signal into electronic form and transmitting in optical form onto all the required outgoing links. Here, by default nodes are considered to have wavelength conversion capability. However in our work, we assume that the intermediate routers forward the optical signal without converting it into electronic form as mentioned in [2]. A node having both splitting and wavelength conversion capabilities is called a Virtual Source (VS). Such a node can transmit every incoming message to any number of output links on any wavelength. The benefit of VS node is discussed in [3].

In this paper, we dealt with protecting multicast sessions from single link and single node failure. We consider a network with nodes having different capabilities. We assume the lightpaths with wavelength continuity constraint and the wavelength conversion at some nodes may happen in the optical domain. The restoration schemes to protect against network components failures are broadly classified into reactive and proactive methods. In a reactive method, when an existing link in primary multicast tree fails, a search is initiated to find a new multicast tree, which does not use the failed links. In proactive method, backup tree is identified and resources are reserved along the backup tree at the time of establishing primary tree itself. By doing so, this method yields 100% restoration guarantee.

In literature, some proactive methods to achieve fault-tolerant multicast routing are proposed for a network with full splitting and wavelength conversion capabilities. They are link-disjoint, arc-disjoint, segment-based and path-based protection schemes [4], [5]. Full splitting and full wavelength conversion at every node is achieved by converting the optical signal into electronic form. The signal arriving at the input fiber link of a node is electronically converted and replicated to as many outgoing ports as required. One copy may be dropped at the local node. The algorithms proposed in [4], [5], generate the multicast trees by either pruned Prim's heuristic or minimum cost-path heuristic. These two heuristics assume splitting capability at all nodes. Hence to apply these heuristics to a sparse splitting network, we may either modify the generated tree or these heuristics need to verify the splitting capability at the nodes while generating the tree. However, these methods if applied to a network with sparse splitting may require more resources as mentioned in [3]. Also, in the algorithms proposed in [4], [5], the cost of links, which are used for backup path, are made zero to implement backup multiplexing. Since, all nodes are not having splitting and wavelength conversion capabilities, these set of links may not be used. To incorporate the backup multiplexing it is necessary to verify the splitting and wavelength conversion capabilities. Hence, the algorithms proposed in [4], [5] require modifications to extend them to a network with sparse splitting and wavelength conversion capabilities.

The rest of the paper is organized as follows. Section 2 explains our proposed algorithms for protecting multicast sessions in a network with sparse splitting and sparse wavelength conversion. Section 3 explains performance study of our algorithms. Section 4 concludes the paper.

2 Our Work

In this section we present our algorithms for generating backup trees for a network with sparse splitting and wavelength conversion. We use the heuristic mentioned in [3] for generating primary multicast trees, which exploits various capabilities of optical nodes. We assume that all nodes in the network have the DaC capability. Some nodes may have both splitting capability and wavelength conversion capabilities (VS nodes), whereas some other nodes have only splitting capability (Split nodes). Our algorithms LFLD (LinkFailureLinkDisjoint), LFAD (LinkFailureArcDisjoint), LFSD (LinkFailureSegmentDisjoint), and LFPD (LinkFailurePathDisjoint) deal with providing protection to multicast sessions from link failures. The definitions mentioned in [4], [5] for link, arc, segment, and path disjointness are also used in our paper. However, our algorithms aim at a network with sparse splitting and sparse wavelength conversion and made use of special capabilities such as DaC. Split, and VS. Due to space limitation, we present here only LFPD algorithm. We also propose NFND (NodeFailureNodeDisjoint), NFPD (NodeFailurePathDisjoint), and NFCB (NodeFailureCapabilityBased) algorithms to provide protection from node failures. Due to space limitation we present here only NFCB algorithm.

2.1 LFPD (LinkFailurePathDisjoint) Algorithm

In a sparse splitting network only a few nodes have split capability. These split capable nodes need to be used to generate primary and backup paths. Hence all special capable nodes are maintained as a list ($setN$), so that they can be used while generating the tree. This list contains all split capable nodes and also the DaC nodes that are not used for extending the tree. . The backup tree is computed as the least cost path among the following paths: Shortest path from source to destination and shortest paths from VS or DaC node of $setN$ the destination.

Algorithm

- Create a primary tree by considering VS and DaC nodes.
- Find all DaC and VS nodes in the primary tree and add them to $setN$.
- For every destination node of the session, repeat the following steps.
- Compute a link-disjoint shortest path between the source and destination node.
- Compute a link-disjoint shortest path from every node in $setN$ to destination node.
- Select the least cost path from the above computed paths as backup path.
- Find all DaC nodes and VS nodes in the backup path and add them to $setN$.

2.2 NFCB (NodeFailureCapabilityBased) Algorithm

Here, the restoration of various nodes is done based on their capabilities. For example, if a DaC node is failed then only one path needs to be restored. This is because, a DaC node can be used to send optical signal to only one node. If a Split node is failed, then all paths that are passing through the Split node need to be restored. The failure of a VS node is dealt in a similar way, but paths passing through a VS node may use different wavelengths. Hence, different methods need to be used to restore the sessions affected due to failure of nodes with different capabilities. The algorithm that takes care of capabilities of the nodes while restoring the sessions is given below:

Algorithm

- Compute a primary tree by considering VS and DaC nodes.
- Find all DaC and VS nodes in the primary tree and add them to $setN$.
- For every node in the primary tree repeat the following steps.
- Remove a node F from the tree.
- If the node F is a DaC node, use NFND algorithm to
    - Compute shortest paths from upstream VS or source node of node F to the immediate down stream node of node F.
    - Compute a path from every node of $setN$ to the immediate down stream node of node F.
    - Select the least cost path from the above computed paths as backup path.
- If the node F is a VS node, then use NFND algorithm to
    - Compute shortest path from upstream VS or source node of node F to the every immediate down stream node of node F.
    - Compute a path from every node of $setN$ to the every immediate down stream node of node F.
    - Select the least cost path from the above computed paths as backup path.
- Find all DaC and VS nodes in the backup path and add them to $setN$.

3 Performance Study

The performance of our link failure protection algorithms and node failure protection algorithms are studied and compared. Extensive simulation experiments are conducted on NSFNET. The network is assumed to have nodes with splitting and/or wavelength conversion capabilities distributed uniformly and randomly. The sessions are generated randomly and with a single source and a set of destinations. Every node is equally likely to be a destination for a session. A node may be the source in more than one session. The destination set is also chosen randomly according to the cardinality G which is a fraction of nodes in the network. We studied the effect of group size (G) and number of Virtual Source (VS) nodes on the number of wavelength channels for a session (bandwidth consumed). To find the effect of G, we consider 30% of nodes as VS nodes.

Figure 1 depicts the number of wavelength channels required for various group sizes (G) when both VS and DaC nodes are present in the network for LFSD and LFPD algorithms. As the group size increases the difference in wavelength channel requirement of LFSD and LFPD algorithms also increases.

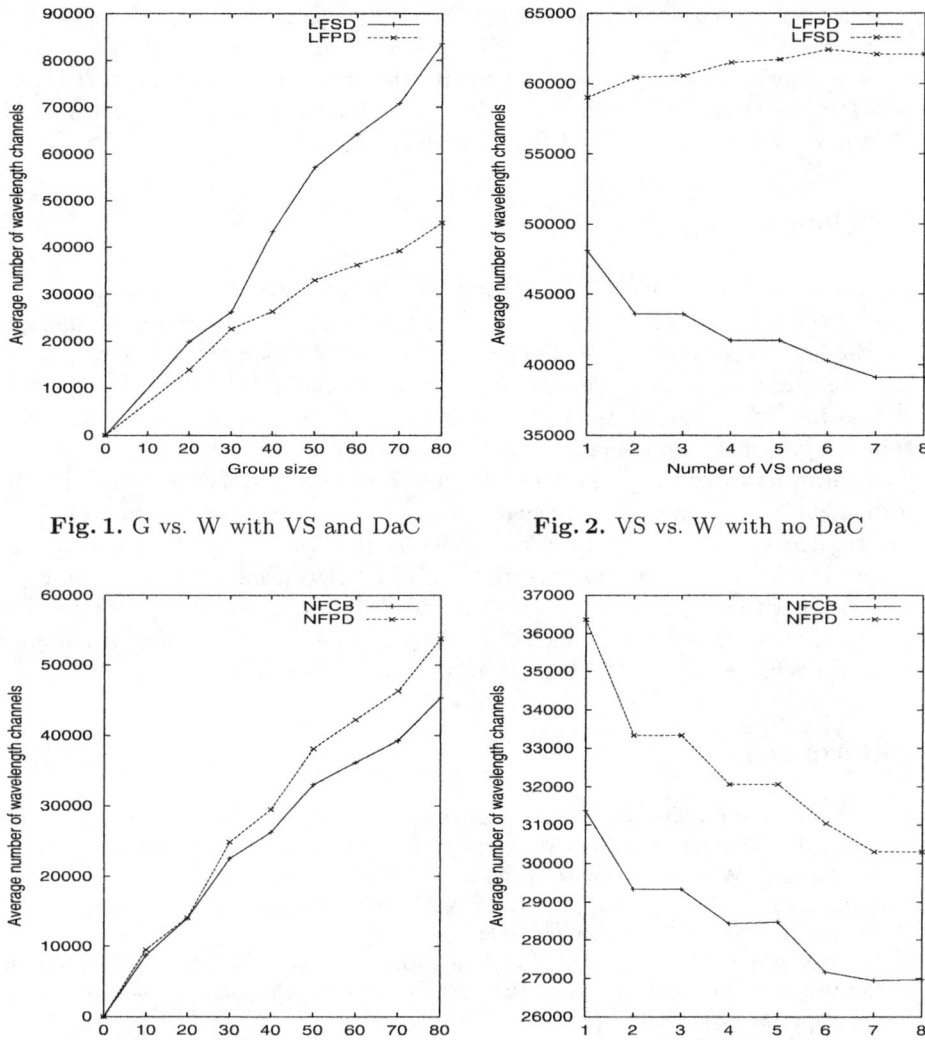

Fig. 1. G vs. W with VS and DaC

Fig. 2. VS vs. W with no DaC

Fig. 3. G vs. W with VS and DaC

Fig. 4. VS vs. W with no DaC

Figure 2 depicts the number of wavelength channels required for varying number of VS nodes for LFSD and LFPD algorithms. Since the results are taken when no DaC nodes are present, it explains the effect of VS nodes on LFSD and LFPD algorithms. As the number of VS nodes increases, number of wavelength channels increases for LFSD algorithm whereas it decreases for LFPD algorithm.

Figure 3 depicts the number of wavelength channels required for various group sizes (G) when both VS and DaC nodes are present in the network for NFPD and NFCB algorithms. NFCB algorithm shows better performance than that of NFPD algorithm.

Figure 4 depicts the the number of wavelength channels required for varying number of VS nodes for NFPD and NFCB algorithms. Since the results are taken when no DaC nodes are present, it explains the effect of VS nodes on NFPD and NFCB algorithms. As the number of VS nodes increases, number of wavelength channels decreases for both NFPD and NFCB algorithms.

4 Conclusions

In this paper, we proposed algorithms for protecting multicast sessions in a wavelength routed WDM network with sparse splitting and sparse wavelength conversion. These algorithms differ from the earlier protection schemes mainly in, considering split and wavelength conversion capabilities while constructing the backup tree. Our multicast protection algorithms are suitable for both full splitting and sparse splitting networks and deals with both link and node failures. The performance of the proposed algorithms are compared based on the amount of bandwidth (number of wavelength channels) consumed by the primary and backup trees. The performance of LFPD to restore the sessions due to link failure requires less resources than that of LFSD algorithm. The performance of NFCB algorithm to restore the sessions failed due to node failure requires less resources than that of NFCB algorithm. At present we are developing distributed algorithms for generating fault tolerant multicast sessions.

References

1. R. Ramaswami, "Multiwavelength Lightwave Networks for Computer Communication", IEEE Communications Magazine, vol. 31, no. 2, pp. 78-88, February 1993.
2. X. Zhang, J. Wei, and C. Qiao, "Constrained Multicast Routing in WDM Networks with Sparse Light Splitting", IEEE/OSA Journal of Lightwave Technology, vol. 18, no. 12, pp. 1917-1927, December 2000.
3. N. Sreenath, G. Mohan, and C. Siva Ram Murthy, "Virtual Source Based Multicast Routing in WDM Optical Networks," Photonic Network Communications, vol. 3, no. 3, pp. 217-230, 2001.
4. N.K. Singhal, L.H. Sahsrabuddhe, and B. Mukherjee, "Provisioning of Survivable Multicast Sessions Against Single Link Failures in Optical WDM Networks", IEEE/OSA Journal of Lightwave Technology, vol.21, no.11, November 2003.
5. N.K. Singhal and Biswanath Mukherjee, "Algorithms for Provisioning Survivable Multicast Sessions Against Link Failures in Mesh Networks," in the Proceedings of 5th International Workshop on Distributed Computing (IWDC 2003), Lecture Notes in Computer Science, Springer-Verlag, Vol. 2918, pp. 361-371, December 2003.

Oasis: A Hierarchical EMST Based P2P Network

Pankaj Ghanshani and Tarun Bansal

Department of Electronics and Computer Engineering,
Indian Institute of Technology Roorkee,
Roorkee 247667, India
`{yomanuec, yahoouec}@iitr.ernet.in`

Abstract. Peer-to-peer systems and applications are distributed systems without any centralized control. P2P systems form the basis of several applications, such as file sharing systems and event notification services. P2P systems based on Distributed Hash Table (DHT) such as CAN, Chord, Pastry and Tapestry, use uniform hash functions to ensure load balance in the participant nodes. But their evenly distributed behaviour in the virtual space destroys the locality between participant nodes. The topology-based hierarchical overlay networks like Grapes and Jelly, exploit the physical distance information among the nodes to construct a two-layered hierarchy. This highly improves the locality property, but disturbs the concept of decentralization as the leaders in the top layer get accessed very frequently, becoming a performance bottleneck and resulting in a single point of failure. In this paper, we propose an enhanced m-way search tree (EMST) based P2P overlay infrastructure, called Oasis. It is shown through simulation that Oasis can achieve both the decentralization and locality properties along with high fault tolerance and a logarithmic data lookup time.

1 Introduction

In recent years, peer-to-peer (P2P) systems have been the burgeoning research topic in large distributed system. Gnutella [1] and Napster [2] are the most famous peer-to-peer file-sharing systems, but both of them have the scalability problem. All such unstructured networks lead to a common problem of wastage of network resources due to heavy flooding. Peer-to-peer networks like CAN [3], Chord [4], Pastry [5], Tapestry [6] try to address this problem by using Distributed Hash tables (DHT). Although each of them has different location and routing algorithms, all of them use consistent hashing (like SHA-1) to let the participant nodes and objects be distributed uniformly in their virtual space. These systems can achieve fairly good load balancing. But the primitive DHT schemes have a significant disadvantage that they may violate the locality property. During the locating and routing process, the next hop is chosen without considering the physical topology information. This produces inefficient effects in response time and overall physical path length for lookup service.

To address this problem, the DHT based approaches should take into consideration the relative physical position among the participant nodes. Grapes [7] provide a hierarchical virtual network infrastructure using physical topology information. It has a two-layered overlay network, the upper layer called super-network, the lower layer called sub-network; in both layers, any DHTs routing algorithm can be used. Each sub-network has a leader that forms a part of the super-network and manages the sub-network. The physically nearby nodes construct the sub-network. Because the physical distance of any node pairs in sub-network is short, it reduces the lookup distance. Although Grapes can highly improve the locality property of DHTs, it disturbs the decentralization property. The leader has to route all the queries of its sub-network and has to manage super-network routing too, thus becoming a performance bottleneck. Suppose a query has to go to Canada from India, it may first go to Pakistan, then to Europe and then to Canada. This route may get even longer, leading to inflated look-up latency. So a multi-hop route in the super-network is a great disadvantage.

We propose a scheme- Oasis that *solves the problem of decentralization* by distributing the network traffic between multiple hosts. Every node is a cluster of hosts dividing the traffic load among them and saving the network from a single point of failure. Oasis also *increases the fault tolerance* of the system by sending multiple copies of a query through different paths so as to increase the probability of a query reaching its destination. Although the total load on the network gets increased, this does not affect performance because the load is already extensively divided (Section 2.3). Further, the nodes in the super-network are directly connected to each other i.e. the query would go directly from India to Canada, *reducing the look up latency* drastically.

In this paper we discuss the design of Oasis, a self organizing hierarchical network. The rest of this paper is organized as follows. Section 2 describes the design of Oasis. Section 3 presents the Oasis protocol. Section 4 gives the simulation results and comparison. Finally Section 5 gives the conclusion and directions for future work.

2 The Design of Oasis

In this section we describe the basic structure of our system. The overlay has two structures: the nodes having physical proximity constitute a sub-network. Each sub-network is an Enhanced m-way search tree (EMST, explained in Section 2.1). The super-network is composed of the leaders of all the sub-networks. Both these networks can use any of the standard hashing schemes (such as SHA-1) for locating and routing purposes. Fig. 1 shows the super-network where each node is the root node of the sub-network below it and is connected to all other nodes in the super-network so as to get fast transmission over large distances. Again, every node is a cluster of hosts.

While a host inserts an object into the system, it sends a request to its sub-network leader. The leader first inserts the object into its own sub-network by the hashed key

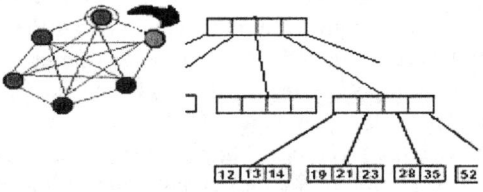

Fig. 1. The structure of Oasis

of that object. After that, it finds the associated leader in the super-network's virtual space by its key. Finally, that leader inserts the object into the corresponding position in its sub-network, completing the inserting process of that object.

While a host looks for an object, it first searches the sub-network by the key of that object. If it fails, it searches the super-network through its leader. After the leader finds the object outside its sub-network, it caches the object in its own sub-network. Consequently, a host will find the object in its sub-network with high probability.

2.1 The Fundamental Hierarchy: The Enhanced m–Way Search Tree (EMST)

The basic structure of our network is an EMST. It is basically an m-way search tree with a restriction that a node can have children only after it has m-1 elements. In other words, a branch of the tree will not grow in height until the capacity of the branch with the given height is fully utilized.

To insert in an EMST at first, we search for the element to be inserted. If the cluster at which the search terminates already has m-1 hosts then the new host is inserted as a child. For deletion we replace the leaving host by the host with the largest key in its left sub-tree or by the host with smallest key in its right sub-tree. Sometimes there might be a need of a rearrangement at the leaf level so as to complete the deletion process (Fig. 2a and Fig. 2b). If we delete the element with key 95, it is replaced by the largest element in its left sub-tree (70). Now the cluster which had 70 as an element will have to do a rearrangement to get 61 at its position. Also, for the purpose of intra-cluster management like insertion and deletion we have a leader in each cluster.

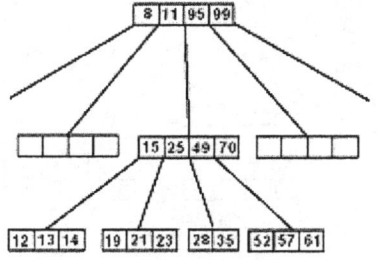
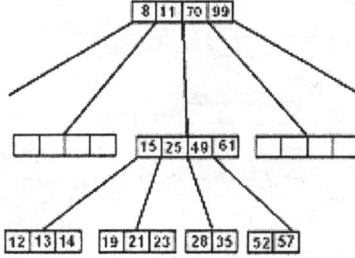

Fig. 2a. The initial state of the tree   **Fig. 2b.** The rearrangement after the deletion

2.2 The Parent Child Relationship

All the children of a certain node are divided equally among the hosts of that node. "Divided" here is in terms of queries and maintenance. As shown in the figure 3, the first host of the parent node maintains the first hosts of its children nodes, the second maintains the second hosts and so on. Every host in the network maintains an address book with the information of (1) the sub-network leader, (2) brothers (and their children), (3) children (and their helpers), (4) parent and (5) helpers (Section 2.3). Information here refers to {IP, Key (and query traffic limit for children and their helpers)}.

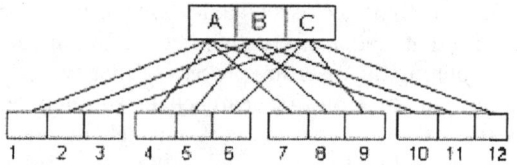

Fig. 3. Division of Children

2.3 Total Decentralization

At first, the load gets divided due to the presence of multiple hosts in a given node. Now there is a possibility that even after this division the host is unable to handle the query load. In that case a host from a leaf node (that would be free most of the time) is requested to share the load (the *helper host*). At the same time, the loaded host informs its parent that if the network traffic crosses a certain limit then the extra queries should be sent to the helper host. This limit depends on the capacity of the host's available bandwidth, processing power etc. e.g. if host B can handle at most 10 queries/sec, then on getting overloaded, it requests host X (helper) to share the network load (Fig. 5), and informs the parent (A) to forward extra queries to X.

Now host X also starts acting as a level h+1 host and will forward queries to level h+2 nodes (note that this will not interfere in the EMST key distribution). This help will also relieve the hosts at the root node i.e., the leader hosts. This decentralization also assures that the network does not have a single point of failure.

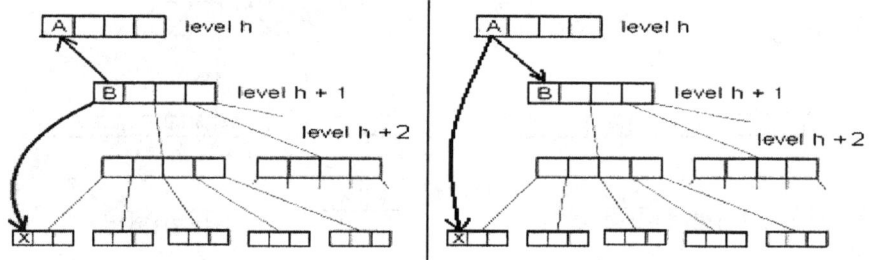

Fig. 4. B informs its parent about X and also sends its address book to X

The above procedure can be carried out again until the traffic load on hosts becomes bearable. This provision also gives a liberty to the user about how much bandwidth (above a certain minimum) does he want to allocate for network service.

2.4 Query Replication

A host makes a query by sending it to more than one host in the leader node of its sub-network. Now each host forwards this query to its child in the relevant node. In this way, the query gets passed to the relevant child node but to multiple hosts in the same node and in this way it reaches the node which contains the host being searched.

Consider an example (Fig. 5): host X (91) generates a query for '25'. First it sends his query to three hosts (23, 46 and 59) of the sub-network leader. These hosts then find the appropriate child node (25 lies between 23 and 46 – node no. 2) and pass it to their respective children in that node. Again these hosts pass it to the relevant child node. Finally, when the query reaches the destination node, brothers 27, 26 pass it to 25. The above scheme shows that the query fails only when at least one host on each of the paths fail simultaneously. This mechanism greatly reduces the probability (detailed analysis in Section 4.4) of a fault.

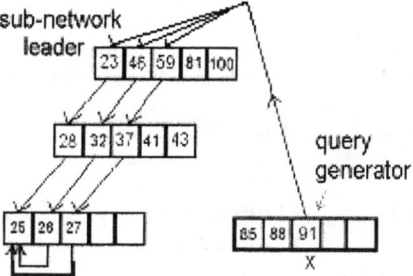

Fig. 5. The flow of a query. 91 originates a query for 25 which follows the above path.

3 The Oasis Protocol

In this section, we discuss the various algorithms and the entire procedures of insertion, deletion and routing in Oasis.

3.1 Host Insertion

Whenever a new host, 'H' joins the network, the bootstrap provides it with an address of any sub-network leader ('nxtldr'). Host H keeps on checking its physical distance from 'nxtldr' and if it finds a suitable leader('suitableldr') i.e. a leader with physical distance less than the distance threshold('dist_thresh') it inserts into that sub-network. In case there is no such sub-network then it inserts into the super-network forming a new sub-network without any sub-nodes. Finally, after forming a new sub-network it informs all other sub-network leaders about its arrival.

When a new host has to be inserted into a sub-network, a query is made for its own key to find its proper position in the EMST which may take O (Log (N)) time (*find-pos*). The node on which the query terminates ('tmnode') informs its cluster leader that a new host has to be inserted. Now the cluster leader sends an invitation to the new host to join as a child or a brother depending on whether the node capacity is full or not respectively. This is done in order to prevent multiple hosts in a node from inviting new hosts at the same time. The join requests are handled by the leader one by one. If the new host H joins as a child, it becomes the cluster leader of its new cluster with one host and stores the address of its parent host. Otherwise, if it joins as a brother, it stores all the information about its node (including cluster leader and addresses of its brothers) and informs all its brothers about its arrival (*inform_arr*).

The following function gives a pseudo code for the mentioned procedure:

```
insert_host (host H) {
  nxtldr = bootstrap.subnetldr;
  suitableldr = NULL;

  while (suitableldr==NULL && nxtldr!=NULL) {
    d = distance(H, nxtldr);
    if (d<dist_thresh) suitableldr =  nxtldr;
    else nxtldr   = nxtsub(nxtldr,H);
  }

  if (suitableldr == NULL) { //no subnet found
    H = new subnetldr;
    H.inform_arr(H, all subnetldrs);
    return;   //insertion process complete
  }
  tmnode = suitableldr.findpos(H.key);
  if (tmnode.full())   {//insert as a child
    H.store(parenthost.info);
    H   = new clusterldr;
  }
  else{         //insert as a brother
    H.store( node.info );
    H.inform_arr(H, all_brothers);
  }
  H.inform_arr(H, parentnode);
}
```

3.2 Host Deletion

When a host 'H' logs off the network, it carries out the following procedure: it informs all the brothers and the parent host about its departure (inform_dep). In case the leaving host is the cluster-leader it appoints a new leader (which is the host with the smallest key). Now the cluster-leader searches for a replacement 'R' for the leaving host 'H' (from a leaf node, no replacement is required if the leaving host is already in a leaf node). The replacement host, R before leaving its old node informs all its relatives about its departure (rearrangement is done at the leaf level, if required). R takes its new address book from H and finally informs its new relatives about its arrival.

3.3 Host Failure

If a host H goes off the network without informing any other host, such a situation is referred to as 'host failure'. In such a situation the host which discovers its failure first, X takes the responsibility of informing all other related hosts. The brothers and the parent of a host ping it at regular intervals so a failure is either discovered by a brother or its parent. Under the first possibility, the brother X informs H's relatives and then the leader of the node. If H was the leader then the host with the smallest key becomes the new cluster leader. If X is the parent of H then it informs all hosts in H's node. Now, the cluster-leader of H's node carries out all the operations as in the case of host deletion (finding a replacement and then giving it all the information about H). The total time that is required from the point of failure to the moment when finally the replacement R informs everyone about its arrival is called recovery period. The fault tolerance of the network is directly dependent on this parameter. The higher the time it takes to replace the failed node higher will be the probability of a query getting lost or stuck somewhere in the path.

```
Host_failure (host H) {
  if ( X.ping(H) == fail) {
    if ( H == X.brother ) {
      X.inform_dep(H, {all brothers, parent node});
      if ( H == X.clusterldr)
      X.clusterldr = X.node->firsthost;

      Host R = X.clusterldr.findreplacement(H);
      R.inform_dep(R, {all brothers, parent host});
      if (R.children()) rearrangement(R);
      R.store ( X.clusterldr.info(H) );
    }

    else    { //H == X.child
      X.inform_dep(H, all children);
      Host R= X.child.clusterldr.findreplacement(H);
      R.inform_dep(R, {all brothers, parent host});
      if(R.children()) rearrangement(R);
      R.store ( X.child.clusterldr.info(H) );
    }
    R.inform_arr(R, {all brothers, parent host});
    R.inform_arr(R, {children});
}
```

3.4 Query

The originator ('Orig') of a query sends it to 'r' (replication factor) hosts in the leader node of its own sub-network. Every host on receiving a query checks its brothers and forwards it to him if his key matches the search otherwise forwards it to the relevant child. While forwarding a query to any host in its child node, a host checks if it has already forwarded more queries than the child's bandwidth limit (the child is loaded). If it is so, it sends the query to the helper host in the leaf node. Otherwise, it will simply forward it to the child host. The query searching mechanism is the same as that in an m-way search tree, but the query proceeds through r parallel paths. This query is first searched in the sub-network and on failing to get a positive response from the sub-network; the leader then forwards the query to the super-network.

```
RecvQuery(Host Orig, key) {
  if(storedkeys(key)==true) {
    sendreply(Orig);
    return;
  }
```

```
   for i = m-1 downto 1
     if(key== brother[i].key ) {
       queryforward(brother[i], Orig, key)
       return;
     }
   i = find_app_child(key);
   if (i ==-1 ) sendreply(Orig);
   if (child[i].traffic_limit() == true)   {
     j = 0;
     while(!child[i].helper[j].traffic_limit()) j++;
     queryforward(child[i].helper[j],Orig,key);
   }
   else queryforward(child[i], Orig, key);
}
```

In the next section we discuss the simulation and performance analysis of Oasis.

4 Simulation

The Oasis simulation software was implemented in C++. We used the following metrics to evaluate Oasis:

1. Data lookup time
2. Path Length
3. Decentralization
4. Fault tolerance in terms of data look up failures

While conducting experiments on the simulation the following parameters were taken into account:

Number of hosts: (N): This is a parameter which shows the scalability of the network. For the analysis we made 128 sub-networks with varying N.

Cluster Size: (m-1): This is a crucial parameter which can significantly affect the performance, especially the path length and consequentially the look up latency. Also, the fault tolerance of the system gets affected by this parameter.

Replication Factor: (r): This factor indicates the number of copies of a query that is originally sent to the sub-network leader.

Threshold (distance_threshold): When the new node joins Oasis, the threshold determines whether the node is inserted to one's sub-network or super-network. In the following simulation, we fixed the threshold at 100ms (the ping interval).

4.1 Data Lookup Time

The data look up time in Oasis comes out to be logarithmic in nature which is as good as other DHT based network schemes (Fig. 6).

The different curves for varying cluster size come out to be a straight line parallel to the x-axis indicating O(logN) complexity of the metric. The look up latency reduces as the cluster size is increased but at the same time the network overhead also increases because of the increased size of the address book and thus higher number of

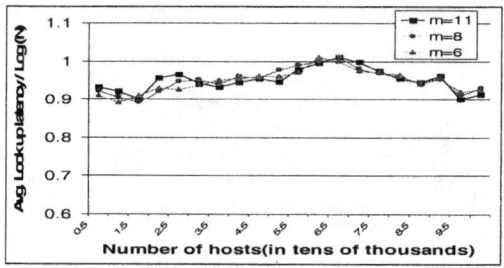

Fig. 6. (Average data look up latency)/Log (N) vs. number of hosts in the network

hosts will have to be communicated with. This leads to a trade off and thus the cluster size can be chosen according to the requirements and capabilities of the network.

4.2 Path Length

Consider a network with 's' no. of sub-networks. Assume that the height of the EMST is 'h' and the queries are uniformly distributed over the network, we have the following average path length in terms of the above parameters.

avg<path_len> = 1/s(avg<path_len> $_{local\ query}$) + (1-1/s)(avg<path_len>$_{local\ query}$ + 1)

A query going into the super-network will have one hop extra for the forwarding between sub-network leaders; this is why we have one added in the expression with (1-1/s). Now, avg<path_len>$_{local\ query}$ = ((m-1) / N) * (1 + 2m + 3m^2 + 4m^3 +..+ hm^{h-1}) = h − 1/(m-1), which implies avg<path_len> = h + 1 -1/(m-1) − 1/s. Now, the total no. of hosts in a complete EMST with a height 'h' is m^h − 1, which is N. Thus, h = Log $_m$ (N + 1). Assuming 1/s to be small,

avg<path_len> = Log $_m$ (N + 1) + 1 -1/(m-1)

Fig. 7a and Fig. 7b show path length characteristics for the network size of 10000 hosts. By increasing the cluster size, the path length of a query reduces significantly. The curve resembles Log (N), the system being basically a network of search trees.

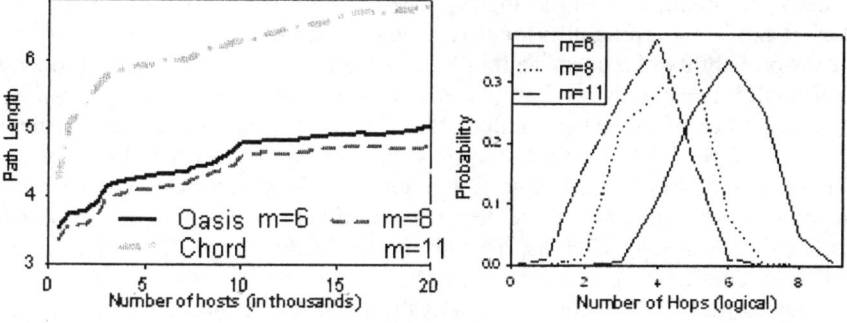

Fig. 7a. The Path Length for Chord and Oasis

Fig. 7b. The Path-Length Probability Distribution with m = 6, m=8 and m=11

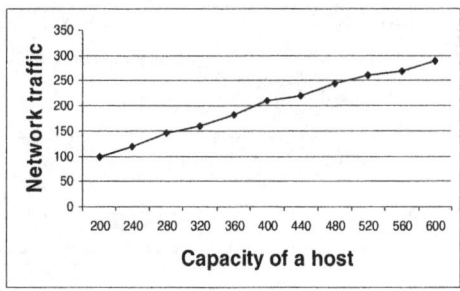

Fig. 8. Query load versus capacity of a host in terms of the bandwidth availability

The above curve also shows that the path length decreases as the cluster size is increased. Also, it is visible that the path length of Oasis is considerably less than that of CHORD. In Grapes, routing within the sub-network does not significantly add to the look up latency but multiple hops in the super-network is costly thus Oasis uses a fully connected super-network.

4.3 Decentralization

The most important and distinctive feature of Oasis is its property of decentralization along with a proper structure for exploiting locality and at the same time giving logarithmic search time. With the concept of a helper it seems quite obvious that no host will have to handle traffic load which is above its capacity. Also the network gets saved from a single point of failure. For N= 10000 and 500,000 uniformly distributed queries, Figure 8 shows a curve between query load (for highly loaded hosts, mostly the sub-network leaders) and the capacity of a host in terms of bandwidth availability.

4.4 Fault Tolerance

Next we evaluated the impact of a massive failure on Oasis's performance and on its ability to perform correct lookups. Once the network becomes stable, each host is made to fail with probability 'f'. We can safely assume that the average path length for a network having N hosts is log(N). Then for a query passing through log(N) number of hosts, the probability of it reaching the destination becomes $(1-f)^{\log(N)}$. Hence the probability of a query getting failed becomes $1 - (1-f)^{\log(N)}$. The probability of a successful query is the probability of at least one query reaching its destination i.e. 1- probability of all getting failed. Probability of a successful query becomes $p = 1 - (1 - (1-f)^{\log(N)})^r$. Table 1 shows the percentage of successful look ups under varying probability of failure, 'f' and at the same time 50000 queries were generated.

Table2 gives the summary of the performance of various existing peer to peer networks together with that of Oasis, d being the no. of dimensions in CAN. It can be seen that the data look up complexity of Oasis is log(N) which is as good as other DHT schemes like CAN, Chord etc. but at the same time exhibits locality property. Grapes, Jelly [8] have the locality property but do not have decentralization and suffer from the problem of a single point of failure, where as Oasis is decentralized and is robust to host failures.

Table 1. Percentage of successful data lookups as a function of the size of the network, the probability of host failure for r = 2 and for r = 3, cluster size (m-1) = 5

Number of hosts	r = 2			r = 3		
	Probability of host failure, f			Probability of host failure, f		
	0.10	0.25	0.50	0.10	0.25	0.50
60000	99.51	94.21	79.67	99.95	98.70	90.98
80000	98.95	94.10	78.14	99.89	98.57	89.78
100000	98.89	93.17	76.82	99.88	98.21	88.84

Table 2. Performance comparison (## : depends on the DHT used like CAN, Chord, Pastry etc)

Network Design	Hops	Locality	Fault Tolerance	Decentralization
CAN	$D(N)^{1/d}$	No	No	Yes
Chord	Log N	No	No	Yes
Pastry	Log N	No	Yes	Yes
Grapes	##	Yes	##	No
Jelly	##	Yes	##	No
Oasis	Log N	Yes	Yes	Yes

5 Conclusion and Future Work

Fault tolerance and decentralization are two important requirements of a peer to peer network. In this paper we have proposed a self organizing hierarchical topology based network which exploits the proximity between hosts without any centralized support of a single host and also provides fault tolerance through query replication. Geographically closer hosts form the sub-network. We propose the concept of an enhanced m-way search tree (EMST) for constructing the sub-network. Use of multiple hosts at each node, distributes the network load between hosts and hence an appreciable degree of decentralization is achieved. Further, the concept of helper also ensures total decentralization among participant hosts. Also, query replication and its passage through different paths results in a high degree of fault tolerance. We are considering designing an adaptive network hierarchy with reduced overhead and higher flexibility in terms of the size of the cluster and intra-cluster communication.

Acknowledgments

The authors would like to thank Dr (Mrs). K. Garg without the help of whom this work would have been very difficult. They also thank Dr. Manoj Misra and Dr. Sumit Gupta for their valuable guidance. Finally, they also thank the anonymous reviewers whose suggestions benefited the paper.

References

1. Gnutella http://www.gnutella.com
2. Napster http://www.napster.com
3. S. Ratnasamy, P. Francis, M. Handley, R. Karp, and S. Schenker: A Scalable Content-Addressable Network. Proceedings of SIGCOMM 2001, ACM
4. I. Stoica, R. Morris, D. Karger, M F. Kaashoek and H. Balakrkshnan: Chord: A Scalable-Peer-to-peer Lookup Service for Internet Applications. Proceedings of SIGCOMM 2001, ACM
5. A. Rowstron and P. Druschel: Pastry: Scalable Distributed Object Location and routing for Large-scale Peer-to-peer Systems. Proceedings of IFIP/ACM Middleware (2001)
6. B. Y. Zhao, J. D. Kuibiatowicz, and A. D. Joseph: Tapestry: An Infrastructures for Fault-tolerant Wide-area Location and Routing. Technical Report UCB/CSD-01-1141, UC Berkeley, EECS (2001)
7. K. Shin, S. Lee, G. Lim, H. Yoon, and J. S. Ma: Grapes: Topology-based Hierarchical Virtual Network for Peer-to-peer Lookup Services. Proceedings of the International Conference on Parallel Processing Workshops (ICPPW' 02)
8. R. Hsiao and S. Wang: Jelly: A Dynamic Hierarchical P2P Overlay Network with Load Balance and Locality. Proceedings of the 24th International Conference on Distributed Computing Systems Workshops (ICDCSW'04)

GToS: Examining the Role of Overlay Topology on System Performance Improvement

Xinli Huang, Yin Li, and Fanyuan Ma

Department of Computer Science and Engineering,
Shanghai Jiao Tong University, Shanghai, P.R. China 200030
{huang-xl, liyin, ma-fy}@cs.sjtu.edu.cn

Abstract. Gnutella's notoriously poor scaling led some to propose distributed hash table solutions to the wide-area file search problem. Contrary to that trend, in this paper, we advocate retaining Gnutella's simplicity while proposing *GToS*, a *G*nutella-like *T*opology-*o*riented *S*earch protocol for high-performance distributed file sharing, by examining the role of overlay topology on system performance improvement. Building upon prior research [10], we propose several modifications as enhancements and then refine these novel ideas, with the aim of trying to remedy the "mismatch" between the logical overlay topology and its projection on the underlying network. We test our design through extensive simulations and the results show a significant system performance improvement.

1 Introduction

1.1 Motivations

The most dominant application currently in use on peer-to-peer (P2P) networks is still large-scale distributed file sharing [1], and such systems are usually designed as *unstructured* networks (e.g., BearShare, LimeWire based on Gnutella [2], Kazaa based on FastTrack [3]). Unlike *structured* P2P networks (e.g., Chord [4], CAN [5], Pastry [6], and Tapestry [7]) where both the data placement and the overlay topology are tightly controlled, unstructured P2P systems do not have any association between the content and the location where it is stored, thereby eliminating the complexity of maintaining such an association in a dynamic scenario, adapt well to the transient activity of peers with very little management overhead, and allow users to perform more elaborate queries. These properties make such systems more suitable for applications of large-scale distributed file sharing. A major limitation and also the key challenging open-question of current unstructured P2P systems lie, however, in their "blind" and constrained broadcast search algorithms, which results in fatal scaling problems in two important ways: first, poor search performance, and second, heavy traffic load of underlying networks. The main difficulty in designing such algorithms is that currently, very little is known about the nature of the network topology on which these algorithms would be operating [8]. The end result is that even simple protocols, as in the case of Gnutella, result in complex interactions that directly affect the overall system's performance.

In this paper, we focus on Gnutella-like decentralized and unstructured P2P file-sharing systems. The main objective of this work is to develop techniques to render the search process more efficient and scalable with high network utilization, by examining the role of overlay topology on the performance improvement of such systems.

1.2 Overview and Contributions

In an earlier paper [10], we present ToA^3, a novel P2P file-sharing system, focusing on surmounting the limitation of Gnutella-like unstructured P2P networks by utilizing topology-oriented adaptability, availability and underlying-network-awareness. In our current work, we refine those ideas and present an extended design (which we call *GToS*) by incorporating several significant modifications as enhancements.

While GToS does build on these previous contributions, it is, to our knowledge, the first open design that (a) recognizes the intrinsic topological properties, like small-world characteristics and power-law degree distributions [8, 9], and further more adapts its protocols to account for these properties, (b) considerss the viewpoints on how to remedy the mismatch between the logical overlay topology and its projection on the underlying network, (c) differentiates the proximity of neighbor nodes and applies different search strategies on them, (d) takes into account not only the search process but also the large-sized file download process, and most importantly, (f) deliberately synergizes these various design features to achieve total system performance improvements.

1.3 Paper Organization

The rest of this paper is organized as follows: we discuss related work in Section 2, some significant inspirations and guidelines from Gnutella topology in Section 3. Based on this knowledge, we then detail the GToS design in Section 4. Section 5 describes the methodology used for the evaluation of GToS, and the simulation results. Finally, we conclude the paper and outline our future work in the last section.

2 Related Work

There have been numerous attempts to leverage aspects of the Gnutella design [1]. The authors in [11] reported, perhaps a little too bluntly, that the fixed "TTL-based mechanism does not work". They argued that by making better use of the more powerful peers, Gnutella's scalability issues could be alleviated. Instead of its flooding mechanism, they used *random walks*. Their preliminary design to bias random walks towards high capacity nodes did not go as far as the ultra-peer proposals in that the indexes did not move to the high capacity nodes. Adamic et al. in [12] suggested that the random walk searches be directed to nodes with higher degree, that is, with larger numbers of inter-peer connections. They assumed that higher-degree peers are also capable of higher query throughputs. However without some balancing design rule, such peers would be swamped with the entire P2P signaling traffic. In addition to the above approaches, there is the "*directed breadth-first*" algorithm [13]. It forwards queries within a subset of peers selected according to heuristics on previous performance, like the number of successful query results.

Another algorithm, called *probabilistic flooding* [14], has been modeled using percolation theory. The authors in [15], propose *Gia*, a P2P file-sharing system extended from Gnutella, by focusing on strong guarantees of the congruence between high-capacity nodes and high-degree nodes. But they do not consider neighbors' proximity in underlying networks and assume that high-degree nodes certainly process high capacity and be more stable than the average, which is in fact not the truth in highly dynamic and transient scenario of P2P networks. In [16], the authors introduce *Acquaintances* to build interest-based communities in Gnutella through dynamically adapting the overlay topology based on query patterns and results of preceding searches. Such a design, because of no feasible measures to limit the explosive increase of node degree, could quickly become divided into several disconnected sub-networks with disjoint interests. The authors in [17], explore various policies for peer selection in the *GUESS* protocol, and conclude that a "most results" policy gives the best balance of robustness and efficiency. However, they only concentrated on the static network scenario.

In summary, these Gnutella-related investigations are characterized by a bias for high degree peers and very short directed query paths, a disdain for flooding, and concern about excessive load on the "better" peers. Generally, the analysis and utilization of intrinsic topological properties for dynamic networks remains open.

3 Inspirations and Guidelines from Gnutella Topologies

We develop this section by introducing the following three questions and then exploring the answers to them step by step:

1. What are the intrinsic properties stemmed in the topologies of Gnutella?
2. What kind of inspirations can be taken from the impacts of such properties on behaviors and performance of these networks?
3. How can these inspirations guide us for the design of GToS?

Many studies, through modeling and network simulations, verify the existence of such intrinsic properties of Gnutella-like topologies as: (a) "small-world" properties, (b) power-law degree distributions [8, 9], (c) heterogeneity and hierarchy that arise entirely from the nature of degree distributions [18], and (d) a significant mismatch between logical overlay and its projection on the underlying network [19].

The existence of the above topological properties in Gnutella-like P2P networks presents significant inspirations for us when designing new, more efficient and scalable application-level protocols.

First of all, dynamical systems with small-world coupling display enhanced signal-propagation speed, computational power, and synchronization, which provide useful cues for efficient navigation of distributed algorithms such as routing and searching in large-scale information networks.

Second, power-law degree distributions play a crucial role in the effectiveness of searching. The basic principle behind the discovery of short paths is that in such a graph the expected degree of a node following an edge is much larger than the average degree, which means most nodes are connected to a few high-degree nodes and whereby have many second neighbors. Most of the second neighbors would be local in a small range but a finite fraction would be randomly distributed throughout the network. Since there would be so many second neighbors, with high probability one of those randomly placed ones would be located close to the target [20].

Third, the "mismatch" between Gnutella logical overlay and its projection on the underlying network indicates that, when building desirable topologies, it is a beneficial idea to take into account the nature of underlying-network-awareness.

Finally but not the least, due to the hierarchical nature and heterogeneity in Gnutella, queries in search process should be forwarded towards deliberately-chosen neighbors. That means, a more intelligent neighbor selection strategy is also a must.

The above inspirations taken from intrinsic topological properties provide us with several significant guidelines as the design rationale for GToS:

1. *Algorithm design should be topology-oriented.* This guideline is on the level of overlay topology. The topologies with desirable properties should be the ones that possess low diameter, large clustering, and are constructed obeying power-law distributions using just degree-focused local knowledge.
2. *Message duplication should be minimized.* This guideline is on the level of search mechanism. Duplicated receiving and forwarding of messages makes major overhead in flooding-based search [11]. In this sense, the key to scalable searches in unstructured networks is to cover the right number of nodes as quickly as possible and with as little overhead as possible. As for small-world-like topologies, the right nodes may mean the next neighbors on the characteristic path. As for high heterogeneity and dynamism, the right nodes should be identified as those with high availability, not just those with high capacity. Besides, adaptive termination is also very important.
3. *Being underlying-network-aware.* This guideline is on the level of underlying network. A proper search algorithm, if being aware of physical network, can speed up query process and improve network utilization without much reducing the success rate.

4 GToS Design

We begin this section with a brief introduction to our previously proposed ToA^3 system first and then present the key components of GToS, focusing on the enhancements extended for ToA^3.

4.1 A Brief Introduction to ToA^3

ToA^3 is a novel P2P file-sharing system [10], built upon Gnutella-like unstructured overlay networks. The key idea of ToA^3 is to generate an overlay topology with *desirable* properties, adapt peers towards *better* neighbors dynamically, and direct queries towards *right* next hops with as few duplicated messages as possible. To achieve this goal, ToA^3 introduces several innovative techniques such as: (a) a dynamic topology adaptation algorithm with self-sustaining power-law degree distributions, (b) simply but efficient utilization of peer-to-peer network heterogeneity, (c) a proper implementation of the underlying-network-awareness, and (d) Smart Search, a biased search algorithm designed special for ToA^3.

4.2 Dytopa

As an extended and enhanced version, our GToS in this work mainly consists of three key components: (a) *Dytopa*—an extended dynamic topology adaptation protocol, (b) *SSplus*—an enhanced search algorithm coupled with several novel mechanisms for optimizations, and (c) *BigDownload*—a unique solution designed for large-sized file download process.

Dytopa is the core component that connects the GToS node to the rest of the network. Building upon the prior inspirations and guidelines, we then focus on constructing topologies with desirable properties by introducing novel techniques detailed as follows.

1. *Self-sustaining power-law degree distribution and its resultant small-world properties.* We prefer keeping such distributions and utilizing their resultant "small-world" properties. We achieve the goal by adding and deleting links in a way that the out-degree of each node is conserved (see Fig.1). We choose a node A at random, build a link from this node to a new node B chosen by a certain metric, and then immediately delete an existing link say with C to conserve links at A. By increasing the fraction of links rewired we get the required low diameter: If the fraction of links deleted and rewired is p, then for very small p the average path length $L(p)$ comes down by orders of magnitude and is close to that of a random graph whereas the clustering coefficient $C(p)$ is still much large similar to that of a regular graph [20]. This is just what we desire: "small-world" properties.

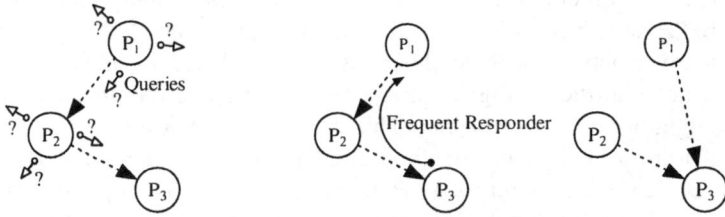

Fig. 1. The way a GToS node self-sustains its out-degree during dynamic topologic changes

2. *Proximity-based neighbor classification.* To realize the underlying-network-awareness, links of a GToS node to its neighbors are divided into two categories: short links and long links. The fraction of links that are short, called the proximity factor α, is a key design parameter that governs the overall structure of the topology. A node with out-degree d has αd short links and $(1-\alpha)d$ long links. α takes values from 0 to 1, inclusive: $\alpha=0$ corresponds to all-long-links (like a random graph) and $\alpha=1$ corresponds to all-short-links (like a regular graph). Different values of α let us span the spectrum of this class of overlay topologies. In between these two ends of the spectrum, we foresee that the topologies, with many short links and few long links, have desirable properties: they not only have low diameter, large search space and connectedness, but are also

aware of the underlying network. We aim to find a suitable balance between these advantages by simulation through populating the range of α value. An appropriate metric for distance (e.g., latency used in GToS) in the underlying network defines the closeness δ of neighbors. Given the dynamic conditions of P2P networks, nodes periodically evaluate the distance to their neighbors and replace them if necessary to maintain the invariant ratio of short/long links. Besides, we introduce α for another purpose: deploying biased searches for two kinds of neighbors respectively to obtain further performance improvement (to be addressed in Section 4.3).

3. *Availability-based better neighbor selection strategy.* We prefer high-availability as a proper measure of better neighbors, different from (but much better than) just the high node capacity that is used in [15]. The availability can characterize P2P network dynamics and heterogeneity more accurately than just node capacity [1], and we propose *MaxDocRtd* as a proper metric for high-availability in the GToS design. A *MaxDocRtd* node is defined as the responder that has returned the maximum relevant results most frequently in the near past. Indeed, a peer that has consistently and frequently returned good results is actually the most available node to the requester and is likely to serve a large number of files it requires in the future. Moreover, the metric of *MaxDocRtd* can also help to realize the locality in interest-based semantic naturally, which is really a positive by-product.

4. *Dynamic neighborhood maintenance based on lease and migration.* The greedy fashion of high-availability-based neighbors selection and replacement may result in such a problem: according to the above rules, if more and more queries issued by P are successfully responded by long-range but high-availability nodes, many existing local neighbors will be replaced by these remote ones, which means that the average diameter (with respect to the physical proximity) of P's neighborhood is increasing. This case is not what we desired. As a remedy, we propose a dynamic neighborhood maintenance strategy based on the concept of lease and migration. P re-computes the average diameter of its neighborhood at regular intervals of time. The interval between successive recomputations is a tunable parameter T. whenever the recomputed average diameter of P's neighborhood increases beyond a pre-configured threshold Δ, P chooses one of its neighbors (say L) and tags it with a *lease*, a random number drawn uniformly from $[T, 2T]$. Then all messages that pass through P, including both the incoming and the outgoing, are migrated to L. L contacts P only when its lease expires. At that time, it informs P about the changes of the physical proximity. If achieving a satisfied gain, P assigns it with a new lease; otherwise, P takes over its job and looks for another target in case the mentioned situation continues.

The above four techniques designed for Dytopa is to ensure that high-availability nodes are indeed selected as better neighbors and that the neighborhood of a peer should evolve itself towards underlying-network-awareness. Below we give the pseudo code of Dytopa, showing how to achieve these goals.

```
Variables:
    NbrsList: Ordered list of neighbors, ordered by Я
    CandList: List of candidate neighbors, ordered by Я
    short_NbrsList, long_NbrsList: List of short/long neighbors
    Я(P): the availability of node P, measured by the number of rele-
          vant results returned successfully by P in the near past
    α: Proximity factor, the fraction of links that are short
    β: Aging factor, with value in (0,1)
    δ: closeness between two nodes in the underlying network
    T: the interval of time, used for dynamic neighborhood maintenance

// Upon a successful query from the requester Pr answered by Pa
WHILE (min(Я(Pi, ∀Pi∈NbrsList)) < max(Я(Pj, ∀Pj∈CandList))) DO
    {NbrsList←cand_node_max; CandList←nbr_node_min}
age all nodes in NbrsList and CandList by a factor β;
Я(Pa) ++;
IF (Pa ∈ NbrsList)        // Pa is an existing neighbor
    do nothing; return;
IF (Я(Pa) > min(Я(Pi, ∀Pi∈NbrsList)))//Pa is a candidate or a new node
    {NbrsList←Pa; CandList←nbr_node_min;
     examine whether needs dynamic neighborhood maintenance;
    }
ELSE
IF (Pa ∉ CandList)
    {CandList←Pa; return}

// Upon a neighbor, say Py, leaving the network
IF (CandList != ∅)
    {NbrsList←cand_node_max;
     examine whether needs dynamic neighborhood maintenance;
    }
ELSE
initiate K peers in CandList randomly by means of existing neighbors;
enforce a neighbor randomly chosen from CandList;

// Ranking nodes in NbrsList by δ incrementally, build short_NbrsList
// and long_NbrsList by α for further utilization by SSplus
short_NbrsList←first α·N peers of all the N nodes in NbrsList;
long_NbrsList←the remaining peers of NbrsList;
```

4.3 SSplus

The deliberate combination of availability-focused better neighbor selection (whereby peers take more available and more relevant nodes as neighbors) and proximity-based neighbors classification (whereby the system is aware of the underlying network) ensure that increasing requests can be answered by neighbor nodes or by their nearby nodes on the overlay, and that many such answerers may be close to the requester. Based on such a design, SSplus conducts a bi-forked and directed search strategy as follows: rather than forwarding incoming queries to all neighbors (the typical way of Gnutella) or randomly chosen neighbors (the way of random walks), the algorithm forwards the query to:

1. all short neighbors using scoped-flooding with a much smaller *TTL* value;
2. *k* long neighbors using random walks coupled with the mechanisms of adaptive termination-checking and duplication-avoiding.

In order to further improve the search efficiency and the network utilization, we also incorporate into SSplus a novel load balancing solution based on free availability of nodes and an intelligent 2-level replication scheme, addressed as follows.

A Novel Load Balancing Solution based on Free Availability. In our previously proposed ToA^3 system, a peer that has many neighbors could quickly become a hot-spot, not only because it receives more queries, but also because it typically sends more files to requesting peers. To avoid overloading these nodes, we use the following mechanism to better balance the traffic load. Before successfully answering a query, a peer first checks whether any of its neighbors also possesses the queried file. If YES, it delegates the responsibility for answering the query to the peer among those serving the file that has the *highest free availability*. Otherwise, it sends the file itself. Then the question is how to identify free availability of a node? In the SSplus algorithm, the free availability of a node is denoted as the remaining number of queries it can still process and is provided by the node itself as a variable observed by other peers. Based on the design principles of GToS in this paper, there is a good probability that some of the neighbors of a peer also have the same files. Therefore, we force the less loaded peer to assume part of the load.

An Intelligent 2-Level Replication Scheme. To improve the search efficiency, we also introduce a novel intelligent replication scheme into the SSplus algorithm. Each GToS node actively maintains an index of the content of each of its neighbors. These indices are exchanged when neighbors establish connections to each other, and periodically updated with any incremental changes. Thus, when a node receives a query, it can respond not only with matches from its own content, but also provide matches from the content offered by all of its neighbors. When a neighbor is lost, either because it leaves the system, or due to topology adaptation, the index information for that neighbor gets flushed. This ensures that all index information remains mostly up-to-date and consistent throughout the lifetime of the node. It should be noted that this kind of replication is just at the level of *index* of files, not the files themselves. That means the download process for popular files may still overload the provider of these files if this provider is not the node with high availability. To make high-availability peers surely store more files, especially more popular files, we then introduce another kind of replication scheme that is at the level of *content* of files themselves (rather than simple pointers to files) [15]. In the SSplus algorithm, this is implemented in an on-demand fashion where the high-availability nodes replicate content only when they receive a query and a corresponding download request for that content.

4.4 BigDownload

If all of the above efforts we made could really solve the insurmountable scaling problems of Gnutella-like unstructured P2P file-sharing systems, we conjecture that the next bottleneck limiting scalability is likely to be the *file download process*. This will be particularly true if, as recent measurement studies indicate, increasing files in networks are large-sized (e.g., multimedia files) [21]. This situation also underscores the significance of distributed multimedia sharing applications. In order to take into account this factor, we couple the GToS system with another unique technique named *BigDownload* based on mechanisms of resources booking and reservation. It should be noted that, although this technique is mainly related to the file download process, it can also contribute significantly to improving the success rate of search, as well as the acceptance rate of incoming queries. In most proposed Gnutella flow-control mechanisms [22], which are reactive in nature: receivers drop packets when they start to become overloaded; senders can infer the likelihood that a neighbor will drop packets based on responses that they receive from the neighbor, but there is no explicit feedback mechanism. As a remedy, we advocate that the overloaded receivers respond to the senders via a message like *"Query hit, try to fetch it after an interval τ"* as a delayed but positive confirmation, rather than the above mentioned rejection of just dropping it. To detail the idea in an algorithmic perspective, a node P maintains a data structure variable of *Overloading_Window*, with its size $sizeOW(P)$ set according to P's capacity of processing queries, and its values recording the first $sizeOW(P)$ incoming queries that arrive just after P reaches its capacity limit. In this case, the senders of these queries (named $S_1, S_2, ..., S_{sizeOW(P)}$ for convenience) are considered having booked the availability of P and can access P after an given interval τ_i (increased incrementally from 1 to $sizeOW(P)$). This is what we call, resources booking, which is expected to improve the network utilization. As for the other mechanism resources reservation, once a request for file-download has been accepted, the related resources, such as available network bandwidth, will be kept reserved during the download process, in order to support some kind of QoS (Quality of Service) that is often required in multimedia sharing applications. The detail design of these mechanisms is omitted due to the space limitations.

5 Performance Evaluation

In this section, we use simulations to evaluate GToS, mainly focusing on the performance gains when at the presence and absence of the above proposed modifications and enhancements.

We consider a P2P network made of 4,096 nodes, which corresponds to an average-size Gnutella network [8]. We rely on the *PLOD*, a power-law out-degree algorithm, to generate an overlay topology with desired degree distribution over the P2P network simulator [23]. In the simulations, 1,000 unique files with varying popularity are introduced into the system. Each file has multiple copies stored at different locations chosen at random. The number of copies of a file is proportional to their popularity. The count of file copies is assumed to follow a Zipf distribution with 2,000 copies for the most popular file and 40 copies for the least popular file. The queries that search for these files are also initiated at random hosts on the overlay topology. Again the number of queries for a file is assumed to be proportional to its popularity.

We evaluate GToS by referring to the following four models:

1. *FG*: Search using TTL-limited *F*looding over *G*nutella. This represents the classic Gnutella model.
2. *RR*: Search using *R*andom walks over uniform *R*andom topologies. This represents the recommended search suggested by [11] against the flooding search.
3. ToA^3: using Smart Search on the ToA^3 topologies [6].
4. *GToS*: the protocol suite proposed in this paper, using the Dytopa topology adaptation procedure and the SSplus search algorithm.

We use the following performance metrics for evaluation:

1. *Pr(success)*: defined as the probability of finding the queried object before the search terminates. This is a metric of user aspect.
2. *avg. #msgs per node*: defined as the average number of search messages each node in the P2P network has to process. This is a metric of average load.
3. *D* and *stress*: D is defined as the average distance in the underlying network to the nearest results, showing whether the protocol is underlying-network-aware; *stress* is one of the most common definitions of traffic load in overlay networks [19], defined as the number of logical links whose mapped paths include the underlying link. These two metrics examine the network utilization.

Fig. 2 plots the success rate of query as a function of the average number of hops needed, showing that both GToS and ToA3 get a much higher success rate than the other two models, with the former performing a little better than the latter. To illustrate the performance gains of our modifications to the search algorithm, we plot Fig.3 and Fig.4, concentrating on the comparisons between *Smart Search* used by ToA3

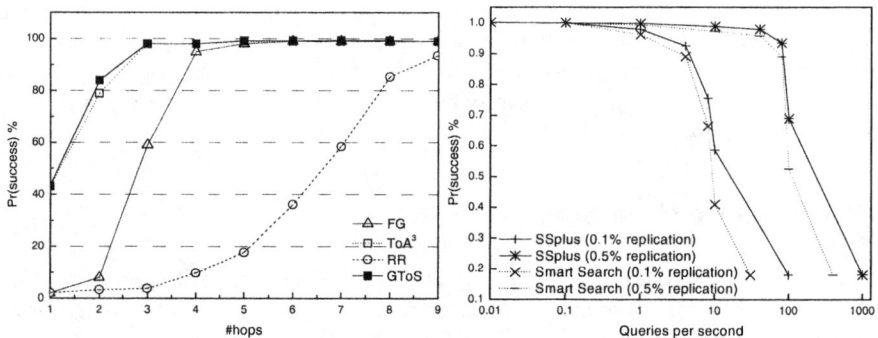

Fig. 2. Success rate of queries as a function of the average hops number

Fig. 3. Success rate as a function of increasing query load

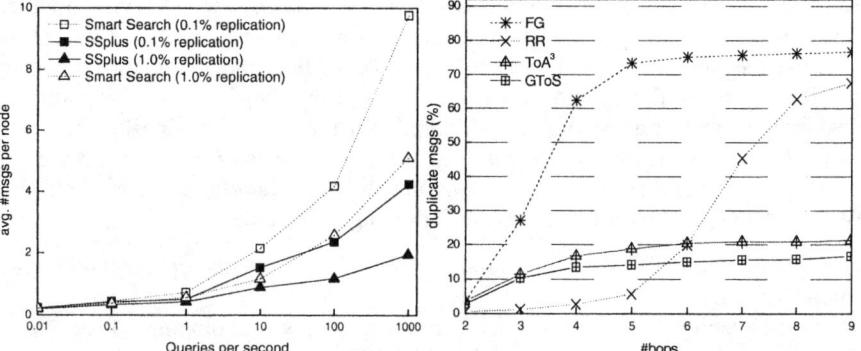

Fig. 4. The average number of messages per node as a function of increasing query load

Fig. 5. The percentage of duplicate messages as a function of the average hops number

and *SSplus* used by GToS. Both the results indicate that, in the case of SSplus over GToS topology, we can achieve a higher success rate of query and distribute the query load more evenly across the network, which also verifies the success of our modifications. In addition, from Fig.5 we can see that, GToS and ToA^3 generate much lower duplicate messages (they are pure overhead!) than the other models, especially after going through several hops. This is mainly due to the intelligent better neighbor selection strategy and the deliberate combination of related optimizations.

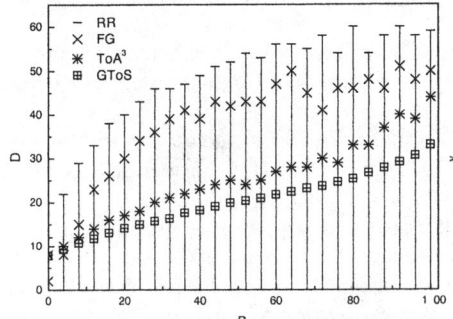 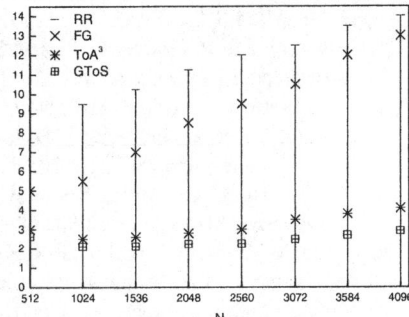

Fig. 6. The distance to search result (D) as a function of variable file popularities (P)

Fig. 7. The variation of mean stress (v) as a function of increasing node population (N)

As for the aspect of the network utilization, we can see from both Fig.6 and Fig.7 that our solution can make better use of the knowledge of underlying network, by dynamically optimizing the neighborhood quality to reduce the distance to search result, and by mapping more logical links to local physical links. These results further verify the significant performance gains of our solution.

6 Conclusions

In this paper, we propose *GToS*, a *G*nutella-like *T*opology-*o*riented distributed *S*earch protocol, by extending our previously proposed *ToA^3* protocol to include several novel techniques for optimizations and enhancements, with the aim of trying to remedy the "mismatch" between the logical overlay topology and its projection on the underlying network. Our simulations suggest that these modifications provide significant performance gains in both the search efficiency and the network utilization: while making search process much more scalable, the design also has the potential to improve the system's file download process by more fully distributing the load. In addition, the improved performance is not due to any single design innovation, but is the result of the synergy of various modifications. Further optimizations to the Dytopa procedure and the SSplus algorithm, such as the considerations of query resilience and more intelligent replication strategies, are orthogonal to our techniques and could thus be used to improve the system performance of GToS.

References

1. J. Risson et al, "Survey of Research towards Robust Peer-to-Peer Networks: Search Methods", Technical Report UNSW-EE-P2P-1-1, University of New South Wales, 2004
2. "http://rfc-gnutella.sourceforge.net"
3. "http://www.fasttrack.nu"
4. I. Stoica, R. Morris, D. Karger, M. Kaashoek, and H. Balakrishnan, "Chord: A scalable peer-to-peer lookup service for internet applications," in ACM SIGCOMM, Aug. 2001
5. S. Ratnasamy, M. Handley, R. Karp, and S. Shenker, "A scalable content addressable network," in ACM SIGCOMM, Aug. 2001
6. A. Rowstron and P. Druschel, "Pastry: Scalable, distributed object location and routing for large-scale peer-to-peer systems," in IFIP/ACM International Conference on Distributed Systems Platforms (Middleware), Nov. 2001
7. B.Y. Zhao, J. Kubiatowicz, and A.D. Joseph, "Tapestry: An infrastructure for fault-tolerant wide-area location and routing," Tech. Rep. UCB/CSD-01-1141, Computer Science Division, University of California, Berkeley, Apr. 2001
8. M.A. Jovanovic, F.S. Annexstein. "Scalability issues in large peer-to-peer networks - a case study of Gnutella". Technical Report, University of Cincinnati, 2001
9. M.A. Jovanovic, F.S. Annexstein, and K.A. Berman. "Modeling Peer-to-Peer Network Topologies through Small-World Models and Power Laws", in Proc. of IX Telecommunications Forum Telfor, Belgrade, November 2001
10. X. Huang, Y. Li, F. Liu, and F. Ma. "ToA^3: Beyond the Limit of Unstructured P2P Networks", to appear in Proc. of ICAS&ICNS'2005, Tahiti, French Polynesia, Oct. 2005
11. Q. Ly, P. Cao, E. Cohen, K. Li, and S. Shenker. "Search and replication in unstructured peer to peer networks", In Proc. of the 16th international conference on super-computing, Jun. 2002
12. L. Adamic, R. Lukose, A. Puniyani and B. Huberman. "Search in power-law networks", Physical review E, The American Physical Society 64(046135), 2001
13. B. Yang and H. Garcia-Molina. "Efficient Search in Peer-to-Peer Networks", in Proc. of the 22nd International Conference on Distributed Computing Systems, Vienna, July 2002
14. F. Banaei-Kashaniand C. Shahabi. "Criticality-based analysis and design of unstructured peer-to-peer networks as 'complex systems'", in Proceedings of the 3rd IEEE/ACM International Symposium on Cluster Computing and the Grid: pp351-358, 2003
15. Y. Chawathe, S. Ratnasamy, L. Breslau, N. Lanham, and L. Breslau, "Making Gnutella-like P2P systems scalable", in ACM SIGCOMM, Aug. 2003
16. V. Cholvi, P. Felber, and E.W. Biersack, "Efficient Search in Unstructured Peer-to-Peer Net-works", in European Transactions on Telecommunications, Special Issue on P2P Networking and P2P Services, Volume 15, Issue 6, 2004
17. B. Yang, P. Vinograd and H. Garcia-Molina. "Evaluating GUESS and Non-Forwarding Peer-to-Peer Search", The 24th International Conference on Distributed Computing Systems (ICDCS 2004), Tokyo University of Technology, Hachioji, Tokyo, Japan, Mar. 2004
18. H. Tangmunarunkit, et al. "Network Topologies, Power Laws, and Hierarchy", Tech Report USC-CS-01-746
19. M. Ripeanu, et al, "Mapping the Gnutella Network: Properties of Large Scale Peer-to-Peer Systems and Implications for System Design", IEEE J. on Internet Computing, 2002
20. A.R. Puniyani, R.M. Lukose, and B.A. Huberman, "Intentional Walks on Scale Free Small Worlds", Technical paper, http://arXiv.org/abs/cond-mat/0107212

21. S. Saroiu, K.P. Gummadi, R.J. Dunn, S.D. Gribble, and H.M. Levy, "An Analysis of Internet Content Delivery Systems", In Proc. of the Fifth Symposium on Operating Systems Design and Implementation, Boston, MA, Dec. 2002
22. S. Osokine, "The Flow Control Algorithm for the Distributed Broadcast-Route Networks with Reliable Transport Links", http://www.grouter.net/gnutella/flowcntl.htm, 2001
23. C.R. Palmer and J.G. Steffan, "Generating Network Topologies That Obey Powers", in Proc. of Globecom'2000, San Francisco, November 2000

Churn Resilience of Peer-to-Peer Group Membership: A Performance Analysis*

Roberto Baldoni, Adnan Noor Mian, Sirio Scipioni, and Sara Tucci-Piergiovanni

DIS, Universitá di Roma *La Sapienza*, Via Salaria 113, Roma, Italia

Abstract. Partitioning is one of the main problems in p2p group membership. This problem rises when failures and dynamics of peer participation, or churn, occur in the overlay topology created by a group membership protocol connecting the group of peers. Solutions based on Gossip-based Group Membership (GGM) cope well with the failures while suffer from network dynamics. This paper shows a performance evaluation of SCAMP, one of the most interesting GGM protocol. The analysis points out that the probability of partitioning of the overlay topology created by SCAMP increases with the churn rate. We also compare SCAMP with DET – another membership protocol that deterministically avoids partitions of the overlay. The comparison points out an interesting trade-off between (i) reliability, in terms of guaranteeing overlay connectivity at any churn rate, and (ii) scalability in terms of creating scalable overlay topologies where latencies experienced by a peer during join and leave operations do not increase linearly with the number of peers in the group.

1 Introduction

Peer to peer (p2p) systems are rapidly increasing in popularity. Their interest stems from the fact that a peer-to-peer system is a distributed system without any centralized control. Thus, there is no need of a costly infrastructure for direct communication among clients. Another specific characteristic of these systems concern peer participation that is each peer joins and leaves the system at any arbitrary time. Indeed, the dynamics of peer participation, or *churn* (the continuous arrival and departure of nodes) is an inherently property of a p2p system. The peers communicate through application-level multicast protocols over an overlay network formed by the peers themselves [12], [6]. Due to churn, the overlay continuously changes. This implies that the group membership management protocols are crucial to the success of multicasting. Two issues are usually taken into account by such group membership protocols: (i) scalability, that is, the operational overhead will not grow linearly with the size of the network and (ii) reliability which is the capacity to keep the overlay network connected in face of network dynamics.

Epidemic or gossip-based protocols [9], [10] are considered good candidates to cope with the issues of scalability and reliability. However, these kind of protocols emerged in fairly static systems [10] and their behavior in systems with high churn rates has

---

* The work described in this paper was partially supported by the Italian Ministry of Education, University, and Research (MIUR) under the IS-MANET project.

received little attention in the literature. Recently this issue has been addressed in [1] which shows, through analytical model, that the probability of partitioning increases with the increase of churn rate.

The contribution of this paper is in understanding the churn resilience of group membership protocols in p2p systems in terms of reliability and scalability. More specifically we compare SCAMP [10] against DET [2]. SCAMP is one of the most interesting GGM protocols, it is an adaptive GGM in the sense that with the changing the size of the group, it maintains a reasonable overhead for each node and a certain degree of reliability. DET is a protocol which deterministically maintains the overlay connectivity assuming that a certain threshold of failures holds. In the experimental comparison we evaluate (i) reliability by calculating the proportion of nodes reached by an application-level multicast, and (ii) scalability by analyzing the overlay topology w.r.t. join/leave latencies. Experimental results confirm that SCAMP suffers from churn in terms of reliability, while it scales for any churn rates. Specifically, under a churn rate equal to 1 membership change (a join or a leave) per second the proportion of nodes reached by the multicast is the 80%. However, the node degree (for every node) remains always equal to the logarithmic size of the group. The analysis of DET shows that the overlay remains connected for any churn rate. This determinism is at the cost of an increase of latency for join and leave operations (note that these operations are instantaneous in SCAMP). In particular, it is shown that if either (i) DET protocol adopts policies to maintain low latencies for join/leave operations, or (ii) the churn rate increases, then the overlay converges to a star topology, thus resulting in an overloading of one node completely – this means that scalability is compromised in any case.

The paper in Section 2 presents a brief description of SCAMP and DET protocols. In Section 3 experimental results are shown. Section 4 discusses the related work and Section 5 concludes the paper.

2 Group Membership Protocols

In the following we briefly describe the two protocols, namely SCAMP [10] and DET [2], which are evaluated in the simulations. Before that let us introduce the system model.

2.1 System Model

The system consists of an unbounded set of nodes Π (Π is finite). Any node may fail either by crashing or by leaving the system without using the defined protocol. A node that never fails is correct. The system is asynchronous: there is no global clock and there is no timing assumption on node scheduling and message transfer delays. Each pair of nodes p_i, p_j may communicate along point-to-point *unidirectional fair lossy links*[4].

Each node $p_i \in \Pi$ may subscribe (join) and unsubscribe (leave) from the group G. The set of nodes constituting the group G at a certain point of time is a subset of Π with size unbounded and finite. The rules defining the membership of G are the following: (i) a node $p \in \Pi$ becomes a member of G immediately after the completion of the subscription operation, (ii) a node p ceases to be member of G immediately after the completion of the unsubscription operation.

2.2 The SCAMP Probabilistic Protocol [10]

Scamp is a gossip-based protocol, which is fully decentralized and provides each node with a partial view of the membership. It is adaptive w.r.t. a-priori unknown size of the group, by resizing partial views when necessary.

Data Structures. Each node maintains two lists, a PartialView of nodes it sends gossip messages to, and an InView of nodes that it receives gossip messages from, namely nodes that contain its node-id in their partial views.

Subscription Algorithm. New nodes join the group by sending a subscription request to an arbitrary member, called a contact. They start with a PartialView consisting of just their contact. When a node receives a new subscription request, it forwards the new node-id to all members of its own PartialView. It also creates c additional copies of the new subscription (c is a design parameter that determines the proportion of failures tolerated) and forwards them to randomly chosen nodes in its PartialView. When a node receives a forwarded subscription, provided the subscription is not already present in its PartialView, it integrates the new subscriber in its PartialView with a probability $p = 1/(1 + sizeof PartialView_n)$. If it doesn't keep the new subscriber, it forwards the subscription to a node randomly chosen from its PartialView. If a node i decides to keep the subscription of node j, it places the id of node j in its PartialView. It also sends a message to node j telling it to keep the node-id of i in its InView.

Unsubscription Algorithm. Assume the unsubscribing node has ordered the id's in its PartialView as $i(1), i(2), ..., i(l)$ and the id's in InView as $j(1), j(2), ..., j(l)$. The unsubscribing node will then inform nodes $j(1), j(2), ..., j(l-c-1)$ to replace its id with $i(1), i(2), ..., i(l-c-1)$ respectively (wrapping around if $(l-c-1) > l$). It will inform nodes $j(l-c), ..., j(l)$ to remove it from their list but without replacing it by any node id.

Recovery from isolation. A node becomes isolated from the graph when all nodes containing its identifier in their PartialViews have either failed or left. In order to reconnect such nodes, a heartbeat mechanism is used. Each node periodically sends heartbeat messages to the nodes in its PartialView. A node that has not received any heartbeat message in a long time resubscribes through an arbitrary node in its PartialView.

Indirection. This mechanism lets new subscriptions to be targeted uniformly at existing members. This is done by forwarding the newcomer's subscription request to a node that is chosen approximately at random among existing members. The interested reader may refer to [10] for further details.

Lease mechanism. Each subscription has been given a finite lifetime called its lease. When a subscription expires, every node holding it in its PartialView removes it from the PartialView. Each node re-subscribes at the time that its subscription expires. Nodes re-subscribe to a member chosen randomly from their PartialView. Re-subscriptions differ from ordinary subscriptions in that the partial view of a re-subscribing node is not modified.

2.3 A Deterministic (DET) Protocol [2]

The DET algorithm deterministically avoids the partition of the overlay. In particular, it provides each member with a partial view of at least $2f + 1$ members, where f is the number of tolerated failures. The other important feature of the algorithm consists in imposing a partial order on nodes to manage concurrent leaves that potentially may cause a partition.

Data Structures. Each node p_i maintains two sets $sponsors_i$ and $sponsored_i$. The union of these two sets is the partial view of the nodes p_i sends messages to. An integer variable $rank_i$ gives an indication of the position of p_i in the overlay, inducing a partial order on nodes. A boolean variable $leaving$ is initialized to $\bot$.

Initialization of the group. A set of nodes $\{p_1, ...p_{3f+1}\} \subseteq \Pi$ totally interconnected and defined in the initialization phase instantiates the group. All these nodes have rank $rank_i = 0$. They are special nodes having the property that they never leave the group.

Subscription Algorithm. Each node p_i joins the group by sending [1] a subscription request to an arbitrary set of members, called $contacts$. When p_i receives $2f + 1$ acknowledgments: (1) p_i includes in $sponsors_i$ all the senders; (2) it sets $rank_i = max(rank_k, \forall senderp_k)+1$. The subscription operation locally returns. When p_i receives a subscription request from p_j and p_i is already a member: (1) p_i inserts p_j in $sponsored_i$; (2) it sends an acknowledgment to p_j along with its own rank $rank_i$. At the end of subscription operation, a newly joined member has $2f + 1$ members around itself. Note that, differently from SCAMP, the newly node becomes a group member only after $2f + 1$ connections to current members have been established.

Unsubscription Algorithm. Each node p_i leaves the group by setting $leaving_i = \top$ and by sending an unsubscription request to $sponsors_i$ along with (i) its own rank $rank_i$ and (ii) nodes is responsible for ($sponsored_i$). When p_i receives a majority of acknowledgments from its sponsors the unsubscription operation locally returns. When p_i receives an unsubscription request from p_j and $rank_i < rank_j$ and $leaving_i = \bot$ (p_i is not concurrently leaving): (1) p_i inserts the nodes p_j that was responsible for in $sponsored_i$; (2) it sends an acknowledgment to p_j and (3) sends a notification to all nodes previously sponsored by p_j to notify that p_j has been replaced by itself. When p_i receives a notification from p_j it replaces the old sponsor with p_j.

3 Simulations

Simulation is conducted by using Ns-2 simulator [14][2]. Let us remark that the aim of the simulation is to evaluate the real impact of the churn (joins and leaves/sec) on the

---

[1] Each message is sent through a fair lossy link, the send primitive embeds a retransmission mechanism that ends to retransmit until an acknowledgment is received. The send primitive is supposed non-blocking.

[2] The choice of Ns-2 was mainly due to the possibility of testing our protocol at the application level by using the full protocol stack. But, also as remarked in [1], due to the exponential nature of the phenomena it was only possible to simulate for small view size and/or high churn rates.

SCAMP and DET behavior. Thus, we conducted simulations in which no failure is simulated but only join and leave.

3.1 Simulation Framework

Each simulation involves a global number of nodes n_{tot}. Each simulation is divided in four intervals: the *bootstrap* interval, the *perturbation* interval, the *transitory* interval and the *measurement* interval.

The bootstrap interval. The bootstrap interval Δ_b is intended as the phase in which the group grows (until a desired value is reached) and no leave occurs. In the bootstrap interval the group starts at time t_0 with n_0 bootstrap nodes. At the end of the bootstrap interval (time t_1) the group contains n_1 nodes. This means that the membership changes in the bootstrap interval consist in n_1 joins.

The perturbation interval and the transitory interval. The perturbation interval Δ_p is intended as the interval in which all membership changes (joins and leaves) are injected in the system. The transitory interval Δ_t is intended as the interval in which all membership changes injected in the perturbation interval take effect. In each simulation the group starts the perturbation interval at t_1 with a number of nodes n_1 obtained after the bootstrapping and it ends the transitory interval at t_3 with a number of nodes $n_f = \frac{1}{2}n_{tot}$. In the perturbation interval we have a total number of leaves equal to $\frac{1}{2}n_{tot}$ and a number of joins equal to $n_{tot} - n_1$.

The measurement interval. In the measurement interval Δ_m all measures are taken. In particular we test for both protocols (i) the proportion of nodes reached by a set of (data) messages sent by each node during the measurement interval, (ii) the average node degree and its distribution, where the node-degree is the number of active connections per-node [3]. The first and second metrics are related to the level of reliability shown by the protocols. Moreover, the second metrics shows the overhead of the protocol. In the case of our protocol we also test the average latency of leaves, i.e. the average time between the leave invocation and the actual departure of the node from the group.

All measures are taken by varying the dynamics rate in the perturbation interval. To characterize the dynamics we use the *churn rate* metrics. The churn rate is the ratio between the number of membership changes, i.e. joins and leaves, and the duration of the perturbation interval. By considering a fixed number of joins and leaves, the churn rate varies by varying the duration of Δ_p. In particular for each simulation Δ_p varies from $5sec$ to $200sec$. Arrivals and up-times follow an exponential distribution.

Simulated Scenarios. All the following simulations have $n_{tot} = 160$. We have compared the two protocols in a scenario in which no bootstrap occurs. The following Table resumes this simulated scenario. In this scenario, at the beginning of Δ_p, the starting node has a partial view which contains only itself.

---

[3] Active connections of a node p_i are intended as the pairs (p_i, p_j) such that in the testing interval p_j is in the group and belongs to the p_i's partial view.

	n_0	n_1	joins during Δ_p	leaves during Δ_p	n_f
$\Delta_b = 0$	-	1	160	80	80

We have also evaluated SCAMP in another scenario to study the impact of bootstrapping. Due to the lack of space the interested reader can find these experimental results in [3] [4].

Each point in the plots has been computed as an average of 40 simulation distinct runs. For each point all the results of these runs were within 4% each other, thus variance is not reported in the plots.

Protocols parameters. As no failure is simulated, we consider for DET $f = 0$ and for SCAMP $c = 0$. Even if we do not consider failures we have implemented for SCAMP a heartbeat mechanism to avoid isolation due to leaves [5]. The heartbeat mechanism we figured out forces a node to re-subscribe if it has not received any heartbeat from its InView in 2.5 seconds. In some plot we have implemented the lease mechanism for SCAMP with a lease duration equal to 50secs.

The determination of the contacts in DET. In this version of DET we use the following mechanism to join the group: each joining node sends a message to a list of *contacts* in which the node with rank 0 is always comprised and the other nodes are arbitrary. This allows to always get an acknowledgment in a short time (from the node with rank 0) when other contacts are not in the group. When the joining node gets more than one acknowledgment it selects as its sponsor the node with highest rank. Since *all* contacted members add the joining node in their partial views even if this node will select only one sponsor among all contacts, extra-messages are needed to purge non-necessary connections [6]. More sophisticated mechanisms can be considered for the join operation (as pointed out in [2]) at the cost of a high latency upon join/leave operations.

The determination of the contacts in SCAMP. For SCAMP the contact is only one and there is no special node that always belongs to the group (as the node of rank 0 in DET). For this reason, in order to augment the probability of finding an active contact we have implemented an extra mechanism in which the joining node broadcasts a message to Π and chooses its contact inside the list of active nodes that have replied to the broadcast. Clearly, for high churn rates this node may choose a contact that has become inactive immediately after the reply. Note that for SCAMP once a subscription is sent, the node is logically a member of the group. Then, an inactive contact is a real problem that affects reliability. Note that even the indirection mechanism does not solve this problem as it is a mechanism that works well in fairly static systems [10].

---

[4] These experiments point out that SCAMP, in the bootstrapping interval, builds a cluster of nodes which are very-well connected. But this cluster remains poorly connected to nodes that join during the perturbation interval. At this point reliability depends on "who leaves the system", i.e. if all the nodes forming the cluster leave, then the reliability of the overlay will be low, leading to partitions and nodes isolations.

[5] Isolation may occur since a contact leaves the system without giving a notice to nodes which joined through it.

[6] Mechanisms to purge non-necessary connections are discussed in [2].

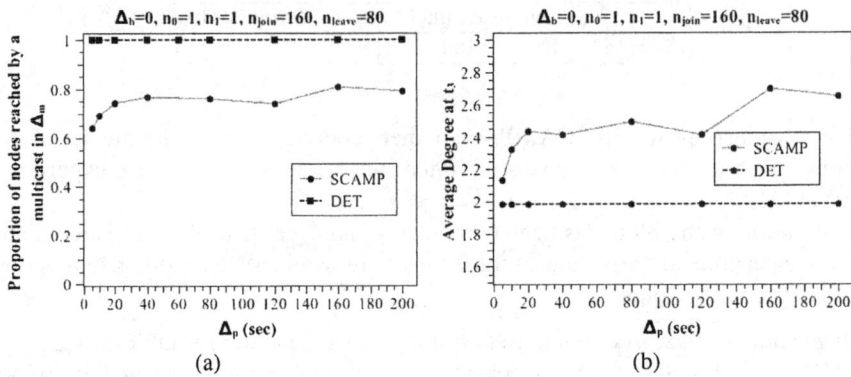

Fig. 1. SCAMP vs DET after a Δ_p with churn rate equal to $240/\Delta_p$

3.2 Experimental Results

Evaluating Reliability of the Topology generated by SCAMP & DET. In the measurement interval the group is freezed in a certain configuration. Thus, members at the beginning of this interval remains in the group till the end of the simulation and no new member is added.

The plot in Fig. 1(a) shows that DET is able to guarantee that each message sent by a group member in the measurement interval is delivered by every group member independently of the churn rate suffered during the perturbation. On the other hand, SCAMP is sensitive to different churn rates suffered in the perturbation interval. In particular, only with churn rates lower than 1.5 membership changes (joins or leaves) per-second the proportion of nodes reached by a multicast is the 80% of nodes. Plot in Fig.1(b) shows as the poor reliability of SCAMP is due to a small average degree (from 2.1 to 2.7). This degree is ever less than the threshold of $log(n_f = 80) = 4.38$ to be reached for a successful working of SCAMP. The average node-degree of DET only points out that the built topology is a tree, it has no direct relation with reliability. In the next paragraph we discuss scalability of DET considering the node-degree distribution.

Evaluating Scalability of the Topology generated by SCAMP and DET. To evaluate the scalability for SCAMP and DET it is necessary to examine the structure of the topology that they build.

In particular for DET the size of the contact list has a huge impact on the overlay topology since a small *contacts* list contributes to keep the message overhead small but the obtained topology converges to a star topology with the node of rank 0 in the middle. In the Fig. 2 plots showing the distribution of the node-degree in case of a contact list with size equal to n_{tot} (Fig. 2(a)) and equal to $n_{tot}/10$ (Fig. 2(b)). Note that the size of the contact list is a predominant parameter with respect to the churn rate. In particular, if the contact list is small the topology converges to a star even for low churn rates (Fig. 2(b) curve for $\Delta_p = 200sec$). With a large contact list the topology converges to a star (more properly, the topology shows a set of hubs) only for high churn rates (Fig. 2(a) curve for $\Delta_p = 5sec$), but the tree become deeper for low churn rates (Fig. 2(a) curve

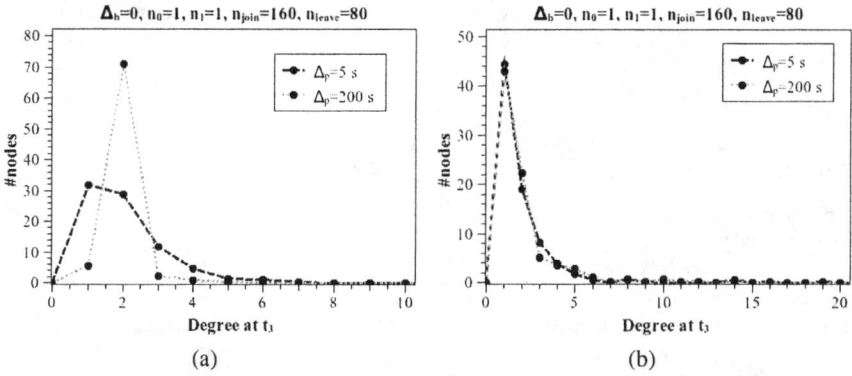

Fig. 2. DET: Degree Distribution at t_3 for $|contacts| = n_{tot}$ and $|contacts| = n_{tot}/10$

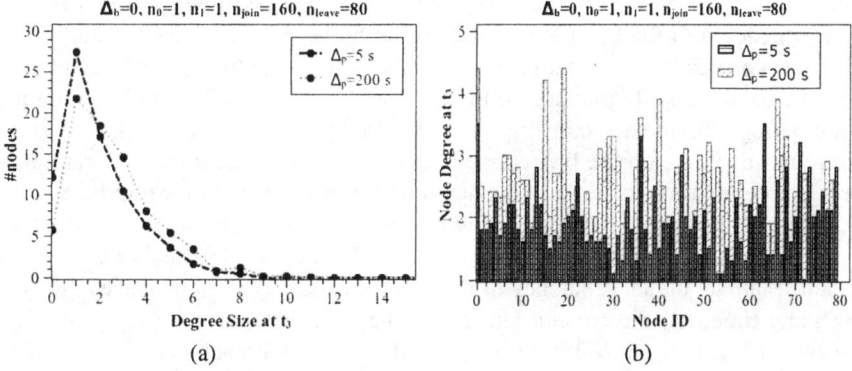

Fig. 3. Degree distribution and node-degree of each node for SCAMP

for $\Delta_p = 200 sec$). Thus, it is confirmed for DET that the faster joins (a join takes a time equal to the maximum round-trip time between contacts links) the least scalable is the topology [7]. On the contrary, SCAMP is always able to balance the degree for each node (see Fig. 3(b)) showing a great scalability. The churn rate impacts only on the average degree and in the distribution degree (see Fig. 3(a)) affecting reliability.

Evaluating Leave Latency for DET. We evaluate leave latency as the average time that passes from the invocation of a leave (the sending of an unsubscription message) to the actual departure of the node from the group (the receiving of an acknowledgment). Note that for each node of rank 1, this time is equal to the round-trip time on the link connecting the node with the node with rank 0. As the rank increases the latency may increase as well. In the worst case a node with rank i may concurrently invoke its leave with all nodes with lower rank belonging to its branch. In this case the latency

---

[7] Note that the more sophisticated mechanisms to join, pointed out in [2], try to maintain a small contact list and a scalable generated topology at the same time. However, these mechanism with high churn rates may lead to unpredictable latency of join/leave operations.

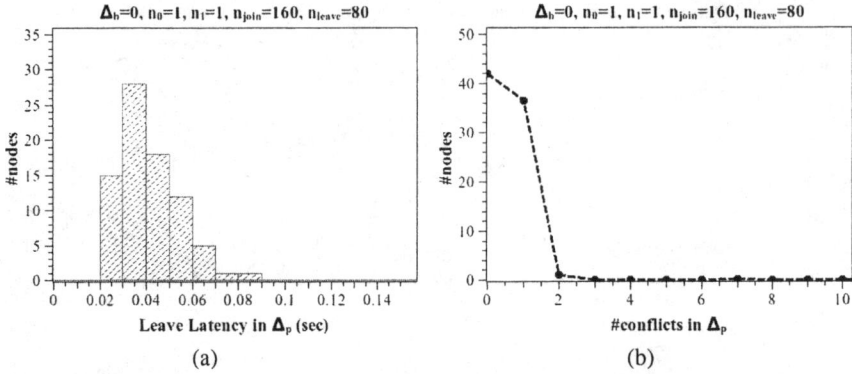

Fig. 4. Leave latency and conflicts number for a scalable topology built by DET

becomes proportional to the rank of a node. Three factors influence the latency of a leave (i) the depth of the tree (deeper trees bring higher latency), (ii) the rate of leaves (higher rates brings higher latency) and (iii) link delays. The first factor depends on the size of the contact list. In practice, with a contact list very small (as pointed out in the previous paragraph) the tree converges to a star. In this case the average latency is equal to the round-trip time on the link connecting the node with the node with rank 0. The third factor depends on the underlying network behavior, then it may unpredictable. To avoid that an unexpected network behavior biases our analysis we consider (only for this particular evaluation) that all links have a RTT equal to 0.02ms [8]. Then, we have chosen to evaluate the leave latency in the case in which (i) all node leaves the system at the same time, (ii) the contact list is very large ($|contacts| = n_{tot}$) and (iii) the churn rate is low ($\Delta_p = 200s$). In this way the tree is a branch and we can evaluate the worst case for leave latency but the best case for scalability of the topology. In Fig. 4 the latency distribution and the number of conflicts for each node, i.e. the number of unsubscription messages received by a node when it was leaving, is shown.

This behavior confirms that the most scalable topology for DET is at the cost of latency of leaves and joins (as pointed out in the previous paragraph). For this reason DET provides reliability at the cost of scalability (either in terms of a not scalable topology or in terms of join/leave latency).

The impact of the lease mechanism in SCAMP. The plots in Fig. 5 shows as the lease impacts the reliability of SCAMP under churn. In practice, the lease mechanism does not influence in the average the reliability of SCAMP. What the lease produces is a high clustering of the group, i.e. most of the nodes are very-well connected and some nodes are isolated. To point out this behavior see (i) Figure 6(a) in which the average number of isolated nodes is in SCAMP higher than in SCAMP without lease and (ii) Fig.6(b) in which not isolated nodes have an average degree higher than in the case of SCAMP without lease.

The reason underlying this behavior is that the lease mechanism forces even a connected node to re-subscribe contacting an arbitrary member of its partial view. With

---

[8] This value has been chosen so small only for convenience.

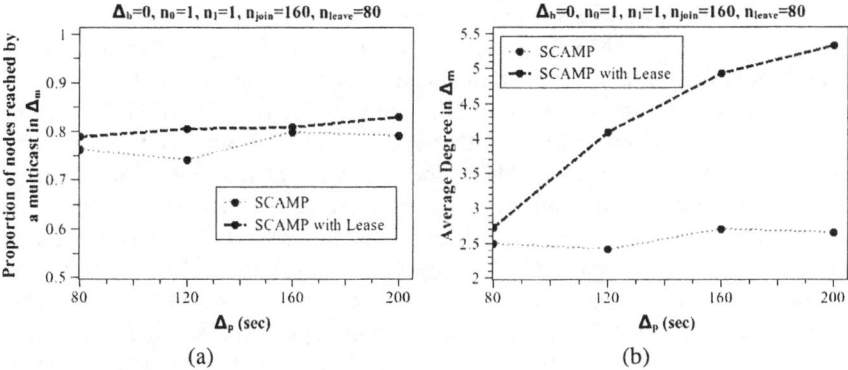

Fig. 5. The impact of the lease in SCAMP on reliability and on the average degree

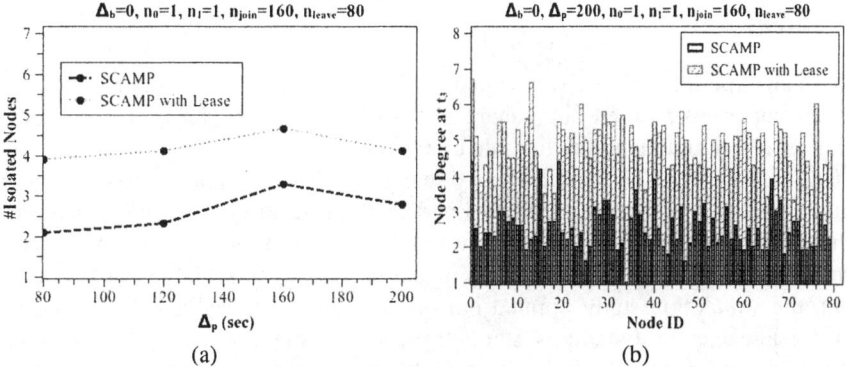

Fig. 6. Node isolation and degree for SCAMP with lease and SCAMP without lease

high churn this member may be inactive. Even if the lease is repeated the same scenario may occur. On the other hand, for those nodes that find an active node upon the re-subscription, there is a new dissemination in the system of their node identifiers that enlarges partial views. It is clear that in fairly static systems (systems with very low churn rates) the lease mechanism has a valuable impact as shown in [10].

4 Related Work

The group membership problem has been extensively studied, and many specifications and implementations exist in literature ([8], [5], [7] just to name a few). These group membership mechanisms ensure greater consistency of group views at the expense of latency and communication overhead.

Probabilistic gossip-based algorithms are being widely studied now. While gossip protocols are scalable in terms of the communication load imposed on each node, they usually rely on a non-scalable membership algorithm. This has motivated work on distributing membership management [9], [10] in order to provide each node with a random partial view of the system, without any node having global knowledge of the membership. However, Jelasity et al. in [11], through an extensive and valuable experimental analysis (not comprising SCAMP), point out the inability of GGMs to make a uniform sampling of peers. Allavena, Demers and Hopcroft have recently proposed a new scalable gossip based protocol[1] for local view maintenance without requiring the assumption of uniformly random views but based on a so-called reinforcement mechanism. They have also given theoretical proofs regarding the connectivity of the graph under churn. They prove that all GGM protocols that does not enjoy a reinforcement mechanism converge to star topology under churn. Liben-Nowell et al. [13] has given a theoretical analysis of structured p2p networks under churn. They define the half-life metric which essentially measures the time for replacement of half the nodes in the network by new arrivals. This metrics is coarser than churn rate and useful when the size of the network is fixed.

5 Conclusion

Through an experimental analysis which compares two p2p group membership protocols, this paper has pointed-out a sharp trade-off between reliability of the generated overlay topology and its ability to scale under churn.

In particular, maintaining an overlay scalable under high churn rates and without sacrificing reliability, latencies of joins and leaves operations become unpredictable. On the other hand, keeping latencies reasonably small (at least predictable) under high churn rates without sacrificing reliability means obtaining not-scalable overlays as stars. In fact, the simulation study pointed out that to obtain overlay scalability and small join/leave latencies in dynamic systems, reliability is compromised. On the contrary, to obtain overlay reliability and join/leave latencies predictable in dynamic systems, overlay scalability is compromised.

References

1. André Allavena, Alan Demers, John E. Hopcroft. *Correctness of a Gossip Based Membership Protocol*, Proceedings of ACM Conference on Principles of Distributed Computing (2005)
2. Roberto Baldoni and Sara Tucci Piergiovanni, *Group Membership for Peer-to-Peer Communication*, Technical Report May 2005, available on http://www.dis.uniroma1.it/$\sim$midlab/publications
3. Roberto Baldoni, Adnan Noor Mian, Sirio Scipioni and Sara Tucci Piergiovanni, *Churn Resilience of Peer-to-Peer Group Membership: a Performance Analysis*, Technical Report May 2005, available on http://www.dis.uniroma1.it/~midlab/publications
4. Anindya Basu, Bernardette Charron-Bost, Sam Toeug: *Simulating Reliable Links with Unreliable Links in the Presence of Process Crashes*. Proceedings of the 10th International Workshop on Distributed Algorithms: 105 - 122 (1996)
5. Kenneth Birman and Robert van Renesse: Reliable Distributed Computing with the Isis Toolkit. IEEE Computer Society Press (1994).
6. Miguel Castro, Peter Druschel, Anne-Marie Kermarrec, Antony Rowstron.*Scribe: A Large-scale and Decentralized Application-level Multicast Infrastructure*. IEEE Journal on Selected Areas in communications (2002)

7. Gregory Chockler, Idit Keidar, Roman Vitenberg. *Group Communication Specifications: a Comprehensive Study*. ACM Comput. Surv. 33(4): 427-469 (2001)
8. Flaviu Cristian. *Reaching Agreement on Processor Group Membership in Synchronous Distributed Systems*. Distributed Computing, 4(4):175-187, April 1991.
9. Patrick Th. Eugster, Rachid Guerraoui, Sidath B. Handurukande, Petr Kouznetsov, Anne-Marie Kermarrec.*Lightweight Probabilistic Broadcast*. ACM Trans. Comput. Syst. 21(4): 341-374 (2003)
10. Ayalvadi J. Ganesh, Anne-Marie Kermarrec, Laurent Massoulié: *Peer-to-Peer Membership Management for Gossip-Based Protocols*. IEEE Trans. Computers 52(2): 139-149 (2003)
11. Márk Jelasity, Richard Guerraoui, Anne-Marie Kermarrec and Maarten van Steen. *The Peer Sampling Service: Experimental Evaluation of Unstructured Gossip-Based Implementations*, Middleware 2004, volume 3231 of Lecture Notes in Computer Science, pages 79-98. Springer-Verlag, (2004)
12. John Jannotti, David K. Gifford, Kirk L. Johnson, M. Frans Kaashoek, James W. O'Toole. *Overcast: Reliable Multicasting with an Overlay Network*. Proceedings of 4th Symposium on Operating System Design and Implementation (2000)
13. David Liben-Nowell, Hari Balakrishnan, David Karger: *Analysis of the Evolution of Peer-to-Peer Systems*. Proceedings of ACM Conference on Principles of Distributed Computing (2002)
14. Ns-2 simulator. http://www.isi.edu/nsnam/ns.

Uinta: A P2P Routing Algorithm Based on the User's Interest and the Network Topology*

Hai Jin[1], Jie Xu[1], Bin Zou[2], and Hao Zhang[1]

[1] Cluster and Grid Computing Lab School of Computer Science and Technology,
Huazhong University of Science and Technology, 430074 Wuhan, China
{hjin, jiexu, haozhang}@hust.edu.cn
[2] School of Mathematics and Computer Science,
Hubei University, 430062 Wuhan, China
zoubin0502@hubu.edu.cn

Abstract. Peer-to-peer (P2P) overlay networks, such as CAN, Chord, Pastry and Tapestry, lead to high latency and low efficiency because they are independent of underlying physical networks. A well-routed lookup path in an overlay network with a small number of logical hops can result in a long delay and excessive traffic due to undesirably long distances in some physical links. In these DHT-based P2P systems, each data item is associated with a key and the key/value pair is stored in the node to which the key maps, not considering the data semantic. In this paper, we propose an effective P2P routing algorithm, called Uinta, to adaptively construct a structured P2P overlay network. Uinta not only takes advantages of physical characteristics of the network, but also places data belonging to the same semantic into a cluster and employs a class cache scheme to reduce the lookup routing latency. Simulations make some comparisons between Chord and our Uinta algorithm all running on the GT-ITM transit stub topology. The results show Uinta routing algorithm significantly improves P2P system lookup performance.

1 Introduction

A *peer-to-peer* (P2P) network is a specialized distributed system at the application layer, where each pair of peers can communicate with each other through the routing protocol in the P2P layer. Routing algorithm is the key component of P2P networks. It nearly determines the total performance of P2P networks.

P2P systems can be classified into two main categories, namely unstructured and structured. Unstructured systems like Gnutella [1], KazaA [2] and Freenet [3] are composed of peers joining the network with some loose rules, without any prior knowledge of topology. It is easier to build and maintain. Typically, new peers randomly connect to existing alive nodes in the network and the searching process for data is flooding across the overlay with a limited scope. However, the flooding-based searching mechanism consumes too much bandwidth to be suitable for large systems.

---

* This work is supported by National Science Foundation of China under grant No.60433040.

Another type of P2P systems named structured P2P systems [4][5] [6][7] follow some predetermined structures. These structures need to be maintained by participant peer nodes. Such structured P2P systems use *Distributed Hash Table* (DHT) as a substrate, Data object (or value) location information is placed deterministically at the peers with identifiers corresponding to the data object's unique key, which makes the routing mechanism more efficient. However, these systems are constructed in overlay networks at the application layer without taking physical network topologies into consideration. Therefore it is possible to result in high lookup delays and unnecessary wide-area network traffic when a routing hop takes a message to a peer with a random location in the Internet.

In order to reduce lookup delays, some researchers have proposed several DHT-based virtual network infrastructures using physical topology information [8][9][12], which map the overlay logical identifier onto the physical network so that neighboring nodes in the logical space are close in the physical network. But all of these systems ignore the user's interest and not consider the data semantic.

The primary contribution of our work is that we propose an overlay network named Uinta to address both the user's interest and the physical topology. All peers are divided into several clusters based on the physical topology of network, which makes peers in the same cluster have small link latency and peers in the different cluster have long link latency. Because users always retrieve data of the same semantic with their interests, we store the data information based on the data semantic, which makes data belonging to the same semantic content be placed in the same cluster. A cache scheme is also employed to reduce the routing cost. Not only data searched recently but also their category information are cached. So it can use the information of the cache table directly if the user searches data of this category next. It is obvious that P2P system workload has temporal and spatial localities just as that in the web traffic [10]. For example, a user who retrieves a song is likely to retrieve other songs in subsequential requests. A high hit rate for this cache schema can be expected, thus a reduced average number of routing hops and lower routing network latency can be achieved.

Uinta is a two-layer overlay network in which peers are organized in different clusters. Routing messages are routed to the destination cluster through the inter-cluster overlay first, and then routed to the destination peer using an intra-group overlay. We take a torus overlay structure in Chord system to construct Uinta for both layers because the ring geometry allows the greatest flexibility, and hence achieves the best resilience and proximity performance [11].

The remainder of the paper is organized as follows. Section 2 provides the method to construct Uinta overlay network. Section 3 shows an overview design of Uinta routing algorithm and the theoretical analysis of algorithm. Our experimental results are described in Section 4. Related work is discussed in Section 5. Section 6 concludes the paper and gives future works.

2 Uinta Overlay Network

In this section, we show how to incorporate the underlying topological information and the data semantic in the construction of Uinta overlay to improve the routing performance.

2.1 Construction of Uinta Overlay Network

Construction of Uinta overlay network involves three major tasks: 1) forming peer clusters based on the physical topology of network; 2) assigning an identifier to a peer or a key to locate a peer in the peer cluster; 3) constructing an overlay network across peer clusters.

1) Cluster formation: The goal of our clustering scheme is to have a set of peers partitioned into several clusters so that peers within a cluster are closer to one another than to ones in a different cluster. So peers should be organized into clusters based on the physical topology of network. Because the cluster formation strategy has great impact on Uinta efficiency, it must be simple and fast with minimal overhead. Also, it must be approximately accurate and can group the close peers into the same cluster.

A simple and relatively accurate topology measurement mechanism is the distributed binning scheme proposed by Ratnasamy and Shenker [12]. In this scheme, a well-known set of machines are chosen as landmark nodes, and system peers are partitioned into disjoint bins so that peers that fall within a given bin are relatively closer to each other in terms of network link latency. Although the network latency measurement method (ping) is not very accurate and determined by many uncertain factors, it is adequate for Uinta to use the method similar with [12] for cluster formation.

Table 1 shows 6 sample nodes A, B, C, D, E and F in Uinta system with measured network link latencies to 3 landmark nodes L1, L2, and L3. We might divide the range of possible latency values into 3 levels: level 0 for latencies in the range [0,100] ms, level 1 for latencies between [100,200] ms and level 2 for latencies greater than 200ms. The cluster name is created according to measured latencies to the 3 landmark nodes L1, L2, and L3, and this information is used

Table 1. Sample peers in a Uinta system with three landmark nodes

Peer	Dist-L1	Dist-L2	Dist-L3	Cluster Name
A	110ms	150ms	240ms	112
B	22ms	135ms	235ms	012
C	285ms	264ms	45ms	220
D	260ms	244ms	67ms	220
E	30ms	120ms	220ms	012
F	28ms	115ms	225ms	012

| P_1 | .... | P_m | S_1 | .... | S_n |

Fig. 1. The binary format of identifier in Uinta

for peers clustering. For example, Peer A's landmark information is 112. Peers C and D have the same information: 220, so they are in the same cluster named 220. The other nodes belong to the cluster named 012.

2) Assignment of identifier: This task also includes four subtasks: assignment of the peer id, the cluster id, the key id and the class id. In Uinta, the $(m+n)$-bit identifier for each peer, each cluster and each key is composed of two parts: the m-bit prefix and the n-bit suffix. For the peer id, the m-bit prefix is assigned to the identifier of a cluster that the peer belongs to and the n-bit suffix is assigned to the identifier chosen by hashing the peer's IP address. For the key id, the m-bit prefix is assigned to the identifier of a class that the key belongs to and the n-bit suffix is assigned to the identifier chosen by hashing the key. For the cluster id (or the class id), the m-bit prefix is assigned to the identifier generated by hashing the cluster name (or the class name) and the n-bit suffix is assigned to 0. The consistent hash function such as SHA-1 [13] is used to avoid the possible identifier duplication problem.

The binary format of identifier in Uinta is shown in Fig.1, in which P_i and $S_j (i = 1, 2, \ldots, m$, and $j = 1, 2, \ldots, n)$ are assigned to 0 or 1. $P_1 \ldots P_m$ is referred to the prefix of an identifier that is marked as P, which is the identifier of cluster or class. $S_1 \ldots S_n$ is referred to the suffix of an identifier that is marked as S, which is hashed by the peer's IP address or the key. So the identifier is equal to $D = P * 2^n + S$.

3) Uinta overlay network construction: To construct the overlay, each peer p in Uinta system maintains two finger tables: the c-finger table and the l-finger table, and a class cache table. Let D_p be the identifier of peer p and $D_p = P_p * 2^n + S_p$. The ith entry in the c-finger table with m entries at peer p contains the identifier of first-joined peer q in the cluster that succeeds $P_p * 2^n$ by $2^{i-1} * 2^n$ on the inter-cluster identity circle, i.e., $q = c\text{-}successor((P_p + 2^{i-1}) \bmod 2^m * 2^n)$, where $1 \leq i \leq m$. We call peer q the ith c-finger of peer p, and denote it by $p.c\text{-}finger[i]$. The ith entry in the l-finger table with n entries at peer p contains the identifier of peer q whose suffix identifier S_q succeeds S_p by 2^{i-1} on the intra-cluster identity circle, i.e., $S_q = l\text{-}successor((S_p + 2^{i-1}) \bmod 2^n)$ and $q = P_p * 2^n + S_q$, where $1 \leq i \leq n$. We call peer q the ith l-finger of peer p, and denote it by $p.l\text{-}finger[i]$. A class cache table entry includes both the class identifier of data searched recently and the identifier of peer at which data information stores (see Fig.2).

Besides the three tables above, in Uinta, each peer uses the landmark table to maintain the information of landmark nodes. It simply records IP addresses of all landmark nodes, which can help a peer joining decide in which cluster it should be located.

	Notation		Definition
c-finger table	c-finger[i]	identifier	$(P_p + 2^{i-1}) \bmod 2^m * 2^n$
		node	the firstly-joined peer in the cluster that succeeds $(P_p + 2^{i-1}) \bmod 2^m * 2^n$, where $1 \le i \le m$
	c-successor		the firstly-joined peer in the next cluster
	c-predecessor		the firstly-joined peer in the previous cluster
l-finger table	l-finger[i]	identifier	$(S_p + 2^{i-1}) \bmod 2^n$
		node	peer $q = P_p * 2^n + S_q$ in the same cluster, where $S_q = l$-successor $((S_p + 2^{i-1}) \bmod 2^n)$ and $1 \le i \le n$
	l-successor		the next peer in the same cluster
	l-predecessor		the previous peer in the same cluster
class cache table	class identifier		the class identifier of data searched recently
	node		a peer in the cluster data information stores

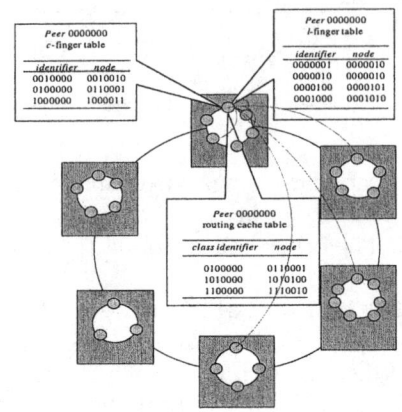

Fig. 2. Definition of data structures for peer p, using the $(m+n)$-bit identifier

Fig. 3. An illustrative example of Uinta

Fig. 3 shows an example of Uinta (with $m=3$, $n=4$). As shown in the figure, the search space is partitioned into 6 clusters after a series of peers join and leave. Peer 0 in cluster 0 maintains three tables: the c-finger table, the l-finger table and the class cache table. The first l-finger of peer 0 points to peer 2 because peer 2 is the first node that succeeds peer 0 within cluster 0. Similarly, the first c-finger of peer 0 points to peer 18 because peer 18 is the first-joined peer of the first cluster that succeeds cluster 0. The class cache table can be established after searching. From this table, we know the entry of class 5 is peer 84, which can not get from the c-finger table directly. Tables of other peers are not shown here for clarity of presentation.

2.2 Peer Operation

1) Peer joins: When a new peer p joins the system, it sends a join message to a nearby peer q that is already a member of system. This process can be done in different methods. We simply assume it can be done quickly (this is the same assumption as in other DHT algorithms). Then peer p gets the information of landmark nodes from this nearby peer q and fulfills its own landmark table. It then decides the distance between landmark nodes and itself and then uses the distributed binning scheme to determine the suitable cluster P_p it should join. The identifier D_p of peer p can be gotten by $D_p = P_p * 2^n + S_p$ (S_p is the hash value of IP address of peer p). Consequently, peer p connects peer p' in the cluster P_p through the c-finger table of peer q and then is located in the cluster based on the suffix S_p. In the following step, it creates routing data structures: the c-finger table and the class cache table that are the same as that of peer p' and the l-finger table. The mechanism used in Chord [4] can be introduced without modification.

If P_p is among the prefix of $c\text{-}finger[i].identifier$ and the identifier prefix of $c\text{-}finger[i].node$ denoted as peer x, peer p will form a new cluster with identifier prefix P_p. Peer p acquires peer x as its c-successor and peer q as its c-predecessor which is the c-predecessor of peer x. Every peer in the cluster where peer x located, when notified by peer p, acquires peer p as its c-predecessor. When the peer whose origin c-successor is peer x next runs of stabilize [4], which is periodically to learn about newly joined nodes, it asks its origin c-successor (for example peer x) for its c-predecessor (peer p now); then this peer acquires peer p as its c-successor. Because peer p already knows one peer x nearby the cluster in the system, it can learn its c-fingers table by asking peer x to look them up in the whole P2P overlay network. The detailed process is described in [4]. All data structures of l-finger table point to itself. Keys between $P_p * 2^n$ and $X_p * 2^n$ are moved form cluster $X_p * 2^n$ to cluster $P_p * 2^n$. Peer p joins the system successfully.

2) Peer leaves or fails: To increase robustness, each Uinta peer maintains an l-successor list of size r containing the first r successors of peer in the same cluster, a c-successor list of size r containing first-joined peers in the first r successor clusters and a cl-successor list of size r containing r peers in the cluster that the c-successor of peer locates in. If a peer's immediate c-successor or l-successor or cl-successor does not respond, the peer can substitute the second entry in its c-successor list or l-successor list or cl-successor list.

The method for a peer leaving or failure is similar with that for a peer leaving in Chord. We do not give the detail description here any more.

2.3 Cache Scheme

The caching scheme is one of the most important aspects which distinguishes Uinta from other P2P systems. OceanStore [14] and CFS [15] also use cache to improve the system performance, where files are cached along the routing path. Because of the large storage requirement for caching files and blocks, an individual node can not cache many files or blocks, thus they can not anticipate a high cache hit rate. Such a caching scheme is not very efficient, especially in a large-scale dynamic system with a large amount of files being shared. In Uinta, it caches the information about classes of data rather than data, and therefore we can hold a large amount of routing information with a relative small cache space and achieve a high cache hit rate. The foundation for using the class cache scheme is that the P2P system workload has temporal and spatial localities. The user tends to search data he is interested in, which always have the same semantic and belong to the same class. For example, a user who retrieves a song is likely to retrieve other songs in subsequential requests. Thus, the user can know which cluster it stores at directly from the class cache table for the next request to search another song, and then a significant fraction of searching will be intra-cluster transfers, which can bypass inter-cluster transfers and generate a more efficient routing algorithm.

3 Uinta Routing Algorithm and Theoretical Analysis

3.1 Routing Algorithm

1) When a peer p wants to obtain the file associated with key k and its class c, it gets the class identifier P_k of file hashed by SHA-1 with c;

2) Check whether exists an entry $(P_k * 2^n, q)$ for the class identifier P_k in the class cache table; if does, jump to peer q directly, then to 6); otherwise, to 3);
3) Check whether P_k falls between the P_p of p and the P_q of its c-successor q; if does, jump to q, then to 6); otherwise, to 4);
4) $x = p$;
repeat
Search peer x's c-finger table for peer q whose prefix of identifier P_q immediately precedes P_k;
$\qquad x = q$;
until P_k falls between the P_x of x and the P_q of its c-successor q;
5) Jump to peer q;
6) Find a peer d through the l-finger table of peer q so as to make the suffix of key identifier S_k hashed by SHA-1 with k fall between the S_x of x and the S_d of its l-successor d;
7) Return the identifier of peer d and $(key, value)$ pair searched to peer p, and join $(P_k * 2^n, d)$ to the class cache table of peer p.

3.2 Theoretical Analysis

In this section, we analyze the routing latency for Uinta. We suppose that there are N peers in both Chord and Uinta and let M be the number of clusters in Uinta. Assuming the average network latency for each hop (hop latency) in Chord is $L_{Chord-hop}$, thus the average routing latency in Chord is:

$$L_{Chord} = \frac{1}{2} * \log_2 N * L_{Chord-hop} \qquad (1)$$

While in Uinta, assuming the average network latency for each hop between the clusters in Uinta is $L_{Uinta-inter}$, the average network latency for each hop within the cluster in Uinta is $L_{Uinta-intra}$ and there are N_i peers in cluster i, thus the average routing latency in Uinta is:

$$L_{Uinta} = \frac{1}{2} * \log_2 M * L_{Uinta-inter} + \frac{1}{M} * \frac{1}{2} * \log_2 \prod_{i=1}^{M} N_i * L_{Uinta-intra} \qquad (2)$$

In our simulations, we find the inter-cluster hop latency in Uinta is nearly the same or slightly larger than the hop latency in Chord, the intra-cluster hop latency is much smaller and $N \gg M$. Thus, we have

$$L_{Chord-hop} \approx L_{Uinta-inter} \qquad (3)$$
$$L_{Chord-hop} > L_{Uinta-intra} \qquad (4)$$

and

$$(N_1 N_2 \cdots N_m)^{\frac{1}{M}} \leq \frac{1}{M}(N_1 + N_2 + \cdots + N_m) \qquad (5)$$

Then we get

$$
\begin{aligned}
L_{Uinta} &\leq \tfrac{1}{2} * \log_2 M * L_{Uinta-inter} + \tfrac{1}{M} * \tfrac{1}{2} * \log_2(\tfrac{N}{M})^M * L_{Uinta-intra} \\
&= \tfrac{1}{2} * \log_2 M * L_{Uinta-inter} + \tfrac{1}{2} * \log_2 \tfrac{N}{M} * L_{Uinta-intra} \\
&< \tfrac{1}{2} * (\log_2 M + \log_2 \tfrac{N}{M}) * L_{Uinta-inter} \\
&\approx \tfrac{1}{2} * \log_2 N * L_{Chord-hop} \\
&= L_{Chord}
\end{aligned}
\quad (6)
$$

From above discussions, we can expect a routing reduction by using the routing algorithm in Uinta. Supposing in a P2P system with 2^{20} nodes, the average latency per hop in Chord is 100ms and the average latency between the clusters in Uinta is 108ms. The average routing latency in Chord algorithm is 1000ms. Assuming all the peers are formed 2^{10} clusters in Uinta system, the average latency within the cluster is only half of the latency between the clusters which is 54ms each hop, thus the average routing network latency in Uinta is approximately to 810ms. The average system routing latency reduces by 19%. If we consider the cache scheme used in Uinta and assuming the hit ratio is P, we get

$$
\begin{aligned}
L_{Uinta-cache} &\leq P(1 * L_{Uinta-inter} + \tfrac{1}{2} * \log_2 \tfrac{N}{M} * L_{Uinta-intra}) \\
&+ (1-P)(\tfrac{1}{2} * \log_2 M * L_{Uinta-inter} + \tfrac{1}{2} * \log_2 \tfrac{N}{M} * L_{Uinta-intra}) \\
&= [P + \tfrac{1}{2}(1-P) \log_2 M] * L_{Uinta-inter} + \tfrac{1}{2} * \log_2 \tfrac{N}{M} * L_{Uinta-intra} \quad (7) \\
&\leq \tfrac{1}{2} \log_2 M * L_{Uinta-inter} + \tfrac{1}{2} * \log_2 \tfrac{N}{M} * L_{Uinta-intra} \\
&< L_{Chord}
\end{aligned}
$$

So we can reduce more routing latency using the cache scheme. Assuming P is 40%, $L_{Uinta-cache}$ in Uinta is less than 637 ms, which reduces the latency by 36%.

4 Performance Evaluations

4.1 Simulation Methodology and Performance Metrics

In our simulation, we use the GT-ITM [16] transit stub topology generator to generate the underlying network, the number of system nodes is varied from 1000 to 10000. As far as the logical overlay is concerned, we build Uinta based Chord simulator. Each peer in the overlay is uniquely mapped to one node in the IP layer. We choose 4 landmarks placed at random and there are three levels for the latency from the landmark to the peer. $100 * N$ pseudo fields that are classified into 100 categories are generated and distributed across all the peers in the simulated network. For each experiment, 100000 randomly generated routing requests (including fields and their types) are executed. We choose Chord as the platform because the ring geometry allows the greatest flexibility. However, Uinta can also be easily deployed in other structured P2P systems such as CAN and Pastry.

We consider three metrics to verify the effectiveness of Uinta: (1) Routing hop; (2) Routing latency; (3) Latency stretch: the ratio of the average latency on the overlay network to the average latency on the physical network.

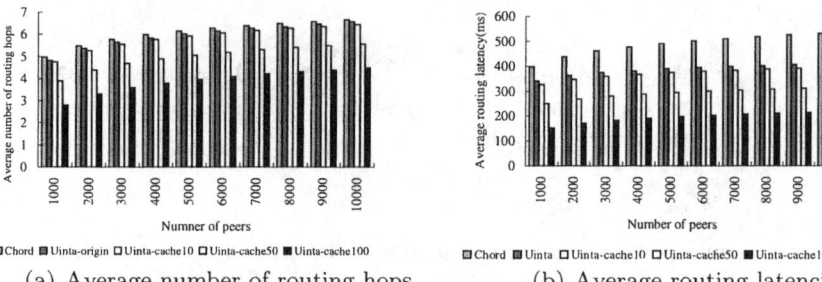

Fig. 4. Uinta and Chord routing performance comparisons

4.2 Routing Cost Reduction

The primary goal of Uinta algorithm is to reduce the routing cost in the P2P system. Fig.4 shows results of routing cost evaluation. In this simulation, we compare routing performances of Uinta-origin, Uinta-cache10, Uinta-cache50, Uinta-cache100 with that of Chord under different network sizes. Uinta-origin is referred to the Uinta algorithm without the cache scheme, while Uinta-cache$n$ is referred to the Uinta algorithm with n cache entries.

Fig.4(a) shows the routing performance comparison result measured with the average number of routing hops. Uinta, Uinta-cache10, Uinta-cache50, Uinta-cache100 and Chord have good scalability: as the network size increases from 1000 nodes to 10000 nodes, average numbers of routing hops only increase around 25%, 27%, 26%, 30%, 38% respectively. Obviously, with the introduction of class cache scheme, the routing cost in Uinta is reduced significantly. For the original Uinta system, the average number of routing hops is a little smaller than that of Chord, which only gets a 2.2% reduction. Using the class cache scheme, the average number of routing hops drops significantly. The more entries in the class cache table, the more performance gain achieved. With a 10-entry class cache table, the average number of routing hops drops by 3.8%. As the number of entries increases to 50, Uinta can get 19.3% reduction. As the number of entries increases to 100, the average number of routing hops decreases 38.5%.

As a proximate metric, the average number of routing hops cannot represent the real routing cost. The actual routing latency highly depends on the average latency for each hop. Fig.4(b) shows the measured results of average routing latency in Uinta, Uinta-cache10, Uinta-cache50, Uinta-cache100 and Chord. Although the original Uinta has the nearly equal average number of routing hops with that of Chord, it has the smaller average routing latency. For the original Uinta and Uinta-cache10, average routing latencies get 20.1% and 23.1% reduction respectively compared with that of Chord. As the number of entries increases to 50, Uinta can get 39.5% reduction. As the number of entries increases to 100, the average routing latency decreases 59.9%.

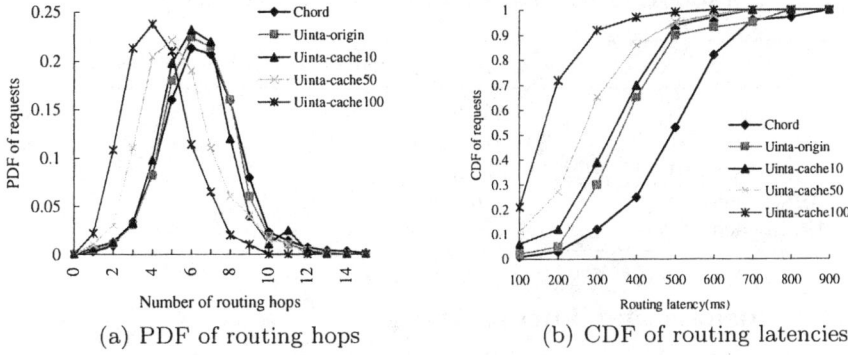

Fig. 5. Performance comparisons in case of a 10000-peer network

4.3 Routing Cost Distribution

In this section, we measure the probability density function (PDF) distribution of average number of routing hops and the cumulative density function (CDF) distribution of average routing latency to analyze the performance of Uinta algorithm.

Fig.5(a) plots the PDF of average routing hops for a network with 10000 peers. The maximum numbers of routing hops for Chord, Uinta-origin, Uinta-cache10, Uinta-cahce50 and Uinta-cache100 are 15, 13, 12, 11, and 9, respectively. The average numbers are 6.64, 6.57, 6.42, 5.56, and 4.47 , respectively. Routing hops for four Uinta algorithms get 1.1%, 3.3%, 16.3%, and 32.7% decreasing for a 10000 peer network, respectively. Fig.5(b) plots the CDF of average routing latency for a network with 10000 peers. Average routing latencies for Chord, Uinta-origin, Uinta-cache10, Uinta-cahce50 and Uinta-cache100 are 531.51ms, 412.36ms, 395.50ms, 316.60ms, and 217.88ms, respectively. Average routing latencies for four Uinta algorithms decrease 22.4%, 25.6%, 40.4%, 59%, respectively compared with that of Chord. In Uinta, routing hops is divided into two parts: inter-cluster hops and intra-cluster hops. The latency for inter-cluster hops is more than that for intra-cluster hops, therefore the latency have more decreasing even though the decreasing of routing hops is little.

4.4 Stretch Reduction

The latency stretch is referred to the ratio of the average latency on the overlay network to the average latency on the IP network, which can be used to characterize the match degree of the overlay to the physical topology. Table 2 summarizes stretch statistics in the case of a 10000-peer network. According to it, we know that the stretch is reduced significantly using Uinta with the cache scheme. This shows that using the topology-aware and semantic-aware overlay construction with the cache scheme, we can achieve significant improvements in the lookup performance.

Table 2. Latency stretch result for Chord and Uinta

Algorithm	Average routing latency	Latency stretch
Chord	531.51ms	4.40
Uinta	412.36ms	3.51
Uinta-cache10	395.30ms	3.19
Uinta-cache50	316.60ms	2.69
Uinta-cache100	217.88ms	1.75

5 Conclusions and Future Work

We propose an overlay network named Uinta, in which peers are clustered according to the physical topology and data information with similar semantics into the same cluster. The user's interest is taken into consideration, and we employ the class cache scheme. From our simulation, we conclude that Uinta offers significant improvements versus random overlay networks. We believe that Uinta can help improve the lookup performance of current and future P2P systems where data information is naturally clustered and the physical topology and users' interests are taken into account. In the future, we plan to explore how to express the data semantic instead of the method now in which the users give the category of data. Load balanced placement of data information is also our next consideration.

References

1. Gnutella, http://www.gnutellaforums.com/.
2. KazaA, http://www.kazaa.com.
3. Freenet, http://freenet.sourceforge.net/.
4. I. Stoica, R. Morris, D. Karger, M. F. Kaashoek, and H. Balakrishnan: Chord: A scalable peer-to-peer lookup service for internet applications. In Proceedings of the ACM Special Interest Group on Data Communication (SIGCOMM), August 2001.
5. S. Ratnasamy, P. Francis, M. Handley, R. Karp, and S. Shenker: A scalable content-addressable network. In Proceedings of the 2003 ACM Special Interest Group on Data Communication (SIGCOMM), Auguest 2001.
6. A. Rowstron and P. Druschel: Pastry: Scalable, distributed object location and routing for large-scale peer-to-peer systems. In Proceedings of IFIP/ACM International Conference on Distributed Systems Platforms (Middleware), November 2001.
7. B. Y. Zhao, L. Huang, J. Stribling, S. C. Rhea, A. D. Joseph, and J. Kubiatowicz: Tapestry: A resilient global-scale overlay for service deployment. IEEE Journal on Selected Areas in Communications, 22 (2004): 41–53
8. K. Shin, S. Lee, G. Lim, H. Yoon, and J. S. Ma: Grapes: Topology-based Hierarchical Virtual Network for Peer-to-peer Lookup Services. In Proceedings of the International Conference on Parallel Processing Workshops (ICPPW'02), 2002.
9. Z. Xu, R. Min, and Y. Hu: HIERAS:A DHT-Based Hierarchical Peer-to-Peer Routing Algorithm. In Proceedings of the 2003 International Conference on Parallel Processing (ICPP'03), pp.187-194, October 2003.

10. A. Mahanti: Web proxy workload characterisation and modelling. Master's Thesis, Department of Computer Science, University of Saskatchewan, September 1999.
11. K. Gummadi, R. Gummadi, S. Gribble, S. Ratnasamy, S. Shenker, and I. Stoica: The impact of DHT routing geometry on resilience and proximity. In Proceedings of ACM SIGCOMM, 2003.
12. S. Ratnasamy, M. Handley, R. Karp, and S. Shenker: Topologically-aware overlay construction and server selection. In Proceedings of IEEE INFOCOM'02, New York, NY, June 2002.
13. D. R. Karger, E. Lehman, F. Leighton, M. Levine, D. Lewin, and R. Panigrahy: Consistent hashing and random trees: Distributed caching protocols for relieving hot spots on the World Wide Web. In Proc. 29th Annu. ACM Symp. Theory of Computing, El Paso, TX, pp.654-663, May 1997.
14. J. Kubiatowicz, D. Bindel, P. Eaton, Y. Chen, D. Geels, R. Gummadi, S. Rhea, W. Weimer, C. Wells, H. Weatherspoon, and B. Zhao: OceanStore: An architecture for global-scale persistent storage. In Proceedings of the 9th International Conference on Architectural Support for Programming Languages and Operating Systems (ASPLOS'00), pp.190-201, Cambridge, MA, Nov. 2000.
15. F. Dabek, M. F. Kaashoek, D. Karger, R. Morris, and I. Stoica: Wide-Area Cooperative Storage with CFS. In Proceedings of the 18th ACM Symposium on Operating Systems Principles (SOSP'01), pp.202-215, Banff, Alberta, Canada, Oct. 2001.
16. E. W. Zegura, K. Calvert, and S. Bhattacharjee: How to model an internet work. In Proceedings of IEEE INFOCOM, 1996.

Optimal Time Slot Assignment for Mobile Ad Hoc Networks

Koushik Sinha

Honeywell Technology Solutions Lab, Bangalore, 560076 India
sinha_kou@yahoo.com

Abstract. We present a new approach to find a collision-free transmission schedule for mobile ad hoc networks (MANETs) in a TDM environment. A hexagonal cellular structure is overlaid on the MANET and then the actual demand for the number of slots in each cell is found out. We assume a 2-cell buffering in which the interference among different mobile nodes do not extend beyond cells more than distance 2 apart. Based on the instantaneous cell demands, we propose optimal slot assignment schemes for both homogeneous (all cells have the same demand) and non-homogeneous cell demands by a clever reuse of the time slots, without causing any interference. The proposed algorithms exploit the hexagonal symmetry of the cells requiring $O(\log \log m + mD + n)$ time, where m is the number of mobile nodes in the ad hoc network, n and D being the number of cells and diameter of the cellular graph.

1 Introduction

In a time division multiplexed (TDM) environment, the existing solutions to time slot assignment in a MANET attempt to assign a globally unique time slot to each node in the network, usually through graph coloring techniques [13, 14, 15], or by finding an appropriate set of partitions of the set of nodes and then assigning a unique time slot to each of these partitions [7, 10], so that no two nodes transmit during the same slot. The algorithms described in [6, 7, 10] need more slots (non-optimal assignment) than the optimal solution and also the number of slots increases rapidly with increase in the maximum node degree of the network graph, although the average node degree may be very small. [15] uses a *maximal independent set* of the nodes to generate a self-organizing TDMA schedule.

In this paper, we introduce a novel strategy for assigning time slots to the nodes in an ad hoc network based on the location information of the individual nodes. The proposed solution significantly improves slot utilization by an elegant technique of *re-using* the time slots by sufficiently distant nodes, avoiding any collision during transmission. For this, we first partition the deployment zone into regular hexagonal cells, similar to the cellular networks. Using the location information of the nodes, the number of active nodes and hence, the actual *demand* of each *cell* at that instant of time is computed. We use this cell demand information to assign time slots to each mobile node by a clever re-use of the time slots which exploits the hexagonal symmetry of the imposed cellular structure, and avoids interference among the nodes. The proposed technique ensures an *optimal collision-free* assignment for every node of the network in $O(m)$ time, m

being the number of nodes in the network. We term this problem of finding an *optimal time slot assignment schedule* for the ad hoc network as the *Slot Assignment Problem* (SAP). The slot assignment algorithm presented here supersedes the existing algorithm in [10, 13] with respect to optimality, and require $O(\log \log m + mD + n)$ time to determine an optimal, collision-free slot assignment schedule for the entire network, n being the number of cells in the overlaid cellular graph and D being the diameter of the ad hoc network.

Mobility of the nodes is also considered by invoking the assignment algorithm whenever a node moves from one cell to an adjacent cell. Appropriate protocols for identifying such a situation through the use of special *control slots* and broadcasting the *id* of the leader of every cell to all nodes within that cell during these control slots, have been presented.

2 System Model

We assume the pre-existence of a partitioning of the MANET deployment area into a number of disjoint cells. The nodes in the network are assumed to possess location information which are either GPS enabled or able to use the network infrastructure to determine their locations relative to the deployment zone [4, 9]. A mapping is used to convert the geographical region to hexagonal grid cells [5, 8]. The nodes need to be synchronized in time. GPS can provide highly accurate and synchronized global time, besides accurate location information.

3 Preliminaries

We first consider the static model of the slot assignment problem, where the number of slots required for each cell is known a priori. The available time space is partitioned into equal length time slots and are numbered $0, 1, 2, \ldots$ from the lower end. The interference between two assigned time slots is represented in the form of *co-slot constraints*, due to which the same slot is not allowed to be assigned to certain pairs of cells simultaneously.

We consider a *2-cell buffering* slot assignment problem (similar to 2-band buffering in [1, 2, 3]) for a hexagonal cellular network overlaid on an ad hoc network, in which a slot can be reassigned to a cell more than distance 2 away. Following the notations in [1, 2], let s_0, s_1 and s_2 be the minimum slot separations between assigned slots in the same cell, in cells at distances one and two apart respectively. In our case of slot assignment in a TDM environment, $s_0 = s_1 = s_2 = 1$. A *cellular graph* is a graph $G = (V, E)$, where each cell of the hexagonal grid is represented by a node and an edge exists between two nodes if the corresponding cells are adjacent to each other, i.e., they share a common cell boundary. Cells i and j are distance-k apart if the minimum number of hops it takes to reach node i from j in G is k. All edges are assumed to be symmetrical.

Figure 1 shows a cell a and its six adjacent cells. The diagram on the right models this scenario as a *hexagonal cellular graph* of seven nodes. The notation $N_i(u)$ denotes the set of all cells that are at a distance $\leq i$ from cell u.

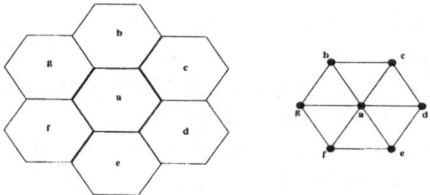

Fig. 1. Conversion of a hexagonal grid to a hexagonal cellular graph

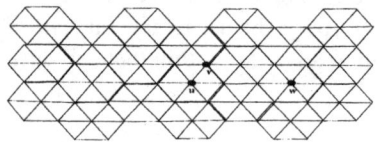

Fig. 2. A hexagonal cellular graph

Definition 1. *Suppose $G = (V, E)$ is a cellular graph. A subgraph $G' = (V', E')$ of G is said to be a distance-k clique, if every pair of nodes in G' is connected in G by a path of length at most k and V' is maximal.*

Definition 2. *A distance-2 clique of 7 nodes in a hexagonal cellular network is defined as a complete distance-2 clique. The node that is at a distance-1 from all other nodes in the complete distance-2 clique is termed as its central node or central cell and the remaining nodes are termed as its peripheral nodes or peripheral cells.*

In a 2-cell buffering environment, the co-slot interference may extend up to cells at distance 2 apart. In view of this, we define a *cellular distance-2 clique* as follows.

Definition 3. *A cellular distance-2 clique $G_2 = (V_2, E_2)$ is a graph generated from a complete distance-2 clique G_1 by adding edges to G_1 between every pair of nodes that are at a distance two in G_1.*

Figures 3(a) and 3(b) illustrate a complete distance-2 clique and the corresponding cellular distance-2 clique. Cell 0 is the central node of the graph. The dashed edges in the cellular distance-2 clique are the edges joining the distance-2 neighbors.

Definition 4. *If G_1 is a cellular distance-2 clique with node u as the central node, then a cellular distance-2 clique G_2 is said to be adjacent to G_1 iff, i) u is a peripheral node of G_2, ii) the central node of G_2 is also a peripheral node of G_1, and iii) G_1 and G_2 have a total of 4 nodes in common, including the central nodes of G_1 and G_2.*

4 Minimum Slot Requirement for Cellular Networks

Let $\mathcal{D}_7^{(2)}(G)$ be the sum of demands of all cells of a cellular distance-2 clique, $G = (V, E)$, where the cardinality of V, $|V| \leq 7$. Then, $\mathcal{D}_7^{(2)}(G) = \sum_{i \in G} w_i$, where w_i is the demand from the cell i.

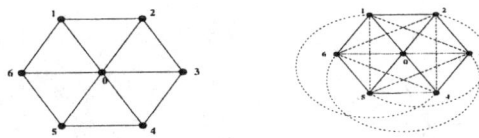

(a) A complete distance-2 clique   (b) A cellular distance-2 clique

Fig. 3. A complete distance-2 clique and the corresponding cellular distance-2 clique

Definition 5. *A 7-node cellular distance-2 clique G or its subgraph is called a critical block, CB_7, which is composed of a maximum of 7 cells, such that the sum of the demands of the cells in CB_7 is maximal over all possible cellular distance-2 cliques in the network.*

We denote the demand of a critical block by $\mathcal{D}_7^{(2)*}$. Thus $\mathcal{D}_7^{(2)*} = \max_{\forall G} \mathcal{D}_7^{(2)}(G)$. Note that there may be more than one such cellular distance-2 clique. We first consider the simpler case of homogenous cell demand, where all cells have the same demand.

4.1 Homogeneous Cell Demand

Let w represent the homogeneous demand for all cells in the network. For $w = 1$, the critical block demand $\mathcal{D}_7^{(2)*}$ would be 7 time slots. Referring to figure 4(a), we see that due to structural symmetry, any distance-2 clique can be chosen as the critical block. Without any loss of generality, let the cellular distance-2 clique $abcdefg$ be designated as the 7-node critical block, with node g as the central node. Considering now the cellular distance-2 clique $gbpqrdc$, centered at c, we note that, node p can be assigned the same time slots as those of nodes e and f, node r can be assigned the same time slots as those of nodes a and f, while node q can be assigned the time slots as those of nodes a, e and f. Thus, we find that the demand of the cellular distance-2 clique $gbpqrdc$ can be satisfied completely by the time slots assigned to the critical block. Figure 4(b) depicts a possible assignment scheme for the cellular graph of figure 4(a). We now state the following results.

Lemma 1. *For any given unsatisfied node u, adjacent to one or more satisfied cellular distance-2 cliques, it is always possible to find a satisfied node v at a distance-3 from u such that the slot assigned to v is unused within a distance two of u.*

We now extend the results of homogeneous demand with $w = 1$ to the general case of $w > 1$ by simply assigning blocks of w consecutive slots to each node, instead of a single slot, leading to the following result.

Lemma 2. *The optimal number of slots required for a cellular graph with homogeneous demand of w slots per cell is $7w$ time slots.*

For all positive and negative integer values (including 0) of m and n, we define the operation $(m, n) \bmod k$ as returning the slot numbers starting from $m \bmod k$ to $n \bmod k$, (including both m and n). The algorithm to handle w slots per cell demand is presented below, which uses only the optimal number of required slots.

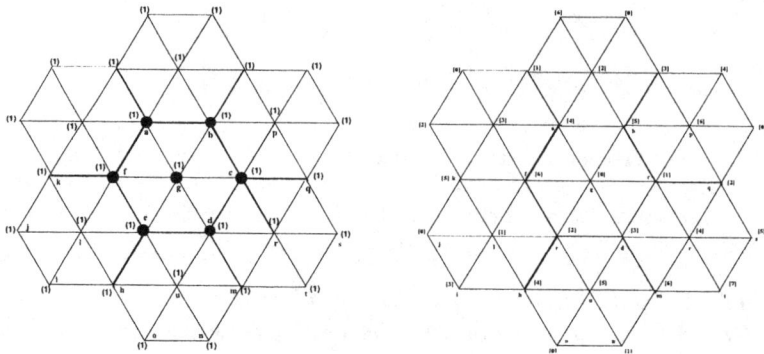

(a) Homogeneous demand of unit slot (b) An optimal assignment scheme

Fig. 4. Slot assignment for a cellular graph with homogeneous unit demand

Algorithm homogeneous_slot_assignment

Step 1 : Assign slot numbers $(0, w - 1)$ to the central cell of the critical block.

Step 2 : Assign slot numbers $(iw, (i + 1)w - 1) \mod 7w$, $i \geq 1$ to the ith cell to the right of the central cell along a particular direction, say along the horizontal line as shown in figure 4(b). That is, we assign the increasing order slot numbers $(0, w - 1), (w, 2w - 1), \ldots, (6w, 7w - 1)$ repeatedly to the cells to the right of the central cell along the horizontal direction.

Step 3 : Assign slot numbers $(-iw, -(i - 1)w - 1)) \mod 7w$, $i \geq 1$ to the ith cell to the left of the central cell. That is, we assign the decreasing order slot values $(7w - 1, 6w), (6w - 1, 5w), \ldots, (w - 1, 0)$ repeatedly to the cells to the left of the central cell.

Step 4 : For rows below the central cell, shift the $(0, w - 1)$ slot value 3 cells to the *left* and then repeat steps 2 and 3 to obtain a slot assignment for each such row.

Step 5 : For rows above the central cell, shift the $(0, w - 1)$ slot value 3 cells to the *right* and then repeat steps 2 and 3 for each such row.

4.2 Heterogenous Cell Demand

We now consider the general case of SAP, where cells have different demands,i.e., $\exists w_i, w_j, i \neq j$, such that $w_i \neq w_j$. The 7-node critical block is insufficient to determine the optimal number of slots of the cellular graph, as demonstrated below.

Example 1. Consider the cellular graph as shown in figure 5. The numbers in parentheses beside each cell denotes the demand of the cell. The cellular distance-2 clique $abfihde$ has a demand of 62 slots. The subgraphs $bcgjief$ and $abcef$ have demands of 61 and 62 time slots respectively. Thus, we see that there are two candidate critical blocks in the network : either subgraph $abfihde$ or subgraph $abcef$. We arbitrarily choose the subgraph $abfihde$ as our 7-node critical block. For the distance-2 clique $bcgjief$ adjacent to the critical block, cells g and j can have their demands satisfied from the slots assigned to the cells a and d. However, the demand of cell c ($w_c = 12$)

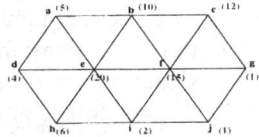

Fig. 5. Heterogeneous demand - 7 node CB fails to give minimum number of slots

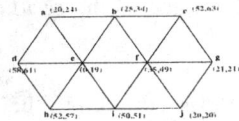

Fig. 6. An assignment scheme requiring 64 slots

is greater than the slots assigned to its two distance-3 neighbors, d and h of the 7-node critical block. The demand sum of cells d and h, $w_d + w_h = (4 + 6) < w_c = 12$. Hence, it is necessary to assign slots in addition to those assigned to the critical block to satisfy the demand of cell c. Thus, we see that for heterogeneous demand, in general, the 7-node critical block will not always give the optimal number of slots of the cellular network. Figure 6 shows a possible slot assignment scheme for the graph in figure 5. The 2-tuple beside each cell denotes the slots assigned to that cell - (m, n) indicates the slots in the range m to n, both inclusive.

The 7-node critical block fails to give the optimal number of slots as it is possible for one of the nodes adjacent to a node of the critical block but not a part of it, to have a demand that exceeds the sum of the demands of its distance 3 neighbors in the critical block. From the cellular graph we see that for every peripheral node of the critical block, there are three neighbors which are at a distance 3 from some other peripheral node of the critical block. Consider for example the node f in figure 5 with neighbors c, g and j. Node d can contribute to satisfying the demands of all of these three nodes while the node a can only satisfy the demands of j and g, and node h can only satisfy the demands of c and g. Hence, each of these three neighbors is a potential source of excess demand over that of the $\mathcal{D}_7^{(2)^*}$, either individually or in combination with the others. This suggests that it is necessary to include all of these three nodes in computing the optimal number of slots. Using a 8-node or 9-node critical block would also fail to obtain a lower bound on the number of slots for the same reasons as for a 7-node critical block. So, we consider a 10-node block consisting of a 7-node distance-2 clique and three other nodes outside this distance-2 clique which are neighbors of a peripheral node of this distance-2 clique. We thus get the following result.

Lemma 3. *For a cellular network with a heterogeneous demand vector, to find the optimal bandwidth requirement of the network, it is necessary to consider a 10 node critical block, as using a critical block with fewer than 10 nodes would not be sufficient to compute the minimum slot requirement of the network.*

In order to compute the demand of the 10-node critical block for which the number of slots will be maximum among all such 10-node blocks, let $C = (V, E)$ be a

cellular distance-2 clique. Let $free_u$ denote the number of slots of node $u \in C$ that can be used by a node which is at a distance three from u and not a part of C, and $used_u(j)$ be the number of slots assigned to $u \in C$ that are reused by node $j \notin C$ and at a distance three from u. Noting that $N_3(u)$ is the set of all distance-3 neighbors of node u, we define *residual demand* res_j of node $j \in N_1(i)$, $i \in C, j \notin C$ as,
$res_j = \max(0, w_j - \sum_{u \in N_3(j) \cap V} used_u(j))$. For $i \in C$, the sum of residual demands of $N_1(i)$ which are not in C will be termed as the *residual sum* of neighbors of i and is defined as $Res_i = \sum_{j \in N_1(i), j \notin C} res_j$.

We demonstrate the procedure for computing the 10-node critical block with the help of the following example.

Example 2. Consider the cellular graph shown in figure 7. Let $abcdefg$ be a candidate critical block. Without any loss of generality, we consider the three neighbors x, y and z of node c. Initially, $free_a = w_a$, $free_f = w_f$ and $free_e = w_e$. The computation of Res_c would be as follows,

Step 1 : Assign slots to node x using maximum number of slots from node e, and the rest, if any, from the node f.
$used_e(x) = min(w_x, free_e); \quad free_e = free_e - used_e(x)$
$used_f(x) = min(w_x - used_e(x), free_f); \quad free_f = free_f - used_f(x)$
$res_x = \max(0, w_x - (used_e(x) + used_f(x)))$

Step 2 : Assign slots to node z using maximum number of slots from node a, and the rest, if any, from the node f.
$used_a(z) = min(w_z, free_a); \quad free_a = free_a - used_a(z)$
$used_f(z) = min(w_z - used_a(z), free_f); \quad free_f = free_f - used_f(z)$
$res_z = \max(0, w_z - (used_a(z) + used_f(z)))$

Step 3 : Assign slots to y using available number of slots from nodes e, a and f.
$res_y = \max(0, w_y - (free_a + free_e + free_f))$

Step 4 : Sum the residual demands of x, y and z, i.e., $Res_c = res_x + res_y + res_z$.

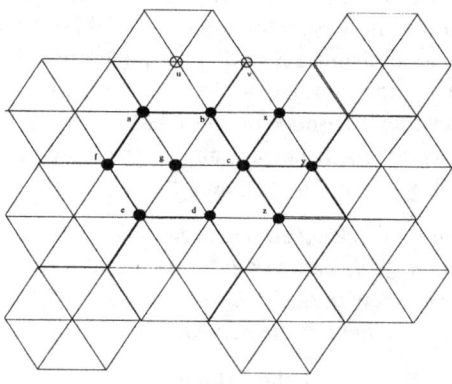

Fig. 7. A 10 node critical block

Let $Res_{max}(\mathcal{C}) = \max_{i \in \mathcal{C}}[Res_i]$. Referring to figure 7, let $\mathcal{D}_{10}^{(2)}(\mathcal{G})$ represent the demand of the 10-node subgraph, $\mathcal{G} \equiv abcdefgxyz$, where $\mathcal{D}_{10}^{(2)}(\mathcal{G}) = \mathcal{D}_{7}^{(2)}(\mathcal{C}) + Res_{max}(\mathcal{C})$. The demand of the 10-node critical block, $\mathcal{D}_{10}^{(2)*}$ is then defined as the demand of a 10-node subgraph that has the maximal $\mathcal{D}_{10}^{(2)}(\mathcal{G})$ in the network, i.e., $\mathcal{D}_{10}^{(2)*} = \max_{\forall \mathcal{G}}[\mathcal{D}_{10}^{(2)}(\mathcal{G})]$

Let $\mathcal{R}_{\mathcal{C}}$ represent the set of nodes that are outside $\mathcal{C}$, but adjacent to some peripheral node of $\mathcal{C}$, corresponding to $Res_{max}(\mathcal{C})$. We call $\mathcal{R}_{\mathcal{C}}$ as the *maximum residual set* of $\mathcal{C}$.

Theorem 1. *The demand sum $\mathcal{D}_{10}^{(2)*}$ is the optimal bandwidth requirement of a hexagonal cellular network having a heterogeneous demand vector.*

Proof. We established from lemma 3 that it is necessary to consider at least a 10 node critical block in order to compute the minimum slot requirement of a cellular network. We now prove that the demand of a 10-node critical block is necessary and sufficient to compute the optimal bandwidth requirement of a hexagonal cellular network.

Let CB_{10} denote the 10-node critical block in a cellular network. Suppose the subgraph $abcdefgxyz$ in figure 7 is our critical block. Let $G = abcdefg$ be the cellular distance-2 clique of the 10-node critical block. Let $\mathcal{R}_G$ denote the maximum residual set of G. Thus, $\mathcal{R}_G = \{x, y, z\}$ in figure 7. We note that our 10-node subgraph for a hexagonal cellular network is actually composed of two adjacent cellular distance-2 cliques.

To establish theorem 1, consider an assignment scheme which proceeds in a spiral, layer by layer fashion, starting with the 10-node critical block. Layer 0 is composed only of CB_{10}, layer 1 composed of all unassigned cellular distance-2 cliques adjacent to CB_{10}. Layer 2 includes all unassigned distance-2 cliques adjacent to the distance-2 cliques in layer 1, and so on. Once the demand of CB_{10} has been satisfied, we first start with the unassigned distance-2 clique in layer 1 that includes all the nodes of $\mathcal{R}_G$ and then move in an anti-clockwise spiral order. Call this distance-2 clique C_1. Now in figure 7, the nodes c, x, y and z of C_1 are already satisfied. For the remaining three unassigned nodes in C_1, the nodes a, f, g and e can be used to satisfy their demands. As the slots assigned to CB_{10} are from slot 0 to $\mathcal{D}_{10}^{(2)*} - 1$, if the remaining three nodes in C_1 were to require slots beyond $\mathcal{D}_{10}^{(2)*} - 1$, it would imply that $\mathcal{D}_{10}^{(2)}(C_1) > \mathcal{D}_{10}^{(2)*}$, which is a contradiction. For the remaining distance-2 cliques adjacent to CB_{10}, we see that for any such distance-2 clique, $\mathcal{C}$, there can be maximum of three unassigned nodes in $\mathcal{C}$. The assigned nodes are a part of CB_{10}. To prove that $\mathcal{D}_{10}^{(2)*}$ slots are sufficient to satisfy their demands, we partition the set of the remaining distance-2 cliques adjacent to CB_{10} into two sets:

1. Set of distance-2 cliques which has at least one but not all unassigned nodes within distance 2 of the nodes in $\mathcal{R}_G$.
2. Set of distance-2 cliques whose unassigned nodes are all at a distance 3 from any node in $\mathcal{R}_G$.

We first consider the scenario when there is at least one unassigned node within distance two of $\mathcal{R}_G$. Without any loss of generality, let u and v be two unassigned nodes of the distance-2 clique $C = auvxcgb$ within distance two of $\mathcal{R}_G$ as shown in figure 7. Node u is 2-hop and v is 1-hop away from x. From figure 7 it is apparent that any such C would have to be adjacent to CB_{10}. Now, $\mathcal{D}_{10}^{(2)^*}$ for a cellular network would not be optimal if node u or v would require slots beyond that required by CB_{10}. Suppose, without any loss of generality, u requires slots beyond that assigned to CB_{10}. This implies that res_u must be greater than $res_y + res_z$, or else these two nodes could additionally be used along with the nodes d and e from the subgraph $abcdefg$ of CB_{10} to satisfy the demand of u. Now, if $res_u > res_y + res_z$

$\Rightarrow res_u + res_x > res_x + res_y + res_z \Rightarrow res_u + res_v + res_x > res_x + res_y + res_z$

This would imply that the nodes u, v and x form the set $\mathcal{R}_G$ of the distance-2 clique $abcdefg$. In other words, the clique $abcdefg$ and the three nodes u, v and w would form the 10-node critical block, which would be a contradiction to the original assumption that the nodes x, y and z form the set $\mathcal{R}_G$ for the cellular distance-2 clique G.

Considering now the second scenario of a distance-2 clique C such that all its unassigned nodes are no less than distance 3 from all nodes of $\mathcal{R}_G$. If C is adjacent to G, then the demand of any unassigned node $u \in C$ can be satisfied using all nodes of $\mathcal{R}_G$, in addition to the nodes in G that are at a distance three from u. If the slots from 0 to $\mathcal{D}_{10}^{(2)^*} - 1$ were not sufficient to satisfy the demand of node u, then arguing as before, if the residue demand of an unassigned node $u \in C$, res_u is greater than $Res_{max}(G)$, then it implies that, $res_u > res_x + res_y + res_z$, which would again be a contradiction to our original assumption that $\mathcal{R}_G = \{x, y, z\}$ represents the maximum residue set of the distance-2 clique G. Thus, it is possible to satisfy the demands of all distance-2 cliques adjacent to CB_{10}, using the slots from 0 to $\mathcal{D}_{10}^{(2)^*} - 1$.

Using a similar assignment procedure and argument as above, we can show that the slots from 0 to $\mathcal{D}_{10}^{(2)^*} - 1$ are sufficient to satisfy the demands of all unassigned distance-2 cliques in layer 2 that are adjacent to satisfied distance-2 cliques in layer 1. The process can be repeated for distance-2 cliques in layer 3, 4, 5, ..., to obtain an assignment scheme that requires only slot values from 0 to $\mathcal{D}_{10}^{(2)^*} - 1$. Hence, $\mathcal{D}_{10}^{(2)^*}$ is the optimal required bandwidth for a cellular network. □

Note that the cellular distance-2 clique of a 10-node critical block may not be a 7-node critical block, as may be seen from figure 8. In figure 8 we see that the 7-node critical block demand is 62 slots, while the 10-node critical block demand is 65 slots. Subgraphs $abfjide$ and $abcef$ both have demands of 62 slots (corresponding to a 7-node critical block), while the subgraphs $pqrstuvwxy$ and $pqwvrstu$ both have demands of a 10-node critical block. If, $abfjide$ ($abcef$) is chosen as the 7-node critical block, then the demand of the 10-node subgraph $abfjidecgk$ ($abcefdij$) would be 64 time slots.

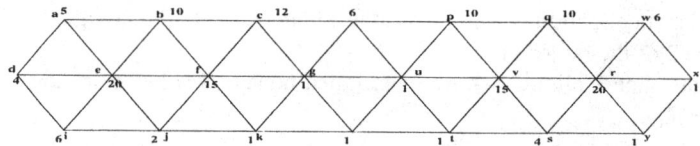

Fig. 8. A 10 node critical block not formed by a 7 node critical block

The algorithm for finding an optimal slot assignment for a cellular network with heterogeneous demand, while satisfying the *2-cell buffering* constraint is as follows :

Algorithm heterogeneous_slot_assignment

Step 1 : For each cell i of the network, construct a cellular distance-2 clique, $\mathcal{C}$ with i as the central node. Compute the demand sum, $\mathcal{D}_7^{(2)}$, of the cells belonging to $\mathcal{C}$.
Step 2 : For each peripheral node $j \in \mathcal{C}$, compute the *residual sum set, Res_j*.
Step 3 : The *maximum residual sum set*, $\mathcal{R}_\mathcal{C}$ corresponds then to the set of neighbors of a peripheral node $k \in \mathcal{C}$ such that $Res_k = Res_{max}(\mathcal{C}) = \max_{j \in \mathcal{C}}[Res_j]$. Let $\mathcal{G}$ denote the 10-node subgraph corresponding to central node i of $\mathcal{C}$. Then, $\mathcal{G} = \mathcal{C} \cup \mathcal{R}_\mathcal{C}$.
Step 4 : Compute the demand of $\mathcal{G}$, $\mathcal{D}_{10}^{(2)}(\mathcal{G}) = \mathcal{D}_7^{(2)} + Res_{max}(\mathcal{C})$.
Step 5 : Repeat step 1 to 4 to obtain the demand $\mathcal{D}_{10}^{(2)}(\mathcal{G})$ of all 10-node subgraphs in the network. The maximum of these demands is the 10-node critical block demand. $\mathcal{D}_{10}^{(2)^*} = \max_{\forall \mathcal{G}}[\mathcal{D}_{10}^{(2)}(\mathcal{G})]$
Step 6 : Now arbitrarily choose one of the 10-node candidate critical blocks as the 10-node critical block of the cellular network.
Step 7 : Satisfy the demand of the nodes of CB_{10} under the *2-cell buffering* constraint.
Step 8 : Satisfy the demands of all distance-2 cliques in layer 1, adjacent to CB_{10}. Begin with the one formed by the nodes of maximum residual set of CB_{10}.
Step 9 : Continue the process of assigning slots to distance-2 cliques in layer 2, layer 3 and so on, in a spiral, layer by layer fashion as described in theorem 1.

5 A Centralized Optimal Slot Assignment Algorithm (COSA)

We present in this section a centralized slot allocation algorithm for assigning slots as per demand of each cell in the cellular network, while utilizing the minimum number of slots required for generating a collision-free transmission schedule that satisfies the *2-cell buffering* constraint, $s_0 = s_1 = s_2 = 1$.

Each MT is assigned a unique identifier (id) from the set $\{1, 2, 3, \ldots, m\}$, where m is the total number of mobile terminals. Initially, each mobile terminal (MT) knows its positional co-ordinates.

In order to handle mobility of the mobile terminals, each cell keeps a few slots for transmitting control messages and some unused slots for handling new MTs joining the network and hand-off scenarios. In general, a cell i computes its demand w_i as the sum of the number of mobile terminals in the cell, the number of slots allocated for control messages and an additional few unused slots. We assume the number of unused slots to be some fraction f of the number of mobile terminals currently in the cell. If m_i is the number of MTs currently in cell i and c slots are used for control purpose, then the demand, w_i of cell i is, $w_i = m_i + c + \max(1, \lceil fm_i \rceil)$, $0 \leq f \leq 1$.

5.1 Algorithm COSA

The steps of the algorithm are as follows :

Step 1 : Elect an MT as the *network leader* through some leader election protocol [11, 12] and call this MT as L.

Step 2 : L broadcasts the mapping to convert the geographical region into a hexagonal grid structure to all the nodes of the network. Each node, on receiving this message, appends it with its own location co-ordinates to be known to all other nodes. An MT i transmits its message in i^{th} slot to avoid collision during this step.

Step 3 : For each cell i, a *cell leader* L_i is elected from the MTs residing in cell i, based on some metric such as remaining battery power, load, location, etc. [12].

Step 4 : The demand of each cell i, w_i is communicated by each cell leader L_i to the network leader L. L produces an optimal, collision-free transmission schedule by executing either *homogeneous_slot_assignment* or *heterogeneous_slot_assignment* algorithm.

Step 5 : L broadcasts the slot assignment schedule of the network to each cell leader. The slot assignment schedule details the slots assigned to each cell i, which had demanded w_i slots. Once a cell leader L_i of cell i receives the information about the slots assigned to it from L, it generates a transmission schedule for the MTs in the cell i and does a periodic local broadcast of this schedule within the cell i.

Due to space constraints, we briefly describe the handling of various dynamic situations like joining/leaving of mobile terminals and hand-off.

- **New mobile terminal joining the network :** When a new MT joins the network in some cell i, it first waits to hear a *cell status message* broadcast by the cell leader, L_i and then tries to join the network by sending a request to L_i. A recomputation of global slot assignment by L is required if not enough free slots exist in cell i.
- **Mobile terminal leaving cell or network:** If a cell (network) leader leaves a cell then a new cell (network) leader is elected from the remaining MTs (cell leaders).
- **Hand-off of mobile terminals :** The process of hand-off is treated in the same way as a new MT u joining cell j, from cell i, with an additional message from L_j to L_i to indicate the new cell in which u can be found.

5.2 Complexity Analysis

The leader election process in step 1 of *algorithm COSA* takes $O(\log \log m)$ time [11, 12]. Steps 2 and 5 each takes $O(mD)$ time for round-robin broadcast, assuming $\forall i, d_i \ll D$ and $w_i = O(m)$ for step 5 of *algorithm COSA*. Step 3 of *algorithm COSA* takes $O(1)$ time. Computation of an optimal slot assignment schedule by either algorithm *homogeneous_slot_assignment* or algorithm *heterogeneous_slot_assignment* takes $O(n)$ time, n being the number of cells in the cellular network. Thus, step 4 takes $O(mD) + O(n)$ time. Hence the complexity of our proposed *algorithm COSA* is $O(\log \log m + mD + n)$ time.

6 Conclusion

We have presented a novel approach to the problem of generating a collision-free transmission schedule for mobile terminals in a mobile ad hoc network. Our proposed algorithm overlays a MANET with a hexagonal cellular grid structure and then generates a collision-free transmission schedule with the minimum number of time slots, while satisfying the *2-cell buffering constraint* using a low overhead. Due to the absence of collisions in the network and use of optimal number of time slots, the proposed scheme provides smaller network latency, higher network throughput and increased battery life of the mobile terminals.

References

1. Ghosh, S.C., Sinha, B.P., Das, N.: Channel Assignment using Genetic Algorithm based on Geometry Symmetry. IEEE Trans. Vehi. Tech.,Vol. 52 (July 2003) 860–875
2. Ghosh, S.C., Sinha, B.P., Das, N.: A New Approach to Efficient Channel Assignment for Hexagonal Cellular Networks. Int. J. Found. Comp. Sci., Vol. 14 (June 2003) 439–463
3. Ghosh, S.C., Sinha, B.P., Das, N.: Coalesced CAP: An Efficient Approach to Frequency Assignment in Cellular Mobile Networks. Proc. Int. Conf. Adv. Comp. Comm., India (Dec. 2004) 338–347
4. Zangl, J., Hagenauer, J.: Large Ad Hoc Sensor Networks with Position Estimation. Proc. 10th Aachen Symp. Signal Theory. Aachen, Germany (2001)
5. Liao, W.-H., Tseng, Y.-C., Sheu, J.-P.: GRID: A Fully Location-aware Routing Protocol for Mobile Ad Hoc Networks. Telecom. Systems, Kluwer Acad. Pub., Vol. 18 (2001) 37–60
6. Sinha, K., Srimani, P.K.: Broadcast Algorithms for Mobile Ad Hoc Networks based on Depth-first Traversal. Proc. Int. Workshop Wireless Inf. Sys., Portugal (Apr. 2004) 170–177
7. Sinha, K., Srimani, P.K.: Broadcast and Gossiping Algorithms for Mobile Ad Hoc Networks based on Breadth-first Traversal. Lecture Notes in Computer Science, Vol. 3326. Springer-Verlag (Dec. 2004) 459–470
8. Tseng, Y.-C., Hsieh, T.-Y.: Fully Power-aware and Location-aware Protocols for Wireless Multi-hop Ad Hoc Networks. Proc. IEEE Int. Conf. Comp. Comm. Networks (2002)
9. Capkun, S., Hamdi, M., Hubaux, J.P.: GPS-free Positioning in Mobile Ad-hoc Networks. Proc. 34'th Hawaii Int. Conf. System Sciences (HICSS) (January 2001)
10. Basagni, S., Bruschi, D., Chlamtac, I.: A Mobility Transparent Deterministic Broadcast Mechanism for Ad Hoc Networks. IEEE Trans. Networking, Vol. 7 (Dec. 1999) 799–807
11. Nakano, K., Olariu, S.: Randomized Initialization Protocols for Ad-hoc Networks. IEEE Trans. Parallel and Distributed Systems, Vol. 11 (2000) 749–759
12. Nakano, K., Olariu, S.: Uniform Leader Election Protocols for Radio Networks. IEEE Trans. on Parallel and Distributed Systems, Vol. 13, Issue 5 (2002) 516–526
13. Perumal, K., Patro, R.K., Mohan, B.: Neighbor based TDMA slot Assignment Algorithm for WSN. Proc. IEEE INFOCOM (2005)
14. Pittel, B., Weishaar, R.: On-line Coloring of Sparse Random Graphs and Random Trees. J. on Algorithms (1997) 195–205
15. van Hoesel, L.F.W., Nieberg, T., Kip, H.J.,Havinga, P.J.M.: Advantages of a TDMA based, Energy-efficient, Self-organizing MAC Protocol for WSNs. Proc. IEEE Vehi. Tech. Conf., Italy (2004)

Noncooperative Channel Contention in Ad Hoc Wireless LANs with Anonymous Stations[*]

Jerzy Konorski

Gdansk University of Technology,
ul. Narutowicza 11/12, 80-952 Gdansk, Poland
`jekon@eti.pg.gda.pl`

Abstract. Ad Hoc LAN systems are noncooperative MAC settings where regular stations are prone to "bandwidth stealing" by greedy ones. The paper formulates a minimum-information model of a LAN populated by mutually impenetrable groups. A framework for a noncooperative setting and suitable MAC protocol is proposed, introducing the notions of verifiability, feedback compatibility and incentive compatibility. For Random Token MAC protocols based on voluntary deferment of packet transmissions, a family of winner policies called RT/ECD-Z is presented that guarantees regular stations a close-to-fair bandwidth share under heavy load. The proposed policies make it hard for greedy stations to select short deferments, therefore they resort to smarter strategies, and the winner policy should leave the regular stations the possibility of adopting a regular strategy that holds its own against any greedy strategy. We have formalized this idea by requiring evolutionary stability and high guaranteed regular bandwidth shares within a set of heuristic strategies.

1 Introduction

In the field of medium access control for single-channel AD Hoc wireless LANs, a wide class of protocols prescribes random deferment of packet transmissions upon detection of the beginning of a protocol cycle. This is meant to avoid packet collisions, while retaining the simplicity of distributed contention. The prevailing approach is to synchronize deferments to a global slotted time axis, with each slot spanning at least the LAN's maximum end-to-end propagation delay, and each deferment being a slot multiple. The generic term *Random Token* (RT) subsumes a class of deferment-based MAC mechanisms where the duration of a deferment (counted in slots) is drawn at random from some finite range of integers. A typical condition for a LAN station to access the medium in the present protocol cycle – i.e., its deferment being extreme among the contending stations – is in that case not unlike

---

[*] Effort sponsored by the Air Force Office of Scientific Research, Air Force Material Command, USAF, under grant FA8655-04-1-3074. The U.S Government is authorized to reproduce and distribute reprints for Governmental purpose notwithstanding any copyright notation thereon. The views and conclusions contained herein are those of the author and should not be interpreted as necessarily representing the official policies or endorsements, either expressed or implied, of the AFOSR or the U.S. Government.

capturing a unique token that visits stations at random rather than along a logical ring. RT mechanisms have been described in a pure form in [2]. They are part of the leading standard solutions, cf. the CSMA/CA technique of IEEE 802.11 [6] (where the shortest deferment wins) and the elimination phase of HIPERLAN/1 [3] (where the longest deferment, advertised as elimination burst, wins). Reference RT mechanisms and their suitability for AD Hoc systems are discussed in Sect. 2.

An RT-type MAC protocol exemplifies an election process with deferments representing elective actions. Thus, the protocol breaks up into:

- an (*election*) *strategy*, entirely within a station's discretion, dictates elective actions in successive protocol cycles, and
- a distributed *winner policy*, common to all stations, defines the feasibility of selected actions (whether they fit into a feasible action range) and defines winning actions in each protocol cycle, producing one winner or none.

An interesting line of research deals with distributed communication mechanisms in a noncooperative setting in which adherence to the common rules cannot be counted on for global optimization [13]. In the context of RT-like MAC, two types of stations can be envisaged, regular and greedy. Regular stations are the cooperative type: they use regular strategies e.g., based on a predefined probability distribution over the action space, optimized with a view to improve global performance indices such as bandwidth utilization and fairness (e.g., uniform in IEEE 802.11 DCF or truncated geometric in HIPERLAN/1). Greedy stations are free to adopt any greedy strategies to self-optimize their bandwidth share to the detriment of regular stations. There is a strong motivation for stations to become bandwidth-greedy on account of the growing volumes of offered traffic; enter advanced chip technology offering increasingly tailor-made and self-programmable station interfaces [12]. In choosing more sophisticated strategies, greedy stations only have to keep their complexity within reason and adhere to the winner policy for synchronization; otherwise they may reasonably hope to get away with the "bandwidth stealing" they commit. This is particularly true in AD Hoc systems given the inherent station mobility (meaning that a station's actions are difficult to trace down, enforce or prevent), and anonymity (e.g., stations' identities may be temporary and/or unavailable at MAC level). Still, most studies of noncooperative MAC settings, unlike ours, assume that stations' identities are recoverable [1], [10], [12]. Few exceptions include [8], [11].

One should guarantee regular stations a fair bandwidth share regardless of the greedy stations' behavior, especially at heavy load. We advocate self-regulatory rather than administrative measures, an appropriate approach for AD Hoc systems and one that promises more flexibility at less cost. A framework for a noncooperative setting is proposed in Section 3 with a focus on preventing certain brute-force "bandwidth stealing" strategies; in this context, the notion of *verifiability* is discussed. Leaving greedy strategies to backstage designers, we focus on the design of a winner policy enabling some regular strategies to hold their own against any greedy strategies.[1]

---

[1] Alternatively, a regular strategy might induce a predictable learning process in greedy stations, drawing on the rich theory of learning in games [5]. For example, "aggressiveness" might be responded to in kind (cf. the backoff freeze mechanism of IEEE 802.11). However, it is difficult to distinguish other stations' "aggressive" play from a traffic increase, leading to poor bandwidth utilization [9].

A framework for a reasonable winner policy and greedy strategy is proposed in Section 4. In Section 5 we describe a family of winner policies called *RT/ECD-Z*, and in Section 6 evaluate them via simulation against a reference RT-type winner policy, assuming a number of heuristic election strategies. The idea behind the evaluation is that a good winner policy should admit a clear candidate for a standard election strategy; we are especially after strategies that exhibit a form of evolutionary stability and fare well when played against any other strategy. Section 7 concludes the paper.

2 Random Token Winner Policies for Ad Hoc Systems

An Ad Hoc LAN uses a wireless medium, has no fixed communication infrastructure and little administration. For simplicity we assume that all stations remain within the hearing range of one another, use a single channel and perceive a common slotted time axis. We adopt a *minimum-information model* whereby a station

- is free to join and leave without prior notification, and change location and/or identity at will, and
- relies on binary per-slot channel feedback i.e., can only distinguish an empty slot from a carrier one, except that recipients of a successful (non-colliding) transmission are also able to interpret the slot's content; non-recipients perceive successful and colliding transmissions alike as just bursts of carrier.

We thus envisage a wireless LAN populated by *mutually impenetrable groups* (Fig. 1). Stations of each group know one another, may use a full packet encryption scheme and need not exchange any user or control data with other groups, whose presence they only perceive as bursts of carrier reducing the available bandwidth.

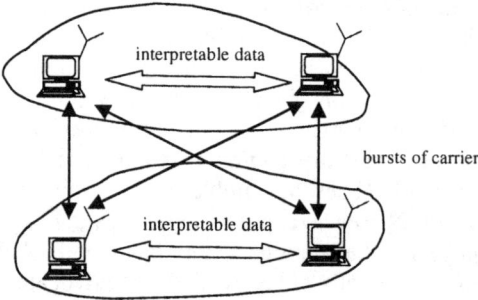

Fig. 1. Perception of transmission in mutually impenetrable groups

RT-like winner policies employ CSMA/CA [2]. To further suppress collisions, a two-phase policy we refer to as RT/CA-Y (cf. HIPERLAN/1's EY-NPMA [3]), in the elimination phase has a station willing to transmit a packet defer its transmission for a random number of slots from the range $[0, E-1]$. Then, unless the channel is sensed busy, the station transmits a 1-slot burst of carrier called *pilot* to discourage stations that have selected longer deferments. Finally, along with other stations that transmitted their pilots in the same slot, it enters a yield phase where a deferment is selected at random from the range $[0, Y-1]$ ($Y=1$ produces pure CSMA/CA).

Our reference policy is called *RT with Extraneous Collision Detection* (RT/ECD). The winners of the elimination phase each transmit an interpretable 1-slot pilot containing the addresses of the intended packet transmission's recipient(s), and await reaction in the following slot (Fig. 2). On sensing a successful pilot, a recipient issues a *reaction* burst of carrier, while refraining from reaction if a collision of pilots is sensed. The presence of reaction prompts the (single) winner to start its packet transmission in the ensuing slots, whereas the absence of reaction prompts the winners to back off, thereby starting a new protocol cycle (Although similar to the RTS and CTS of IEEE 802.11 [6], pilots and reactions serve to ensure verifiability, discussed further, rather than cope with hidden stations.)

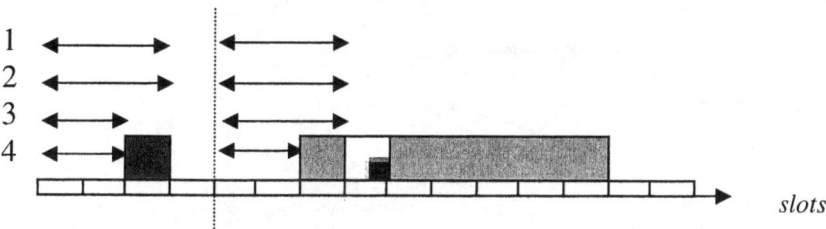

Fig. 2. RT/ECD: stations 3 and 4 transmit pilots, no reaction follows and a new protocol cycle begins in which station 4 transmits pilot successfully, reaction follows and station 4 starts packet transmission

RT/ECD outperforms RT/CA-Y in terms of bandwidth utilization. Let the actions (deferments) be drawn from a probability distribution (p_l, $l = 0,...,E - 1$). Suppose the stations transmit packets of constant size L slots; denote by O the average scheduling overhead per protocol cycle (number of slots not devoted to packet transmission), and by W the probability of exactly one winner per protocol cycle. Then, if all stations are always ready to transmit packets, the total bandwidth utilization, U, equals $W \cdot (1 + O/L)$ for RT/CA, and $1/(1 + O/(W \cdot L))$ for RT/ECD. Calculation of O and W given the above description is a simple exercise in probability. Fig. 3 plots U against E for RT/ECD and RT/CA-Y, assuming $N = 10$, $L = 50$, $Y = 7$, and $p_l = \text{const.} \cdot q^l$ with $q = 2$, 1 or 0.5. These three values of the parameter q typify, respectively, "gentle," "moderate," and "aggressive" behavior. Proper choice of q ensures that RT/ECD is distinctly superior to RT/CA-Y regardless of E and N, as is RT/CA-Y to pure CSMA/CA. The benefits of extraneous collision detection are thus tangible. Unfortunately, under both RT/CA-Y and RT/ECD, straightforward greedy strategies exist that consist in selecting "shorter-than-random" deferments. To prevent frequent collisions with other greedy stations using similar strategies, a greedy station may draw its deferments from a probability distribution biased toward 0.

3 Framework for a Non-cooperative MAC Setting

It might seem unnatural that a greedy station should commit "bandwidth stealing" given that typically it is willing to both transmit and receive packets. Within our model of mutually impenetrable groups, however, a station is indistinguishable from – and thus an adequate model of – a group of stations. What the outsiders perceive as a sequence of actions (elimination deferments) of a station can in fact be produced by a group of stations that have reached an intra-group agreement as to how to take turns at transmitting pilots. Thus more *transmission* opportunity for a greedy station models more *communication* opportunity for a group. A noncooperative setting will henceforth be modeled as one with N stations, of which G are greedy ($0 \leq G \leq N$).

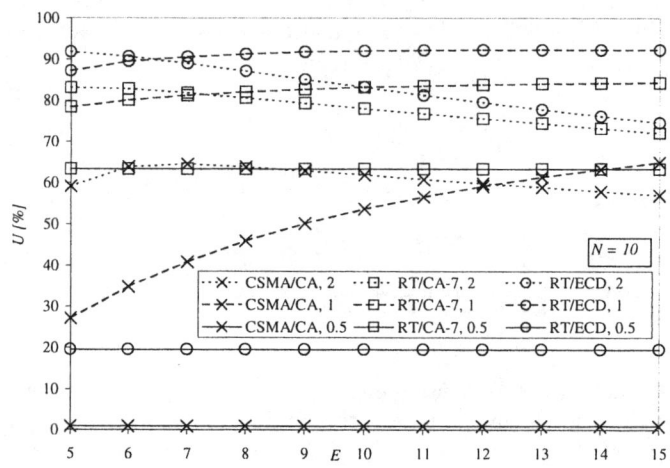

Fig. 3. Bandwidth utilization under CSMA/CA, RT/CA-Y and RT/ECD (values of q indicated)

Brute-force strategies should be prevented that consist in deviations from the MAC protocol being used. E.g., a station under RT/CA-Y may join in the yield phase having issued no pilot; a station under RT/ECD may jam any pilot it senses. (While the former strategy is rational, the latter is not.) Under RT/ECD, a greedy station might also start its packet transmission claiming to have sensed a reaction, or refrain from reaction on the claim that channel errors corrupted the pilot into a perceived collision. (Again, the former strategy is rational, while the latter is not, as it prevents reception of data.) Deviations such as the above raise the issue of a winner policy's verifiability.

A conceptual *verifier* (meant as a deterrent but not necessarily deployed) can be thought of as an extra station complete with a directional and an omnidirectional antenna. It is able, which the greedy stations are aware of, to lock the directional antenna upon a station and, upon detection of a deviation, impose predefined sanctions e.g., jam all that station's pilots. A *verifiable* winner policy defines relevant actions so that any rational deviation from the MAC protocol is verifier detectable. For example, pure CSMA/CA does not qualify: starting a packet transmission immediately is a rational but not detectable deviation (may pass as drawing a 0-slot deferment). It is advisable that elective actions consist in transmission of some physical signals; a rational deviation on the part of a station then involves making false claims as to sensing or not sensing carrier on the channel. Such behavior will not go unnoticed if a verifier has locked its directional antenna upon that station, while using its omnidirectional antenna to correctly perceive the signals of other stations.

4 Framework for a Winner Policy and Greedy Strategy

Recall that a station having a packet ready to transmit selects its elective action from the range $[0,...,E-1]$. Selecting an action a means transmitting a pilot after an a-slot deferment. Let c_a be the number of stations that have selected action a in the current protocol cycle, thus the vector $C = (c_0,...,c_{E-1})$ reflects the actions selected by all the stations. A winner policy defines a winning action (or a no-winner contention) by specifying a binary-valued *payoff function* $u_a(C)$, with $u_a(C) = 1$ naming a as the winning action. It also defines feasible actions for each station given its recent behavior. In a *plausible* winner policy,

- $u_a(C) = 1$ implies that $c_a = 1$ and $u_x(C) = 0$ for all $x \neq a$, and
- for any a there exists a C such that $c_a > 1$ and $u_x(C) = 1$ for some $x \neq a$.

The latter condition, related to the notion of protectiveness [13], precludes "fail-safe" actions that render any other action non-winning, as well as trivial strategies based on repeatedly taking such actions in order to discourage other stations (note that neither CSMA/CA nor RT/ECD qualifies, the action 0 being "fail-safe").

Let $F(C)$ be the observable channel feedback upon a set of actions reflected by C. For example, take Fig. 2 and assume that $E = 4$. In the first protocol cycle, $C = (0, 0, 2, 2)$ and $F(C) = $ (empty, empty, carrier, empty) i.e., no station selects 0 or 1 (two empty slots), next a pilot collision in slot 3 (a carrier slot) is followed by no reaction (an empty slot). In the second protocol cycle, $C = (0, 0, 1, 3)$ and $F(C) = $ (empty, empty, carrier/successful, carrier) i.e., the non-recipients of the pilot in slot 3 perceive a carrier slot, whereas the recipients perceive a successful slot and react thus producing another carrier slot. Denote $C_a = (c_0,...,c_a)$. A winner policy should be

- *feedback compatible* i.e. (with a slight abuse of notation), $u_a(C) = u_a[F(C_a)]$,
- *incentive compatible* i.e., if $c_a = 1$ and $u_x[F(C_x)] = 0$ for all $x < a$ such that $c_x > 0$, then $u_a[F(C_a)] = 1$, and
- *verifiable* i.e., a station selecting an infeasible action a or attempting a rational deviation from the protocol by generating a channel feedback F such that $u_a(F) = 1$ is verifier detectable.

Feedback compatibility ensures that all stations perceive the same winner based on the observed channel feedback and that each station is able to determine its payoff immediately upon the action it has selected. E.g., this rules out hash-based policies [9] whereby winning actions are only decided upon gathering the whole C. Incentive compatibility ensures that no action is dismissed a priori as non-winning based on the channel feedback observed so far – otherwise stations might be unwilling to take any actions or certain slots would be unused. This rules out an RT-like policy whereby a second-shortest deferment wins or one whereby a shortest deferment only wins if it is "sufficiently large." Checking a station for action feasibility should be based on recent past since a verifier may be unable to track a station for long.

A regular station calculates (and a greedy station also self-optimizes) its bandwidth share based on its payoffs in a number of protocol cycles. Let the respective shares be U_r and U_g, and let U_{rc} correspond to a cooperative MAC setting ($G = 0$). We seek a winner policy that is both *fair*, in that U_r is comparable with U_{rc}, and *efficient*, in that U_{rc} is comparable with U_{rc} under RT/ECD. Fairness and efficiency are not a winner policy's features; rather, they depend on the class of permissible greedy strategies. Assuming verifiability, the only viable greedy strategy consists in selecting "shorter-than-random" deferments. A permissible greedy strategy is *isolated* i.e., not colluding with other greedy stations (whose number and status it has no means of knowing), and *rational* i.e., aiming to maximize U_g and not to just diminish U_r at the price of self-damage; this implies that stations currently without packets to transmit select no action, and that a greedy strategy may revert to regular if $U_g < U_r$ or $U_g < U_{rc}$.[2]

5 RT/ECD-Z Winner Policy

Intuitively, a smart enough greedy strategy quickly "learns the game" against a simple regular strategy based on randomization. This it does by systematically selecting "shorter-than-random" deferments. In view of feedback and incentive compatibility, discrimination of short deferments is not possible via the payoff function alone. The idea of the proposed family of policies, called *RT/ECD with Collision Count and Penalties*, is to combine a suitable payoff function and recent behavior-based definition of action feasibility to create a tension between the immediate gain from a short deferment and a diminished performance in near future. Given a parameter Z from the range $[0,...,E-1]$, put $u_a(C) = 1$ if $c_a = 1$ and

- there is no $x < a$ such that $c_x = 1$ i.e., a yields the first successful pilot, and
- the number of distinct x's such that $x < a$ and $c_x > 0$ is less than Z i.e., C_a yields fewer than Z pilot collisions.

If no such a exists, a no-winner contention is perceived; in that case, let x_Z be the maximum deferment followed by a pilot from any station (reaction slots not counting) i.e., x_Z is the Zth smallest x such that $c_x > 0$. Action feasibility is checked based on penalties a station self-imposes, motivated by the possibility that a verifier has locked upon it and is tracing the intervals between successive pilots. An action a is feasible if $a \geq b$, where b is the current penalty self-imposed by the station. If the previous protocol cycle ended with a packet transmission then $b = 0$; otherwise

$$b = \begin{cases} E - a' - 1, & \text{if } a' \leq x_Z \\ \max\{0, b' - x_Z\}, & \text{if } a' > x_Z \end{cases} \quad (1)$$

where a' and b' are the station's selected action and self-imposed penalty in the previous protocol cycle. In particular, if a no-winner contention was perceived and

---

[2] However, not knowing N or the other stations' identities, a greedy station cannot reliably detect either. Gradient-based search for a higher U_g may not help if the current play is close to a Nash equilibrium [5]. Thus an ill-designed greedy strategy may lead to a lose-lose situation where both U_g and U_r are low. For a discussion of rational behavior see e.g., [4].

$a' = 0$ then $a = E - 1$, if $a' = 1$ then $a \geq E - 2$ etc., whereas stations that had no chance to transmit their pilots reduce their penalties, a mechanism resembling backoff freezing in IEEE 802.11. The above specification will be referred to as RT/ECD-Z.

Fig. 4 illustrates a possible scenario. Each elimination slot containing a pilot is followed by a reaction slot. Stations whose pilots collide and thus are not reacted to perceive themselves as non-winners and back off, while the rest may take their actions later. The protocol cycle continues until a successful pilot is reacted to and followed by a packet (in which case the penalties become irrelevant), or Z pilot collisions occur (and the penalties are recalculated), or E elimination slots elapse.

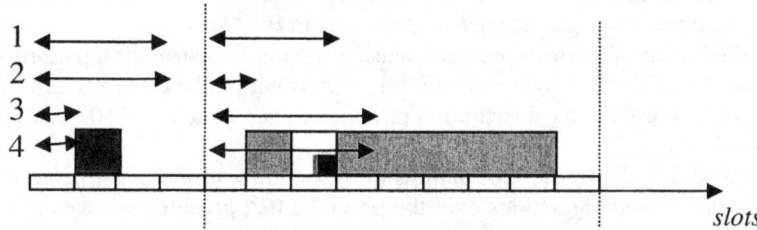

Fig. 4. RT/ECD-1, $E = 4$ (penalties at station i are denoted b_i); initially, $b_1 = b_2 = 2$, $b_3 = 1$, $b_4 = 0$; stations 3 and 4 select deferment 1; no reaction and no-winner contention after pilot collision ($x_Z = 1$); in the next protocol cycle, $b_1 = b_2 = 2 - 1 = 1$, $b_3 = b_4 = 4 - 1 - 1 = 2$ enable station 2 to select deferment 1 and win (note that active deferments are frozen during reaction slots)

The choice of Z is a compromise between no-winner contentions and penalty relevance: for $Z = E - 1$ penalties are irrelevant, but no-winner contentions are rare. ($Z = 1$ combined with $b_i \equiv 0$ yields RT/ECD.) We summarize Sects. 4 and 5 as follows.

Proposition: RT/ECD-Z is plausible, feedback compatible, incentive compatible, and verifiable.

Note that any rational deviation should consist in either disregarding the penalty, or a packet transmission not preceded by a pilot, or transmitting more than one pilot in one protocol cycle, or finally, jamming other stations' pilots and subsequently transmitting one's own. All these deviations are verifier detectable. At most $2E$ slots of continuous lock on a particular station are required on the part of a verifier. Moreover, it need not distinguish successful pilots from pilot collisions: refraining from a reaction upon the former or issuing one upon the latter is not rational.

6 Performance Evaluation

In a series of simulation experiments under heavy load, various strategies were used to obtain U_r and U_g for RT/ECD-Z against the backdrop of RT/ECD. Simulation imitated the slot-wise channel state evolution as exemplified in Figs. 2, 3, and 5. Runs were repeated until the 95% confidence intervals shrank to 10% of the sample averages. In each run, $N = 8$, $E = 10$, $L = 50$, and G were fixed. Escalation of "aggressiveness" (suggested in the footnote in Sect. 1) was found to lead to poor efficiency. Each of the eight heuristic strategies briefly described below was adopted in all regular stations and played against itself and each of the other seven, employed in all greedy stations, producing 36 regular vs. greedy strategy scenarios. Strategies 1 and 2 are better suited for regular stations because of their simplicity, while strategies 3 through 8 are better suited for greedy stations as they employ reinforcement learning [4]. The latter define an *update period* (UP) spanning a number of recent protocol cycles (20 except for the initial UP, which was of random length to make the learning asynchronous across the stations). The experimented strategies featured:

1. uniform probability distribution of actions (designated "neutral" in Fig. 3),
2. truncated geometric probability distribution of actions with parameter $q = 0.5$ i.e., biased toward 0 (designated "aggressive" in Fig. 3),
3. adjustment of the truncated geometric probability distribution parameter based on the comparison of own and winning actions within the previous UP,
4. uniform probability distribution of actions over a subset of $\{0,...,E-1\}$ adjusted similarly,
5. probability distribution of actions corresponding to the constructed histogram of *fictitious* winning actions over the previous UP; given C, a is a fictitious winning action if $c_a = 0$ and $u_a(C') = 1$, where C' coincides with C except that $c'_a = 1$,
6. cyclic sequence of actions within UP e.g., 1, 2, 3, 4, 1, 2, ..., with length and starting point adjusted based on own payoffs over the previous UP (this strategy is supposed to mimic token passing among a set of anonymous stations),
7. schedule of actions within an UP adjusted based on a technique similar to simulated annealing [7]: an action yielding the lowest sum of payoffs in the previous UP is tentatively replaced by another one whose sum of payoffs over the next UP, k, determines the probability of its final admittance into the schedule according to the formula $Pr[\text{admittance}] = 1/(1 + \exp(-k))$, and
8. schedule of actions within an UP adjusted based on somewhat modified simulated annealing, with an action admitted similarly as above except that k is the sum of payoffs and the number of no-winner contentions over the next UP.

Ideally $U_r = U_g = 1/N$ of the available bandwidth, an "ideally fair and efficient" share. Scheduling overhead causes it to drop even in a cooperative MAC setting (at $G = 0$), whereas "bandwidth stealing" (at $G > 0$) may bring about discrepancies between U_r and U_g and a further decrease in U_r. We take the viewpoint of a regular station and examine U_r (normalized with respect to $1/N$) as a function of G, the winner policy parameter Z and adopted election strategies.

Sample results are plotted in Fig. 5 for RT/ECD-Z with $Z = 1, ..., 4$ (since $N = 8$, the maximum number of pilot collisions per protocol cycle is 4). In Fig. 5a, regular strategy 1 was played against greedy strategy 2. It can be seen that strategy 1 completely fails for RT/ECD, but copes with strategy 2 for RT/ECD-Z with $Z > 1$ regardless of G. Unfortunately, as seen in Fig. 5b, should greedy stations adopt strategy 7, U_r can fall as low as 30% to 40% of the "ideally fair and efficient" bandwidth share for intermediate values of G unless $Z = 1$. Interestingly, the smart strategy 7 appears a little capricious: for small Z and large G it fails to "learn the game" and permits U_r in excess of 60%; however, in other cases strategy 1 is distinctly cut off from the channel. When greedy stations adopt any of the strategies 4, 5, or 6, we get a similar picture. It turns out from further experiments that strategy 7

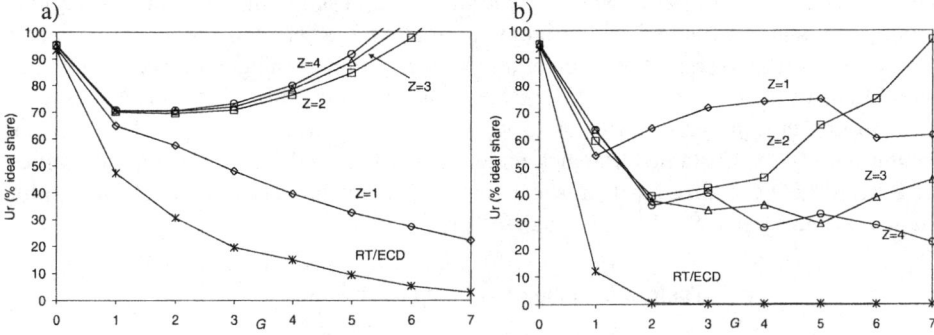

Fig. 5. Strategy 1 performance under RT/ECD-Z and RT/ECD vs. a) strategy 2, b) strategy 7

copes well with any other strategy regardless of G and Z; being somewhat capricious, it requires more research in order to be standardized as a regular strategy. The other strategies fare better or worse depending on Z and the strategy they play against.

A more systematic approach to winner policy evaluation is possible given an exhaustive set S of conceivable strategies within the framework of Sec. 4. Since the above eight strategies do not constitute such a set, although they do cover a wide range of common-sense heuristics, our further considerations are only indicative of results obtainable with a broader set of strategies. A conjecture based on research into a number of heuristic election strategies other than 1,...,8 is that there is little chance of finding a strategy which exhibits a qualitatively different behavior.

Let $U_r(s, t; G)$ denote the regular bandwidth share when a regular strategy $s \in S$ plays against a greedy strategy $t \in S$, there being G greedy stations. Define the *guaranteed* regular bandwidth share $U(s, t) = \min_{1 \leq G \leq N-1} U_r(s, t; G)$. In search for good candidates for a standard regular strategy, an important consideration is related to the notion of *evolutionary stability* [5, 14]. Informally, a standard regular strategy s should be among the best opponent strategies to s, and for any best opponent $t \neq s$, s should be the single best opponent to t. This precludes any rational deviations from s from being regarded "as good as the standard" and thus from initially being adopted at some stations while most stations adopt s, subsequently competing with s and finally supplanting s in a process that models natural evolution. We shall modify this notion with reference to the set S and considering that estimation of the obtained bandwidth share may be subject to error $\varepsilon > 0$. Thus a strategy s will be called *evolutionarily* (S, ε)-*stable* if it fulfills the following two conditions:

$$\forall t \in S \ \ U(t, s) \leq U(s, s) \tag{2}$$

$$\forall t \neq s \ [\text{if } U(t, s) \geq (1 - \varepsilon)U(s, s) \text{ then } \forall \ s' \neq s \ \ U(s', t) < U(s, t)] \tag{3}$$

Strategies fulfilling the "if" condition in (7) may be called (S, ε)-*best opponents* of strategy s. Furthermore, it is natural to require of an evolutionarily (S, ε)-stable strategy s that both $U^*(s) = U(s, s)$ and $U^{**}(s) = \min_{t \in S} U(s, t)$ be large. The former represents the regular bandwidth share in a cooperative setting when all the stations adopt strategy s i.e., U_{rc}, whereas the latter represents the guaranteed regular bandwidth share achieved by strategy s against the *hardest* opponent strategy (possibly itself) and should be comparable with U_{rc}. In designing a winner policy, one should ensure that a strategy fulfilling the above requirements exists (ideally exactly one, so that no ambiguity arises as to which regular strategy to adopt). Table 1 lists evolutionarily (S, ε)-stable strategies under RT/ECD-Z and the corresponding values of Z and U_{rc} (normalized with respect to the "ideally fair and efficient" bandwidth share), assuming that $S = \{1,...,8\}$ and $\varepsilon = 0.1$.

Table 1. Evolutionarily (S, 0.1)-stable strategies

strategy	Z	U_{rc} (% of ideal share) w.r.t. each Z value
1	1	93.8
2	1	65.2
5	1, 2, 3, 4	97.5, 97.5, 97.5, 97.5
7	1, 2, 3, 4	78.2, 88.7, 88.2, 87.9
8	1	87.9

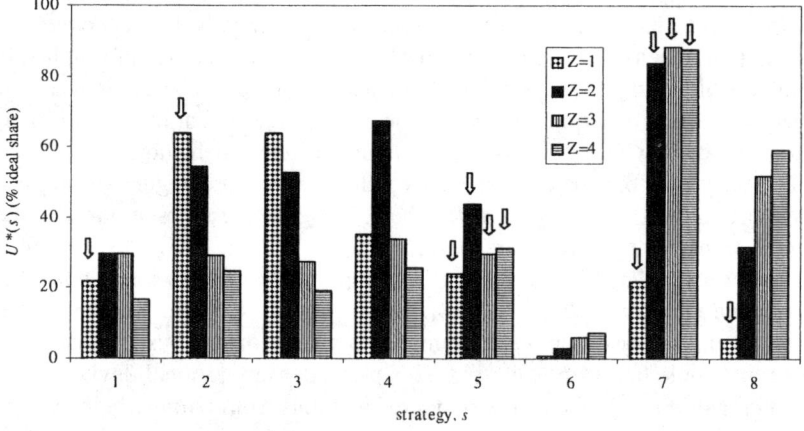

Fig. 6. Minimum guaranteed bandwidth share against hardest opponent

Considering evolutionary stability alone may be misleading since in general a strategy s may fare worse against a strategy t that is not its (S, ε)-best opponent than against one that is. For example, greedy stations may come up with "not too rational" a strategy or may be unable to compute a (S, ε)-best opponent to s. Take $Z = 2$, and $s = 7$. We have $U(7, 7) = 88.7\%$ and yet it turns out that $U(7, 5) = 67.1\% < U(7, 7)$ even though strategy 5 is not strategy 7's (S, 0.1)-best opponent ($U(5, 7) = 43.9\%$). In view of this, $U^{**}(s)$ is a more conservative measure. Fig. 6 depicts $U^{**}(s)$, with evolutionarily (S, 0.1)-stable strategies indicated by arrows. Note in passing that the supposedly token passing-like strategy 6 fares poorly against any opponent strategy, apparently failing to establish a valid token ring in our anonymous setting. Of the two strategies that remain evolutionarily (S, 0.1)-stable for all Z, strategy 7 yields a higher $U^{**}(s)$ for $Z > 1$, its relatively high complexity and capriciousness notwithstanding. For $Z = 1$, the simple strategy 2 makes a good candidate, yielding a distinctly lower $U^{**}(s)$, however. In our experiments, $Z = 3$ looks like optimal design, rendering only strategies 7 and 5 evolutionarily (S, 0.1)-stable, the former clearly superior with respect to $U^{**}(s)$. The fact that a distinctly superior candidate emerges and that $Z = 3$ is an intermediate rather than extreme value confirms the usefulness of RT/ECD-Z. In view of the discussion related to Fig. 5 it is obvious that RT/ECD is not satisfactory.

7 Conclusion

Designing contention mechanisms for anonymous stations is a reasonable approach, as it gives some "safety upper bounds" for mechanisms relying on permanent station identities. A framework for a noncooperative MAC setting, winner policy and greedy election strategy has been proposed. For a class of Random Token MAC protocols based on voluntary deferment of packet transmissions, a new family of winner policies under the name RT/ECD-Z has been presented that guarantees regular stations a close-to-fair share of the available bandwidth under saturation load. The proposed policies make it hard for a greedy station to decide a priori on short deferments, which are advantageous under existing policies. Therefore greedy stations resort to smarter strategies and the task of the winner policy is to enable a regular strategy that holds its own against any greedy strategy. We have formalized this idea by requiring evolutionary stability and high guaranteed regular bandwidth shares within a set of heuristic strategies. Directions for future research include extensions to multihop wireless topologies, more complex traffic environments and QoS issues.

References

1. Cagalj, M., Ganeriwal, S., Aad, I., Hubaux, J. -P.: On Cheating in CSMA/CA Ad Hoc Networks, Proc. IEEE INFOCOM 2005, Miami FL, March 2005
2. Chlamtac, I., Ganz, A.: Evaluation of the Random Token Protocol for High-Speed and Radio Networks, IEEE J. Select. Areas of Comm. JSAC-5 (1987) 969-976
3. ETSI TC Radio Equipment and Systems: High Performance Radio Local Area Network (HIPERLAN); Services and Facilities; Version 1.1, RES 10 (1995)
4. Friedman, E. J., Shenker, S.: Synchronous and Asynchronous Learning by Responsive Learning Automata, Mimeo (1996)
5. Fudenberg, D., Levine, D. K.: The Theory of Learning in Games, MIT Press 1998
6. IEEE 802.11 Standard: Wireless Media Access Control (MAC) and Physical Layer (PHY) Specifications (1999)
7. Ingber, L.: Simulated Annealing: Practice versus Theory, Mathl. Comput. Modelling 18 (1993) 29-57
8. Konorski, J.: Packet Scheduling in Wireless LANs – A Framework for a Non-cooperative Paradigm, Proc. IFIP Int. Conf. on Personal Wireless Comm., Kluwer (2000) 29-42
9. Konorski, J., Kurant, M.: Application of a Hash Function to Discourage MAC-Layer Misbehaviour in Wireless LANs, J. Telecomm. and Inf. Technology 2 (2004) 38-46

10. Kyasanur, P., Vaidya, N. H.: Detection and Handling of MAC Layer Misbehavior in Wireless Networks, Proc. Int. Conference on Dependable Systems and Networks, San Francisco CA, June 2003
11. MacKenzie, B., Wicker, S. B.: Selfish Users in ALOHA: A Game-Theoretic Approach, Proc. Vehicular Technology Conference Fall 2001, Atlantic City NJ, Oct. 2001
12. Raya, M., Hubaux, J. -P., Aad, I.: DOMINO: A System to Detect Greedy Behavior in IEEE 802.11 Hotspots, Proc. MobiSys 2004, Boston MA, June 2004
13. Shenker, S.: Making Greed Work in Networks: A Game-Theoretic Analysis of Switch Service Disciplines, Proc. SIGCOMM'94, London UK, June 1994
14. Yao, X.: Evolutionary Stability in the n-Person Iterated Prisoners' Dilemma, BioSystems 39 (1996) 189–197

A Power Aware Routing Strategy for Ad Hoc Networks with Directional Antenna Optimizing Control Traffic and Power Consumption

Sanjay Chatterjee[1], Siuli Roy[1], Somprakash Bandyopadhyay[1],
Tetsuro Ueda[2], Hisato Iwai[2], and Sadao Obana[2]

[1] Indian Institute of Management, Calcutta 700104, India
sc@iimcal.ac.in
[2] ATR Adaptive Communications Research Laboratories, Kyoto 619-0288, Japan
http://www.acr.atr.jp/acr/top-e.html

Abstract. This paper addresses the problem of power aware data routing strategies within ad hoc networks using directional antennas. Conventional routing strategies usually focus on minimizing the number of hops or route errors for transmission but they do not usually focus on the energy depletion of the nodes. In our proposal, if a node in the network has depleted its battery power, then an alternative node would be selected for routing so that not only the power is used optimally but there is an automatic load sharing or balancing among the nodes in the network. The usage of directional antenna in this scheme has some key advantages outperforming the omni-directional counterpart. The space division multiple access, range extension capabilities and power requirement of the directional antenna is itself a reason for its choice. We illustrate how directional antenna can be combined with the power aware routing strategy and using simulations, we quantify the energy benefits and protocol scalability.

1 Introduction

In an ad hoc network mobile, hosts depends on the assistance of the other nodes in the network to forward a packet to the destination in case the destination node is multi-hop away from the source. Thus each node may also act as a router.

One of the major concerns here is how to decrease the power usage or battery depletion level of each node among the network so that the overall lifetime of the network can be stretched as much as possible. In conventional routing schemes, the same node may be selected repeatedly, thereby causing severe depletion in its energy level. In our proposal, if a node in the network has heavily depleted its battery power, then an alternative node would be selected for routing so that not only the power of each node is used optimally but there is an automatic load sharing or balancing among the nodes in the network. The usage of directional antenna has some key advantages which outperforms the omni-directional counterpart. The space division multiple access and the range extension capabilities of the directional antenna is itself a reason for its choice. The power requirement of the directional antenna is also much less than that of the omni-directional version covering the same range. A salient feature of directional antenna is that it doesn't overhear the nodes outside its own cone of coverage and allows simultaneous communication without interference. This additionally helps to reduce power depletion of nodes. We illustrate how directional antenna can be combined with the power aware routing strategy and, using simulations, we quantify the energy benefits and protocol scalability. Our initial evaluation offer encouraging results, indicating the potential benefits of power aware routing using directional antenna.

2 Related Work

A survey of power optimization techniques for routing protocols in wireless networks can be found in [1]. Suresh Singh et al. [2] presented five power aware metrics. The protocol is based on the original MACA protocol with the addition of a separate signaling channel. The manner in which nodes power themselves off in this scheme does not influence the delay or throughput characteristics of the protocol. However, the power balancing among the nodes cannot be guaranteed, thereby causing non-uniform power conservation characteristics of nodes.

An online approximation algorithm for power-aware message routing has been proposed in [3]. An algorithm that requires accurate power values for all the nodes in the system at all times. They further proposed a second algorithm which is hierarchical, known as Zone-based power-aware routing partitioning the ad-hoc network into small number of zones. Each zone can evaluate its own power level. These power estimates are then used as weights for the zones. A local path for the message is computed so as not to decrease the power level of the zone too much moreover, formation of hierarchical zone and its maintenance is a serious problem in dynamic ad hoc networks.

In our proposed strategy, each node knows the approximate battery power status of the other nodes and topology information. This is done through periodic propagation of power status along with topology information. To minimize the power usage, directional ESPAR antennas [4] have been used. We illustrate how directional antenna can be combined with the power aware routing strategy using a modified version of the Mac protocol, developed in our earlier work [5].

3 System Description

In order to fully exploit the capabilities of directional antenna, all the neighbors of a source and destination should know the direction of communication so that they can initiate new communications in other directions, thus preventing interference with ongoing data communication between source and destination. Thus, it becomes imperative to have a mechanism at each node to track the direction of its neighbors and get some vital information like power status and neighborhood information. A model of an ESPAR antenna, a low-cost, low-power, small-sized smart antenna, has been used in our simulation experiments.

3.1 Location Tracking and MAC Protocol

In our framework, each node waits in omni-directional-sensing mode while idle. Whenever it senses some signal above a threshold, it enters into rotational-sector-receive-mode. In rotational-sector-receive mode, node n rotates its directional antenna sequentially in all directions at 30 degree interval, covering the entire 360 degree space in the form of the sequential directional receiving in each direction and senses the received signal at each direction. After one full rotation, it decides the best possible direction of receiving the signal with maximum received signal strength. Then it sets its beam to that particular direction and receives the signal. We have used three types of broadcast (omni-directional) control packets: Global Link State Table (GLST), RTS (Request to send) and CTS (clear to send) for medium access control. Data packets and the control packet ACK is a directional control packet. A detailed description of directional MAC is illustrated in [6].

3.2 Information Percolation Mechanism in the Network

The purpose of information percolation mechanism is to make each node aware of the approximate topology and the power depletion status of each node in the network. The objective here is to get accurate local, but approximate global perception of the network information. This awareness would be helpful to implement both MAC and a power-aware routing protocol using directional antennas.

3.3 Global Link-State Table (GLST)

It contains the global network topology information as well as the battery power status of the corresponding nodes as perceived by a node n at that instant of time. Each node broadcasts a beacon at a periodic interval, say T_A. When a node n receives a beacon from all or any of its neighbors (say node i, j and k), node n forms the GLST(n) to include node i, j and k as its neighbors and records the best possible direction of communicating with each of them and even their battery power status. Initially when the network commences, all the nodes are just aware of their own neighbors and are in a don't-know-state regarding the other nodes in the system. Periodically, each node broadcasts its GLST as update to its neighbors thereby slowly updating the entire network about the topology [6].

4 Power Aware Routing Strategy

A lot of effort is currently going on to reduce the power consumed in a mobile device within the ad hoc network and our power aware routing strategy can ensure optimal usage of battery power of each node. It is to be noted that our proposed strategy not only balances the battery usage of each node extending the network life but it also ensures network traffic balancing when the congestion is high.

When following only the shortest path algorithm it will be observed that source and intermediate nodes will deplete their power much more early then their neighbors. Consider the following topology, as shown in Fig 1.

Here, packets are to be sent from node 1 to node 3. Let us assume that the shortest path algorithm selects 1 -> 6 -> 3 as the best path. Disregarding the source node 1 and destination node 3 (which are fixed in this case), it will be observed that the intermediate node 6 will suffer heavy depletion in its battery power because only node 6 is selected repeatedly as intermediate node by the shortest path algorithm.

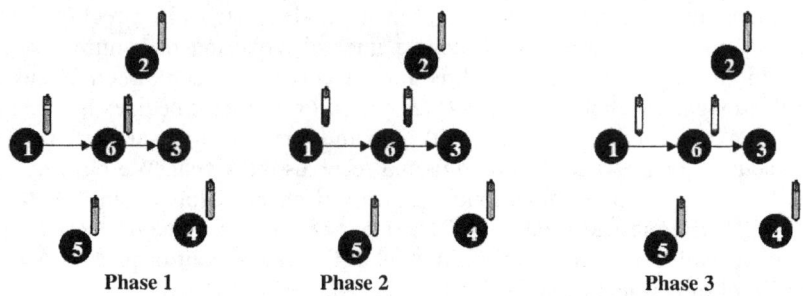

Fig. 1. Battery Status without Power Aware Routing with Shortest Path Algorithm

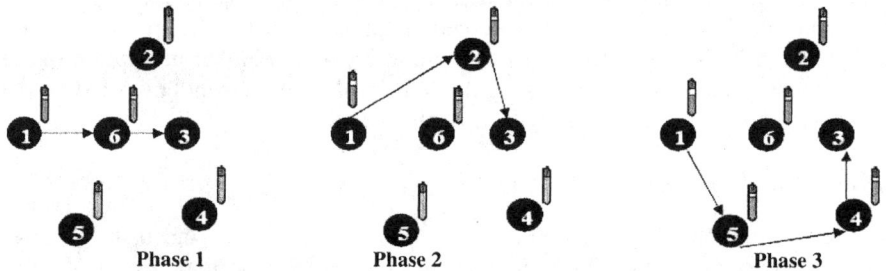

Fig. 2. Battery Status with Power Aware Routing together with Shortest Path Algorithm

Now let us shift our focus on our proposed algorithm for route selection using residual power aware routing strategy. Fig 2 represents the case where data packets are forwarded using this strategy from the same source to destination. After phase 1, the battery of intermediate node 6 has depleted by 10 % (say) and so in phase 2, node 6 will not be considered. An alternate path 1 -> 2 > 3 will be selected (say next shortest path), since node 6 has less battery power than that of node 2.

Now let us consider phase 3 in Fig 2. Both node 2 and 6 have depleted their power by 10% (say). For transmission of next set of data packets, both the intermediate nodes would be rejected and intermediate nodes 5 and 4 will be selected (1 -> 5 -> 4 -> 3), since they have their battery power much higher than node 2 and 6. It is to be noted that not only the power is used optimally but there is an implicit property of the algorithm to automatically balance the network traffic and distribute it in an even fashion choosing different best paths from source to destination.

5 Performance Evaluation

The simulations are conducted using QualNet 3.1 network simulator using the ESPAR antenna model. 60 nodes are placed over 1000 x 1000 sq. meter area using the grid topology with transmission power of 10dBm. Nodes are randomly chosen to

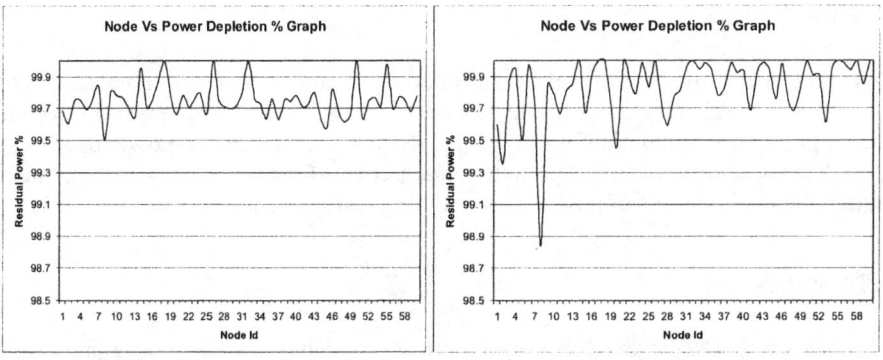

Fig. 3a. With Power Control **Fig. 3b.** Without Power Control

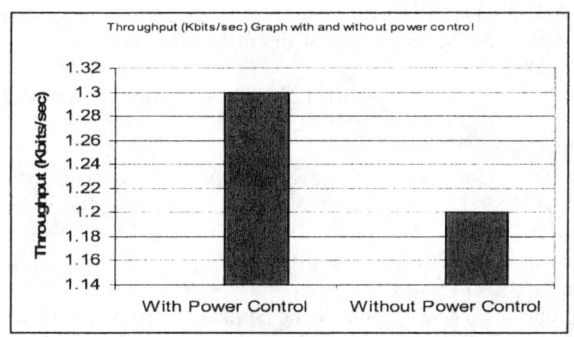

Fig. 4. Throughput in Static scenario

be CBR (constant bit rate) sources, each of which generates 512 bytes data packets to a randomly chosen destination at a rate of 2 to 500 packets per second. The entire simulation period is of 7 minutes with 4 pairs of CBR traffic.

Fig 3a and 3b shows the power depletion graphs in a static scenario. Fig 3a represents the nature of power depletion characteristics among the nodes when our power aware routing strategy is used. Fig 3.b on the other hand shows the power depletion characteristics without our power aware routing strategy but using only the shortest path algorithm. A close study reveals the fact that some nodes in Fig 3.b suffer heavy depletion, although most of the nodes have nearly the same initial power. These results in early die out of some nodes in the network and thus the entire network may get partitioned into two or more sub networks. In other words, multi-hop communication would be restricted to a great extent because the intermediate nodes have died out much earlier than the neighbors which still have more battery power. Now we shift our focus on Fig 3.a which shows the power depletion graph characteristics when our power aware routing strategy is used. This graph represents a uniform power depletion curve, leading to increased life-time of the network. Fig 4. represents the throughput of the network in a static scenario. The underlying reason for improved throughput with power-aware routing is the automatic load balancing nature of the algorithm, as illustrated in Section 4.

6 Conclusion

This strategy mainly optimizes the power depletion and maintains a more or less uniform power usage among all the nodes in the network while maintaining effective throughput. In our simulation, we observe a sharp performance and power usage gains using the proposed algorithm. Our initial evaluation offer encouraging results, indicating the potential benefits of power aware routing using directional antenna.

References

1. S. Lindsey, K. Sivalingam, C.S Rahgavendra: Power optimization in routing protocols for wireless and mobile networks. Wireless Networks and Mobile Computing Handbook, Stojmenovic I (ed.), John Wiley & Sons: 2002; 407-424
2. Suresh Singh, Mike Woo, CS Raghavendra: Power-Aware Routing in Mobile Ad Hoc Networks. MOBICOM 1998
3. Qun Li, Javed Aslam, Daniela Rus: Online power-aware routing in wireless Ad-hoc Networks. Proceedings of the Seventh Annual International Conference on Mobile Computing and Networking. 2001
4. T. Ohira and K.Gyoda: Electronically Steerable Passive Array Radiator (ESPAR) Antennas for Low-cost Adaptive Beam forming. IEEE International Conference on Phased Array Systems, Dana Point, CA May 2000
5. Siuli Roy, Dola Saha, Somprakash Bandyopadhyay, Tetsuro Ueda, Shinsuke Tanaka.: A Network-Aware MAC and Routing Protocol for Effective Load Balancing in Ad Hoc Wireless Networks with Directional Antenna. Proc. of the Fourth ACM International Symposium on Mobile Ad Hoc Networking and Computing (MobiHoc 2003) Annapolis, Maryland, USA, June 1-3, 2003
6. Tetsuro Ueda, Shinsuke Tanaka, Dola Saha, Siuli Roy, Somprakash Bandyopadhyay: A Rotational Sector-based, Receiver-Oriented mechanism for Location Tracking and medium Access Control in Ad Hoc Networks using Directional Antenna. Proc. of the IFIP conference on Personal Wireless Communications PWC 2003. September 22-25, 2003 - Venice – ITALY

Power Aware Cluster Efficient Routing in Wireless Ad Hoc Networks

Sanjay Kumar Dhurandher[1] and G.V. Singh[2]

[1] Division of Computer Engineering, Netaji Subhas Institute of Technology, New Delhi
dhurandher@rediffmail.com
[2] School of Computer & Systems Sciences, Jawaharlal Nehru University, New Delhi
gvs10@hotmail.com

Abstract. In Ad Hoc networks a routing protocol is either proactive or reactive. The former maintains consistent up-to-date routing information from each node to every other node in the network, whereas the latter creates route to the destination only when desired by the source node using "flooding". In flooding packets are broadcast to all destinations with the expectation that they eventually reach their intended destination. This proves to be very costly in terms of the throughput efficiency and power consumption. For reactive protocols, researchers have tried to enhance the throughput efficiency and reduce power consumption using techniques that cut down flooding. In this paper we propose a routing protocol called Power Aware Cluster Efficient Routing (PACER) protocol for multi-hop wireless networks. In PACER, the network is dynamically organized into partitions called clusters with the objective of maintaining a relatively stable effective topology. The protocol uses the Weight Based Adaptive Clustering Algorithm (WBACA), developed by us for cluster formations. The main objective is to significantly reduce the number of overhead messages and the packet transfer delay. We demonstrate the efficiency of the proposed protocol with respect to average end-to-end delay, control overheads, throughput efficiency and the number of nodes involved in routing.

1 Introduction

Ad Hoc networks are peer-to-peer, multi-hop mobile wireless store-and-forward packet transfer networks. The low resource availability in these networks necessitates their efficient utilization; hence the motivation for optimal routing in mobile Ad Hoc networks (MANETs). With an increase in the size of the networks flat routing schemes do not scale well in terms of performance. The routing tables and topology information in the mobile stations also gets tremendously large. Routing schemes such as DSR [3] that perform well for small networks results in low bandwidth utilization in large networks because of high load and longer source routes. To solve this problem some kind of organization is required in large mobile Ad Hoc networks. The nodes in the network are grouped into easily manageable sets known as *clusters* [7]. Certain nodes, known as *clusterheads*, are responsible for the formation of clusters and maintenance of the topology of the network.

In this paper we are proposing a power aware cluster efficient routing (PACER) protocol that is highly efficient in terms of control overhead and delay in communication. The performance evaluation of the various routing algorithms is done in terms of achievable efficiency. The rest of this paper is organized as follows. The related work done in the area of routing is reviewed in Section 2, which includes an overview of AODV and DSR algorithms. The proposed routing algorithm is described in detail in Section 3. In Section 4, the simulation results demonstrating the efficiency of the proposed algorithm are presented. Finally, Section 5 concludes this paper.

2 Related Work

Routing in a MANET depends on many factors, such as modeling of the topology, selection of routers, initiation of request, and specific underlying characteristics that could serve as a heuristic in finding the path efficiently. The existing routing protocols can be classified either as proactive or reactive [5]. Proactive protocols attempt to evaluate continuously the routes within the network, so that when a packet needs to be forwarded, the route is already known and can be immediately used. The family of distance vector protocols such as Destination-Sequenced Distance Vector (DSDV) [2] routing is an example of a proactive protocol. Reactive protocols, on the other hand, invoke a route determination procedure only on demand. The family of classical flooding algorithms belongs to the reactive group of protocols. Some examples of reactive Ad Hoc network routing protocols are Ad Hoc On-Demand Distance Vector (AODV) protocol [1], Dynamic Source Routing (DSR) protocol, etc.

3 Proposed Routing Protocol

This section describes the proposed *Power Aware Cluster Efficient Routing (PACER)* protocol. This routing algorithm uses our previously developed Weight Based Adaptive Clustering Algorithm (WBACA) [11]. PACER is based on the concept of clustering and is a highly adaptive, loop-free, on-demand routing protocol. The key design concept of PACER is the minimization of control messages by limiting them to a very small set of nodes. To accomplish this, nodes need to maintain information about adjacent one-hop and two-hop nodes, which is obtained at the time of cluster formation. The route discovered for the destination node is stored at the clusterheads only that lie on the discovered path and not at the other intermediate nodes. This protocol performs two basic functions: route creation and route maintenance. The steps in the routing process are:

STEP1. Nodes in the network are identified as clusterheads, gateways and ordinary nodes using WBACA.

STEP2. When a packet is to be transmitted, the node checks if the destination node is present in its neighbor table.

STEP3. If the destination node is found in the neighbor table, the source node directly transmits the packet to the destination.

STEP4. If the destination node is not found in the neighbor table, then the source node checks its two-hop neighbor table. If the destination node is found there, then the transmission takes place through the intermediate node.

STEP5. If the entry of the destination node is not found in the two-hop neighbor table, then
  (a) If the node is an ordinary node, the node initiates a route discovery by sending a Route Request (RREQ) packet to its clusterhead.
  (b) If the node is a clusterhead, then it initiates a route discovery by sending a RREQ packet to all its gateway nodes.
  (c) If the node is a gateway, the node initiates a route discovery by sending a RREQ packet to its clusterhead.

In the case of an intermediate node, the node is either a clusterhead or a gateway.
  (a) If the node is a clusterhead, it stores the path list from the source node up to the current node in its route cache table and, then forwards the RREQ packet to all its gateway nodes.
  (b) If the node is a gateway, it forwards a RREQ packet to all one-hop clusterhead neighbors, leaving the clusterhead from which it received the RREQ packet. If the gateway node is not having any clusterheads as its neighbors, but has other gateway nodes as its neighbors, then the RREQ packet is forwarded to these gateway nodes.

STEP6. Each intermediate node appends itself in the path list. Whenever a clusterhead is encountered in the route, the clusterhead stores the path list.

STEP7. This process (i.e. steps 5, and 6) is continued till a route to the destination is found.

STEP8. Once the RREQ packet reaches the node, which has the destination node present in its two-hop neighbor table, it responds by unicasting a Route Reply (RREP) packet to the source node using the path list.

The route maintenance procedure is accomplished through the use of *route update*, *route modify* and *route error* messages. Steps involved in the route maintenance are:

STEP1. If the next-hop node on the route has moved or is not reachable, the current node generates a *Route Update* (RUPDT) packet and sends it to all the nodes in the path list up to the source node.

STEP2. The current node then tries to find if it can reach the next-hop node, by consulting its two-hop neighbor table.
  (a) If the current node finds the next-hop node in its two-hop neighbor table, it modifies the route in the path list and generates a *Route Modify* (RMOD) packet. This message is then sent to all the nodes up to the source node and each node then modifies its path list accordingly.
  (b) If the current node does not find the next-hop node in its two-hop neighbor table, it checks if it can reach the node next to the next-hop node in the path list by consulting its one-hop and two-hop neighbor tables. If found, it modifies the route in the path list and generates a

RMOD packet. This message is then sent to all the nodes up to the source node and each node then modifies its path list accordingly.

STEP3. In case step 2 fails, the current node starts the route creation procedure. On receiving a RREP, it modifies the route in the path list and generates a RMOD packet. This message is then sent to all the nodes up to the source node and each node then modifies its path list accordingly.

STEP4. In case both step 2 and step 3 fail, the current node generates a *Route Error* (RERR) packet and sends it to all the nodes up to the source node. This results in a new route creation procedure by the source node.

4 Simulation Study

The simulation experiments conducted for the performance evaluation were implemented in the Global Mobile Information System Simulator (GloMoSim) library [9]. GloMoSim is a scalable simulation environment for large wireless and wireline communication network systems using the parallel discrete-event simulation language called PARSEC [10]. The IEEE 802.11 [6] is used as the MAC layer. The roaming space considered is 2000x2000 meters square. Nodes move according to the random waypoint model [4].

To determine the efficiency of the proposed PACER protocol, we monitored four parameters: the control packet overhead, the average end-to-end delay, the number of nodes involved in routing, and the throughput. The control packet overhead is computed by counting the total number of control packets transmitted during the simulation period.

Figure 1 shows the average end-to-end delay for the three routing protocols as a function of the number of nodes in the network. The graph shows that PACER gives better performance than the other two protocols. DSR has the largest end-to-end delay.

Figure 2 shows the control overheads for the three routing protocols as a function of the number of nodes in the network. The larger the number of control packets, more is the power consumed in routing the data. Here, we observe that the control overheads increase with the increase in the number of nodes. It is found that PACER performs very well.

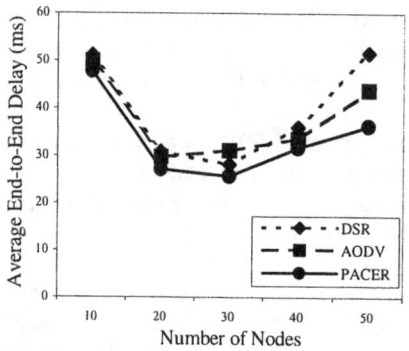
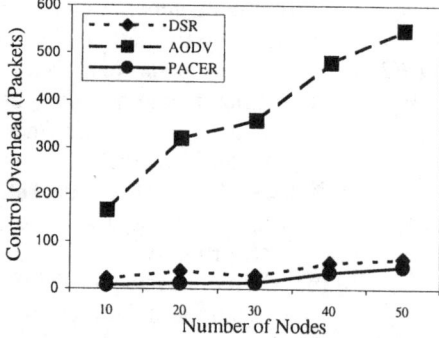

Fig. 1. Avg. End-to-End Delay vs. No. of Nodes

Fig. 2. Control Overhead vs. No. of Nodes

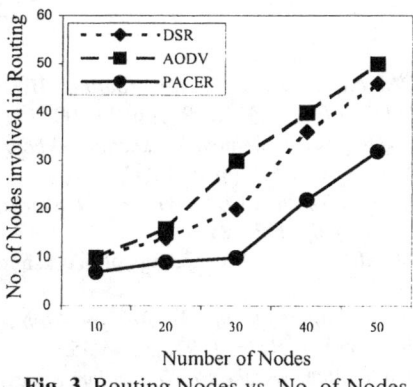 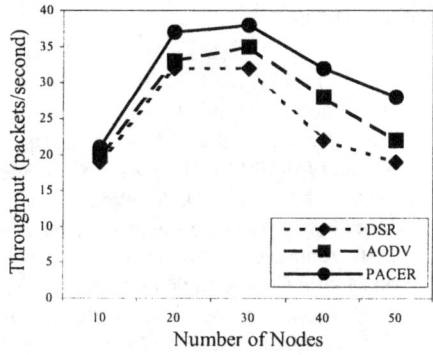

Fig. 3. Routing Nodes vs. No. of Nodes   **Fig. 4.** Throughput vs. No. of Nodes

Figure 3 illustrates the total number of nodes involved in the routing process. More number of nodes leads to more power dissipation in the network. As can be seen from the graph, PACER performs best. AODV gives the worst performance. AODV has almost all the nodes involved in routing. This is due to the flooding of packets in route discovery.

Figure 4 demonstrates the throughput achieved in case of the three routing protocols. PACER achieves better throughput than AODV and DSR. For a small number of nodes, the three protocols give almost the same performance. But, for a large number of nodes PACER is found to be the best. DSR is seen to have the lowest throughput.

5 Conclusion

In this paper, we have shown how routing can be applied with clustering in wireless mobile Ad Hoc networks. The proposed on-demand Power Aware Cluster Efficient Routing (PACER) is one such routing protocol, which can adapt itself to the changing topology of the network. The simulation experiments show that the proposed PACER protocol outperforms the existing AODV and DSR protocols with respect to power consumption, control overhead, throughput, number of nodes involved in routing and the average packet transfer delay. Currently, we are in the process of conducting simulation experiments for comparing PACER protocol with the Cluster Based Routing Protocol (CBRP) [8]. Our study till now shows that the PACER performs better than the CBRP.

Acknowledgement

This work has been supported by the research project funded by the All India Council for Technical Education (AICTE) at the School of Computer and Systems Sciences, Jawaharlal Nehru University (Grant No. 8020/RID/TAPTEC-53/2001-2002).

References

1. C. E. Perkins and E. Royer, *Ad Hoc On-Demand Distance Vector Routing*, IEEE Workshop on Mobile Computing Systems and Applications, Vol. 3, 1999, pp. 90-100
2. C. E. Perkins and P. Bhagwat, *Highly Dynamic Destination-Sequenced Distance-Vector Routing for Mobile Computers*, Computer Comm. Review, 1994, pp. 234-244
3. D. B. Johnson and D. A. Maltz, *Dynamic Source Routing in Ad Hoc Wireless Networks*, Mobile Computing, Kluwer Academic Publishers, 1996, pp. 153-181.
4. D. B. Johnson, *Routing in Ad Hoc Networks of Mobile Hosts*, in Proceedings of Workshop on Mobile Computing and Applications, Dec. 1997.
5. E. Royer and C. K. Toh, *A Review of Current Routing Protocols for Ad Hoc Mobile Wireless Networks*, IEEE Personal Communications, Vol. 7, No. 4, 1999, pp. 46-55.
6. IEEE Computer Society LAN MAN Standards Committee, *Wireless LAN Medium Access Protocol (MAC) and Physical Layer Specification*, IEEE Std. 802.11-1997.
7. M. Gerla and J. Tsai, *Multicluster, mobile, multimedia radio network*, ACM-Baltzer Journal of Wireless Networks, Vol.1, No.3, 1995, pp. 255-265.
8. M. Jiang, J. Li and Y. C. Tay, *Cluster Based Routing Protocol (CBRP) Functional Specification Internet Draft*, draft-ietf-manet-cbrp.txt, June 1999.
9. M. Takai, L. Bajaj, R. Ahuja, R. Bagrodia and M. Gerla, *GloMoSim: A Scalable Network Simulation Environment*, Technical report 990027, UCLA, 1999.
10. R. Bagrodia, R. Meyer, M. Takai, Y. Chen, X. Zeng, J. Martin, and H. Y. Song, *PARSEC: A Parallel Simulation Environment for Complex Systems*, IEEE Computer, Vol. 31, No. 10, 1998, pp.77-85.
11. S. K. Dhurandher and G. V. Singh, *Weight Based Adaptive Clustering in Wireless Ad Hoc Networks*, IEEE ICPWC, New Delhi, January 2005, pp. 95-100.

A New Routing Protocol in Ad Hoc Networks with Unidirectional Links*

Deepesh Man Shrestha and Young-Bae Ko

Graduate School of Information & Communication,
Ajou University, South Korea
{deepesh, youngko}@ajou.ac.kr

Abstract. Most of the proposed algorithms in ad hoc networks assume homogeneous nodes with similar transmission range and capabilities. However, in heterogeneous ad hoc networks, it is not necessary that all nodes have bidirectional link with each other and hence, those algorithms may not perform well while deployed in real situations. In this paper, we propose a scheme for an ad hoc on-demand routing protocol which utilizes the unidirectional links during the data transmission. Simulation shows that it is not only possible to use unidirectional links but it is also better in terms of performance metrics we defined in different situations.

1 Introduction

Ad hoc networks have emerged as a solution for the type of network where no infrastructure exists and various types of devices communicate with each other in a self-organizing fashion. Military scenarios, disaster relief situations are the examples where diverse communication equipments communicate in multi-hop fashion without any infrastructure. Since devices vary in types and capabilities, heterogeneity prevails in such network scenarios. However, many proposed algorithms assume homogeneous nodes with similar transmission radius and capabilities [1], and hence may not perform well while deployed in real situations.

A unidirectional link arises between a pair of nodes in a network when a node can send a message to another but not vice versa. Let us consider two nodes A and B. If A has the higher transmission range compared to B and the distance between them is greater than the transmission range of B, acknowledgement from B cannot be received by A. In this case both will assume that the link does not exist between them. One of the major causes for the existence of such links is the variation in transmission range of nodes. These links also arise due to collision or noise, which however does not persist for a long time.

The detection of unidirectional links provides two options for routing protocols: (1) either avoid the route or (2) utilize it for current data transmission.

---

* This work was supported by the Korea Research Foundation Grant funded by the Korean Government (R05-2003-000-10607-02004) and and also supported by the MIC (Ministry of Information and Communication), Korea, under the ITRC support program.

Avoiding the path with such links incur higher cost of route re-discovery and also lead to network partitions. On the other hand, utilization may cause variability in path affecting upper layer protocols. In this paper, we propose to utilize these links resulting from the disparity of transmission range due to heterogeneity. Using such links has an advantage of retaining the connectivity and using the shortest path route. We show that the routing protocols can effectively use it for data transmission without having to restart route discovery process.

In the performance analysis, proposed scheme is compared with AODV-EUDA [1] using random mobility and static model. We show that the proposed scheme is better based on metrics implying that using unidirectional links for on-demand ad hoc routing protocol is not only possible but also better in terms of efficiency. For the sake of readability, we refer to [1] as the AODV-EUDA.

In the next section we briefly describe research efforts that is close to our work. In Section 3 we present our scheme. In Section 4 we present performance analysis and finally conclude in Section 5.

2 Related Works

Problems encountered due to unidirectional links are uncommon as many routing protocols cannot function normally in such conditions. Unidirectional links affect AODV protocol [2] by causing route discovery failures even in presence of alternate bidirectional paths between source and destination. This is due to the occurrence of such links in the shortest path, where route replies fail to reach the source and re-discovery process recurrently attempts to find the path through same set of nodes. This problem is well illustrated in [1] and [3]. Some of the schemes that handles unidirectional links are studied in [4], [6], [7]. All of these previous approaches avoid the path containing unidirectional links. Our paper extends upon the recently proposed algorithm that detects unidirectional links called AODV-EUDA [1]. In AODV-EUDA detection is immediately done when it receives a RREQ packet during route discovery process. A node embeds its power information either in RREQ or a MAC frame. Each receiving node calculates the distance between itself and the RREQ sender from the parameters in RREQ and compares with its maximum transmit range. The link is unidirectional if its transmit range is shorter than the computed distance and hence discards that RREQ and waits for other RREQs from other bidirectional links. Unlike avoiding unidirectional links detected in AODV-EUDA, in our scheme, we utilize unidirectional links for data packet delivery.

3 Routing with Unidirectional Link

For the purpose of utilizing unidirectional links our scheme requires two steps. In first step, a node detecting a unidirectional link (as in AODV-EUDA) initiates election mechanism for selecting a monitor node. A *monitor node* is a node in a routing path that has a bidirectional link with both sender and receiver. In second step we utilize unidirectional link for successfully transmitting data by

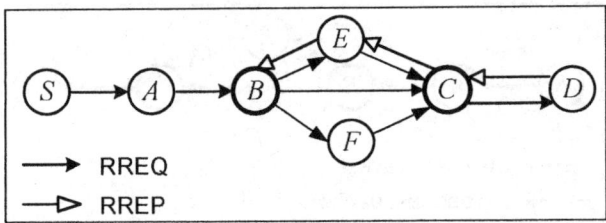

Fig. 1. E and F sends RREQ to C that decides one as a monitoring node. Here, E (monitor) replies with ACK to B.

local broadcast and receive acknowledgement from the monitor node. Detailed operation of our scheme is presented below.

3.1 Election of Monitoring Node

During route discovery process, while RREQ is being forwarded from the source to destination, node that detects the unidirectional link buffers them instead of forwarding immediately to other nodes for some time period. During this period, if it receives RREQs from the node that has a bidirectional link with itself and the sender, it selects a monitor node from which the first RREQ is received. Note that the collected RREQs must be the ones from the sender node with which the receiver has a unidirectional link. A sender is made aware about the monitor node when RREP is received back from the receiver.

In Fig 1, both E and F send RREQ to C and have bidirectional links with both B and C. C does not immediately forward the RREQ that was received from B, unless other RREQs are received from E and F. So assuming that E's RREQ is received earlier then F, C will select E as the monitor. In the process of sending RREP back to the source, sender B receives RREP from the monitor and hence is informed about the unidirectional link with C.

3.2 Utilizing Unidirectional Links

A sender node aware of unidirectional link needs to locally broadcast data packets so that they can be received by its neighbor nodes. A receiver node with a path further unicast these data packets towards the destination.

A monitor node in between receives passive acknowledgement through overhearing and passes it over to the sender. From this indirect acknowledgement, sender with the outgoing unidirectional link gets confirmation about the proper delivery of the data. This mechanism is illustrated in Fig. 2. Following from the previous example B is aware of the unidirectional link with C. First, when B receives the packets from A, it is locally broadcasted so that both E and C will receive the packet. C delivers this packet to D and at the same time, passive acknowledgement is received by E (a monitor node) through overhearing. Finally, the acknowledgement is sent from E to B. This ensures the proper delivery of the packet through unidirectional link.

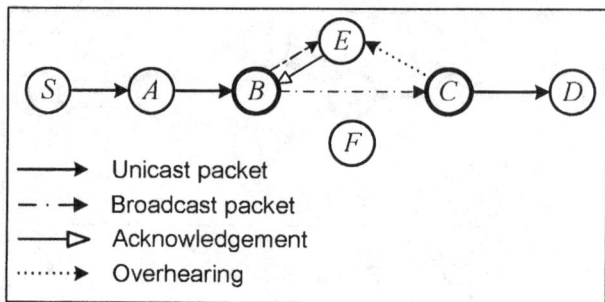

Fig. 2. Local broadcasting of data packets from node, Overhearing by and receiving acknowledgement from E

It is also possible that monitor node can change its location due to mobility and may not be reachable for overhearing. For example, E shifts its position from the current location and becomes unreachable from B. In such cases B will try to re-transmit the data packet three times, and if not successful it will send the route failure error back to source S for route re-discovery. In another situation, if the monitor node is not present in the scene, our protocol subsumes to AODV-EUDA.

4 Performance Evaluation

4.1 Simulation Environment

In this section, our scheme is compared with AODV-EUDA. We performed a simulation using the network simulator ns-2 in static and random mobility model with 100 nodes. In random mobility model all nodes move around a rectangular region of size $1500 \times 300 m^2$. Speeds ranging from 0m/s to 20m/s are used without pause. Total simulation time is 900 sec and each scenario is repeated ten times. Traffic pattern consists of 10 CBR connections running on UDP generating four 512-byte data packets per second. In static model we linearly increased unidirectional links from 1 to 5, around the rectangular region of size $2000 \times 300 m^2$ with a simulation time of 300sec.

4.2 Simulation Results

In our experiments, we capture the performance based on packet delivery ratio, delay and energy consumption for both protocols. Fig. 3(a) shows that the packet delivery ratio of the proposed scheme and AODV-EUDA is similar in static model. Both algorithms achieve route on the first attempt by the source, for AODV-EUDA (at least if one bidirectional link is available) and for the proposed scheme even if unidirectional link is present. Fig. 3(b) shows the packet delivery ratio as a function of variation of the maximum speed of nodes. As the mobility of node increases, our proposed scheme shows weaker performance

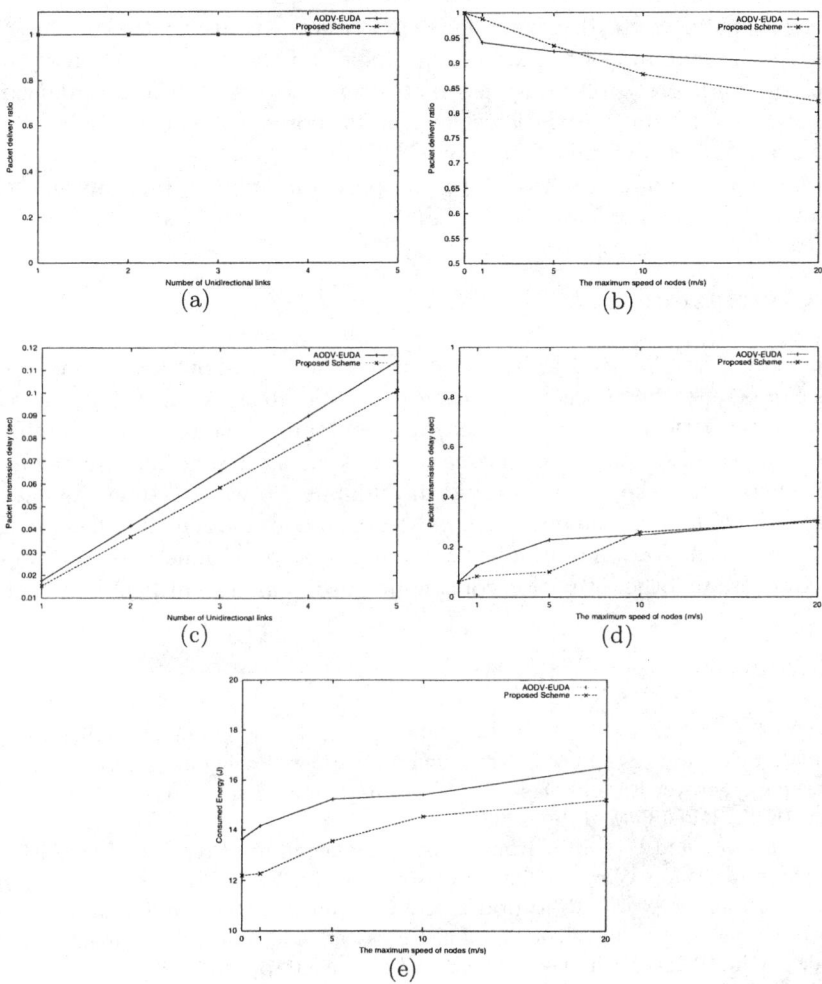

Fig. 3. Results on (a) Packet delivery ratio in static model (b) Packet delivery ratio in mobility model (c) Delay in static model (d) Delay in mobility model (e) Energy consumption

than AODV-EUDA. By analyzing the traces, we found that the stability of unidirectional links becomes poor with the increase in the mobility of nodes. Next, in Fig. 3(c) and (d) we report average end to end delay in static and mobile scenario respectively. Our scheme provides better shortest path in using unidirectional links, and hence shows lesser delay than AODV-EUDA. However, if the mobility of nodes becomes high and the route break occurs more frequently, the route re-discovery time is added to the end-to-end delay. Fig. 3(e) shows the normalized consumed energy per node of the two protocols as a function of the maximum speed of nodes. We can see that, AODV-EUDA consumes more energy then the proposed scheme. It is due to the fact that the number of nodes participating in route discovery decreases when we utilize the unidirectional links. As the mobility of nodes becomes high and the number of control packet increases, both protocols consume more energy. However, the normalized consumed energy is consistently lower for the proposed scheme as it is affected by total bytes (or bits) of data transmitted by nodes. As the amount of successfully delivered packet dominate total bytes, despite of high mobility, proposed scheme consumes less energy than AODV-EUDA.

5 Conclusion

In this paper, we have described a novel scheme that shows how unidirectional links can be effectively used by routing protocols. Results show that our scheme shows better performance in many cases as compared with protocols running over bidirectional links. Our protocol consistently selects the shortest route, consumes lesser energy and shows comparable throughput. So we conclude that utilizing unidirectional link can be beneficial in heterogeneous mobile ad hoc networks. In this research, utilization of unidirectional links has been done over AODV protocol, however any other situation routing protocols can also utilize this technique.

References

1. Ko Y-B, Lee S-J, Lee J-B: Ad Hoc Routing with Early Unidirectionality Detection and Avoidance. Personal Wireless Communications, Springer, (2004).
2. Perkins C., Royer E., and Das S.: Ad-hoc on-demand distance vector (AODV) routing. IETF, RFC 3561, July (2003).
3. Marina M.K. and Das S.R.: Routing performance in the Presence of Unidirectional Links in Multihop Wireless Networks. Proc. of the 3rd ACM International Symposium on Mobile Ad Hoc Networking and Computing (MOBIHOC), Jun. (2002).
4. Prakash R.: A routing algorithm for wirelss ad hoc networks with unidirectional links. ACM/Kluwer Wireless Networks, Vol.7, No.6, pp. 617-625.
5. Johanson P. and Maltz D.: Dynamic source routing in ad hoc wireless networks. Mobile Computing, Kluwer Publishing Company, (1996), ch. 5, pp. 153-181.
6. Ramasubramanian V., Chandra R. and Mosse D.: Providing Bidirectional Abstraction for Unidirectional Ad Hoc Networks. Proc. of the 21st IEEE INFOCOM, Jun. (2002).
7. Bao L. and Garcia-Luna-Aceves J.J.: Link state routing in networks with unidirectional links. Proc of IEEE ICCCN, Oct. 1999 Jun. (2002).

Impact of the Columbia Supercomputer on NASA Science and Engineering Applications

Walter Brooks, Michael Aftosmis, Bryan Biegel, Rupak Biswas,
Robert Ciotti, Kenneth Freeman, Christopher Henze, Thomas Hinke,
Haoqiang Jin, and William Thigpen

NASA Advanced Supercomputing (NAS) Division,
NASA Ames Research Center, Moffett Field, CA 94035
wbrooks@mail.arc.nasa.gov

Abstract. Columbia is a 10,240-processor supercomputer consisting of 20 Altix nodes with 512 processors each, and currently ranked as one of the fastest in the world. In this paper, we briefly describe the Columbia system and its supporting infrastructure, the underlying Altix architecture, and benchmark performance on up to four nodes interconnected via the InfiniBand and NUMAlink4 communication fabrics. Additionally, three science and engineering applications from different disciplines running on multiple Columbia nodes are described and their performance results are presented. Overall, our results show promise for multi-node application scaling, allowing the ability to tackle compute-intensive scientific problems not previously solvable on available supercomputers.

1 Introduction

During the summer of 2004, NASA began the installation of Columbia, a 10,240-processor SGI Altix supercomputer, at its Ames Research Center. Columbia is a constellation comprised of 20 nodes, each containing 512 Intel Itanium2 processors and running the Linux operating system. In October of 2004, the machine achieved 51.9 Tflop/s on the Linpack benchmark. According to the June 2005 Top500 supercomputing list, Columbia is ranked as the third fastest system in the world. The system increased NASA's total high-end computing capacity ten-fold, and helped put the U.S. back on the technology leadership track. Through unprecedented collaboration between government and industry partners, this world-class system was conceived, designed, built, and deployed in a mere 120 days. Since its installation, Columbia has garnered worldwide interest among scientists, industry, academia, and the public.

The system currently has over 650 users solving problems across many scientific and engineering disciplines. In this paper, we give a detailed system description and examine the performance characteristics of its 2,048-processor capability subsystem. Through benchmarking tests and real-world applications in the areas of large-scale molecular dynamics, computational fluid dynamics in aerospace design, and high-resolution global ocean modeling, we demonstrate Columbia's current and potential impact on science and engineering applications.

Fig. 1. The 10,240-processor Columbia constellation

2 Columbia Overview

The Columbia system and its supporting infrastructure (see Fig. 1) are housed at the NASA Advanced Supercomputing (NAS) Division in California. Beyond the physical facility upgrades for power and cooling, significant upgrades were made to the mass storage system, local area network, and security perimeter. During installation, NAS computer scientists conducted extensive benchmark tests to further understand the performance characteristics and to grasp the magnitude of the computational capabilities of this massive Altix system. Upgrades to NASA's wide area network to 10-gigabit Ethernet (10 GigE) are underway.

2.1 System Description

Columbia is a 10,240-processor constellation comprised of 20 nodes, each consisting of 512 Intel Itanium2 processors employing single system image (SSI) technology and running the Linux operating system. Twelve nodes are SGI Altix 3700 and eight are Altix BX2 (doubled processor count in rack from 32 to 64). All 512 processors within a node are interconnected via NUMAlink (SGI's proprietary non-uniform memory access advanced interconnect technology for clusters). In turn, all of the nodes are connected together via five networks: InfiniBand (IB) (high performance, switched fabric interconnect standard for servers), 10 GigE, and three GigE. Four of Columbia's BX2 nodes are linked via NUMAlink4, making a 2,048-processor SMP (symmetric multiprocessing) system with a peak of 13.1 Tflop/s.

The Columbia storage array consists of 16 RAID racks, eight of which each have 20 TB of FibreChannel (FC) storage; the other eight each have 35 TB of Serial ATA (SATA) storage. Each RAID array is quad-connected to two 128-channel FC switches and each node has two to four FC dual-ported Host Bus Adapters connecting between the two switches. SGI's CXFS shared file system is currently being installed to allow sharing of file systems among groups of nodes. These file systems provide users with temporary scratch storage available for the duration of a computation. In addition, users have assigned permanent storage provided through a network file system (NFS) via GigE connections. The Columbia tape robot mass storage system enables storage of up to 200 GB of data per tape, with a total theoretical capacity of 10 PB. This StorageTek system holds data from several NASA centers and takes approximately 20 seconds to mount data from the tape robots, creating a transparent process to the user.

The physical cable plant for Columbia consists of patch panels with Category 5e Unshielded Twisted Pair (UTP), Multi-Mode Fiber (MMF), and Single-Mode Fiber (SMF) for each node in centrally located cabinets along with Ethernet, IB, and FC switches. SMF primarily supports connections to storage servers in remote locations, while MMF and UTP are heavily used to provide GigE, 10 GigE, CXFS, and MetaData Server (MDS) interconnects between nodes. Patch panels are key to addressing the dynamic configurations, with approximately 15 added, moved or removed connections per week.

The overall Columbia perimeter protection system includes the Secure Front Ends (SFE), Secure Unattended Proxies (SUP), and Perimeter Enforce and Controller, which collectively serve as the security reference monitor for access to systems located within the Columbia enclave. The SFE mediates all interactive accesses to the enclave and is the point at which a user must be identified and authenticated using RSA's SecurID authentication. The SUP supports unattended file transfers (where the user is not present to perform two-factor authentication) by allowing the use of SecurID to acquire a ticket based on public key technology.

SGIs Linux Environment 7.2 with the SGI ProPack kernel enables a single system image on each 512-processor node. Programming paradigms available on Columbia include MPI, OpenMP, multi-level parallelism (MLP), and hybrid (MPI across nodes and OpenMP/MLP within a node).

2.2 Altix Architecture

The 64-bit processors used in the 3700 architecture run at 1.5 GHz and can issue two MADDs (multiply and add) per clock, with a peak performance of 6 Gflop/s. These processors are grouped in sets of four—each set is called a "C-brick." All 128 C-bricks within a 3700 node are connected via SGI's NUMAlink3 (a high-performance network with fat-tree topology). Each brick has 8 GB of local memory and two SHUBs, a proprietary Application Specific Integrated Circuit (ASIC) designed by SGI. Peak bandwidth between bricks in a single 3700 node is 800 MB/s per processor.

Being twice the density of a 3700, each C-brick in a BX2 node contains eight processors, for a total of 64 bricks. Each of these 64 bricks is interconnected via NUMAlink4, yielding twice the bandwidth of that between bricks in a 3700. Each brick in a BX2 has a 16 GB memory capacity and four SHUBs. The 64-bit processors used in the BX2 architecture run at 1.6 GHz and can issue two MADDs per clock with a peak performance of 6.4 Gflop/s.

The memory hierarchy of the Itanium2 processor consists of 128 floating point registers and three on-chip data caches: 32 KB of L1; 256 KB of L2; and 6 MB of L3. The memory hierarchy of the processors in the BX2 nodes is identical except for a larger 9 MB L3 cache. As a Cache Coherent Non-Uniform Memory Access (CC-NUMA) system, local cache-coherency is maintained between processors on the Front Side Bus (FSB) in both the 3700 and BX2 architectures. Global cache coherency is implemented via a SHUB chip and is a refinement of the protocol used in the DASH computing system (a scalable shared-memory multiprocessor developed at Stanford University).

2.3 Benchmark Performance

Several microbenchmarks, low-level benchmarks, computational kernels, and real applications for various regression testing, verification, validation, and planning purposes are employed to enable scientists and administrators to research, design, and develop an optimized and tuned computing system. Here we present some performance data using a subset of the NAS Parallel Benchmarks (NPB) [6, 12]; detailed characterization results can be found in [2].

Figure 2 shows the per-processor Gflop/s rates reported from NPB runs, a horizontal line indicating linear scaling. The four graphs on the left show MPI and OpenMP results on three types of the Columbia nodes: 3700, BX2a with 1.5 GHz CPUs and 6 MB caches, and BX2b with 1.6 GHz clock and 9 MB caches. Results demonstrate that the double density packing for BX2 produces shorter latency and higher bandwidth in NUMAlink access. The effect of doubled network bandwidth of BX2 on OpenMP is evident; it is less profound on MPI performance until communication starts to dominate. A bigger cache in the BX2b produces substantial performance improvement for MPI codes on large processor counts when the data can fit into local cache. However, no significant

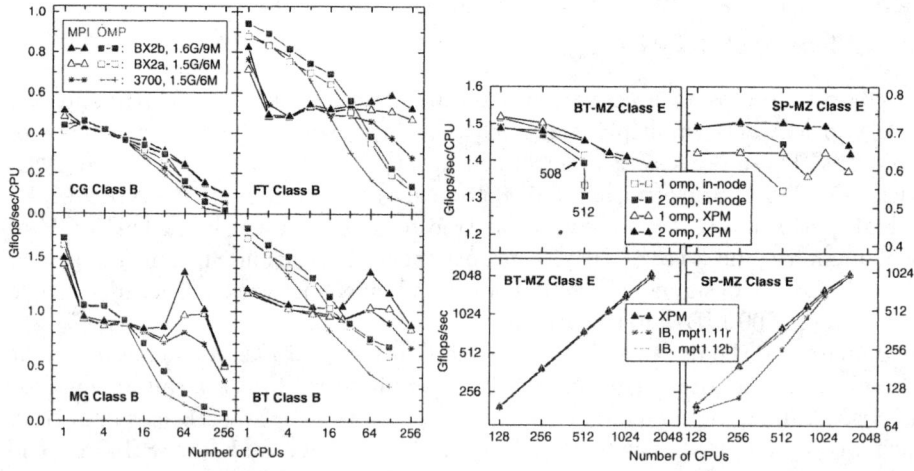

Fig. 2. NPB performance comparison on three types of the Columbia nodes (left), and under three different interconnects (right)

difference is observed for OpenMP codes because the cost of accessing shared data from each thread increases substantially with the number of processors. In the case of MPI, the falloff from the peak is due to increased communication-to-computation ratio as this is a strong scaling test. The slightly larger processor speed of the BX2b brings only a marginal performance gain.

The four graphs on the right of Fig. 2 show performance of the hybrid MPI+OpenMP codes of the NPB multizone benchmarks. These were tested across four Columbia nodes connected with both the NUMAlink4 network and the IB switch. The Class E problem size (4096 zones, 1.3 billion grid points) was used for these experiments. The top two graphs compare multi-box NUMAlink4 results with those from a single BX2b node. For 512 or fewer CPUs, multi-node performance is comparable to or even better than single-node results. The bottom two graphs compare runs using NUMAlink4 with those using IB, taking the best process-thread combinations. The IB results are only 7% worse; however, performance is sensitive to a few SGI runtime environment parameters that control how MPI accesses its internal message buffers.

3 Applications

The following applications are examples of compute-intensive work being performed on Columbia, all of which have been scaled beyond a 512-processor node.

3.1 Large-Scale MD Simulations

There is growing interest in large-scale molecular dynamics (MD) simulations [13] involving several million atoms, in which interatomic forces are computed quantum mechanically [3] to accurately describe chemical reactions. Such large reactive MD calculations provide the requisite coupling of chemical reactions, atomistic processes, and macroscopic materials phenomena, to solve a wide spectrum of science and engineering problems. One example of technological significance is that of energetic nanomaterials used to boost the impulse of rocket fuels in which chemical reactions sustain shock waves (see Fig. 3). Petaflops-scale computers could potentially extend the realm of quantum mechanics to macroscopic scales, but only if scalable simulation technologies were developed. A multidisciplinary team of physicists, chemists, materials scientists, and computer scientists at NASA and several academic institutions are working toward solving this challenging problem. They have developed a scalable parallel computing framework for reactive atomistic simulations, based on data locality principles.

Density functional theory (DFT) has reduced the exponentially complex quantum mechanical (QM) N-body problem to $O(N^3)$, by solving N one-electron problems self-consistently instead of an N-electron problem [7]. Unfortunately, DFT-based MD simulations [3] are rarely performed for $N > 10^2$ atoms because of the excessive computational complexity, which severely limits their scalability. Over the past few years, two promising approaches have emerged toward achieving million-to-billion atom simulations of chemical reactions.

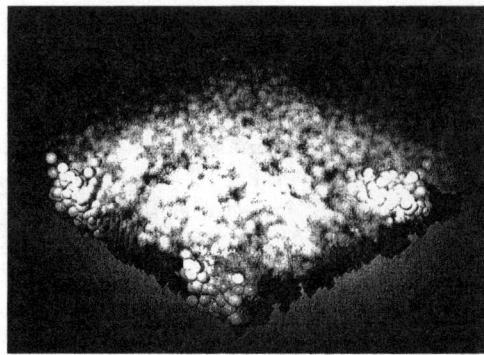

Fig. 3. Reactive force-field MD simulation of shock-initiated combustion of an energetic nanocomposite material (nitramine matrix embedded with aluminum nanoparticles)

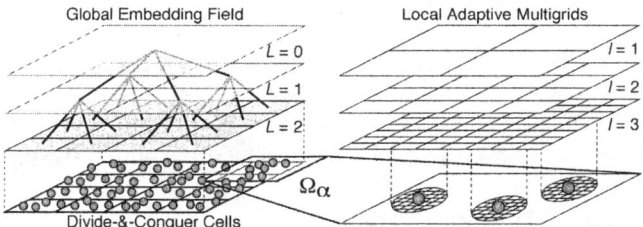

Fig. 4. Schematic of an embedded divide-and-conquer (EDC) algorithm

One approach is to perform a number of small DFT calculations on-the-fly to compute interatomic forces quantum mechanically during an MD simulation. The team has recently designed an embedded divide-and-conquer DFT algorithm (EDC-DFT) and used it to simulate a 1.4 million-atom problem. An alternative to this concurrent DFT-MD approach is a sequential DFT-informed MD strategy, which employs environment-dependent interatomic potentials to describe charge transfers, and chemical bond formation and breakage. A first principles-based reactive force-field method (ReaxFF) where parameters in the interatomic potentials are trained to best-fit many DFT calculations on small ($N\sim 10$) clusters of various atomic-species combinations has been developed. A new $O(N)$ parallel implementation of ReaxFF enabled a 0.56 billion-atom MD simulation of chemical reactions.

Linear-Scaling EDC Algorithms. The embedded divide-and-conquer (EDC) algorithms, based on data locality principles, solve spatially localized subproblems in a global embedding field, which are then efficiently computed with tree-based methods. Examples of the embedding field are the electrostatic field in MD simulations and the self-consistent Kohn-Sham potential in DFT. A suite of these linear-scaling EDC algorithms developed by the team solves multiresolution MD

(MRMD) based on a many-body interatomic potential model; environment-dependent ReaxFF MD; and QM calculation based on DFT.

Figure 4 shows a schematic of an EDC algorithm. In the left panel, the physical space is subdivided into spatially localized cells, with local atoms (spheres) constituting subproblems that are embedded in a global field (shaded) solved with a tree-based algorithm. To solve the subproblem in domain Ω_α in the EDC-DFT algorithm, coarse multigrids (shaded in right panel) are used to accelerate iterative solutions on the original real-space grid (corresponding to the grid refinement level, $l = 3$). Fine grids are adaptively generated near the atoms to accurately operate the ionic pseudopotentials on the electronic wave functions.

Performance Results. Major design parameters for MD simulations of materials include the number of atoms in the system and the methods to compute interatomic forces (classically in MRMD, semi-empirically in P-ReaxFF, or quantum-mechanically in EDC-DFT). Figure 5 shows parallel performance for each of the three algorithms on Columbia and a design-space diagram on 1,920 processors. Execution and communication times are shown per MD step. The largest benchmark tests include 18,925,056,000-atom MRMD, 557,383,680-atom P-ReaxFF, and 1,382,400-atom EDC-DFT calculations. Results demonstrate excellent linear scaling for all three algorithms, spanning five orders of magnitude in problem size. The only exception is P-ReaxFF below 100 million atoms, due to the high communication-to-computation ratio. Parallel efficiency on 1,920 processors is 0.87, 0.91, and 0.76 for MRMD, P-ReaxFF, and EDC-DFT, respectively. Further code optimizations are currently underway to understand and eliminate the jumps in timings at and beyond 480 processors.

3.2 High-Fidelity Aerospace Applications

Computational fluid dynamics (CFD) techniques have been applied to aerospace analysis and design problems since the advent of the supercomputer; however, their historical impact on the vehicle design process has been limited. Platforms like Columbia now promise to unlock the full potential of these simulation systems both by producing more optimal designs and by permitting parametric analyses that examine a vehicle's performance over the complete flight envelope. The large-scale parallel hardware improves accuracy in all phases of the process both by enabling simulations employing grids with one or two orders of magnitude higher resolution, and simultaneously permits tens of thousands of runs to be made as part of design optimization or parametric performance studies.

NASA's Cart3D is a high-fidelity simulation package aimed at design and aero-performance prediction for vehicles with complex geometry. It is in widespread use both within NASA, and throughout other government agencies and industry. The package is based upon the solution of the Euler equations of fluid motion on locally adapted Cartesian grids with embedded boundaries. This approach permits fully automated mesh generation for extremely complex geometries and gives it the ability to dynamically re-mesh configurations when control surfaces are deployed, or when the underlying CAD geometry is significantly modified by a shape optimizer [11].

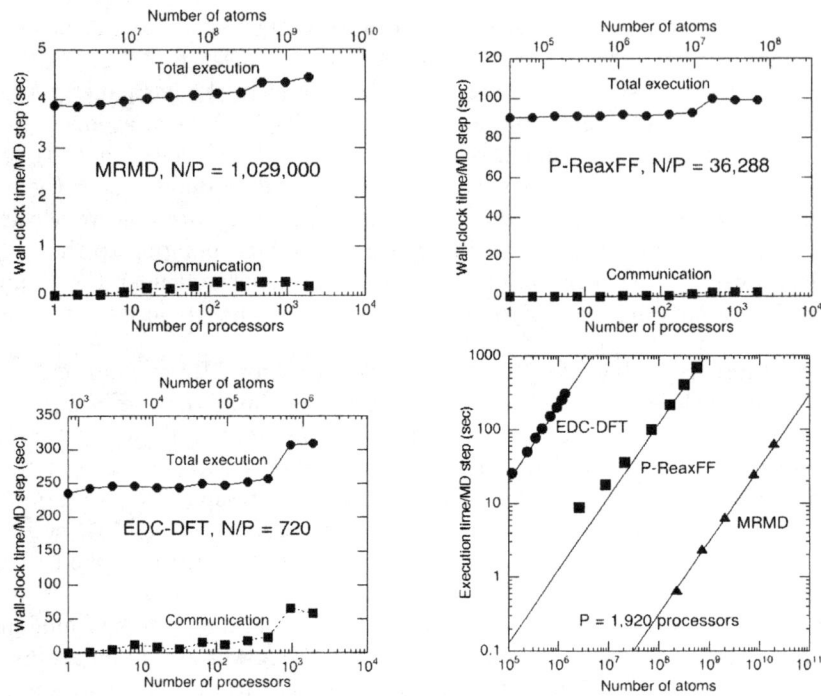

Fig. 5. Total execution and communication times, and design space diagram for three linear-scaling MD algorithms: MRMD, P-ReaxFF, and EDC-DFT

Parallel Implementation. Cart3D employs several techniques to enhance its efficiency on distributed parallel machines. It uses multigrid for convergence acceleration and employs a domain-decomposition strategy for subdividing the global solution among the many processors of a parallel machine [1]. The mesh coarsener and the partitioner in Cart3D take advantage of the hierarchical nesting of adaptively refined Cartesian meshes. This structure permits the efficient use of Space Filling Curves (SFCs) both for domain decomposition and mesh coarsening. The same SFC that partitions the fine mesh is also used to partition the coarser meshes. This approach produces meshes with generally good overlap between coarse and fine mesh partitions; however, they are not perfectly nested. Thus, while most of the communication for multigrid restriction and prolongation in a particular subdomain will take place within the same local memory, these operators will incur some degree of off-processor communication. This approach favors workload balancing on each mesh in the hierarchy at the possible expense of increased communication [1, 10].

Performance Results. Several performance experiments were devised to examine Cart3D's scalability for a typical large grid case based on the full Space Shuttle Launch Vehicle (SSLV) shown in Fig. 6. For scalability testing, the mesh density was increased to 25 million cells, with approximately 125 million degrees-

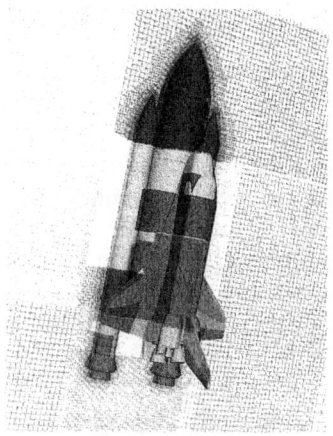

Fig. 6. Cartesian mesh (left) and pressure contours (right) around full SSLV configuration. Mesh color indicates 16-way decomposition of 4.7 million cells using the SFC partitioner, while pressure contours are at Mach 2.6 and 2.3° angle-of-attack.

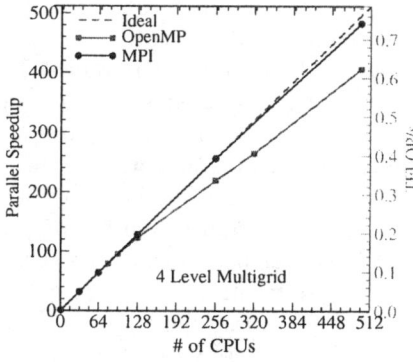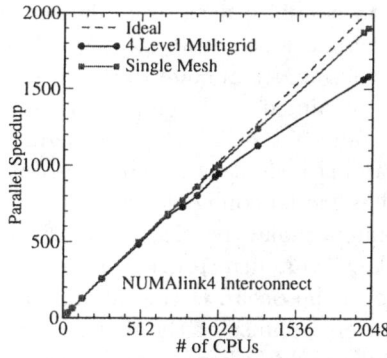

Fig. 7. Parallel scalability of Cart3D solver for SSLV using a 25 million cell mesh on one node (left) and four nodes (right) of Columbia

of-freedom. An aerodynamic performance database and virtual-flight trajectories using this configuration were presented in [11].

Cart3D's solver module can be built against either OpenMP or MPI communication libraries. On Columbia, cache-coherent shared memory is not maintained between nodes; thus, pure OpenMP codes are restricted to a single box. The left panel in Fig. 7 shows scalability for the test problem using both OpenMP and MPI on a single Altix node. In calculating parallel speedup, perfect scalability was assumed on 32 CPUs. Performance with both programming libraries is very nearly ideal; however, the OpenMP results display a break near 128 processors. Beyond this point the curve is again linear, but with a slightly reduced slope. This degradation is probably attributable to the routing scheme used within the Altix nodes. They are built of four 128-processor double cabinets; within any one of these, addresses are dereferenced using the complete pointer. More distant addresses are dereferenced by dropping the last few bits of the address. On average, this translates into slightly slower communication when addressing distant memory. Since only the OpenMP version uses the global address space, the MPI results are not impacted by this pointer swizzling.

The graph on the right in Fig. 7 examines parallel speedup for the problem spread across four nodes of Columbia using the NUMAlink4 interconnect. Simulations were run using one and four grids in the multigrid hierarchy, and reducing the number of multigrid levels clearly de-emphasizes communication (relative to floating-point performance) in the solution algorithm. Scalability for the single grid scheme is nearly ideal, but deteriorates at around 688 processors for multigrid because the coarsest mesh in the sequence has only about 16 cells per partition when using 2016 CPUs. Given this relatively modest decrease in performance, it appears the bandwidth demands of the solver are not greatly in excess of that delivered by NUMAlink4. Detailed performance results are in [10].

3.3 High-Resolution Global Ocean Model

Finally, we describe how we are using Columbia's 2,048-processor SMP subsystem to simulate ocean circulation globally at resolutions up to 5km ($\approx \frac{1}{16}^\circ$). The simulations employ the M.I.T. General Circulation Model (MITgcm), a finite volume ocean code that can scale efficiently to large processor counts. The study is aimed at developing a clearer understanding of the physical processes that underly the skill improvements that eddy resolving ocean models show, and at gaining insights into what resolution is sufficient for a particular purpose.

The model configurations employed are significant in that, at the resolutions Columbia makes possible, numerical ocean simulations begin to truly represent the key dynamical process of oceanic meso-scale turbulence. Meso-scale turbulence in the ocean is the analog of synoptic weather fronts in the atmosphere. However, because of the density characteristics of seawater, the length scale of turbulent eddy phenomena in the ocean is around 10 or less kilometers. In contrast, in the atmosphere, where the same dynamical process occurs, it has length scales of thousands of kilometers. Although it has been possible to resolve ocean eddy processes well in regional ocean simulations [5] for some time, global scale simulations that resolve or partially resolve the ocean's energetic eddy field are still rare [8, 9] because of the immense computational challenge they represent.

Altix Implementation. The MITgcm algorithm is rooted in the incompressible form of the Navier-Stokes equations for fluid motion in a rotating frame of reference [4]. The equations are discretized in time and stepped forward explicitly using an Adams-Bashforth procedure that is second order accurate. The equations are discretized in space using a finite volume technique yielding a solution procedure that requires at each time step explicitly evaluated local finite volume computations and an implicit two-dimensional elliptic inversion.

Our parallel formulation takes a global finite volume domain with $N_x \times N_y \times N_z$ cells in three dimensions, and decomposes it into $N_{sx} \times N_{sy}$ sub-domains each of size $(S_{nx} + 2 \times O_x) \times (S_{ny} + 2 \times O_y) \times N_z$ such that $S_{nx} \times N_{sx} = N_x$

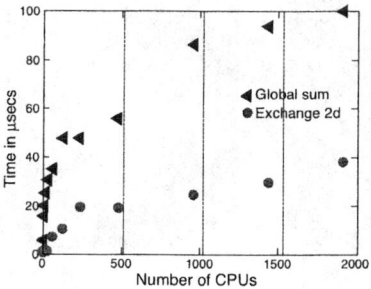 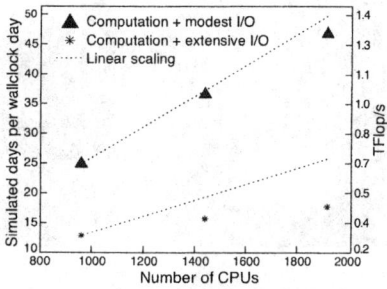

Fig. 8. Performance of key primitives used on the $\frac{1}{16}°$ resolution simulation of 1.25 billion grid cells: exchange times for a sub-domain of size 96×136 with $O_x = O_y = 3$ (left), and overall scaling and performance on 960, 1440, and 1920 processors (right)

and $S_{ny} \times N_{sy} = N_y$. The O_x and O_y values are overlap region finite volume cells that are added to the boundaries of the subdomains to hold replicated data from neighboring subdomains. Each computational process integrating forward the MITgcm is then given a static set of one or more subdomains.

A single time-step is split into a series of *Compute*, *Exchange*, and *Sum* phases. *Compute* contains only local computations (predominantly arithmetic and associated memory loads/stores) and I/O operations. Performance is sensitive to the volume of I/O and computation involved, local CPU and memory capabilities of the hardware, and to the system I/O capacity. *Exchange* involves point-to-point communication between neighbor processes. Performance hinges on the interconnect and inter-process communication software stack. *Sum* involves all subdomains collectively combining locally calculated 8-byte floating point values to yield a single global sum. It is sensitive to how system performance for collective communication scales with processor count. Scaling behavior for the *Sum* and *Exchange* phases are shown in the left panel of Fig. 8, and overall scaling, with and without diagnostic I/O, is shown in the right panel.

Performance Results. A series of numerical simulations at $\frac{1}{4}°$, $\frac{1}{8}°$, and $\frac{1}{16}°$ resolutions were performed on the Columbia 2048-processor SMP subsystem. Results in Fig. 9 show significant changes in solution with resolution. The plots capture changes in sea-surface heights due to eddy activity over a single month. The Gulf Stream region at $\frac{1}{4}°$ resolution shows a relatively small area of vigorous sea-surface height changes, but the $\frac{1}{8}°$ and $\frac{1}{16}°$ resolution simulations show more extensive areas of changes. Key behaviors like how tightly waters "stick" to the coast, or how far energetic eddies penetrate the ocean interior, change significantly between resolutions and can be seen in these images.

At first glance, the three different resolution runs show significant differences. There does, however, seem to be a smaller change between the $\frac{1}{8}°$ and $\frac{1}{16}°$ simulations. A next step is to undertake a fourth series of runs at even higher resolution. Formally quantifying the changes between these runs would provide important information on whether ocean models are reaching numerically converged solutions. Performance on Columbia shows it is well suited for

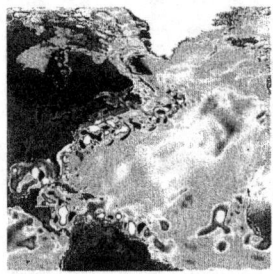 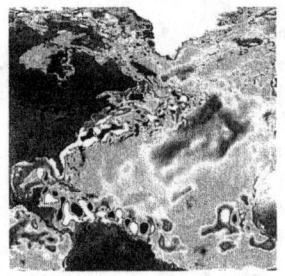 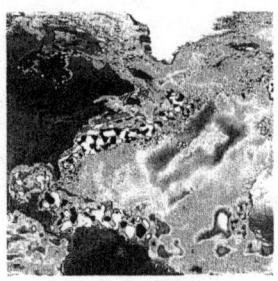

Fig. 9. Gulf Stream region sea-surface height difference plots at different resolutions for one month: $\frac{1}{4}°$ (left), $\frac{1}{8}°$ (middle), and $\frac{1}{16}°$ (right). Color scale -0.125m to 0.125m.

addressing these questions. The code achieved a sustained performance of 12% of peak on 1,920 processors. The scaling across multiple Altix nodes is encouraging and suggests that configurations that span eight or more nodes, and that would therefore enable $\frac{1}{20}°$ and higher resolution simulations, are today within reach.

4 Summary and Conclusions

Through innovative engineering techniques by NASA computer scientists and industry partners, some of today's most computationally challenging problems are being solved on the Columbia supercomputer. It has proven itself to be a valuable national resource, running massive computationally intensive programs in relatively short time periods, and giving scientists and engineers a tool to effectively and efficiently solve the most difficult problems in diverse areas such as materials science, aeronautics, and earth science.

References

1. M.J. Berger and M.J. Aftosmis, Performance of a new CFD flow solver using a hybrid programming paradigm, *J. of Parallel Dist. Comput.*, 65 (2005) 414–423.
2. R. Biswas et al., An application-based performance characterization of the Columbia supercluster, in: *Proc. SC2005* (Seattle, WA, 2005).
3. R. Car and M. Parrinello, Unified approach for molecular dynamics and density functional theory, *Phys. Rev. Lett.*, 55 (1985) 2471–2474.
4. C. Hill and J. Marshall, Application of a parallel Navier-Stokes model to ocean circulation. in: *Proc. Parallel CFD* (1995) 545–552.
5. H.E. Hurlburt and P.J. Hogan, Impact of 1/8 to 1/64 resolution on gulf stream model-data comparisons in basin-scale subtropical atlantic ocean models, *Dynamics of Atmosphere and Oceans*, 32 (2000) 283–329.
6. H. Jin and R.F. Van der Wijngaart, Performance characteristics of the multi-zone NAS Parallel Benchmarks, in: *Proc. IPDPS2004* (Santa Fe, NM, 2004).
7. W. Kohn and P. Vashishta, General density functional theory, *Inhomogeneous Electron Gas* (N. March and S. Lundqvist, eds.), Plenum (1983) 79–184.
8. M.E. Maltrud and J.L. McClean, An eddy resolving global 1/10 ocean simulation, *Ocean Modeling*, 8 (2005) 31–54.

9. Y. Masumoto et al., A fifty-year eddy-revolving simulation of the world ocean, *J. of the Earth Simulator*, 1 (2004) 35–56.
10. D.J. Mavriplis et al., High-resolution aerospace applications using the NASA Columbia supercomputer, in: *Proc. SC2005* (Seattle, WA, 2005).
11. S.M. Murman et al., Automated parameter studies using a Cartesian method, AIAA Paper 2004-5076 (Providence, RI, 2004).
12. NAS Parallel Benchmarks, see URL *http://www.nas.nasa.gov/Software/NPB*.
13. J. Phillips et al., NAMD: Biomolecular simulation on thousands of processors, in: *Proc. SC2002* (Baltimore, MD, 2002).

Hierarchical Routing in Sensor Networks Using k-Dominating Sets

Michael Q. Rieck[1] and Subhankar Dhar[2]

[1] Drake University, Des Moines, Iowa 50311, USA
michael.rieck@drake.edu
[2] San José State University, San José, CA 95192, USA
dhar_s@cob.sjsu.edu

Abstract. For a connected graph, representing a sensor network, distributed algorithms for the Set Covering Problem can be employed to construct reasonably small subsets of the nodes, called k-SPR sets. Such a set can serve as a virtual backbone to facilitate shortest path routing, as introduced in [4] and [14]. When employed in a hierarchical fashion, together with a hybrid (partly proactive, partly reactive) strategy, the k-SPR set methods become highly scalable, resulting in guaranteed minimal path routing, with comparatively little overhead.

1 Introduction

Recent advances in micro-electro-mechanical systems (MEMS) and wireless research led to the development of sensor networks that show a lot of promise for future mobile applications [1]. Research efforts have been made to build low cost micro-sensors that possess processing capability as evidenced in the Smart Dust Project [7], [15], the PicoRadio Project [11] and WINS Project [12]. A large number of wireless sensor networks consist of portable mobile devices with limited battery power. In order to address this limitation, energy-efficient routing algorithms and protocols are a major focus of current research.

In our work, we model sensor networks by a connected weighted graph having bidirectional links. For the sake of simplicity, the network nodes are presumed to be identical in nature and to have the same transmission radii. Edge weights are used as a measurement of the impact on the network of using a given link. These weights will be referred to as "costs", and the exact details of how such costs are assigned will not be important in our discussion.

It will simply be understood that the higher the cost of a link, the less desirable it is to transmit using this link. Costs might be a function of the minimal transmission energy required for the link, and/or the relative impact on the battery levels of the nodes involved in the link. The minimal transmission energy is of course a function of the proximity of the two nodes, as well as any interference. The relative impact on a node's battery energy level is additionally sensitive to the node's current battery level. Ideally, the links at a node with a weak battery should all have a high cost.

2 Our Approach

Our routing strategy is based on special k-dominating sets of nodes, namely k-SPR sets, that generalize similar sets from our earlier work [4], [14]). The nodes in such a set serve as "routers" and play a central role in facilitating route requests. Moreover, the nature of a k-SPR set is such that this guarantees minimal path routing under reasonable assumptions, where minimal path means shortest weighted path based on edge weights.

k-SPR sets can be used in a hierarchical way, based on an increasing finite sequence of numbers k_i, with one of these numbers corresponding to each of the levels of the hierarchy. This leads to an easily maintained and quite natural hybrid hierarchical routing strategy. It too guarantees minimal path routing. We supply detailed algorithms for forming such a hierarchy of k-SPR sets, which we call a $\mathcal{K}$-SPR sequence.

A reasonable choice for these numbers would be $k_i = k^i$, for some fixed integer $k \geq 2$. Since the largest k_i can be assumed not to exceed the diameter of G, the number of hierarchy levels in this case would be bounded by the logarithm of the diameter of G. Consequently, our hybrid routing strategy is highly scalable. Moreover, it is quite unique in its ability to also ensure minimal path routing. Although dominating sets have been used to construct virtual backbones in ad hoc and sensor networks, this is the first attempt to use k-hop connected k-dominating sets for hierarchical routing that is also minimal path routing.

3 Related Work

Routing protocols for sensor networks are active areas of research and several researchers have proposed several protocols/heuristics in this regard. Since our framework for routing is based on minimum connected dominating set, we will here focus on only some of these, ones that are highly relevant to our own approach and that utilize a (k-)dominating set. The nodes in such a set provide a virtual backbone of router nodes, and in general, must be supplied with global routing information.

Span [2] is one of several ad hoc networking protocols based on the notion of a dominating set. In Span, "coordinators" - a group of nodes that form a connected dominating set over the network - do not sleep. Non-coordinator nodes follow a synchronized sleep/wake cycle, exchanging traffic using an algorithm based on the beaconing and traffic announcement methods of IEEE 802.11 IBSS power save. The routing protocol is integrated with the coordinator mechanism so that only coordinators forward packets, acting as a low latency routing backbone for network. Span is intended to maximize the amount of time nodes spend in the sleep state, while minimizing the impact of energy management on latency and capacity.

The algorithm of J. Wu and H. Li is a distributed algorithm [16] that is used to construct a connected dominating set in a connected graph of radius at least two. The set produced by their algorithm is used to form a virtual backbone of a wireless ad hoc network. In [14], the authors generalized the Wu-Li algorithm so as to produce a k-hop connected k-dominating set that work as routers. (See Section IV for definitions.) One of the important aspect of their routing scheme was that it also guaranteed shortest path routing through the network along a path that was guaranteed at any point along the way, to encounter another router node within every k steps. Later the authors modified this algorithm and proposed a number of variations on it [4]. These were largely motivated by the following study of k-hop dominating sets.

In [8], B. Liang and Z. J. Haas proposed a distributed greedy algorithm to produce a small k-dominating set. In order to do so, they reduced the problem to a special case of the Set Covering Problem. A similar but different reduction to this problem was also used in [4]. For a given value of k, though, the latter requires fewer steps than the Liang-Haas method. In addition it produces a set that is not only k-dominating, but is also k-hop connected, and has a special property to facilitate shortest path routing.

Hierarchical routing has gained special attention for sensor networks for their scalability and flexibility. In order to orchestrate hierarchical routing, various clustering algorithms have been developed for this purpose [3]. However, all these clustering strategies do not guarantee shortest path routing.

Low-energy adaptive clustering hierarchy (LEACH) is a hierarchical-based protocol that minimizes energy dissipation in sensor networks [5]. The purpose of LEACH is to randomly select sensor nodes as cluster-heads, so the high energy dissipation in communicating with the base station is spread to all sensor nodes in the sensor network. Clusterhead selection is difficult to optimize in many situations.

The Power-Efficient Gathering in Sensor Information Systems (PEGASIS) [9] is another hierarchical protocol that is an improvement of the LEACH protocol. As opposed to forming clusters like LEACH, PEGASIS first constructs chains consisting of sensor nodes so that each node transmits and receives from a neighbor and only one node is selected from that chain to transmit to the base station (sink). Performance evaluation of PEGASIS indicates that it outperforms LEACH for different network sizes and topologies. However, one of the major drawback of PEGASIS is that it introduces excessive delay for distant node on the chain. Moreover, the single node acting as a leader of the chain can sometimes become a bottleneck.

Hierarchical-PEGASIS [10], which is an extension of PEGASIS, is designed to addresses the delay incurred for packets during transmission to the base station. In order to improve the performance by reducing the delay in PEGASIS, messages are transmitted simultaneously.

4 k-SPR Sets and $\mathcal{K}$-SPR Sequences

The k-SPR sets to be presented are a straightforward generalization of the k-SPR sets defined in [4] (where they are called "d-SPR sets") and essentially introduced in [14]. The generalization is for the purpose of handling graphs that are equipped with link weights. After a discussion of k-SPR sets, sequences of such will be considered and ultimately used to facilitate hierarchical routing. Throughout this discussion, G will denote a finite connected graph representing a sensor network, with positive link weights referred to as "costs".

4.1 Basic Definitions and a Relationship Between These

Given a path in G, the *cost* of the path is the sum of the costs of the links along the path. Given two nodes, u and v, the *cost* $c(u,v)$ between these is the minimum of the costs of the paths connecting these two nodes. A path from u to v is said to be a *minimal path* if its cost is $c(u,v)$. The *radius* of G is the largest number $R \geq 0$ such that for each node u, there exists a node v satisfying $c(u,v) \geq R$. Let V denote the set of nodes of G. Let $N = |V|$.

Some fundamental definitions concerning subsets of V and claims about these required for the routing strategy to be described in the next section will now be presented.

Definition 1. Fix a positive number k. Fix a subset S of the set of nodes in V.

(a) S is *k-dominating* if every node in V is within a cost k of some node in S.
(b) S is *k-hop connected* if, given any two nodes u and v in S, there is a path in G from u to v such that the cost between consecutive elements of S along this path never exceeds k.
(c) S is a *k-SPR* set if, given any two nodes u and v in V satisfying $c(u,v) > k$, there exists some node w in S such that $w \neq u$, $w \neq v$, and $c(u,w) + c(w,v) = c(u,v)$.

The definition of a k-SPR set was formally introduced in [4], and is a central concept in [14] as well. It essentially means that whenever two nodes are sufficiently far apart, there is certain to be at least one node from the k-SPR set lying between them along a minimal path. The three types of subsets of V are related via the following facts, which generalizes [4–Theorem 1], and whose proof is similar.

Theorem 1. *Assume that S is a k-SPR set for G. Then the following are true.*

(a) *Given any two nodes u and v of G, there exists a minimal path connecting u to v such that the set of nodes along this path that are also in $S \cup \{u,v\}$ is k-hop connected.*
(b) *S is k-hop connected.*
(c) *If the radius of G exceeds k, then S is k-dominating.*

4.2 Local Views

When G represents an ad hoc network, [14] and [4] produce a k-SPR set to serve as a virtual backbone for routing purposes. To achieve practical distributed algorithms for finding such a k-SPR set, the following subgraphs of G need to be considered. These generalize similar subgraphs in [14] and [4], but the terminology is altered slightly. A "$(d+1)$-local view" there is called an "extended d-local view" here.

Definition 2. Let v be a node of V. Let $r \geq 0$. The *r-local view* of v is the subgraph induced by all of the nodes within a cost r of v. The *extended r-local view* of v is the subgraph of G obtained by extending the r-local view of v by including also any nodes at a cost greater than r from v that are adjacent to a node in the r-local view, plus the links that realize these adjacencies.

It is clear that the cost from v to another node u in v's r-local view is also the cost between these nodes in G, that is, $c(v, u)$. We will suppose that nodes employ some sort of "extended hello" messages in order that each node be able to learn about its extended r-local view, for some r. It is important for the purposes of shortest path routing to know when the cost between two nodes in some extended r-local view agrees with the corresponding cost in the graph G as a whole. This issue is partly addressed in the first part of [4–Theorem 2]. A somewhat more general claim is the following, which is proved in a similar manner.

Lemma 1. *Let x and y be in the extended r-local view of v. Let c' denote the cost between x and y as measured in this r-local view. If $c(v, x) + c(v, y) + c' \leq 2r$, then $c' = c(x, y)$.*

4.3 A Covering Problem

Another common feature of the routing algorithms to be considered is that they all rely on a bipartite graph $B = B(G)$, based on G, a portion of which is maintained in a data structure by each network node. The bipartite graph B is described as follows.

Definition 3. The nodes of the bipartite graph $B = B(G)$ constitute two sets V and P, each of which is an independent set in B. V is simply the set of all nodes of G. The elements of P are certain unordered pairs of nodes $\{x, y\}$ of G. To describe which, first consider the set $\hat{P}$ of all such pairs satisfying $c(x, y) > k$. Partially order $\hat{P}$ by taking $\{x', y'\} \leq \{x, y\}$ if (after possibly reordering x' and y') $c(x, x') + c(x', y') + c(y', y) = c(x, y)$. (This means that x' and y' lie along some minimal path connecting x and y.) Now P is defined to be the subset of $\hat{P}$ consisting of the minimal elements with respect to this partial order. The description of the bipartite graph B is completed by indicating that $v \in V$ is taken to be adjacent to $\{x, y\} \in P$ if and only if $c(x, v) + c(v, y) = c(x, y)$, but $v \neq x$ and $v \neq y$.

When all the link costs are one, B is the same as the bipartite graph considered in [4]. The following claim is straightforward to check using Definition 2 and part (c) of Definition 1.

Theorem 2. *A subset S of V is a k-SPR set for G if and only if every element of P is adjacent in B to some element of S.*

When this adjacency condition holds, we say that S *covers* P. The second part of [4–Theorem 2] may now easily be generalized to produce the following needed fact.

Lemma 2. *Let e be an upper bound on the link costs of G. Suppose that $u, v \in V$ both cover the pair $\{x, y\} \in P$. Then $c(u, v) \leq k + e$.*

4.4 $\mathcal{K}$-SPR Sequences

The constructs presented in this subsection anticipates the hierarchical nature of the routing strategy to be introduced in the next section. Given a k-SPR set, it will be helpful to consider the following derived link-weighted graph.

Definition 4. *Let S be a k-SPR set for G. Define the link-weighted graph $G[S, k]$ as follows. The node set for the graph $G[S, k]$ is S. Two elements u and v of S are made adjacent in $G[S, k]$ if $c(u, v) \leq k$ (in G). In this case, the link connecting u and v in $G[S, k]$ is assigned the cost $c(u, v)$.*

By part (b) of Theorem 1, this graph is connected. Moreover, the cost between any two nodes in $G[S, k]$ when measured in this graph agrees with the cost between them when measured in G. To accommodate a hierarchical version of k-SPR routing, this derived graph notion will now be used to introduce a generalization of the notion of a k-SPR set.

Definition 5. *Fix a set of positive numbers $\mathcal{K} = \{k_1, ..., k_l\}$ with $k_1 < k_2 < \cdots < k_l$. A $\mathcal{K}$-SPR sequence for G is a collection $\mathcal{S} = \{V_1, ..., V_l\}$ of sets of nodes of G with the following property. Letting $G_0 = G$ and $V_0 = V$, and letting G_i denote $G_{i-1}[V_i, k_i]$ for $i = 1, 2, ..., l$, the set V_i is required to be a k_i-SPR set for the graph G_{i-1}, for $i = 1, 2, ..., l$. The following numbers will also be needed. Let $r_0 = k_1$ and for $i > 0$, let $r_i = k_{i+1} + 2k_i + \cdots + 2k_1$.*

Thus $V = V_0 \supseteq V_1 \supseteq \cdots \supseteq V_l$. Part (a) of Theorem 1 now generalizes as follows, and is proved by induction on k.

Theorem 3. *Let $\mathcal{K} = \{k_1, ..., k_l\}$ be a set of positive numbers with $k_1 < k_2 < \cdots < k_l$. Let $\mathcal{S} = \{V_1, ..., V_l\}$ be a $\mathcal{K}$-SPR sequence for G. Given any two nodes u and v in V, there exists a minimal path p connecting u to v such that, for $i = 1, 2, ..., l$, the set of nodes consisting of the all the nodes along p and belonging to V_i, together with the first and last nodes along p and belonging to V_{i-1}, form a k_i-hop connected set for G. Moreover, V_i is a r_{i-1}-SPR set for G, for $i = 1, 2, ..., l$.*

4.5 An Example

Consider the following example using $k_1 = 3$ and $k_2 = 9$. The graph on the left in Figure 1 is the original graph G. The dashed edges have cost one, while the solid edges have cost two. The dark vertices form a 3-SPR set V_1 for G. The graph on the right is then $G_1 = G[V_1, 3]$. It has two types of edges. The dashed edges have cost two, while the solid edges have cost three. Here V_2 consists of the lone dark vertex in the figure. This is a 9-SPR for G_1. Thus $G_2 (= G_1[V_2, 9])$ would consist only of one vertex, and the process terminates.

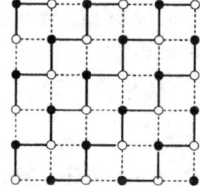 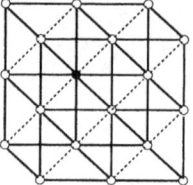

Fig. 1. G and G_1

Now, in Theorem 3, consider the case where u and v are the top-left node and bottom-right node of G, respectively. There are several minimal paths connecting u and v, and we see that their cost is 15. One of these path starts at u, and repeatedly moves down one hop and then right one hop, zigzagging until arriving at v. Call this path p. Notice that it goes through the only node in V_2, which we'll call w. Consider the claim in Theorem 3 when $i = 2$. The first and last nodes along p that belong to V_1 are u and v. The fact that $\{u, v, w\}$ is 9-hop connected in G gives evidence in support of Theorem 3.

Let's try a different choice for u and v, say by taking these to be the top-right node and the bottom-left node, respectively. Now the cost between u and v is only 10 and there is an evident unique minimal path connecting them. Let p now denote this path, which uses only edges of cost one, and which alternates between nodes in V_1 and nodes not in V_1. Letting x and y denote the first and last nodes along the path that belong to V_1, we see that $c(x, y) = 8$. There are no nodes from V_2 along p. So using $i = 2$ again, we now notice that $\{x, y\}$ is 9-hop connected in G, as required.

5 Hierarchical Routing Via $\mathcal{K}$-SPR Sequences

5.1 Establishing a $\mathcal{K}$-SPR Sequence

Let k_0 be an upper bound on the link costs of G. Let $\mathcal{K} = \{k_1, ..., k_l\}$ be a set of positive numbers satisfying $k_0 < k_1 < \cdots < k_l$. The distributed algorithms of [14] and [4] can now be altered to handle graphs with weighted links. By iteratively applying such an algorithm, it then becomes straightforward to obtain a $\mathcal{K}$-SPR sequence for G. Once this has been accomplished, the routing strategies described in the next section can be implemented.

For example, the greedy algorithm approach in [4] is easily adapted to handle a graph with link costs, as will now be outlined. The following algorithm shows how this would proceed at level i, that is, when applied to the graph G_i in order to find a k_{i+1}-SPR set for it. Note however that when $i > 0$, the processing at level i begins locally only after processing at level $i - 1$ has completed locally. The distributed greedy algorithm used here, at each level, does not require strict synchronization though.

Each node in the network has a unique ID number. Each node that becomes a level-i node (element of V_i) begins participating in the process of selecting level-$(i+1)$ nodes (elements of V_{i+1}). Initially it is in the "undecided" state, but ultimately ends up in either the "selected" or "not selected" state after completing the algorithm. The selected nodes are of course the level-i nodes that are selected to become level-$(i+1)$ nodes, that is, the nodes of G_{i+1}. The distributed greedy algorithm is as follows.

Distributed Greedy Algorithm

Step 1: Each node $v \in V_i$ gathers information about its r_i-local view of G_i, which will henceforth be referred to as v's *level-i view*. This requires several rounds of passing local link-state information. Some nodes in this local view may still be actively participating in the greedy algorithm at a lower level. If this happens, then the level-i algorithm must stall until these nodes complete the lower level algorithms.

Step 2: v determines P_v and C_v, where these are defined as follows. P_v denotes the set of all the nodes pairs $\{x,y\}$ covered by v in the bipartite graph B. C_v denotes the set of all the nodes that cover some node pair in P_v. ($v \in C_v$, and by Lemma 2, v is able to "see" the elements of P_v and C_v. Actually, only a $(k_{i+1} + k_i)$-local view is required for this.) v also computes its current *covering number* $|C_v|$ (the size of C_v).

Step 3: v multi-casts a message containing its covering number and its status (undecided, selected or not selected) to each node in C_v. (Note that the first time this step is executed, v is undecided, and the last time it executes this step, it will be in one of the two decided states.)

Step 4: If v has entered one of the two decided states (selected or not selected), then it essentially terminates its participation in this algorithm (at the current level), except to help route messages between other nodes. Otherwise, if it is still undecided, then ....

Step 5: v waits until it receives messages as in Step 3 from each node in C_v. For each such node u that has become decided, v removes u from C_v, and if u has become selected, then v also removes any pairs from P_v that u covers. Accordingly, v recomputes its covering number as necessary.

Step 6: If v's covering number is now zero, then v enters the "not selected" state, and loops back to Step 3. Otherwise....

Step 7: v checks to see if its own *priority* is the highest among all the nodes of C_v. Priority here is defined to be the ordered pair (covering_number, ID), lexicographically ordered (as in [6–Subsection 2.1]). If v has the highest priority, then v enters the "selected" state. In either case, it loops back to Step 3.

Remarks:

1. Once a selected node has terminated the greedy algorithm at level i, it can proceed to initiate its participation in the greedy algorithm at level $i + 1$, where it is of course initially undecided at this level.
2. In Step 3, a node v is obliged to send a message to some of the nodes in its level-i view. This can be handled efficiently by means of "optimal routing trees" and lower level local routing.

3. It is also possible to let the ultimate number of levels be initially unspecified, perhaps until a level is reached consisting of a single node. The set $\mathcal{K}$ would then grow according to some formula, as new levels are constructed.
4. Other algorithms can be used in place of the greedy algorithm. For example, it is possible to adapt the "d-SPR-C method" of [4]. Unlike the greedy algorithm, and assuming that link costs reflect transmission time delays, this algorithm completes in a time period that does not depend on the overall size of the network, but rather only depends on the maximum link cost and maximum node degree.

5.2 Local Unicast Routing at a Given Level

Once a $\mathcal{K}$-SPR sequence has been established up to some level, say i, it is possible for a level-i node v to efficiently route a message to another level-i node u within its level-i view, as follows. Recall that if $j < i$, a level-i node is also a level-j node. Now v can easily discover a minimal path in the level-i view connecting it to u. Let u_{i-1} denote the first node on this minimal path after v. Since $c(v, u_{i-1}) \leq k_i$, the node u_{i-1} is visible to v in its level-$(i-1)$ view. It can then find a minimal path connecting itself to u_{i-1} at this level. Let u_{i-2} be the node after v on this minimal path. And so forth, down to level zero.

Letting $u_i = u$, v can append the sequence $\{u_j\}_{j=1}^{i}$ as routing information to the message, before sending it to its neighbor (in G) u_0. The level-zero views of the nodes along the way now aid to easily route the message to u_1. By similar reasoning, requiring both level-one and level-zero views, the message can then delivered to u_2. And so forth, until it ultimately arrives at u. Moreover, the path (in G) used to route the message from v to u is guaranteed to be a minimal path.

5.3 Special Multicasting to Routers

We now consider a very specific multicasting problem for a network with an established $\mathcal{K}$-SPR sequence. This will be employed for both the proactive and reactive aspects of the hybrid routing scheme proposed in the next subsection. We will need the following definition and lemma.

Definition 6. *Consider an arbitrary node v. For $i \geq 1$, a level-i node v_i will be called a level-i router for v if the only level-i node u satisfying $c(v, u) + c(u, v_i) = c(v, v_i)$ is $u = v_i$.*

Thus a level-i router for v is a level-i node such that any shortest path connecting it to v contains no other level-i nodes. The following is straightforward to establish, using induction on i.

Lemma 3. *A level-i router v_i for v satisfies $c(v, v_i) \leq k_1 + k_2 + \cdots + k_i$.*

The goal now is to allow v to send a message to all of its routers, at all levels. In fact, this goal will be accomplished in such a way that forwarded messages always move along minimal paths, moving away from the source node v. Moreover, there will be no redundancy in the message forwarding, in the sense that no node will receive more than one copy of the message. That is, the message will move along a tree rooted at v, and each path from v in this tree will be a minimal path. This sort of "multicasting to routers" will provide a basis for the hybrid routing scheme described in the next subsection.

To manage the proposed multicasting, it is necessary for a level-i router v_i for v that receives the message along a given minimal path, to decide to which of the level-$(i+1)$ routers for v it must forward the message. As a technical detail, in order for v_i to make this decision, it will be necessary that a list of all the level-i routers for v, along with their costs from v, be included in the header of the message that v_i receives. Under reasonable conditions, this list will not be large. Before v_i forwards the message to level-$(i+1)$ the routers, it will likewise be necessary for it to append a list of all the level-$(i+1)$ routers for v, and their costs from v. However, the level-i router information can be removed from the header at this point.

Now v_i is within a cost $k_1 + \cdots + k_i$ of v, as are all of the level-i routers for v. Moreover, v_i has received a list of these together with their costs from v. Let u denote one such level-i router. Consider a level-$(i+1)$ node w within a cost k_{i+1} of u. Such a node is potentially a level-$(i+1)$ router for v, and all level-$(i+1)$ routers for v fit this description for some u. Now, with w fixed, it turns out that v_i is able to determine which level-i routers u lie along a minimal path in G connecting v to w. In the first place, w is in the level-i view of v_i, which can be seen by considering a shortest possible path from v_i to v, and then to u, and then to w. The cost of this does not exceed r_i, so $c(v_i, w) \leq r_i$. Also, $c(v_i, u) \leq 2(k_1 + \cdots + k_i)$ and $c(u, w) \leq k_{i+1}$. It follows by Lemma 2 that v_i is able to correctly compute $c(u, w)$, using its level-i view (of level-i nodes within a cost r_i). It is now straightforward to see that v_i is able to determine whether or not w is a level-$(i+1)$ router for v. If it is, then v_i is also able to determine any level-i routers for v that lie along a minimal path connecting v and w.

There is one last detail. In order to avoid redundant messages, for each level-$(i+1)$ router v_{i+1} for v, exactly one of the level-i routers for v lying between v and v_{i+1} along a minimal path should be selected to forward the message to v_{i+1}. Each of these routers is aware of the others and so some criterion can be used that they will all agree on in order to make the selection. For example, this decision could be made by using a simple criterion such as choosing the level-i router for v with the largest ID.

5.4 A Hybrid Hierarchical Routing Strategy

The routing strategy that will be developed here has the following theorem as its foundation. (Choose i here as large as possible such that $k_i < c(u, v)$.)

Theorem 4. *Given any two nodes u and v, there exists a minimal path p connecting u and v, and a positive integer i, such that p contains a level-i router u_i for u and a level-i router v_i for v with $c(u_i, v_i) \leq c(u, v) \leq k_{i+1} \leq r_i$.*

During the process of establishing a $\mathcal{K}$-SPR sequence in a sensor network, say by the greedy algorithm method, it is easy to arrange for each node v to be known to all of the nodes within a cost k_1, as well as all of the level-one nodes that are within a cost k_2 of a level-one node that is within a cost k_1 of v, as well as all of the level-two nodes that are within a cost k_3 of a level-two node that is within a cost k_2 of a level-one node that is within a cost k_1 of v, and so forth. In fact, this does not require any additional messages, but rather only the inclusion of more information in the already required selection overhead messages.

It may be assumed that in this way each level-i router v_i of v maintains a list of nodes $\{v = v_0, v_1, v_2, ..., v_i\}$ with the property that there exists a minimal path in G connecting v to v_i such that v_j is a level-j router for v $(j = 1, ..., i)$. In addition, all level-i nodes within a cost k_{i+1} of one of the level-i routers v_i of v will be made aware of v, and we may assume that these too have been provided with routing information to v. If the network is allowed to change dynamically, then any new node that joins the network later would be obliged to announce itself to its routers and to each level-i node within a cost k_{i+1} of one of its level-i routers. This could be managed using a variation of the multicasting to routers method discussed in the previous subsection.

Now, after establishing the $\mathcal{K}$-SPR sequence and the above routing information, suppose that a node u has a need to contact a node v, say to establish a virtual circuit in order to conduct an extended conversation with v. Suppose too that u is currently unaware of where v is in the network, and so has no routing information concerning it, other than the ID number of v or some other identifier such as a unique name. In particular, this would mean that $c(u, v) > k_1 = r_0$.

As a result of Theorem 3 and the assumptions we are making about the local information maintained by each node, at each level, the node u is able to find the node v as follows. u multicasts a request message to its routers, as described in the previous subsection. Eventually some node receiving the request will know about the existence of v, and will know a shortest path to it. This node can then reply by relaying this information back to u along with the information that describes a minimal path from itself to u. It does not need to forward the message to higher level routers. In this way, u learns a path to v, as well as its cost. At least one of the paths thus discovered will be a minimal path from u to v.

Point-to-point communication between u and v can now be effected via routing information placed in the header. However, this only needs to involve a sequence $u = u_0, u_1, \cdots u_i, v_i, \cdots v_1, v_0 = v$ of nodes, where u_j and v_j are level-j routers for u and v, respectively $(j = 1, ..., i)$ and $c(u_i, v_i) \leq k_{i+1}$. The routing between these nodes can be managed by means of the appropriate local views of the various nodes along the way.

References

1. I.F. Akyildiz, W. Su, Y. Sankarasubramaniam, E. Cayirci, Wireless Sensor Networks: A Survey, *Computer Networks*, Vol 38, 2002, pp. 393-422.
2. B. Chen, K. Jamieson, H. Balakrishnan and R. Morris, Span: an energy-efficient coordination algorithm for topology maintenance in ad hoc wireless networks, *Proc. Mobicom*, 2001, pp. 85-96.

3. Y. P. Chen, A. L. Liestman, J. Liu, Clustering Algorithms for Ad Hoc Wireless Networks, in *Ad Hoc and Sensor Networks*, Edited by Y. Xiao and Y. Pan, Nova Science Publisher, 2004.
4. S. Dhar, M. Q. Rieck, S. Pai and E. J. Kim. Distributed Routing Schemes for Ad Hoc Networks Using d-SPR Sets, *Microprocessors and Microsystems, Special Issue on Resource Management in Wireless and Ad Hoc Mobile Networks*, Volume 28, Issue 8, October 2004, pp. 427-437.
5. W. R. Heinzelman, A. Chandrakasan, H. Balakrishnan, Energy-effcient communication protocol for wireless microsensor networks, *IEEE Proceedings of the Hawaii International Conference on System Sciences*, January 2000, pp. 1-10.
6. L. Jia, R. Rajaraman, T. Suel, An efficient distributed algorithm for constructing small dominating sets, *Proc. Annual ACM Symposium on Principles of Distributed Computing*, 2001, pp. 33-42.
7. J.M. Kahn, R.H. Katz, K.S.J. Pister, Next century challenges: mobile networking for smart dust, *Proceedings of ACM MobiCom 99*, August 1999, pp. 271-278.
8. B. Liang, Z. J. Haas, Virtual backbone generation and maintenance in ad hoc network mobility management, *Proc. 19th Ann. Joint Conf. IEEE Computer and Comm. Soc. INFOCOM*, 2000, pp. 1293-1302.
9. S. Lindsey and C. S. Raghavendra, PEGASIS: Power Efficient Gathering in Sensor Information Systems, *Proceedings of the IEEE Aerospace Conference*, Big Sky, Montana, March 2002.
10. S. Lindsey, C. S. Raghavendra and K. Sivalingam, Data Gathering in Sensor Networks using the Energy*Delay Metric, *Proceedings of the IPDPS Workshop on Issues in Wireless Networks and Mobile Computing*, San Francisco, CA, April 2001.
11. PicoRadio: $http : //bwrc.eecs.berkeley.edu/Research/Pico\_Radio.htm.$
12. G.J. Pottie, W.J. Kaiser, Wireless integrated network sensors, *Communications of the ACM*, 43:5, 2000, pp. 51-58.
13. S. Rajagopalan, V. V. Vazirani, Primal-dual RNC approximation of covering integer programs, *SIAM J. Computing*, 28, 1998, pp. 525-540.
14. M. Q. Rieck, S. Pai, S. Dhar, Distributed Routing Algorithms for Multi-hop Ad Hoc Networks Using d-hop Connected d-Dominating Sets, *Computer Networks*, Volume 47, Issue 6, April 2005, pp. 785-799.
15. B. Warneke, M. Last, B. Liebowitz, K.S.J. Pister, Smart dust: communicating with a cubic-millimeter computer, *Computer Magazine*, January 2001, pp. 44-51.
16. J. Wu, H. Li, On calculating connected dominating set for efficient routing in ad hoc wireless networks, *Proc. 3rd Int. Wksp. Discrete Algorithms and Methods for Computing and Communications*, 1999, pp. 7-14.

On Lightweight Node Scheduling Scheme for Wireless Sensor Networks*

Jie Jiang, Zhen Song, Heying Zhang, and Wenhua Dou

School of Computer,
National University of Defense Technology,
410073, Changsha, China
jiangjie@nudt.edu.cn

Abstract. Energy efficient self-organization is a crucial method to prolong the lifetime of wireless sensor networks consisting of energy constrained sensor nodes. In this paper, we focus on a distributed node scheduling scheme to extend network lifespan. We discuss the network coverage performance when sensor nodes are deployed according to Poisson point process and reveal the internal relationship among the required coverage performance, expected network lifetime and the intensity of Poisson point process. Also the impact of uniformly distributed time asynchrony on network coverage performance is analyzed. Simulation results demonstrate that the proposed scheme works well in the presence of time asynchrony.

1 Introduction

Because of advances in micro-sensors, wireless networking and embedded processing, wireless sensor networks (WSN), which consists of a large number of tiny sensor nodes with limited computation, communication capabilities and constrained energy resource, are becoming increasingly available for commercial and military applications, such as environmental monitoring, chemical attack detection, and battlefield surveillance [1],[2].

Energy is the most precious resource in wireless sensor networks. First, sensor nodes are usually supported by batteries with limited capacity due to the extremely small dimensions. Second, it is usually hard to replace or recharge the batteries after deployment, either because the number of sensor nodes is very large or the deployment environment is hostile and dangerous (e.g. remote desert or battlefield). But on the other hand, the sensor networks are usually expected to operate several months or years once deployed. Therefore reducing energy consumption and extending network lifetime is one of the most critical challenges in the design of wireless sensor networks.

One promising approach to extending network lifetime is node scheduling, which only keeps a subset of sensor nodes active and puts other sensor nodes

---

* This work is supported by the National Natural Science Foundation of China under grant number 90104001.

into low-powered sleep status. Most of existing work [3],[6],[7],[11],[12] on node scheduling relies on exact location information, which is expensive and difficult to obtain in large scale wireless sensor networks.

In this paper we propose a distributed node scheduling scheme for random wireless sensor network. The network lifetime can be extended to be about kT_s (T_s is the lifetime of individual sensor node) when sensor nodes are organized into k node disjoint sensor covers and each of these sensor covers is activated in a round-robin manner. In our scheme, sensor nodes randomly select a number i between 1 and k, then joins node set NS_i, works during set NS_i's working shift and sleeps during the rest of time. This scheme is lightweight as it does not require any message communication among sensor nodes and the computation cost is low. It is also location free and does not rely on the expensive localization service in wireless sensor networks. As shown later in this paper, the proposed scheme can achieve good coverage quality if the intensity of node deployment is large enough. Our theoretical analysis also reveals the relationship between expected network lifetime and node deployment intensity. Further, the proposed scheme can work well even in the presence of clock asynchrony among sensor nodes.

2 Related Work

Many research efforts have been made to exploit the inherent coverage redundancy to extend the lifetime of wireless sensor networks. Slijepcevic et al. [3] propose a centralized heuristic solution for the NP-hard problem of finding the maximal number of disjoint sensor sets, where each set can cover the target region completely. Abrams et al. [4] address a variation of the problem, where the objective is to partition the sensors into mutually exclusive covers such that the number of covers that include an area, summed over all areas, is maximized. Ye et al. [5] present a distributed, probing based algorithm to extend network lifetime. Tian et al. [6] propose a distributed node scheduling scheme that exploits the coverage overlap among neighboring sensors to prolong network lifespan. Chen et al. [7] propose a grid-based approach for selecting working nodes in sensor networks. Carbunar et al. [8] propose a distributed algorithm with a view to improving energy efficiency while preserving network coverage. Yan et al. [9] address the issue of providing differentiated surveillance service for various target area. Zhang et al. [10] present a decentralized density control algorithm (OGDC) to choose a minimal set of working sensor nodes while these active sensor nodes can maintain the initial coverage and the communication connectivity. Wang et al. [11] introduce a coverage configuration protocol that aims to maintain both the sensing coverage and the network connectivity when scheduling sleep intervals for redundant sensors. Gupta et al. [12] propose a centralized greedy algorithm to construct a minimal connected sensor cover, which covers the target region completely and forms a connected communication network. The most closely related work is [13] by Liu and Wu, where a similar idea is discussed. Here our work focuses on different node deployment and time asynchrony model.

3 Lightweight Node Scheduling Scheme

3.1 Basic Idea

The work in [3] proposes to organize sensor nodes in node disjoint sensor covers to prolong the network lifetime. It aims to calculate the maximal number of such sensor cover because the network lifetime is proportional to the number of sensor cover. Here we consider another related problem. Given the expected lifetime requirement, kT_s, how to organize sensor nodes into these k disjoint node sets in a distributed, lightweight and location-free manner?

Given the parameter k, in the initial phase each sensor node randomly selects a number between 1 and k with equal probability of $1/k$, and all nodes choosing number i form the i'th node set. In the following working phase, these k node sets work in a round-robin manner and there is only one node set working at any time instance.

3.2 Performance Analysis

A. System Model

We consider static sensor networks in a two-dimensional region. And we use binary sensing model to model sensor node's sensing capability. In binary sensing model, sensor can reliably detect events within the circle centered at the sensor node with radius of sensor's sensing range. Such circle is called sensor node's *sensing disk* and the radius of the sensing disk is called sensor node's sensing radius (denoted by R_s). We assume that the sensor network is homogeneous, i.e., all sensor nodes have the same sensing radius.

We consider the random sensor network where sensor nodes are randomly deployed (e.g., dropped form airplane) according to Poisson point process [14], which has been widely used in researches [15],[16],[17] on random wireless sensor networks. In Poisson point process, the probability of that an region A contains m sensor nodes is given by

$$\Pr\{N(A) = m\} = \frac{(\lambda \|A\|)^m e^{-\lambda \|A\|}}{m!} \quad (1)$$

where $\|A\|$ denotes the area of A, $N(A)$ denotes the number of nodes in region A, and λ is the intensity of Poisson point process.

B. Performance Analysis

Definition 1. Coverage Intensity for a Specific Point [13]

For a given point p in the deployed region, the coverage intensity for this point is $C_p = T_c/T$, where T is any given long time period and T_c is the total time during T when point p is covered by at least one active sensor node.

Definition 2. Network Coverage Intensity [13]

The network coverage intensity, C_n, is defined to be the expectation of C_p: $C_n = E(C_p)$.

Theorem 1. *With the proposed scheduling scheme,*

$$C_n = 1 - exp\left(-\frac{\|\lambda A\|}{k}\right) \qquad (2)$$

where k is the given network lifetime requirement, λ is the intensity of the Poisson point process, and $\|A\| = \pi R_s^2$ is the area of sensor node's sensing disk.

Proof. For any given point p in the deployment region, suppose there are totally N_p sensor nodes that cover point p. Let S_p denote the set of these N_p sensor nodes. Using the proposed scheduling scheme, each node in S_p assigns itself to one of the k node sets with equal probability $1/k$. Let A_i denote the event that the i $(1 \leq i \leq k)$'th node set NS_i does not include any node in S_p, then $Pr\{A_i\} = \left(1 - \frac{1}{k}\right)^{N_p}$, and $Pr\{\overline{A_i}\} = 1 - \left(1 - \frac{1}{k}\right)^{N_p}$.

Let's define an indicator function as follows:

$$I_i = \begin{cases} 1 & \text{if } A_i \text{ not holds} \\ 0 & \text{else} \end{cases}$$

Then $I = \sum_{j=1}^{k} I_j$ is the total number of the node set that can cover point p.

As $E[I] = E\left[\sum_{j=1}^{k} I_j\right] = \sum_{j=1}^{k} E[I_j]$ and $E[I_j] = 1 - \left(1 - \frac{1}{k}\right)^{N_p}$, we have $E[I] = k \times \left[1 - \left(1 - \frac{1}{k}\right)^{N_p}\right]$. Therefore $C_p = \frac{E[I] \times T}{k \times T} = 1 - \left(1 - \frac{1}{k}\right)^{N_p}$. According to the binary sensing model and the definition of Poisson point process,

$$C_n = E[C_p] = 1 - E\left[\left(1 - \frac{1}{k}\right)^{N_p}\right]$$

$$= 1 - \sum_{N_p=0}^{\infty} \left(1 - \frac{1}{k}\right)^{N_p} \times \frac{(\lambda\|A\|)^{N_p} e^{-\lambda\|A\|}}{N_p!}$$

$$= 1 - exp\left(-\frac{\lambda\|A\|}{k}\right)$$

where $\|A\| = \pi R_s^2$. ∎

Corollary 1. *For a given λ, the possible maximal number k of disjoint node sets while the network coverage intensity is at least α is given by $\frac{\lambda\|A\|}{-\ln(1-\alpha)}$.*

Proof. $C_n \geq \alpha \Rightarrow 1 - exp\left(-\frac{\lambda\|A\|}{k}\right) \geq \alpha \Rightarrow \ln(1-\alpha) \geq -\frac{\lambda\|A\|}{k}$
As $0 \leq \alpha < 1$, $\ln(1-\alpha) < 0$, so $k \leq \frac{\lambda\|A\|}{-\ln(1-\alpha)}$. ∎

Corollary 2. *For a given k and a required network coverage intensity α, the lower bound of the intensity of the Poisson point process, λ, is given by $\frac{-k\ln(1-\alpha)}{\|A\|}$.*

Proof. $C_n \geq \alpha \Rightarrow \lambda \|A\| \times \left(\frac{1}{k}\right) \geq -\ln(1-\alpha) \Rightarrow \lambda \geq \frac{-k\ln(1-\alpha)}{\|A\|}$. ∎

These two corollaries, which point out the internal relationship among the network coverage intensity, the expected network lifetime, and the intensity of the Poisson point process, are instructive in practice when determining the largest number of disjoint node sets (k) if the required network coverage intensity (α) and the intensity of Poisson point process (λ) are given *a priori*. Also with given k and α, we can determine the required smallest intensity of Poisson point process.

4 Network Coverage Intensity with Clock Asynchrony

The proposed scheduling scheme organizes sensor nodes into different node disjoint node sets and these node sets work alternately to prolong the network lifetime. This requires that each sensor node should know the starting and the ending time of the working shift of the node set which it belongs to. But exact time synchronization is hard to realize in large scale wireless sensor networks. In this section, we analyze the impact of clock asynchrony on the performance of the proposed scheduling scheme. The analysis here is similar to that in [13]. But we consider different model of time asynchrony under Poisson point process.

Consider any point p in the target region. Assume there are totally N_p sensor nodes that can cover point p initially and $N_p{}^i$ sensor nodes are assigned to node set NS_i. Point p will not be covered during the working shift of node set NS_i only in three situations. First, all $N_p{}^i$ sensor nodes start working ahead of the starting time of NS_i. Then there will be a time interval at the end of the working shift of NS_i when all the $N_p{}^i$ sensor nodes have stopped and p will not be covered. Second, all $N_p{}^i$ sensor nodes start working behind the starting time of NS_i. In this situation, there will be a time interval at the beginning of the working shift of NS_i when all the $N_p{}^i$ sensor nodes haven't waken up and therefore p will not be covered. Third, and finally, a part of $N_p{}^i$ sensor nodes starts working ahead of the starting time of NS_i while the remains are behind the time, and there is a gap period between them. Therefore in this gap period p is not covered by any sensor node.

Note that both the sensor nodes in N_p that are assigned into node set NS_{i+1} and with ahead-of-starting time, and the sensor nodes in N_p that are assigned into node set NS_{i-1} and with behind-of-starting time can help to reduce the uncovered time period during the working shift of node set NS_i. But we ignore these cases in our following analysis because of the complexity induced by the correlation among neighboring node sets. Therefore, the calculated network coverage intensity in the following sections is the lower bound of the actual value. That is, the actual network coverage intensity is larger or at least equal to the theoretical value presented.

We make the following assumptions in our following analysis.

(1) The starting time of each sensor node may not be synchronized precisely with the standard time, but the internal time ticking frequency is accurate. So there will be no accumulation of time drift.

(2) Let T denote the working duration of each node set in one round. We assume that the difference between the starting time of each sensor node and the standard time, Δt, is less than $T/2$. We assume that $\Delta t \geq T/2$ is an extremely rare case and could be ignored. This assumption eliminates the possibility of the third case described above and reduces the complexity of analysis.
(3) The time difference, Δt, is a random variable which is uniformly distributed between $(-T/2, T/2)$, i.e, $\Delta t \sim U(-T/2, T/2)$.

We are interested in the expectation of the length of time when point p is not covered by any of these $N_p{}^i$ sensor nodes during the working shift of node set NS_i. Let $E_{uc}{}^i$ denote this expectation. Obviously, $E_{uc}^i = T$ if $N_p{}^i = 0$. When $N_p{}^i > 0$,

$$E_{uc}^i = \int_0^\infty x f_1(x) dx + \int_{-\infty}^0 -y f_2(y) dy \qquad (3)$$

where $x = \min\{\Delta t_j, 0 \leq j \leq m_i - 1\}$, $y = \max\{\Delta t_j, 0 \leq j \leq m_i - 1\}$ and Δt_j denotes the difference between node j's starting time and the standard time, $f_1(x)$ and $f_2(y)$ are the p.d.f of x and y respectively. The first and the second item in equation (3) correspond respectively with the time interval when point p is not covered due to the first and the second reasons described previously. Since $\Delta t_1, \Delta t_2, \ldots, \Delta t_j$ are independently random variables uniformly distributed in $(-T/2, T/2)$, we can get

$$E_{uc}^i = \int_0^{T/2} x f_1(x) dx + \int_{-T/2}^0 -y f_2(y) dy$$

Since $x = \min\{\Delta t_j, 0 \leq j \leq N_p{}^i - 1\}$,

$$\Pr\{x \geq \alpha\} \Leftrightarrow \Pr\{\forall j \in [0, N_p{}^i - 1], \Delta t_j \geq \alpha\} = [1 - F(\alpha)]^{N_p{}^i}$$

where $F(x)$ is the c.d.f of uniform distribution. Therefore $\Pr\{x < \alpha\} = 1 - \Pr\{x \geq \alpha\} = 1 - [1 - F(\alpha)]^{N_p{}^i}$. Then we can get the p.d.f of x:

$$f_1(x) = N_p{}^i f(x) [1 - F(x)]^{N_p{}^i - 1}$$

where $f(x)$ is the p.d.f of uniform distribution. According to the definition of uniform distribution, we have

$$f_1(x) = \begin{cases} \dfrac{N_p{}^i}{T} \left(\dfrac{1}{2} - \dfrac{x}{T}\right)^{N_p{}^i - 1}, & -T/2 < x < T/2 \\ 0, & \text{otherwise} \end{cases}$$

So by symmetry,

$$E_{uc}^i = 2 \int_0^{\frac{T}{2}} x f_1(x) dx = \frac{T}{2^{N_p{}^i}} \cdot \frac{1}{N_p{}^i + 1}$$

then

$$E_{uc} = E\left(E_{uc}^i\right) = \sum_{j=0}^{N_p} E_{uc}^i \times \Pr\left\{N_p{}^i = j\right\}$$

$$= T \times \left(1 - \frac{1}{k}\right)^{N_p} + \sum_{j=1}^{N_p} E_{uc}^i \times \Pr\left\{N_p{}^i = j\right\}$$

$$= T \times \left(1 - \frac{1}{k}\right)^{N_p} + T \sum_{j=1}^{N_p} \frac{1}{j+1}\binom{N_p}{j}\left(\frac{1}{2k}\right)^j\left(1 - \frac{1}{k}\right)^{N_p-j}$$

$$= T \times \left(1 - \frac{1}{k}\right)^{N_p}$$
$$+ \frac{2kT}{N_p+1}\left[\left(1 - \frac{1}{2k}\right)^{N_p+1} - \frac{N_p+1}{2k}\left(1 - \frac{1}{k}\right)^{N_p} - \left(1 - \frac{1}{k}\right)^{N_p+1}\right]$$

Let $E_c = T - E_{uc}$, then the expectation of the time interval when point p is covered in the working shift of any node set is given by

$$E(E_c) = E(T - E_{uc})$$
$$= T - T \times \sum_{N_p=0}^{\infty}\left(1 - \frac{1}{k}\right)^{N_p} \times \frac{e^{-\lambda\|A\|} \times (\lambda\|A\|)^{N_p}}{N_p!}$$
$$- \sum_{N_p=0}^{\infty} \frac{2kT}{N_p+1} \times \left(1 - \frac{1}{2k}\right)^{N_p+1} \times \frac{e^{-\lambda\|A\|} \times (\lambda\|A\|)^{N_p}}{N_p!}$$
$$+ T \sum_{N_p=0}^{\infty}\left(1 - \frac{1}{k}\right)^{N_p} \times \frac{e^{-\lambda\|A\|} \times (\lambda\|A\|)^{N_p}}{N_p!}$$
$$+ \sum_{N_p=0}^{\infty} \frac{2kT}{N_p+1} \times \left(1 - \frac{1}{k}\right)^{N_p+1} \times \frac{e^{-\lambda\|A\|} \times (\lambda\|A\|)^{N_p}}{N_p!}$$
$$= T - T \times \exp\left(-\frac{\lambda\|A\|}{k}\right)$$
$$- \left\{\frac{2kT}{\lambda\|A\|} \times \left[\exp\left(-\frac{\lambda\|A\|}{2k}\right) - \exp\left(-\frac{\lambda\|A\|}{k}\right)\right] - T\exp\left(-\frac{\lambda\|A\|}{k}\right)\right\}$$

The network coverage intensity with time asynchrony uniformly distributed $C_n{}'$ is:

$$C_n{}' = \frac{k \times E(E_c)}{k \times T} = C_n - \Delta \qquad (4)$$

where

$$\Delta = \frac{2k}{\lambda\|A\|} \times \left[\exp\left(-\frac{\lambda\|A\|}{2k}\right) - \exp\left(-\frac{\lambda\|A\|}{k}\right)\right] - \exp\left(-\frac{\lambda\|A\|}{k}\right)$$

The second item in equation (4), Δ, indicates the impact of the uniformly distributed time asynchrony on network coverage intensity.

5 Simulation

5.1 Simulation Setup

In our simulation, we use the binary sensing model describe in section 3. Based on the information from [18], we set the sensing radius to be 6. This is consistent with other current sensor types, such as Smart Dust (U.C.Berkeley), CTOS dust, Wins (Rockwell) [19], and JPL [20]. And the target region is a square of 50×50. Sensor nodes are randomly distributed in the target region according to the Poisson point process with intensity λ. All simulations are conducted using MATLAB and the simulation of Poisson point process is implemented based on the information from [21]. We are interested in the network coverage intensity with different network lifetime requirement k, different intensity of Poisson point process λ and with or without time asynchrony among sensor nodes. We also investigate the impact of time asynchrony on network coverage intensity when time asynchrony is uniformly distributed. For each simulation scenario, ten runs with different random node distributions are conducted and only the average is presented.

5.2 Simulation Results

Fig. 1 shows how the network coverage intensity varies with the intensity of Poisson point process when the value of k equals to 3, 6, 9, and 12 respectively. From

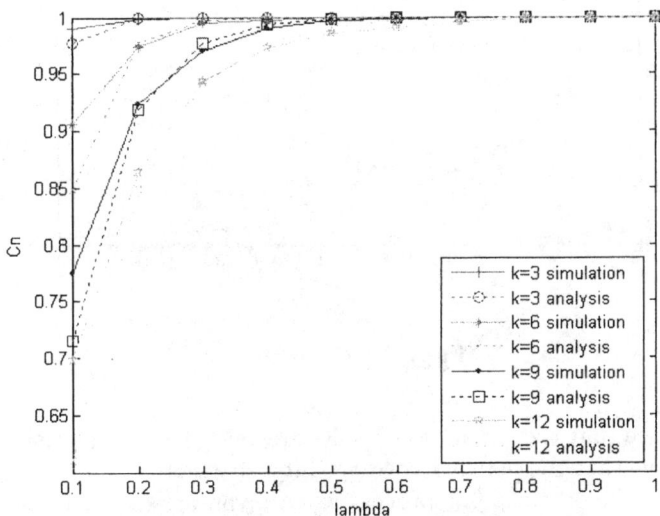

Fig. 1. C_n vs. λ

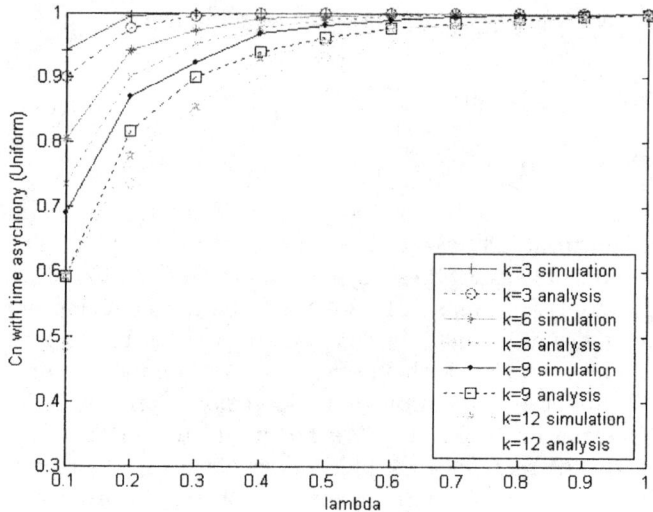

Fig. 2. C_n' vs. λ

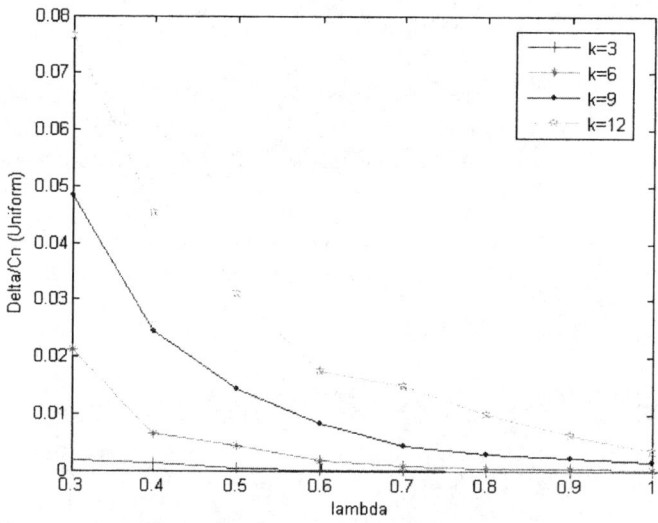

Fig. 3. Δ/C_n vs. λ

this figure, we see that the simulation results are very close to the theoretical results. We observe that the network coverage intensity increases with the increase of the intensity of Poisson point process when given a fixed k. Larger deployment intensity will deploy more sensor nodes in the network and each node set will include more sensor nodes when k is fixed. Therefore the network coverage intensity of each node set is improved. But the network coverage intensity becomes saturated at some node intensity. For example, the network coverage intensity is larger than 99.9% when $\lambda = 0.5$ and $k = 6$. This means that larger node intensity will not benefit the network coverage intensity remarkably, but increase the deployment cost hugely. We also observe that when λ is fixed, smaller k will lead to better network coverage intensity. This is because when the node number is fixed, smaller k means fewer node sets and each node will include more sensor nodes.

Fig. 2 shows how the network coverage intensity varies with the intensity of Poisson point process when sensor nodes are not precisely synchronized and the time difference is uniformly distributed in interval $(-T/2, T/2)$. It can be seen that the simulation curves match the theoretical analysis very well when the value of k is 3, 6, 9, and 12 respectively. Fig. 3 shows how the impact of time asynchrony on the network coverage varies with the intensity of Poisson point. Even for $k = 12$, when the node intensity λ increases up to about 0.5, this ratio of Δ/C_n decreases rapidly to about 0.036. These simulation results demonstrate that the proposed scheduling scheme can work well even in the presence of time asynchrony.

6 Conclusions

In this paper, we discuss a distributed, lightweight and location-free node scheduling scheme that aims to extend the lifetime of wireless sensor networks. This scheme neither incurs any communication overhead nor relies on expensive localization service. Thus it is scalable to large scale sensor networks. We focus on the network coverage performance when sensor nodes are deployed randomly in the target region according to Poisson point process. Theoretical analysis reveals the internal relationship among the required coverage performance, expected network lifetime and the intensity of Poisson point process. We also discuss the impact of time asynchrony on network coverage intensity when the time asynchrony is uniformly distributed. Simulation results demonstrate that the proposed scheme is robust to time asynchrony.

References

1. Elson J. and Estrin D.: Sensor Networks: A Bridge to the Physical World. Wireless Sensor Networks, Kluwer, (2004).
2. Akyildiz I. F., Su W., Sankarasubramaniam Y., and Cayirci E.: Wireless Sensor Networks: A Survey. Computer Networks (Elsevier) Journal,pp.393-422, (2004).
3. Slijepcevic S. and Potkonjak M.: Power Efficient Organization of Wireless Sensor Networks. In Proc. of IEEE ICC'01, Helsinki, Finland, (2001).
4. Abrams Z., Goel A., and Plotkin S.: Set K-Cover Algorithms for Energy Efficient Monitoring in Wireless Sensor Networks. Proc. of Information Processing in Sensor Networks (IPSN), Berkeley, California, USA, (2004).
5. Ye F., Zhong G., Lu S., and Zhang L.: Peas: A Robust Energy Conserving Protocol for Long-Lived Sensor Networks. In Proc. of ICDCS'03, (2003).

6. Tian D. and Georganas N. D.: A Coverage-Preserving Node Scheduling Scheme for Large Wireless Sensor Networt. In Proc. of WSNA'02, Atlanta, Geogia, USA, (2002).
7. Chen H., Wu H., and Tzeng N. Grid-Based Approach for Working Node Selection in Wireless Sensor Networks. In Proc. of IEEE ICC'04, Paris, France, (2004).
8. Carbunar B., Grama A., Vitek J., and Carbunar O.: Coverage Preserving Redundancy Elimination in Sensor Networks. In Proc. of SECON 2004, Santa Clara, CA, USA, (2004).
9. Yan T., He T., and Stankovic J. Differentiated Surveillance Service for Sensor Networks. In Proc. of SenSys'03, Los Angels, CA, USA, (2003).
10. Zhang H. and Hou J. C.: Maintaining Sensing Coverage and Connectivity in Large Sensor Networks. In Proc. of NSF International Workshop on Theoretical and Algorithmic Aspects of Sensors, Ad Hoc Wireless, and Peer-to-Peer Networks, (2004).
11. Wang X., Xing G. et al: Integrated Coverage and Connectivity Configuration in Wireless Sensor Networks. In Proc. of SenSys'03, Los Angeles, CA, (2003).
12. Gupta H., Das S. R., and Gu Q. Connected Sensor Cover: Self-Organization of Sensor Networks for Efficient Query Execution. In Proc. of MobiHoc'03, Annapolis, Maryland, USA, (2003).
13. Liu C., Wu K., and King V. Randomized Coverage-Preserving Scheduling Schemes for Wireless Sensor Networks. In Proc. of IFIP Networking 2005, Waterloo Ontario, Canada, (2005).
14. Okabe A., Boots B., Sugihara K., and Chiu S. N.: Spatial Tessellations: Concepts and Applications of Voronoi Diagram. John Wiley & Sons Press, (1999).
15. Liu B. and Towsley D. A Study of the Coverage of Large-Scale Sensor Networks. In Proc. of The 1st IEEE International Conference on Mobile Ad-hoc and Sensor Systems (MASS'04), Florida, USA, (2004).
16. Kumar S., Lai T. H., and Balogh J.: On K-Coverage in a Mostly Sleeping Sensor Network. In Proc. of ACM MobiCom 2004, Philadelphia, USA, (2004).
17. Zhang H. and Hou J.: On Deriving the Upper Bound of Alpha-Lifetime for Large Sensor Networks. In Proc. of the 5th ACM international symposium on Mobile ad hoc networking and computing (MobiHoc), Roppongi Hills, Tokyo, Japan, (2004).
18. http://www-bsac.eecs.berkeley.edu/shollar
19. http://wins.rsc.rockwell.com
20. http://sensorwebs.jpl.nasa.gov
21. Stoyan D., Kendall W. S., and Mecke J.: Stochastic Geometry and Its Applications. Second Edition. Wiley Series in Probability and Statistics. (1995).

Clique Size in Sensor Networks with Key Pre-distribution Based on Transversal Design

Dibyendu Chakrabarti, Subhamoy Maitra, and Bimal Roy

Applied Statistics Unit, Indian Statistical Institute,
203 B T Road, Kolkata 700 108, India
{dibyendu_r, subho, bimal}@isical.ac.in

Abstract. Key pre-distribution is an important area of research in Distributed Sensor Networks (DSN). Two sensor nodes are considered connected for secure communication if they share one or more common secret key(s). It is important to analyze the largest subset of nodes in a DSN where each node is connected to every other node in that subset (i.e., the largest clique). This parameter (largest clique size) is important in terms of resiliency and capability towards efficient distributed computing in a DSN. In this paper, we concentrate on the schemes where the key pre-distribution strategies are based on transversal design and study the largest clique sizes. We show that merging of blocks to construct a node provides larger clique sizes than considering a block itself as a node in a transversal design.

1 Introduction

A sensor node is a small, inexpensive and resource constrained device that operates in RF (radio frequency) range. It has limitations in different aspects such as communication, computation, power and storage. A DSN (distributed sensor network) is an ad-hoc network consisting of sensor nodes. The sensor nodes are often deployed in an uncontrolled environment where they are expected to operate unattended. In many situations, the DSN is also very large. In either case, though one might try to control the density of deployment, the only deployment option is to randomly scatter the nodes to cover the target area. The consequence is that the location or topology is not available prior to deployment.

Given the various limitations, the security of the DSN hinges on efficient key distribution techniques. Even with the present day technology, public key cryptosystems are considered too computation intensive for DSNs and typically a DSN establishes a secure network by the use of pre-distributed keys. The following four metrics are often used to evaluate key pre-distribution solutions.

1. Scalability: The distribution must allow post-deployment increase in the size of network.
2. Efficiency:
   (a) storage: Amount of memory required to store the keys.
   (b) computation: Number of cycles needed for key establishment

(c) communication: Number of messages exchanged during the key generation/agreement phase.
3. Key Connectivity (probability of key share): The probability that two nodes share one/more keys should be high.
4. Resilience: Even if a number of nodes are compromised, i.e., the keys contained therein are revealed, the complete network should not fail, i.e., only a part of the network should be affected.

One of the challenges in DSNs is to find efficient algorithms to distribute the keys to sensor nodes before they are deployed. The solutions may be categorized as follows:

1. Probabilistic: The keys are randomly chosen from a given collection of keys and distributed to the sensor nodes.
2. Deterministic: The key distribution is obtained as the output of some deterministic algorithm.
3. Hybrid: A combination of deterministic and probabilistic approaches.

A trivial (and obvious) deterministic solution to the problem is to put the same key in all the nodes. However, the moment a single node is compromised, the network fails. To guard against such a possibility, one can think of using distinct keys for all possible pair of nodes in the DSN. The very good resilience notwithstanding, the solution is not viable for even networks of moderate size due to the limited storage capacity of the nodes. If there are N nodes, then there will be $\binom{N}{2}$ keys in total and each node must have $N-1$ many keys. It is not possible to accommodate $N-1$ many keys in a node given the current memory capacity of sensor hardware when N is moderately large, say ≥ 500.

Let us now briefly refer a few state of the art key pre-distribution schemes. The well known Blom's scheme [1] has been extended in recent works for key pre-distribution in wireless sensor networks [5, 7]. The problem with these kinds of schemes is the use of several multiplication operations (as example see [5-Section 5.2]) for key exchange. The randomized key pre-distribution is another strategy in this area [6]. However, the main motivation is to maintain a connectivity (possibly with several hops) in the network. As an example [6-Section 3.2], a sensor network with 10000 nodes has been considered and to maintain the connectivity, it has been calculated that it is enough if one node can communicate with only 20 other nodes. Note that the communication between any two nodes may require a large number of hops. However, only the connectivity criterion (with too many hops) may not suffice in an adversarial condition. Further in such a scenario, the key agreement between two nodes requires exchange of the key indices. The use of combinatorial and probabilistic design (also a combination of both – termed as hybrid design) in the context of key distribution has been proposed in [2]. In this case also, the main motivation was to have low number of common keys.

In [8] transversal design (see Subsection 2.1 for more details) has been used where the blocks correspond to the sensor nodes. In our recent works [3, 4], we have proposed to start from a combinatorial design and then apply a probabilistic extension in the form of random merging of blocks to form the sensor nodes and in this case there is good flexibility in adjusting the number of common keys between any two nodes. In our earlier works [3,4], we dealt with the cases of (i) unconstrained random merging of blocks and (ii) random merging of blocks with the restriction that the nodes are composed of disjoint blocks (do not share common keys among themselves). The computation to find out a shared key under this framework is of very low time complexity [8,3,4], which basically requires calculation of the inverse of an element in a finite field. That is the reason this kind of design becomes popular for application in key pre-distribution.

In the domain of distributed computing, the nodes forming a complete graph is an "ideal situation". As mentioned earlier, one gains a lot in terms of resilience. Moreover, the communication complexity decreases because fewer messages are exchanged between the nodes in order to generate/agree upon a key. In such a scenario, there is no question of "multi-hop" paths and since there is a unique key shared between any two nodes, the computational complexity decreases as well.

Thus, in a DSN, it is important to study the subset of nodes (clique, in graph theoretic terminology) that are connected to each other. By connectivity of two nodes we mean that the nodes share one or more common secret key(s) for secure communication. In this paper we study the basic combinatorial designs [8] and their extensions in terms of merging [3,4] to estimate the cliques of maximum size. We show that if one uses a $(v = rk, b = r^2, r, k)$ configuration, where each block corresponds to a node [8], then the maximum clique size is $r = \sqrt{b}$. We also study the extension of the basic design where a few blocks are merged to get a node [3,4] and show that in such a strategy the clique size becomes considerably larger than what is available in the basic design [8].

2 Preliminaries

2.1 Basics of Transversal Design

Let A be a finite set of subsets (also known as blocks) of a set X. A *set system* or *design* is a pair (X, A). The degree of a point $x \in X$ is the number of subsets containing the point x. If all subsets/blocks have the same size k, then (X, A) is said to be uniform of rank k. If all points have the same degree r, (X, A) is said to be regular of degree r.

A regular and uniform set system is called a $(v, b, r, k) - 1$ design, where $|X| = v, |A| = b$, r is the degree and k is the rank. The condition $bk = vr$ is necessary and sufficient for existence of such a set system. A $(v, b, r, k) - 1$ design is called a (v, b, r, k) configuration if any two distinct blocks intersect in zero or one point.

A (v, b, r, k, λ) BIBD is a $(v, b, r, k) - 1$ design in which every pair of points occurs in exactly λ many blocks. A (v, b, r, k) configuration having deficiency $d = v - 1 - r(k - 1) = 0$ exists if and only if a $(v, b, r, k, 1)$ BIBD exists.

Let g, u, k be positive integers such that $2 \le k \le u$. A group-divisible design of type g^u and block size k is a triple $(X, \mathcal{H}, \mathcal{A})$, where X is a finite set of cardinality gu, $\mathcal{H}$ is a partition of X into u parts/groups of size g, and $\mathcal{A}$ is a set of subsets/blocks of X. The following conditions are satisfied in this case:

1. $|H \bigcap A| \leq 1 \; \forall H \in \mathcal{H}, \; \forall A \in \mathcal{A}$,
2. every pair of elements of X from different groups occurs in exactly one block in $\mathcal{A}$.

A Transversal Design $TD(k, n)$ is a group-divisible design of type n^k and block size k. Hence $H \bigcap A = 1 \; \forall H \in \mathcal{H}, \; \forall A \in \mathcal{A}$.

Let us now describe the construction of a transversal design. Let p be a prime power and $2 \leq k \leq p$. Then there exists a $TD(k, p)$ of the form $(X, \mathcal{H}, \mathcal{A})$ where $X = \mathbb{Z}_k \times \mathbb{Z}_p$. For $0 \leq x \leq k-1$, define $H_x = \{x\} \times \mathbb{Z}_p$ and $\mathcal{H} = \{H_x : 0 \leq x \leq k-1\}$.

For every ordered pair $(i, j) \in \mathbb{Z}_p \times \mathbb{Z}_p$, define a block $A_{i,j} = \{x, (ix + j) \bmod p : 0 \leq x \leq k-1\}$. In this case, $\mathcal{A} = \{A_{i,j} : (i, j) \in \mathbb{Z}_p \times \mathbb{Z}_p\}$. It can be shown that $(X, \mathcal{H}, \mathcal{A})$ is a $TD(k, p)$.

Now let us relate a $(v = kr, b = r^2, r, k)$ configuration with sensor nodes and keys. X is the set of $v = kr$ number of keys distributed among $b = r^2$ number of sensor nodes. The nodes are indexed by $(i, j) \in \mathbb{Z}_r \times \mathbb{Z}_r$ and the keys are indexed by $(i, j) \in \mathbb{Z}_k \times \mathbb{Z}_r$. Consider a particular block $A_{\alpha,\beta}$. It will contain k number of keys $\{(x, (x\alpha + \beta) \bmod r) : 0 \leq x \leq k-1\}$. Here $|X| = kr = v$, $|\mathcal{H}_x| = r$, the number of blocks in which the key (x, y) appears for $y \in \mathbb{Z}_r$, $|A_{i,j}| = k$, the number of keys in a block. For more details on combinatorial design refer to [9, 8].

Note that if r is a prime power, we will not get an inverse of $x \in \mathbb{Z}_r$ when x is not a unit of $\mathbb{Z}_r$ i.e., $\gcd(x, r) > 1$. This is required for key exchange protocol. So basically we should consider the field $GF(r)$ instead of the ring $\mathbb{Z}_r$. However, there is no problem when r is a prime by itself. In this paper we generally use $\mathbb{Z}_r$ since in our examples we consider r to be prime.

2.2 Lee-Stinson Approach [8]

Consider a $(v = rk, b = r^2, r, k)$ configuration. There are $b = r^2$ many sensor nodes, each containing k distinct keys. Each key is repeated in r many nodes. Also v gives the total number of distinct keys in the design. One should note that $bk = vr$ and $v - 1 > r(k-1)$. The design provides 0 or 1 common key between two nodes. The design $(v = 1470, b = 2401, r = 49, k = 30)$ has been used as an example in [8]. The important parameters of the design are as follows.

The expected number of common keys between any two nodes is $p_1 = \frac{k(r-1)}{b-1} = \frac{k}{r+1}$. In the given example, $p_1 = \frac{30}{49+1} = 0.6$.

There is a good proportion of pairs (40%) with no common key, and two such nodes will communicate through an intermediate node. Assuming a random geometric deployment, the example shows that the expected proportion such that two nodes are able to communicate either directly or through an intermediate node is as high as 0.99995.

Under adversarial situation, one or more sensor nodes may get compromised. In that case, all the keys present in those nodes cannot be used for secret communication any longer, i.e., given the number of compromised nodes, one needs to calculate the proportion of links that cannot be used further. The expression for this proportion is $fail(s) = 1 - \left(1 - \frac{r-2}{b-2}\right)^s$, where s is the number of nodes compromised. In this particular example, $fail(10) \approx 0.17951$. That is, given a large network comprising as many as 2401 nodes, if 10 nodes are compromised, almost 18% of the links become unusable.

3 Analysis of Clique Sizes

First we study the maximum clique size where the $(v = rk, b = r^2, r, k)$ configuration is used and each block in the design corresponds to a sensor node, which is the idea proposed in [8].

Theorem 1. *Consider a DSN with b many nodes constructed from a $(v = rk, b = r^2, r, k)$ configuration. The maximum clique in this case is of size r.*

Proof. First we prove that there is a clique of size r. It is known that a key is repeated in r many different blocks. Fix a key. Thus, there are r many distinct blocks which are connected to each other by the fixed key. Hence there is a clique of size r.

Now we prove that there is no clique of size $r + 1$, because that will rule out the possibility of cliques of larger size. Let there be a clique of size $r + 1$. Note that the (v, b, r, k) configuration results from $TD(k, r)$ (see Subsection 2.1). In this case each block is identified by two indices (i, j), $0 \le i, j \le r - 1$. Further two blocks having same value of i (i.e., in the same row) can't have a common key. The moment one chooses $r + 1$ blocks, at least two of the blocks must be from the same row (by pigeon hole principle as there are at most r many rows) and are disjoint, which is a contradiction to the basic assumption of a clique having size $r + 1$. □

It should be observed that the clique size r is exactly the square-root of the number of nodes $b = r^2$. Note that in such a case two nodes/blocks either share a common secret key or not. Consider the graph with b^2 many nodes/vertices where each block corresponds to a node. Now two vertices are connected by an edge if they share a common secret key, otherwise they are not connected. Now a block contains k many distinct keys. For each key, a clique of size r is formed. Thus a vertex/node in this graph participates in k many cliques each of size exactly r.

Given two keys, which never occur together in the same block, will form cliques which are completely disjoint. On the other hand, two keys may occur together at most in a single block. In such a case, the two different cliques generated by them can intersect on a single node/vertex corresponding to the block that contains both the keys.

3.1 The Merging Approach

To overcome certain restrictions in the strategy provided in [8] (explained in the previous subsection), we have provided a strategy to merge certain blocks to construct a sensor node [3, 4]. The basic idea is to start from a $(v = rk, b = r^2, r, k)$ configuration. Then we merge z many blocks to form a single sensor node. Thus the maximum number of sensor nodes available in such a strategy is $\lfloor \frac{r^2}{z} \rfloor$. We have studied a random merging strategy in [3], where randomly chosen z many blocks are merged to get a sensor node. In such a scenario, we found that the number of common keys among any two nodes approximately follows the binomial distribution $\mathcal{B}(z^2, \frac{k}{r+1}))$. The expected number of common secret keys among any two nodes is $\frac{z^2 k}{r+1}$ (see [3–Theorem 1] for more details). It has been shown that this strategy provides favorable results compared to [8]. Note that in [3], the blocks are merged randomly. So it may happen that the blocks being merged may have common secret key(s) among themselves. This is actually a loss, since we really do not need a common key among the blocks that are merged to get a single node. Hence, in [4], we improved the strategy such that only disjoint blocks are merged to construct node. This provides little better parameters compared to [3]. In this paper we will show that our strategy [3, 4] provides better clique size than that of the design presented in [8].

Now we concentrate on the cliques where blocks are merged to get a node [3, 4]. It is worth mentioning that the number of blocks is $\lfloor \frac{r^2}{z} \rfloor$ in this case. From [3–Theorem 1], each key will be present in Q many nodes, where average value of Q is $\hat{Q} = \frac{1}{kr} \left(\lfloor \frac{b}{z} \rfloor \right) \left(zk - \binom{z}{2} \frac{k}{r+1} \right) \approx r$. So cliques of size $\approx r$ are available in the design where merging strategy is employed.

We like to highlight that the value of z is much less than r (as example, $r = 101, z = 4$) though it is not a serious restriction in the proof of our results in the following discussion.

Thus we like to point out the following improvement in the merging strategy over the basic technique.

1. In the basic design, there are r^2 many nodes (each block corresponds to a sensor node) and the maximum clique size is r.
2. Using the merging strategy, there are $\lfloor \frac{r^2}{z} \rfloor$ many nodes (z many blocks are merged to get a sensor node) and the maximum clique size is $\approx r$. Thus there is an improvement by a factor of $\sqrt{z}$ in the size of clique.

Let us present some examples to illustrate the comparison. The design $(v = 1470, b = 2401, r = 49, k = 30)$ has been used as an example in [8]. Hence there are 2401 nodes and the largest clique size is 49. Now consider a $(v = 101 \cdot 7, b = 101^2, r = 101, k = 7)$ configuration and merging of $z = 4$ blocks to get a node. Thus there will be 2550 (we take this value as it is comparable to 2401) many nodes. We have cliques of size ≈ 101 on an average, which shows the improvement.

Next we provide a more improved result by increasing the clique size beyond r. We present a merging strategy where one can get a clique of size $r + z - 1 \geq r$ for $z \geq 1$. The result is as follows.

Theorem 2. *Consider a (v, b, r, k) configuration with $b = r^2$. We merge z many blocks to form each node in achieving a DSN having $N = \lfloor \frac{b}{z} \rfloor$ many sensor nodes.*

Then there exists an initial merging strategy which will always provide a clique of size $r + z - 1$.

Proof. Let's denote the nodes by $\nu_1, \nu_2, \ldots$. Initially choose the first column of the $TD(k,r)$ and place the r blocks (indexed by $(i,0)$ for $0 \leq i \leq r-1$) successively to fill up the first slot (out of the z slots) of the first r nodes $\nu_1, \nu_2, \ldots, \nu_r$. That will obviously yield a clique of size r as any two blocks in a specific column always share a common key.

The rest of the available blocks will always be traversed in column-wise manner. That is the next available block is now the one indexed by $(0,1)$. Let us refer to the next available block by (i,j) for the rest of the present discussion. Once a block is used, we apply the update function on its index to get the next available node. Update (i,j) to $((i+1) \bmod r, j + \delta)$, where $\delta = 0$, if $i < r - 1$ and $\delta = 1$ when $i = r - 1$.

We go on adding new nodes for $t = 1$ to $z - 1$ to generate a clique of size $r + z - 1$ at the end.

To add a new node ν_{r+t}, proceed as follows. Choose the first available block (i,j) and put it in ν_{r+t}. Place the next available blocks in $\nu_1, \nu_2, \ldots, \nu_k$ as long as $i \leq r - 1$. After using the last element of current column, the update function provides the first block of the next column. In that case, we add this new block $(0,j)$ to the node ν_{r+t}. Then again the next available blocks are put into the nodes $\nu_{k+1}, \nu_{k+2}, \ldots$, in the similar manner. Once the blocks in that column gets exhausted, we again add the first block of the next column to ν_{r+t} and the following blocks to the nodes as long as we reach ν_{r+t-1}. Thus it is clear that all the nodes $\nu_1, \ldots, \nu_{r+t-1}$ are connected to ν_{r+t} increasing the size of the clique by 1.

In this strategy, the value of t is bounded above by $z - 1$ as otherwise the number of blocks in a node will increase beyond z. The remaining blocks will be arranged randomly to have z blocks in each node to get $\left\lfloor \frac{r^2}{z} \right\rfloor$ many nodes in completing the merging strategy. □

Now we present an example corresponding to the strategy presented in Theorem 2.

Example 1. Consider the $TD(k, r = 25)$. Let $z = 2$. Consider the 5^2 blocks of the TD arranged in the form of a 5×5 matrix. If we adopt the strategy outlined in the proof of Theorem 2, initially, the following clique is obtained: $\nu_1 \to \{(0,0)\}$, $\nu_2 \to \{(1,0)\}$, $\nu_3 \to \{(2,0)\}$, $\nu_4 \to \{(3,0)\}$, $\nu_5 \to \{(4,0)\}$. Next $(0,1)$ is put in the new node ν_6 and then $(1,1)$ is added to ν_1, $(2,1)$ is added to ν_2, $(3,1)$ is added to ν_3, $(4,1)$ is added to ν_4. As the second column gets exhausted, $(0,2)$ is added to the new node ν_6 and then $(1,2)$ is added to ν_5. Thus we get, $\nu_1 \to \{(0,0), (1,1)\}$, $\nu_2 \to \{(1,0), (2,1)\}$, $\nu_3 \to \{(2,0), (3,1)\}$, $\nu_4 \to \{(3,0), (4,1)\}$, $\nu_5 \to \{(4,0), (1,2)\}$, $\nu_6 \to \{(0,1), (0,2)\}$ and they form a clique of size 6.

Next we observe that the clique size we present in Theorem 2 is not the maximum achievable one. One can indeed find a different merging strategy that provides a clique of larger size. Here is an example.

Example 2. Taking a different arrangement compared to Example 1, we get a clique of size 7 as follows: $\nu_1 \to \{(0,0),(2,1)\}$, $\nu_2 \to \{(1,0),(3,1)\}$, $\nu_3 \to \{(2,0),(4,1)\}$, $\nu_4 \to \{(3,0),(0,2)\}$, $\nu_5 \to \{(4,0),(1,2)\}$, $\nu_6 \to \{(0,1),(2,2)\}$, $\nu_7 \to \{(1,1),(3,2)\}$.

Thus it will be interesting to device a merging strategy which will provide the largest clique size when the (v,b,r,k) configuration and z are fixed.

Note that in the basic (v,b,r,k) configuration or after our merging strategy, the size of cliques are not dependent on the number of keys in each block/node. It is clear that the connectivity of the DSN increases with the increasing number of keys in each node. However, increasing the number of keys is constrained by the limited memory capacity of a sensor node. It is a nice property that the clique size does not increase with number of keys in each node as otherwise one may be tempted to obtain cliques of larger sizes by increasing the number of keys in each node (i.e., by increasing the edges in the graph).

3.2 Configurations Having Complete Block Graphs: Projective Planes

Since we are talking about cliques, we should also revisit the designs where the entire DSN forms a clique. In [8–Theorem 11, 12], it has been pointed out that the block graph of a set system is a complete graph if and only if the set system is the dual design of a BIBD and in particular, there exists a key pre-distribution scheme for a DSN having $q^2 + q + 1$ nodes, in which every node receives exactly $q+1$ keys and in which any two nodes share exactly one key. It is also stated that such designs are not recommendable as a key pre-distribution scheme in large DSNs because of storage limitation in each sensor node. We like to point out that even if the storage space is not a limitation, then also this scheme is not suitable. The reason is as follows.

In this design any two nodes share a common key. However, for better resiliency one may like to have more common keys among any two nodes (this is one important motivation for our merging strategy [3, 4]). Even if one maintains multiples keys against each identifier, the projective planes does not help because compromise of a single node results in discarding the identifiers contained in each node (block) and all the corresponding keys for each identifier also get discarded. Thus the resiliency measure $fail(s)$, (the probability that a given link is affected due to the compromise of s number of randomly chosen nodes) does not improve (i.e., does not reduce).

4 Conclusion

In this paper we consider the DSNs where the key pre-distribution mechanism evolves from combinatorial design. Such schemes provide the advantage of very low complexity key exchange facility (only inverse calculation in finite fields). In terms of distributed computing and communication among the sensor nodes, it is important to study the subset of nodes that are securely connected to each other (clique). In this paper we have studied that in details. We studied the cliques corresponding to the (v, b, r, k) configuration where each block corresponds to a node. Further we study the scenario when more than one blocks are merged to generate a node. We show that the clique size gets improved in such a scenario. An interesting future work in this area is to implement a merging strategy such that one can get cliques of maximum size after the merging.

References

1. R. Blom. An optimal class of symmetric key generation systems. Eurocrypt 84, pages 335–338, LNCS 209, 1985.
2. S. A. Camtepe and B. Yener. Combinatorial design of key distribution mechanisms for wireless sensor networks. Eurosics 2004.
3. D. Chakrabarti, S. Maitra and B. Roy. A Key Pre-distribution Scheme for Wireless Sensor Networks: Merging Blocks in Combinatorial Design. To be presented in *8th Information Security Conference, ISC'05*, Lecture Notes in Computer Science, volume 3650, Springer Verlag.
4. D. Chakrabarti, S. Maitra and B. Roy. A Hybrid Design of Key Pre-distribution Scheme for Wireless Sensor Networks. To be presented at *1st International Conference on Information Systems Security, ICISS 2005*, Jadavpur University, Kolkata, India, December 19-21, 2005. Proceedings to be published in Lecture Notes in Computer Science, Springer Verlag.
5. W. Du, J. Ding, Y. S. Han, and P. K. Varshney. A pairwise key pre-distribution scheme for wireles sensor networks. Proceedings of the 10th ACM conference on Computer and Communicatios Security, Pages 42–51, ACM CCS 2003.
6. L. Eschenauer and V. B. Gligor. A key-management scheme for distributed sensor networks. Proceedings of the 9th ACM conference on Computer and Communicatios Security, Pages 41–47, ACM CCS 2002.
7. J. Lee and D. Stinson. Deterministic key predistribution schemes for distributed sensor networks. SAC 2004.
8. J. Lee and D. Stinson. A combinatorial approach to key predistribution for distributed sensor networks. IEEE Wireless Computing and Networking Conference (WCNC 2005), 13–17 March, 2005, New Orleans, LA, USA.
9. A. P. Street and D. J. Street. Combinatorics of experimental design. Clarendon Press, Oxford, 1987.

Stochastic Rate-Control for Real-Time Video Transmission over Heterogeneous Network

Jae-Woong Yun[1], Hye-Soo Kim[1], Jae-Won Kim[1],
Youn-Seon Jang[2], and Sung-Jea Ko[1]

[1] Department of Electronics Engineering, Korea University, Seoul, Korea
{jyun, hyesoo, jw9557, sjko}@dali.korea.ac.kr
[2] ETRI, 161, Gajeong-dong, Yusong-gu, Daejeon, Korea
ysjang@etri.re.kr

Abstract. In this paper, we propose a stochastic rate control method to provide seamless video streaming for vertical handoff between WLAN and 3G cellular network. In the proposed method, we first estimate the channel rate by using the state transition probabilities that can be found from the relationship between the packet loss ratio (PLR) and the medium access control (MAC) layer parameters. The proposed method performs bit allocation at the frame level using the estimated channel rate, minimizing the average distortion over an entire sequence as well as variations in distortion between frames. Experimental results indicate that the proposed method provides better visual quality than the existing TMN8 rate control method in heterogeneous wireless network.

1 Introduction

The rapid growth of wireless communications and networking protocols, such as 802.11 [1] and 3G cellular network [2],[3], and the combination of wireless technologies, offer the possibility of achieving anywhere, anytime communication, bringing benefits to both end users and service providers. The movement of a user within or among different types of networks is referred to as vertical mobility. One of the major challenges for seamless service with vertical mobility is vertical handoff, where handoff is the process of maintaining a mobile user's active connections with changes in the point of attachment [4].

In recent years, since digitized multimedia applications such as videophone and video conference have intensified, the latest application trends have created an increasing interest in providing practical multimedia streaming systems to meet the needs of mobile computing. A successful video streaming solution is to implement an adaptive multimedia streaming system that allows a mobile user to receive uninterrupted service of the best quality multimedia in any communication environment.

The rate control scheme in TMN8 is optimized for a CBR channel in a wired channel, not for a VBR channel [5]. Unlike a wired channel where the signal strength is relatively constant and the errors at receiver are mainly due to additive noise, the errors in a wireless channel are mainly due to the time varying

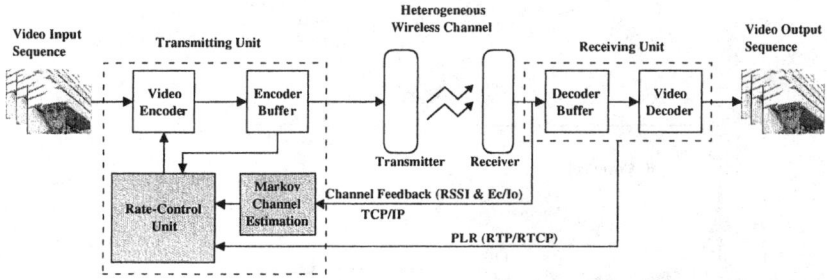

Fig. 1. System block diagram

signal strength caused by the multi-path fading. Wireless radio networks suffer from high bit error rates with channel characteristics that are time varying. Especially, vertical handoff that effects in heterogeneous network must be considered in the rate control system because the rate can be changed dramatically after handoff. Several rate control schemes for wireless channel have been proposed in [6],[8]. In these papers, it is proposed to use the automatic repeat request (ARQ) scheme with an adaptive source rate control that dynamically changes both the number of the intra-coded macroblocks and the quantization scale used in a frame, based on the packet-error-rate in a sliding window. In the ARQ scheme, a lot of retransmissions occur in poor channel conditions and this will increase the delay. Such a retransmission scheme is not good for a real time system. Furthermore, in conventional schemes, the channel status of heterogeneous networks is not considered.

In this paper, we provide an alternative practical solution to allocate the number of bit-budgets adaptively and to determine the rate of channel coder depending on the channel conditions obtained from the stochastic channel information. To enhance the image quality, we propose a stochastic rate control method that exploits the channel rate estimated by using a three-state Markov model to predict the channel condition and dynamically re-allocatates the target number of bits for each frame. This method dynamically changes the target bit rate by using the relation of the RSSI, E_c/I_o and PLR. Fig. 1 shows overall system block diagram. Experimental results indicate that the proposed rate control method provides better visual quality than the existing TMN8 rate control method in heterogeneous wireless network.

This paper is organized as follows. In Section 2, we describe a wireless channel model for vertical handoff. The proposed stochastic rate control scheme is presented in Section 3. Section 4 shows the experimental results. Finally, our conclusions are given in Section 5.

2 Wireless Channel Model for Heterogeneous Network

As described in Section 1, wireless networks suffer from high bit error rates since wireless channel conditions frequently vary over time. In particular, it is

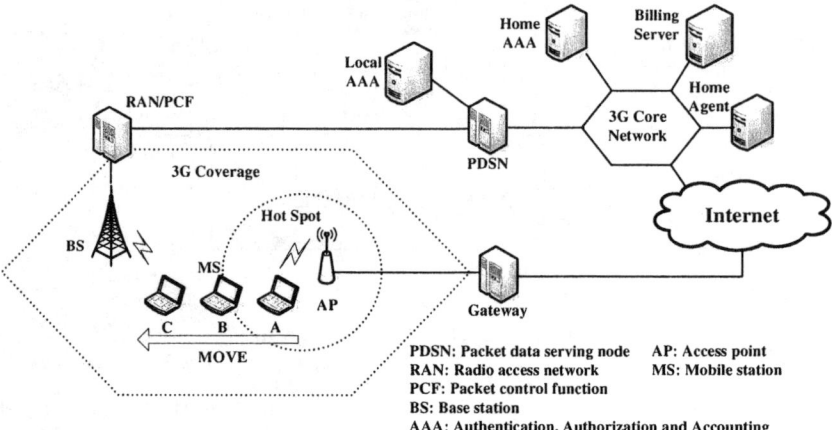

Fig. 2. Upward vertical handoff scenario

essential to monitor the vertical handoff status because the rate can be changed dramatically after handoff. In order to estimate the time varying channel status, we first define a wireless channel model. The wireless channel is modelled as a three-state Markov model considering vertical handoff.

2.1 Vertical Handoff Scenario

A horizontal handoff is defined as a handoff between base stations (BSs) that use the same type of wireless network interface. This is a traditional definition of handoff for homogeneous cellular systems. A vertical handoff is defined as a handoff between BSs that use different wireless network technologies such as WLAN and 3G cellular network. Vertical handoff can be divided into upward vertical handoff and downward vertical handoff. Upward vertical handoff is a handoff from a smaller network with higher bandwidth to a larger network with lower bandwidth. Downward vertical handoff is a handoff from a larger network to a smaller network [9].

Fig. 2 shows the network architecture to integrate WLAN and 3G cellular network. As shown in Fig. 2, WLAN covers a smaller network with higher bandwidth and 3G cellular network covers a larger network with lower bandwidth. In Fig. 2, a upward vertical handoff occurs when a mobile station (MS) moves from location A in WLAN to location C in 3G cellular network. As the MS leaves the access point (AP), the strength of the beacon signal received from the AP weakens. If its strength decrease below a threshold value, the MS tries to connect to 3G cellular network and starts synchronizing with the system to prepare the handoff.

2.2 Channel Rate Estimation for Vertical Handoff

The specific channel under consideration is a wireless channel such as WLAN and 3G cellular network, for a mobile transmission environment, where channel

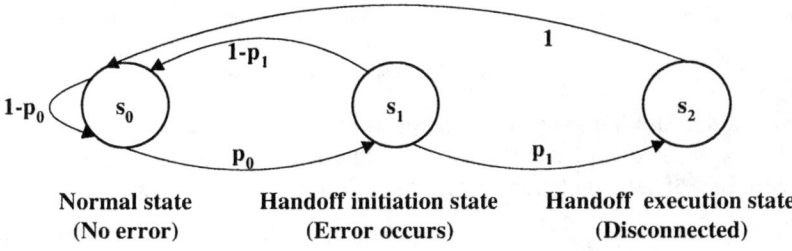

Fig. 3. Three-state Markov model

errors tend to occur in bursts during channel fading periods and vertical handoff. The packet loss results in the quality degradation of streaming video. In order to reduce the video quality degradation in vertical handoff, we first define a wireless channel model. The wireless channel is modelled as a three-state Markov model.

Fig. 3 shows the three-state Markov model of upward vertical handoff. This Markov model has three channel states, s_0, s_1, and s_2 where s_0, s_1, and s_2, respectively, are the "normal state", the "handoff initiation state", and the "handoff execution state". The transition probabilities can be obtained by using the channel characteristic information such as the RSSI and the E_c/I_o measured in our experimental platform. When the channel is in state s_n, $n \in \{0,1,2\}$, the transition of the channel state goes to the next higher state or back to state s_0 based on the channel information. If the channel is in state s_2, it will always transit to state s_0.

Define $p_n = Prob(s_{n+1}|s_n)$ as the transition probability from state s_n to s_{n+1}. The transition probability matrix for the three-state Markov model can be set up as

$$P = \begin{bmatrix} 1 - p_0 & p_0 & 0 \\ 1 - p_1 & 0 & p_1 \\ 1 & 0 & 0 \end{bmatrix}. \qquad (1)$$

We define the state probability $\pi_n(k|S(t))$ as the probability that the channel is in state s_n at time k given the channel state observation $S(t)$. Note that t and k are all discrete values.

$$\vec{\pi}(k|S(t)) = [\pi_0(k|S(t)),\ \pi_1(k|S(t)),\ \pi_2(k|S(t))]. \qquad (2)$$

The initial state probability $\pi_n(t|S(t))$ at time t can be set up as

$$\forall n \in \{0,1,2\}, \quad \pi_n(t|S(t)) = \begin{cases} 1, & \text{if } S(t) = s_n \\ 0, & \text{otherwise.} \end{cases} \qquad (3)$$

In the Markov model, the vector of state probabilities $\vec{\pi}(k|S(t))$ at time k can be derived from the state probabilities $\vec{\pi}(k-1|S(t))$ at the previous time slot and the transition probability matrix P in (1) as

$$\vec{\pi}(k|S(t)) = \vec{\pi}(k-1|S(t)) \cdot P. \qquad (4)$$

The vector of state probabilities at time k can be obtained by using (4) recursively as

$$\vec{\pi}(k|S(t)) = \vec{\pi}(t|S(t)) \cdot \mathrm{P}^{k-t}. \tag{5}$$

We consider the heterogeneous wireless channel, where each bandwidth provides the different data rate. Thus, we define the channel transmission rates $\bar{R}$ as the number of bits sent per second as follows:

$$\bar{R} = \begin{cases} \mathrm{R}_w^{\max}, & \text{for the WLAN,} \\ \mathrm{R}_c^{\max}, & \text{for the 3G network,} \end{cases} \tag{6}$$

where $\mathrm{R}_w^{\max}$ and $\mathrm{R}_c^{\max}$ are the maximum channel rates in WLAN and 3G cellular network. In our channel model, packets are transmitted correctly when the channel is in state s_0, while errors occur when the channel is in any other state s_i, $i \in \{1,2\}$. Therefore, $\pi_0(k|S(t))$ is the probability of correct transmission at time k. Let $C(k)$ be the future channel transmission rate where $k > t$. The expected channel rate $E[C(k)|S(t)]$ given the observation of channel state $S(t)$ can be calculated as

$$E[C(k)|S(t)] = \bar{R} \cdot \pi_0(k|S(t)). \tag{7}$$

Finally, we define the wireless channel rate $\widehat{R}_E$ as follows:

$$\widehat{R}_E = E[C(k)|S(t)]. \tag{8}$$

In this paper, we show how to make use of both a probabilistic model of the channel and observations of the current channel state in the context of this rate-control problem.

3 Improved Frame-Layer Rate-Control Scheme

In this section, we describe the framework of the proposed rate-control scheme to reduce the video quality degradation when the vertical handoff occurs in the heterogeneous mobile network and when the wireless channel state is poor. The frame-layer rate control scheme uses the channel model to estimate the current channel rate and adjusts the frame target bit rate by using the estimated channel rate. The obtained target bit budget is optimally allocated to each frame by using the frame-layer rate control scheme to minimize the average distortion over an entire sequence as well as variations in distortion between frames [10].

Before encoding of the current frame, the encoder buffer will be updated as the number of bits. In the conventional TMN8, if the encoder buffer is larger than, or equal to, some maximum value M, the encoder skips encoded frames until the buffer fullness is below M. For each skipped frame, the buffer fullness is reduced by an additional R/F bits where R is the channel rate and F is the frame rate. In our proposed scheme, R can be replaced by the expected channel rate obtained by the proposed channel model. The number of bits in the encoder buffer, W, is modified as follows:

$$W = \max(W_{prev} + B' - \widehat{R}_E/F, 0). \tag{9}$$

First, we estimate the target bandwidth for video transmission over wireless network. We estimate the target bandwidth for the period that is the time interval between two successive measurements of the link status. Next, the target bit budget is optimally allocated to each frame using the frame-layer rate control method. Fig. 4 shows the basic concept, where the bundle of frames during the time interval is referred to as the temporal frame segment.

For the frame-layer rate control, an empirical data-based frame-layer R-D model is employed using the quadratic rate model and the affine distortion model [11] with respect to the average quantization parameter (QP) in a frame, which is given by

$$\hat{R}(\bar{q}_i) = (a \cdot \bar{q}_i^{-1} + b \cdot \bar{q}_i^{-2}) \cdot MAD(\hat{f}_{ref}, f_{cur}), \tag{10}$$

$$\hat{D}(\bar{q}_i) = a' \cdot \bar{q}_i + b', \tag{11}$$

where $a, b, a', $ and b' are the model coefficients, $\hat{f}_{ref}$ is the reconstructed reference frame at the previous time instant, f_{cur} is the uncompressed image at the current time instant, $MAD(\cdot,\cdot)$ is the mean of absolute difference between two frames, $\bar{q}_i$ is the average QP of all macroblocks in the ith frame, and $\hat{R}(\bar{q}_i)$ and $\hat{D}(\bar{q}_i)$ are the rate and distortion models of the ith frame, respectively. The model coefficients are determined by using the linear regression analysis and the formula consisting of the previous encoding results as follows:

$$a = \frac{\sum_{i=1}^{N} \left(\frac{R_i \cdot \bar{q}_i}{MAD(\hat{f}_{i-1}, f_i)} - b \cdot \bar{q}_i^{-1} \right)}{N}, \tag{12}$$

$$b = \frac{N \cdot \left(\sum_{i=1}^{N} \frac{R_i}{MAD(\hat{f}_{i-1}, f_i)} \right)}{N \cdot \left(\sum_{i=1}^{N} \bar{q}_i^{-2} \right) - \left(\sum_{i=1}^{N} \bar{q}_i^{-1} \right)^2} - \frac{\left(\sum_{i=1}^{N} \frac{R_i \cdot \bar{q}_i}{MAD(\hat{f}_{i-1}, f_i)} \right) \left(\sum_{i=1}^{N} \bar{q}_i^{-1} \right)}{N \cdot \left(\sum_{i=1}^{N} \bar{q}_i^{-2} \right) - \left(\sum_{i=1}^{N} \bar{q}_i^{-1} \right)^2}, \tag{13}$$

$$a' = \frac{\sum_{i=1}^{N} D_i \cdot \sum_{i=1}^{N} \bar{q}_i - N \cdot \sum_{i=1}^{N} D_i \cdot \bar{q}_i}{\left(\sum_{i=1}^{N} \bar{q}_i \right)^2 - N \cdot \sum_{i=1}^{N} \bar{q}_i^2}, \tag{14}$$

$$b' = \frac{\sum_{i=1}^{N} D_i - a' \cdot \sum_{i=1}^{N} \bar{q}_i}{N}, \tag{15}$$

where N is the number frames observed in the past, D_i and R_i are the actual distortion and bit rate of the encoded ith frame, respectively.

A new formulation of frame-layer rate control based on the R-D model is considered as follows: Determine $\bar{q}_i, i = 1, 2, ..., N_k^{SEG}$ to minimize

$$\sum_{i=1}^{N_k^{SEG}} \hat{D}_i(\bar{q}_i) \cdot (\hat{D}_i(\bar{q}_i) - D_{i-1}), \tag{16}$$

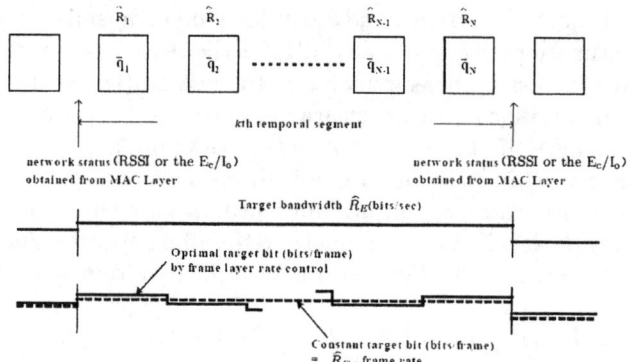

Fig. 4. Bandwidth estimation using network status information and bit allocation for a frame

subject to

$$\sum_{i=1}^{N_k^{SEG}} R_i \leq \widehat{R}_k^{SEG} \cdot T_k^{SEG}, \quad (17)$$

where $\hat{D}_i$ is the estimated distortion of the current frame, D_{i-1} is the actual distortion of the previous frame, N_k^{SEG} is the number of encoding frames in the kth temporal segment, $\widehat{R}_k^{SEG}$ and T_k^{SEG} are the expected channel rate and the time interval of kth temporal segment, respectively. In (16), a formulation is introduced to minimize the average distortion over an entire sequences as well as variations in distortion between frames.

The optimization task in (16) and (17) can be solved using Lagrangian optimization where a distortion term is weighted against a rate term. The Lagrangian formulation of the minimization problem is given by

$$J_i(\bar{q}_i) = \hat{D}_i(\bar{q}_i) \cdot (\hat{D}_i(\bar{q}_i) - D_{i-1}) + \lambda_i \cdot max(\hat{B}_i^{res}, 0), \quad (18)$$

$$\hat{B}_i^{res} = \sum_{j=1}^{i-1} R_j + \hat{R}_i(\bar{q}_i) - \sum_{j=1}^{i} \frac{MAD_k^j}{Ave\_MAD_{k-1}} \frac{\widehat{R}_k^{SEG} \cdot T_k^{SEG}}{N_k^{SEG}}, \quad (19)$$

where the Lagrangian rate-distortion function $J_i(\bar{q}_i)$ is minimized by the particular value of the Lagrange multiplier λ_i for the ith frame, R_j is the used bit-rate for the jth frame, MAD_k^j is the MAD between $(j-1)$th and jth frames of the kth temporal frame segment, and Ave_MAD_{k-1} is the average of MADs of the $(k-1)$th temporal frame segment, respectively. Note that $\hat{B}_i^{res}$ denotes the estimated bit based on the R-D model.

Based on the rate and distortion models, the optimal QP can be determined to minimize the above penalty function. It was shown in [12] that $J_i(\bar{q}_i)$ is a convex function generally. Thus, its optimal solution can be obtained by using the gradient method as described in (20).

$$\bar{q}_i^* = arg \min_{\bar{q}_i} J_i(\bar{q}_i). \quad (20)$$

Note that what is finally needed is not $\bar{q}_i^*$, but $\hat{R}_i(\bar{q}_i^*)$ which is the target bit budget for the ith frame.

The proposed frame-layer rate control algorithm consists of two steps. The first step is to find the optimal bit-rates with the current Lagrange multiplier, and the second step is to adjust the Lagrange multiplier based on residual bit-rates. The properties of the Lagrange multiplier method are very appealing in terms of computation. Finding the best quantizer for a given λ is easy and can be done independently for each coding unit. In order to achieve the optimal solution at the required rate, an optimal λ must be found. Several approaches including the bisection search algorithm [13] are proposed to find a correct λ. However, the number of iterations required in searching for λ can be kept low as long as an exact match of the budget rate is not required. Moreover, since allocations may be performed on successive frames having similar characteristics in video coding, it is possible to adjust λ for a frame using the value achieved for the previous frame. Thus, the adaptive adjustment rule [14] is employed given by

$$\lambda_{i+1} = \lambda_i + \Delta\lambda, \quad \Delta\lambda = \frac{B_i}{B_{target,i}} - 1, \quad (21)$$

where λ_i is the Lagrange multiplier for the ith frame and

$$B_i = \sum_{j=1}^{i} R_j, \quad (22)$$

$$B_{target,i} = \sum_{j=1}^{i} \frac{MAD_k^j}{Ave\_MAD_{k-1}} \frac{\hat{R}_k^{SEG} \cdot T_k^{SEG}}{N_k^{SEG}}. \quad (23)$$

Therefore, the proposed rate control algorithm does not produce encoding time delay. However, a negligible performance loss due to its intrinsic sub-optimality is inevitable in this design.

Once the bit rate is allocated to the frame using the aforementioned frame-layer rate control, the TMN8 macroblock layer rate control algorithm allocates the bit budget to each macroblock with the solution $\hat{R}_i(\bar{q}_i^*)$.

4 Experimental Results

The channel state transition of the proposed wireless channel model is performed by experimental thresholds which are 35 of RSSI and 10.8 of E_c/I_o. The transition probabilities are acquired by using the relationship between the PLR and the MAC layer parameters. Using the relationship in Fig. 5, the transition probability matrix can be found to be p_0=0.8125, p_1=0.6667 in WLAN, p_0=0.9545, p_1=0.4285 in 3G network.

With the proposed wireless channel model, we simulated vertical handoff according to the vertical handoff scenario to show the effectiveness of the proposed video streaming method. Our stochastic rate-control system has been implemented in a H.263+ standard codec. The test video sequences are "FOREMAN",

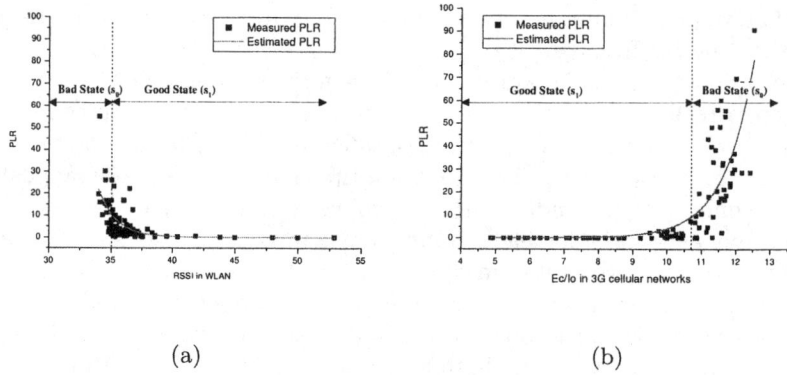

Fig. 5. Channel state determination. (a) PLR vs RSSI in WLAN and (b) PLR vs E_c/I_o in 3G cellular network.

Table 1. Performance comparison of the proposed algorithm with TMN8 in upward vertical handoff (WLAN to 3G cellular network)

Test sequence	Rate-control method	Average PSNR	Frame skipping
FOREMAN	TMN 8	31.14	11
	Proposed method	35.19	5
CARPHONE	TMN 8	34.40	8
	Proposed method	36.23	4
AKIYO	TMN 8	38.76	9
	Proposed method	39.71	4
NEWS	TMN 8	36.47	10
	Proposed method	38.22	5

"CARPHONE", "NEWS", and "AKIYO". The test sequences are encoded to the H.263+ CBR bitstream of 128kbps with 30fps.

The performance of the proposed stochastic rate-control scheme is compared with that of TMN8. For the performance comparison, we show the average PSNR value and the frame skipping reduction in Table 1.

It is clearly seen that the proposed rate control algorithm can reduce the video quality degradation as compared with TMN8. Fig. 6 shows plots associated with the "FOREMAN" sequences as a function of the frame number. Thus, the proposed frame rate control can reduce the quality degradation better than TMN8. The average PSNR results for different channel status are depicted in Fig. 6. It can be seen that the proposed rate control algorithm significantly improves the video quality, especially for the environment that the channel status is not good or the handoff execution status because the proposed algorithm considers channel status. Fig. 6-(b) shows that we obtain better PSNR for QCIF "FOREMAN" sequence in the vertical handoff from WLAN to 3G network.

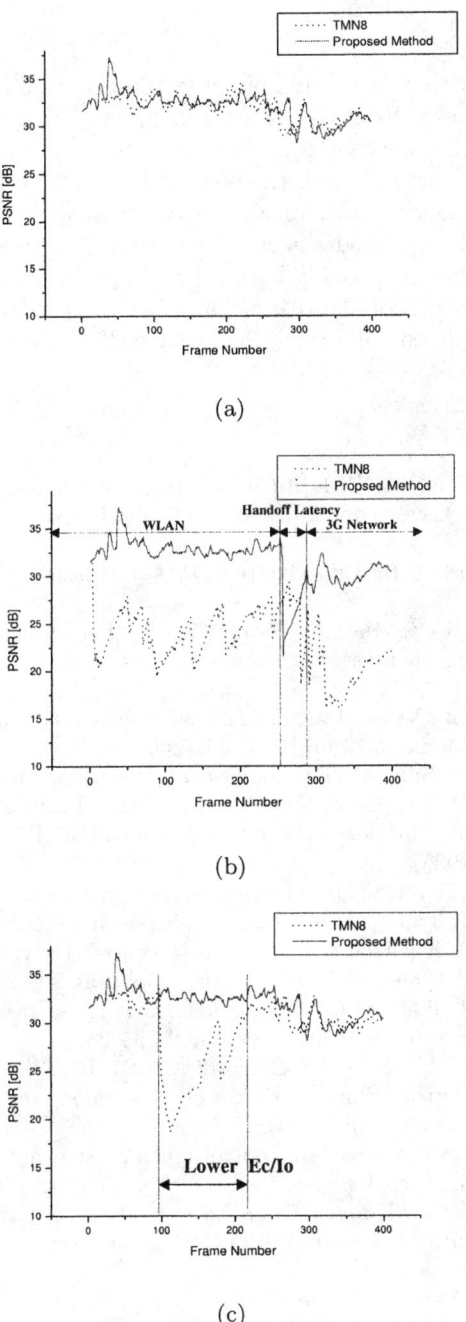

Fig. 6. PSNR comparison: (a) QCIF FOREMAN with the RSSI at 59 in WLAN (b) QCIF FOREMAN with the RSSI at 35 in WLAN (c) QCIF FOREMAN with the E_c/I_0 at 10.8 in 3G cellular network

5 Conclusions

When video streams are transmitted in heterogeneous mobile networks, the compressed video can suffer from the video quality degradation. To reduce degradation of video quality, we have proposed the stochastic rate-control scheme for real-time video transmission in vertical handoff. The experimental results show that the proposed scheme can reduce the video quality degradation even in the vertical handoff. The proposed algorithm has been tested on several sequences, and it has been found to provide better PSNR performance than that of the existing TMN8 rate-control algorithm. Furthermore, the proposed algorithm is robust and can handle channel variations very well.

References

1. ISO/IEC 8802-11 - ANSI/IEEE Std 802.11: Information Technology Part 11: Wireless LAN medium access control (MAC) and physical layer (PHY) specifications. IEEE (1999)
2. TIA/EIA/IS-2000.1-A: Introduction to CDMA2000 standards for spread spectrum systems. (2000)
3. Huang, J., Yao, R.-Y., Bai, Y., Wang, S.-W.: Performance of a mixed-traffic CDMA2000 wireless network with scalable streaming video. Video Technol. **13** (2003) 973-981
4. McNair, J., Fang, Z.: Vertical handoffs in fourth-generation multinetwork environments. Wireless Communication IEEE **11** (2004) 8-15
5. Aramvith, S., Pao, Sun, M.-T.: A rate-control scheme for video transport over wireless channels. IEEE Trans. Circuits Syst. Video Technol. **11** (2001) 569-580
6. ITU-T:Video coding for low bit rate communication. ITU-T Recommendation H.263 version 1 (1995)
7. Corbera, R., Lei, S.: Rate Control in DCT video coding for low-delay video commmunications. IEEE Trans. Circuits Syst. Video Technol. **9** (1999) 172-185
8. Liu, H., Zarko, M.-E.: Adaptive source rate control for real-time wireless video transmission ACM Trans. Mobile Networks Applicant **3** (1998) 49-60
9. Stemm, M., Katz, R.-H.: Vertical handoffs in wireless overlay networks. ACM Trans. Networking and Applications **3** (1998) 335-350
10. Kim, Y., Pyun, J. -Y., Kim, H. -S., Park, S. -H., Ko, S. -J.: Efficient real-time frame layer rate control technique for low bit rate video over WLAN. IEEE Trans. on Consumer Electronics. **49** (2003) 621-628
11. Chiang T., Zhang Y.: A new rate control scheme using quadratic rate distortion model. IEEE Trans. Circuits Syst. Video Technol. **7** (1997) 246-250
12. Lin L. J., Orterga A., Kuo C. C.: Rate control using spline interpolated R-D characteristics. in Proc. of SPIE Visual Communication Image Processing (1996) 111-122
13. Ramchandran K., Vetterli M.: Best wavelet packet bases in a rate-distortion sense. IEEE Trans. Image Processing (1993) 160-175
14. Wiegand T., Lightstone M., Mukherjee D, Campbell T. G., Mitra S. K.: Rate-distortion optimized mode for very low bit rate video coding and emerging H.263 standard. IEEE Trans. Circuits Syst. Video Technol. (1996) 182-190

An Efficient Social Network-Mobility Model for MANETs

Rahul Ghosh, Aritra Das, P. Venkateswaran, S.K. Sanyal, and R. Nandi

Dept. of Electronics & Tele-communication Engineering,
Jadavpur University, Kolkata 700 032, India
rahul3dspace@yahoo.co.uk, mail_artd@yahoo.com
pvwn@yahoo.co.in, s_sanyal@ieee.org,
robnon@hotmail.com

Abstract. An efficient deployment of a mobile Ad Hoc network (MANET) requires a realistic approach towards the mobility of the hosts who want to communicate with each other over a wireless channel. Since Ad Hoc networks are driven by the human requirements, instead of considering the random movement of the mobile nodes, we concentrate on the social desire of the nodes for getting connected with one another and provide here a framework for the mobility model of the nodes based on Social Network Theory. In this paper, we capture the preferences in choosing destinations of pedestrian mobility pattern on the basis of social factor (Ψ_F) and try to find out the essential impact of Ψ_F on the pause time of the nodes. Further, our paper also provides a mobility distribution pattern, and a relative comparison has been done with Random Way-Point (RWP) Model under a certain constrained simulation.

1 Introduction

In an Ad Hoc network, the network topology may be subjected to a rapid change due to frequent link failure and due to the mobility of the nodes. A good number of research works have been published regarding different issues like routing protocols, mobility model, Quality of Service (QoS), bandwidth optimization for mobile Ad Hoc networks (MANETs). However, in the absence of established properties of real mobility patterns, it is not yet clear today, what are the essential parameters to consider while constructing a mobility model. The current scenarios on the available mobility models for MANETs are synthetic models based on simple, homogeneous, random processes [1], [2]. For example, Random Walk Mobility Model is used to represent pure random movements of the entities of a system. A slight enhancement of this, is the Random Way-Point (RWP) Model, in which waypoints are uniformly distributed over the given convex area and the nodes have so called "thinking times" (pause times) before next destination. However, all such synthetic movement models generally do not reflect the real world situations regarding the mobility of nodes. In practice, a mobile user, within a campus or in any geographic location does not roam about in a random manner. Though the present synthetic models are more tractable for mathematical analysis and easy for trace generation, they do not capture the delicate details like time-location dependence and community behavior of pedestrian mobility. Human decisions and socialization behavior play a key role in typical Ad Hoc networking deployment scenarios of disaster relief teams, platoon of soldiers etc.

In this paper, we emphasize on the mobility pattern of individual nodes biased by the strength of social relationships. The reviews of the social networks may be found in [3]. Here, we have systematically developed some social indicators out of the needs of an Ad Hoc environment and then we have transformed them into mathematical domain to formulate key factors. These factors are then mapped to a topographical space to show the distribution pattern for our model. Thus we present the design and analysis of the individual as well as group mobility model based on the social network theory.

The rest of the paper is organized as follows. In Section 2, we give a brief overview of the related works. Section 3 provides the proposed mobility model. Section 4 provides our simulation results and analysis. The conclusion is given in Section 5.

2 Related Works

In [2], an example of realistic mobility model for MANETs, which enables the inclusion of the obstacles in the network simulation, is given. Mathematical models of complex and social networks have been shown to be useful in describing many relationships, including real social relationships [4]. In [5], an approach has been presented towards a mobility model on the relationships of people though the paper lacks a rigorous mathematical representation of the relationship between individuals. The authors of [6] have presented a mobility model based on Social Network Theory from theoretical point of view. Though their work provides a general framework for the mathematical analysis based on the social relationships of the nodes, certain assumptions make their formulations unsuitable for implementation in real world cases.

3 The Proposed Model

Instead of using heuristic approach, we develop our mobility model on the basis of the following assumptions. The assumptions are:

- A_1: The mobile nodes tend to select a specific destination and follow a well-defined path to reach that destination.
- A_2: Path selection process is biased by the social interaction and community demand and it is different at different locations and time.
- A_3: The pause time of the nodes, being a function of social network, is not random instead it follows a specific user oriented distribution at different locations.

With the help of these assumptions, we try to find out the factors controlling the mobility of nodes, and then study the effect of the factors on both the individuals and the groups.

3.1 Different Social Issues Controlling Mobility

We represent a social network using a weighted graph where weights associated with each edge of network are an indicator of the direct interactions between individuals.

We assign a value in the range [0, 1] to signify the degree of social interaction between two people, where '0' indicates no interaction and '1' indicates strongest social interaction. Here, we use a symbolic matrix M, called Interaction Matrix [6] whose diagonal elements are 1 and the generic element 'm_{ij}' represents the interaction between two individuals 'i' and 'j'. For the sake of simplicity, the matrix used in this model is symmetric.

Since, every relation between two mobile nodes is not strong; we introduce here the term connection threshold (CT), which indicates a limit of social connectivity. Contrary to [6] we do not assign an arbitrary value to CT and express it as a function of time, network parameters and social issues. Here, in context, we define the following terms-

- Link Duration [LD (t)]: The average time duration along which a channel is formed between two mobile nodes.
- Frequency of Connectivity [FC]: The number of times a mobile node i is connected to j over a single existing time of Ad Hoc network.

Let us first discuss how CT depends on LD (t) and FC. A high value of link duration between two nodes suggests that the social interaction between them is considerably high. Again frequent connectivity between two nodes through out the life-time of the MANET is indicative of the fact that the nodes prefer specific social relation instead of general social relation involving large amount of nodes. On the basis of above, the connection threshold of a node j denoted by CT_j in a group of 'n' number of nodes can be defined as:

$$CT_j = \frac{\sum_{i=1}^{n} LD_i(t) * FC}{n * T_{total}} \quad (1)$$

where, n = the total no. of nodes present in the current MANET with whom the node j gets connected, and T_{total} = the total time elapsed by the node j in an Ad Hoc environment.

Since the total time elapsed by the node j in an Ad Hoc environment is much greater than the total communication time between two nodes, we can argue that

$$CT < 1 \text{ As } \sum_{i=1}^{n} LD_i(t) * FC < T_{total} \quad (2)$$

Till now, we have considered only a single network topology. However, the social behavior of a node essentially depends on its community behavior; i.e. the involvement of the node to different social scenarios. In this context, we define another parameter called Community Factor (CF), as follows:

$$CF = \frac{\sum_i C_i * NNC}{\sum_i C_i} \quad (3)$$

where, NNC = New Network Coefficient whose value is either 0 or 1, and Ci = Specific grade assigned to a particular social network e.g. battlefield, cafeteria etc.

Here, the term NNC indicates whether it is exposed to a new network or not. Clearly, for a new network, its value is 0, since we do not consider the contribution of a new network to the value of CF.

With the help of these factors, we now try to find out an indicator of the attitude of a node towards the interaction with others. To this end, we introduce Social Factor (Ψ_F), which gives a measure of the degree of interaction between a node and others present in the Ad Hoc network. For a node i, the social factor (Ψ_F) is given as:

$$\Psi_{Fi} = \frac{\sum_{\substack{j=1 \\ j \neq i \\ m_{ij} > CT}} m_{ij} * CF_i * CF_j}{N} \tag{4}$$

where, N = Total no. of social neighbors above the CT level in a social network of i.

From (2), we can state that CT approaches a steady state value less than 1. Since, for a highly social node the value of N is very high compared to the numerical values of CFs, in that case, Ψ_{Fi} also tends to a steady value less than 1.

3.2 Formulation of Pause Time

We explicitly define pause time (PT) for our mobility model as the time elapsed by a node when it meets a social neighbor over a wireless channel, or in a geographic location in a MANET, and try to develop an expression of pause time based on our social issues as in section 3.1. This is being done, because instead of taking a random value of pause time (as in the case of RWP), as we make pause time as a function of social network parameters.

Again, we define another quantity namely, Previous Average Connectivity (PAC), which is the average time of connection with a node i to a social group G_i. Thus, associating all the variables together (including Ψ_F), we give an empirical relation connecting Ψ_F and PT:

$$PT = \Psi_F * GA_i * [1 + PAC(t)] \tag{5}$$

where, GA_i is the individual group attraction force of the node i to the group G_i and has a value in the range [0, 1] i.e. a node may have no pause time at all. The term PAC (t) also serves as a history parameter for different nodes. Thus, instead of using random pause time for the mobile users scattered across a social gathering, we try to find out a node specific pause time.

3.3 Effect of Group Velocity on the Mobile Nodes

For the sake of clarity, we use the basic relationship between the group velocity and the position of the group members as in [6]. But, here we introduce a slight modification such that instead of direct relationship between V_n and V_g, there is also an influence of GA, which is defined in section 3.2. The new position of a mobile node (N_n) is given as:

$$N_n = N_p \pm \int_0^T \frac{\partial Vn}{\partial t} dt \pm \left[\int_0^T \frac{\partial Vg}{\partial t} dt \right] * GA \qquad (6)$$

where, N_p = Previous Node position, T = Total time elapsed by a node in the present group and V_n and V_g are the node and group velocity respectively. It is obvious from (6) that there will be a tendency for the mobile host to change its present group, if a strong group attraction force is exerted on it from an outside group. This is an important issue since, joining a group or leaving a group is analogous to a new link set-up and link failure respectively. Using the same relation, we can also gather information about the social connectivity of the nodes after a period of time.

4 Simulation Results and Analysis

We have considered an Ad Hoc environment in which we have arbitrarily placed a node as a group centre (G_c), velocity of which indicates the overall cluster velocity or group-velocity. The transmission range of G_c has been considered to be 250 meters and other mobile nodes are placed randomly around it with about 80% of the nodes within this range. A node is said to be within the group, if it is within the transmission range of G_c. Now, we have considered an indicator variable (I_v) through out the simulation process, which is defined as:

I_v = 1; if the node is within the range.
  = 0; if the node is out of the group.

Under this scenario, we have placed 100 nodes in an arbitrary fashion with a velocity within the range 1-3 m/s. The group centre has been assigned a velocity within the range 0-1 m/s. Nodes (including G_c) move in a random direction with an angle $\theta \in [0, 2\pi]$ and after a random interval of time, it takes a pause-time generated from (5). Again, a node is connected to a group at a particular time if the value of I_v for the node is 1 at that instant. Readings have been taken at an interval of 5 sec to measure the number of nodes connected to the group.

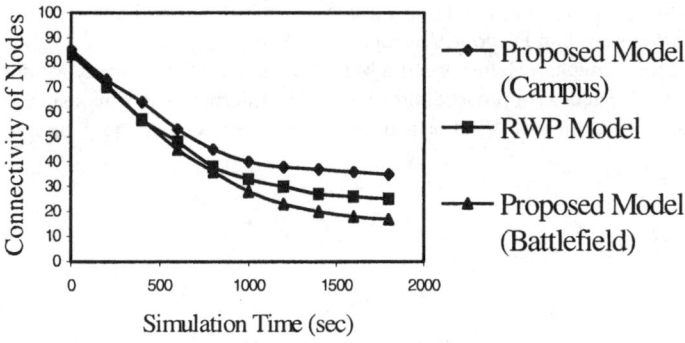

Fig. 1. Percentage of Nodes Connected Vs Simulation Time

From the simulation results, we have extracted the node distribution pattern within an Ad Hoc clustered network. Fig.1 shows a comparison of the proposed model for two scenarios (campus and battlefield) with the RWP model. It is evident from the graph that unlike RWP model, our proposed model is able to capture the time location dependence of mobility distribution for different social scenarios since it does not assume random pause time. Moreover, the degree of connectivity of mobile nodes will suffer a major change for different communities. Thus, our model reflects the near actual pattern of pedestrian mobility distribution.

5 Conclusion

In this paper, we presented a theoretical framework for the mobility distribution of the nodes in a MANET. We have considered the effect of social behavior on the movement of a node which is basically a move and pause type of motion. Instead of assuming random pause-time distribution for the mobile hosts, we have designed a theoretical background for the pause-time formulation. The simulation result of our model shows a marked improvement over the existing RWP model. Finally, we plan to refine our model by considering the presence of obstacles within the transmission range, which is left as a future work.

References

1. F. Bai, N. Sadagopan, and A. Helmy: The Important Framework for Analyzing the Impact of Mobility on Performance of Routing for Ad Hoc Networks. Ad Hoc Networks Journal, Vol. 1, Issue 4, pp. 383-403, Nov (2003).
2. Jardosh, E. M. Belding-Royer, K. C. Almeroth, and S. Suri: Towards Realistic Mobility Models for Mobile Ad hoc Networks. in proceedings of ACM MobiCom, pp.217-229, September (2003).
3. M. E. J. Newman: The structure and function of complex networks. SIAM Review, 19(1):1–42, (2003).
4. D. J. Watts: Small Worlds: the Dynamics of Networks between Order and Randomness. Princeton Studies on Complexity. Princeton University Press, (1999).
5. K. Hermann: Modeling the sociological aspect of mobility in ad hoc networks. In Proceedings of MSWiM'03, San Diego, California, USA, September (2003).
6. Mirco Musolesi, Stephen Hailes, Cecilia Mascolo: An Ad Hoc Mobility Model Founded on Social Network Theory. In proceedings of 7th ACM International Symposium on Modeling, Analysis and Simulation of Wireless and Mobile Systems, Venice, Italy, pp.20-24, (2004).

Design of an Efficient Error Control Scheme for Time-Sensitive Application on the Wireless Sensor Network Based on IEEE 802.11 Standard*

Junghoon Lee[1], Mikyung Kang[1], Yongmoon Jin[1], Gyungleen Park[1], and Hanil Kim[2]

[1] Dept. of Computer Science and Statistics
[2] Dept. of Computer Education, Cheju National University,
690-756, Jeju City, Jeju Do, Republic of Korea
{jhlee, mkkang, ymjin, glpark, hikim}@cheju.ac.kr

Abstract. This paper proposes and analyzes the performance of an efficient error control scheme for time sensitive applications on wireless sensor networks. The proposed scheme divides DCF into HDCF and LDCF without changing PCF, aiming at maximizing the successful retransmission of a packet that carries critical data. While channel estimation obviates the unnecessary polls to the node in channel error during PCF, two level DCF enables prioritized error recovery by making only the high priority packet be retransmitted via HDCF. A good chop value can distribute the retransmission to each period, maximizing recovered weight, or criticality as well as keeping low the possible loss of network throughput. The simulation results show that the proposed scheme can improve recovered weight by 8% while showing 97% successful transmission at maximum for the given simulation parameter.

1 Introduction

In the past few years, smart sensor devices have matured to the point that it is now feasible to deploy a large, distributed network of such sensors [1]. Some mobile devices such as telematics terminals can carry the sensors to the spot of concern. Communication between the sensors and sinks requires wireless networks, and all nodes in the network share one common communication media. Message flows exchanged in a sensor network are mainly periodic and need guaranteed delay for a computing node to make a meaningful and timely decision [2]. Many real-time scheduling and fair packet scheduling algorithms have been developed for wired networks. However, it is not clear how well these algorithms work for wireless sensor networks where channels are subject to unpredictable, location-dependent, and time-varying bursty errors [3].

---

* This research was supported by the MIC Korea under the ITRC support program supervised by the IITA.

The IEEE 802.11 was developed as a MAC (Medium Access Control) standard for WLAN. The standard consists of a basic DCF (Distributed Coordination Function) and an optional PCF (Point Coordination Function). The DCF exploits CSMA/CA (Carrier Sense Multiple Access with Collision Avoidance) protocol for non-real-time messages. While collision-free PCF can provide a QoS guarantee, the error-prone nature of wireless network makes indispensable the error control procedure during DCF. The retransmission should carefully consider the priority of a packet, and network should try to enhance the successful retransmission of higher priority packets [4]. To meet such requirement, this paper proposes and analyzes an error control scheme for sensor data on DCF interval of IEEE 802.11 WLAN, aiming at supporting, though limited, level of priority in recovering the packet transmission error, during DCF. To this end, AP divides the DCF into two subperiods, makes their loads different, and gives more chance to the higher priority message by transmitting it via lower load network.

The rest of this paper is organized as follows: Section 2 introduces the background of this paper, including IEEE 802.11 WLAN standard, and real-time communication on WLAN. Then Section 3 proposes the communication architecture for time-sensitive sensor traffic. After demonstrating the simulation result in Section 4, Section 5 finally concludes this paper with a brief summarization and the description of future works.

2 Background

2.1 IEEE 802.11 WLAN

The wireless LAN operates on both CP (Collision Period) and CFP (Collision Free Period) phases alternatively in BSS (Basic Service Set) as shown in Fig. 1. Each superframe consists of CFP and CP, which are mapped to PCF and DCF, respectively. PC (Point Coordinator) node, typically AP, sequentially polls each station during CFP. In contrast, DCF is the basis of the standard CSMA/CA access mechanism and it uses the RTS (Request To Send)/CTS (Clear To Send) clearing technique to further reduce the possibility of collisions. The PC attempts to initiate CFP by broadcasting a *Beacon* at regular intervals derived from a network parameter of *CFPRate*. Round robin is one of the popular polling policies for CFP, in which every node is polled once in a polling round. Senders expect acknowledgment for each transmitted frame and are responsible for retrying the transmission. After all, error detection and recovery is up to the sender station, as positive acknowledgments are the only indication of success.

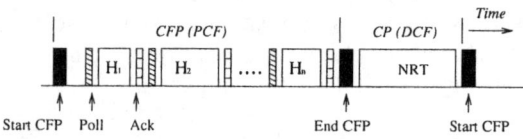

Fig. 1. Time axis of wireless LAN

2.2 Real-Time Communication on WLAN

The traffic of sensored data is typically *isochronous* (or synchronous), consisting of message streams that are generated by their sources on a continuing basis and delivered to their respective destinations also on a continuing basis [5]. In case of a change in the active flow set, bandwidth is to be reallocated or network schedule mode is changed. This paper follows the general real-time message model which has n streams, namely, $S_1, S_2, ..., S_n$, and for each S_i, a message sized less than C_i is produced at the beginning of its period, P_i. Each packet must be delivered to its destination within P_i units of time from its generation or arrival at the source, otherwise, the packet is considered to be lost.

As for the outstanding real-time communication scheme on WLAN, M. Caccamo and et. al. have proposed a MAC that supports deterministic real-time scheduling via the implementation of TDMA (Time Division Multiple Access), in which the time axis is divided into fixed size slots [6]. Unfortunately, to implement implicit contention, each node must schedule all messages in the network and their scheme didn't consider the network error at all. Choi and Shin suggested a unified protocol for real-time and non-real-time communications in wireless networks [2]. To handle location-dependent, time-varying, and bursty channel errors, the channel state can be predicted via channel probing before the packet is transmitted. Adamou and his colleagues have addressed the scheduling problem of achieving fairness among real-time flows with deadline constraints as well as maximizing the throughput of all the real-time flows over a wireless LAN [3]. This scheme is built on the assumption that BS knows which station has messages to retransmit as well as their deadlines, and decides which one to poll among them according to the criteria.

3 Message Scheduling Scheme

3.1 Channel Management

According to the operation of AP, the time axis of WLAN consists of a series of superframes and each of them consists of PCF, H-DCF, and L-DCF. Naturally, each channel can interfere with one another, due to the deferred beacon problem, that is, a beacon message can get delayed and the start of PCF can be put off, if another packet is already occupying the network. The maximum amount of deferment coincides with the maximum length of a data packet, as can be

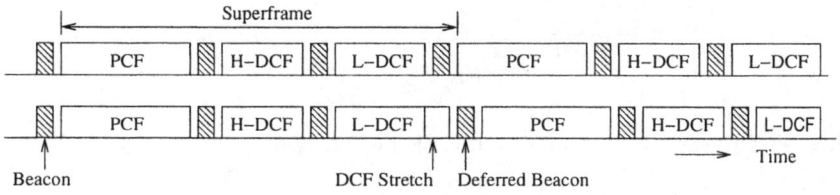

Fig. 2. Time axis of proposed network

inferred in Fig. 2. Additionally, we assume that the length of PCF and that of H-DCF are not reduced even if their starts are delayed. Only L-DCF shrinks its length when its start gets delayed, as shown in the right-hand part of Fig. 2. Each node transmits its message on each poll for the predefined time duration decided by a specific bandwidth allocation scheme. AP polls only those nodes whose channel is estimated to be *good*, since *bad* channel has no possibility to success considering the error characteristics of wireless channel. If a transmission fails or is deferred, the sender moves the packet to the retransmission queue via H-DCF or L-DCF according to its priority.

The 802.11 radio channel is modeled as a Gilbert channel [7]. We can denote the transition probability from state *good* to state *bad* by p and the probability from state *bad* to state *good* by q. The average error probability, denoted by ϵ, and the average length of a burst of errors are derived as $\frac{p}{p+q}$ and $\frac{1}{q}$, respectively. We take the estimation method from Bottiglieno's work [8]. To trace the channel status, AP maintains a state machine, or simply flag, associated to each sensor node. If the ACK/NAK is sent from the receiver to AP as soon as it receives a packet, AP sets the state to *good*. Otherwise, a timeout triggers the state to *bad*. Each *bad* channel has its own counter, and when a counter expires the AP attempts to send a single data frame to check the channel status.

3.2 Bandwidth Allocation

By allocation, we mean the procedure of determining capacity vector, $\{H_i\}$, for the given superframe time, F, as well as message stream set, $\{S_i(P_i, C_i)\}$. Though there have been plenty of bandwidth allocation schemes for the real-time message stream or sensor data stream, we exploit Lee's scheme form which the basic scheduling policy stems [9]. Let δ denote the total overhead of a superframe including polling latency, IFS and the like, while D_{max} the maximum length of a data packet. If P_{min} is the smallest element of set $\{P_i\}$, the requirement for the superframe time, F, can be summarized as follows:

$$\sum H_i + \delta + D_{max} \leq F \leq P_{min} \tag{1}$$

The minimum value of available transmission time, X_i is calculated as Eq. (2).

$$\begin{array}{ll} X_i = (\lfloor \frac{P_i}{F} \rfloor - 1) \cdot H_i & if (P_i - \lfloor \frac{P_i}{F} \rfloor \cdot F) \leq D_{max} \\ X_i = \lfloor \frac{P_i}{F} \rfloor \cdot H_i & Otherwise \end{array} \tag{2}$$

For each message stream, X_i should be greater than or equal to C_i ($X_i \geq C_i$).

$$\begin{array}{ll} H_i = \frac{C_i}{(\lfloor \frac{P_i}{F} \rfloor - 1)} & if (P_i - \lfloor \frac{P_i}{F} \rfloor \cdot F) \leq D_{max} \\ H_i = \frac{C_i}{\lfloor \frac{P_i}{F} \rfloor} & Otherwise \end{array} \tag{3}$$

By this, we can determine the length CFP period (T_{CFP}) and that of CP (T_{CP}) as follows:

$$T_{CFP} = \sum H_i + \delta, \quad T_{CP} = F - T_{CFP} \geq D_{max} \tag{4}$$

3.3 Scheduling of Retransmission

The proposed system has 3 virtual transmission links, PCF link, high-priority DCF link, and low-priority DCF link, while each of them is mapped to PCF, H-DCF, and L-DCF periods, respectively. The lower the load, the higher the probability of successful transmission, so we are to make the load of H-DCF lower than that of L-DCF, actually differentiating the upper bounds of maximum load for two periods. H-DCF transmits those packets whose priority is higher than c. If a packet recovery fails in H-DCF, it can be retried in the L-DCF with a normal CSMA/CA procedure. The value c is a tunable parameter that can be set according to the network load, current error rate, weight distribution, and so on [10]. It ranges from the lowest priority value, W_{min} to the highest one, W_{max}. The optimal value of c which maximizes value of recovered weight, can be found empirically or via analytical model for the given network parameters.

4 Performance Analysis

This section measures the performance of the proposed scheme via simulation using SMPL [11]. With SMPL, we implemented restricted contention protocol based on RTS/CTS mechanism for DCF. The number of active sensors is 5 and their utilization is 0.5. Each packet fits to the length of $0.1F$, being associated to a priority randomly picked from 0 to 19.

The first experiment measures the effect of chop value with fixed error rate, ϵ, set to 0.01, while the length of error duration, denoted as $\frac{1}{q}$ in Gilbert error model, distributes exponentially with average $2.0F$.

The y-axis of Fig. 3 plots the ratio of total weights of recovered packets to those of packets that failed in the first transmission. The gap between the proposed scheme and the non-partitioned DCF through ordinary CSMA/CA protocol is maximized when chop value is 0.55. Fig. 4 exhibits the measurement result of recovered weights according to the ϵ ranging from 10^{-3} to 10^{-2}. As

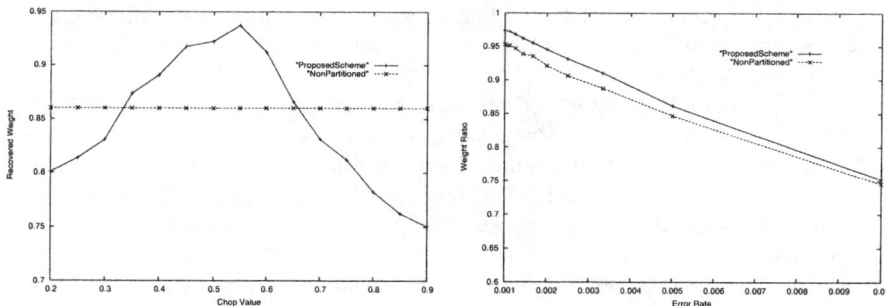

Fig. 3. Recovered weights vs. chop value   **Fig. 4.** Total weights vs. error rate

shown in the figure, the proposed scheme always outperforms the non-partitioned retransmission and achieves almost 97% of success of transmission for the given network and error parameter.

5 Conclusion

In this paper, we have proposed and analyzed the performance of communication architecture capable of efficiently dealing with channel error on the wireless sensor network for the time-sensitive sensor application based on the IEEE 802.11 WLAN standard. The proposed scheme makes AP always estimate channel status between itself and each sensor node, to avoid polling a node whose channel is not in normal condition. Once the packet transmission fails, it should be retried in a best-effort manner within its deadline. After all, it can support the prioritized error recovery by dividing the DCF into two sub-periods and differentiating their loads. The experiment performed via simulation using SMPL shows that the proposed scheme can improve the recovered weight compared with the traditional non-partitioned scheme with a good chop value. For the given parameters, it shows about 8% improvement when the chop value is 0.55. In addition, for the sum of weights of successfully transmitted packets, the proposed scheme always outperforms non-partitioned scheme. As a future work, we will investigate a method to find the optimal chop value for the given importance distribution as well as other real-time communication parameters.

References

1. Madden, S., Franklin, M., Hellerstein, J., Hong, W.: The design of an acquisitional query processor for sensor networks. ACM SINGMOD, (2003).
2. Choi, S., Shin, K.: A unified wireless LAN architecture for real-time and non-real-time communication services, IEEE/ACM Trans. on Networking, pp.44-59, Feb. (2000).
3. Adamou, M., Khanna, S., Lee, I., Shin, I., Zhou, S.: Fair real-time traffic scheduling over a wireless LAN. Proc. IEEE Real-Time Systems Symposium, pp.279-288, Dec. (2001).
4. Vaidya, N., Bahl, P., Gupta, S.: Distributed fair scheduling in a wireless LAN. Sixth Annual Int'l Conference on Mobile Computing and Networking, Aug. (2000).
5. Liu, J.: Real-Time Systems. Prentice Hall, (2000).
6. Caccamo, M., Zhang, L., Sha, L., Buttazzo, G.: An implicit prioritized access protocol for wireless sensor networks, Proc. IEEE Real-Time Systems Symposium, Dec. (2002).
7. Bai, H., Atiquzzaman, M.: Error modeling schemes for fading channels in wireless communications: A survey. IEEE Communications Surveys, Vol. 5, No. 2, pp.2-9,(2003).
8. Bottigliengo, M., Casetti, C., Chiaserini, C., Meo, M.: Short term fairness for TCP flows in 802.11b WLANs. Proc. IEEE INFOCOM, (2004).

9. Lee, J., Kang, M., Jin, Y., Kim, H., Kim, J.: An efficient bandwidth management scheme for a hard real-time fuzzy control system based on the wireless LAN. Accepted to LNCS: Embedded Systems for Ubiquitous Computing, (2005).
10. Gao, B., Garcia-Molina, H.: Scheduling soft real-time jobs over dual non-real-time servers. IEEE Trans. Parallel and Distributed Systems, pp.56-68, Jan. (1996).
11. MacDougall, M.: Simulating Computer Systems: Techniques and Tools. MIT Press, (1987).

Agglomerative Hierarchical Approach for Location Area Planning in a PCSN

Subrata Nandi, Purna Ch. Mandal, Pranab Halder, and Ananya Basu

Department of Computer Science and Engineering,
National Institute of Technology,
Durgapur, WB 713209, India
sn_nitdgp@yahoo.co.in

Abstract. Location area (LA) planning in PCSN is a NP-hard problem. In this paper we modeled it as a clustering problem where each LA is considered to be a cluster. Agglomerative Hierarchical Algorithm (AHA) is applied to form the cell clusters. The algorithm starts assuming each cell as a separate cluster. In successive iterations the clusters are merged randomly in a bottom up fashion based on a total cost function (TCF) till the desired numbers of clusters are obtained. Total Cost Evaluation Metric (TCEM) is proposed to compare AHA with other schemes. Experimental results show that AHA provides better results in most of the cases compared to Greedy Heuristic based approach.

1 Introduction

In Personal Communication Service Network (PCSN) [1] a set of LAs form the Service Area (SA). Each LA consists of a group of cells and is served by a Mobile Switching Center (MSC). The mobile terminals (MT) within each cell are controlled by a Base Station (BS). Each BS is connected to the MSC by a cable. BSs within the same LA communicate with each other through the (MSC) of that LA. If the MT moves from one cell to other within same LA there is no location update, but if MT crosses LA boundary then the handoff invokes a location update (LU). Given a set of cells, MSC/switches and their call handling capacity the problem is to assign the cells to a switch such that it minimizes the total hybrid cost including LU cost due to handoff and cabling cost under the constraint of call handling capacity of the switches. It is known as the static LA planning or cell to switch assignment (C2S) problem and is NP hard [1], [3].

Several Integer Programming based and heuristic based [1-2], [4], [9] approaches have been proposed to solve the C2S problem. Till now the approaches made towards solving the above problem, requires explicitly prior knowledge of MSC location, further none of these has used a common evaluation metric to compare the efficiency of the proposed scheme with others.

The goal of this paper is to propose a common cost evaluation metric and design an algorithm to explore the possibility of composing better solution by applying Agglomerative Hierarchical Clustering Algorithm (AHA) [10]. Clustering technique is used to group the cells among which traffic flow (handoff) is maximum and the distance is minimum. We define an objective function called Total Cost Function (TCF) which contains two factors (a) Handoff cost which is proportional to traffic flow in between the cells (b) Cabling cost which is proportional to distance. AHA starts by initializing each cell SA as a separate cluster. In successive iteration a randomly chosen cluster say c_K, is merged with adjacent cluster c_J for which TCF is optimum. After successive iteration of merging in bottom up fashion desired number clusters are obtained. Experimental results show that AHA gives better result than greedy heuristic algorithm (GHA) [5] in terms of the proposed cost evaluation metric.

2 The Proposed Approach

We consider fixed spatial distribution of inherently adjacent hexagonal cells. The entire SA in modeled as a 2D Graph. Let there be N Cells and M switches. The problem is to form M clusters of cells. All cells belonging to a particular cluster are assigned to the corresponding switch which is assumed to be located at the mean position of each cluster. We have considered single homing i.e. non-overlapping clusters.

If cell i and j are assigned to different switches i.e. different clusters, then cost is incurred every time a handoff occurs between cell i and cell j. Let h_{ij} be the handoff cost between cell i cell j per unit time where i, j = 1,2...N. Obviously, h_{ij} is proportional to the handoff frequency between cell i and cell j which is known before hand form statistics derived from simulation model or vehicular traffic measurement [2]. The amortized fixed cabling cost between cell i and switch k is proportional to the distance between the cell i and switch k. Let λ_i denotes the number of calls that cell i handles per unit time. Let S_k is known to be the call handling capacity of switch k.

The objective is to group the cells into optimal clusters so that total cost including handoff cost and amortized cabling cost per unit time is minimized such that call handling capacity of switches are not exceeded.

2.1 Problem Formulation

To formulate the problem mathematically we consider following notations:
If cell i belong to cluster c_k then $X_{ik} = 1$ otherwise $X_{ik} = 0$. The constraint on call handling capacity of switch k is as follows,

$$\sum_i \lambda_i X_{ik} \leq S_k, \quad \forall i=1,2...N \quad (1)$$

It means the total traffic from all cells belonging to a particular cluster must be less or equal to the call handling capacity of the switch corresponding to the cluster. To find the cost between a pair of clusters c_k and c_l, total cost function (TCF) is defined. The two components of TCF are as follows:

1. Total handoff cost per unit time say, H_{kl} between c_k and c_l. It is defined as the sum of handoff cost of the cells belonging to c_k which are adjacent to c_l :

$$H_{kl} = \sum_{i=1, j=1}^{n} h_{ij} . X_{ik} . X_{jl} \quad (2)$$

2. Cabling cost which is proportional to distance. Distance between mean position of cluster c_k and c_l say, D_{kl}. Let Cord_X_i and Cord_Y_i be the x and y coordinate of cell i respectively and Mean_X_k and Mean_Y_k be the x and y coordinate of mean position of cluster c_k.

$$\text{Mean\_}X_k = (\sum_{i=1}^{N} \text{Cord\_}X_{ik} \cdot X_{ik})/n(c_k); \quad \text{Mean\_}Y_k = (\sum_{i=1}^{N} \text{Cord\_}Y_i \cdot X_{ik})/n(c_k) \quad (3)$$

D_{kl} is obtained using the Euclidian distance metric

$$D_{kl} = \sqrt{(\text{Mean\_}X_k - \text{Mean\_}X_l)^2 + (\text{Mean\_}Y_k - \text{Mean\_}Y_l)^2} \quad (4)$$

We normalize both the components since H_{kl} and D_{kl} are in different scale,

$$\text{norm}(H_{kl}) = H_{lk}/\sum(H_{km}); \quad \text{norm}(D_{kl}) = D_{kl}/\sum(D_{km}), \quad \forall \text{ m of } c_m \text{ adjacent to } c_k \quad (5)$$

TCF is used as the key condition to be checked while merging clusters. A given cluster k will be merged with one of its adjacent cluster l iff norm (H_{kl}) is maximum and norm (D_{kl}) is minimum among all its adjacent clusters. Therefore we define TCF_{kl} as a maximizing function as follows,

$$TCF_{kl} = \text{norm}(H_{kl}) + 1/\text{norm}(D_{kl}) \quad (6)$$

A randomly selected cluster is merged with one of its adjacent cluster in aech iteration such that the objective function TCF in (6) is maximized subject to the constraints in (2). Thus clusters are merged in a bottom up fashion based on TCF till the desired numbers of clusters are obtained. Finally, a Cost Evaluation Metric (CEM) is defined as half of the sum of $TCF_{i,j}$ ($i \neq j$) between each adjacent pair of clusters to compare the final solutions obtained from different schemes i.e.

$$CEM = (\sum_{i,j} TCF_{i,j})/2, \quad \forall \text{ i of } c_i \text{ adjacent to } c_k \text{ and } i \neq j \quad (7)$$

2.2 The AHA Algorithm

Input:

a) Number of switches M to be installed in the SA.
b) Traffic handling capacity of each switch S_k where k=1, 2....M.
c) Call volume of each cell λ_j where i=1, 2...N.

Output:

Set of M clusters with the set of cells in each cluster and CEM for the solution.

Procedure:

a) Initialize each cell as clusters i.e. cluster i={cell $_i$}. Form initial set of clusters CLST_SET={c_i} where i=1,2....N. Make a list of available switches AVAIL_MSC={switch j} where j=1,2....M sorted in descending order of their-call volumes. Initialize list of assigned switches ASSIGN_MSC={NULL}. Finally it contains (switch#, cluster#) tuples.

b) Compute call handling capacity of each cluster as the sum of call volume of the cells corresponding to that cluster i.e. Clust_callvol$_i$=$\sum \lambda_i$, $\forall$ j belonging to c$_i$
c) Iteration:
1) Randomly choose a cluster, say c$_i$ that has not been considered in this iteration.
2) Find set of clusters adjacent to c$_i$, not considered in this iteration, say ADJ_SET$_i$
3) Make a list L$_i$ of TCF$_{ij}$ corresponding c$_i$ for all j adjacent to c$_i$.
4) Sort the list L$_i$ in non-increasing order of TCF.
5) Select a cluster from the list L$_i$ for which TCF is maximum, say c$_k$.
6) Let CV=Clust_callvol$_i$+Clust_callvol$_j$ and call volume of 1st switch in AVAIL_MSC is MSC_callvol. If MSC_callvol>=CV then merge c$_i$ with c$_k$ and set Clust_callvol$_i$ = CV. Mark the c$_i$ as considered. If n(CLUST_SET) > M goto Step 9.
else Select next cluster from TCFij and repeat Step (c6).
7) If c$_i$ is not merged in (c6). Check if there exists a switch that best fits the capacity of c$_i$. Remove c$_i$ from CLST_SET, make an entry in ASSIGN_MSC. Repeat Steps (c1-c7) till atleast two clusters remain unmarked in CLST_SET.
8) If n(CLST_SET) is same after last iteration then split the cluster with maximum call volume into two as it was before merging. Goto Step (c) for the next iteration.
9) Compute cost evaluation metric CEM of the final solution.

3 Results and Discussion

To test the effectiveness of AHA in solving the C2S problem for large SA, we compare the results with Greedy Heuristic Algorithm (GHA)[5]. Comparative results corresponding to a 15 cell SA shown in Fig.1, with two switches are presented in Table1. Results corresponding to a 27 cell SA shown in Fig.1, with 2 switches are given in Table2 along with the same SA with 3 switches. Results are compared by varying switch positions for both the cases with 2 switches within the SA, except the 3 switch case with 27 cells. Location of switches and cells and their call handling capacity is provided as input. Both tables lists the LAs formed with the set of cell identifiers within parenthesis and CEM of each solution in italics. As randomness is involved in selecting a cluster for merging different results are produced in different runs. For each input, the best obtained result out of five runs is tabularized.

Table 1. Results obtained by using GHA and AHA on a 15 cell SA layout of Fig.1 with 2 switches with call volume capacities 33.14 each. CEM is shown in italics.

MSC Location	GHA Output	AHA Output
	LAs formed - Cost of solution (CEM)	LAs formed - Cost of solution (CEM)
3,14	(3,4,6,7,10,11,2,8),(14,15,13,9,12,5,1) -*13.1*	(3,6,7,1,4,5,8,2),(14,13,12,15,9,11,10) -*7.9*
6,11	(5,6,10,13,9,2,1),(11,14,15,12,3,4,8,7) - *4.3*	(6,10,13,9,2,1),(11,14,15,12,3,4,8,7) - *4.3*
9,7	(9,13,5,10,6,2,1,14),(7,4,8,11,3,12,15) - *4.5*	(7,8,4,3,15,14,12,11),(9,10,13,6,2,1,5)- *4.3*
1,15	(1,2,3,5,9,6,13,10),(15,14,11,12,8,7,4) - *5.0*	(7,8,4,3,15,14,12,11),(9,10,13,6,2,1,5)- *4.3*

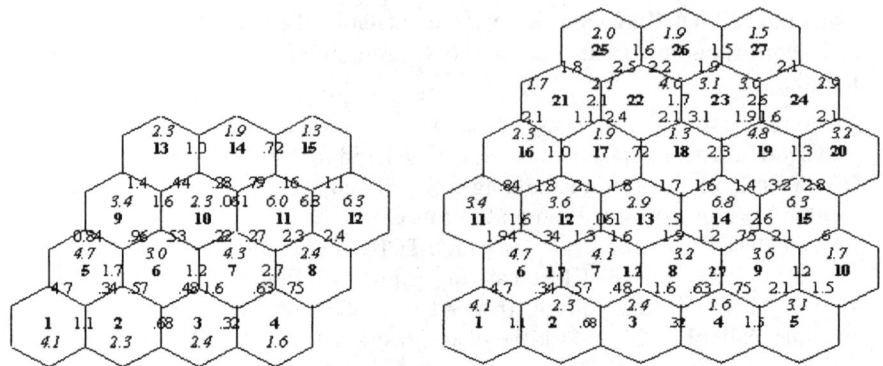

Fig. 1. A sample SA with 15 cell layout and a sample SA with 27 cell layout. Call volume of each cell is written in italics beside the cell in the figures. The handoff cost for each pair of adjacent cells is labeled at the corresponding edge.

Table 2. Results obtained by using GHA and AHA on a 27 cell SA layout of Fig. 1 with 2 switches with call volume capacities 55.0 each for switch locations. The last row gives solution for the same layout with 3 switches with capacities 35.0 each. CEM is shown in italics.

MSC Location	GHA Output	AHA Output
	LAs formed: Cost of solution (CEM)	LAs formed: Cost of solution (CEM)
15,14	Unsuccessful	(12,15,16,17,19,20,21,22,23,24,25, 26), (1,2,3,4,5,6,7,8,9,10,11,13,14,18) - *21.5*
1,24	(1,2,3,4,5,6,7,8,9,11,12,16),(13,14,15,17,18, 19,20,20,22,23,24,25,26,27) - *15.9*	(1,2,3,4,5,6,7,8,9,10,11,12),(13,14,15,16,17, 18,19,20,21,22,23,24,25,26,27) - *12.9*
11,20	(1,2,6,7,11,12,16,17,22,25,26), (3,4,5,8 ,9,10,13,14,15,18,19,20,23,24,27) - *14.2*	(11,12,17,21,16,19,18,24,23,27,26,22,25), (1, 2,3,4,5,6,7,8,9,10,13,14,15) - *14.2*
10,20	(1,2,3,4,5,6,7,8,9,10,11,12),(13,14,15,16, 17,18,19,20,21,22,23,24,25,26,27) – *12.9*	(1,2,3,4,5,6,7,8,9,10,11,12),(13,14,15,16,17, 18,19,20,21,22,23,24,25,26,27) - *12.9*
7,23	(1,2,3,4,5,6,7,8,9,10,11,12,13,16,17),(14, 15,18,19,20,21,22,23,24,25,26,27) – *13.1*	(1,2,3,4,5,6,7,8,9,10,11,12,13,16,17), (14, 15, 18,19,20,21,22,23,24,25,26,27) – *13.1*
5,25	(1,2,3,4,5,6,7,8,910,13,14,15,20),(25,26,27, 24,22,23,21,11,12,16,17,18,19) – *14.2*	(1,2,3,4,5,6,7,8,9,10,11,12,13,16),(14,15,17, 18,19,20,21,22,23,24,25,26,27) – *13.8*
2, 9, 25	(1,2,3,6,7,11,12), (4,5,8,9,10,13, 14,15, 19 20),(16,17,18,21,22,23,24,25,26,27) - *23.1*	(1,2,3,6,7,11,12,13,16,17),(4,5,8,9,10,14,15, 20),(18,19,21,22,23,24,25,26,27) – *20.6*

As observed in most of the cases AHA produces optimal or near optimal results which are better in most cases or atleast equally good compared to GHA. Further, in AHA the CEM remains unchanged in spite of small change in switch location in some cases. Thus, using AHA we can find the best possible switch position within the SA through a series of experiments. The first experiment in Table 2 shows GHA may not provide a solution even if exist because of its greedy nature. But AHA always explores and gives some solution if exist, because it allows backtracking.

4 Conclusion and Future Work

In this paper, we have modeled the C2S problem as a clustering problem and used Agglomerative Hierarchical approach to cluster the BSs. Experiments results have demonstrated the effectiveness of the AHA algorithm. AHA requires several runs, therefore takes more computation time than the GHA but finds much better solution. Computation time is not a major concern because here computation is an offline activity. Some of the results show that change of switch position does not alter the quality of solution, if they are nearer to the center of the LAs. So AHA provides more flexibility as any of these locations can be used to place the switch while designing a new SA. The AHA can be used effectively both for designing new SA and extending existing SA.

We can find the optimal number of switches to be placed in a SA by analyzing the behaviors of cost evaluation metric against number of clusters. Multihoming i.e. assigning boundary cells to more than one switch to reduce location update cost may be implemented if we consider fuzzy clusters.

References

1. Merchant, A., Sengupta, B.: Assignment of cells to switches in PCS networks, IEEE/ACM Trans. Networking, Vol. 3, no. 5, pp. 521–526, Oct. (1995)
2. Saraydar, C. U., Kelly, O., Rose, C.: One-dimensional location area design, IEEE Trans. Vehicular Technology, Vol. 49, pp. 1626–1632, Sept. (2000)
3. Gary, M. R., Johnson, D. S.: Computers and Intractability, A Guide to the Theory of NP-Completeness, New York: Freeman, (1979)
4. Bhattacharjee, P. S., Saha, D., Mukherjee, A.: Heuristics for assignment of cells to switches in a PCSN: A comparative study, Proc. IEEE Int. Conf. Personal Communication, Jaipur, India, Feb. 17–19, pp. 331–334, (1999)
5. Bhattacharjee, P. S., Saha, D., Mukherjee, A.: An Approach for Location area Planning in a Personal communication service Network (PCSN), IEEE Transactions on Wireless Communications, Vol. 3, No. 4, pp. 1176-1187, July (2004)
6. Saha, D., Mukherjee, A., Bhattacharjee, P. S.: A simple heuristic for assignment of cells to switches in a PCS network, Wireless Personal Communication, Amsterdam, The Netherlands: Kluwer Academic, Vol. 12, pp. 209–224, (2000)
7. Saha, D., Mukherjee, A.: Design of hierarchical communication network under node/ link failure constraints, Computer Communication., Vol. 18, no. 5, pp. 378–383, (1995)
8. Gondim, P.: Genetic algorithms and location area partitioning problem in cellular networks, Proc. IEEE Vehicular Technology Conference , Atlanta, GA, Apr., (1996)
9. Mandal, S., Saha, D., Mahanti, A.: A heuristic search for generalized cellular network planning, Proc. IEEE Int. Conf. Personal Communication, New Delhi, India, pp. 105–109, Dec.,(2002)
10. http://www2.cs.uregina.ca/~hamilton/courses/831/notes/itemsets/itemset_prog2.html

A Clustering-Based Selective Probing Framework to Support Internet Quality of Service Routing

Nattaphol Jariyakul[1] and Taieb Znati[1,2]

[1] Department of Information Science and Telecommunications
[2] Department of Computer Science,
University of Pittsburgh, Pittsburgh PA 15260
{njariyak, znati}@cs.pitt.edu

Abstract. Two Internet-based frameworks, IntServ and Differentiated DiffServ, have been proposed to support service guarantees in the Internet. Both frameworks focus on packet scheduling; as such, they decouple routing from QoS provisioning. This typically results in inefficient routes, thereby limiting the ability of the network to support QoS requirements and to manage resources efficiently. To address this shortcoming, we propose a scalable QoS routing framework to identify and select paths that are very likely to meet the QoS requirements of the underlying applications. Scalability is achieved using selective probing and clustering to reduce signaling and routers overhead. A thorough study to evaluate the performance of the proposed d-median clustering algorithm is conducted. The results of the study show that for power-law graphs the d-median clustering based approach outperforms the set covering method. The results of the study also show that the proposed clustering method, applied to power-law graphs, is robust to changes in size and delay distribution of the network. Finally, the results suggest that the *delay bound* input parameter of the d-median scheme should be no less than 1 and no more than 4 times of the average delay per one hop of the network. This is mostly due to the weak hierarchy of the Internet resulting from its power-law structure and the prevalence of the small-world property.

1 Introduction

The Internet has emerged as the most prominent communication infrastructure, carrying an ever broadening range of protocols and applications. The traditional best-effort service of the Internet, however, is inadequate to support diverse characteristics and different Quality-of-Service (QoS) requirements of multimedia applications. Depending upon the application and media type, such requirements may involve stringent temporal constraints. Different multimedia applications are sensitive to different factors and possess a variety of service constraints, including bandwidth, delay bounds and loss bounds. To meet these constraints, two service models, namely IntServ and Differentiated DiffServ, have been proposed to support service guarantees in the Internet. IntServ supports service guarantees on a per-flow basis. The framework, however, is not scalable due to the fact that routers have to maintain a large amount of state information for each supported flow. DiffServ was proposed as an alternate solution to address the lack of scalability of the IntServ framework. DiffServ uses class-based service differentiation to achieve aggregate support for QoS requirements. This approach eliminates the need to maintain per-flow states on a hop-by-hop basis and reduces considerably the overhead routers incur in forwarding traffic.

DiffServ focuses on packet scheduling; as such, it decouples routing from QoS provisioning. This typically results in inefficient routes, thereby limiting the ability of the network to support QoS requirements and to manage resources efficiently. To address this shortcoming, we propose a scalable cluster-based scheme to support QoS routing in Internets. The tenet of our approach is based on seamlessly integrating routing into the DiffServ framework to extend its ability to support QoS requirements. Scalability is achieved using selective probing and clustering to reduce signaling and router overhead, while identifying paths that satisfy a specific constraint, such as delay.

In the proposed cluster-based scheme, nodes whose metrics are highly correlated are clustered together, and the metrics inquiries are performed on a per-cluster basis. In this work, we focus on *delay* as the metric of interest. Therefore, the nodes located in the same cluster are said to share equivalent delay. Furthermore, each cluster is represented by one *anchor* node, usually located at the "center" of the cluster. The QoS metric dissemination and measurements are performed by the anchor node; the delays to the rest of the nodes in the same cluster are estimated to be equal to the anchor delay. The actual delay measured from the anchor, however, may be slightly different from those of the rest of the nodes in the same cluster. This difference is referred to as the *estimation error*, and should be bounded for each cluster. The estimation error determines the accuracy of the scheme.

There is a trade off between scalability and accuracy of the scheme. Suppose the network of n nodes is clustered into k clusters; the routing overhead is then reduced by a factor of (n/k). For scalability, the number of clusters k should be small, which implies large cluster sizes must be used. This approach, however, may result in high estimation errors caused by the highly likely delay diversity among the large number of nodes in the cluster. On the other hand, using small-sized clusters may reduce the estimation error. This, however, can only be achieved at the cost of reduced scalability as the number of clusters in large networks is likely to increase. As a result, a design tradeoff between accuracy and overhead must be carefully considered.

To address this issue, the paper proposes a delay-based clustering approach, referred to as d-median, which efficiently clusters large-scale networks, based on delay such that scalable routing can be achieved, while maintaining the routing accuracy to an acceptable level. A thorough study to evaluate the performance of the proposed d-median clustering algorithm is conducted. The results show that the d-median algorithm outperforms the existing approach and the clustering results are robust to the changes in network topologies. We also observe that a range of very small cluster sizes, in terms of delay, must be used due to the loosely hierarchical nature of the Internet.

The rest of the paper is organized as follows. Section 2 reviews work related to clustering in computer networks. Section 3 discusses the proposed clustering and probing framework. Section 4 describes the d-median clustering approach. Section 5 defines the methodology for evaluating the d-median. Section 6 discusses the results of the performance evaluation study. Finally, Section 7 concludes the work.

2 Related Work

Clustering is widely used to solve a diverse set of problems in the area of computer networks. Typically, the models proposed are often referred to as discrete location models or facility location models. These models deal with optimally locating a set of facilities in order to satisfy one or more requirements, e.g., to minimize the number of facilities used to cover the entire network, or to minimize the average distance from every node to its nearest facility. In this paper, we use the terms *facility* and *distance* to represent *anchor* and *delay*, respectively. Discrete location problems can be formulated as Integer Programming problems and are known to be *NP-hard*. Therefore, approximation algorithms are generally required to obtain near-optimal solutions.

For the past several years, discrete location models were used in network design to solve problems such as placement of Internet routers or cache servers [12], [8] or replication of web server in Content Distribution Networks (CDN) [9].

In [13], a scheme is proposed to determine the location of web server replicas in CDNs. The approach formulates the problem as k-median problem. Various algorithms for solving the k-median problem were proposed and evaluated. The evaluation was performed on various network configurations. Results indicated that the greedy-based algorithm outperforms other approaches in terms of accuracy and robustness. In [3], an overlay network scheme, referred to as Iso-bar, is proposed for distance monitoring and estimation in the Internet. The framework divides an overlay network into clusters and estimates the distance (delay) between any pair of nodes using both distance between clusters and distance within clusters. The Iso-bar scheme clusters the network using three discrete location models, namely set covering, k-center, and k-median.

Set covering is one of the simplest models used in discrete location models. The objective of the set covering problem is to find a minimum number of facilities from among a finite set of candidate facilities so that every demand node is covered by at least one facility. The set covering problem in a general graphs is *NP-hard* [6]. Despite the intensive studies on the set covering problem, the best approximation algorithm known is greedy-based [11]. In this algorithm, the approximation factor is $ln(n)$ and the running time is proportional to n^2, where n is the number of nodes in the network. In practice, a comparative study of nine different approximation algorithms for the set covering problem was conducted on 60 randomly generated problem sets, for which the optimal solutions were known [7], [2]. The greedy-based algorithms (both in the case of randomized and deterministic variants) yield the best results. The solutions obtained from a greedy-based algorithm deviate only by 5%, in average, from the optimum.

The k-median approach uses the concept of the linear cost function to locate k facilities in the network so that the *total cost*, in terms of distance, is minimized. This results in three constraints: the first that each node is connected to exactly one facility, the second ensures that this facility must be available, and the third ensures that the number of facilities does not exceed k. The k-median problem in general graphs is *NP-hard* [6]. Approximation algorithms are generally required. A simple greedy-based algorithm for k-median has been proposed in [4]. The running time of this algorithm is $O(kn^2)$, where k is the maximum number of clusters and n is the number of nodes in the network. A major shortcoming of the greedy-based approach is that it has no guaranteed approximation factor. However, the algorithm was run against 40 problem sets, for which the optimal solutions are known [2]. The results show that, in the worst case, the solution obtained by the algorithm deviates from the optimum one by less than 5% [10].

The k-median based approach has a desirable property in that it tries to minimize the delay between every node and its nearest anchor. However, the model cannot guarantee the maximum delay bound from an anchor to the farthest node in its cluster. Similarly, a set covering based approach is inadequate to address our clustering criteria. To address this shortcoming, the d-median scheme takes the coverage distance (maximum delay bound), d_c, as an input and determines the number of clusters, k. The d-median scheme tries to locate the minimum number of anchors such that the sum of the connection cost is minimized and the maximum delay of every cluster does not exceed the delay bound input, d_c.

3 Clustering and Probing Framework

Scalability and efficiency of the QoS routing architecture may be achieved using efficient *network clustering* and *selective probing*. Network clustering reduces the number of nodes which participate in routing information dissemination and path selection. Nodes, whose delay variations are bounded by a network-wide specified delay value, d_c, are said to be in same *class of equivalence*. These nodes are grouped to form a *cluster*. A cluster can be viewed as a logical node, called *meta-node*. The topology, derived from the physical connectivity of the meta-nodes, represents a *meta-graph*.

Once the network is clustered into meta-graph, selective probing can be used for metric acquisition and dissemination among meta-nodes, on a per-cluster, as opposed to a per-node, basis. For each cluster, an *anchor node* representative of its equivalence class is selected to *probe* its peers in other clusters and exchange QoS metric information. Once the QoS metric information has been exchanged between meta-nodes, path computation can be undertaken to locate appropriate paths that satisfy the QoS requirements of the underlying applications. Note that the process of metric acquisition and estimation, and the process of path selection can be done periodically or upon request. Network may also be re-clustered to update the meta-graph topology following significant changes in the underlying physical topology or after a long period of time. In the following section, we describe the clustering approach used in the proposed framework to minimize the number of anchors, while guaranteeing the maximum delay bound within a cluster.

4 The d-Median Based Strategy

The main objective of the proposed d-median strategy is to keep the estimation error bounded while minimizing the signaling overhead. Before we describe the clustering scheme, we need to define the desirable properties of the clustering strategy. First, the clustering method must minimize the number of clusters k to reduce the signaling overhead in routing process. The smaller number of clusters k results in increased performance of the clustering method. Second, the coverage distance d_c of every cluster must be bounded to limit the effect of estimation error. The coverage distance d_c is referred to as the *delay bound input* of the clustering method. Finally, the average delay for each node to reach its nearest anchor (the connection cost) should be small in order to reduce the effect of estimation error. Based on the above, the Integer Programming formulation of the d-median can be expressed as follows:

MINIMIZE $\quad \sum_{i \in F, j \in C} d_{ij} x_{ij}$

SUBJECT TO: $\quad \forall j \in C: \quad \sum_{i \in F} x_{ij} = 1$

$\forall i \in F, j \in C: \quad y_i - x_{ij} \geq 0$

$\forall j \in C: \quad d_c \geq d_{ij} x_{ij}$

$\forall i \in F, j \in C: \quad x_{ij} \in \{0,1\}$

$\forall i \in F: \quad y_i \in \{0,1\}$

In the above formulation, F is a set of facilities and C is a set of all nodes in the network. d_{ij} denotes the connection cost associated with node j and facility i. d_c denotes the coverage distance. y_i and x_{ij} are the decision variables. y_i has its value set to 1 if and only if an anchor i is selected. Similarly, x_{ij} has its value set to 1 if and only if node j is served by the anchor i. Based on this formulation, each cluster is bounded with coverage distance d_c and the total connection cost is minimized. The d-median problem is NP-hard. To solve this problem, we propose an approximation algorithm based on the k-median heuristic. The algorithm is described in Figure 1.

Note that, in the set covering problem, a node is said to be covered if $d_{ij} \leq d_c$. Once a node is covered, no weight or cost associated with d_{ij} is taken into further consideration. However, in the proposed framework, a smaller d_{ij} indicates a smaller delay and, hence, more accuracy in metrics estimation; larger values of d_{ij}, however, implies less accuracy and therefore is less desirable. This suggests that we should include the weight associated with d_{ij} when deciding about locations. The simplest way to achieve this is by treating d_{ij} as a linear cost function.

The d-median approximation algorithm uses an inverse approach of the original k-median counterpart. Note that each algorithm takes either d_c or k as an input and determines the other. Therefore, we can say that d_c determines k, and vice versa. Given a network topology, the d-median algorithm and the k-median algorithm will produce the same clustering results, when the appropriate values for d_c and k are used. Consequently, the accuracy of the algorithm is identical to that of k-median. However, the running time of the d-median algorithm is $O(n^3)$.

1. Set $F = \phi$
2. If every node has a connection cost less than or equal to d_c, go to (5)
3. For each node $i \notin F$
   a. Calculate the total connection cost with the set of facilities $F \cup \{i\}$ (assuming that each node connects to the nearest facility)
4. Select i that yields the **minimum** connection cost
   a. $F := F \cup \{i\}$
   b. Go to (2)
5. Return F.

Fig. 1. Approximation algorithm for the d-median algorithm: greedy-based approach

5 Evaluation Methodology

To evaluate the performance of the d-median and the set covering clustering methods, we simulate their corresponding approximation algorithms on a variety of network topologies. In the following, we first discuss the network topologies and a set of performance metrics used in our evaluation.

5.1 The Internet Topology

In this work, we consider the performance of the clustering methods over large-scale Internets. Recent work has shown that the node degree in the Internet induced graph exhibits power law properties [5], [14]. Several algorithms have been proposed to generate power-law graphs. It is widely accepted, however, that the degree-based network topology generators are superior to structural generators in generating graphs with power-law degree distributions [15]. In this study, the degree-based network topology generator INET 3.0 was used to generate the Internet topologies [16].

In this work, we study the behavior of the two clustering methods, using various network sizes, namely 3,037, 3,500, 4,000, and 4,500 nodes. The reason behind choosing these network sizes stems from the fact that power laws hold only for large data sets. Furthermore, the power laws properties of the Internet were first discovered when the number of Internet nodes was 3,037.

In general, the Internet topology consists of routing nodes and connectivity information associated with these nodes. Therefore, metrics, such as hop-count, can be easily derived. However, our work is based on the delay metric. Unfortunately, due to the high variability of this metric in the Internet, neither the generated Internet topologies nor the measured Internet topologies supply this information [1]. To overcome this problem, the delay associated with each link in a simulated power-law graph is assigned based on one of the following standard distributions: Uniform, Normal, Exponential, and Heavy-tailed.

5.2 The Performance Metrics

In this section, we introduce the performance metrics that we will use as the tools to study the behavior and performance of set covering and d-median clustering methods.

Table 1 lists each performance metric used in this study and provides a brief description of its meaning.

The *number of clusters* is used for performance comparison between the two clustering schemes. More specifically, for a given delay bound, it is assumed that clustering method, which produces the smaller number of clusters, is considered to yield superior performance. However, it was observed that, in several cases, a large portion of the clusters contained only one node, thereby resulting in inefficient clustering. To address this concern, the concept of *effective clusters* was introduced. Based on this concept, single node clusters are not considered. Both number of clusters and number of effective clusters are shown in the percentage of clusters to the total number of nodes in the network. The *average delay* denotes the average delay between a node and its nearest anchor. A small value of the average delay indicates a small estimation error. The last performance metric is the *average cluster size*, which represents the expected number of nodes in the clusters.

Table 1. List of the main performance metrics used in this study

Performance Metrics	Descriptions
Number of clusters (%)	Total number of clusters
Number of effective clusters (%)	Total number of clusters consisting of more than 1 node
Average delay	Average delay from each node to the nearest anchor
Average cluster size	Average number of nodes in each cluster

6 Results and Evaluation

To evaluate and compare the performance of the d-median and the set covering clustering approaches, the corresponding approximation algorithms were executed using the same range of delay bounds, against a variety of synthetic Internet topologies. In the following, we report on the performance of these experiments.

6.1 Performance Comparison

In the preliminary analysis, several experiments were conducted. Each experiment is dedicated to one of four performance metrics, namely the number of clusters or effective clusters expressed as a percentage of the total number of nodes, the average cluster delay and the average cluster size. Furthermore, for each experiment the link delay distribution and the network size were varied. Both d-median and set covering require the delay bound d_c as an input. We normalized the d_c unit so that one unit equals the mean of the delay assigned to every link in the network. This normalized unit is hereafter referred to as *mean-hop-delay*. The optimal or near-optimal clustering results of each clustering method, computed over a specific set of network topologies and delay bounds, are then computed. The results show that for both set covering and d-median, the number of clusters decreases as we increase the delay bound input. This is due to the fact that the larger coverage-area clusters cover more nodes, and hence the number of clusters required to cover entire network is reduced. This behavior holds for all network topologies.

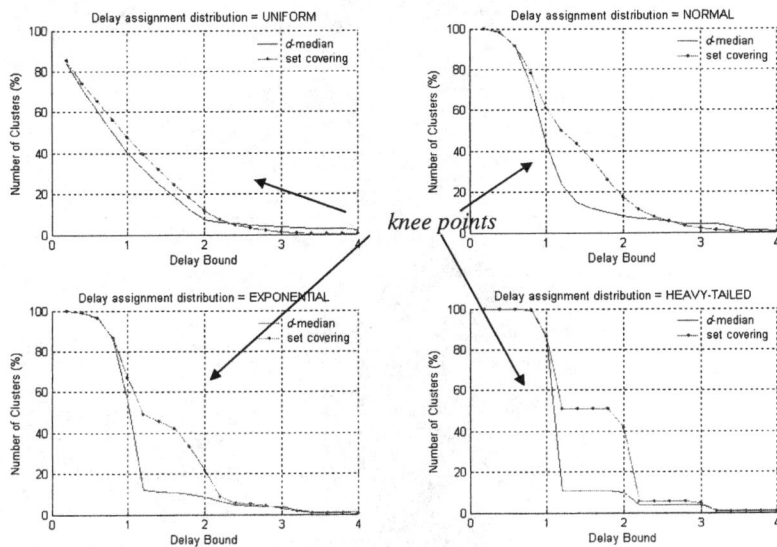

Fig. 2. The number of clusters produced by the d-median and set covering heuristics

Fixing the network size to 4,500 nodes, an experiment was carried out to determine the number of clusters produced by each method for different delay distributions. The results are as shown in Figure 2. In most cases, we observe that d-median yields smaller number of clusters than set covering for any given delay bound input. In general, a smaller number of clusters imply a smaller amount of signaling exchange in the network. This suggests the performance of d-median is better than that of set covering.

Also note that, in the case of the d-median, the number of clusters decreases rapidly in the beginning and becomes *stable* after 1 or 2 mean-hop-delays. We named the point where the steep slope ends and the graph becomes stable the *knee point*, as indicated in Figure 2. We will discuss the importance of these knee points in Section 6.3.

As mentioned previously, considering only the number of clusters may be misleading. One possible reason is that many of these clusters are *one-node clusters*, as nodes may be located in remote areas. A one-node cluster may also occur because of inefficient clustering. In this case, the clustering method fails to identify and avoid one-node clusters, thereby increasing the total number of clusters.

Considering only the number of effective clusters, the results show both the d-median and set covering start with a steep ascent to reach a peak before the number decreases as delay bound increases. When the delay bound is relatively small, most of the nodes scattered in the network form their own one-node clusters. As the delay bound increases, the one-node clusters merge with other clusters in their vicinity, thereby increasing the number of effective clusters. As the boundaries of clusters grow larger, at the total number of clusters required to cover the network is reduced, and so is the number of effective clusters, as shown in the plot. In all cases, however, the results show that the d-median always yields larger number of effective clusters than set covering. This is depicted in Figure 3 which presents the ratio of the number of

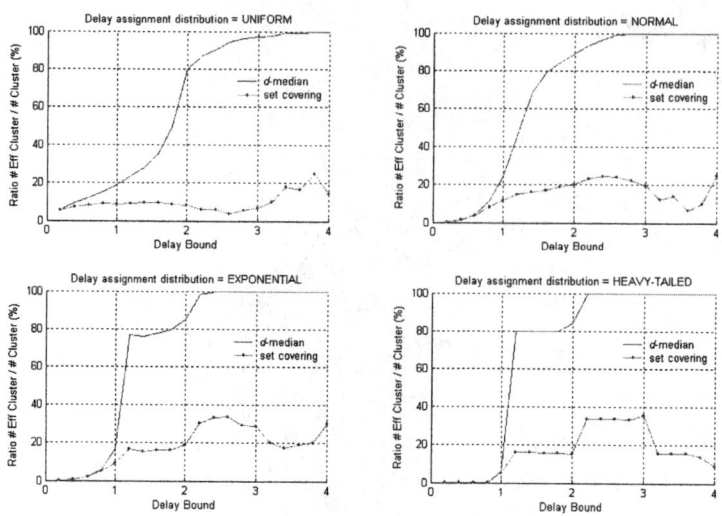

Fig. 3. Ratio of the number of effective clusters to the total number of clusters

effective clusters to the total number of clusters, for the case of 4,500 nodes. We can see that d-median can reach the point where every cluster is a non-one-node cluster, for a delay bound around 2 to 4 mean-hop-delay. However, set covering does not exhibit such a performance, thereby failing to eliminate unnecessary one-node clusters.

With respect to the *average delay*, defined as the estimated delay, in mean-hop-delay units, necessary for each node to reach its nearest anchor, results show that set cover exhibits in some cases smaller average delays than the d-median. Theoretically, the d-median's objective function is to minimize the overall delay; as such it should yield smaller average delay in every case. A closer look at the results, however, reveals that the portion of one-node clusters obtained by set covering is high. These one-node clusters have zero delay and consequently artificially reduce the average delay. We conclude that, based on our performance metric, d-median outperforms set covering as it produces a smaller number of clusters, a larger ratio of effective clusters, and a smaller average delay.

6.2 Sensitivity Analysis

The sensitivity analysis to the network sizes is performed to study the effect of changes in network sizes on the behavior of the two clustering approaches. To our surprise, the two clustering approaches exhibit a high degree of similarity. Specifically, the correlation coefficient ranges from 0.8376 to 1.0000 for the case of d-median and from 0.9014 to 1.0000 for the case of set covering. The high correlation coefficient indicates that the proposed scheme leads to acceptable performance as the network size increases, assuming a power-law topology. Sensitivity to delay, however, is more subtle. Unlike the impact of network sizes, changes in delay assignments show direct impact on the clustering results. The dissimilarities among the results produced by each clustering method for different delay assignments are noticeable, as indicated by a correlation coefficient that can be as low as 0.3382 in the worst case. Furthermore, the results show that for all delay assignments, d-median always yields smaller number of clusters than set covering, around the knee points, and a higher number of effective clusters. Finally, it was observed that the average cluster size obtained by d-median is always smaller. This confirms that, overall d-median outperforms set covering.

6.3 The Delay Bound Input

Both d-median and set covering take the delay bound as an input. The delay bound is the maximum allowable delay from an anchor to the rest of the nodes in the cluster. We believe that it is beneficial to find a range of *practical* delay bounds for which the d-median clustering scheme performs efficiently, as a small delay bound may result in unnecessary one-node clusters, while a large delay bound may result in exceedingly large clusters and consequently high estimation errors. A careful analysis of the results show that when the delay bound is around 5 mean-hop-delays, the network is dominated by a single cluster (average cluster size equal to network size and the number of clusters is one). Actually, the single-cluster domination starts around 4 mean-hop-delays when the average cluster size starts to rise abruptly and the number of clusters is reduced to a few clusters. The upper bound of the delay bound input of the d-median clustering approach should, therefore, be no larger than 4 mean-hop-delays.

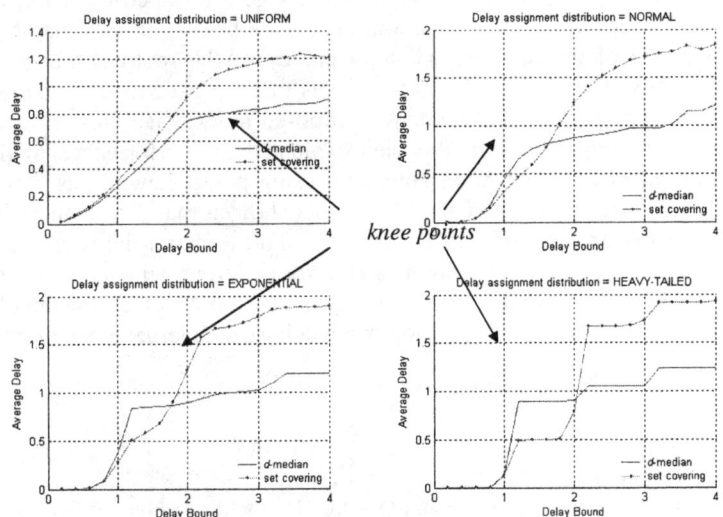

Fig. 4. Average delay for d-median and set covering heuristics

Results also show that inefficient clustering also occurs when the delay bound input is very small, as small delay bounds cause the number of clusters either to become exceedingly high or in some cases even equal to the number of nodes in the network. Now consider the average delay obtained for the 4,500-node network, as depicted in Figure 4. In this case also, *knee points* can be identified. Note that these knee points are located in exactly the same locations as in Figure 2. The knee points represent the points at which clustering results become stable, i.e., the number of clusters and the average delay do not change considerably as the delay bound increases. We propose using these knee points as the lower bound of the delay bound. Therefore, the lower bound should be around 1 and 2 mean-hop-delays.

It is also important to note that a very small delay bound input can efficiently cluster various network topologies, independently of their size and delay assignments. Recall that the mean-hop-delay is an average delay on one hop in the network. Therefore, in most cases, the cluster size is bounded by a few hops away from its anchor. In particular, if we cluster the network using d-median, the average delay for each node to reach its nearest anchor is merely around 1 mean-hop-delay, as shown in Figure 4. This is a remarkable result since it implies that the large scale topology of the Internet can be efficiently clustered, where the nodes within a cluster are located only a few hops away from each other. This is due to the loosely hierarchical nature of the Internet as mentioned in [14]. This fact is also confirmed by the works of [5] and [14], which estimate that the *diameter* of the Internet is between 4 and 5 hops.

7 Conclusion

In this work, we considered the clustering-based metrics acquisition scheme that aims to reduce the routing overheads in the Internet, where the metrics acquisitions are done on a per-cluster basis, rather than on a per-node basis. Our two major concerns are the scalability of the scheme and the accuracy of the routing information. We considered two existing discrete location models that are used to solve the problems of network clustering in the literature. We proposed a d-median clustering approach and its approximation algorithm. We then evaluated the performance of d-median approach, compared to set covering approach, using power-law graphs with various network sizes and delay assignments. The results showed that d-median outperforms set covering based on our performance metrics. Furthermore, the results show that the behavior of the clustering scheme is stable for different network sizes and delay assignments. The results suggest that the delay bound input to the d-median heuristic should be around 2 to 4 times the per-hop mean delay, for Internet clustering.

References

1. http://mscmga.ms.ic.ac.uk/info.html
2. Chen, Y., Lim, K. H., Katz R. H., and Overton, C.: On the stability of network distance estimation. ACM SIGMETRICS Performance Evaluation Review, Vol. 30 Issue 2 (2002).
3. Daskin, M. S.: Network and discrete location model, algorithms, and applications. John Wiley & Sons, Inc. (1995).
4. Faloutsos, M., Faloutsos, P., and Faloutsos, C.: On power-law relationships of the Internet topology. ACM SIGCOMM Computer Communication Review, Vol. 29 Issue 4 (1999).
5. Garey, M. R., and Johnson, D. S.: Computer and intractability: A guide to the theory of NP-completeness. W.H. Freeman (1979).
6. Grossman, T., and Wool, A.: Computational experience with approximation algorithms for the set covering problem. European Journal of Operational Research (1997) 81-92.

7. Guha, S., Meyerson, A., and Munagala, K.: Hierarchical placement and network design problems. In Proceedings of 41st Annual Symposium on Foundations of Computer Science (2000) 603 – 612.
8. Jamin, S., Jin, C., Jin, Y., Raz, D., Shavitt, Y., and Zhang, L.: On the placement of Internet instrumentation. In Proceedings of IEEE INFOCOM (2000).
9. Jariyakul, N.: A Clustering-based selective probing framework to support Internet quality of service routing. Master's Thesis, University of Pittsburgh (2004).
10. Johnson, D.: Approximation algorithms for combinatorial problems. Journal of Computer and System Sciences. Vol. 9 (1974) 256-278.
11. Li, B., Golin, M. J., Italiano, G. F., Deng, X., and Sohraby, K.: On the optimal placement of web proxies in the Internet. In Proceedings of IEEE INFOCOM. Vol. 3 (1999) 1282 – 1290.
12. Qiu, L., Padmanabhan, V. N., and Voelker, G. M.: On the placement of web server replicas. In Proceedings of IEEE INFOCOM. Vol. 3 (2001) 1587-1596.
13. Siganos, G., Faloutsos, M., Faloutsos, P., and Faloutsos, C.: Power laws and the AS-level Internet topology. IEEE/ACM Transactions on Networking. Vol 11 Issue 4 (2003) 514 – 524.
14. Tangmunarunkit, H., Govindan, R., Jamin, S., Shenker, S., and Willinger, W.: Network topology generators: degree-based vs. structural. ACM SIGCOMM Computer Communication Review, Vol 31 Issue 4 (2002).
15. Winick, J., and Jamin, S.: University of Michigan Technical Report CSE-TR-456-02, http://topology.eecs.umich.edu/inet/

A Fair and Reliable P2P E-Commerce Model Based on Collaboration with Distributed Peers*

Chul Sur[1], Ji Won Jung[2], Jong-Phil Yang[1], and Kyung Hyune Rhee[3]

[1] Department of Computer Science, Pukyong National University,
599-1, Daeyeon3-Dong, Nam-Gu,
Busan 608-737, Republic of Korea
{kahlil, bogus}@mail1.pknu.ac.kr
[2] Department of Information Security,
Pukyong National University
forji@mail1.pknu.ac.kr
[3] Division of Electronic, Computer and Telecommunication Engineering,
Pukyong National University
khrhee@pknu.ac.kr

Abstract. In this paper we present a fair and reliable e-commerce model for P2P network, in which communication parties can buy and sell products by P2P contact. In particular, we focus on a fair exchange protocol that is based on collaboration with distributed communication parties and distinguished from the traditional fair exchange protocols based on a central trusted authority. This feature makes our model very attractive in P2P networking environment which does not depend on any central trusted authority for managing communication parties.

1 Introduction

Recently Peer-to-Peer (P2P) networking paradigms and its applications offer opportunities for new services over both Internet and Mobile Ad-hoc Networks (MANETs). Specially, mobile devices such as mobile phones and PDAs are already used widespread, and functionality and performance of these devices are improved day by day. Due to the rapid growth of these technologies, mobile devices are expected to have capability to provide various services beyond the request of desired services. Hence, new services have appeared in P2P network, in which contents are bought and sold among parties by using mobile devices. Moreover, P2P network encourages an efficient model for contents distribution among communication parties. Since each communication party in P2P network does not depend on any central trusted authority for management, it is inherently scalable to implement communication models. Therefore, designing an e-commerce model in P2P network is a promising challenge which we have never met before in Internet environment.

---

* This research was supported by University IT Research Center Project, MIC, Korea.

However, due to the lack of the central trusted authority, P2P network does not efficiently provide all the services required by e-commerce transaction such as reliability and fairness. In particular, guaranteeing *fairness* is a major challenge in e-commerce model. Moreover, since the dynamic nature of P2P network implies that the consecutive connectivity between communication parties is not provided, it is more difficult to guarantee fairness for e-commerce transaction in P2P network.

Our Contribution. In this paper we design a new e-commerce model for guaranteeing fairness and reliability in P2P network, in which communication parties can buy and sell digital contents by P2P contact. Especially, we focus on an optimistic fair exchange protocol based on collaboration with distributed communication parties and distinguished from the traditional optimistic fair exchange protocols relied on a central trusted authority. Moreover, the proposed fair exchange protocol provides desirable property such as *availability* for P2P e-commerce model since we consider the threshold cryptography to design the protocol.

The rest of the paper is organized as follows. The next section identifies the security requirements for the P2P e-commerce services we have considered and describes cryptographic tools to induce the motivation of the paper. We outline the proposed e-commerce model suitable for P2P network in Section 3. An optimistic fair exchange protocol with Distributed TTP that provides fairness and reliability for the model is presented and analyzed in Section 4. Finally, we have a conclusion in Section 5.

2 Preliminaries

2.1 Security Requirements for P2P e-Commerce Service

Not all the P2P services are offered with a robust central server, and collaboration among peers in P2P commercial transaction is performed under ad-hoc and temporal connection. Therefore, these characteristics result in formidable challenge as far as providing the security services required by e-commerce service such as *confidentiality, authentication, integrity, non-repudiation*. Furthermore, the following requirements are desirable in e-commerce service:

- **Fairness** : No party should be able to interrupt or corrupt the protocol to force an outcome to his or her own advantage. The protocol should terminate with either party having obtained the desired information, or with neither one acquiring anything useful.
- **Effectiveness** : If no messages are lost, both parties behave according to the protocol and do not abandon the exchange, then both parties receive the desired items.
- **Timeliness** : It guarantees that both parties will achieve their desired items in the exchange within finite time.

Specially, *fairness* is the most considerable requirement in e-commerce service. Consequently, it is crucial that the protocol guarantees fairness between communication parties in P2P e-commerce model.

2.2 Cryptographic Tools

Threshold Cryptography. *Threshold cryptography* distributes the ability to provide a cryptographic service such as signing or decryption[3][10]. In a t out of n threshold scheme, any subset of greater than t peers (out of a total of n peers) can compute the desired functionality while any subset of less than or equal to t peers cannot. It offers better fault tolerance than non-threshold cryptography: even if some peers are unavailable, others can still perform the desired functionality. Threshold cryptography also provides better security since no single peer is entrusted to perform the desired functionality in its entirety. Consequently, it seems like an ideal choice to provide security services, such as reliable and fair exchange in P2P network.

Fair Exchange Protocol. *A fair exchange protocol* ensures that, at the end of the exchange, either each party receives the item it expects or neither party receives any information about the other's item. The classical solution to the fair exchange problem is based on the idea of *gradually* exchanging small parts of the items. Works in this approach generally rely on the unrealistic assumption that the two parties have equal computational power or require many rounds to execute properly.

The practical approach to resolve the problem is to use a trusted third party(TTP) as arbitrator. Specifically, this approach can be classified as *on-line* protocol and *optimistic* protocol according to their involvement of TTP[1][8][11]. On-line protocol requires the presence of the TTP as a delivery channel, intervening in each transaction. As the TTP is always involved in every transaction, this protocol considerably implies the communication and computational bottleneck. In optimistic protocol the TTP is not used during the transaction when the communication parties behave correctly, but is involved only in case of disputes with one of the parties. Since the TTP is mostly off-line, this protocol reduces the communication and computational overhead of the TTP.

3 A Fair and Reliable P2P e-Commerce Model

3.1 System Components and Communication Model

In this section we describe the proposed P2P e-commerce model, in which communication parties can buy and sell their products. The proposed model consists of peers who play both roles of a seller and a buyer, and DTTP (Distributed TTP) which manages the service key of a peer community. The description of system components is as follows:

- **Peer** : An entity who plays either role of a seller or a buyer according as the demand that it desires.

- **DTTP(*Distributed TTP*)** : DTTP is composed of a set of n special peers($n \geq 3t + 1$) which are called *master peers*, each runs on a separate device in a network. Each master peer has the service secret key share ss_i of a peer community and performs threshold cryptographic operations for assuring fairness and reliability between commercial transaction parties in the peer community.

In addition, we introduce *an adversary* who can easily steal or otherwise compromise all peers including master peers. Thus, our adversary model includes active(or Byzantine) adversary who can compromise some bounded fraction of peers in the network. However, we assume that fewer than or equal to 1/3 of the master peers are corrupted or malicious during the entire lifetime of the shared service secret key. This means that at least $2t + 1$ master peers are *available* at any time.

Generally, the quality of communication channels can be classified as *reliable*, *resilient* and *unreliable*. Previously proposed fair exchange protocols[1][2][8][11] assume that communication channel between the party and the TTP is *resilient* in order to resolve the dispute, because it is impossible to guarantee fairness without at least resilient channel between those parties. However, resilient channel assumption is not sufficient for our model since P2P network implies that no robust central servers are offered and the consecutive connectivity is not provided between communication parties including master peers. Therefore, to clarify our communication model for real P2P networking environment, we employ the idea used in Byzantine environment[4] with respect to communication channel between a peer and an available master peer.

Definition 1 (Fair Communication). *A communication channel between two correctly behaving parties is **fair** if no part of the network becomes permanently unavailable, given sufficient number of retransmissions, every message is delivered eventually.*

Consequently, in our model we assume that communication channel between peers who carry out e-commerce transaction is *unreliable* by the nature of P2P network, and that communication channel between a peer and an available master peer is *fair* by the definition above. Upon taking into consideration of the nature of P2P network, our qualitatively weaker communication model is very reasonable and realistic. Finally, we assume that communications is carried over confidential and broadcast channels.

3.2 Initialization of Peer Community

In the initial phase, each peer who wants e-commerce transactions of its contents constitutes a peer community as a virtual market. Every peer community has a service public key and a corresponding service private key for guaranteeing fairness and reliability for e-commerce transaction in the peer community. The high-level description of initialization is as follows:

1. To provide fairness and reliability for e-commerce transaction, master peers are chosen from the constructed peer community. The master peers can be chosen by the peer community founder, or can be the participants at the beginning of the peer community.
2. Each master peer obtains his service secret key share ss_i for obvious reasons and service public key of the peer community from a centralized dealer or by collaborative computation among master peers using a t out of n threshold scheme. For example, the threshold scheme described in [3] provides share distribution by collaboration among master peers, while the threshold scheme presented in [10] supports share distribution by a trusted dealer.
3. Each master peer publishes his identity and the service public key. After obtaining the identity of master peer and the service public key, a peer who wishes to buy or sell its own digital contents, of course including master peers, performs membership enrollment protocol presented in the subsequent section to affiliate himself with the peer community.

3.3 Notations

We use the following notations to describe our protocols:

- B, S : the identities of buyer and seller, respectively.
- MP_i : the identity of i-th Master Peer, where $1 \leq i \leq n$.
- f : a flag that indicates the purpose of a message.
- $item_X$: an item of the peer X.
- pay_X : a payment information of the peer X.
- $desc_{item_X}, desc_{pay_X}$: the description of the item and the payment of the peer X, respectively.
- t_X : the local timestamp value of the peer X.
- com_X : a randomly chosen commitment value by the peer X.
- $DTTP$: a set of Master Peer's identities.

$$DTTP := \{MP_1, \cdots, MP_n\}$$

- PH : the protocol header, which contains relevant information such as the identities of the peers involved, the description of the desired item and payment.

$$PH := \{B, S, DTTP, desc_{item_X}, desc_{pay_X}\}$$

- $H()$: a collision resistant one-way hash function.
- K : a randomly chosen secret key for symmetric-key encryption function.
- $E_K()$: a symmetric-key encryption function under secret key K.
- $C := E_K(item_X)$: the cipher of $item_X$ under secret key K.
- $Sig_X()$: a signature function under X's private key.
- $PU_X()$: an asymmetric-key encryption function under X's public key.
- $PD_X()$: an asymmetric-key decryption function under X's private key.
- $X \rightarrow Y$: m : message m is sent from a peer X to a peer Y.
- $X \rightarrow \forall Y_i$: m : message m is broadcasted from a peer X to every peer Y_i, where $1 \leq i \leq n$.
- $\forall X_i \rightarrow Y$: m : message m is sent from every peer X_i to a peer Y, where $1 \leq i \leq n$.

3.4 Membership Enrollment Protocol

Every peer who wishes to buy and sell digital contents in a peer community needs to affiliate himself with the peer community. Figure 1. describes the detailed steps of the protocol.

Step 1. A prospective peer P_{new} who wishes to perform e-commerce transaction generates his own public key/private key pair, and constitutes a membership credential request message to enroll in the peer community. Then the prospective peer broadcasts the credential request message to all the master peers.

$$[\text{E-1}] \quad P_{new} \rightarrow \forall MP_i \; : \; Sig_{P_{new}}(f_{EnrollReq}, P_{new}, t_{P_{new}}, PK_{P_{new}})$$

Step 2. Each master peer verifies the received [E-1]. Each MP_i who want to approve enrollment of the peer community for the prospective peer computes a partial signature $Sig_{ss_i}(f_{Enrolled}, P_{new}, t_{P_{new}}, PK_{P_{new}})$ with its service secret share ss_i, then sends confirmation of enrollment to the prospective peer.

$$[\text{E-2}] \quad \forall MP_i \rightarrow P_{new} \; : \; Sig_{MP_i}(Sig_{ss_i}(f_{Enrolled}, P_{new}, t_{P_{new}}, PK_{P_{new}}))$$

Step 3. To generate a valid membership credential, the prospective peer needs at least $t+1$ correct partial signatures. Hence, the prospective peer chooses $t+1$ correct partial signatures, and finally obtains the membership credential $Cre_{P_{new}} = Sig_{DTTP}(f_{Enrolled}, P_{new}, t_{P_{new}}, PK_{P_{new}})$ that can be used to prove admission of the peer community. Finally, the peer broadcasts its own credential to all master peers

---

Fig. 1. Membership Enrollment Protocol

After becoming a member of the peer community, the peer who plays the role of seller can broadcast the information of its digital contents and its membership credential to all other peers of the peer community at any time.

Finally, common issues associated with peer community that we have to consider are a peer community policy, an advertisement of digital contents and payment mechanisms. However, it remains beyond the scope of this work.

4 Optimistic Fair Exchange Protocol with Distributed TTP

In this section, we present and analyze *an optimistic fair exchange protocol with Distributed TTP*, which is used for guaranteeing the fairness and the reliability in our P2P e-commerce model.

The proposed protocol is composed of three sub-protocols: *the main protocol, the abort protocol, the recovery protocol*. The main protocol consists of messages exchanged directly between a buyer and a seller. In case of problematic happening during this main protocol, two possibilities are offered to the parties.

Either the buyer can execute the abort protocol in order to cancel the exchange, or the buyer(or the seller) can launch the recovery protocol to complete the exchange.

4.1 Main Protocol

We assume that a buyer has already obtained the description of the desired item and all parties agree on the DTTP to be possibly invoked in case of conflict. When a buyer wishes to receive the desired item from a seller against a payment of the item, the buyer can launch the main protocol. The detailed steps are described in Figure 2.

Step 1. A buyer who wants to perform e-commerce transaction constitutes a protocol header PH. The buyer also selects a commitment value com_B and a timestamp value t_B, then computes $H(pay_B), H(com_B), PU_{DTTP}(pay_B)$. The buyer configures a purchasing message including all above parameters and signs the purchasing message, then sends it with her credential to the seller as [M-1].

$$[\text{M-1}] \quad B \rightarrow S \ : \ Sig_B(PH, H(pay_B), H(com_B), t_B, PU_{DTTP}(pay_B)), Cre_B$$

Step 2. The seller who receives [M-1] checks whether the signature of purchasing message is valid. If the check is invalid, the seller *quits* the exchange. Otherwise the seller constitutes the protocol header $\overline{PH}$, then chooses a random secret key K and computes $C, H(item_S, K), PU_{DTTP}(K)$. The seller forms a selling message and signs the selling message, then sends it to the buyer as [M-2].

$$[\text{M-2}] \quad S \rightarrow B \ : \ Sig_S(\overline{PH}, H(item_s, K), C, PU_{DTTP}(K))$$

Step 3. After having checked the validity of the received message in step 2, the buyer sends $PU_S(pay_B, com_B)$ together with its signature on those information to the seller as [M-3]. If the validity of [M-2] is not satisfied, or the buyer gives up receiving the [M-2] message, then the buyer runs *the abort protocol*.

$$[\text{M-3}] \quad B \rightarrow S \ : \ Sig_B(PU_S(pay_B, com_B))$$

Step 4. The seller checks the validity of [M-3]. If the check is valid, the seller obtains the desired payment information pay_B. The seller sends the encrypted secret key $PU_B(K)$ to the buyer together with its signature. If any problem occurs in above process, the seller may *quit* the protocol.

$$[\text{M-4}] \quad S \rightarrow B \ : \ Sig_S(PU_B(K))$$

Step 5. After receiving the [M-4] message from the seller, the buyer verifies the signature and obtains the desired item by using the secret key K. If the validity of the received message is incorrect or the buyer gives up finishing the protocol, then launches *the recovery protocol*.

Fig. 2. Main Protocol

The protocol headers are constituted of both parties, PH and $\overline{PH}$, contain not only the identities of the parties involved, but also the description of the desired item and payment, respectively. Hence, each protocol header has to be checked, by both parties, to confirm the correctness of information relevant to the protocol.

The use of the commitment com_B, in steps 1 and 3, prevents a malicious seller from launching the recovery protocol without sending the second message to a buyer. Unless receiving commitment com_B, the DTTP does not run the recovery protocol to resolve the conflict.

Timestamp t_B is used to identify the execution for buyer requests. Timestamps for buyer's requests are totally ordered such that later requests have higher timestamps than earlier ones, e.g., the timestamp could be the value of the buyer's local clock when the request is issued.

4.2 Abort Protocol

If the seller does not send the second message of the main protocol, the buyer can collaborate with DTTP in order to abort the protocol. The detailed steps are described in Figure 3.

By using fair communication, the buyer periodically repeats step 1 until it receives sufficient [A-2] messages as the response to its abort request. In fact, the buyer can try to compute the abort token as soon as it has received $t+1$ partial signatures from master peers. So, the buyer has to wait for more partial signatures only if some partial signatures it received are incorrect.

Our protocol has been designed by considering threshold RSA schemes because threshold schemes based on discrete logarithms may require an agreement upon random number to generate partial signature. Furthermore, threshold RSA scheme can be applicable to threshold decryption. Since the validation of partial

---

Step 1. The buyer broadcasts an abort request and her credential to all the master peers.

[A-1] $B \to \forall MP_i \ : \ Sig_B(f_{AbortReq}, t_B, [\text{M-1}]), Cre_B$

Step 2. Each master peer verifies the received [A-1]. If [A-1] is correct, each master peer computes partial signature $Sig_{ss_i}(f_{Aborted}, t_B, [\text{M-1}])$ with its service secret share ss_i, then sends an abort confirmation to the buyer.

[A-2] $\forall MP_i \to B \ : \ Sig_{MP_i}(Sig_{ss_i}(f_{Aborted}, t_B, [\text{M-1}]))$

Step 3. To generate a valid signature of DTTP, the buyer needs at least $t+1$ correct partial signatures. Hence, the buyer chooses $t+1$ correct partial signatures, and computes an abort token $Sig_{DTTP}(f_{Aborted}, t_B, [\text{M-1}])$. This abort token can be used to guarantee the fairness in case of potential dispute.

Fig. 3. Abort Protocol

signature depends on the underlying threshold scheme, the buyer can check the validation of partial signature by means of applying threshold RSA schemes that provide the *robustness*[5][10] to our protocol.

4.3 Recovery Protocol

If the seller does not send her final message of the main protocol, the buyer can launch the recovery protocol by means of collaborating with DTTP in order to complete the exchange. Figure 4. describes the detailed steps of the recovery protocol.

Since the recovery protocol is performed in the same manner as the abort protocol by using fair communication, the buyer periodically repeats step 1 until it receives sufficient [R-2-B] messages. Also, each master peer who intervenes in the recovery protocol periodically resends the recovery information to the seller until it receives the acknowledgment of [R-2-S] from the seller.

Step 1. The buyer broadcasts the received [M-1],[M-2] and her commitment com_B along with her signature to all the master peers.

$$[\text{R-1}] \quad B \to \forall MP_i \;:\; [\text{M-1}], [\text{M-2}], Sig_B(f_{RecoverReq}, t_B, com_B)$$

Step 2. Each master peer checks all the validity of received [R-1]. If the check is valid, each master peer performs the followings:
- To complete the exchange for the buyer, each master peer generates partial decryption $PD_{ss_i}(PU_{DTTP}(K))$ of the secret key with its service secret share ss_i, then sends recovery information to the buyer.

$$[\text{R-2-B}] \quad \forall MP_i \to B \;:\; Sig_{MP_i}(f_{Recovered}, t_B, PD_{ss_i}(PU_{DTTP}(K)))$$

- Also, each master peer computes partial decryption $PD_{ss_i}(PU_{DTTP}(pay_B))$ of the payment information with its service secret share ss_i, then sends corresponding information to the seller.

$$[\text{R-2-S}] \quad \forall MP_i \to S \;:\; [\text{M-1}],[\text{M-2}],$$
$$Sig_{MP_i}(f_{Recovered}, t_B, com_B, PD_{ss_i}(PU_{DTTP}(pay_B)))$$

Step 3. Finally, Each buyer and seller performs the followings, respectively.
- To generate the secret key K, the buyer chooses $t+1$ correct partial decryptions, and computes the secret key K. Therefore, the buyer can obtain the desired item by using secret key K.
- The seller selects $t+1$ correct partial decryptions, then obtains the desired payment with respect to her item. Then the seller sends $Sig_S(f_{Recovered}, t_B, PD_{ss_i}(PU_{DTTP}(pay_B)))$ as acknowledgment of [R-2-S] to all master peers corresponding to received message.

Fig. 4. Recovery Protocol

The seller does not engage in the recovery protocol with DTTP in the main protocol, basically the seller needs not launch the recovery protocol for assuring fairness. However, the seller is able to recognize the activity of recovery caused by receiving [R-2-S] message when the buyer runs the recovery protocol. Thus, if the seller does not receive sufficient information to generate the desired payment information in desired amount of time, the seller can launch the recovery protocol together with commitment com_B,[M-1],[M-2] within [R-2-S] for assuring her fairness.

4.4 Analysis

Here we give an analysis of our fair exchange protocol, checking the requirements described in Section 2, and then we discuss additional desirable property provided by our protocol. Our claim is as follows:

Claim. *The optimistic fair exchange protocol with distributed TTP is a fair exchange protocol which provides fairness, timeliness, effectiveness, authentication, confidentiality, integrity, and non-repudiation.*

Proof Sketch. Clearly our protocol provides authentication, non-repudiation, and integrity by means of the signatures of each communication parties on the exchanging messages and the hash values of $H(pay_B)$ and $H(item_S, K)$ in [M-1] and [M-2], respectively. Furthermore, these hash values can be used for potential dispute resolution.

Regarding confidentiality, it is sufficient to prove that: any master peer which belongs to DTTP cannot open $PU_{DTTP}(pay_B)$ or $PU_{DTTP}(K)$ while intervening in the exchange. Since any master peer has not entire service secret key, but has service secret key share ss_i through the threshold scheme, it is possible for any master peer to open $PU_{DTTP}(pay_B)$ or $PU_{DTTP}(K)$ if and only if it must conspire with at least $t+1$ other master peers.

It is obvious that both parties obtain the expected items if the main protocol is executed without errors. Therefore, our protocol provides effectiveness.

Before proving the fairness and the timeliness of our protocol, let us consider the *availability* of the entire DTTP in terms of fair communication model that is applied to a peer and an available master peer. In contrast to previously presented fair exchange protocols that assume a robust central TTP in terms of resilient communication model between a party and the TTP, communication channel among all parties is *really unreliable* and *no robust central server* are offered in P2P network. To overcome the nature of P2P network, our protocol is based on collaboration with distributed communication parties for guaranteeing fairness by the use of threshold scheme. This feature inherently implies that any single party is not wholly entrusted to guarantee the desired fairness. Hence, regarding the availability of the DTTP, it is sufficient to show that: a peer who wishes to contact the DTTP should eventually receive enough information from any available subset of the DTTP and stop retransmitting its own requests. Since we have assumed that the entire DTTP contains at least $2t+1$ *available* master peers at any time, all peers are able to eventually contact at least $2t+1$ master peers among DTTP, and further, obtain the desired information through sufficient number of retransmissions. Consequently, the entire DTTP is *always available* in terms of fair communication model.

Now let us prove the fairness and the timeliness of our protocol. When regarding timeliness, we consider three situations:

1. The main protocol ends up successfully without any time-out.
2. The buyer aborts the protocol and receives the abort confirmation signed by DTTP within a time period which may be arbitrarily long, yet finite amount of time.
3. The buyer(or if necessary, the seller) has the ability to launch the recovery protocol to complete the exchange, and eventually receives the desired item in a finite period of time.

Therefore, our protocol provides timeliness.

Finally, let us show the fairness of our protocol for both the seller and the buyer. We start by proving the fairness of the seller.

1. In the main protocol the seller does not basically need to engage in both the abort protocol and the recovery protocol for assuring fairness, because the seller sends the secret key to the buyer after receiving the desired payment information.
2. Also, if the buyer starts the recovery protocol to complete the exchange, the seller can recognize the activity of the recovery. In this case, the seller may receive sufficient information to generate the desired payment information from DTTP, otherwise he can launch the recovery protocol to complete the exchange.

For the fairness of the buyer, we analyze the following case in which the buyer does not obtain the desired item $item_S$.

1. If the seller stops the main protocol after receiving the [M-3] message, the buyer can perform the recovery protocol with collaborating DTTP in order to compute the secret key K. All information sent to the buyer by DTTP may be eventually arrived as our communication model.
2. If the seller does not send the [M-2] message to the buyer, the buyer can launch the abort protocol through collaborating with DTTP to obtain the abort token which can be used in case of potential conflict.
3. Also, we note that the seller can not perform the recovery protocol without the commitment com_B as discussed earlier. The seller can launch the recovery protocol to complete the exchange if and only if the buyer has launched the recovery protocol in advance. So, in this case, it will never happen that the seller gains pay_B while the buyer does not receive $item_S$.

Therefore, our protocol provides fairness. □

Finally, our protocol provides additional interesting property that a seller does not basically need to engage in both the abort protocol and the recovery protocol in order to guarantee her fairness. Therefore, the seller does not need to maintain state information regarding the transaction in the main protocol. This feature makes our protocol more practical in e-commerce environments in which seller would be prefer to involve in commercial transactions rather than being involved by buyer.

5 Conclusion

In this paper, we have presented a fair and reliable e-commerce model suitable for P2P network, in which communication parties can buy and sell digital contents by P2P contact. In particular, we have proposed and analyzed a new optimistic fair exchange protocol with distributed TTP which is used to guarantee the fairness and the reliability for presented P2P e-commerce model.

Compared with the traditional fair exchange protocols that are required a central trusted authority for providing fairness and reliability, our protocol does not require any central trusted authority since it guarantees fairness and reliability by means of collaboration with distributed community parties. Consequently, our protocol is very attractive in P2P networking environment which does not naturally depend upon any central trusted authority for managing communication parties.

References

1. N. Asokan, V. Shoup, and M. Waidner: Asynchronous protocols for optimistic fair exchange. In Proceeding of the IEEE Symposium on Research in Security and Privacy, May (1998).
2. N. Asokan, V. Shoup, and M. Waidner: Optimistic fair exchange of digital signatures. In Proc. Eurocrypt'98, LNCS 1403, pp. 591-606, (1998).
3. D. Boneh, M. Franklin: Efficient generation of shared RSA keys. In Proceedings Crypto'97, pp.425-439, (1997).
4. M. Castro and B. Liskov: Practical Byzantine fault tolerance. In Proc. the 3rd USENIX OSDI'99, pp.173-186, (1999).
5. R. Gennaro, S. Jarecki, H. Krawczyk, and T. Rabin: Robust and efficient sharing of RSA functions. In Proc. Crypto'96, LNCS 1109, pp.157-172, (1996).
6. R. Housley, W. Ford, W. Polk, D. Solo: Internet X.509 Public key infrastructure certificate and CRL profile, RFC 2459. January (1999).
7. T. Iwao, Y. Wada, S. Yamasaki, M. Shiouchi, M. Okada, and M. Amamiya: A Framework for the Next Generation of E-Commerce by Peer-to-Peer Contact. IEEE WET ICE 2001, (2001).
8. O. Markowitch and S. Saeednia: Optimistic Fair Exchange with Transparent Signature Recovery. In Proc. Financial Cryptography 2001, LNCS 2339, pp. 339-350, (2002).
9. Tal Rabin: A Simplified Approach to Threshold and Proactive RSA. Advances in Cryptology-CRYPTO'98, LNCS 1462, pp. 89-104, (1998).
10. Victor Shoup: Practical threshold signatures. In Proc. Eurocrypt 2000, LNCS 1807, pp.207-220, (2000).
11. Holger Vogt: Asynchronous Optimistic Fair Exchange Based on Revocable Items. In Proc. Financial Cryptography 2003, LNCS 2851, pp. 193-207, October (2003).

An Efficient Access Control Model for Highly Distributed Computing Environment

Soomi Yang

The University of Suwon,
Kyungki-do Hwasung-si Bongdam-eup Wau-ri san 2-2,
445-743, Korea
smyang@suwon.ac.kr

Abstract. For a secure highly distributed computing environment, we suggest an efficient role based access control using attribute certificate. It reduces management cost and overhead incurred when we change the specification of the role. In this paper, we grouped roles and structured them into the role group relation tree. It results in secure and efficient role updating and distribution. For scalable role specification certificate distribution, multicasting packets are used. We take into account the packet loss and quantify performance enhancements of structuring role specification certificates.

1 Introduction

Traditional access control mechanisms are inherently centralized and existing attempts to distribute the functionality suffer from problems of scalability. Our access control is a new distributed access control paradigm designed for a highly distributed computing environment. It defines a hierarchical access control mechanism, which relies exclusively on role based access control using specific attribute certificate. It is particularly designed to operate in un-trusted environments where the lack of global knowledge and control are defining characteristics. Due to the lack of central control, the autonomous entities form trust relations [3]. They can be chained to represent recommendations and the propagation of trust.

For scalability, we use multicast for group communication. It makes distribution of role specifications faster. In the experimental section, we will show the performance enhancements gained.

This paper is organized as follows. Section 2 gives a brief overview of related work. Section 3 describes the secure role group model with group communication. Section 4 shows the performance of our method. Section 5 concludes this paper.

2 Related Work

In [1], D. Ferraiolo et al. modeled RBAC (Role Based Access Control) as combinations of user, role, permission, administrator and others. They also gave it the priority relation. Following their research, many variants were suggested. However they dealt with the group of subjects only. No research considers group of roles. In [6], J. Joshi et al.

introduces a temporal privilege delegation. It provides flexible permission delegation for dynamically changing environments. However it did not consider the group of roles or multicasting. Our method distributes the role specifications according to the levels of access. It accords with the characteristics of the distributed environments and sometimes it is inevitable. So our method is different from the privilege delegation. It can be thought of as the distribution of privileges in groups of roles.

In relation to security of highly distributed computing, multicasting packets are used mainly for distribution of cryptographic keys [7]. We applied the ideas for distribution of attribute certificates.

3 Secure Role Group Model

The secure role group is an extended version of the secure group [7]. It consists of a finite and nonempty set of role groups, a finite and nonempty set of permissions and there exists a binary relation between the set of role group and the set of permission.

According to the ITU-T X.509 Recommendation (ISO/IEC 9594-8) [2] Attribute Certificate (AC) is composed of version, holder, issuer, signature, serialNumber, attrCertValidityPeriod, attributes, issuerUniqueID, extensions. IETF RFC 3281[4] defines AC similarly. AC fields match PKC (Public Key Certificate) fields which are composed of version, serialNumber, signature, issuer, validity, subject, subjectPublicKeyInfo, issuerUniqueIdentifier, extensions. AC and PKC should be related through holder and subject. You can find specific descriptions of each field in [2], [4]. We need to make it simple for terse explanation by including the related fields only. So in the following explanations, we will use the abbreviated figures as shown in Fig. 1 and 2.

The holder field conveys the identity of the attribute certificate's holder. It should match to the subject field of PKC. In the roles model of PMI (Privilege Management Infrastructure), role name shall also appear in the holder field of the role specification certificate.

The roles model [1], [2], [4] provides a means to indirectly assign privileges to individuals. Individuals are issued role assignment certificate that assign one or more roles to them through the role attribute contained in the certificate. Specific privileges are assigned to a role name through role specification certificate. This level of indirection enables the privileges assigned to a role to be updated, without impacting the certificates that assign roles to individuals.

In role extensions field, if a certificate is a role assignment certificate, a privilege verifier needs to be able to locate the corresponding role specification certificate. So the role name used as a role specification certificate identifier would be the same as that in the holder component of the role specification certificate being referenced by this extension. Role certificate serial number or role certificate locator can do the same function. We propose the extension that role extension fields can be included in role specification certificate as well as in role assignment certificate. It makes a chain of role specification certificates.

We group roles and structure in a role specification certificate. The role groups are different from the subject groups. And its structure differs from the delegation of roles. It gathers common roles and builds the trust structure similar to [3]. The chain of role specification certificates can incur the overhead when a subject is going to use some privileges. The problem can be solved through the use of coherent caching of role specification certificates [5]. In highly distributed environments the distribution of the specifications of roles is inevitable. We consider the case of updating the roles, specifically changing the role specification certificates.

Attribute certificates or public key certificates can be used as role assignment certificates. When the public key certificates are used, the extensions field should have the information about role specification certificates.

On the other hand if the attribute certificates are used, the attribute certificate should have the contents as shown in Fig. 1.

In other words holder field has pkc subject and attribute field has roles and extensions field has the information for the role specification certificates. According to Fig. 1 role specification certificate should have the structures as shown in Fig. 2.

For the role specification certificates shown in Fig. 2, extensions field can have another role specification certificate information repeatedly such as role name or serial number. It forms the tree structure as shown in Fig. 3.

Field Name	Holder	attributes	extensions
Content	pkc subject	role information	role specification certificate information

Fig. 1. Contents of a role assignment certificate

Field Name	holder	attributes	extensions
Content	role name	role information	role specification certificate information

Fig. 2. Contents of a role specification certificate

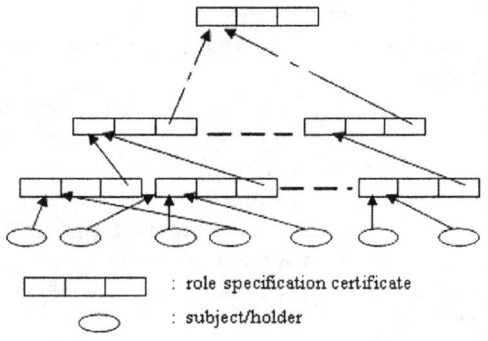

Fig. 3. Role grouping of a role assignment certificate

We call the node that corresponds to the role specification certificate having child role specification certificate as role group. Although there should be overhead incurred when privileges are applied, it can be overcome by the use of caching. But if the nodes are distributed geographically, the performance enhancements gained when the role specifications should be changed are overwhelming. We are going to show the performance gains quantitatively in Section 4.

If role group notion is not used, in Fig. 3, role holder should possess all the upper level role specifications. In that case, the application of the role can be done directly, but each holder/subject should have all the role specification certificates required and the small memory devices used in ubiquitous computing environment cannot afford it.

Updated role specification certificates are delivered by the multicast communication. The distribution of updated role specification certificates of our method is modeled as having R roles, G role groups constructing the tree structure of height h and degree d. In general, the roles are included in subset of role groups. Thus, an unnecessary role group creation can be avoided by determining the proper value of h. From the viewpoint of the reliable delivery, a role specification certificate at level l of the tree structure has to be delivered to d^{h-l} receivers. If the roles are grouped, it needs to be delivered to only d members. Let M be the number of times a role specification certificate will need to be transmitted in order to successfully deliver it to all the related receivers. The probability that one of the receivers will not receive the updated role specification if it is transmitted once is equal to the probability of packet loss, p, for that receiver, since all the packet loss events for some receiver, including replicated packet and retransmissions, is mutually independent and is geometrically distributed. Thus the probability that the role specification certificate is delivered successfully within m packet transmissions is $1-p^m$. Thus the expected number of packet transmission is $1/(1-p)$. Since lost packet events at different receivers are independent of each other, the probability that all the receivers will receive the packet within m transmissions is $(1-p^m)^{\#receivers}$. Thus the average expected role specification packet transmission time is $\sum_{m=1}^{\infty}(1-(1-p^{m-1})^{\#receivers})$. We can compute it by truncating the summation when the mth value falls below the threshold.

4 Performance Evaluation

For each given packet loss p, we examine the average packet transmission for the various values of threshold. We used Visual Studio and Gnuplot. Fig. 4 shows the impact of packet loss p on the average packet transmission E when $m=10$. When roles are not grouped (ung-diff-pm2-m10.dat) E results in higher value. However when roles are grouped (g-diff-pm2-m10.dat) E results in lower value.

In Fig. 5 we plot the expected packet transmission E for packet loss p and the degree difference $(h-l)$. For better readability we plot two dimensional graph for the case of $p=0.04, 0.1, 0.2$. In Fig. 5 we can see that E shows great increase when the roles are not grouped (ung-*.dat) and it shows very little increase when the roles are grouped (g-*.dat). So we can see that when the quality of network is more inferior (so p is greater) the performance enhancements obtained through role grouping becomes greater.

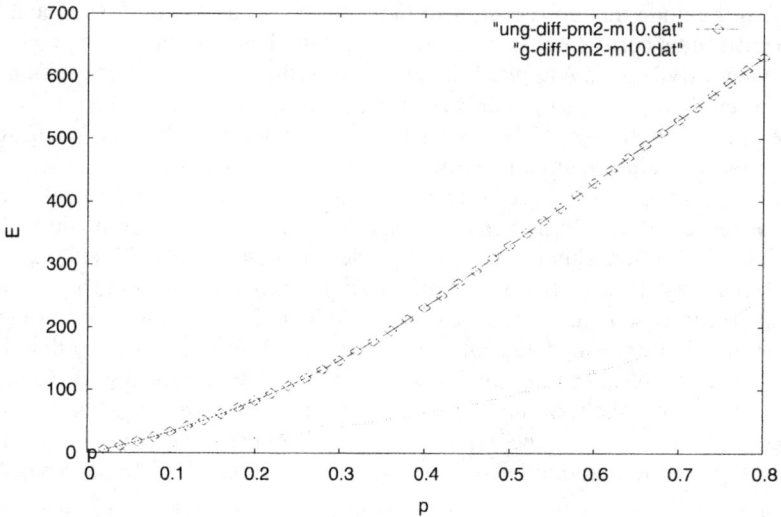

Fig. 4. A comparison of the expected packet transmission as a function of p and m

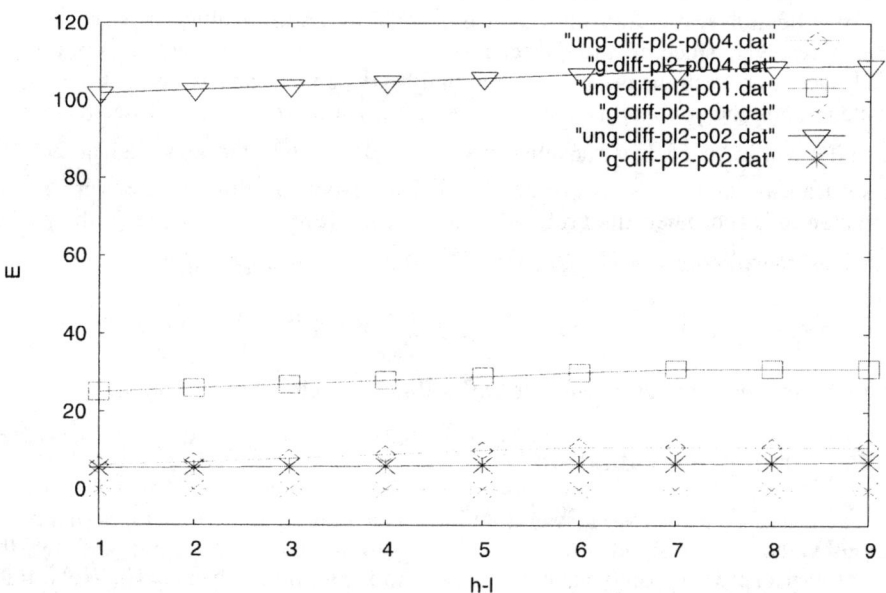

Fig. 5. A comparison of the expected packet transmission as a function of p and h-l

5 Conclusion

For efficient access control considering the characteristics of highly distributed computing environment, we adopt the trust model. As an efficient access control using attribute certificate, we use the technique of structuring role specification certificates. It can reduce the management cost and overhead incurred when changing the specification of the role. Highly distributed computing environments such as ubiquitous computing that cannot have global knowledge and control, need another attribute certificate management technique. Therefore we group roles and make the role group relation tree. It results in secure and efficient role updating and the distribution of role specification certificates. For scalable role specification certificate distribution, multicasting packets are used. We took into account the packet loss to some large values of unreliable network and quantified performance enhancements. And we showed that our scalable access control technique improved the existing access control techniques.

References

1. D. Ferraiolo, R. Sandhu, S. Bavrila, D. Kuhn and R. Chandramouli: Proposed NIST Standard for Role-Based Access Control, ACM Trans. Info. and Syst. Security, 4 (3), (2001)
2. ITI, Role Based Access Control ITU/T. Recom. X.509 I ISO/IEC 9594-8, ITOSI-The Directory: Public-Key and Attribute Certificate Frameworks (2003)
3. C. English, P. Nixon, S. Terzis, A. McGetrtrick and H. Lowe: Dynamic Trust Models for Ubiquitous Computing Environments, Workshop on Security in Ubiquitous Computing (UBICOMP 2002)
4. S. Farrell and R. Housley: An Internet Attribute Certificate Profile for Authorization, IETF RFC 3281, (2002)
5. S. Yang: Role Based Access Control Supporting Coherent Caching of Privilege Delegation Which Utilizes Group Key. The Journal of Suwon Information Technology, 3 (2004)
6. J. Joshi, E. Bertino, A. Ghafoor: Temporal hierarchies and inheritance semantics for GTRBAC, Proc. of the 7th ACM Symp. Access control models and technologies (2002)
7. C. Wong, M. Gouda and S. Lam: Secure Group Communications Using Key Graphs, IEEE/ACM Trans. Networking 8 (1) (2000)

Cryptanalysis and Improvement of a Multisignature Scheme*

Manik Lal Das[1], Ashutosh Saxena[1], and V.P. Gulati[2]

[1] Institute for Development and Research in Banking Technology,
Castle Hills, Road No.1, Masab Tank, Hyderabad 500057, India
{mldas, asaxena}@idrbt.ac.in
[2] Tata Consultancy Services, Software Units Layout,
Madhapur, Hyderabad 500081, India
vp.gulati@tcs.com

Abstract. A multisignature scheme for implementing safe delivery rule in group communication systems (MSGC) was recently proposed by Rahul and Hansdah. In this paper we show that the MSGC scheme is insecure against forgery attack and signature integrity attack. We propose an improved scheme that resists the weaknesses of MSGC scheme.

1 Introduction

A multisignature is a digital signature that allows multiple signers to generate a signature in sequential and/or parallel manner. For example, an approval requires signatures in a sequential manner, whereas, signing a contract by two or more parties is an example of parallel multisignature. In 1983, Itakura and Nakamura [4] first introduced the notion of multisignature. Since then, several schemes and improvements have been proposed [2], [3], [8], [9] for multisignatures; however, a formal security model on multisignature was absent until the work by Micali et al. [7]. Afterwards, Lin et al. [5] and Boldyreva [1] generalized the security notion of multisignatures.

Recently, Rahul and Hansdah [10] proposed a multisignature scheme for implementing safe delivery rule in group communication systems (MSGC). They claimed that the MSGC scheme can be used in client-server model for safe delivery rule in group communication systems. In this paper, we show that the MSGC scheme is insecure against forgery and signature integrity attacks. We give two scenarios where an adversary can easily forge individual partial signatures or multisignature without the knowledge of signers' private key. Moreover, the adversary can modify partial signatures or multisignature without being detected by the verifier. We present an improved scheme that resists the weaknesses of MSGC scheme. The rest of the paper is organized as follows. In the next Section, we review the MSGC scheme and show its vulnerability in Section 3. We present an improved scheme in Section 4. Finally, we conclude the paper in Section 5.

---

* The research was supported in part by the Ministry of Communications and Information Technology, Govt. of India, under the grant no. 12(35)/05-IRSD:18/Jan/2005.

2 The MSGC Scheme [10]

The following are the notation used in the rest of the paper.

pq	product of two primes p and q.
g	$g < pq$ and has a large order.
α_i	private key of the party M_i.
$x_i \equiv g^{\alpha_i} \bmod pq$	public key of the party M_i.
$H(\cdot)$	a collision–resistant hash function [6].

The public parameters are $pq, g, x_i, H(\cdot)$, and the private key is α_i. The scheme consists of two phases, namely the partial signature generation phase and the partial signature aggregation phase. The phases work as follows.

2.1 Partial Signatures Generation

Suppose the group consists of n members. For $i = 1, 2, \cdots, n$, the member M_i creates partial signature on message m as follows:

1. Select a random number r_i
2. Compute $s_i = \alpha_i + H(m)r_i$
3. Compute $t_i = g^{r_i} \bmod pq$
4. The tuple (s_i, t_i) is M_i's partial signature on m.

Now, the member M_i sends (s_i, t_i) to the group communication system (GCS). Then, the GCS verifies the partial signature by whether $g^{s_i} \equiv x_i \cdot t_i^{H(m)} \bmod pq$.

2.2 Partial Signatures Aggregation

The GCS combines the partial signatures of the group members into a multisignature (s, t) as follows:

1. Compute $s = \sum_{i=1}^{n} s_i$
2. Compute $t = \prod_{i=1}^{n} t_i \bmod pq$.

The GCS sends (s, t) to the sender of m. The validity of (s, t) on m is verified by checking whether $g^s \equiv x \cdot t^{H(m)} \bmod pq$, where $x = \prod_{i=1}^{n} x_i \bmod pq$.

2.3 Correctness of Multisignature

$$g^s \equiv g^{\sum_{i=1}^{n} s_i} \bmod pq \equiv \prod_{i=1}^{n} g^{s_i} \bmod pq \equiv \prod_{i=1}^{n} g^{\alpha_i + H(m)r_i} \bmod pq$$
$$\equiv \prod_{i=1}^{n} x_i \cdot t_i^{H(m)} \bmod pq$$
$$\equiv x \cdot t^{H(m)} \bmod pq.$$

3 Cryptanalysis of the MSGC Scheme

In this section, we show that the MSGC scheme [10] suffers from forgery attack and signature integrity attack.

3.1 Forgery Attacks

In the MSGC scheme, a valid partial signature on message m is the tuple (s_i, t_i), where s_i is computed by the signer's private key α_i. The verification algorithm needs signer's public key x_i to prove that the signature was created by the signer.

Forgery attack on partial signatures: An adversary[1] can forge any member's (say M_i) partial signature by the following computations:

1. Select a random number R
2. Compute $\sigma_i = H(m)R$
3. Compute $T_i = (x_i)^{-\frac{1}{H(m)}} \cdot g^R \bmod pq$
4. The partial signature on m is the tuple (σ_i, T_i).

Now, the adversary sends (σ_i, T_i) to the GCS. The GCS validates the partial signature by verifying whether $g^{\sigma_i} \equiv x_i \cdot T_i^{H(m)} \bmod pq$.
We note that the signature was generated without signer's private key and the verification still holds good by the following checks:

$$g^{\sigma_i} \equiv (g^R)^{H(m)} \bmod pq \equiv (x_i^{\frac{1}{H(m)}} T_i)^{H(m)} \bmod pq \equiv x_i \cdot T_i^{H(m)} \bmod pq.$$

Forgery attack on multisignature: The above attack is also applicable to the combined partial signatures (i.e., multisignature). The attack works as follows:

1. Select a random number R
2. Compute $\sigma = H(m)R$
3. Compute $T = (x)^{-\frac{1}{H(m)}} \cdot g^R \bmod pq$
4. The tuple (σ, T) is the multisignature of m.

Correctness: $g^{\sigma} \equiv (g^R)^{H(m)} \bmod pq \equiv (x^{\frac{1}{H(m)}} T)^{H(m)} \bmod pq \equiv x \cdot T^{H(m)} \bmod pq.$

3.2 Attack on Signature Integrity

Here we show that if an adversary modifies a valid partial signature (s_i, t_i), the verification algorithm is unable to detect the modified signature. Thus, the signature does not posses signature integrity. The attack works as follows:

1. Select a random number R
2. Compute $\sigma_i = s_i + H(m)R$
3. Compute $T_i = t_i \cdot g^R \bmod pq$
4. The partial signature on m is the tuple (σ_i, T_i).

---

[1] The adversary could be any third party or any malicious group members.

The adversary sends (σ_i, T_i) to the GCS. The GCS validates the partial signature by verifying whether $g^{\sigma_i} \equiv x_i \cdot T_i^{H(m)} \mod pq$. The verification holds as follows:

$$g^{\sigma_i} \equiv g^{s_i} \cdot g^{R \cdot H(m)} \mod pq \equiv x_i \cdot t_i^{H(m)} g^{R \cdot H(m)} \mod pq \equiv x_i \cdot T_i^{H(m)} \mod pq.$$

We note that the same attack is also applicable to the combined partial signatures (i.e., multisignature).

4 An Improvement and Analysis

We present an improved scheme which can resist the weaknesses of the MSGC scheme. The improved scheme is based on Schnorr's signature [11], where the security assumption is based on the hardness of the discrete logarithm problem. The improved scheme works as follows.

For $i = 1, 2, \cdots, n$, the group member M_i selects a random number r_i, computes $t_i = g^{r_i} \mod pq$ and broadcasts t_i.

Then, all members and GCS compute $t = \prod_{i=1}^{n} t_i \mod pq$. Now, M_i computes his partial signature on message m as $s_i = r_i \cdot t + \alpha_i \cdot H(m, t)$.

The tuple (s_i, m) is the partial signature of M_i on m. The member M_i sends (s_i, m) to the GCS. The GCS validates the partial signature by verifying whether

$$g^{s_i} \equiv t_i^t \cdot x_i^{H(m,t)} \mod pq.$$

After validation of all partial signatures, the GCS computes $s = \sum_{i=1}^{n} s_i$ and sends multisignature tuple (s, t) to the sender of m. The validity of (s, t) is checked by whether

$$g^s \equiv t^t \cdot x^{H(m,t)} \mod pq. \quad (1)$$

If it holds, the signature is valid, else invalid.

Correctness:

$$g^s \equiv g^{\sum_{i=1}^{n} s_i} \mod pq$$
$$\equiv \prod_{i=1}^{n} g^{s_i} \mod pq$$
$$\equiv \prod_{i=1}^{n} g^{r_i \cdot t + \alpha_i \cdot H(m,t)} \mod pq$$
$$\equiv \prod_{i=1}^{n} t_i^t \cdot x_i^{H(m,t)} \mod pq$$
$$\equiv t^t \cdot x^{H(m,t)} \mod pq.$$

4.1 Analysis

We show that the improved scheme withstands the weaknesses of the MSGC scheme.

Security against Forgery attack: In order to forge any M_i's partial signature, the adversary has to perform the following computation in line with the attack mentioned in Section 3.1 without the knowledge of the private key α_i:

1. Select a random number R
2. Compute $\sigma_i = H(m,t)R$
3. Compute $T_i = (x_i)^{-\frac{1}{H(m,t)}} g^R \mod pq$.

The signature tuple (σ_i, m) fails to validate the verification algorithm eq. (1) of Section 4, because

$$g^{\sigma_i} \equiv g^{H(m,t)R} \mod pq \equiv (T_i \cdot x_i^{\frac{1}{H(m,t)}})^{H(m,t)} \mod pq \equiv T_i^{H(m,t)} \cdot x_i \mod pq,$$

which is incorrect for a valid partial signature. For successful forgery, the adversary needs to select R such that $t_i^{\frac{t}{H(m,t)}} x_i \equiv g^R \mod pq$, but this leads to solve the discrete logarithm problem (DLP) which is computationally hard. Thus, the adversary can not forge partial signatures. With the same argument, we articulate that the adversary can not forge the multisignature, as the verification algorithm is similar for both partial signature and multisignature.

Security against Signature Integrity: If an adversary modifies a valid partial signature (s_i, m), the verification algorithm can detect it. Suppose an adversary modifies the signature as follows:

1. Select a random number R
2. Compute $\sigma_i = s_i + H(m,t)R$
3. Compute $T_i = t_i \cdot g^R \mod pq$

The signature tuple (σ_i, m) fails to validate the verification algorithm eq. (1) of Section 4, because $g^{\sigma_i} \equiv g^{s_i + H(m,t)R} \mod pq$
$\equiv g^{s_i} \cdot g^{H(m,t)R} \mod pq$
$\equiv T_i^{T_i} \cdot x_i^{H(m,t)} (\frac{T_i}{t_i})^{H(m,t)} \mod pq,$

which is incorrect for a valid partial signature. Thus, any fraudulent attempts on signature tampering is detected in the verification stage.

The improved scheme is based on Schnorr's scheme [11] and the underlying security is based on the hardness of DLP. The adversary who wants to forge the multisignature of a message m requires the knowledge of individual group members private keys. Without the knowledge of group member's private key, a party can not forge the member's signature as well as the multisignature, as it leads to break Schnorr's signature which is proven to be secure as long as DLP is hard.

5 Conclusion

In this paper we studied a recently proposed MSGC scheme [10] and showed that the scheme is vulnerable to forgery and signature integrity attacks. The design of the MSGC scheme was so weak that it neither requires the signer's private key to create valid signature nor detects the integrity of the signature. We proposed an improved scheme that can successfully resist the weaknesses of the MSGC scheme, where the security of the improved scheme is based on the discrete logarithm problem.

Acknowledgements

The authors would like to acknowledge Dr. Deepak B Phatak for his suggestions and encouragements.

References

1. Boldyreva, A.: Efficient threshold signature, multisignature and blind signature schemes based on the gap Diffie-Hellman group signature scheme. In Proceedings of Public Key Cryptography'03, LNCS 2567, Springer-Verlag (2003) 31–46.
2. Boyd, C.: Digital multisignatures. Cryptography and Coding, Oxford University Press (1989) 241–246.
3. Harn, L.: Group-oriented (t, n) threshold digital signature scheme and digital multisignature. IEE Proc. Computers and Digital Techniques **141** (1994) 307–313.
4. Itakura, K., Nakamura, K.: A public-key cryptosystem suitable for digital multisignatures. NEC Research & Development **71**, (1983) 1–8.
5. Lin, C. Y., Wu, T. C., Zhang, F.: A structured multisignature scheme from the gap Diffie-Hellman group. Cryptology ePrint Archive, Report no. 90, (2003).
6. Menezes, A., van Oorschot, P. C., Vanstone, S.: Handbook of Applied Cryptography. CRC Press (1996).
7. Micali, S., Ohta, K., Reyzin, L.: Accountable-subgroup multisignatures. In Proceedings of ACM Computer and Communications Security, ACM press (2001) 245–254.
8. Mitomi, S., Miyaji, A.: A general model of multisignature schemes with message flexibility, order flexibility and order verifiability. IEICE Transactions on Fundamentals **E84-A** (2001) 2488–2499.
9. Ohta, K., Okamoto, T.: Multisignature schemes secure against active insider attacks. IEICE Transactions on Fundamentals **E82-A** (1999) 21–31.
10. Rahul, S., Hansdah, R. C.: A multisignature scheme for implementating safe delivery rule in group communication systems. In Proceedings of International Workshop on Distributed Computing (IWDC'04), LNCS 3326, Springer-Verlag (2004) 231–239.
11. Schnorr, C.: Efficient signature generation by smart cards. Journal of Cryptology **4** (1991) 161–174.

Key Forwarding: A Location-Adaptive Key-Establishment Scheme for Wireless Sensor Networks

Ashok Kumar Das, Abhijit Das, Surjyakanta Mohapatra,
and Srihari Vavilapalli

Department of Computer Science and Engineering,
Indian Institute of Technology, Kharagpur 721 302, India
{akdas, abhij}@cse.iitkgp.ernet.in,
surjyakanta@gmail.com, hvpalli@yahoo.com

Abstract. In this paper we propose an improved alternative for the path key establishment phase of bootstrapping in a sensor network. Our scheme lets the network adapt to the deployment configuration by secure transmission of predistributed keys. This results in better connectivity than what path key establishment can yield. The communication overhead for our scheme is comparable with that for path key establishment. Moreover, the assurance of good connectivity allows one to start with bigger key pools, thereby improving resilience against node capture.

1 Introduction

Sensor networks are widely deployed in a variety of applications ranging from military to environmental and medical research. Chiefly for military applications, data collected by sensor nodes need be encrypted before transmission. Due to resource limitations in sensor nodes, it is not feasible to use public key routines. A symmetric cipher (like DES, RC5, IDEA, or AES) is the only viable option for encryption or decryption of secret data. However, setting up symmetric keys among communicating nodes continues to remain a challenge. Pairwise key establishment between neighboring sensor nodes in a sensor network is done by using a protocol which is popularly known as the *bootstrapping* protocol. A bootstrapping protocol involves several steps. In the key-predistribution phase, each sensor node is loaded with a set of pre-distributed keys. This is done before the deployment of the sensor nodes in a target field. After deployment, a direct key establishment (shared key discovery) phase is performed by the sensor nodes in order to establish direct pairwise keys between them. Path key establishment phase is an optional stage and, if executed, adds to the connectivity of the network. When two physical neighbors fail to establish a direct key during the shared key discovery phase, they attempt to find out a secure path to transmit a new pairwise key.

Key predistribution in sensor networks has received considerable research attention in recent years [1], [2], [3], [4], [5], [6]. Eschenauer and Gligor [1] proposed the first basic random key predistribution called the EG scheme. Chan et al. [2] proposed several modifications of the EG scheme. Liu and Ning's polynomial-pool based key predistribution scheme [3] and the matrix-based key predistribution proposed by Du et al. [5] improve security considerably.

In this paper, we propose a modification of the existing bootstrapping framework. We introduce the concept of *key forwarding* as an alternative to the path key establishment phase. Our technique yields better connectivity at a cost comparable to (if not better than) that associated with path key establishment, and does not degrade the security of the network.

2 Location-Adaptive Key Forwarding Scheme

The deployment topology of a sensor network cannot usually be determined before the actual deployment of the nodes. If, however, an approximate deployment configuration is known a priori, a host of modifications can be incorporated in the key predistribution schemes so as to achieve substantially improved connectivity and high resilience against node captures. Such *location aware* schemes [4], [5] lose their performance enhancements as the error between the actual and the expected deployment locations of the sensor nodes increases. For sufficiently large errors, a location aware scheme essentially degrades to a random scheme without a priori knowledge of deployment configuration.

A *location adaptive* scheme, on the other hand, may or may not start with prior knowledge of the deployment configuration, but adapts to the geography of deployment, thereby improving local connectivity in the sensor network. The path key establishment phase is a location adaptive feature in the bootstrapping process. We propose an alternative to the path key establishment scheme, namely the *key forwarding scheme*, which leads to considerably better connectivity than the path key establishment scheme. Our scheme works on any geographic distribution of sensor nodes in the deployment area.

The key forwarding scheme is motivated by the following consideration. Consider the basic scheme (EG scheme) with each node capable of storing m (say 200) keys. Assume also that each node has at most d (for example, 100) physical neighbors. Even when a node is connected securely to all of these neighbors, at least $(m-d)$ keys remain unused in the node. Loading a key ring with more keys than the neighborhood size is necessitated by the desire to achieve decent local connectivity. Now imagine a situation where a node v is in the physical neighborhood of two other nodes u and w. Suppose that u and v share a predistributed key and so also do u and w, but not v and w. The nodes u and w may or may not be in the physical communication ranges of one another. The node u then forwards the key k shared between u and w to node v. Since u and v have a secure link between them, k can be forwarded securely. Once v receives k, a secure link between v and w is established by using either k or any other pairwise key set up using this key k. Each round of the key forwarding phase involves the following steps:

Algorithm KeyForwarding

1. *for* each node v in the network:
2.   *for* each physical neighbor w of v with which v does not share a key:
3. v broadcasts a request whether any of its neighbors shares a key with w
4.   *if* a neighbor u of v responds affirmatively and if u and v share a key,
5.   *then* u securely forwards to v a key k shared between u and w.
6. v generates a new pairwise key k', encrypts k' with k, and sends the encrypted key and the id of u to w.
7. w retrieves k' by decrypting using k.
8.   both v and w record k' for future communication between them.
9. v deletes k from its memory, if k happens to occupy large space (like polynomial shares).

The above steps are to be carried out after the shared key discovery phase and can be repeated multiple times. In order to reduce communication overhead, the number of rounds of the key forwarding stage may be restricted to 2 or 3.

The security of the key forwarding stage is based on the assumption that bootstrapping is done securely, i.e., no nodes are captured during the initial key establishment phase. Incidentally, this is the assumption inherent in the path key establishment phase too.

Here we shall analyze our scheme applied only to the EG scheme [1] and the poly-pool scheme [3].

2.1 Network Connectivity of Key Forwarding Under the EG Scheme

Let M be size of the key-pool, m the number of keys pre-distributed in each node, and p the probability that two physical neighbors share one or more keys in their key rings. It is easy to deduce (see [1]) that $p = 1 - \prod_{i=0}^{m-1} \frac{M-m-i}{M-i}$. Let us now calculate the theoretical probability p_r that a secure link exists between two physical neighbors v and w after r rounds of key forwarding. Let d denote the (average) physical neighborhood size of each node. After the direct key establishment phase, we have: $p_0 = p$.

For the derivation of p_1, let us take two physical neighbors v and w that do not share a key. A new pairwise key is established between v and w if there exists a neighbor u of v sharing a key with both v and w. The probability that a physical neighbor u of v has this property is p^2. So the probability that neither of the d neighbors of v can help to establish a secure v-w link is $(1-p^2)^d$. Thus among the $(1-p)d$ neighbors of v with whom v does not share a key, about $d(1-p)(1-p^2)^d$ links remain insecure. We then have $p_1 = 1 - (1-p)(1-p^2)^d$. This analysis can be repeatedly generalized as: $p_r = 1 - (1-p_{r-1})(1-pp_{r-1})^d$ for all $r \geq 1$.

The probabilities p_r are plotted in Figure 1 for $M = 100000, m = 100$ (so that $p = 0.0953$) and for several values of d. From the figure, it is clear that when the average number of neighbors increases, the connectivity also increases. This is expected, since the probability that two unconnected nodes v and w can

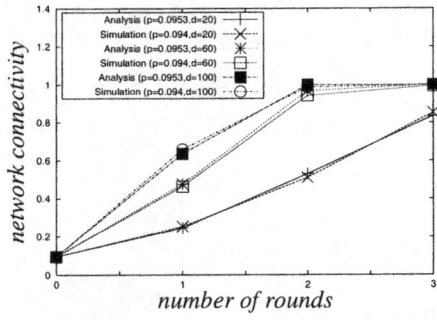 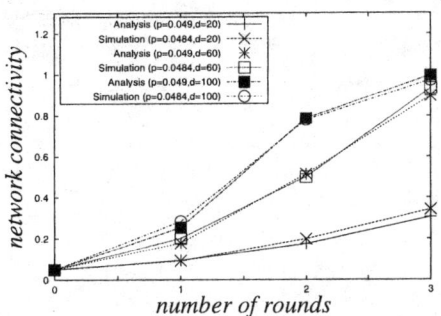

Fig. 1. Analysis and simulation of network connectivity for key forwarding under the EG scheme ($n = 10000$, $d = 20, 40, 60, 80, 100$, $M = 100000$, $m = 100$)

Fig. 2. Analysis and simulation of network connectivity for key forwarding under the poly-pool scheme ($n = 10000$, $d = 20, 40, 60, 80, 100$, $s = 500$, $s' = 5$)

establish a pairwise key between them increases with the number of nodes that can help in this process. The figure also illustrates that one obtains high network connectivity after two rounds of key forwarding.

2.2 Network Connectivity of Key Forwarding Under the Poly-Pool Scheme

Let s be the polynomial pool size, and s' the number of polynomial shares given to each node. Analogous to the EG scheme, the local connectivity p can be computed as (see [3]) $p = 1 - \prod_{i=0}^{s'-1} \frac{s-s'-i}{s-i}$.

The probability p_r of two sensor nodes sharing a key after r rounds of key forwarding can be derived analogously as before and can be given by the equations: $p_0 = p$, $p_1 = 1 - (1-p)(1-p^2)^d$, $p_r = 1 - (1-p_{r-1})(1-pp_{r-1})^d$ for all $r \geq 1$.

For $s = 500$ and $s' = 5$, we have $p = 0.0492$, that is, the network is likely to remain disconnected with high probability after shared key discovery. From Figure 2, it is clear that after executing two to three rounds of key forwarding we expect to achieve high network connectivity.

3 Simulation Results

3.1 Connectivity Measurement

For the EG scheme, we have taken the parameters $n = 10000$, $M = 100000$, $m = 100$, $d = 20, 40, 60, 80, 100$. The theoretical and simulated connectivity probabilities are plotted in Figure 1. For the poly-pool scheme, we have considered the parameters: $s = 500$, $s' = 5$, $n = 10000$, $d = 20, 40, 60, 80, 100$. The theoretical and simulated probabilities are plotted in Figure 2.

In Figures 3 and 4 we compare simulated connectivity between key forwarding and path key establishment. Key forwarding is found to clearly outperform

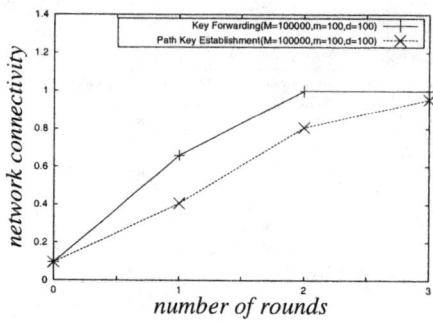 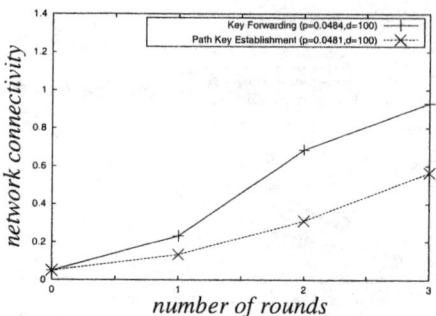

Fig. 3. Comparison of connectivity between key forwarding and path key establishment under the EG scheme ($n = 10000$, $d = 100$, $M = 100000$, $m = 100$)

Fig. 4. Comparison of connectivity between key forwarding and path key establishment under the poly-pool scheme ($n = 10000$, $d = 100$, $s = 500$, $s' = 5$)

path key establishment, particularly for the poly-pool scheme. In fact, key forwarding may render an initially disconnected network connected, whereas path key establishment can never achieve this.

3.2 Resilience Measurement

Following conventional practice, we measure the resilience of the network against node capture by the fraction of compromised links among uncaptured nodes and express this resilience as a function of the number of nodes captured. We assume that bootstrapping is done securely, i.e., no nodes are captured during bootstrapping. If the adversary also does not intercept any transmission during bootstrapping, the resilience of the network against node capture becomes the same as that of the original EG or poly-pool scheme under the given parameters. Since considerable connectivity is guaranteed by key forwarding, we can start with parameters leading to extremely high resilience.

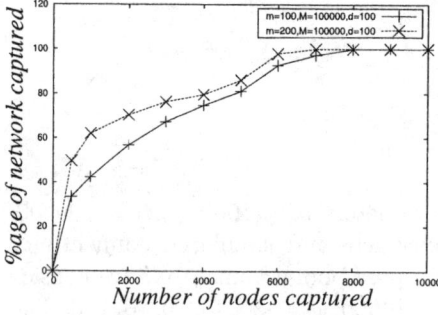 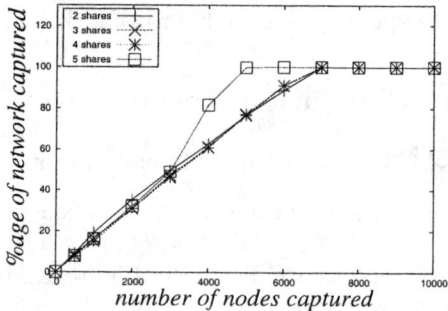

Fig. 5. Resilience measurement of key forwarding under the EG scheme ($m = 100, 200$, $M = 100000$, $n = 10000$, $d = 100$)

Fig. 6. Resilience measurement of key forwarding under the poly-pool scheme ($n = 10000$, $d = 100$, $s = 500$, $s' = 2, 3, 4, 5$)

So we assume now that an eavesdropper does not capture any node during bootstrapping but records every transaction made during bootstrapping. Later the eavesdropper manages to capture some nodes. The record of bootstrapping transactions reveals to the eavesdropper the following secret information: (i) All the pairwise keys resulting from the initial key predistribution based on captured keys or polynomial shares, (ii) All the pairwise keys established using forwarded keys or polynomial shares that are captured, (iii) For the poly-pool scheme, if more than t shares of a polynomial f are captured, any pairwise key established using any share of f during both shared key discovery and key forwarding.

Simulation results for resilience measurement under the EG scheme are shown in Figure 5 for various parameter values. Results for resilience measurement under the poly-pool scheme are shown in Figure 6 for various parameter values.

4 Conclusion

In this paper, we have proposed an alternative to the path key establishment phase of bootstrapping in a sensor network. Our scheme offers markedly better connectivity compared to path key establishment. We have corroborated this claim both theoretically and by running simulations. Better connectivity lets one start with bigger networks and/or bigger pool sizes, both leading to better resilience against node captures. The extra communication overhead incurred by key forwarding is comparable with, if not better than, that associated with path key establishment.

References

1. Eschenauer, L., Gligor, V.D.: A key management scheme for distributed sensor networks. In: 9th ACM Conference on Computer and Communication Security. (2002) 41–47
2. Chan, H., Perrig, A., Song, D.: Random key predistribution for sensor networks. In: IEEE Symposium on Security and Privacy, Berkely, California (2003) 197–213
3. Liu, D., Ning, P.: Establishing pairwise keys in distributed sensor networks. In: Proceedings of 10th ACM Conference on Computer and Communications Security (CCS), Washington DC (2003) 52–61
4. Liu, D., Ning, P.: Location-based pairwise key estalishments for static sensor networks. In: ACM Workshop on Security in Ad Hoc and Sensor Networks (SASN '03). (2003)
5. Du, W., Deng, J., Han, Y.S., Chen, S., Varshney, P.K.: A key management scheme for wireless sensor networks using deployment knowledge. In: 23rd Conference of the IEEE Communications Society (Infocom'04), Hong Kong, China (2004)
6. Du, W., Deng, J., Han, Y.S., Varshney, P.K.: A pairwise key pre-distribution scheme for wireless sensor networks. In: ACM Conference on Computer and Communications Security (CCS'03), Washington DC, USA (2003) 42–51

New Anonymous User Identification and Key Establishment Protocol in Distributed Networks*

Woo-Hun Kim[1] and Kee-Young Yoo[2]

[1] Department of Information Security, Kyungpook National University,
Daegu 702-701, Republic of Korea
whkim@infosec.knu.ac.kr
[2] Department of Computer Engineering, Kyungpook National University,
Daegu 702-701, Republic of Korea
yook@knu.ac.kr

Abstract. Recently, Wu and Hsu showed that Lee and Chang's anonymous user identification and key establishment protocol was insecure with regard to two attacks and proposed an improved protocol, called the WH protocol. In this paper, we show that the WH protocol is still vulnerable to an unknown-key share attack. Then, we propose an improved protocol to address this problem by applying a mutual authentication method.

1 Introduction

We will demonstrate an unknown-key share scenario in the following: This scenario was first described by Diffie, van Oorschot and Wiener [1]. Let B denote a bank branch and A denote an account holder. Suppose that the protocol for an electronic deposit of funds is to exchange a key with a bank branch via an authenticated key agreement with a key confirmation protocol. At the end of the protocol run, A sends encrypted funds to B. Suppose that no further authentication is done in the encrypted message which is needed to save the bandwidth. If the unknown-key share attacks are successfully launched, the deposit will be made to the adversary's account instead of B's.

In 2000, Lee-Chang proposed a anonymous user identification and key establishment protocol based on the security of the factoring problem and the one-way hash function, called LC protocol [2]. In 2004, however, Wu-Hsu showed that the LC protocol suffers from two weaknesses [3]. They also proposed a modified version to overcome these weaknesses, called WH protocol.

In this paper, we show that the WH protocol is still vulnerable to an unknown-key share attack. In addition, we will propose an improved protocol to overcome this weakness.

---

* This research was supported by the MIC (Ministry of Information and Communication), Korea, under the ITRC (Information Technology Research Center) support program supervised by the IITA (Institute of Information Technology Assessment).

2 Review of the WH Protocol

The WH protocol consists of two phases: Key generation phase and an anonymous user identification phase. The key generation phase that is performed in the WH protocol is described as follows:

U_i		P_j
	Service Request $\longrightarrow$	choose k
choose t	$z = g^k S_j$	compute $z = g^k S_j$
	$\longleftarrow$	
compute $a = z^e / ID_j$		
$x = S_i f(a^t \| T)$		
$y = g^{et}$		
	x, y, T	check T and verify
	$\longrightarrow$	$ID_i \stackrel{?}{=} (x/f(y^k \| T))^e$
$k_{ij} = a^{tx} = g^{ektx}$		$k_{ji} = y^{kx} = g^{ektx}$

Fig. 1. Illustration of the WH protocol

Step 1 The Smart Card Producing Center (SCPC) first chooses two large primes p and q, computes $N = pq$, picks an element $g \in Z_N^*$ and a hash function f, and selects e and d such that $ed = 1 \mod \phi(N)$, where $\phi(N)$ is the Euler totient function. After that, $N, e, g,$ and f are considered published and d, p and q are kept secret by the SCPC.

Step 2 With a secure channel, the SCPC then sends each user U_i (or service provider P_i) a secret token $S_i = ID_i^d \mod N$, where ID_i is the identity of U_i (or P_i).

The anonymous user identification phase that is performed in the WH protocol is described as follows.

Step 1 In order to request service from the service provider P_j, U_i first submits a service request to P_j.

Step 2 After receiving the request, P_j chooses a random number k and computes $z = g^k S_j \mod N$, which is then sent to U_i.

Step 3 U_i in turn, randomly chooses a number t and computes $a = z^e / ID_j \mod N$, $x = S_i f(a^t \| T) \mod N$, $y = g^{et} \mod N$, where T is the time-stamp. After that, U_i sends (x, y, T) to P_j.

Step 4 Finally, P_j checks T and verifies the equality $ID_i \stackrel{?}{=} (x/f(y^k \| T))^e \mod N$. If it holds for some ID_i that exists on the identity list, U_i is accepted as an authorized user and the service request will be granted.

Note that the user and the service provider share one common session key after the identification protocol as $k_{ij} = a^{tx} = y^{kx} = g^{ektx} \mod N$, which can be used in subsequent communications for confidentiality. The illustration of the WH protocol is provided in Fig. 1.

3 Weakness of the WH Protocol

In this section, we show that the WH protocol is vulnerable to an unknown-key share attack. An illustration of the unknown-key share attack scenario is given in Fig. 2.

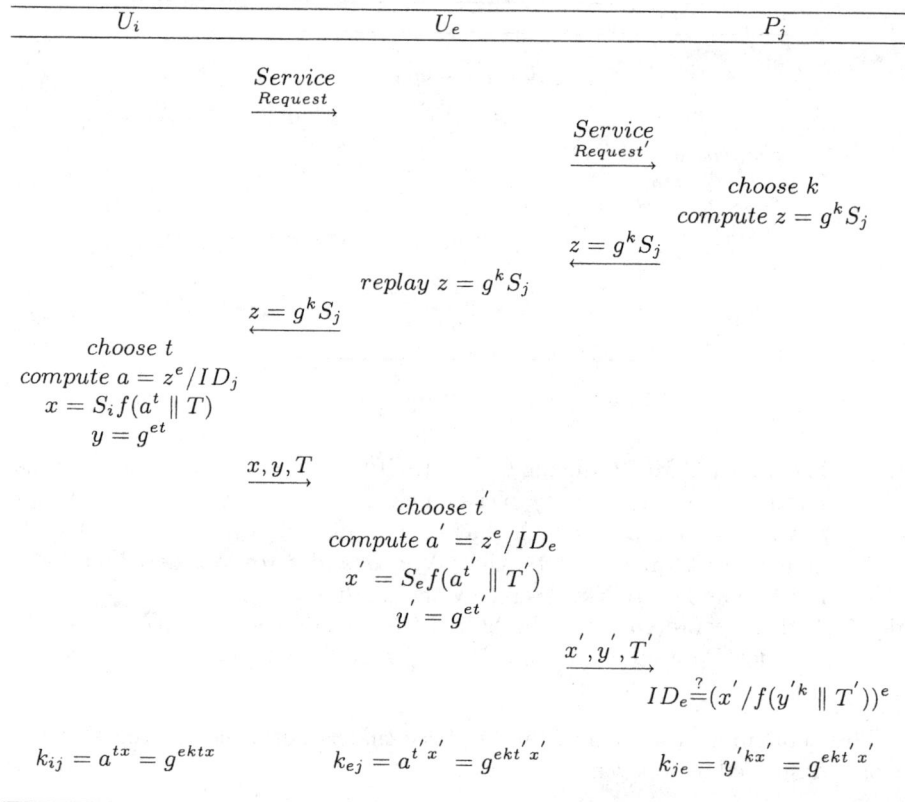

Fig. 2. Illustration of an unknown-key share attack scenario

Suppose that U_e intercepts the communication between U_i and P_j. An unknown-key share attack scenario in the WH protocol is described as follows:

Step 1 U_i sends a *Service Request* to P_j.
Step 2 U_e intercepts the message from the network during the legal user U_i and sends the *Service Request* to the legal service provider P_j.
Step 3 After intercepting the *Service Request*, U_e sends the *Service Request* to the legal service provider P_j.
Step 4 After receiving the *Service Request*, P_j chooses a random number k and computes $z = g^k S_j \mod N$, which is sent to U_e.
Step 5 U_e pretends to be the service provider P_j in order to re-send the replayed value $z = g^k S_j \mod N$ to user U_i.

Step 6 U_i in turn, randomly chooses a number t and computes $a = z^e/ID_j$ mod N, $x = S_i f(a^t \parallel T)$ mod N, $y = g^{et}$ mod N, where T is the time-stamp. After that, U_i sends (x, y, T) to U_e.

Step 7 Then U_e computes his own $x' = S_e f(a^{t'} \parallel T')$ mod N, $y' = g^{et'}$ mod N, T' where T' is the time-stamp. After that, U_e sends (x', y', T') to P_j.

Step 8 Finally, P_j checks T' and verifies the equality $ID_e \stackrel{?}{=} (x'/f(y'^k \parallel T'))^e$ mod N. If it holds for ID_e that exists in the identity list, U_e is accepted as an authorized user.

Then, U_e and the service provider P_j share a common session key $k_{ej} = a^{t'x'} = y^{kx'} = g^{ekt'x'}$ mod N. While in fact, U_i believes that the session key $k_{ij} = a^{tx} = y^{kx} = g^{ektx}$ mod N is shared between P_j and U_i. Hence, the WH protocol can not achieve the security requirement of an unknown-key share resilience.

4 The Proposed Protocol

Also, the proposed protocol can be divided into two phases, as are those in the WH protocol. The key generation phase is also the same as those in the WH protocol. In the following, we describe the anonymous user identification phase of the proposed protocol. Note that all values are generated in modulus N.

U_i		P_j
choose t		choose k
compute $S \cdot R = g^t S_i$	$\xrightarrow{S \cdot R}$	compute $u = (S \cdot R)^e / ID_i$
		$z = g^k S_j$, $v = f(g^{ke} \parallel u)$
	$\xleftarrow{z, v}$	
compute $a = z^e / ID_j$		
and verify $f(a \parallel g^{te}) \stackrel{?}{=} v$		
$x = S_i f(a^t)$	$\xrightarrow{x}$	verify $ID_i \stackrel{?}{=} (x/f(u^k))^e$
$k_{ij} = a^{tx}$		$k_{ji} = u^{kx}$

Fig. 3. Illustration of the anonymous user identification phase of the proposed protocol

The anonymous user identification phase that is performed in the proposed protocol, is described as follows:

Step 1 In order to request service from the service provider P_j, U_i chooses a random number t and submits a $S \cdot R = g^t S_i$ to P_j.

Step 2 After receiving the request, P_j chooses a random number k and computes $z = g^k S_j$, $u = (S \cdot R)^e / ID_i$ and $v = f(g^{ke} \parallel u)$. Then z and v are sent to U_i.

Step 3 U_i in turn computes $a = z^e/ID_j$. After that, U_i verifies the equality $f(a \parallel g^{te}) \stackrel{?}{=} v$. If it holds, U_i assures that z and v came from the legal service provider P_j. After verifying the service provider P_j, U_i computes $x = S_i f(a^t)$. After that, U_i sends x to P_j.

Step 4 Finally, P_j verifies the equality $ID_i \stackrel{?}{=} (x/f(u^k))^e$. If it holds for some ID_i that exists in the identity list, U_i is accepted as an authorized user and the service request will be granted.

An illustration of the proposed anonymous user identification phase is provided in Fig. 3. After the identification phase, the user and the service provider share a common session key $k_{ij} = a^{tx} = u^{kx} = g^{ektx}$. In order to avoid unknown-key share attacks, we applied a mutual authentication method in our proposed protocol.

5 Security Analysis

In this section, we analyze our proposed protocol under the difficulty of a factoring problem, a discrete logarithm problem and the intractability of the one-way hash function.

Theorem 1. *The proposed protocol provides user anonymity.*

Proof: If the adversary knows the service provider's random number k, he/she can obtain the participant's identity from the following equation $ID_i=(x/f(u^k))^e$. In order to compute k from the public value, however, this is the equivalent to solving the discrete logarithm problem.

Theorem 2. *The proposed protocol resists a user impersonation.*

Proof: An adversary tries to impersonate U_i by forging the $S \cdot R$ and x. It, however, is impossible to compute $S \cdot R = g^t S_i$ and $x = S_i f(a^t)$ without the legal user's secret token S_i, which is stored in each user's smart card.

Theorem 3. *The proposed protocol resists a service provider impersonation.*

Proof: An adversary tries to impersonate the service provider P_j by replaying the previously captured messages or forging messages. It, however, is impossible to compute $z = g^k S_j$ and $v = f(g^{ke} \parallel u)$ without a legal service provider's secret token S_j.

Theorem 4. *The proposed protocol resists a session key compromise.*

Proof: An adversary wishes to derive the session key from the transmitted messages of an identification phase. The adversary can obtain $M = \{S \cdot R, z, v, x\}$. In order to compute the session key $k_{ij} = k_{ji} = g^{ektx}$, the adversary can obtain a service provider's random value k and user's random value t.

6 Comparisons

Table 1 summarizes the main features of the previously-proposed protocol and our proposed protocol.

Table 1. A comparison of the previously proposed protocols and our proposed protocol

Features	LC protocol	WH protocol	The proposed
Number of rounds	3	3	3
Time-stamp	o	o	×
Challenge-response	×	×	o
Providing user anonymity	×	o	o
Mutual authentication	×	×	o
Unknown-key share attack resilience	×	×	o

7 Conclusions

This paper has demonstrated the weakness of the WH protocol, the anonymous user identification and the key distribution protocol using smart card. Their protocol use one-way user authentication, in that only a service provider can verify with whom he is communicating. Their protocol is vulnerable to unknown-key share attacks. In order to overcome this weakness, we have proposed a new anonymous user identification and key distribution protocol.

References

1. W. Diffie, P. van Oorschot, M. Wiener, Authentication and authenticated key exchanges, Designs, Codes and Cryptography 2 (1992) 107-125.
2. Lee WB, Chang CC, User identification and key distribution maintaining anonymity for distributed computer network, Computer Syst Sci Eng 15-4 (2000) 211-214.
3. Tzong-Sun Wu, Chien-Lung Hsu, Efficient user identification scheme with key distribution preserving anonymity for distributed computer networks, Computers and Security 23 (2004) 120-125.

Semantic Overlay Based Services Routing Between MPLS Domains

Chongying Cao[1,2], Jing Yang[3], and Guoqing Zhang[1]

[1] Institute of Computing Technology, Chinese Academy of Sciences,
P.O. Box 2704, Beijing 100080, China
[2] Graduate School of the Chinese Academy of Sciences, Beijing, China
[3] UTStarCom Company, China
caocy@ict.ac.cn, Jamesy@utstarcom.com, gqzhang@ict.ac.cn

Abstract. Service routing across different MPLS and access networks domains is a critical issue in next generation networks. In spite of being a de facto inter domain routing standard, BGP can not fulfill the inter domain traffic engineering needs and it does not take into account service metrics, which results in the low routing efficiency of BGP. We set up a system model by decomposing the problem. Based on the model, we set up a service network reduction algorithm. At the initial stage we use a grid service instance to map service paths of different MPLS domains, then we construct service network graph, and set up index for service routing based on ontology. At the last stage, we get a single service metric to set up service routing by applying the service ontology metric function.

1 Introduction

Next Generation Network (NGN) is expected to support services across different networks with diverse requirements. It brings great challenges to the MPLS networks, which will be the core networks in NGN. Service routing is an overlay network routing, which provides corresponding service capability. For example, we often need to set up VPN services across different MPLS network domains. The objective of service routing is three-fold: to find a feasible path for each service; to optimize the usage of the network by balancing the load; and to satisfy different MPLS domain's requirements.

Since the Border Gateway Protocol (BGP) [1] is a de facto standard for inter domain routing, it can also be used as a inter domain routing protocol between MPLS domains. As a path vector routing protocol, BGP is similar to any other distance vector routing protocol that doesn't take into account service metrics and its criteria for selecting the best path are based on the length of AS path, which result in the failure on the part of BGP to support different service routing between MPLS domains. Each domain, owned by different network providers, has different service providing methods and policies. Different MPLS domains lack common languages to understand each other. At the same time, in order to set up service routing across different MPLS domains, multi service constraints must be considered in the service routing algorithm. This is a NP hard problem. Due to these reasons, service routings across different network domains are difficult to be set up only through BGP.

To address these problems, we introduce an innovative idea — service semantic P2P overlay network to organize the resources of service routing across different network domains based on ontology. In this paper, we propose a new model and some related algorithms for the problems.

In Section 2, we introduce the related research works. Based on the above, we analyse and decompose the problem to set up a model in Section 3. In Section 4, we set up service routing based on the model. In Section 5, we do experiments to evaluate our model and algorithm. Finally, a conclusion is reached.

2 Related Work

There are many researches to improve BGP protocol [2], [3]. Many heuristics also have been proposed due to its NP-completeness [4], [5]. Due to the complexity imposed on the router, their improvements are limited. High complexity prevents their practical applications. Some algorithms only suit a specific network, and can't support different service routing.

To reduce the online overhead, A. Orda proposed a considerable reduction through precompution [6]. Recently, overlay network is proposed to decompose the functions from routers [7], [8]. However, these solutions have not considered different service routing requirements and search complexity is still high. To expose network capability through open APIs, Parlay/OSA [10] has been proposed. In Parlay, MPLS traffic engineering capability can be mapped up. All kind of traffic engineering requirements can be computed in all kinds of Parlay service capability servers. However, service convergence and virtualization is the key to reduce online overhead for establishing routing across heterogeneous MPLS domains. Parlay is tightlyly coupled and can't provide such kind of capability. OGSA (Open Grid Services Architecture) [11] is proposed to provide such kind of service convergence and virtualization platform. It is loosely coupled. At the same time, it can manage network resource status which can not be supported by Parlay and web services. It provides a feasible way to reduce overhead for setting service routing.

Scalability is also very important for MPLS inter domain service routing. P2P routing can provide such scalability. Xiaohui Gu of University of Illinois gave a framework to compose the QoS-aware service for large-scale peer-to-peer systems [12]. However, current P2P routing algorithms are based on single key words, which don't solve multi services problem according to traffic engineering.

In order to let different MPLS domains understand each other, semantic network can be used. Semantic Overlay Networks (SONs) [13] appear to be a way to group peers sharing the same schema information. This approach facilitates routing, since a peer can easily identify relevant peers instead of broadcasting query requests on the networks.

3 System Model

The system model is to expose the capability of different MPLS domains and optimize placement of service routing across the traffic trunks provided by different MPLS domains. BGP is not enough to expose the capability of MPLS domains. Computational time complexity and space complexity is very large as the size of network becomes large. In fact, it will be NP-hard for any optimal problem whose dimensions are higher than two in large network systems. Current routing architectures and algorithms force all computation in a component, which makes it hard to reduce the problem space. To solve the problem, we formulate a model and setup a realization model for it.

3.1 Formulation of Model

We can define a service overlay network $G = (V, E, C)$ for service routing. We denote the set of service nodes in the network as V, the set of service links in the network as E and the set of constraints as C. C is a vector set of service routing constraints.

We decompose the problem space by decomposing network functions into different components. In order to do this, we can formulate the problem by decomposing it into three sub problems. One sub problem is to expose the capability and map service into single service ontology according to the related constraints based on traffic engineering. The second sub problem is to setup a single service metric according to the service ontology. The last sub problem is to optimize the service routing according to the single service metric. Through such kind of decomposition, we reduce the problem dimension to 1 for service routing in large networks, while addressing complex computing problems in local systems. Then the three sub problems can be formulated as the following:

1) Service Ontology Mapping Function:

$$S_{ij} = M(S, x_{ij}, t) \tag{1}$$

2) Service Ontology Metric Function:

$$C_{ij}(S) = F(R(S), x_{ij}, t) \tag{2}$$

3) Service Routing Optimization Function:

$$\text{Min} \sum_{i,j} C_{ij}(S) * x_{ij}(S) \tag{3}$$

where S is a kind of service ontology, x_{ij} is the network path between node i and j, t is the time for the service routing, $C_{ij}(S)$ is service ontology metric, $x_{ij}(S)$ is the service path between node i and j for service ontology S, $R(S)$ is the resource requirement of service ontology S.

Ontology [14] is a formal, explicit specification of a shared conceptualization. Service Ontology Mapping Function in formula (1) maps the underlying MPLS traffic trunk into the service ontology according to the requirement of traffic engineering and different MPLS management domains. Through ontology, we can get a single service metric for a service routing. Service Ontology Metric Function in formula (2) is to map local resource status to a single service metric according to the requirement of traffic engineering. Service Routing Optimization Function in formula (3) is to optimize service routing based on the single service metric.

3.2 Realization Model

To implement the model, we use semantic P2P overlay network in system architecture as shown in Fig. 1, which will be used to set up service routing across different MPLS networks and access network domains. In the model, resources in MPLS domains are mapped by Parlay Gateway. Every MPLS service will be mapped to service ontology. Using service ontology mapping, we can solve the first problem and the second problem. The Services Semantic P2P Overlay Network forwards according to the service ontology. Due to P2P function, we can solve the last problem.

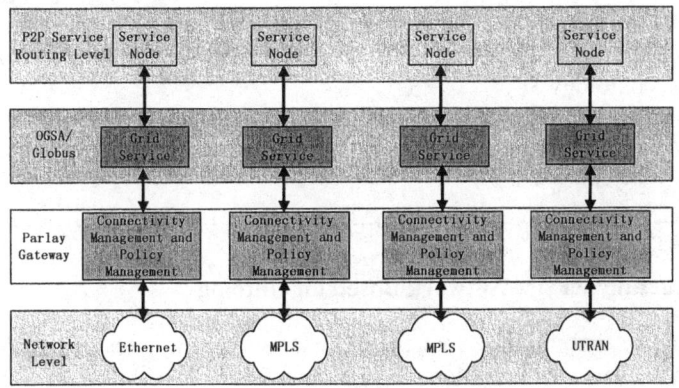

Fig. 1. System Architecture

At first, we need to use Parlay API to open the MPLS network capability, which includes connectivity management [15] and policy management [16]. To get network resource status between different MPLS domains, we use common network measurement platform. These capabilities are wrapped as Grid Services, which can effectively manage MPLS network resources and their states. On the foundation, we set up a single semantic image — ontology. Through ontology, we can get, describe, express and compose services to set up service routing across different MPLS domains and other networks. At the same time, different service relationships can be set up through semantics. In semantic grid level, different MPLS domains and other access networks can understand each other through ontology, and become a transparent homogeneous network. In such a homogeneous network, we can use semantics to set up P2P service routing.

4 Setup of Service Routing

Actually, the above model reduces the network graph into a service network graph of the same ontology. Instead of trying to find multiple metrics, the single metric can be used to move new load applied to lightly loaded paths in the *reduced graph*. We can use the Service Network Graph Reduction (SNGR) algorithm (in Table 1), which executes two consecutive actions on the path set G to construct a service network based on ontology and applies selection criteria on the resulting subset to set up service routing. The time complexity of general routing is $O(mn^2)$, where n is number of network nodes and m is the number of service constraints. However, SNGR's time complexity is only $O(kn)$, where k is the number of service ontology based on the index structure below.

Table 1. Service Network Graph Reduction algorithm

SNGR( )
{
   Step 1: Transform the directed graph G into $G' = (V', E', C')$, $V' \in V$, $E' \in E$, $C' \in C$. We select only those service paths that can bear service ontology S.
   Step 2: Compute $P(a, d, S)$ according to a single service metric, and let it be the set including one minimum-cost path from each source to each destination.
}

4.1 Constructing Service Network Based on Ontology

We can conceive that an inter domain MPLS service path across inter domain paths consists of a sequence of intra domain segments. Using semantic grid, we can set up grid services to represent these paths. Each grid service wraps service pipes exposed by Parlay Gateway. It is mapped based on service ontology. The inter domain service routing consists of a list of composable grid services, which is connected into a service path. According to such a model, we can setup service network graph and its index structure.

4.1.1 Service Network Graph Construction

In the model, the service network graph (SNG) (Fig. 2) represents a "snap-shot" of the P2P resource requirement and availability. A, B, C, D are MPLS domains, A^s, B^s, C^s, D^s are service network domains. There are several attributes of traffic engineering. S is the service vector of the service pipe which is provided by parlay gateway. Formally, we define the vectors S as follows: $S = [s_1, s_2, ..., s_n]$. These vectors represent the service capability of the service ontology. These capabilities include a set of attributes associated with service paths which collectively specify their behavioral characteristics and a set of attributes associated with resource which constrain the placement of service paths through them. These can also be viewed as topology attribute constraints. The SNG is defined as follows:

(1) *SNG nodes*: The service node of SNG represents the border node of a MPLS domain. In service network domain B^s, both a^s and b^s are the border service nodes, which is mapped from a and b, but c isn't the border node.

(2) *SNG edges:* Edges from source service border node to destination service border node within a MPLS domain. In domain B^s, $<a^s, b^s>$ is the service edge which is mapped from underlay network edge.
(3) *SNG service instances:* In a SNG edge, a grid service instance wraps a service pipe across an underlay physical path. Its resource requirement vector S^{req} (a, b), can be satisfied by the current availability of the corresponding service pipe provided by the Parlay gateway. In domain B, S_1 and S_2 are two grid service instances, which represent different service pipe. S_1 is across the physical path (a, b), S_2 is across the physical path (a, c) and (c, b).

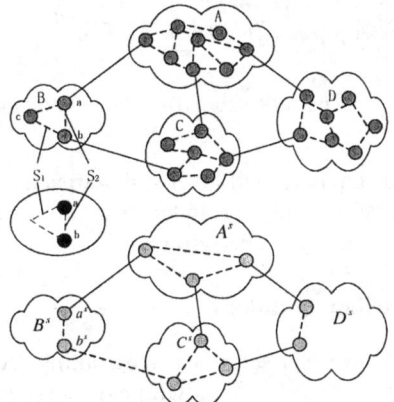

Fig. 2. Service Network Graph

4.1.2 Index Structure Based on Ontology

Service network graph will become very complex when the size of network becomes larger. Maintaining the global view of the service network graph is difficult. So we adopt super node P2P architecture. Semantic grid built on Parlay gateway is a super peer, which is a node that acts both as a server to a set of clients, and as an equal in a network. A peer group is based on the same ontology to be set up.

Based on ontology, we can summarize the service relationships within a MPLS domain which is managed by Parlay gateway. We have employed three levels of summarization in our framework (Fig. 3). The lowest level is *Parlay Gateway level*. The second level is named as *service ontology peer level*; all information owned by a peer is summarized according to service ontology. In this level, peers are the service instances that are grid services which wrap service pipes provided by Parlay gateway. Finally, in the third level, named as *semantic grid level*, all information contained by a peer group is registered. Each semantic grid maintains two pieces of summaries: the super level summaries of its group and its neighboring groups, and peer level summaries of its group. According to ontology, a super peer can determine which peer group is relevant. To further improve the efficiency of the system, we maintain indexes on the ontology information. We name the three indexes for Parlay gateway,

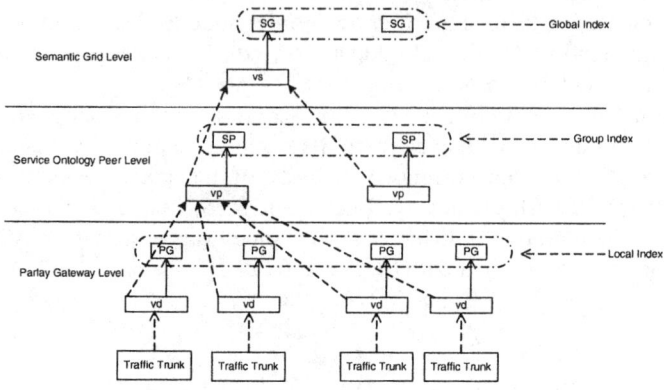

Fig. 3. Index Structure Based on Ontology

service ontology peer, and semantic grid level summaries as *local index, group index*, and *global index*, respectively. Using the index structure based on ontology, we can find the entry to the service ontology peer group.

4.2 Service Routing Based on Ontology

Based on the above work, we can setup service routing. At first, different service routing will have different strategies, which will be discussed first. Then we will setup a single service metric and service routing algorithm.

4.2.1 Local Strategies
At first, we need to discuss the service instances finding by using local strategies. Our algorithm must effectively balance the network load among different MPLS domains and efficiently utilize the network resources. Suppose C_{ij} is the required capability of service routing (i, j) and DC_x is the capability of MPLS domain which service routing (i, j) goes across. Then Overhead Rate in the domain can be defined as:

$$\text{OverheadRate}_{ij} = C_{ij}/DC_x \qquad (4)$$

Our algorithm is based on the following criteria. If the capability DC_x of the domain x is larger than the capability DC_y of domain y, then we require that the probability of choosing DC_x should be larger than DC_y. The overhead rate will be used to select the service instance group.

When we select service instance in a service group, we also need to consider the resource utility of the service path. The resource utility can be defined as in formula (5), where *Bandwidth*$_{ij}$ is the used bandwidth of the service instance, and B_{ij} is the total bandwidth of the link which is represented by the service instance.

$$ResourceUtility_{ij}(S) = Bandwidth_{ij}/B_{ij} \qquad (5)$$

In order to balance the load, we select the service instance among the service instance group according to the largest resource utility rate as in formula (6), where $\{S_{ij}\}$ is the service instance set between service node i and j.

$$Max(ResourceUtility_{ij})\{S_{ij}\} \tag{6}$$

4.2.2 Service Ontology Metric Computation

The service ontology allows different MPLS management domains to understand each other. Different domains have different constraint conditions for service ontology. We must compute service ontology metric for a single peer selection in which the current peer needs to choose the next hop peer according to its local resource information to decide the metric.

Table 2. Service Routing Algorithm

```
SRA(a, d, S) {
    FindServiceNetworkGraphEntry(a, d, S).
    Bellman-Ford(G, w, c, a, d)
}
Bellman-Ford(G, w, c, a, d) {
    For i = 0 to |N(G)|-1
        PATH(i) = Φ
    PATH(a) = { 0⃗ }
    For i = 0 to |N(G)|-1
        For each edge (u, v) ∈ E(G)
            FindServiceInstance(u, v, S, G)
            If there are service instance in the edge
                Evaluate(u, v, w)
    For each w(p) in PATH(d)
        If (w(p) < c) then return "yes"
    Return "no"
}

Evaluate (u, v, w) {
    For each w(p) in PATH(u)
        flag = 1
        For each w(p) in PATH(v)
            If (w(p) + w(u, v) >= w(q))
                flag = 0
            Else
                Remove w(q) from PATH(v)
        If (flag = 1)
            Add w(p) + w(u, v) to PATH(v)
}
```

The weight of a service path vector as defined in (2) is a vector sum. As in linear algebra, the length of a service path vector requires a vector norm to be defined. The definition of the service path length is needed to be able to compare paths since the path weight components all reflect different service ontology metrics with specific units. We propose a straightforward choice of a linear path length to compute service ontology metric in formula (7), where w_i is the service constraint factors

$$C_{ij}(S) = \sum_{i=1}^{k} a_i \cdot w_i \qquad (7)$$

Vector $a = (a_1, a_2, \ldots, a_{m+1})$ satisfy forma (8), where a_i are positive real numbers.

$$\sum_{i=1}^{m+1} a_i = 1 \qquad (8)$$

4.2.3 Service Routing Algorithm

Using the set $\{C_{ij}(S)\}$ of service routing of service ontology S which is pre-computed, we can compute the single metric path problem by using service routing algorithm as shown in Table 2. In the inter domain service routing, scalability must be considered, so we use dynamic programming to optimize service routing. Using the service ontology S and service nodes a and d, we use the index structure based ontology to get the entry of a reduced service network graph which is suitable for the service routing. This greatly reduces the search space for the routing. In the algorithm, FindServiceNetworkGraphEntry(a, d, S) uses the index based on service ontology to find the service network graph entry. There are many service instances in a service edge. FindServiceInstance(u, v, S, G) is used to find service instances. $w(p)$ is the service ontology metric, which will be computed in service ontology metric function.

5 Experiments

In this section, we evaluate the performance of the SNRG model by simulation.

5.1 Experimental Set Up

In the experiment, a semantic overlay is built over a simulated MPLS network. The inter domain topology is generated based on power-law model [15], and the intra domain topology is generated based on the Waxman model [9]. The number of domains is 100, while the number of paths in a MPLS domain is randomly selected from 10 to 100. The degree of a domain is defined as the total number of inter domain links adjacent to the border nodes of the domain. 15% of the domains have a degree of one, and the degrees of the other domains follow the power law. Every path is randomly assigned initial resource constraints S = [ServiceConstraints, Bandwidth, Delay] from [1, 100, 1] to [10, 1000, 10]. ServiceConstraints refers other service constraint. The service instances are created over these paths. Every service instance is assigned initial resource constraints according to service ontology definition.

We select three kinds of services. Service 1 needs to consider only delay across different MPLS domains; we use the delay as the cost. Service 2 needs to consider bandwidth, delay is not an important factor; we use a threshold to control bandwidth (Bandwidth $\{S_{ij}\} < Th$, where Th is the threshold of packet loss rate). Service 3 needs to consider both bandwidth and delay. At the same time, we define a set of traffic attributes as the service vector.

5.2 Experimental Results and Discussions

1) Success ratio:
A feasible request may be rejected due to imperfect approximation of service routing requirement. Success ratio is used to measure quantitatively how well an algorithm finds feasible paths and it is defined as:

$$\frac{\text{total number of accepted feasible requests}}{\text{total number of requests}} \qquad (9)$$

The dividend represents all connection requests that are accepted by inter domain service routing in MPLS network. The divider is the total number of feasible requests (not the total number of requests). Therefore, in our simulation, success ratio measures the performance with respect to the optimal performance.

Fig. 4 shows the performance of SNGR algorithm for three kinds of services, and compares it with BGP. AS0, AS1, and AS2 represent three services of SNGR. B0, B1 and B2 represent three services of BGP. From the figure, we can see the success ratio of SNGR algorithm is higher than BGP. The constraint condition of service 3 is more stringent than other services. The success ratio of the service is lower than other services. As the request number increases, the network resource will decrease, and we can see that the success ratio is also decreasing. The decreasing rate of success ratio of service 3 is also higher than that of other services.

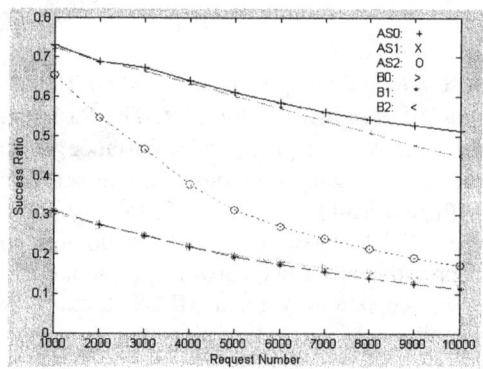

Fig. 4. Success Ratio

2) Balancing Service Path Capacity among different MPLS Domains:
SNGR also needs to balance the service routing overhead among different MPLS domains. It can achieve this goal by balancing the residual capacity of different MPLS domains. We use service network domain load rate deviation (*SNLRD*)

(formula (10)) to evaluate the performance of traffic balancing among different MPLS domains. SC_i is the load rate of the i th service network domain; SC is the average value of load rate, while N_{SND} is the number of service network domain.

$$SNLRD = \sqrt{\frac{\sum_{i=1}^{N_{SND}}(SC_i - SC)^2}{N_{SND}}} \qquad (10)$$

Fig. 5 shows the service network domain load rate deviation of SNGR algorithm and BGP algorithm. It shows that the balancing capability of the SNGR algorithm is better than that of BGP algorithm. At the same time, we find that the deviation increases with the increase in request number. This is because as the request number increases, the network resource becomes more disordered. This makes the balancing more difficult.

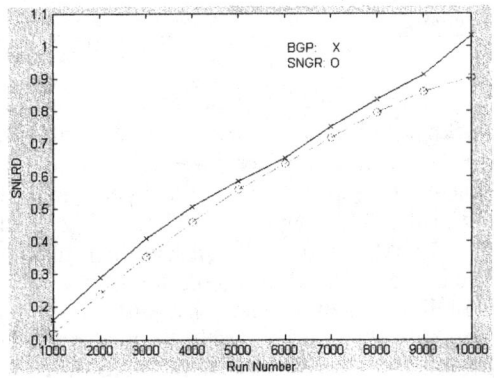

Fig. 5. Service network domain load rate deviation

6 Conclusion

In this paper, we propose a model using semantic P2P overlay service network to set up service routing across different MPLS domains. The Parlay gateway abstracts the capability of MPLS domain. We use grid service instance to represent traffic trunk provided by Parlay gateway. By using ontology, we can set up a transparent service network for different MPLS domains.

We also propose the SNGR algorithm and the following things: 1) using grid service instances to set up reduced service network graph based on ontology, 2) using index to organize the service ontology for a MPLS domain which is managed by semantic grid, and 3) set up a service routing algorithm based on the reduced service network graph.

References

1. Rekhter, Y., Li, T.: A Border Gateway Protocol 4 (BEP-4) (RFC 1771), March, 1995
2. Griffin, T. G., Shepherd, F. B., Wilfong, G.: The Stable Paths Problem and Inter domain Routing, IEEE/ACM Transactions on Networking, Vol.10, No.2, April 2002

3. Gao, L., Griffin, T. G., Rexford, J.: Inherently Safe Backup Routing with BGP", IEEE INFOCOM 2001
4. Garey, M. S., Johnson, D. S.: Computers and Intractability: A Guide to the Theory of NP-Completeness, W.H. Freeman, New York, 1979
5. Korkmaz, T., Krunz, M.: Multi-Constrained Optimal Path Selection, IEEE INFOCOM 2001
6. Orda, A., Sprintson, A.: QoS Routing: the precomputation perspective, IEEE INFOCOM 2000
7. Chimento, P., et al.: QBone Signalling Design Team, QBone Bandwidth Broker Architecture, Work in progress http://sss.advanced.org/bb/bboutline2.html.
8. Agarwal, S., Chuah, C., Katz, R. H.: OPCA: Robust Interdomain Policy Routing and Traffic Control, OPENARCH 2003
9. Waxman, B. M.: Routing of multipoint connections, IEEE J. Select. Areas Commun., vol. 6, Dec, 1988.
10. Moerdijk, A., Klostermann, L.: Opening the Networks with Parlay/OSA: Standards and Aspects Behind the APIs, IEEE Network, 2003,17(3): 58 ~ 64
11. Foster, I., Kesselman, C., et.al.: Grid Services for Distributed System Integration. IEEE Computer, 35(6), 2002, pp: 37 ~ 46
12. Gu, X., et al.: QoS-aware service composition for large-scale peer-to-peer systems
13. Crespo, A., Garcia-Molina, H.: Semantic Overlay Networks, 2003, http://www-db.standford.edu/~crespo/publication/op2p.pdf
14. Studer, R., Benjamins, V. R., Fensel, D.: Knowledge Engineering, Principles and Methods, Data and Knowledge Engineering, 1998, 25(1-2)
15. Faloutsos, M., Faloutsos, P., Faloutsos, C.: On power-law relationships of the internet topology, in Proc. ACM SIGCOMM, 1999.

Effective Static Task Scheduling for Realistic Heterogeneous Environment

Junghwan Kim[1], Jungkyu Rho[2], Jeong-Oog Lee[1], and Myeong-Cheol Ko[1]

[1] Department of Computer Science, Konkuk University,
Danwol-dong 322, Chungju si, Chungbuk 380-701, Korea
{jhkim, ljo, cheol}@kku.ac.kr
[2] Department of Computer Science, Seokyeong University,
Jungneung-dong 16-1, Sungbuk-gu, Seoul, Korea
jkrho@skuniv.ac.kr

Abstract. Effective task scheduling is crucial for achieving good performance in high performance computing. Many scheduling algorithms have been devised for heterogeneous computing, but most of algorithms have not been considered in realistic heterogeneous environments which are not arbitrarily heterogeneous but have locality in communication. In this paper we present new scheduling algorithms by considering the locality. It is thought that critical-path tasks are often important in reducing schedule length, however one of the previous scheduling algorithms, CPOP (Critical-Path-On-a-Processor) does not show good result against to expectation. Our first heuristic uses a cluster of processors for critical-path tasks while a single processor is used in the CPOP. This heuristic well exploits realistic computing environments in which communication costs are not arbitrarily heterogeneous. In an additional heuristic the critical-path tasks are considered to finish (or start) as early as possible when even non critical-path tasks are scheduled. For a performance study five scheduling algorithms are compared by experimenting on three different environments. The experimental results show our scheduling algorithm outperforms the others in the realistic heterogeneous environments.

1 Introduction

The task scheduling problem includes assigning the tasks of an application to processors and determining the execution order of tasks on the processors. The goal of task scheduling is usually to minimize the schedule length (makespan). Most scheduling problems are known to be NP-complete except a few restricted cases [1], [2]. Many sub-optimal algorithms have been devised to reduce the schedule length because of its importance on performance.

The scheduling algorithms could be classified into *static* or *dynamic* scheduling according to whether it is done at compile time (static scheduling) or on-the-fly (dynamic scheduling) [3]. In static scheduling an application is usually represented by DAG (Directed Acyclic Graph).

Our research focuses on a comparative study of several static DAG scheduling algorithms including our own algorithms in realistic heterogeneous environments. In this paper five scheduling algorithms, HEFT, CPOP, CPOP_E, CPOC and CPOC_E will be described and compared in three different environments: realistic, non-clustered and unrealistic heterogeneous environments. The first two algorithms HEFT and CPOP have been proposed by Topcuoglu et al. [4], and the others by us.

Even though it seems critical paths are important to reduce schedule length, the CPOP has lower performance than the HEFT and the HEFT gives good quality of schedules with low cost. However, our scheduling algorithm which was modified from the CPOP to adapt to more realistic environments can give more efficient schedules than the HEFT.

In the CPOP critical-path tasks are allocated on a single processor, however, we allocate those tasks on a cluster of processors among which the communication cost is relatively low. We call it CPOC (Critical-Path-On-a-Cluster). As we exploit a cluster instead of a processor for critical-path tasks, we can get more chance to reduce computation time of critical path while the communication cost still keeps low. Our ultimate scheduling algorithm, CPOC_E(Enhanced CPOC) of which processor selection rule is different from the CPOC shows better performance than the others in the realistic environment, although the CPOC_E does not show better performance than the HEFT in the unrealistic or the non-clustered environments.

The rest of the paper is organized as follows. In section 2 we describe related work about task scheduling and in section 3 five scheduling algorithms including ours are described and compared. In section 4 the experimental results are presented and discussed. Finally section 5 gives concluding remarks.

2 Related Work

The heuristic-based algorithms can be classified into a few categories: TDB (Task-Duplication Based) [5], UNC (Unbounded Number of Clusters) [6], [7], [8], and other scheduling algorithms. The other groups are further classified into BNP (Bounded Number of Processors) [9] and APN (Arbitrary Processors Network) [10], [11] scheduling algorithms. While the BNP assumes contention-free network and no routing strategies, the links are not contention-free in the APN. The idea of TDB algorithms is to reduce the communication overhead by duplicating tasks and allocating them on multiple processors redundantly. In each step of the UNC algorithms some clusters (or tasks) are merged to reduce the completion time. The UNC needs an additional step for mapping the clusters onto the available processors.

Most popular scheduling technique is *list scheduling*. The common idea of list scheduling heuristics is to make a scheduling list and schedule tasks from the front of the list. So it consists of two phases: a *task prioritizing phase* and a *processor selection phase*. On the task prioritizing phase priority of each task is computed to make a ready list, and the most appropriate processor is selected for the current highest-priority task on the processor selection phase.

For task prioritizing *t-level* (top level) and *b-level* (bottom level) are often used (they could be comparable to *downward rank* and *upward rank* respectively).

The upward rank, $rank_u$, is based on average computation time and average communication time. The upward rank of a task i is defined by

$$rank_u(i) = \overline{w}_i + \max_{j \in succ(i)} (\overline{c}_{ij} + rank_u(j)) \ . \tag{1}$$

where $succ(i)$ is the set of immediate successors of task i, $\overline{c}_{ij}$ is the average communication cost of edge (i, j) and $\overline{w}_i$ is the average computation cost of task i.

The *downward rank* of task i, $rank_d(i)$ is defined by

$$rank_d(i) = \max_{j \in pred(i)} (rank_d(j) + \overline{w}_j + \overline{c}_{ji}) \ . \tag{2}$$

where $pred(i)$ is the set of immediate predecessors of task i.

3 Scheduling Algorithms

In this section we describe and compare five scheduling algorithms. Those algorithms use $rank_u$ or $rank_u+rank_d$ on task prioritizing phase, and use one or more of the following heuristics on processor selection phase.
- H1: earliest-finish-time-first
- H2: assigning critical-path tasks to a processor
- H3: assigning critical-path tasks to a cluster
- H4: not earliest-finish-time-first for immediate predecessors of critical-path tasks.

H4 means when immediate predecessors of critical-path tasks are scheduled earliest-finish-time of critical-path tasks (not immediate predecessors itself) is pursued. H1 and H2 were proposed by Topcuoglu et al.[4] while H3 and H4 are proposed by us.

3.1 HEFT (Heterogeneous Earliest Finish Time)

The HEFT was proposed by Topcuoglu et al. [4], which uses $rank_u$ for task priority and H1 for processor selection. On the processor selection phase the task with the highest priority is picked for allocation and a processor which ensures the earliest-finish-time is selected. The earliest-finish-time is considered especially using *insertion-based* allocation. If we allocate tasks in the insertion-based way, a task can be inserted at the point of previously allocated task as long as there is free time slot.

The HEFT scheduling algorithm reflects two heuristics. First, a task which has higher upward rank is more important and preferred to other tasks. Intuitively, the upward rank of a task reflects the average remaining cost to finish all tasks after that task starts up. Second, simple but effective idea is earliest-finish-time-first approach (H1). However, as it pursues the earliest-finish-time-first of the current task, it may fall into local optima like a greedy method.

3.2 CPOP (Critical-Path-On-a-Processor)

The CPOP scheduling algorithm was introduced in the same literature as the HEFT. It differs from the HEFT in not only the priority of a task but also processor selection for the tasks with the highest priority (critical-path tasks). The priority of a task i is given by $rank_u(i) + rank_d(i)$. H1 and H2 are used for processor selection.

The task with the highest priority should be on a critical path, and in that case the priority is thought of as critical path length, |CP|. Intuitively, the upward rank of a task is the expected time to finish the exit task after the task starts, and the downward rank of a task is the expected elapsed time (not including its computation time) after the entry task. So if the summation of the upward rank and the downward rank of a task is the highest, the task is thought of as being on a critical path.

For the set of tasks which are on a critical path, CP, the algorithm finds a processor, p_{CP}, which minimizes the sum of computation time of all CP tasks. On the processor selection phase each task of a ready queue is selected for allocation and if the selected task is on a critical path, the processor p_{CP} is used; otherwise a processor which ensures the earliest-finish-time of the task is used.

3.3 CPOC (Critical-Path-On-a-Cluster)

The CPOP algorithm schedules all CP tasks on a processor which minimizes overall computation cost of the critical path. Since only one processor is used for the critical path, obviously there is no communication cost among CP tasks. However, we think that communication cost among processors which are located in the same local area network are relatively negligible. So we use a cluster of processors to execute CP tasks instead of using a single processor. This heuristic is profitable since it is possible to choose any processor within the cluster for each CP task. It gives more chance to minimize the sum of computation time of CP tasks.

```
                      <task prioritizing phase>
1   Compute rank_u(i) and rank_d(i) for each task i.
2   Assign rank_u(i) + rank_d(i) to each task i as priority.
                      <processor selection phase>
3   Find a set CP of tasks having the largest value of
    rank_u(i) + rank_d(i).
4   c_CP = find_critical_path_cluster(CP)
5   ReadyQ.insert(k) where k is the entry task.
6   while (not ReadyQ.empty()) {
7      i = ReadyQ.delete()
8      if (task i is in CP)
9         schedule i on a processor within cluster c_CP to minimi
          ze EFT(i).
10     else
11        schedule i on a processor to minimize EFT(i).
12     for each immediate successor k of task i,
13        if (every immediate predecessor of task k has been alr
          eady scheduled)
14           ReadyQ.insert(k)
15  }
```

Fig. 1. CPOC scheduling algorithm

Fig. 1 shows the CPOC algorithm, which uses $rank_u + rank_d$ for task prioritizing like CPOP, and uses H1 and H3 for processor selection. On the processor selection phase, for CP tasks the algorithm tries to find a cluster, c_{CP}, which minimizes the sum of computation time of all CP tasks. Each task of a ready queue is considered in turn for allocation. If the selected task is on a critical path, the cluster c_{CP} is used and a processor is selected within the cluster to ensure the earliest-finish-time of the task. Otherwise a processor is selected among all processors.

3.4 Enhancement of Algorithms: Earliest Start Time for CP Tasks

The processor selection scheme based upon earliest-finish-time-first reflects the heuristic that the current task is most important now and should finish as early as possible. In the HEFT scheduling algorithm the task having the highest priority is always selected for allocation at each step, so the earliest-finish-time-first approach matches with it well. In the CPOP or the CPOC algorithms, however, tasks will not be selected in order of priority because only ready tasks can be selected and the tasks having higher priority may not be ready.

Since the current ready task is not of the highest priority among unscheduled tasks, our goal should not be completion of the current task as early as possible. Instead, we should try to shorten the CP length, in other words, make the CP tasks finished as early as possible. So we could pursue the earliest start (or finish) time for the CP tasks when we even schedule other than CP tasks. The goal is to make the CP tasks start as early as possible when we schedule the immediate predecessors of the CP tasks. This processor selection heuristic is H4.

Enhanced CPOP. The scheduling algorithm (CPOP_E) differs from original CPOP in allocating the immediate predecessors of CP tasks. It uses heuristics, H1, H2 and H4 for processor selection instead of H1 and H2.

Enhanced CPOC. The scheduling algorithm (CPOC_E) differs from original CPOC in allocating the immediate predecessors of CP tasks. It uses heuristics, H1, H3 and H4 for processor selection instead of H1 and H3. Just three lines are modified for the CPOC_E algorithm compared with line 10 of Fig. 1 (see Fig. 2).

```
10-1   else if (let j = succ(i) and j is in CP)
10-2      schedule i on a processor to minimize EST(j).
10-3   else
```

Fig. 2. CPOC_E scheduling algorithm

4 Experiments

In this section we would describe experimental results over randomly generated task graphs and network graphs. Five scheduling algorithms will be compared in three different heterogeneous environments.

4.1 Task Graph Generation

Task graphs are randomly generated by a task graph generator which has been developed in this work. Most of input parameters for the task graph generator are similar to those of the HEFT. The followings are the list of the parameters:

- d: Out-degree.
- α: Shape of a task graph. $\sqrt{v}/\alpha$ is the mean of height of the task graph where v is the number of tasks.
- β: Heterogeneity of processor speed. If $\beta=0$, it is exactly homogeneous.
- γ: Heterogeneity of processor architecture. If $\gamma=0$, it is exactly homogeneous.
- ccr: Communication to computation ratio. If $ccr>>1$, it represents communication cost is very high compared to computation cost.
- π : Ratio of non-terminal tasks.

Among the above parameters γ and π are newly introduced in this work. Also, β has somewhat different meaning compared to Topcuoglu's. While all heterogeneity is represented only by β in Topcuoglu's, we use both β and γ. We can generate more various task graphs with both parameters. If $\gamma=0$, there would be no architectural difference among processors which mean there can be only heterogeneity of processor speed.

4.2 Network Graph Generation

Three different types of networks have been used for experiments: *realistic, non-clustered* and *unrealistic*. For the first two types a network generator has been used[12]. The generator could generate not only the realistic global networks which range from WAN to LAN but also just LANs. A LAN just consists of a single cluster, so it is denoted by *non-clustered* in this paper. Fig. 3 shows examples of the two types. The third type, *unrealistic*, can not be visualized, in which arbitrarily random communication costs are assigned to each pair of processor-to-processor.

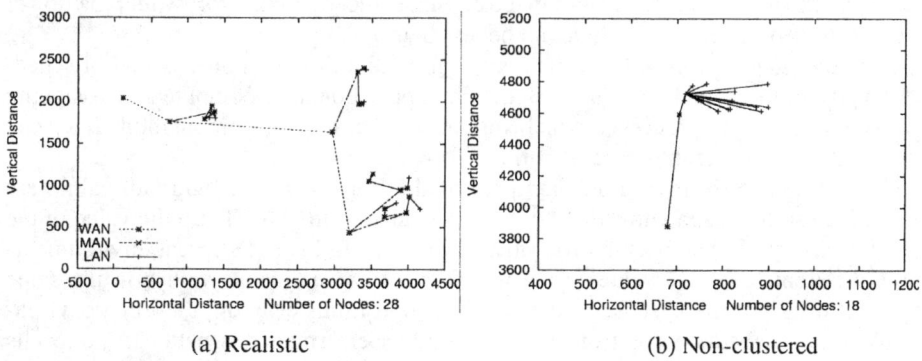

Fig. 3. Examples of networks

As shown in Fig. 3, the network generator gives node-to-node distances. For our experiments, we convert it into latency and bandwidth matrices. The latency matrix can be computed by multiplying the distance by the unit delay and adding the fixed delay. Every node-to-node latency is determined as the smallest value among the values of all possible paths using the Floyd-Warshall shortest path algorithm. The bandwidth matrix can be obtained from the smallest value along the node-to-node path. Table 1 shows parameters for computation of the latency and the bandwidth matrices.

Table 1. Bandwidth and delay parameters

	bandwidth(MB/s)	fixed delay(ms)	delay / unit distance(ms/unit)
WAN-to-WAN	1000	15	0.1
WAN-to-MAN	100	100	0.01
MAN-to-MAN	100	10	0.1
MAN-to-LAN	100	5	0.01
LAN-to-LAN	1	1	0.01

4.3 Performance Results

To compare the schedule of each algorithm, we would use two performance metrics. Though performance goal is to minimize schedule length, we need a normalized metric since each task graph has various schedule length. The first metric, SLR(Schedule Length Ratio) is defined by the following[4].

$$SLR = \frac{schedule\_length}{\sum_{i \in CP_{MIN}} \min\{w_{ik}\}} \quad . \tag{3}$$

$\min\{w_{ik}\}$ is the minimum value among computation costs of task i when it is assigned to processor k, and CP_{MIN} is the set of tasks on a critical path assuming every task has the minimum computation cost. Since the denominator would be lower bound of schedule length, SLR cannot be less than 1.

The other metric is speedup which is given by dividing the upper bound of schedule length by the actual schedule length. The upper bound is computed by assigning all tasks to a single processor which minimizes the sum of computation time, *i.e.*, ensures the best sequential execution time.

Comparative performance data are presented by Fig. 4. In the figure all values are normalized so that each value is relative to the value of the HEFT. So the value of the HEFT is always 1. The reason of normalization is as follows. Different environments show different levels of values, so it is difficult to visualize all values in the same figure. It does not make sense to compare the algorithms crossing the three environments from an absolute point of view. Generally, performance is better if the results present the smaller SLR or the higher speedup.

In cases of 'Unrealistic' or 'Non-clustered' the HEFT algorithm shows the best performance. We think that the HEFT generally gives good schedules in arbitrarily heterogeneous computing environments. However, as the environments reflect more realistic configuration which may be biased by constituting several clusters, the winner becomes the CPOC_E algorithm. The CPOP and the CPOP_E algorithms show poor performance in 'Unrealistic' or 'Non-clustered', but the performance is greatly improved in 'Realistic'.

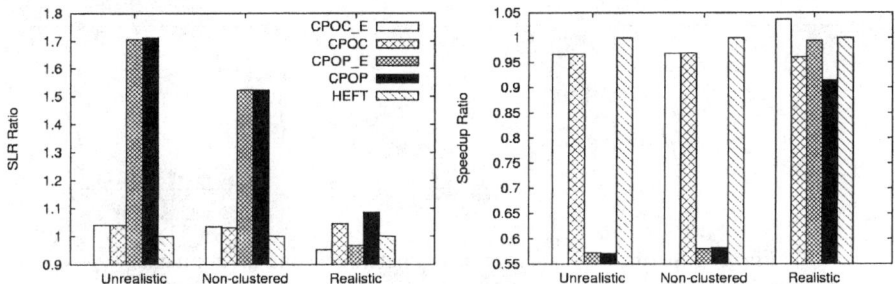

Fig. 4. Performance comparison of HEFT, CPOP, CPOC, CPOP_E and CPOC_E

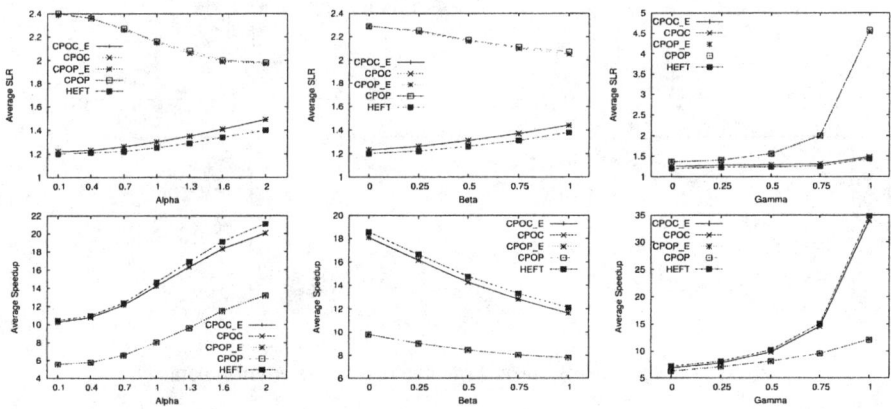

Fig. 5. Experiments on *unrealistic* heterogeneous environments

Fig. 5, 6 and 7 show results for α, β and γ parameters in three different heterogeneous environments. Note that there is very little difference between the CPOP and CPOP_E, and also between the CPOC and the CPOC_E in 'Unrealistic' or 'Non-clustered' environments. However, the enhanced algorithms show better performance in the 'Realistic' environment. This accounts for effectiveness of heuristic H4. H4 pursues earliest-start-time-first of CP tasks instead of that of its immediate predecessors when we schedule the immediate predecessors. It means that the heuristic considers communication cost between CP tasks and its immediate predecessors as well as computation time of the immediate predecessors. The communication cost does not affect performance if it would be given in arbitrarily random style. However, the heuristic is effective in a biased environment that some communication costs are

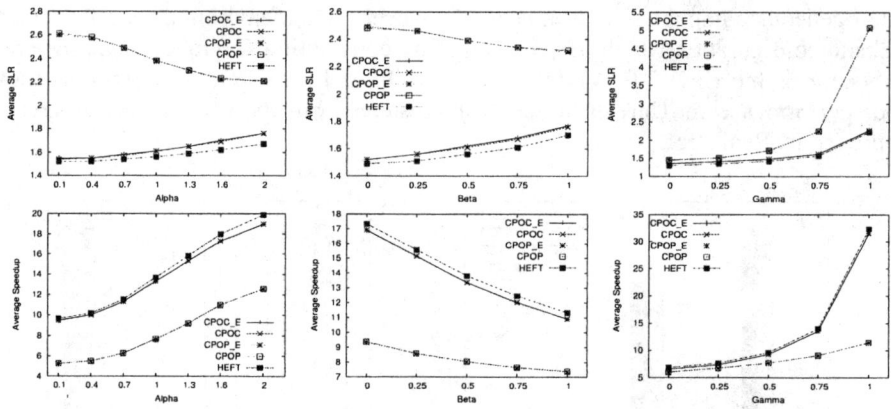

Fig. 6. Experiments on *non-clustered* heterogeneous environments

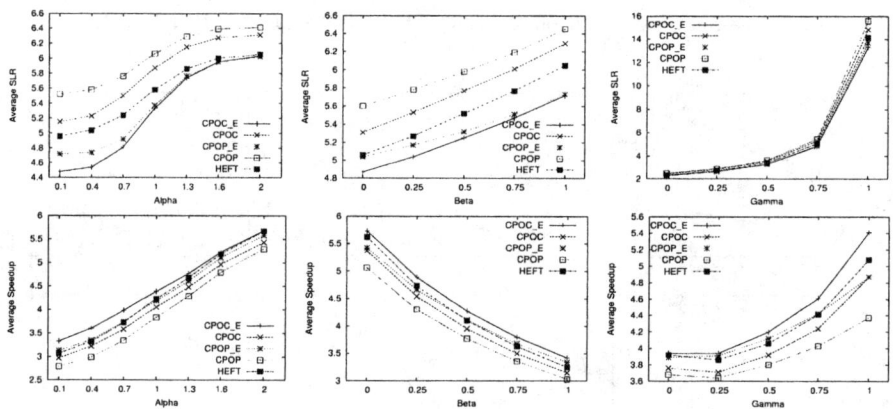

Fig. 7. Experiments on *realistic* heterogeneous environments

relatively higher than others. Although 'Realistic' environments ranged from WAN to LAN in the experiments, we think that H4 is still effective in any environments in which communication costs are asymmetric and biased.

Generally, as the SLR decreases the speedup increases, and vice versa. However, in cases of α or γ, the results are not accordance with the general tendency.

As α increases the lower bound rapidly decreases compared to the actual schedule length. So SLR would increase as α increases. However, the upper bound hardly changes over α. Undoubtedly the shape of a task graph does not have any correlation to the upper bound. Since the schedule length can be reduced for increment of α, speedup may increase. As shown in Fig. 5 and 6, the SLRs of CPOP and CPOP_E present different tendency from the others. If α is very small *i.e.*, height of a task graph is very high, critical path would be very long. Since only one processor is used for the critical path in cases of CPOP and CPOP_E, SLR could be very high. As α increases this situation would be mitigated.

γ represents heterogeneity of architecture, so the higher γ gives a chance to obtain less computation time for each task. It means very rapid decrement of the lower bound, which results in rapid increment of SLR. Though it is obvious that architectural heterogeneity gives more chance to shorten the length of each task, when we use a single-processor assignment it hardly affects the upper bound. So speedup would quickly increases.

5 Conclusions

In this paper we presented two new heuristics (H3 and H4) for processor selection and devised new scheduling algorithms, CPOC and CPOC_E by applying those heuristics. Five scheduling algorithms including ours have been discussed with four heuristics and compared experimentally in three different environments.

Compared to HEFT scheduling algorithm, the CPOC_E does not have good performance in arbitrarily (unrealistic) heterogeneous environments. However, the CPOC_E generates better schedules than the others in realistic heterogeneous environments.

From a comparative study it is confirmed that the heuristics H3 and H4 are effective in the realistic heterogeneous environments. This is mainly due to that locality of communication can be exploited in the realistic environments. Especially the communication latencies are possibly not arbitrarily heterogeneous. So the heuristic H3 uses a cluster to allocate CP tasks instead of a single processor. The heuristic H4 also reflects the realistic heterogeneous environments. It differs from H2 in that it pursues the earliest-start-time-first of CP tasks instead of the immediate predecessor's own earliest-finish-time-first.

The experimental results show CPOC_E outperforms CPOC and CPOP_E outperforms CPOP in the realistic environments. This accounts for effectiveness of H4. However, in the unrealistic environments there is very little difference between the CPOC and the CPOC_E, and also between the CPOP and the CPOP_E. The results also show CPOC outperforms CPOP, and this accounts for effectiveness of H3.

References

1. H. El-Rewini and H. H. Ali: Task Scheduling in Multiprocessing Systems, IEEE Computer, pp.27-37, Dec. 1995.
2. Y. Kwok and I. Ahmad: Static Scheduling Algorithms for Allocating Directed Task Graphs. ACM Computing Surveys, Vol. 31, No. 4, pp.407-471, Dec. 1999.
3. T. L. Casavant and J. G. Kuhl: A Taxonomy of Scheduling in General-Purpose Distributed Computing Systems," IEEE Trans. on Software Engineering, Vol. 14, No. 2, pp.141-154, Feb. 1988.
4. H. Topcuoglu, S. Hariri and M. Wu: Performance-Effective and Low-Complexity Task Scheduling for Heterogeneous Computing. IEEE Trans. on Parallel and Distributed Systems, Vol. 13, No. 3, pp.260-274, March 2002.
5. I. Ahmad and Y. Kwok: On Exploiting Task Duplication in Parallel Program Scheduling. IEEE Trans. on Parallel and Distributed Systems, Vol. 9, No. 9, pp.872-892, 1998.

6. M. Wu and D. D. Gajski: Hypertool: A Programming Aid for Message-passing Systems. IEEE Trans. on Parallel and Distributed Systems, Vol. 1, No. 3, pp.330-343, 1990.
7. T. Yang and A. Gerasoulis: DSC: Scheduling Parallel Tasks on an Unbounded Number of Processors. IEEE Trans. on Parallel and Distributed Systems, Vol. 5, No. 9, pp.951-967, 1994.
8. Y. Kwok and I. Ahmad: Dynamic Critical-path Scheduling: An Effective Technique for Allocating Task Graphs to Multiprocessor. IEEE Trans. on Parallel and Distributed Systems, Vol. 7, No. 5, pp.506-521, 1996.
9. J. Hwang, Y. Chow, F. D. Anger and C. Lee: Scheduling Precedence Graphs in Systems with Interprocessor Communication Times. SIAM J. Comput., Vol. 18, No. 2, pp.244-257, April 1989.
10. G. C. Sih and E. A. Lee: A Compile-time Scheduling Heuristic for Interconnection-constrained Heterogeneous Processor Architectures. IEEE Trans. on Parallel and Distributed Systems, Vol. 4, No. 2, pp.75-87, Feb. 1993.
11. H. El-Rewini and H. H. Ali Scheduling Parallel Program Tasks onto Arbitrary Target Machines. Journal of Parallel and Distributed Computing, Vol. 9, No. 2, pp.138-153, 1990.
12. M. B. Doar: A Better Model for Generating Test Networks. IEEE Global Telecommunications Conference, Nov. 1996.

ehSTCP: Enhanced Congestion Control Algorithm of TCP over High-Speed Networks[*]

Young-Soo Choi, Hee-Dong Park, Sung-Hyup Lee, and You-Ze Cho

School of Electrical Engineering and Computer Science,
Kyungpook National University, Korea
{yschoi, tenetshlee, yzcho}@ee.knu.ac.kr, hdpark@pohang.ac.kr

Abstract. Current TCP congestion control can be inefficient and unstable in high-speed wide area networks due to its slow response with a large congestion window. Several congestion control proposals have already been suggested to solve these problems and two properties have been considered: TCP friendliness and scalability, to ensure that a protocol does not take away too much bandwidth from TCP, while utilizing a bandwidth of high speed networks efficiently. In this paper, we propose a new variant of TCP for a high-speed network which combines delay-based congestion control with loss-based congestion control. Our simulation results show that proposed scheme performs better than the existing high-speed TCP protocols in terms of fairness, stability and scalability, while providing friendliness at the same time.

1 Introduction

TCP has already been widely adopted as a data transfer protocol for the Internet. The demand for high-speed applications such as bulk-data transfer, multimedia web streaming, high energy and nuclear physics, astronomy, bioinformatics, earth sciences, storage area network, and grid networking has increased. However, it has been reported that as the bandwidth-delay product continues to grow, TCP underutilizes the bandwidth and it will eventually become a performance bottleneck itself [1]-[5]. For example, according to [2], for TCP to increase its window to a full utilization of 10Gbps with 1500-byte packets, it will require over 83,333 RTTs. With 100ms RTT, this would take approximately 1.5 hours, and for full utilization in a steady state, the loss rate cannot exceed 1 loss event per 5,000,000,000 packets, which is less than the theoretical limit of the network's bit error rate. Consequently, it is impossible to achieve such a large throughput with TCP, mainly because TCP decreases its congestion window too drastically when packet losses occur, yet only it increases congestion window very slightly when there is no packet loss.

Recently, various adaptive schemes have been designed to offer more flexibility, wider bandwidth scalability, and fairer competition with the standard TCP.

---

[*] This work was supported in part by the Information & Communication Fundamental Technology Research Program and the ITRC of the Ministry of Information and Communication (MIC), Korea.

Such schemes include eXplicit Control Protocol (XCP) [1], High Speed TCP (HSTCP) [2], Scalable TCP (STCP) [3], FAST TCP [4], and Binary Increase TCP (BIC) [5]. XCP generalizes the Explicit Congestion Notification (ECN), enabling more (or explicit) information to be sent about the degree of congestion in the network. XCP resembles the Explicit Rate (ER) allocation algorithm in the available bit rate (ABR) service in ATM networks. XCP provides predominant efficiency, fairness, and stability. However, since XCP requires XCP senders, routers, and receivers to be deployed, deployment issues still remain. In this paper, the above-mentioned protocols such as HSTCP, STCP, and BIC are referred to as high-speed TCP protocols. As an alternative TCP implementation, delay-based congestion avoidance algorithms such as TCP Vegas [6] and most recently FAST TCP have been proposed. To detect and avoid congestion, TCP Vegas estimates the backlog which is the number of buffered packets inside the network. Many studies have shown that TCP Vegas outperforms TCP Reno in terms of link utilization, jitter, and packet loss rate. But if the network congestion occurs in the backward path, TCP Vegas underutilizes the bandwidth on the forward path [7]. FAST TCP is a high-speed version of TCP Vegas for fast, long distance networks.

In this paper, we propose a new variant of HSTCP, called as eHSTCP (enhanced HSTCP), which is a hybrid scheme between loss-based congestion control and delay-based congestion control. First, by using the TCP timestamp option, eHSTCP estimates the effective RTT (eRTT) that is a delay that may be measured if there is no queueing delay along the backward path. Thus eHSTCP avoids the effect of backward path congestion. Second, instead of using Vegas' backlog to prevent packet loss proactively, eHSTCP refines the additive increase multiplicative decrease (AIMD) mechanism of HSTCP to enhance scalability, TCP friendliness, stability, and fairness.

The remainder of this paper is organized as follows: Section 2 briefly discusses some high-speed TCP congestion control algorithms and Vegas-like delay-based congestion control algorithms. Section 3 describes the eHSTCP protocol. Section 4 presents the simulation results and conclusions are given in Section 5.

2 Related Work

The importance of congestion control is now widely acknowledged and extensive research has already been done to enhance the performance of TCP. TCP congestion control is composed of two major algorithms: slow-start and congestion avoidance algorithms which allow TCP to increase the data transmission rate without overwhelming the network. TCP uses a variable called congestion window ($cwnd$) and cannot inject more than $cwnd$ segments of unacknowledged data into the network. The TCP congestion avoidance algorithm is called AIMD and it is the basis for steady state congestion control. In the congestion avoidance phase, TCP increases the congestion window by one packet for each RTT and halves the congestion window in the event of a packet loss. The TCP congestion control is briefly explained in Table 1.

Table 1. TCP congestion control in congestion avoidance

Regular TCP	ACK: $w \leftarrow w + \alpha/w$ LOSS: $w \leftarrow w - \beta \times w$	$\alpha_{TCP} = 1, \beta_{TCP} = 0.5$
HSTCP	ACK: $w \leftarrow w + \alpha_{HSTCP}/w$ LOSS: $w \leftarrow w - \beta_{HSTCP} \times w$	if $w > Low\_Window(= 38)$ $\alpha_{HSTCP} = 0.1578 \times w^{0.8024}$ $\times \beta_{HSTCP}/(2 - \beta_{HSTCP})$ $\beta_{HSTCP} = -0.052 \ln w + 0.6892$ else regular TCP congestion control
STCP	ACK: $w \leftarrow w + \alpha_{STCP}$ LOSS: $w \leftarrow w - \beta_{STCP} \times w$	if $w > Low\_Window(= 16)$ $\alpha_{STCP} = 0.01, \beta_{STCP} = 0.125$ else regular TCP congestion control
BIC	ACK: if $\alpha_{BIC} < S_{max}$ $\quad w \leftarrow w + \alpha_{BIC}/w$ else $\quad w \leftarrow w + S_{max}/w$ LOSS: $max\_w \leftarrow w$ $w \leftarrow w - \beta_{BIC} \times w$ $min\_w \leftarrow w$ $target \leftarrow (max\_w + min\_w)/2$	if $w > Low\_Window(= 14)$ $S_{max} = 32, \beta_{BIC} = 0.125$ $\alpha_{BIC} = max\{(target - w), 0.001\}$ else regular TCP congestion control

HSTCP was introduced by S. Floyd in [2] as a modification of the TCP congestion control mechanism, to improve the performance of TCP in fast, long delay networks. HSTCP is designed to have a different response in an environment with a very low congestion event rate, and have the regular TCP response in an environment with a packet loss rate of maximum 10^{-3}. HSTCP introduces a new relation between the average congestion window w and the steady state packet drop rate p. For simplicity, HSTCP response function provides a straight line on a log-log scale. The HSTCP response function is specified using three parameters: *Low_Window*, *High_Window*, and *High_P*. The *Low_Window* is used to establish a point of transition and to ensure TCP friendliness. The HSTCP response function uses the same response function as the regular TCP when the current *cwnd* is at most *Low_Window*, and uses the HSTCP response function when the current congestion window is greater than the *Low_Window*. Meanwhile, *High_Window* and *High_P* are used to specify the upper end of the HSTCP response function, where *High_P* is the specific drop rate needed in the HSTCP response function in order to achieve a *High_Window* as the average congestion window. The HSTCP response function is represented by a new additive increase and multiplicative decrease parameters. These parameters modify both the increase and decrease parameters according to the *cwnd*.

STCP was described by T. Kelly in [3]. Instead of using an additive increase, the increase is exponential and the multiplicative decrease factor β_{STCP} is set to 0.125. Here, the congestion avoidance algorithm of the STCP is MIMD (Multiplicative Increase and Multiplicative Decrease).

[5] revealed that HSTCP and STCP have a fairness problem when multiple flows with different RTTs are competing. Also, [5] introduced BIC that attempts to correct the RTT unfairness. BIC regards congestion control as a searching problem in which the system can give binary feedback through packet loss as to whether the current congestion window is larger than the network capacity. BIC uses a binary search scheme to quickly find an estimated equilibrium window size, and then slowly increases the congestion window.

The congestion avoidance algorithms of HSTCP, STCP, and BIC are briefly expressed in Table 1. For more details, see [2], [3], and [5], respectively. Since most existing schemes concentrate on bandwidth scalability and TCP friendliness, the performance of high-speed TCP and the impact of its use on the present implementation of TCP have been highlighted. As such, the results show that existing high-speed TCP schemes can relieve bandwidth scalability problem to a certain extent. However further studies in TCP friendliness and fairness are still needed.

TCP and most high-speed TCP protocols are loss-based congestion avoidance protocol. The problem with loss-based congestion avoidance scheme is that a TCP sender keeps increasing its congestion window until it causes a buffer overflow. These "self-induced" packet losses cause increased loss rate, decreased throughput, and significant jitter. Delay-based congestion avoidance protocols attempt to control the congestion window based on RTT measurements. If the congestion window is large enough to saturate the available bandwidth, queueing delay will increase at the congested node, and thus the RTTs will increase. So, the TCP sender decreases the congestion window when the RTTs start increasing to prevent packet loss proactively. There are several variations of the delay based congestion avoidance schemes, including TCP Vegas [6] and most recently, FAST TCP [4]. TCP Vegas performs better regular TCP in terms of link utilization, stability, fairness, packet loss. FAST TCP is a stabilized version of TCP Vegas for high-speed networks. As opposed to constant increase/decrease factor of TCP Vegas, FAST fully exploits delays as a congestion measure and window increments depend upon the current window size and RTT decrements and vice versa.

When a TCP Vegas competes with a TCP Reno, TCP Vegas does not receive a fair share of bandwidth due to its conservative congestion avoidance mechanism. That is, while TCP Vegas tries to maintain a smaller queue, TCP Reno keeps increasing congestion window until a packet loss is detected. Therefore, the performance of TCP Vegas degrades significantly when TCP Vegas coexists with TCP Reno [7], [8]. FAST TCP is still vulnerable to this problem. Choosing a large target backlog can make Vegas-like protocols compatible with Reno-like protocols [9]. However, it is a challenge to choose suitable backlog parameter, since the correct choice will depend on the number of flows and the buffer size of the bottleneck router.

In addition, the correlation between increased delays (or RTTs) and congestive losses has recently been challenged [10], thereby raising serious doubts as to the effectiveness of DCA algorithms given that their main assumption is that RTT measurements can be used to predict and avoid network congestion.

3 eHSTCP Protocol: Mechanisms and Deployment

In this section we propose eHSTCP (enhanced HSTCP), a new variant of TCP for high-speed networks that provides high utilization, stability, and fairness. In contrast to TCP Vegas, eHSTCP uses effective RTT to avoid the effects of reverse path congestion. And rather than preventing packet loss as in TCP Vegas, for additive increase mechanism, eHSTCP uses a backlog as a binary feedback to determine whether the network is fully utilized. Also, effective RTT is used to refine multiplicative decrease mechanism of HSTCP to achieve high link utilization, while guaranteeing TCP friendliness comparable to that of HSTCP.

3.1 Delay Measurement

If network congestion occurs in the backward path, TCP Vegas-like protocols may overestimate RTT and unnecessarily decrease congestion window. By using the TCP timestamp option, our mechanism obtains samples of queueing delay on the forward and backward paths separately. Note that the sender and receiver clocks do not have to be synchronized since we are only interested in the relative time difference. By distinguishing the direction in which congestion occurs, eHSTCP is robust in the case of backward congestion.

To remove the effect of reverse path congestion, we redefine the effective RTT (eRTT) as

$$eRTT = RTT - d_{b,q} \tag{1}$$

$$d_{b,q} = d_b - min(d_b) \tag{2}$$

where RTT is a newly measured round trip time, $d_{b,q}$ is the backward queueing delay, d_b is a measured backward delay, and $min(d_b)$ is the minimum of all measured backward delays. Consequently, the $eRTT$ indicates a round trip time when there is no backward path congestion. We compute the smoothed eRTT by using an exponential weighted moving average (EWMA), with a delay smoothing parameter of 1/8. This value is typically used for computing the smoothed RTT for TCP.

3.2 Congestion Control Based on Effective RTT

Since previous research shows that HSTCP provides acceptable bandwidth scalability and friendliness [2], [5], we modified HSTCP's AIMD mechanisms as follows:

Additive Increase Algorithm: The TCP Vegas estimates a proper amount of extra data to be kept in the network pipe (i.e. backlog) and controls the congestion window size accordingly. The amount is between the two thresholds α and β, as shown in the following:

$$\alpha \leq N = (Expected - Actual) \times RTT_{min} \leq \beta \qquad (3)$$

where $Expected$ is the current congestion window size divided by RTT_{min} (the minimum of all measured RTTs), and $Actual$ represents the current congestion window size divided by the newly measured RTT. According to Little's Law, N represents the backlog at the bottleneck router queue. Thus, TCP Vegas tries to keep at least α packets, but no more than β packets queued in the network.

For our scheme, first, to prevent throughput degradation from the reverse cross-traffic, we redefine N and $Actual$ as follows:

$$Actual' = cwnd/eRTT \qquad (4)$$

$$N' = (Expected - Actual') \times RTT_{min} = cwnd \times d_{f,q}/eRTT \qquad (5)$$

where $d_{f,q}$ is the forward queueing delay. Consequently, N' and $Actual'$ represent the backlog and $Actual$, respectively, if there is no backward queueing delay.

Since random noise in the RTT measurements (due to time resolution, OS interrupts, etc) cannot be avoidable in practice, FAST-like congestion control, which fully exploits delay to congestion control, seems unfeasible in most cases. Nevertheless, a delay is still valuable information as the indication of network congestion. Therefore, we use N' as a binary feedback signal of whether the network is fully utilized or not. More specifically, if measured backlog N' is lower than N^*, we assume the bottleneck is underutilized and uses HSTCP's congestion control algorithm, where N^* is the target backlog. If N' is higher than N^*, we use TCP Reno's congestion control algorithm. When the network is fully utilized, eHSTCP behaves equally to TCP Reno, so that the eHSTCP can stay in this region longer. This mechanism leaves buffer space for other traffic and thus makes eHSTCP TCP friendly. Unchanging or decreasing the congestion window size like TCP Vegas are not considered because TCP Vegas can not keep fairness with TCP Reno and to prevent convergence stalling. Also, it is important to note that flows with small RTTs do not gain a competitive advantage over flows with long RTTs. Proposed additive increase mechanism can significantly correct the RTT fairness problem as compared with other high-speed TCP protocols.

Multiplicative Decrease Algorithm: After a packet loss, TCP Reno halves the congestion window. If we size the router buffer to match the delay-bandwidth product, this mechanism ensures that the buffer does not underflow and goes empty. However, it is generally impractical; because there is no clear way to get average RTT information (even if it exists). Moreover, in high-speed networks large buffers are problematic for both technical as well as cost reasons. HTCP suggests a backoff scheme which makes a more informed decision by using minimum and maximum RTTs [11]. The rationale of the HTCP's backoff scheme is as follows: When congested, the total throughput through the link is given by

$$Throughput^- = \sum_{i=1}^{n} \frac{cwnd_i}{RTT_{max,i}} \qquad (6)$$

where n is the number of flows and RTT_{max},i is the maximum RTT experienced by the i'th source. After the backoff, the throughput is given by

$$Throughput^+ = \sum_{i=1}^{n} \frac{(1 - \beta_i) \times cwnd_i}{RTT_{min,i}} \qquad (7)$$

To ensure the buffer is empty while preventing buffer underflow, HTCP sets $1 - \beta_{HTCP}$ as RTT_{min}/RTT_{max}. eHSTCP, in contrast to HTCP, uses:

$$1 - \beta_{eHSTCP} = \frac{RTT_{min}}{eRTT} \qquad (8)$$

$$w \leftarrow w - w \times min(\beta_{eHSTCP}, \beta_{HSTCP}) \qquad (9)$$

By inspecting the raw data from our simulation results, we found that the measured RTTs are frequently smaller than the maximum RTT when a packet loss occurs. The main reason behind this phenomenon is TCP burstiness. In previous work [12], we pointed out that since the congestion window achieved by a high-speed TCP flow can be quite large, there is a strong possibility that the sender may send a large burst of packets in response to a single acknowledgement. Since the bursty behavior of high-speed TCP can lead to bursty traffic flows in high speed networks, actual measured RTTs are lower than the maximum RTT. This mismatching violates HTCP's assumption and causes link underutilization. Thus, we use RTT instead of RTT_{max}. Also, to exclude reverse path congestion, we use effective RTT.

From the equation 9, eHSTCP reduces the congestion window by a smaller size than HSTCP. Reducing the congestion window less drastically improves utilization and throughput fluctuation but it hurts convergence speed and TCP friendliness since larger window flows give up their bandwidth slowly. To provide comparable TCP friendliness and bandwidth scalability of HSTCP at least while to avoid drastic decreasing congestion window, we employ the following algorithm. After packet loss, if β_{eHSTCP} is smaller than β_{HSTCP}, eHSTCP reduces its congestion window using β_{eHSTCP} and enters a safety check phase. At the same time, w_desg is calculated using β_{HSTCP}. During this safety check phase, eHSTCP does not increase its congestion window and monitors the backlog. If N' exceeds N^* in the safety check phase, eHSTCP assumes that β_{eHSTCP} is too aggressive and reduces its congestion window again to w_desg. Therefore, it takes one RTT time for eHSTCP to decrease its window size to the size of HSTCP used. Otherwise, after safety check phase, eHSTCP enters the additive increase phase. Below, we present the pseudo-code of eHSTCP.

ACK:
```
if (low_window > w)
    w ← w + 1/w;                    // regular TCP congestion control
else
    if(InSafetyCheckPhase)
        if (N' > N*)                // backlog exceeds threshold during SafetyCheckPhase
            w ← w_desg;
        else
            w ← w;
    else
        if (N' < N*)                // not fully utilized → HSTCP congestion control
            w ← w + α_HSTCP/w;
        else                        // fully utilized → regular TCP congestion control
            w ← w + 1/w;
```
LOSS:
```
if (low_window <= w){
    if (InSafetyCheckPhase){        // another packet loss during SafetyCheckPhase
        w ← w_desg − β_eHSTCP×w_desg;
    } else{
        if (β_eHSTCP < β_HSTCP){
            w_desg ← w − β_HSTCP×w;
            w ← w − β_eHSTCP×w;
            InSafetyCheckPhase ← 1;
        }else
            w ← w − β_HSTCP×w;
    }
} else
    w ← w×0.5;                      // regular TCP
```

4 Simulation Results and Discussion

In this section, we compare the simulated performance of eHSTCP with that of HSTCP, STCP, and BIC. Unless explicitly stated, the same amount of background traffic is used for all experimental runs. eHSTCP is implemented into the ns [13] simulation code for TCP SACK. The TCP timestamp option is used to obtain accurate RTT samples. The topology used for the simulation experiments is shown in Fig. 1. Various bottleneck capacities and delays are tested. The bottleneck router uses FIFO scheduling and a drop tail buffer management scheme. By default, the buffer size at the bottleneck router is set to 2000 packets. All high-speed TCP flows use the forward direction. To reduce the phase effect and synchronized feedback, a significant amount of background traffic is

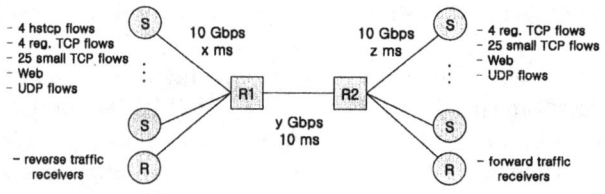

Fig. 1. The network topology for the simulation

used in both directions, along with randomized RTTs and starting times. For background traffic, web traffic, 25 small TCP flows with a limited congestion window size under 64, and 4 long lived TCP flows are created in both directions for all simulations, unless otherwise specified. The packet size is 1000 bytes. In our experiments, we use $N^* = 10$ and the safety check phase $= 5 \times RTT_{max}$.

4.1 Utilization, Fairness, and Stability

In this experiment, RTT of all flows is around 40ms and the bottleneck bandwidth is 2.5Gbps. To evaluate bandwidth scalability, we measure link utilization and the average packet loss rate of the link between router R1 and R2. We also measure the fairness using Jain's fairness index among high-speed TCP flows. And the sample standard deviation normalized by the average throughput is used to evaluate stability. From Table 2, it can be seen that link utilization of eHSTCP is relatively comparable to that of STCP. Also, eHSTCP shows the best performance among all protocols under packet loss rate evaluation criterion. It is found that for HSTCP, BIC, and eHSTCP, the fairness index is approximately equal to 1 and STCP has some fairness issues. eHSTCP stays at the fully utilized region longer and proposed multiplicative decrease mechanism avoids unnecessarily drastic decreasing of congestion window. Therefore, we observe that eHSTCP shows the best stability.

Table 2. Comparison of utilization, fairness, packet loss rate, and stability

	HSTCP	STCP	BIC	eHSTCP
Link Utilization	0.92	0.99	0.95	0.99
Packet loss rate(%)	0.0197	0.1281	0.0206	0.0065
Fairness index	0.99	0.91	0.99	0.99
stddev	0.148	0.149	0.107	0.047

4.2 RTT Fairness

In this experiment, two high speed flows with a different RTT are used. The RTT of flow 1 is 40ms, while the RTT of flow 2 is computed for 120ms and 240ms. The bottleneck bandwidth is 1Gbps. Table 3 depicts the throughput ratio of the two high-speed flows. In Table 3, we see the bias against connections with long RTT. As predicted in [5], there is a serious fairness problem with flows of different RTTs

Table 3. The throughput ratio of two high-speed flows over various RTT ratios under 1Gbps

	HSTCP	STCP	BIC	eHSTCP
Ratio = 3	42.46	111.45	12.03	3.88
Ratio = 6	197.80	341.65	84.65	4.77

for HSTCP and STCP. HSTCP and STCP tend to starve long RTT flows under high bandwidth environments, since short RTT flows quickly dominate the link bandwidth, starving out the other flows. eHSTCP's RTT fairness outperforms HSTCP, STCP and BIC.

4.3 TCP Friendliness

Fig. 2. shows the percentage of the bandwidth shared by each flow type. For 20Mbps, all high-speed TCP protocols show similar TCP friendliness. As the bandwidth gets larger than 300Mbps, the share of bandwidth taken by the web, small TCP, and long-lived TCP flows is substantially reduced due to the TCP bandwidth scalability problem. STCP achieves higher throughput for various scenarios but also, STCP shows the worst TCP friendliness followed by BIC and HSTCP in most cases. eHSTCP utilizes the link bandwidth as efficiently as HSTCP. That is, eHSTCP consistently gives good friendliness relatively to all high-speed TCP protocols for all bandwidths while providing with the bandwidths scalability of HSTCP at least. Note that, under 2.5Gbps, eHSTCP flows consume more bandwidth than HSTCP flows. The increase in eHSTCP bandwidth shares can be accounted by the reduction in unused bandwidth. Although eHSTCP consumes more bandwidth than HSTCP, it does not take bandwidth away from TCP, but instead from the unused bandwidth.

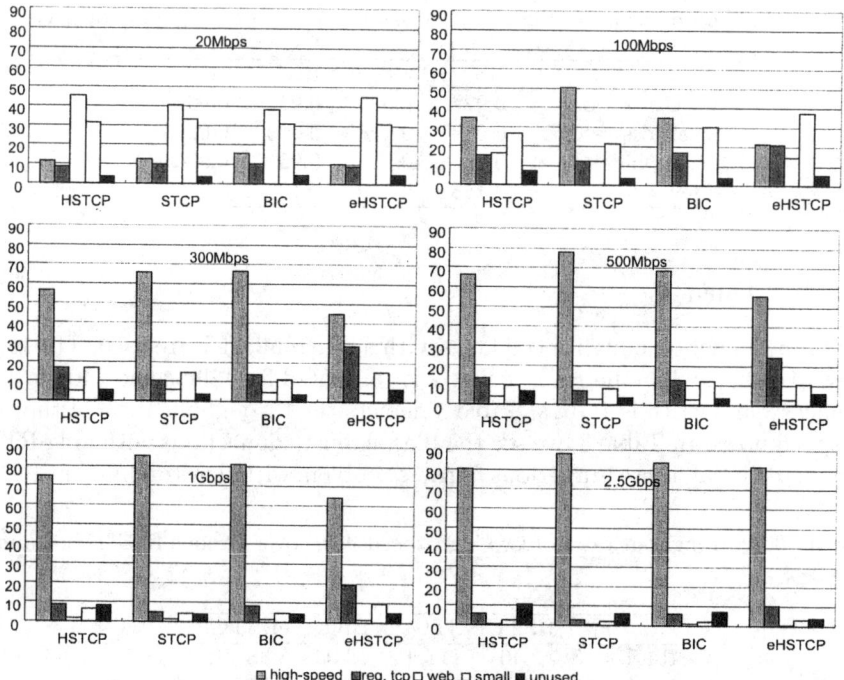

Fig. 2. TCP friendliness for various bandwidth networks

4.4 More Dynamic Scenario

In this scenario, we add 100 UDP flows with ON and OFF times drawn from a heavy-tailed distribution. The mean ON and OFF time is 1 second and the mean OFF time is also 1 second, with each source sending at 5Mbps during an ON time. Table 4 shows the percentage of the bandwidth shared by each flow type. And table 5 shows the average loss rate and the sample standard deviation normalized by the average throughput. BIC searches the equilibrium congestion window size by using loss history. In a dynamic scenario, loss history might be out of date, and thus unused bandwidth increases in BIC scenario. Note that the bandwidth scalability of eHSTCP is comparable to that of STCP and also, eHSTCP is the friendliest protocol of all the high-speed TCP protocols. To summarize, eHSTCP provides good TCP friendliness for all bandwidths while providing bandwidth scalability, which is comparable to STCP in high-speed environments.

Table 4. Comparison of the bandwidth shared by each flow type

	HSTCP	STCP	BIC	eHSTCP
high-speed flows	71.70	77.39	72.66	74.35
regular TCP	4.54	2.51	4.78	5.24
web	0.59	1.50	0.59	0.66
small TCP	2.42	1.82	2.34	2.8
UDP	9.69	10.31	9.86	10.50
unused	11.06	6.47	9.76	6.45

Table 5. Comparison of packet loss rate and stability

	HSTCP	STCP	BIC	eHSTCP
loss rate(%)	0.0375	0.167	0.0514	0.0272
stddev	0.1441	0.1061	0.1073	0.0581

5 Conclusion

In this paper, we propose a new variant of TCP for high-speed network which combines delay-based congestion control with loss-based congestion control. Although existing high-speed TCP schemes solve bandwidth scalability to some degree, there are still problems with fairness, friendliness, and stability. We define the effective RTT as the RTT that may be measured if there is no backward queueing delay along the path. Then, we refine HSTCP's AIMD mechanism. Since delay is error-prone, proposed additive increase algorithm uses effective RTT as binary feedback signal as to whether a network is full utilized. Proposed additive increase mechanism provides enhanced stability, reduced packet loss rate, and TCP friendliness. To guarantee the comparable TCP friendliness and scalability of HSTCP at least while to avoid drastic decreasing congestion window, proposed multiplicative decrease algorithm uses the effective RTT and deploys the safety check phase. We have shown through simulations that the proposed scheme outperforms other high-speed TCPs in terms of fairness, friendliness, and stability, while utilizing a link bandwidth efficiently.

References

1. Katabi D., Handley M., and Rohrs C.: Internet Congestion Control for High Bandwidth-Delay Product Networks. In Proceedings of the ACM SIGCOMM, pp. 89-102, (2002).
2. Floyd S.: HighSpeed TCP for Large Congestion Windows. RFC3649, (2003).
3. Kelly T.: Scalable TCP: Improving Performance in Highspeed Wide Area Networks. ACM SIGCOMM Computer Communication Review, vol.33, pp. 83-91, (2003).
4. Jin C., Wei D. X. and Low S. H.: FAST TCP: motivation, architecture, algorithms, performance. In Proceedings of the IEEE Infocom, vol. 4, pp. 2490-2501, (2004).
5. Xu L., Harfoush K., and Rhee I.: Binary Increase Congestion Control for Fast, Long Distance Networks. In Proceedings of IEEE Infocom, vol. 4, pp. 2514-2524, (2004).
6. Brakmo L. and Peterson L.: TCP Vegas: End to End Congestion Avoidance on a Global Internet. IEEE Journal on Selected Areas in Communication, vol. 13, no. 8, pp. 1465-1480, Oct. (1995).
7. Hengartner U., Bolliger J., and Gross T.: TCP Vegas Revisited. In Proceedings of IEEE Infocom, vol. 3, pp. 1546-1555, (2000).
8. Mo J., La R., Anantharam V., and Walrand J.: Analysis and Comparison of TCP Reno and Vegas. In Proceedings of IEEE Infocom, vol. 3, pp. 1546-1555, (2000).
9. Feng W. and Vanichpun S.: Enabling compatibility between TCP Reno and TCP Vegas. In Proceedings of Symposium on Applications and the Internet, pp. 301-308, (2003).
10. Martin J., Nilsson A., and Rhee I.: Delay Based Congestion Avoidance for TCP. IEEE/ACM Transactions on Networking, vol. 11, no. 3, pp. 356-369, (2003).
11. Shorten R. and Leith D.: H-TCP: TCP for high-speed and long-distance networks. In Proceedings of the PFLDnet, (2004).
12. Choi Y., Lee K., and Cho Y.: Performance Evaluation of High-Speed TCP Protocols with Pacing. Lecture Notes in Computer Science, vol. 3332, pp. 322-329, (2004).
13. The Network Simulator ns2, http://www.isi.edu/nsnam/ns/

Programming Paradigms for Networked Sensing: A Distributed Systems' Perspective*

Amol Bakshi and Viktor K. Prasanna

Department of Electrical Engineering,
University of Southern California, Los Angeles, CA 90089, USA
{amol, prasanna}@usc.edu

Abstract. Research in embedded networked sensing has primarily focused on the design of hardware architectures for sensor nodes and infrastructure protocols for long lived operation of resource constrained sensor network deployments. There is now an increasing interest in the programming aspects of sensor networks, especially in the broader context of pervasive computing. This paper provides a brief overview of ongoing research in programming of sensor networks and classifies it into layers of abstraction that provide the application developer with progressively higher level primitives to express distributed, phenomenon-centric collaborative computation. As a specific instance of a macroprogramming methodology, we discuss the data driven Abstract Task Graph (ATaG) model and the structure of its underlying runtime system. ATaG separates the application functionality from non-functional aspects, thereby enabling end-to-end architecture-independent programming and automatic software synthesis for a class of networked sensor systems. A prototype visual programming, software synthesis, functional simulation and visualization environment for ATaG has been implemented.

1 Introduction

Distributed sensor networks allow intelligent, dense monitoring and control of physical environments and have a wide range of applications such as home and office automation, habitat monitoring, intruder detection, etc. Advances in VLSI technology have enabled the integration of sensing, computation, and wireless communication capabilities into small, inexpensive hardware platforms. Ad hoc wireless sensor networks (WSNs) comprised of such untethered nodes provide embedded sense-and-response capability. The unprecedented degree of access to information about the physical world could provide context awareness to other applications, making WSNs an integral part of the vision of pervasive, ubiquitous computing – with the long term objective of seamlessly integrating this fine grained sensing infrastructure into larger, multi-tier systems. A comprehensive overview of state of the art in wireless embedded sensing can be found in [1].

---

* This work is supported by the National Science Foundation, USA, under grant number IIS-0330445.

With continuing advancements in sensor node design and increasingly complex applications, interest in automatic synthesis of sensor network applications is inevitable, given the fact that ease of programming is perhaps the single most important determinant of the ubiquity and acceptance of a computing platform. In other words, there should be a well-defined methodology to translate high level intentions of the programmer expressed in a suitable formalism into an executable specification for the underlying deployment.

In this paper, we focus on programming of networked sensor systems from a distributed systems perspective. We assume that protocols and services for the basic communication and collaboration infrastructure are already available for the target platform. The job of the programming model is to suitably abstract these existing services and define a model of computation for the distributed system that is useful for application development. A collection of autonomous sensor nodes passing messages through a communication network fits the definition of a distributed computing system. However, some of the fundamentally new characteristics of networked sensing systems that differentiate them from traditional parallel and distributed computing are as follows:

- **Transformational vs. reactive processing:** Most of the traditional parallel and distributed applications are transformational systems characterized by a function that maps input data to output data. The main purpose of parallelism for such systems is to reduce the overall latency of computation and to provide robustness through replication [2]. A networked sensor system is not transformational but is primarily reactive in that it has to continuously respond to external and internal stimuli. An event of interest in the environment triggers computation and communication in one or more nodes of the network, usually in the immediate vicinity of the event.

- **Nature of input data:** In transformational distributed systems, a given set of input data is statically and/or dynamically distributed among various computing nodes in order to perform the 'transformation' with lowest latency. In sensor networks, however, most the data is continuously created in the network through the act of sampling the sensor interfaces. The time and location of origin of a particular piece of data influences the processing performed on it. The typical untethered wireless sensor node is energy constrained. It is desirable to process the data as close to the source as possible, and collaborative, in-network data processing is hence an important consideration in networked sensing.

- **Spatial awareness:** From the end users' perspective, an embedded sensor network ultimately represents a discrete sampling of a continuous physical space. Instead of specifying applications in terms of sensor nodes and the network connectivity, behaviors can be naturally specified using spatial abstractions. For instance, the exact placement of sensor nodes will probably be of incidental interest as long as the set of sensing tasks mapped onto a subset of those nodes at any given time collaboratively ensure the desired degree of coverage.

Macroprogramming of sensor networks broadly refers to an application development methodology – supported by a suitable programming model, compiler, and runtime support – that liberates the programmer from having to compose the complex control, coordination, and state maintenance mechanisms at the individual node in order to accomplish the desired global behavior.

Low level optimizations especially related to the networking layer are important for long lived operation of untethered resource-constrained networks, and protocols for positioning, time synchronization, etc., provide the basic infrastructure for distributed computing in the sensor network. The challenge in defining high level macroprogramming models is achieving the right balance between long lived operation through low level optimizations and ease of application development by hiding most of the low level details from the programmer.

In the next section, we analyze the layers of programming abstractions that naturally emerge from the ongoing research in programming models in the sensor network community. Section 3 discusses our macroprogramming model called the Abstract Task Graph (ATaG) [17] that builds upon the core concepts of the data driven computing paradigm to allow domain experts to develop sensor network applications. ATaG provides support for reactive processing, mechanisms to concisely indicate distributed, in-network collaborative computation, and support for space awareness both for expressing collaborative computation in terms of spatial neighborhoods, and to express notions such as spatial density of task placement that can be used to provide a desired degree of sensing coverage. Details of the ATaG programming model and the runtime system can be found in [3] and [4] respectively. We conclude in Section 4 with a discussion of our broader vision and related work in the context of design automation for sensor networks.

2 Layers of Programming Abstraction

Figure 1 depicts our view of the emerging layers of programming abstraction for networked sensor systems. Many protocols have been implemented to provide the basic mechanisms for efficient infrastructure establishment and communication in ad hoc deployments. These include energy-efficient medium access, positioning, time synchronization, and a variety of routing protocols such as data centric and geographic routing that are unique to spatial computing in embedded networked sensing. Ongoing research, such as MiLAN [5] is focusing on sensor data composition as part of the basic infrastructure. A sensor data composition framework delegates the responsibility of interfacing with physical sensors and aggregating the data into meaningful application-level variables to an underlying middleware instead requiring its incorporation as part of the application-level logic [6].

2.1 Service-Oriented Specification

To handle the complexity of programming heterogeneous, large-scale, and possibly dynamic sensor network deployments and to make the computing substrate

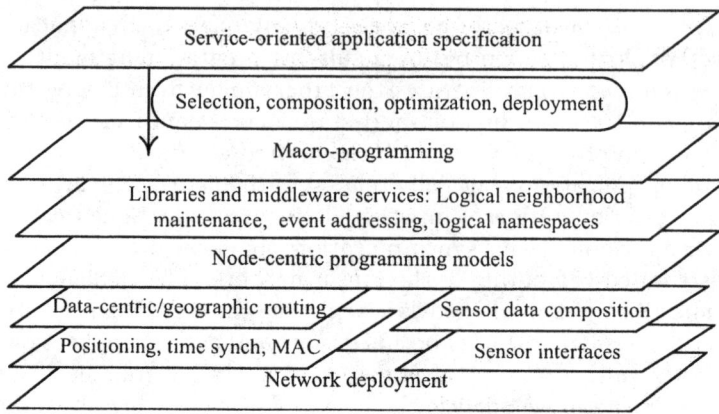

Fig. 1. Layers of abstraction for application development on WSNs

accessible to the non-expert, the highest level of programming abstraction for a sensor network is likely to be a purely declarative language. The Semantic Streams markup and query language [7] is an example of such a language that can be used by end users to query for semantic information without worrying about how the corresponding raw sensor data is gathered and aggregated. The basic idea is to abstract the collaborative computing applications in the network as a set of services, and provide a query interpretation, planning, and resource management engine to translate the service requirements specified by the end user into a customized distributed computing application that provides the result. A declarative, service-oriented specification allows dynamic tasking of the network by multiple users and is also easier to understand compared to low level distributed programming.

2.2 Macroprogramming

The objective of macroprogramming is to allow the programmer to write a distributed sensing application without explicitly managing control, coordination, and state maintenance at the individual node level. Macroprogramming languages provide abstractions that can specify aggregate behaviors that are automatically synthesized into software for each node in the target deployment. The structure of the underlying runtime system will depend on the particular programming model. While service-oriented specification is likely to be invariably declarative, various program flow mechanisms - functional, dataflow, and imperative - are being explored as the basis for macroprogramming languages. Regiment [8] is a declarative functional language based on Haskell, with support for region-based aggregation, filtering, and function mapping. Kairos [9] is an imperative, control-driven macroprogramming language for sensor networks that allows the application developer to write a single centralized program that operates on a centralized memory model of the sensor network state.

ATaG [3] (discussed in more detail in the next section) explores the data flow paradigm as a basis for architecture-independent programming of sensor network applications.

2.3 Node-Centric Programming

In node-centric programming, the programmer has to translate the global application behavior in terms of local actions on each node, and individually program the sensor nodes using languages such as nesC [10], galsC [11], C/C++, or Java. The program accesses local sensing interfaces, maintains application level state in the local memory, sends messages to other nodes addressed by node ID or location, and responds to incoming messages from other nodes. While node-centric programming allows manual cross-layer optimizations and thereby leads to efficient implementations, the required expertise and effort makes this approach insufficient for developing sophisticated application behaviors for large-scale sensor networks.

The concept of a logical neighborhood – defined in terms of distance, hops, or other attributes – is common in node-centric programming. Common operations upon the logical neighborhood include gathering data from all neighbors, disseminating data to all neighbors, applying a computational transform to specific values stored in the neighbors, etc. The usefulness and ubiquity of neighborhood creation and maintenance has motivated the design of node-level libraries [12], [13] that handle the low level details of control and coordination and provide a neighborhood API to the programmer.

Middleware services [5], [14], [15] also increase the level of programming abstraction by providing facilities such as phenomenon-centric abstractions. Middleware services could create virtual topologies such as meshes and trees in the network, allow the program to address other nodes in terms of logical, dynamic relationships such as leader-follower or parent-child, support state-centric programming models [16], etc. The middleware protocols themselves will typically be implemented using node-centric programming models, and could possibly but not necessarily use communication libraries as part of their implementation.

3 Data Driven Macroprogramming with the Abstract Task Graph

The Abstract Task Graph (ATaG) [3], [17] seeks to raise the level of programming abstraction by (a) allowing the architecture-independent specification of application behavior through a mixed imperative-declarative program specification, and (b) transferring the responsibility of low level coordination, communication, and optimization to an underlying runtime system, thereby allowing the application developer to focus on high level behavioral aspects.

Macroprogramming broadly refers to the collaborative tasking of sensor nodes as opposed to configuring individual node behaviors. ATaG support macroness at the application level by allowing the programmer to define and manipulate information at the desired level of abstraction without worrying about how the information is created. ATaG also supports macro-ness at the architecture level by allowing concise specification of common patterns of in-network distributed processing such as neighbor-to-neighbor, many-to-one, tree-based, etc.

3.1 Objectives and Key Concepts

ATaG is designed to support intuitive expression of reactive processing, spatial awareness, network awareness, architecture independence, and composability. The first three are the functional objectives that allow concise and intuitive expression of the behavior of a networked sensing application. The non-functional objectives – architecture independence and composability – are motivated by software development concerns such as ease of programming and code reusability.

To accomplish these objectives, ATaG employs a *data driven programming model* and *mixed imperative-declarative program specification* for separation of concerns. Tasks are defined in terms of their input and output data objects. An underlying runtime system manages task scheduling and inter-task communication. Availability of operands triggers task execution, subject to firing rules. This model is attractive for computing in distributed systems for programming convenience, and the modularity and extensibility of the programs. Also, a sensor network can be viewed as a system for domain-specific transformation of sensor data and many applications can be naturally expressed as a set of transformations on raw and processed sensor readings.

The mixed imperative-declarative specification separates the 'when and where' of processing from the 'what'. The same program can be compiled for a different network size and topology by interpreting the declarative ('when and where') part in the context of that network architecture, while the imperative ('what') part remains unchanged. The ATaG programmer, who writes only the task implementations, is free to focus on application-level design without being concerned about low level details of the sensor node platform and the specifics of a particular deployment.

3.2 Illustrative Example

Figure 2 is a complete ATaG program for a sensor network application with two distinct behaviors. The first is to periodically sample the temperature at each node (through a temperature sensor), continuously compute the average reading, and log it at a designated root node. The second is to detect an object in the network through acoustic sensors, and report the current location of the target to a designated root node.

The figure shows the declarative part of the ATaG program, which consists of the set of abstract tasks (ovals), abstract data items (square rectangles), I/O dependencies or channels (arrows), and annotations (shaded rectangles). The imperative part consists of the user-supplied code associated with each of the five abstract tasks in the program, and the user-supplied data structures that represent the four abstract data items. Abstract tasks represent the types of processing in the application, abstract data items represent the types of application-specific

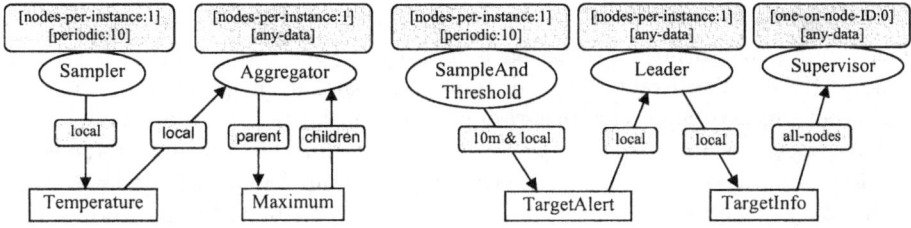

Fig. 2. An ATaG program for temperature monitoring and object tracking

data exchanged between instances of abstract tasks, and the abstract channels denote the I/O relationship between tasks and data. Task annotations govern the placement of task instances and the firing conditions of each instance. Channel annotations govern the scope of dissemination (collection) of instances of a particular data item produced (consumed) by an instance of the abstract task.

The dependencies and annotations in this program specify that the Sampler and Aggregator tasks are to be instantiated on each node. The Sampler periodically produces a Temperature reading which is routed to the Aggregator task on the same node. The Aggregator receives temperature readings from its own node and from its child nodes in a logical tree structure maintained by the runtime. The Aggregator is fired whenever an instance of Temperature is produced on its node or on its child nodes. The aggregated reading is conveyed up the tree. The object tracking algorithm depicted here is based on the one discussed in [12]. Briefly, each node determines whether the object is in its vicinity by periodically sampling and thresholding the reading from its acoustic sensor. If a target is detected (a TargetAlert is produced), all the nodes that detect the target broadcast their readings to all other nodes that might have detected the same target - in this example, the assumption is that a 10m radius includes this set of nodes. The Leader task on each node receives all such readings, including its own. The Leader task on the node that has the highest reading calculates the target position and transmits the TargetInfo to the designated Supervisor on the root node.

The ATaG program has data driven semantics. A particular task instance is scheduled for execution when the firing rules of the abstract task are satisfied. A task can be specified as periodic with a specified period of execution, or its execution can be predicated on the occurrence of either (any-data) or all (all-data) of its input data items. This paradigm intuitively supports *reactive processing* because abstract data items can represent the occurrence of events such as the detection of an intruder, in addition to carrying information about the occurrence such as the location of the intruder. Each execution of a task may not necessarily result in the production of each of its output data items; depending on the (application-specific) semantics of the abstract data items, the output can be produced only when certain conditions are satisfied. For instance, in this example, SampleAndThreshold produces a TargetAlert only if the sampled reading exceeds a specific threshold and not otherwise.

Architecture independence is evident in the fact that both the task and channel annotations are independent of a particular network deployment. Task annotations indicate requirements such as density of placement and can be generic (e.g., instantiate task on each node) or specific (e.g., instantiate task on node ID 0). The exact physical node which hosts an instance of the task will be determined when this program is compiled for a particular deployment. *Spatial awareness and network awareness* is also supported through channel annotations that allow a task to control the scope of input and output of data items. For instance, when an instance of the SampleAndThreshold task produces an instance of TargetAlert, it is disseminated by an underlying runtime system to all nodes within 10m of the producer. Similar annotations can be used to specify the neighborhood in terms of nodes, e.g., 'k-hop'. The application developer need not worry about how the neighborhood information is maintained at that node, what routing protocols are used for the communication, etc.

Finally, the data driven paradigm makes ATaG programs highly composable. The ATaG program in Figure 2 actually consists of two disjoint abstract task graphs, and can be considered as a larger application that is composed by concatenating the ATaG programs of two smaller applications, corresponding to temperature averaging and object tracking respectively. Composability is also enabled by the fact that the only methods available to a task for producing and consuming data items are the put() and get() methods respectively. Similar to the communication orthogonality of tuple spaces, these methods do not require the producer and consumer to know each other's identity. This enables distributed sharing of data both in space and in time. Also, since tasks are not coupled to each other, there is a high degree of code reuse since a new task can be added to the application without modifying the code associated with existing tasks.

3.3 Application Development Methodology

Figure 3 depicts the process of application development using ATaG. The input to the process is an ATaG program and a description of the target deployment

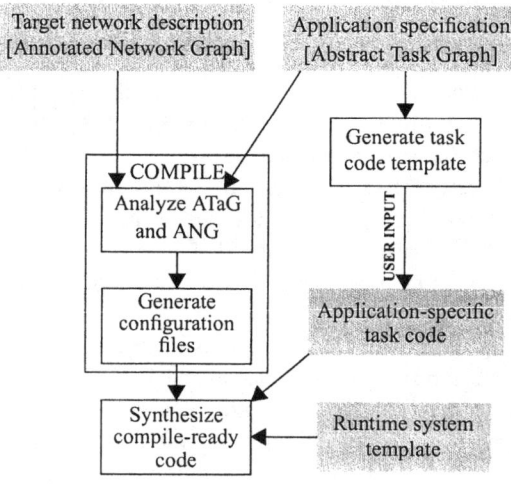

Fig. 3. Application development with ATaG

in the form of an annotated network graph (ANG), which is not discussed in this paper. The ANG contains information such as the number of nodes, the co-ordinates of each node, network connectivity, etc.

The graphical interface to the programming and synthesis environment is through a configurable graphical tool suite called the Generic Modeling Environment (GME) [18]. The declarative part of the ATaG program which consists of the various declarations and their annotations is specified visually. GME stores the model defined by the user in a canonical format. Tools called *model interpreters* can read from and write to this model database. In our case, model interpreters were written for the components represented by unshaded boxes in Figure 3.

4 Towards Design Automation: System-Level Support for Macroprogramming

In the context of programming methodologies for sensor networks, *design automation* refers to the automatic customization of the underlying system level support for a high level language. As depicted in Figure 1, the highest level of abstraction is a declarative specification that expresses the desired semantic information to be extracted from the system. This specification will ideally be translated into a macroprogram after suitable identification, selection, and composition of the individual behaviors that collaborative provide the desired service. The macroprogram in turn is compiled into a distributed software system that includes the application-level functionality as well as the mechanisms for control and coordination within a node and between nodes in the system.

The design of the underlying runtime system is critical for design automation because a well designed runtime system can (i) greatly simplify the compilation and code generation process, and (ii) allow plug and play integration of the various low level protocols and services whose choice could be influenced at compile time by the performance requirements of the end user. The data-driven ATaG runtime (DART) [4] is designed to separate application-independent mechanisms for control and coordination from application-specific configuration information to customize the individual node behavior.

Figure 4 is a high level overview of the modular structure of the data driven ATaG runtime called DART. Each module offers a well-defined interface to other modules in the system, and has complete ownership of the data and the protocols required to provide that functionality. This reduces interactions and dependencies among modules, and hiding the module implementation allows an entirely different set of protocols to be used within a module as long as the interface is not affected. We briefly summarize the purpose of each module - details can be found in [4]. The *ATaGManager* stores the information from the user-specified ATaG program that is relevant to the particular node. This information includes task annotations such as firing rule and I/O dependencies, and the annotations of input and output channels associated with the data items that are produced or consumed by tasks on the node. *Datapool* is responsible

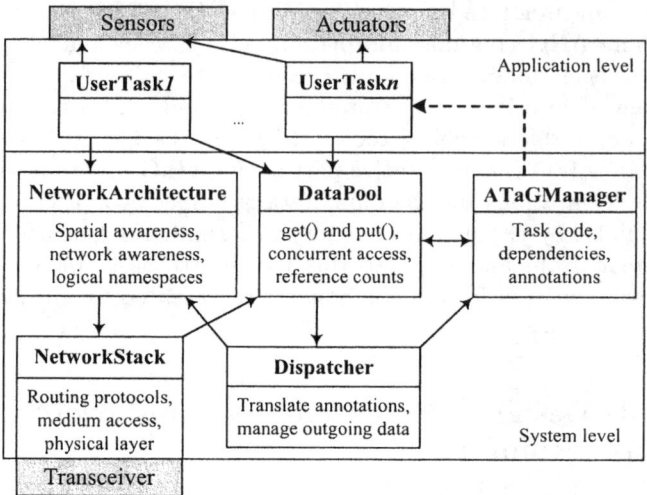

Fig. 4. The structure of the DART runtime system

for managing the instances of abstract data items produced or consumed at the node. *NetworkArchitecture* is responsible for maintaining all information about the real and virtual topology of the network. *NetworkStack* is in charge of communication with other nodes in the network, and manages the routing, medium access, and physical layer protocols. *Dispatcher* is responsible for disseminating data items that are produced on the node to other nodes in the network as specified in the ATaG program. In addition, a *Startup* module is responsible for initializing node-level services such as the transceiver functionality, the protocols for topology discovery, etc., and then starting the initial set of application-level tasks. The remainder of the execution is driven by the side-effects of the get() and put() calls made by the tasks, and the data items arriving over the network interface for addition to the data pool.

During the normal course of application execution, there are three main events that can occur: (i) a get() invocation by a user task, (ii) a put() invocation by a user task, or (iii) a put() invocation by the receiver thread when a data item arrives from another node. When a get() invocation occurs, *DataPool* merely decrements the reference count of the data item in question. When a local task invokes a put(), *DataPool* first checks if the corresponding data item is inactive before adding the newly produced data instance to the pool. This check ensures that all currently scheduled tasks that have been triggered by the production of a particular data instance get a chance to consume the data before it is overwritten by the same or different producer. *DataPool* the informs *ATaGManager* about the production of the data. *ATaGManager* determines the list of tasks that depend on this data item, checks their firing rules, and schedules the eligible tasks for execution. *DataPool* then notifies *Dispatcher* and finally returns control to the user task. *Dispatcher* interacts with *ATaGManager*,

NetworkArchitecture, and *NetworkStack* to send the data item to other nodes as indicated by the ATaG program. When the third type of event - an invocation of put() by the receiver thread of the *NetworkStack* - occurs, it is handled in much the same way as a local invocation, except that *Dispatcher* is not part of the loop.

There are three classes of APIs available to the ATaG programmer: (i) get()/put() calls to the data pool, (ii) the network-awareness and spatial-awareness API that allows a task instance to determine the composition of its neighborhood, and (iii) the API to the sensor interface. Since the runtime system does not know the access pattern of each task to the sensing interface, it cannot optimize resource usage. To enable resource management by the runtime requires, there should be a way for tasks to specify their sensor data requirements at a high level and leave the details of interfacing with sensors to the runtime system. In future work, we plan to extend the ATaG model with a special class of abstract data items to represent readings (scalar values, images, etc.) from the sensing interface(s). A set of annotations will be defined for the abstract 'sensor data' items, to indicate the type of sensing interface and other parameters such as spatial coverage and temporal coverage. This extension will allow the runtime a greater flexibility in task placement and resource management. An important problem in this context is resource allocation in face of conflicting requests from application tasks. The challenge is to develop a robust and scalable mechanism and a common utility scale to arbitrate across disparately developed ATaG libraries that are combined into a larger application. The key challenge in extending the basic model to handle such scenarios is to maintain the core design objectives - especially application neutrality - while enabling the expression of increasingly sophisticated behaviors.

5 Conclusion

The complexity of programming large scale embedded networked sensor systems has stimulated interest in high level programming paradigms that ease the task of application development. Many of the concepts from traditional distributed computing such as different program flow mechanisms (control driven, data driven, and demand driven) and coordination structures (distributed shared memory, tuple spaces, etc.) are applicable at various levels of abstraction in the "programming stack" for sensor networks.

The Abstract Task Graph was discussed in this paper as a demonstration of the applicability of data driven computing for code modularity, reuse, and extensibility, and of mixed imperative-declarative programming for separation of concerns to the programming of networked sensor systems. Programming languages such as ATaG will ultimately act as the intermediate representations that are generated from high level service-oriented specifications and synthesized into deployable software for a target system.

References

1. Iyengar, S.S., Brooks, R.R., eds.: Distributed Sensor Networks. Chapman & Hall/CRC (2004)
2. Bal, H.E., Steiner, J.G., Tanenbaum, A.S.: Programming languages for distributed computing systems. ACM Computing Surveys **21** (1989) 261–322
3. Bakshi, A., Prasanna, V.K., Reich, J., Larner, D.: The abstract task graph: A methodology for architecture-independent programming of networked sensor systems. In: Workshop on End-to-end Sense-and-respond Systems (EESR). (2005)
4. Bakshi, A., Pathak, A., Prasanna, V.K.: System-level support for macroprogramming of networked sensing applications. In: Intl. Conf. on Pervasive Systems and Computing (PSC). (2005)
5. Heinzelman, W., Murphy, A., Carvalho, H., Perillo, M.: Middleware to support sensor network applications. IEEE Network (2004)
6. Cohen, N.H., Purakayastha, A., Turek, J., Wong, L., Yeh, D.: Challenges in flexible aggregation of pervasive data. In: IBM Research Report RC 21942. (2001)
7. Whitehouse, K., Zhao, F., Liu, J.: Semantic streams: a framework for declarative queries and automatic data interpretation. Technical Report MSR-TR-2005-45, Microsoft Research (2005)
8. Newton, R., Welsh, M.: Region streams: Functional macroprogramming for sensor networks. In: 1st Intl. Workshop on Data Management for Sensor Networks (DMSN). (2004)
9. Gummadi, R., Gnawali, O., Govindan, R.: Macro-programming wireless sensor networks using kairos. In: Intl. Conf. Distributed Computing in Sensor Systems (DCOSS). (2005)
10. Gay, D., Levis, P., von Behren, R., Welsh, M., Brewer, E., Culler, D.: The nesC language: A holistic approach to networked embedded systems. In: Proceedings of Programming Language Design and Implementation (PLDI). (2003)
11. Cheong, E., Liu, J.: galsC: A language for event-driven embedded systems. In: Proceedings of Design, Automation and Test in Europe (DATE). (2005)
12. Whitehouse, K., Sharp, C., Brewer, E., Culler, D.: Hood: a neighborhood abstraction for sensor networks. In: 2nd Intl. Conf. on Mobile systems, applications, and services. (2004)
13. Welsh, M., Mainland, G.: Programming sensor networks using abstract regions. In: First USENIX/ACM Symposium on Networked Systems Design and Implementation (NSDI). (2004)
14. Liu, T., Martonosi, M.: Impala: A middleware system for managing autonomic, parallel sensor systems. In: ACM SIGPLAN Symposium on Principles and Practice of Parallel Programming. (2003)
15. Yu, Y., Krishnamachari, B., Prasanna, V.K.: Issues in designing middleware for wireless sensor networks. IEEE Network **18** (2004)
16. Liu, J., Chu, M., Liu, J., Reich, J., Zhao, F.: State-centric programming for sensor-actuator network systems. In: IEEE Pervasive Computing. (2003)
17. Bakshi, A.: Architecture-independent programming and software synthesis for networked sensor systems. PhD thesis, University of Southern California (2005)
18. Generic Modeling Environment, http://www.isis.vanderbilt.edu/projects/gme

Deadlock-Free Distributed Relaxed Mutual-Exclusion Without Revoke-Messages

Sukhamay Kundu

Computer Science Dept, Louisiana State University,
Baton Rouge, LA 70803, USA
kundu@csc.lsu.edu

Abstract. The revoke mechanism in generalized relaxed distributed mutual exclusion algorithm GRME [1] for eliminating a potential deadlock can cause extensive unnecessary revoke actions by the nodes which fail to receive the required granted-replies to their resource requests within a certain predefined time limit. It may happen that there is actually no deadlock present or that only a few nodes need to revoke some of their resource requests to eliminate the deadlock. We show that if the interference graph G is triangle-free (no three nodes are mutually adjacent), then we can choose the request-sets R_i in GRME in such a way that deadlocks are prevented altogether and there is no need to use revoke-messages, while keeping the resource-use decisions fully distributed and allowing non-interfering nodes to use the resource simultaneously.

1 Introduction

The distributed Generalized Relaxed Mutual Exclusion algorithm (GRME) was introduced in [1] to address the dynamic frequency allocation problem to the base stations in a mobile telephone system. Here, each frequency is a separate shared resource for which the base stations compete with each other, and thus there is one mutual exclusion problem for each frequency. A frequency can be shared by two non-interfering base stations, i.e., when they are physically separated by a certain minimum distance. Since there is a limited number of frequencies available to a given wireless service provider and the number of customers far exceeds the number of those frequencies, the sharing of frequencies among the base stations (for their customers) plays a key role in improving the service-performance by reducing the call blocking, which occurs when a base station cannot be assigned a suitable frequency in response to one of its customer's attempt to place a call. (No two customers within a base station cell can share a frequency at any given moment.) The term "relaxed" in GRME refers to the fact that several mutually non-interfering nodes (base stations) can simultaneously use a resource, and the term "generalized" refers to the use of the abstract information structure approach introduced in [6].

In the simplest distributed mutual exclusion algorithm by Ricart and Agrawala [2], a node sends a request-message to all other nodes for permission to use the resource and enters the critical section when it receives all $N-1$ permissions (N being the number of nodes in the network). It uses only $2(N-1)$ messages per resource-access compared to $3(N-1)$ in Lamport's algorithm [3]. The algorithm by Suzuki and Kasami [4] uses $N-1$ request-messages and one granted-message (which is called privilege-message in [4]) per resource access, but the granted-message has N pieces of information (one per node). Here, the node which is currently using the resource determines single-handedly the next node to get the resource and sends that node the granted-message. The granted-message contains for each node i the most recent resource-access number r_i, which is increased to $r_i + 1$ when node i completes a resource-use before it sends granted-message to the next node to access the resource. Maekawa [5] uses a new approach, where a node sends request-message only to a subset of nodes and it accesses the resource when granted permission by each of those nodes; it uses only $O(\sqrt{N})$ messages. In [6], Sanders generalizes Maekawa's idea to an information-structure-based approach. In each of [2-6], at most one node can use the resource at a time, whereas in GRME several non-interfering nodes can access the resource simultaneously.

The resource (frequency) sharing problem is modeled in [1] by an *interference graph* $G = (V(G), E(G))$ whose nodes are the processes (base stations) and whose links (i, j) indicate that process i and process j interfere, i.e., they cannot access the resource at the same time (mutual exclusion). Two nodes i and j can use the resource simultaneously if and only if $(i, j) \notin G$. In GRME algorithm [1], which generalizes Sanders' algorithm [6], each node i sends a request-message to a subset of nodes R_i, and when each node $j \in R_i$ grants the request to i, node i can access the resource. Likewise, when node i is done with resource, it informs a set of nodes $I_i \subseteq R_i$. The behavior of GRME is completely determined by the choice of the sets R_i and I_i, except for the variation of message arrivals due to arbitrary transmission delays. We first review GRME algorithm and show that we must have $I_i = R_i$ for all i in order for GRME to be a meaningful solution to the mutual exclusion problem.

The initial GRME algorithm, which can easily give rise to deadlocks, is modified further in [1] using an abort mechanism that allows a node to withdraw all its pending resource requests using a revoke-message if it does not get the resource within a predetermined time. The later is simply interpreted as a deadlock (there being no explicit test for deadlock detection) although there may not be an actual deadlock. If there is an actual deadlock, it may happen that all nodes involved in the deadlock will exercise the abort mechanism although it might be possible to eliminate the deadlock by aborting only some of the resource requests of one or a few deadlocked nodes. The approach used in [1] has therefore the potential of a significant and unnecessary increase in message load, and indeed the system remains vulnerable to the same deadlock once again. We address the deadlock avoidance problems by focusing on the appropriate selection of the sets R_i so that no deadlock can occur and hence no need to use the revoke-messages. We show this can be done for triangle-free interference graphs G, while keeping the workload fairly distributed among all nodes.

2 Review of GRME Algorithm

Fig. 1 shows a finite-state model (FSM) of the algorithm GRME in [1], where the algorithm was presented simply as a set of rules. The finite-state model shows more clearly the actions available at a node i at various stages. Each transition in Fig. 1 is labeled with a "condition/action" pair. If node i is currently in state σ, then each transition in state σ whose condition-part is satisfied becomes *enabled* and one of the enabled transitions is selected at random for execution. The action-part of the selected transition is then carried out and the head of the transition becomes the new state of node i. The empty-condition, denoted by "–", means that it is always satisfied; the empty-action, also denoted by "–", means no action is performed. Each node i in the interference graph G acts according to this FSM. Table 1 shows the three kinds of messages (M1)-(M3) used in GRME for coordinating the resource allocation to achieve mutual exclusion. The list L_i at node i contains the messages at i that are yet to be processed (or need further processing). The FSM in Fig. 1 is non-deterministic because at each state $\sigma \neq$ start-state one may have several transitions enabled at a given moment. We also do not specify any particular order in which messages are to be selected from L_i for processing. We write $R'_i = \{j: i \in R_j\}$, the nodes which send request-message to node i.

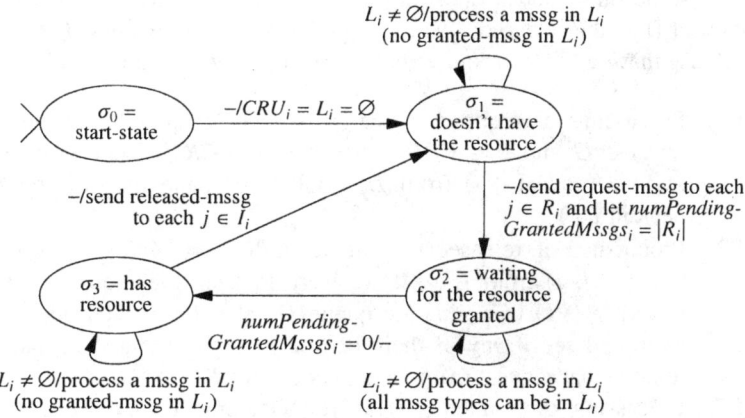

Fig. 1. The finite-state machine for node i in algorithm GRME

Table 1. Messages use in GRME algorithm

(M1) Resource *request* message: A node i sends request(i, j) message to a each node j in a fixed set R_i = request-set(i) when it currently does not have the resource and wishes to acquire it.

(M2) Resource *granted* message: A node $j \in R_i$ sends granted(j, i) message to node i as a response to a request(i, j) message from i if certain conditions hold.

(M3) Resource *released* message: A node i sends released(i, j) message to each node j in a fixed set I_i = inform-set(i) when it is done with the current use of the resource. The released-messages to nodes in I_i as a whole can be thought of as the response of node i to the set of granted-messages it received (from nodes in R_i) as a whole.

It is convenient to define the digraph G_R which has the same nodes as G and the arcs $\{(i, j): j \in R_i, 1 \leq i \leq |V(G)|\}$. Thus, R_i is the set of nodes that are *adjacent from* node i and R'_i is the set of nodes that are *adjacent to* node i. As will be seen, the behavior of GRME in regard to mutual exclusion and deadlock is completely determined by G_R (in spite of its non-deterministic nature) when $R_i = I_i$ for all i, which is the case of primary interest.

2.1 Message Processing

We denote a message from i to j by mssg(i, j), where mssg = request, granted, or released. Each node i maintains, in addition to its message-list L_i, the set CRU_i of nodes j to which it has sent the message granted(i, j) and from which it has not received the corresponding message released(j, i). Thus, $j \in CRU_i$ means that node j is either currently holding the resource or it is waiting to receive granted-message from some other nodes or it may have released the resource but i does not know about it (i.e., the released(j, i)-message has not been processed at i); it may even be the case that i will never know about it because $i \notin I_j$. In particular, $CRU_i \subseteq R'_i$.

(P1) Processing of request(i, j) at j: If $i \notin CRU_j$ and there is no node k such that $(i, k) \in G$ and $k \in CRU_j$, then add i to CRU_j, send granted(j, i) to i, and remove request(i, j) from L_j. (Otherwise, the message request(i, j) is put back in L_j.)

(P2) Processing of released(i, j) at j: $CRU_j = CRU_j - \{i\}$ and remove released(i, j) from L_j. (If we were to keep the request(i, j) that was postponed in (P1) in a separate queue D_j at node j, then one could go through them and see if any of them can be granted now and in that case we would remove that request from D_j and add i to CRU_j.)

(P3) Processing of granted(i, j) at j: Decrement numPendingGrantedMssgs$_j$ (which is initialized to $|R_j|$ in the transition from σ_1 to σ_2 in Fig. 1) by 1 and remove granted(i, j) from L_j.

The states σ_1 and σ_3 in Fig. 1 differ only in that node i sends out different messages (to nodes in R_i or I_i, respectively) in the transitions (σ_1, σ_2) and (σ_3, σ_1); none of these transitions changes L_i. The processing of a particular kind of message does not depend on the state, but the content of L_i can depend on the state.

Example 1. Consider the interference graph G in Fig. 2 and let $R_i = \{4\}$ and $I_i = \{2\}$ for each i. On completion of processing the messages in the sequence ⟨request(1,4), request(3,4), granted(4,1), granted(4,3)⟩, nodes 1 and 3 will have the resource simultaneously and we have $CRU_4 = \{1, 3\}$, $L_1 = L_3 = L_4 = \emptyset$, and numPendingGrantedMssgs$_1$ = numPendingGrantedMssgs$_3$ = 0. (In contrast, for any message-sequence we have $CRU_1 = CRU_2 = CRU_3 = \emptyset$, with L_2 having zero or more messages.) Node 4 cannot now grant future request for resource from nodes 1 and 3 and thus nodes 1 and 3 cannot access the resource again. Also, a request from node 2 or from node 4 for the resource will not be granted by node 4, even after both nodes 1 and 3 have sent released-message to node 2. However, had we chosen $I_1 =$

Fig. 2. A simple interference graph G (shown on the left) and its G_R (shown on the right) where $R_i = 4$ for each i

$\{2, 4\} \supseteq R_1$, say, only node 1 can access the resource again and again, and the same is true for node 4, although both nodes 1 and 4 cannot have the resource at the same time. Note that we can remove 2 from an I_i without having any impact on the accessibility of the resource for that node or any other node.

2.2 Analysis

A node i cannot send its $(n + 1)$th request-message, $n > 0$, until its nth request was granted by every one in R_i and node i has subsequently sent released-message to every node in I_i when it finished the nth use of the resource. It follows that L_j can contain at most one copy of the message request(i, j) because the previous copies of request(i, j) were removed from L_j when node j issued the corresponding granted(j, i). Likewise, the list L_i can contain at most one copy of granted(j, i) for a given j. In regard to release-message, if $j \in I_i - R_i$, then i never enters the set CRU_j and L_j may contain multiple copies of released(i, j). Other than the multiple copies of released(i, j) in L_j, the removal of j from I_i does not prevent or provide access to the resource for node i or any other node for any message sequence. Thus, we can assume (as in [1]) that $I_i \subseteq R_i$ for all i. In that case, there can be at most one released(i, j) in L_j for $j \in I_i$ because for the case $I_i \subset R_i$ node i cannot even get access to the resource more than once and for the case $I_i = R_i$ node i cannot send $(n + 1)$th released-message until its nth released-message was processed by each $j \in I_i = R_i$. Also, both request(i, j) and released(i, j) cannot occur simultaneously in L_j because that would require $j \in R_i \cap I_i$, but then request(i, j) would have been deleted from L_j as node j sent granted(j, i) before node i could send released(i, j). This suggests the following lemma.

Lemma 1. If $I_i \subseteq R_i$ for all i, then $|L_i| \leq |R'_i| + numPendingGrantedMssgs_i$ ($\leq |R'_i| + |R_i|$).

Proof. Suppose $R'_i = \{j_1, j_2, \cdots, j_n\}$ and currently $CRU_i = \{j_1, j_2, \cdots, j_m\}$, where $0 <= m \leq n$ ($m = 0$ means $CRU_i = \emptyset$). Then, L_i can contain at one most one request(j, i) for $j = j_k$, $m + 1 \leq k \leq n$. Let $I'_i = \{j : i \in I_j\} \subseteq R'_i$; these are the nodes j who can send released-message to node i. Since only the nodes that have received granted-message can send released-message, there is at most one released(j, i) for $j = j_{k'}$, $1 \leq k' \leq m$. The first term in the bound for $|L_i|$ then gives the number of request(j, i) and released(j, i) messages in L_i. The second term gives the bound for the number of granted(j, i) messages in L_i, with $j \in R_i$. Since R_i and R'_i may be disjoint, the lemma is proved. □

2.3 Two Simplifications

First, we argue that we can assume $(i, i) \notin G$ for each i (in [1], it is assumed that $(i, i) \in G$ for each i). The only role (i, i) can play is in processing request(i, j) at a node $j \in R_i$ if $i \in CRU_j$. This condition can arise only after a request for resource from node i was granted by each node in R_i and i has already released that instance of the resource (and informed the nodes in R_i), but $j \notin I_i$ and node i making the next request for resource. However, by removing (i, i) from G we can only allow more access to resource at node i, without violating mutual exclusion. Note that if $I_i = R_i$, then we cannot have request$(i, j) \in L_j$ and $i \in CRU_j$.

Next, we argue that we can assume $I_i = R_i$ for each i (Theorem 1 in [1], which we will show to be wrong, assumes that $I_i \subseteq R_i$ for each i). We begin with the simple observation that removing a node $j \in I_i - R_i$ does not affect the resource allocation to node i or any other node; this can be seen from Example 1 with $j = 2$, which does not belong to any R_i. Now, suppose $j \in R_i - I_i$ for some i. Then, following the processing of the first request(i, j) at j, we have $i \in CRU_j$ permanently. There are two cases to consider. If we did not eliminate $(i, i) \in G$, then node i will not ever receive granted(j, i) from j and hence will not receive the resource more than once, leading to starvation. On the other hand, if we have eliminated (i, i) from G, then other problems can arise when there is a node k such that $(i, k) \in G$ and $j \in R_k$. In that case, node k cannot receive the granted(j, k) message and hence cannot get access to the resource following the first access to resource by node i, which is again a case of starvation. On the other hand, if there is no such k, then removing j from R_i will not lead to any new cases of failure of mutual exclusion that were not there before. This shows that we can assume $I_i = R_i$.

Lemma 2. A necessary and sufficient condition to assure that each node i can access the resource infinitely often, assuming a first-come-first-serve processing of the messages from each L_i, is $I_i = R_i \neq \emptyset$. □

Henceforth, we assume that $I_i = R_i \neq \emptyset$ for each node i.

2.4 Corrected Necessary and Sufficient Condition

The purpose of Theorem 1 in [1] is only to assure the relaxed mutual exclusion, i.e., two nodes i and j can get simultaneous access to the resource if and only if $(i, j) \notin G$. It is not intended to prevent deadlock or starvation. We can achieve the relaxed mutual exclusion property by simply taking each $R_i = \{1\}$, say, and each $I_i = \emptyset$. If we do not want empty sets for I_i, then we can take each $R_i = V(G)$ and let each I_i be arbitrarily assigned to $\{1\}$ or $\{2\}$, say. But this shows that the condition "$[(i \in I_i \cap R_j) \wedge (j \in I_j \cap R_i)] \vee [I_i \cap I_j \neq \emptyset]$ for each $(i, j) \in G$" in [1]] is not necessary. The corrected necessary and sufficient condition for RME is then given by the following Theorem. The condition given in [1] reduces to this if we assume that $I_i = R_i$ for each node i.

Theorem 1. If $I_i = R_i$ for each node i, then a necessary and sufficient condition for GRME to provide mutual exclusion is $R_i \cap R_j \neq \emptyset$ for $(i, j) \in G$. □

To prevent starvation of node i, node j should not repeatedly process request(i', j), $i' \neq i$, and send granted(j, i') message while there is a pending request$(i, j) \in L_j$. To see this, consider G with three nodes and the links $(1, 2)$ and $(2, 3)$ and $R_i = I_i = \{j\}$ for some j. It may happen that first node 1 is granted the resource, then request$(2, j)$ is not granted because of $(1, 2) \in G$ and subsequently request$(3, j)$ is granted. Now before node 3 releases the resource it may happen that node 1 has released the resource and has its next request$(1, j)$ granted while node 2 is still waiting with its request$(2, j) \in L_j$ pending. In this fashion, as long as one of nodes 1 and 3 has the resource, node 2's request is not granted, leading to its starvation. This can be easily prevented by each node j keeping a separate local count$(i) =$ the number of requests granted to other nodes while a previous request$(i, j) \in L_j$ for each $i \in R'_j$; initially, each count$(i) = 0$. When a count(i) exceeds a predetermined maximum limit $M_j = |R'_j| - 1$, say, node j does not grant a request to any node that interferes with node i. Moreover, when node j processes a released(i', j)-message, count(i) for each node i which interferes with node i' is increased by one, indicating its higher preference for granting resource by node j. In other words, if there are several request(i, j) messages in L_j, node j will select a random i with the largest count values $> M_j$ (if any, and hence currently having no interfering node granted its request by node j) for granted(j, i) message.

Definition 1. We say that a link $(i, j) \in G$ is *supervised* by a node k if $k \in R_i \cap R_j$, i.e., $\{i, j\} \subseteq R'_k$. The node k then prevents nodes i and j getting access to the resource at the same time. (A given link (i, j) may be supervised by more than one node.)

It is clear that for one successful access-use-release cycle to the resource for node i, it sends $|R_i|$ request-messages, receives that many granted-messages, and finally sends that many released-messages, giving a total of $3|R_i|$ message load. Thus it is important to keep the sets R_i as small as possible satisfying Theorem 1.

Lemma 3. If the sets R_i minimize the total message load, then $R_i \subseteq \bigcup \{R_j : (i, j) \in G\}$ for each i.

Proof. If $k \in R_i - \bigcup \{R_j : (i, j) \in G\}$, then removal of k from R_i cannot destroy the property "$R_i \cap R_j \neq \emptyset$ for $(i, j) \in G$" needed for the mutual exclusion. Thus, we can replace each R_i by $R_i - \bigcup \{R_j : (i, j) \in G\}$, say in the order $i = 1, 2, \cdots, N = |V(G)|$ and repeat the process till none of them changes in a cycle. The finiteness of the various sets implies that the process will terminate (in no more than N^2 cycles). The final values of R_i satisfy the lemma and the intersection property "$R_i \cap R_j \neq \emptyset$ for $(i, j) \in G$" in Theorem 1. □

3 Deadlock Prevention

It is clear that if we select a granted or a released message from L_i at node i for processing, then it can always be processed successfully. Till now we have not imposed any restriction like first-in-first-out for selection of messages in L_i for processing, and this worked fine because the only messages from a node j to i for which node i can influence node j's behavior in terms of node j's access to the resource (or other nodes' access to the resource who depend on granted-message from j) are request(j, i) and released(j, i) and only one of them can be in L_i at any time. We will continue to assume that any message can be selected from L_i for processing at a node i, except for "busy-looping" situation where node i attempts unsuccessfully to process the same request(j, i) message again and again without being able to send granted(i, j) to j, or cycling in that fashion through a set of request-messages for different j.

A deadlock in the present context means that there is a non-empty subset of nodes $D \subseteq V(G)$ which are unable to receive additional granted-message and progress further towards having access to the resource (after having received some granted-messages). The nodes in D may, however, continue to process other messages sent to them, including sending granted-messages to nodes not in D. A minimal message-scenario in a deadlock is therefore one where the only messages present in the system are the unfulfilled (and unprocessable) request-messages from nodes in D. The following definition is a refinement of an observation in [1].

Definition 2. An *alternating cycle* in G_R is a cycle $C = \langle s_1, t_1, s_2, t_2, \cdots, t_m, s_1 \rangle$, where the conditions (1)-(3) below hold. In particular, C has an even number of arcs, which are traversed alternately in the forward direction and the backward direction.

(1) The nodes s_i's are distinct from each other and so are the arcs (s_i, t_i) and (s_{i+1}, t_i); in contrast, the nodes t_j's need not be distinct and, moreover, we may have $s_i = t_j$ for some i and j, including the cases $j = i$ or $i + 1$.

(2) Each arc $(s_i, t_i) \in G_R$ corresponds to a granted request, and

(3) Each arc $(s_{i+1}, t_i) \in G_R$ is a blocked request due to the granted request (s_i, t_i) and $(s_i, s_{i+1}) \in G$ (with $s_{m+1} = s_1$).

Note that the arcs (s_i, t_i) and (s_{i+1}, t_i) form an Eulerian graph on the nodes in an alternating cycle C in the sense that each node in C has an even ($2p \geq 0$) number of arcs to it and an even ($2q \geq 0$) number of arcs from it, with $p + q > 0$ for each node s_i and t_j. This gives a simple method of showing the absence of a possible alternating cycle in G_R without being concerned with which arcs represent granted requests and which arcs represent blocked requests. Each alternating cycle $C = \langle s_1, t_1, s_2, t_2, \cdots, t_m, s_1 \rangle$ in G_R has associated with it the cycle $\langle s_1, s_2, \cdots, s_m, s_1 \rangle$ in G and thus the existence of C depends on both G and the sets R_i, i.e., G_R.

Example 2. Figs. 3(i)-(iii) show an interference graph G, a possible graph G_R, and a deadlock state. Here, node 3 has sent granted(3, 1) to node 1 and node 4 has sent granted(4, 2) to node 2, which now prevent node 3 to send granted(3, 2) and also prevent node 4 to send granted(4, 1). Since neither of nodes 1 and 2 can proceed further, we have a deadlock with $D = \{1, 2\}$. In this case, the deadlock is caused due to the fact that the interference (1, 2) is being supervised by both nodes 3 and 4. A deadlock can occur even if each interference is supervised by exactly one node. This is shown in Figs. 3(iv)-(vi). In each case, we have an alternating cycle C; $C = \langle 1, 3, 2, 4, 1 \rangle$ in Fig. 3(iii) and $C = \langle 1, 2, 5, 1, 4, 5, 3, 4, 2, 3, 1 \rangle$ in Fig. 3(iv). Note that if we remove (5, 1) from G in Fig. 3(iv) and remove the arcs (1, 2) and (5, 2) from G_R in Fig. 3(v) because node 2 is the supervisor of (5, 1) $\in G$, then (1, 3) $\in G_R$ is not part of an alternating cycle since this is the only arc from node 1 and this in turn means (2, 3) $\in G_R$ is not in part of alternating cycle since this is now the only arc to node 3, and so on. It follows that the reduced $G - (5, 1)$ and its associated reduced $G_R - \{(1, 2), (5, 2)\}$ is deadlock-free. (If we apply Theorem 3 to $G - (5, 1)$, then we get a different G_R which is also deadlock-free; we can also get a deadlock-free family of R_i for the interfernce graph G in Fig. 3(iv) using Theroem 3.)

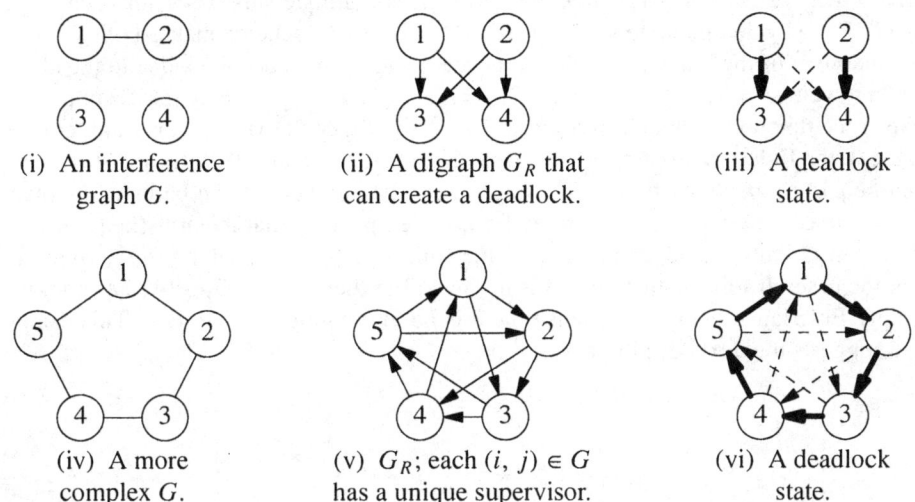

(i) An interference graph G.

(ii) A digraph G_R that can create a deadlock.

(iii) A deadlock state.

(iv) A more complex G.

(v) G_R; each $(i, j) \in G$ has a unique supervisor.

(vi) A deadlock state.

Fig. 3. Illustration of deadlock state; solid arcs show requests granted and dashed arcs show requests blocked

Lemma 4. A given family of sets R_i $(= I_i)$ allows a deadlock in GRME algorithm [1] if and only if there is an alternating cycle in G_R.

Proof. We prove the "only if" part, the "if-part" being trivial. Consider a minimal set of nodes D involved in a deadlock. Each node $s_i \in D$ has therefore at least one of its requests granted that causes at least one request in some other node in D to be blocked, because otherwise we could remove the requests from s_i (i.e., pretend they did not take place), including those that have been granted to s_i, and $D - \{s_i\}$ would still be deadlocked, contradicting the minimality of D. Let t_i be a node that granted a request to s_i and is blocked from granting a request to $s_{i+1} \in D$, $s_{i+1} \neq s_i$. Starting with a node $s_1 \in D$, we can successively traverse the links (s_1, t_1), (t_1, s_2), (s_2, t_2), etc as in an alternating cycle till we have either (t_m, s_{m+1}) where $s_{m+1} = s_i$, $i \leq m$, or we have (s_m, t_m) where $t_m = t_i$ for $i < m$. In either case, we get an alternating cycle involving s_j and t_j, for $i \leq j \leq m$. By minimality of D, it follows that $i = 1$ and $D = \{s_1, s_2, \cdots, s_m\}$. □

Theorem 2. If the sets R_i give a deadlock-free mutual exclusion with GRME and $\pi(x)$ is a permutation of the nodes $V(G)$, then the sets $R_i^\pi = \{\pi(j): j \in R_i\}$ also give a deadlock-free mutual exclusion with GRME.

Proof. Immediate from Theorem 1 and Lemma 4. □

Example 3. Figs. 4(i)-(ii) illustrate a simple method for choosing the sets R_i so that the resulting GRME provides mutual exclusion property in a deadlock-free fashion. For each link $(i, j) \in G$, we simply let the larger node $\max\{i, j\}$ be a supervisor for the link. Note that if G does not contain three nodes which are mutually adjacent to each other, i.e., G is triangle-free, then this gives a unique supervisor for each $(i, j) \in G$. For a cycle on the nodes $i < j < k$ in G, the above scheme makes both j and k a supervisor of the link (i, j) in G and hence it can create a deadlock due to the alternating cycle $C = \langle i, k, j, j, i \rangle$, where $s_1 = i$, $t_1 = k$, and $s_2 = j = t_2$ (cf. Example 2). We argue that no deadlock can occur in Fig. 4(ii). Since $(4, 4)$ is the only arc leaving node 4, the Eulerian property of an alternating cycle C implies that the arc $(4, 4)$ cannot be part of an alternating cycle. Thus, the only way node 4 can be in an alternating cycle C is that it acts in the role of a t_i node implying that the arcs $(2, 4)$ and $(3, 4)$ are in C, but then $(2, 3)$ must be of the form (s_i, s_{i+1}) and belong to G, which is not the case. It follows that node 4 is not in C, but then clearly $G_R - \{4\}$ has no nonempty Eulerian subgraph and hence there is no alternating cycle in G_R. This example suggests the next definition.

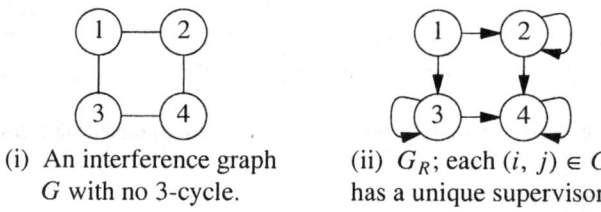

(i) An interference graph G with no 3-cycle.

(ii) G_R; each $(i, j) \in G$ has a unique supervisor.

Fig. 4. Illustration of a deadlock-free scheme R_i when there are no triangles in G

Definition 3. Let $R_i = \{j: (i, j) \in G \text{ and } i < j\}$ if there is no link $(i', i) \in G$ for $i' < i$; otherwise, let $R_i = \{j: (i, j) \in G \text{ and } i < j\} \cup \{i\}$. In particular, $R_i = \emptyset$ if and only if i is an isolated node (is not adjacent to any node). We call this the *max-scheme* for defining the sets R_i or equivalently the digraph G_R, whose arcs are now given by $\{(i, j), (j, j): (i, j) \in G \text{ and } i < j\}$.

Lemma 5. If G is triangle-free and the sets R_i are defined by the max-scheme, then they satisfy the property in Theorem 1 and the property in Lemma 3.

Proof. First, if $(i, j) \in G$ and $i < j$, then $j \in R_i \cap R_j$ and hence the sets R_i satisfy Theorem 1. Now we show that the sets R_i satisfy Lemma 3. For an isolated node i, $R_i = \emptyset$ and the union $\bigcup\{R_j: (i, j) \in G\}$ in Lemma 3 is also $\emptyset$. Now consider a node i which is not isolated and is not adjacent to any node $i' < i$; let the nodes adjacent to i in G be $A(i) = \{j_1, j_2, \cdots, j_m\}$, where $m \geq 1$ and each $j_k > i$. Then, $R_i = \{j_1, j_2, \cdots, j_m\}$ and each $R_{j_k} \supseteq \{j_k\}$, which shows $R_i \subseteq \bigcup\{R_j: (i, j) \in G\}$. For any other node i, there is at least one node $i' < i$, $i' \in A(i)$; there may be other nodes smaller than i which are in $A(i)$. Let $\{j_1, j_2, \cdots, j_m\}$, $m \geq 0$, be the nodes in $A(i)$ that are larger than i. Then, $R_i = \{i, j_1, j_2, \cdots, j_m\}$ and once again $R_i \subseteq \bigcup\{R_j: (i, j) \in G\}$ because $R_{i'} \supseteq \{i\}$ and $R_{i'}$ is included in the union and each $R_{j_k} \supseteq \{j_k\}$. □

Theorem 3. For each triangle-free interference graph G, the request-sets R_i defined by the max-scheme gives a deadlock-free GRME and degree$(i) \leq |R_i| + |R'_i| \leq 2 +$ degree(i), where degree(i) = the number of links at i in G.

Proof. We may assume that G is a connected graph because otherwise G_R can be constructed separately for each component, with each R_i contained in the component that contains node i. Suppose G has N nodes $\{1, 2, \cdots, N\}$, $N \geq 2$. From Def. 3, it follows that $R'_1 = \emptyset$, $R_N = \{N\}$, $|R_1| + |R'_1| = $ degree(1), and degree$(i) \leq |R_i| + |R'_i| = 2 +$ degree(i) for $i > 1$.

The proof for the deadlock-freeness follows a similar argument as in Example 3 together with the induction on the number of nodes N in G. We first show that node N does not participate in an alternating cycle C. Since there is only one arc (N, N) leaving node N in G_R, this node can participate in C only as a t_i-node. But then the corresponding $(s_i, s_{i+1}) \in G$ and also (s_i, t_i) and $(s_{i+1}, t_i) \in G_R$ because of the max-scheme. Since G is triangle-free, this means $s_i = N = t_i$ and s_{i+1} is a node adjacent to N in G, say, the node k. It follows that there is no t_{i-1} with $(s_i, t_{i-1}) \in G_R$ so that it can be traversed backwards in C and it is distinct from (s_{i+1}, t_i). This shows that N is not part of an alternating cycle. By induction $(G - \{N\})_R = G_R - \{N\}$ has no alternating cycle and hence G_R has no alternating cycle. □

The bounds degree$(i) \leq |R_i| + |R'_i| \leq 2 +$ degree(i) in Theorem 3 shows that the message-load $2(|R_i| + |R'_i|) = $ #(messages sent and received by node i for one access to the resource by each node) for node i is proportional to the number of nodes with which it interferes. In this sense, the scheme used in Theorem 3 is fair. Of course, the choice $R_i = \{N\}$ for each i is also deadlock-free, but it has the disadvantage that node N has a disproportionate message-load compared to the other nodes. If G is a tree (and hence triangle-free), with each node $i \neq N$ adjacent to node N, then the max-scheme gives each $R_i = \{N\}$. In this case, we can reduce max$(|R_i| + |R'_i|)$ from $N+2$ to 3 if we interchange the node labels 1 and N. Note that we can combine Theorem 3 with Theorem 2 and get many alternative family of sets $\{R_i: i \in V(G)\}$ to provide a deadlock-free GRME for a triangle-free interference graph G.

4 Conclusion

We have provided a simple sufficient condition "triangle-free property of the interference graph G" and a method for selecting the request-sets R_i to achieve distributed generalized relaxed mutual exclusion, where no deadlock can happen and where the processing load is fairly distributed among the nodes.

References

1. J.J. Jiang, T.-H. Lai, N. Soundarajan, On distributed dynamic channel allocation in mobile cellular networks, IEEE Trans. Parallel and Distributed Systems, **13** (2002) 1024-1037.
2. G. Ricart and A.K. Agrawala, An optimal algorithm for mutual exclusion in computer networks, Communications of ACM, **24** (1981) 9-17.
3. L. Lamport, Time, clocks, and the ordering of events in a distributed system, Communications of ACM, **7** (1978) 558-565.
4. I. Suzuki and T. Kasami, A distributed mutual exclusion algorithm, ACM Trans. Computer Systems, **3** (1985) 344-349.
5. M. Maekawa, A $\sqrt{N}$ algorithm for mutual exclusion, ACM Trans. on Computer Systems, **3** (1985) 145-159.
6. B.A. Sanders, The information structure of distributed mutual exclusion algorithms, ACM Trans. on Computer Systems, **5** (1987) 284-299.
7. M. Singhal, A taxonomy of distributed mutual exclusion, J. Parallel and distributed Computing, **18** (1993) 94-101.

Fault Tolerant Routing in Star Graphs Using Fault Vector

Rajib K. Das

Tezpur University, Tezpur 784028, Assam, India
rkd@tezu.ernet.in

Abstract. In this paper a fault tolerant routing algorithm for unicasting on star graph is proposed. The routing algorithm does not involve back tracking and uses *fault-vectors*. Each node in an n-star has a fault-vector of $\lfloor \frac{3(n-1)}{2} \rfloor$ bits. The kth bit of a node's fault-vector is a measure of its routing ability to nodes which are at distance k from itself. The fault-vector of each node can easily be calculated through $\lfloor \frac{3(n-1)}{2} \rfloor$ rounds of information exchanges among neighbor nodes. For a given source destination pair (u, v), the routing algorithm finds a path of length $d + h$ where d is the length of the shortest path between u and v in a fault-free star graph and $h = 0$, 2 or 4. The space requirement for storing the fault vector is $O(n)$ in each node. Simulation results show that the proposed algorithm far outperforms the routing algorithm based on *safety vectors* [13].

1 Introduction

The star graph originally proposed in [1] has emerged as an attractive interconnection network for distributed memory systems. Both star graph and hypercube fall under one class of graphs called Cayley graphs [3]. Cayley graphs have many desirable properties like node symmetry, edge symmetry, recursive substructure etc. But compared to hypercubes the star graphs have smaller degree and diameter for the same number of nodes. The n-star has $n!$ number of nodes with degree $n - 1$ and diameter $\lfloor \frac{3(n-1)}{2} \rfloor$.

Interprocessor communication plays an important role on the performance of a multicomputer. Several routing problems like unicasting, single node broadcast, all-to-all broadcast have been studied in the context of star graphs [3], [8]. Also, adaptive and fault-tolerant routing algorithms have been the subject of extensive research in recent years. These routing algorithms can be broadly classified as i) with restricted number of faults and ii) where the number of faults is unrestricted.

The routing algorithm proposed in [10] falls under the first category. The star graph is shown to be maximally fault tolerant, i.e., the n-star remains connected in presence of up to $n - 2$ faults. The routing algorithm is based on finding $n - 1$ node-disjoint paths between the source and the destination. So, if the number of faults is restricted to $n - 2$ at least one fault-free path is available. The length of the path found by routing algorithm is $d + e$ where d is the minimum distance between source and destination, and e is 0, 2, or 4.

Those routing algorithms which do not require the number of faults to be less than $n - 1$ again can be put into 3 classes as follows:

Local information based method [6], [7]: Each node knows only the status of its neighbors. This method is based on depth first search and backtracking is required when all forward links are blocked by faulty nodes or links. Hence, the length of the routing path is unpredictable. The algorithm presented in [7] is better than the one in [6] in the sense that it guarantees liveness and deadlock free transmission. Even then this backtracking algorithm has some shortcomings. It incurs some heavy penalty on the length of the path. Even with the number of faults $\leq n - 2$ in an n-star, the penalty can be as large as 12 in a 10-star. When the number of faults is more, a message is routed from source to destination after at most $2(n! - 1) - h_0$ hops, where h_0 is the number of hops needed to route the message on an optimal path in the absence of faults.

Global information based method [9] : Each node knows the status of all the nodes in the network. But a separate process is needed to collect the global information. So, this method may be too costly with respect to time and storage space.

Limited global information based method : Each node knows the exact fault information within distance d and the fault information about nodes that are outside distance d is coded in a special way. This approach requires a relatively simple process to gather and maintain fault information and can be more cost effective than the ones based on global information or local information. Examples of this approach are safety vector ($d = 1$) and extended safety vector ($d = 2$), which are quite successful in hypercube [11], [12]. The safety vector approach was used in star graph where path patterns are taken into account [13]. But the performance of the routing algorithm presented in [13] degrades drastically as the number of faults increase.

The proposed approach is simpler than the one in [13] as we do not take the path patterns into consideration but it still outperforms the one based on safety vector by a great extent. Here, each node is associated with a fault-vector of $\lfloor \frac{3(n-1)}{2} \rfloor$ bits. For the safety vector model the size of the safety vector associated with each node is $\sum p_i$ where p_i is the number of distinct path patterns for paths of length i. For example, when $n = 7$, the length of the fault vector is 9, and the length of the safety vector is 30.

The organization of the paper is as follows. Section 2 describes some basic properties of star graph relevant to our routing algorithm. Section 3 introduces the fault vector and presents the fault tolerant routing algorithm. Section 4 presents some experimental results and Section 5 concludes the paper.

2 Preliminaries and Basic Properties

The n-star graph denoted as S_n consists of $n!$ nodes, each of degree $n-1$. Each node is identified by a permutation of digits 1 through n. Two nodes u and v are connected by an edge if and only if the node label of v can be obtained from the node label of u by interchanging the first symbol with any other symbol. If u is obtained from v by switching the ith symbol with the first, then we write $g_i(u) = v$, where g_i is a generator of S_n. So, the degree of each node in S_n is $n-1$ corresponding to the generators $g_2, g_3, \cdots, g_n$.

The 3-star and 4-star graphs are shown in Fig. 1 and Fig. 2 respectively.

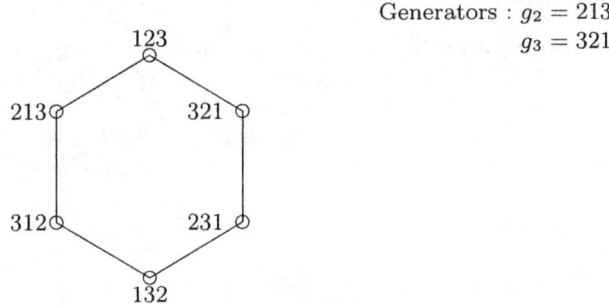

Fig. 1. The 3-star

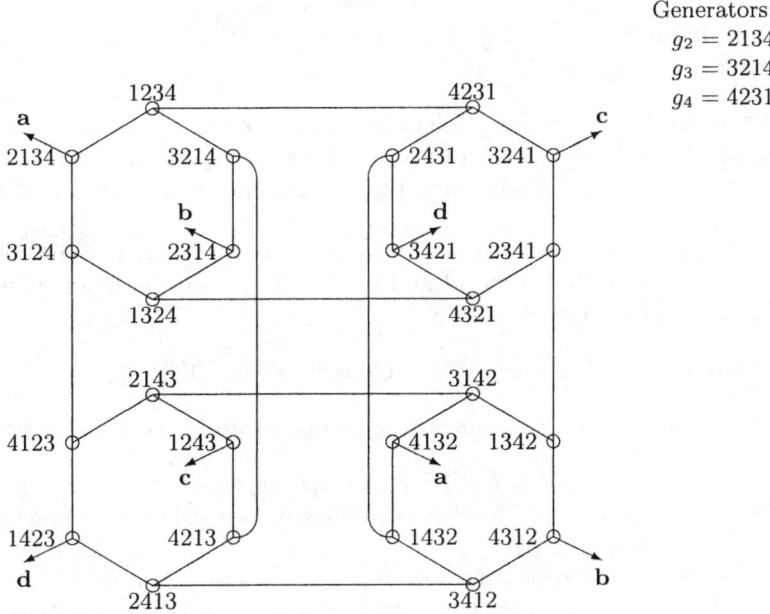

Fig. 2. The 4-star

2.1 Properties of Shortest Path

In this paper we focus on the communication problem which is called unicasting, i.e, sending a message from a particular source to a particular destination. The path from a node s to another node t can be identified as a sequence of generators $g_{i_1}, g_{i_2}, \ldots g_{i_m}$ where m is the length of the path. For example in a 4-star, the path from 1234 to 4321 can be represented by the sequence $g_4 g_2 g_3 g_2$.

Any path from a node u to another node v can be represented by a generator sequence $S = g_{i_1} g_{i_2} \ldots g_{i_m}$, $g_{i_j} \in \{g_2, g_3, \ldots g_n\}$. In such case we write $S(u) = v$.

Definition 1. *Two paths S^1 and S^2 are equivalent if $S^1(u) = S^2(u)$.*

The following lemmas are taken from [5] with a little modification.

Lemma 1. *Let $S = g_{i_1} g_{i_2} \ldots g_{i_n} g_{i_1}$, where g_{i_1}, g_{i_2}, $\ldots$ g_{i_n} are all distinct. Then S is equivalent to a set of $n-1$ node-disjoint paths listed as follows :*

$$g_{i_2} g_{i_3} \ldots g_{i_n} g_{i_1} g_{i_2}$$
$$g_{i_3} g_{i_4} \ldots g_{i_1} g_{i_2} g_{i_3}$$
$$\vdots$$
$$g_{i_n} g_{i_1} \ldots g_{i_{n-1}} g_{i_n}$$

Such a path is denoted by $C(g_{i_1} g_{i_2} \ldots g_{i_n})$ and is called a path of type C.

Definition 2. *Let $S = g_{i_1} g_{i_2} \ldots g_{i_n}$ where all g_{i_j} are distinct. Such a path is called a path of type O.*

Let $X(S)$ denote the set of generators any of which can be the first generator in the shortest path from u to $S(u)$. If S is of type O then the shortest path is unique and $X(S)$ contains only the first generator in S. If S is of type C, then by lemma 1 $X(S) = \{g_{i_1}, g_{i_2}, \ldots, g_{i_n}\}$ where $S = g_{i_1} g_{i_2} \ldots g_{i_n} g_{i_1}$.

Any path between two nodes u and v can be written as $S = AC_1 C_2 \ldots C_k$, where A is a path of type O and C_i, $1 \leq i \leq k$ are paths of type C. Paths of either type can be absent in S.

Lemma 2. *[5] If $S = AC_1 C_2 \ldots C_k$ then $X(S) = \cup_{i=1}^{k} X(C_i) \cup X(A)$.*

Definition 3. *The neighbor of u along generator g_i is denoted by u^i.*

The set $NS(u) = \{u^i | g_i \in X(S)\}$ is called the set of preferred neighbors of u for the destination $S(u)$. The other neighbors of u are called non-preferred neighbors for the destination $S(u)$.

Let S give a shortest path from u to $S(u)$ and $g_i \in X(S)$. Then the shortest path from u^i to $S(u)$ is denoted by $S - g_i$. If S is of type O then $S - g_i$ is obtained by removing the first generator g_i from S. If S is of type C and $S = g_{j_1} g_{j_2} \ldots g_{j_n} g_{j_1}$ where $g_i = g_{j_k}$ then $S - g_i = g_{j_{k+1}} g_{j_{k+2}} \ldots g_{j_n} g_{j_1} \ldots g_{j_k}$.

If $S = AC_1C_2\ldots C_k$ and $g_i \in X(S)$ then $S - g_i$ is obtained as follows : If $g_i \in X(A)$ then remove g_i from the beginning of S. If $g_i \in X(C_j)$ then remove C_j from S and append $C_j - g_i$ to the beginning of S.

An algorithm which finds the generator sequence S for a source destination pair (u, v) in the form given in lemma 2 is presented below. The algorithm is a straight forward adoption of the method described in [3].

Algorithm to generate S such that $S(u) = v$
Let $u = u_1u_2\ldots u_n$ and $v = v_1v_2\ldots v_n$
$t = u$; Let $t = t_1t_2\ldots t_n$; $S = \epsilon$;
while $(t \neq v)$ do
if $(t_1 \neq v_1)$
    find j such that $t_1 = v_j$
    $t = g_j(t)$;
    append g_j to S.
else /* $t_1 = v_1$ */
    find j such that $t_j \neq v_j$
    $t = g_j(t)$;
    append g_j to S
end while

Next we give the algorithm to find $X(S)$ from S based on lemma 2.

Let $S = S_1S_2\ldots S_m$; $X(S) = \epsilon$
while scanning S from left to right
if $S_i = S_j$ for some i, j, $i < j$
    add $S_i, S_{i+1},\ldots S_{j-1}$ to $X(S)$
    mark the generators in S from S_i to S_j
end while
Append the first unmarked generator in S
/* generator from path of type O */ to $X(S)$

3 Fault Tolerant Routing

Before giving the idea of fault-vector we describe briefly the safety vector defined in [13]. In the star graph, routing can be represented as a permutation and the permutation can be represented as the product of p-cycles [7]. A product structure is called a *pattern*. Let A denote the first symbol of the destination and X denote any other symbol in the destination. Then, a pattern is denoted by a product structure with one A and some Xs. For example, for the source nodes 2143 and 3412 and the destination node 1234, their product structures are $(21)(34)$ and $(31)(24)$ respectively. They both belong to the same pattern $(XA)(XX)$. For source 1342 and destination 1234, the permutation is (234) and the pattern is (XXX).

If α is the pattern for the source destination pair (u, v) and u^i is a preferred neighbor of u and α' is the pattern for the pair (u^i, v), then α' is a preferred pattern of α. The set of preferred patterns of α is denoted by $p(\alpha)$. The number of preferred neighbors (of pattern α') of a node (of pattern α) is denoted by $n(\alpha, \alpha')$.

Definition 4. *If one node is at one end of a faulty link, then it treats the other end as a faulty node. The safety vector $u[]$ of a node u is defined as follows :*
If u is a faulty node, then let $u[\alpha] = 0$ for all α. If u is not a faulty node, then let

$$u[(XA)] = \begin{cases} 0, & \text{if } u \text{ is on one end of a faulty link,} \\ 1, & \text{otherwise.} \end{cases}$$

Then, for all α, $\alpha \neq (XA)$, let

$$u[\alpha] = \begin{cases} 1, & \text{if } \exists \alpha' \in p(\alpha), |\Omega_{u,\alpha'}| < n(\alpha, \alpha') \\ 1, & \text{if } |\cup_{\alpha' \in p(\alpha)} \Omega_{u,\alpha'}| < \sum_{\alpha \in p(\alpha)} n(\alpha, \alpha'), \\ 0, & \text{otherwise.} \end{cases}$$

where $\Omega_{u,\alpha} = \{i | 2 \leq i \leq n, u^i[\alpha] = 0\}$.

Although the above definition of safety vector is quite complicated compared to the definition of safety vector in hypercube [11], we see that the routing algorithm based on safety vector performs poorly in star graphs. So, we design a heuristic based on a much simpler idea of fault vector. The fault-vector for a node u is a d length vector $(u_1, u_2, \ldots u_d)$, where d is the diameter of the star-graph. ($d = \lfloor \frac{3(n-1)}{2} \rfloor$ for an n-star) For the definition, we use a threshold $n - f(k)$ where the function $f(k)$, for integer k, is defined as follows.

Definition 5. $f(k) = \begin{cases} \frac{2k}{3}, & \text{if } k \mod 3 = 0 \\ 2\lfloor \frac{k}{3} \rfloor + 1, & \text{otherwise} \end{cases}$

It can be seen that for an n-star, the value of $n - f(k)$ is greater than 0 for $k = 1, 2, \cdots d$, where d is the diameter of the n-star.

Let $u^i = g_i(u), i \in \{2, 3 \cdots n\}$. where g_i is a generator of the star graph. Then the fault-vector is defined recursively as follows : If a node u is faulty then its fault-vector is $(0, 0, \cdots 0)$. If node u is an end node of a faulty link, the other end node will be registered with a fault-vector of $(0, 0, \cdots 0)$ at node u.

For a non-faulty node u,

$$u_1 = \begin{cases} 0, & \text{if } u \text{ is an end-node of a faulty link} \\ 1, & \text{otherwise} \end{cases}$$

for $k = 2, 3 \cdots d$,

$$u_k = \begin{cases} 0, & \text{if } \sum_i u^i_{k-1} < n - f(k) \\ 1, & \text{otherwise} \end{cases}$$

Example of fault-vector is given in Fig. 3, where in a 4-star two nodes and a single link are faulty.

The fault tolerant routing algorithm (named as FTR) takes help of the fault-vector which is computed at every node after $\lfloor \frac{3(n-1)}{2} \rfloor$ rounds of communication among nodes of the n-star. For routing from a source node u to destination node v first the generator sequence S such that $S(u) = v$ is computed at the source node u. Let $k = |S|$, i.e., k is the length of the shortest path from u to v. Then

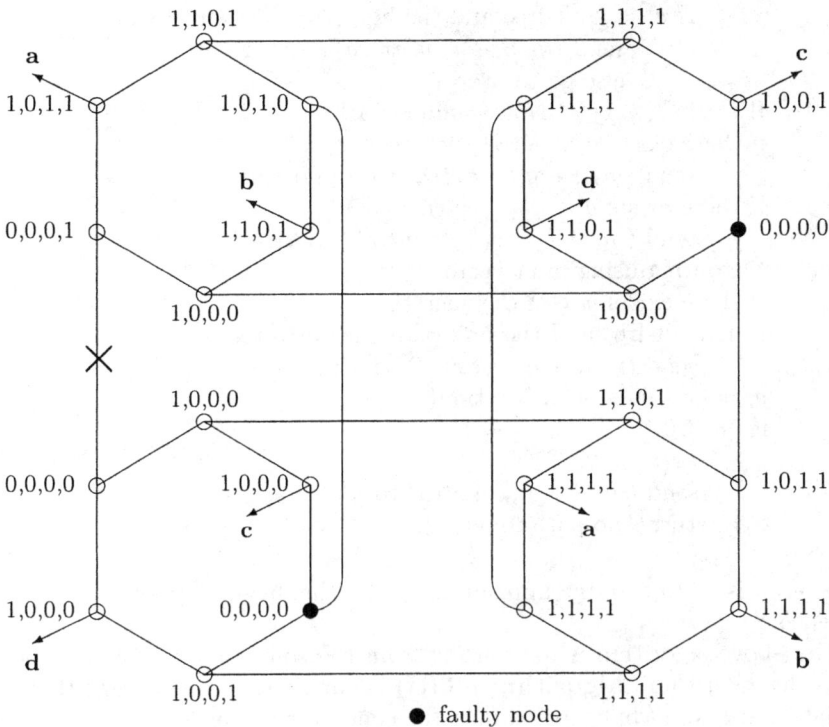

Fig. 3. Fault-vectors in a 4-star

$X(S)$ gives the set of generators any of which can be the first generator in the shortest path from u to v. The algorithm first checks if any preferred neighbor of u has the $(k-1)^{th}$ bit of fault vector set as 1, and if so send the message to such a node. If there are no such nodes the algorithm checks if any preferred neighbor of u has the kth bit of fault-vector set as 1, and if so sends the message to such a node. Otherwise, the message is sent to any preferred fault-free node u^i provided that the link (u, u^i) is also fault-free. If no such node is found, the message is sent to a non-preferred neighbor chosen by the function *best-suboptimal-option*. This function finds among the fault-free non-preferred neighbors of u, a neighbor u^i such that $X(S(u^i, v))$ is of maximum size and returns g_i. If no such neighbor is found it returns 0.

Routing algorithm for the source node:
Algorithm FTR(u, v, m) /* u: source, v : destination, m : message */
Step 1: Get the generator sequence S such that $S(u) = v$.
         $k \leftarrow |S|$; $extra = 0$;
Step 2: Get $X(S)$
Step 3: If $(|X(S)| = 1)$ /* Unique path */
             Let $X(S) = \{g_i\}$.

            If u^i is fault-free and the link (u, u^i) is fault-free
                send $(m, S - g_i, v, extra)$ to u^i; return;
                else go to Step 7.
Step 4:   If $(|X(S)| > 1)$ /* Not a unique path */
            if there exists $g_i \in X(S)$ such that $u_{k-1}^i = 1$
                send $(m, S - g_i, v, extra)$ to u^i; return;
Step 5:   If there exists $g_i \in X(S)$ such that $u_k^i = 1$
                send $(m, S - g_i, v, extra)$ to u^i; return;
Step 6:   /* routing using fault vector fails */
            If there exists $g_i \in X(S)$ such that
            u_i is fault-free and the link (u, u^i) is fault-free
                send $(m, S - g_i, v, extra)$ to u^i; return;
Step 7: $g_i = $ best-suboptimal-option(u, S, v);
            if $(g_i \neq 0)$
                $extra = 2$;
                send $(m, S + g_i, v, extra)$ to u^i
            else return("no path found ");

$S + g_i$ is obtained by appending g_i to the beginning of the generator sequence S.

The above algorithm is for routing from the source node. For intermediate nodes the algorithm (**Algorithm FTRI**) is almost the same except that intermediate nodes won't have to compute the generator sequence S, as it will receive it along with the message. If the message is sent from u along a sub-optimal path to a node x, while routing from x, u becomes a preferred neighbor. In order to ensure that the message is not again sent back to u, we keep a variable *back* which is the link along which the message is received. We also restrict the length of the suboptimal path to at most 4 more than the length of the shortest path. Since, star graphs do not have any odd cycle, every time we use a suboptimal path the path length is increased by 2. We use the variable *extra* to keep track of this overhead.

Routing algorithm for intermediate node:
Algorithm FTRI$(m, S, t, extra)$
/* m : message , S: generator sequence received, t : destination,
extra : number of extra hops in the path;
s : current node
back : g_j, if the message is received from s^j; */
Step 1:   If $(s = t)$ Stop else go to Step 2
Step 2:   If $(|X(S)| = 1)$ /* Unique path */
            Let $X(S) = \{g_i\}$.
            If $g_i \neq back$ and s^i is fault-free and the link (s, s^i) is fault-free
                send $(m, S - g_i, t, extra)$ to s^i; return;
                else goto Step 6;
Step 3:   If $(|X(S)| > 1)$ /* Not a unique path */
            if there exists $g_i \in X(S) - \{back\}$ such that $s_{k-1}^i = 1$

	send $(m, S - g_i, t, extra)$ to s^i; return;
Step 4:	If there exists $g_i \in X(S) - \{back\}$ such that $s_k^i = 1$
	send $(m, S - g_i, t, extra)$ to s^i; return;
Step 5:	/* routing using fault vector fails */
	If there exists $g_i \in X(S) - \{back\}$ such that
	s_i is fault-free and the link (s, s^i) is fault-free
	send $(m, S - g_i, t, extra)$ to s^i; return;
Step 6:	If $(extra = 4)$ return("no path found");
	else
	$\quad g_i =$ best-suboptimal-option(s, S, t);
	$\quad$ if $(g_i \neq 0)$
	$\quad\quad extra = extra + 2$;
	$\quad\quad$ send $(m, S + g_i, t, extra)$ to s^i
	$\quad$ else return("no path found ");

4 Experimental Results

We simulated our fault tolerant routing algorithm on 6-star and 7-star (the algorithm is applicable to star-graphs of any size though) with number of faults ranging from 10 to 120. We compare the routing capability of the proposed algorithm using fault-vector with the one using safety vector. We consider 3 cases, i) when all the faults are node faults, ii) When half of the faults are node faults and iii) all the faults are link faults.

For a fixed number of faults we have taken a random fault distribution and the routing algorithm is applied for different source destination pairs. For a given fault distribution we have taken 100,000 source destination pairs randomly and

Table 1. In a 6-star, when all faults are node faults

	Fault-vector			Safety vector			Optimum		
#of faults	shortest path	+2	+4	shortest path	+2	+4	shortest path	+2	+4
10	97.26	2.68	.05	71.23	15.45	2.33	98.14	1.86	0.0
20	94.61	5.10	.22	51.69	11.43	1.62	96.31	3.67	.005
30	91.89	7.45	.50	38.99	8.45	.86	94.48	5.50	.016
40	89.19	9.66	.83	31.25	6.83	.58	92.76	7.21	.027

Table 2. In a 6-star, when half the faults are node faults

	Fault-vector			Safety vector			Optimum		
#of faults	shortest path	+2	+4	shortest path	+2	+4	shortest path	+2	+4
10	97.96	1.83	.20	64.18	14.73	2.54	98.59	1.40	.000
20	95.88	3.59	.46	42.07	9.49	1.12	97.16	2.83	.009
30	93.66	5.37	.82	30.22	6.89	.62	95.74	4.24	.016
40	91.47	7.07	1.14	24.26	5.50	.44	94.43	5.55	.024

noted the % of cases where the routing algorithm is able to find the shortest path or a sub-optimal path (path of length 2 or 4 more than the shortest path). We compare this value with the optimum value which is obtained by breadth first search. Although no simulation results were presented in [13] for the safety vector model, we have implemented the routing algorithm presented in [13] and obtained the results. Finally, we found the average value over 25 different fault

Table 3. In a 6-star, when all faults are link faults

#of faults	Fault-vector			Safety vector			Optimum		
	shortest path	+2	+4	shortest path	+2	+4	shortest path	+2	+4
10	98.62	.99	.38	58.01	13.70	2.42	99.016	.980	.002
20	97.21	1.93	.82	35.52	8.27	.89	98.037	1.954	.009
30	95.65	2.93	1.29	25.41	5.97	.51	97.066	2.918	.016
40	94.05	3.90	1.77	19.69	4.52	.37	96.109	3.865	.025

Table 4. In a 7-star, when all faults are node faults

#of faults	Fault-vector			Safety vector			Optimum		
	shortest path	+2	+4	shortest path	+2	+4	shortest path	+2	+4
20	99.02	.98	.00	59.63	13.07	2.02	99.38	.62	.00
40	98.06	1.91	.03	40.34	8.86	1.62	98.76	1.24	.00
60	97.12	2.80	.06	29.56	6.19	1.04	98.16	1.84	,00
80	96.16	3.70	.11	22.70	4.51	.53	97.55	2.45	.00
100	95.18	4.61	.17	18.16	3.61	.32	96.93	3.07	.00
120	94.15	5.56	.24	15.17	3.08	.24	96.33	3.67	.00

Table 5. In a 7-star, when all the faults are link faults

#of faults	Fault-vector			Safety vector			Optimum		
	shortest path	+2	+4	shortest path	+2	+4	shortest path	+2	+4
20	99.60	.30	.09	45.37	10.06	1.81	99.74	.25	.00
40	99.22	.58	.19	26.72	5.68	.92	99.49	.51	.00
60	98.82	.88	.29	18.09	3.78	.38	99.23	.77	.00
80	98.39	1.18	.41	13.67	2.96	.25	98.98	1.02	.00
100	97.93	1.52	.53	11.09	2.47	.19	98.73	1.27	.00
120	97.44	1.86	.66	9.30	2.11	.16	98.48	1.52	.00

Table 6. In a 7-star, when half the faults are link faults

#of faults	Fault-vector			Safety vector			Optimum		
	shortest path	+2	+4	shortest path	+2	+4	shortest path	+2	+4
20	99.31	.64	.05	52.04	11.48	1.98	99.56	.44	.00
40	98.65	1.24	.10	32.47	7.05	1.31	99.13	.87	.00
60	97.97	1.86	.16	22.69	4.65	.59	98.70	1.30	.00
80	97.25	2.50	.23	17.00	3.50	.32	98.25	1.75	.00
100	96.50	3.15	.31	13.77	2.93	.23	97.83	2.17	.00
120	95.72	3.84	.39	11.58	2.52	.19	97.41	2.59	.00

distributions with same number of node-faults or link-faults. The experimental results are shown in Table 1 to 6. The second column in each table corresponds to the case where shortest paths are found using fault-vector. Percentage of sub-optimal routing are also reported (in the third and fourth column). The corresponding values for safety-vector model are given in column 5 to 7. The last 3 columns give the optimal values.

From the tables it is evident that the heuristic using fault-vector is much superior to the one using safety vectors. Also, the percentage of cases where a sub-optimal path is found is very small compared to the cases where shortest path is found. The shortest paths and sub-optimal paths found by the proposed heuristic covers more than 99% of all the cases.

5 Conclusion

A fault-tolerant algorithm for routing in faulty star graph has been presented in this paper. First, each node computes its fault-vector after $O(n)$ rounds of information exchanges and then the routing algorithm uses these fault-vectors as navigation tool. The algorithm does not involve back-tracking and can tolerate large number of node as well as link faults. It is superior to the one based on safety-vector with regard to 3 factors i) The fault-vector requires less storage space than safety vector; ii) Computing the fault-vector is simpler; and iii) Routing algorithm based on fault-vector performs much better both for small and large number of faults.

References

1. Akers, S. B., Harel, D., Krishnamurthy, B.: The star graph : An attractive alternative to the n-cube. In: Proceedings of International Conference on Parallel Processing (1987) 393-400.
2. Akers, S. B., B. Krishnamurthy, B.: The fault tolerance of star graphs. In: Proceedings of International Conference on Supercomputing. San Francisco, CA, (1987) 270-276.
3. Akers, S. B., Krishnamurthy, B.: A group theoretic model for symmetric Interconnection Networks. IEEE Transaction on Computers, **38** (1989) 555-566.
4. Rouskov, Y., Srimani, P. K.: Fault diameter of star graph. Information Processing Letters. **48** (1993) 243-251.
5. Misic, J., Jovanovic, Z.: Routing function and deadlock avoidance in a star graph. Journal of Parallel and Distributed Computing. (1994) 216-228.
6. Sur, S., Srimani, P. K.: A fault tolerant routing algorithm in star graph interconnection networks. In Proc. International Conference on Parallel Processing. (1991) 267-270. is
7. Bagherzadeh, N., Nassif, N., Latifi, S.: A routing and broadcasting scheme on faulty star graphs. IEEE Trans. on Computers. **42** (1993) 1398-1403.
8. Mendia, V. E., Sarkar, D.: Optimal broadcasting on the star graph. IEEE Trans. on Parallel and Distributed Systems. **3** (1992) 389-396.
9. Gu, Q. P., Peng, S.: Node-to-node cluster fault routing in star graph. Information Processing letters. **56** (1995) 29-35.

10. Day, K., Ayyoub, A. E. A. I.: Reliable Communication in faulty star. In Proc. International Parallel and Distributed Processing Symposium, IPDPS (2002)
11. Wu, J.: Adaptive fault-tolerant routing in cube-based multicomputers using safety vectors. IEEE Transaction on Parallel and Distributed Systems. **9** (1998) 321-334.
12. Wu, J., Gao, F., Li, Z., Min, Y.: Optimal fault-tolerant routing in hypercubes using extended safety vectorsr. In Proc. Seventh International conference on Parallel and Distributed Systems, ICPADS. (2000) 267-271.
13. Yeh, S-I., Yang, C-B., Chen, H-C.: Fault-tolerant routing on the star graph with safety vectors. In Proc. of International Symposium on Parallel Architectures, Algorithms and Networks, ISPAN. (2002) 301-309.

Optimistic Concurrency Control in Firm Real-Time Databases

Anand S. Jalal, S. Tanwani, and A.K. Ramani

School of Computer Science,
Devi Ahilya University, Indore, India
jalalanand@yahoo.com
{stanwani.scs, ramani.scs}@dauniv.ac.in

Abstract. Concurrency control algorithms for real-time database systems satisfy not only consistency requirements but also meet transaction-timing constraints. Two Phase Locking (2PL) is used often in traditional database systems. However, it has some inherent problems such as the possibility of deadlocks as well as long and unpredictable blocking times. Optimistic concurrency control protocols are non-blocking and deadlock free, but they have the problems of late conflict detection and transaction restarts. Other Concurrency Control techniques, such as Dynamic Adjustment of Serialization Order (DASO) have been found to be better at reducing number of transaction restarts. In this paper, we propose a new optimistic concurrency control algorithm based on DASO using firm deadline in order to effectively reduce number of unnecessary restarts. Since firm real time transaction imparts no value to the system once its deadline expires, therefore in our algorithm, we adjust the timestamp intervals of all conflicting active transactions only after the validating transaction is guaranteed to meet its deadline during the validation phase. A simulator is designed to verify the effectiveness of the proposed method. The simulation results show that the proposed method can significantly reduce the number of unnecessary restarts and thereby improve the miss ratio, commit ratio.

1 Introduction

In real-time database systems (RTDBS), the correctness of a result depends on not only the logical results and functional behavior of the execution, but also the temporal behavior, i.e. the time when the result is delivered [1]. Effective concurrency control algorithms are needed to ensure predictable and timely response in these systems. Most concurrency control algorithms for RTDBS are based on one of the two basic concurrency control mechanism: locking [1], [2] or optimistic concurrency control (OCC) [4], [5], [7].

Optimistic concurrency control [3], [9] is based on the assumption that conflict is rare, that it is more efficient to allow transactions to proceed without delays. There are three phases to an optimistic concurrency control method. During the read phase, the transaction reads the values of all data items it needs from the database and stores them in local variables. Concurrency control scheduler stores identity of these data items to a *read set*. However, writes are applied only to local copies of the data items kept in the transaction workspace. Concurrency control scheduler stores identity of all written data items to *write set*. In the validation phase it is ensured that all the committed transactions are in a serializable fashion. For read only transactions, this consists of checking that the data values read are still the current values for the corresponding data items. For transaction with write operations, the validation consists of determining whether the current transaction has executed in a serializable manner. The third phase, called write phase, follows the successful validation phase for transaction including write operations. During the write phase, all changes made by the transaction are permanently stored into the database.

Optimistic concurrency control protocols have properties of being non-blocking and deadlock free. However, these protocols have two problems: late conflict detection and transactions restart. So, it is important to design new methods to minimize the number of transaction restarts. One efficient way is to avoid unnecessary restarts. The OCC-DA [4], [5], OCC-TI [3] and OCC-DATI [5], [6] concurrency control protocols based on dynamic adjustment of serialization order [4], [5], [6] can avoid some unnecessary restarts. So the number of transaction restarts with these protocols is smaller than with other optimistic concurrency control protocols, such as OCC-BC, OCC-WAIT, and WAIT-X [1], [3], [10].

In this paper, we propose a new optimistic concurrency control algorithm based on DASO using firm deadline with an objective of further reducing unnecessary restarts. Since firm real time transaction imparts no value to the system once its deadline expires, therefore, in our approach, we adjust timestamp intervals of all conflicting active transactions only after the validating transaction is guaranteed to meet its deadline.

2 Proposed Optimistic Concurrency Control Algorithm

This section presents an optimistic concurrency control method named Revised OCC-DATI. It is based on forward validation. The number of transaction restarts is reduced by dynamic adjustment of serialization order using timestamp interval same as in OCC-DATI [5]. In our approach we consider the deadline of validating transactions. Since OCC-DATI is used for firm real time system. As we know, firm real time transactions impart no value to the system once their deadlines expire. In OCC-DATI, it may be possible that after adjustments of timestamp interval in conflicting transactions, the validating transaction misses its deadline. In such a situation, the adjustment is wasteful. Therefore in our approach, we adjust the timestamp intervals of all conflicting active transactions only after the validating transaction is guaranteed to meet its deadline during the validation phase.

Our algorithm resolves conflicts using the timestamp intervals of the transactions. Every transaction is bound to execute within a specific time interval. When an access conflict occurs, it is resolved using the read and write sets of the conflicting transactions together with the allocated time interval. The timestamp interval is adjusted during the validation phase of transaction. In revised OCC-DATI every transaction is assigned a timestamp interval (TI). At the start of the transaction, the timestamp interval of the transaction is initialized as [0, ∞]. This time stamp interval is used to record a temporary serialized order during the validation of the transaction.

At the beginning of the validation, the final timestamp of the validating transaction TS (Tv) is determined from the timestamp interval allocated to the transaction Tv. The timestamp intervals of all other concurrently running and conflicting transactions are adjusted to reflect the serialization order. The final validation timestamp TS (Tv) of the validating transaction Tv is set to be the current timestamp if it belongs to the timestamp interval TI (Tv), otherwise TS (Tv) is set to be the maximum value of TI (Tv).

The adjustment of timestamp intervals is done for Read set (RS) and Write set (WS) of a validating transaction. First, it is checked that the validating transaction has read from the committed transactions. This is done by checking the each data item's read and write timestamp. These values are fetched when the first read and write to the current data item is made. Then, the Read set (RS) and Write set (WS) of validating transaction is compared with the Read set (RS) and Write set (WS) of active conflicting transactions. When access is been made to the same data item both in the validating transaction and in the active transaction, the temporal time interval of the active transaction is adjusted.

Therefore timestamp intervals of all conflicting active transactions are adjusted after the validating transaction is guaranteed to meet its deadline. If the validating transaction misses its deadline, no adjustments are done. Non-serializable execution is detected when the timestamp interval of an active transaction becomes empty. If the timestamp interval of a transaction becomes empty, the transaction is restarted.

3 Performance Evaluation

This section describes the simulation model which is used to test the new optimistic algorithm. We simulate an RTDBS Model to evaluate the performance of different optimistic concurrency control methods (OCC-TI, OCC-DATI, Revised OCC-DATI). The simulator model is developed in C++ and consists of four major components: Source, Transaction Manager, Concurrency controller, Resource Manager. *Source* - It generates the workload of the system with a specified arrival rate. It creates a transaction and put the transaction into the ready queue. *Transaction Manager* - This module receives generated transactions from the source and coordinates their execution.

Concurrency controller - This module manages the data conflict detection and resolution between transactions through one of the optimistic algorithms presented earlier. It checks the read and write set of the validating transaction with read and write set of other concurrently executing transactions. *Resource Manager* - It models system resources, such as buffers, CPUs, disks, and their associated queues.

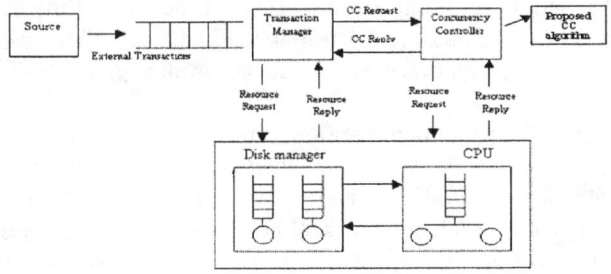

Fig. 1. Simulation Model

The primary performance metrics used in the experiments are as follows:

Miss Ratio - The miss ratio is the ratio of user transactions that miss their deadlines and total number of entered transactions. Miss ratio = $N_{miss} / (N_{miss} + N_{succeed})$, Where N_{miss} is the number of transactions that miss deadlines and $N_{succeed}$ is the number of transactions that succeed.

Restart Ratio - Let Norestart and Nosubmitted represent the number of user transactions restart due to conflict and number of transaction submitted to the system respectively. Then Restart ratio is defined as: Restart ratio = Norestart / Nosubmitted*Commit Ratio* - Let #timely and #submitted represent the number of user transactions committed within their deadlines and number of user transactions submitted to the system respectively. The Commit ratio is defined as: Commit ratio = #timely / #submitted.

4 Experimental Results and Discussions

The simulator described in section 3 is used for experimentation. The number of aborted transaction attempts could be any number greater than or equal to zero. The simulation run was carried out by varying the load from 10 to 100 transactions per second. Load variation was done by changing inter arrival rate of user transactions. The following experiments were conducted to compare the miss ratio, restart ratio, commit ratio of OCC-TI, OCC-DATI and Revised DATI algorithms. The results obtained through simulation are shown in the figures 2, 3 and 4.

Figure 3 shows the Restart ratio. The results show that the Restart ratio increases as the load increases but in Revised DATI it was drastically reduced after a certain load increment. The reason is that in Revised DATI, as load increases, more number of transactions miss their deadline during validation phase. So, the timestamp interval adjustment of other active conflicting transactions is discarded, resulting in reduced restart ratio.

As show in Figure 4, it is observed that as the load is increased, the Commit ratio of revised DATI is greater as compared to other algorithms as more number of transactions is committed.

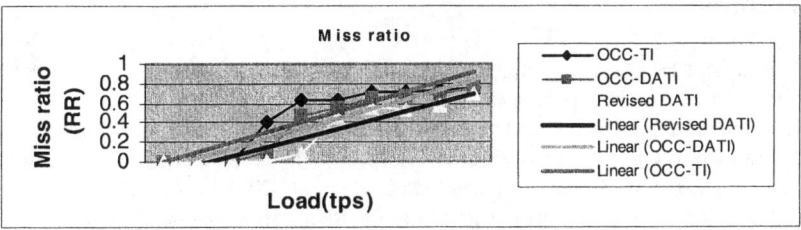

Fig. 2. Application Load vs. Miss ratio

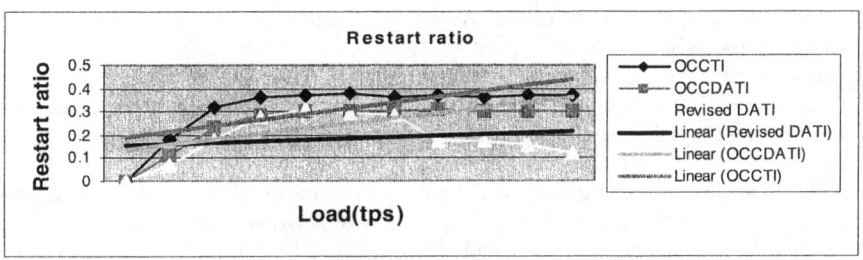

Fig. 3. Application Load vs. Restart ratio

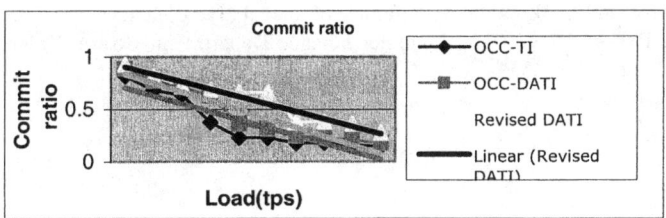

Fig. 4. Application Load vs. Commit ratio

5 Conclusions

A new optimistic concurrency control algorithm based on DASO using firm deadline is proposed in this paper. Since firm real time transaction imparts no value to the system once its deadline expires, therefore we have adjusted the timestamp intervals of all conflicting active transactions only after the validating transaction is guaranteed to meet its deadline during the validation phase. Experimental studies prove that the number of transactions restart with this algorithm is considerably lower as compared to other algorithms. We intend to extend this work by studying the behavior of the proposed algorithm over a replicated real-time database.

References

1. B. Kao and H. Garcia-Molina: An overview of real-time database systems. Advances in Real-Time Systems, S.H. Son (Ed.), Prentice Hall, 1995, pp. 463-486
2. K. Ramamritham: Real-Time Databases. Distributed and Parallel Databases, Vol. 1, 1993.
3. J. Lindstrom: Optimistic Concurrency control Methods in Real Time Database Systems. Lincentiate Thesis, Series of publications C, Report C-2001-9, Helsinki, Feb 2001.
4. J. Lee: Concurrency Control Algorithms for Real Time Database Systems. PhD thesis, faculty of the school of Engineering and Applied Science, University of Virginia, January 1994.
5. J. Lindstrom and K. Raatikainen: Dynamic adjustment of serialization order using time-stamp intervals in real-time databases. Proceedings of the 6th International Conference of Real-Time Computing Systems and Applications, IEEE Computer Society Press, Hong Kong, China, 1998, pp. 13-20.
6. J. Lindstrom and K. Raatikainen: Using importance of transactions and optimistic concurrency control in firm real-time database. Proceedings of the 7th International Conference of Real-Time Computing Systems and Applications, IEEE Computer Society Press, Cheju Island, South Korea, 2000, pp. 463-467.
7. A. Datta and S. H. Son: A study of concurrency control in real-time active database systems. Tech. Report, department of MIS, university of Arizona, Tucson, 19996.
8. A. Datta, I. R. Viguier, S. H. Son, and V. Kumar: Is a bird in the hand worth more than two in the bush? Limitations of priority cognizance in conflict resolution for firm real-time database systems. IEEE Transactions on Computers, 49(5):482-502, May 2000.
9. P. S. Yu, K. L. Wu, K. J. Lin, K. J. and S. H. Son: On Real-time Databases: Concurrenc-Control and Scheduling. *Proc. of the IEEE*, 82(1):140-157, 1994.
10. A. Bestavros and S. Braoudakis: A family of speculative concurrency control algorithms: Technical Report TR-92-017, Computer Science Department, Boston University, Boston, MA, July 1992.

Stochastic Modeling and Performance Analysis for Video-On-Demand Systems

Vrinda Tokekar, A.K. Ramani, and Sanjiv Tokekar

Devi Ahilya University, Indore, MP 452017, India
{vrindatokekar, ramaniak, sanjivtokekar}@yahoo.com

Abstract. This study deals with the VOD system having limited number of channels and servicing of requests for mixed movies of different durations and popularity. The performance measures such as channel utilization, number of channels in use, blocking probability of channels, etc. are quantified in terms of various design parameters like, total number of channels, users request arrival rate, batching interval and request service rate etc.

1 Introduction

The design of VOD systems faces significant challenges to support large number of concurrent customers. Having a hierarchy of servers can reduce the load on the storage servers and network channels [1][2]. In a typical two level hierarchy of such servers, the local servers placed near to the user clusters distribute requested video locally. These servers stream through the desired video from the central repository server using network channels. Such systems can also further efficiently utilize storage and network resources by incorporating various batching and caching techniques.

The different service classes can be characterized in terms of throughput and delay that a network will guarantee. These guarantees are expressed as quality of service (QOS) parameters. The network users must select the appropriate service class and QOS with well-defined pricing structure for their application. Thus, the designed VOD systems in this environment have to be evaluated on two factors: received video quality and pricing of resources [3].

The VOD system proposed in [1] is analyzed for three different caching schemes, and the tradeoff for each scheme has been presented. The work in [4] focuses on analyzing enterprise media server workloads in terms of access pattern, locality and content evolution. The authors in [5] proposed a model for handling mixed workloads of long and short movies and a Generalized Interval Caching (GIC) scheme. Similarly, Wong and Chan in [6] proposed a model of VOD in broadband networks, assuming that sufficient number of network channels is available. However, in practice the network channels are leased at vary high cost, and are available in limited number. In this study, we present the development of a general model to analyze a VOD system which can playback several video movies of varying durations using limited number of network channels. The model is generalized for M movies and N channels. The system is evaluated using number of system design parameters and performance parameters such as blocking probability, channel utilization, average number of users served in a VOD service. The analysis results in the optimum number of network channels to be leased out for the desired values of performance parameters.

2 Model Development

The channels in the VOD system may be leased by the video content service provider and are assumed to be always available. In such environment, the users request for playback of videos of their choice at any time. Before playing the video, a logical channel is allocated to that request. This corresponds to the set of resources required for the playback of a stream. If multiple users request the same video within a short time interval then these requests can be batched together so that single channel is used to satisfy all users. This short interval is called batching period. As the batching period increases the channel utilization increases, thereby user's start up delay also increases. If the start up delay exceeds the users waiting tolerance then user may leave or renege. Therefore, the way batching period is chosen, the usage cost is amortized while allowing an acceptable user loss. In this batching scheme when the first request arrives, the batching interval starts, during this time all the requests for the same movie are grouped and serviced together. The movie is shown after batching interval time (say W). A request arriving beyond the batching interval is serviced in the next batch. The batching interval is kept small enough, so that the user never reneges.

In the proposed system m movies of varying durations are permitted on the available channels. However, due to popularity difference in the demand between the movies, the batching interval is kept variable for optimum utilization of the channel capacity. The behavior of such system is modeled as Continuous Time Markov Chain (CTMC) process, discussed further. At any instance of time the system is capable of serving movies of variable durations simultaneously. Further the different movies are classified among type-1 to type-M categories based upon its characteristics. The objective here is to quantify the number of users served, blocking probability and channel utilization. For this analysis, following system operating assumptions are made:

- Total Number of repository channels which can be allocated for the movies are N.
- Maximum numbers of movie types in the system can be requested by users are M.
- User's requests arrive in Poisson stream at the rate λ_i for movie m_i. Its value depends upon the popularity of the movie and modeled after Zipf's law. If all M movies are arranged in decreasing order of their popularity and let m_1 be the most popular movie whose request rate is λ_1 then according to Zipf's law, the request rate of m_i movie is given by λ_1/i.
- Service rate for a movie m_i is μ_i. Thus duration of the m_i movie is $1/\mu_i$.
- Batching interval for the movie m_i is w_i, thus channel request rate for the same movie is $\alpha_i = 1/(w_i + 1/\lambda_i)$. Number of users requested per batching interval for the movie m_i is Ur_i and given by $Ur_i = \lambda_i/\alpha_i$.

Fig. 2 shows CTMC for M=2 movies, namely m_1, m_2 and N=3 channels. A state (m,n) in the CTMC represents that channels m and n out of N are currently allocated to m_1 and m_2, respectively. Various transitions from a state (m,n) may occur in the following ways :

- When requests for channel for m_1 movie arrive with channel request rate α_1, the transition occurs to state (m, n+1). When requests for channel for m_2 movie arrive with channel request rate α_2, the transition occurs to state (m+1, n).
- When requests for channel for m_1 movie serviced with rate $n\mu_1$, the transition occurs to state (m, n-1). When requests for channel for m_2 movie serviced with rate $m\mu_2$, the transition occurs to state (m-1, n).

There is no further transition from a state if m + n = N. This condition represents that allocation of channels for any two types of movies cannot be more than the total number of channels.

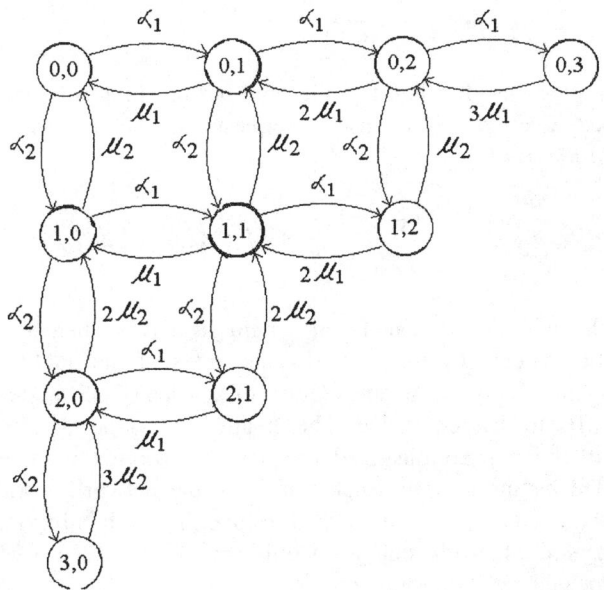

Fig. 1. CTMC model for N=3 & M = 2

3 Analysis

In a generalized Markov model of M movies ($m_1, m_2..m_M$) and N channels, channel request rates of movie m_1 to m_M are α_1 to α_M and service rate are μ_1 to μ_M. A state ($i_1, i_2...i_M$) in the model represents that i_1 channels are allocated to movie m_1 and i_m channels are allocated to movie m_M such that $i_1+i_2+i_3...i_M \leq N$. The steady state probability $\pi_{i_1...i_m}$ of the system staying in state ($i_1, i_2....i_M$) is given by the product formula

$$\pi_{i_1\ldots i_M} = \frac{\dfrac{\left(\dfrac{\alpha_1}{\mu_1}\right)^{i_1} \times \ldots \times \left(\dfrac{\alpha_M}{\mu_M}\right)^{i_M}}{(i_1 \times i_2 \times \ldots \times i_M !)}}{\displaystyle\sum_{j_1=0}^{N} \ldots \sum_{j_m=0}^{N-(j_1+j_2\ldots+j_{M-1})} \left(\left(\dfrac{\alpha_1}{\mu_1}\right)^{j_1} \times \ldots \times \left(\dfrac{\alpha_M}{\mu_M}\right)^{j_M} \bigg/ (j_1 \times j_2 \times \ldots \times j_M !) \right)} \quad (1)$$

The average number of channels in use in the system can be expressed as the expected reward rate in the steady state. These expected reward rates can be obtained by attaching suitable weights to the steady state probabilities which is dependent upon the total number of channels in use, therefore average number of channels in use combining all the movies is given by

$$C_{use} = \sum_{i_1=0}^{N} \ldots \sum_{i_M=0}^{N-(i_1+i_2\ldots i_{m-1})} \pi_{i_1\ldots i_M} \times (i_1 + i_2 \ldots + i_M) \quad (2)$$

The user's requests are batched together and per batching interval one channel is allocated. Hence, average no of user's requests served at any instant of time combining all M movies is

$$R_c = \sum_{i_1=1}^{N} \ldots \sum_{i_M=0}^{N-(i_1+i_2\ldots i_{M-1})} \pi_{i_1\ldots i_M} \times (U_{r1} \times i_1 + \ldots U_{rM} \times i_M) \quad (3)$$

When all the channels are allocated and a request arrives, then it is rejected. The amount of traffic rejected by the system is an index of the quality of the service offered by the system. This is termed as Grade of Service (GOS) and is defined as the ratio of lost traffic to offered traffic. The smaller the value GoS the better is the service. When all channels are busy and a request is generated by the Poisson process then it is rejected by the system. Such traffic on the network is known as Erlang traffic [6] and the Grade of Service is equal to blocking probability P_B. The blocking probability is the sum of steady state probabilities of all those states where number of channels are allocated N. Therefore,

$$P_B = \sum_{i_1=1}^{N} \ldots \sum_{i_m=0}^{N-(i_1+i_2\ldots i_{M-1})} \pi_{i_1\ldots i_M} \qquad \forall (i_1+i_2+\ldots+i_M) = N \quad (4)$$

Utilization of channel can be expressed as

Channel Utilization = Avg. number of channels in use / Total number of channels (5)

$$= C_{use} / N \quad (6)$$

4 Results and Discussions

The optimal values of various system design parameters such as total number of channels, batching interval and request arrival rate can be found out, and several results obtained are plotted, but due to space limitation few graphs are shown through Figs. 2-4, and discussed below:

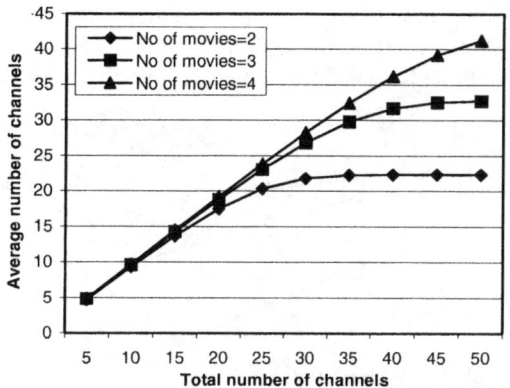

Fig. 2. Average number of channels in use Vs Total number of channels (lemda0=2req/min, mue=120min., window size=10min)

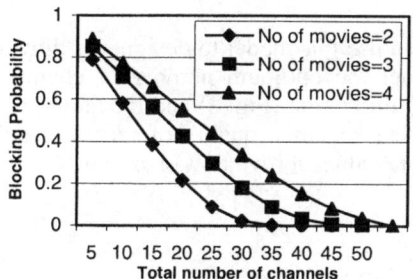

Fig. 3. Blocking Probability Vs Total number of channels (lemda0=2rew/min, mue=120 min., window size=10min)

Fig. 2 shows the variation in average number of channels in use against the number of channels for different number of movie types 2-4. The various input parameters such as request rate of most popular movie λ_1 is kept 2 req./min, service time of movies is assumed to be 120 minutes and batching period is kept 10 minutes. The parameter average channels in use vary exponentially with N and saturates for N > 25. This graph gives the estimate about number of channels needed in the system for given set of input parameters. Provision of any additional channels will only increase the cost. It can also be seen that as the number of movie types increase the number of channel needed also increase. The parameter blocking probability gives the estimate about how many requests are dropped because of the unavailability of free channels.

Figs.3 illustrates the variation in blocking probability versus total number of channels for the same input parameter values as mentioned above. The blocking probability decreases fast as the total number of channels needed increase in the system similar variation is observed with window size (figure not shown due to space limitation). The blocking probability should be sufficiently low (< 0.005) accordingly total number of channels and window size may be deployed in the system for different number of movie types. Fig.4 plots channel utilization against window size. The utilization decreases as the window size increases. The utilization should be sufficiently high (50% or more) accordingly the value for window should be chosen. Thus a tradeoff is obtained among blocking probability, channels utilization and window size.

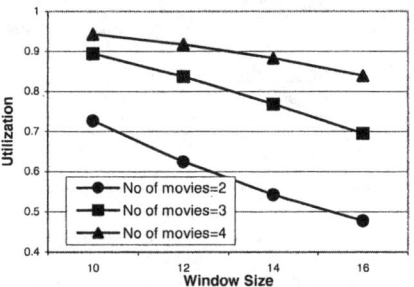

Fig. 4. Utilization Vs Window size (lemda0=2req/min, mue=120 min,)

The VOD operators can use this model to design and dimension their systems. For example, using this model the optimum number of channels can be obtained for desired values of channel utilization (Fig. 2) and blocking probability (Fig. 3). While designing a system to meet specified quality of service requirements (e.g., channel utilization), the appropriate values for window size (Fig. 4) can be chosen.

5 Conclusions

The study has dealt with the development of a Continuous Time Markov Chain model to analyze a Video-On-Demand system, where movies of different types are allocated channels appropriately to yield profit. Various design and performance parameters have been identified based on which the proposed system is analyzed. The various performance parameters such as channel utilization, number of channels in use, blocking probability, etc., are statistically quantified in terms of various design parameters.

References

1. S.H. Gary Chan and Foud Tobagi "Distributed Servers Architecture for Networked Video Services," IEEE/ACM Transactions on Networking, vol. 9, No 2, pp.125-136, April 2001.
2. S.H. Gary Chan and Foud Tobagi "Caching Schemes for Distributed Video Services," Proceedings IEEE, ICC'99, Vancouver, Canada, pp. 988-994, June 1999.

3. 3. C.E. Luna, L.P. Kondi and Aggelos K. Katsaggelos, "Maximizing User Utility in Video Streaming Applications," IEEE Transactions on Circuits and Systems for Video Technology, vol. 13, no. 2, pp. 141-148, Feb. 2003.
4. L. Cherkasova and M. Gupta, "Analysis of Enterprise Media Server workloads: Access Patterns, Locality, Content Evolution and Rates of Change," IEEE/ACM Transactions on Networking, vol. 12, No.5, October 2004.
5. Asit Dan and Dinkar Sitaram, "A generalized Interval caching Policy for Mixed Interactive and Long Video Workloads," IBM Research report, RC 19347, Yorktown Heights, NY, 1993.
6. E.W.M. Wang and S.C.H. Chan, "Performance modeling of Video-on-Demand Systems in Broadband Networks," IEEE Transactions on Circuits systems for video Technology, vol. 11, No.7, pp. 141-148, July 2001.

A Memory Efficient Fast Distributed Real Time Commit Protocol

Udai Shanker, Manoj Misra, and A.K. Sarje

Department of Electronics & Computer Engineering,
Indian Institute of Technology, Roorkee, Uttaranchal, India
udaigkp@gmail.com,{manojfec, sarjefec}@iitr.ernet.in

Abstract. Most of the past researches [1], [2], [3] investigate the behavior of distributed real time commit protocols either under update or blind write model. The effect of both types of models has not been investigated collectively. These protocols also require a considerable amount of memory for maintaining temporary objects (data structure) created during execution of transactions and block the WORKDONE message if cohort is dependent. This paper presents an optimized distributed real time commit protocol (MEFCP) based on new locking scheme and write operation divided into update and blind write. The proposed protocol optimizes the memory required for maintaining the transient information of lender & borrower [1]. It also sends the WORKDONE message if borrower has locked the data in mode 2 only. We also compared MEFCP with PROMPT and 2SC commit protocols through simulation.

1 Introduction

Many real time applications handle large amount of data and an intensive transactions processing by using disk resident database because the amount of data they store is too large (and too expensive) to be stored in the non volatile main memory. The buffer is used to store the execution code, copies of files & data pages, and any temporary objects produced. The buffer manager does this in the main memory [4]. Before the start of execution of a transaction, buffer is allocated to the transaction. When memory is running low, a transaction may be blocked from execution. The amount of memory available in system thus limits the number of concurrently executable transactions [4] in large scale distributed real time database systems (DRTDBS). The execution of transaction will be significantly slowed down. This problem needs to be taken into account to ensure that transactions receive their required resources in time to meet their deadlines. So, it is important for the database designer to develop memory efficient and fast protocols, so that more number of transactions can be executed concurrently at any instant. In this paper, we design an optimized distributed real time commit protocol which optimizes memory usage and processing time. Several commit protocols such as 2PC and its variants have been proposed in past. Soparkar et al. have proposed a protocol that allows individual sites to unilaterally commit. Gupta et al. proposed optimistic commit protocols. Enhancement has been made in PROMPT proposed in [2], which allows executing transactions to borrow data in a controlled manner from the healthy transactions in their commit phase. The technique proposed by Lam et al. [3] maintains three copies of each modified data item for resolving execute-commit conflicts. Biao Qin and Yunsheng Liu [1] proposed a protocol, double space commit (2SC), which classifies the dependencies between lender and borrower cohorts into two types; commit and abort. All above protocols consume a considerable amount of memory for temporary records which, in turn, create additional workload on the system. MEFCP, based on update and blind write model, and new locking scheme reduces the number of temporary records needed, and thus relieves the system from additional load. It also sends the WORKDONE message, if borrower has locked the data in mode 2 only.

Section 2, discusses data access conflict resolving strategies and pseudo code of the protocol, whereas Section 3 describes memory optimization achieved by MEFCP. Section 4 discusses our simulation model and results. Section 5 concludes the paper with future directions.

2 Data Access Conflicts Resolving Strategies

A flag is attached with each data item. The flag is set in any one of three modes given below, if a data item is being locked by a cohort at the time of its arrival.

Mode 1: If a cohort want to use a data item and it is not locked by any other cohort, it sets the flag of data item in Mode 1.

Mode 2: If a cohort T2 wants to update a data item read by another cohort T1 in its committed phase, it convert the flag of data item in Mode 2 from Mode 1. T2 is not allowed to commit until T1 is committed. However, if T1 aborts, T2 does not abort.

Mode 3: If a cohort T2 reads/writes an uncommitted data item written by another cohort T1, it converts the flag of data item in Mode 3 from Mode 1. Here, T2 is not allowed to commit until T1 is committed. However, if T1 aborts, T2 also aborts.

Each site S_i maintains a list which contains the following information.
List $(S_i) : \{(T_j, D) | T_j$ is borrower and has locked the dirty data D$\}$
Let T1 be the committing cohort holding lock on data item X. T2 be the cohort requesting lock on same data. X is in mode 1. Six possible cases of data conflict are:

Case 1: Read-Write(Blind OR Update) Conflict: If cohort T2 requests a Write (Blind OR Update) - Lock while cohort T1 is holding a Read-Lock, the flag associated with data item is set in Mode 2 from Mode 1.

Case 2: Write (Blind) –Write (Blind) Conflict. If cohort T2 requests a Write (Blind)-lock while cohort T1 is holding Write (Blind)-Lock, the flag associated with data item is set in Mode 2 from Mode 1.

Case3: Write (Update)–Write (Update) Conflict: If both locks are Write (Update)–Locks, then flag associated with data item is set in Mode 3 from Mode 1.

Case 4: Write (Update)–Write (Blind) Conflict. If cohort T2 requests a Write (Blind)-Lock while cohort T1 is holding a Write (Update)-Lock, flag associated with data item is set in Mode 2 from Mode 1.

Case 5: Write (Blind)–Write (Update) Conflict. If cohort T2 requests a Write (Update) -Lock while cohort T1 is holding Write (Blind)´-Lock, flag associated with data item is set in Mode 3 from Mode 1.

Case 6: Write (Blind OR Update)-Read Conflict. If cohort T2 requests a Read-Lock while cohort T1 is holding a Write (Blind OR Update)-Lock then flag associated with data item is set in Mode 3 from Mode 1.

Three Possible Cases (when T2 has accessed the data item locked by T1):

1 T1 receives decision before T2 has completed its local data processing:
1. If global decision is to commit, T1 commits.
- Execute all cohorts using the data items locked by T1 whose flag is in either Mode 2 or Mode 3 as usual.
- Flag either in Mode 2 or Mode 3 of data items locked by T1 is set to Mode 1.

2. If the global decision is to abort, T1 aborts.
All cohorts using the data items whose flag is in Mode 2 and already locked by T1 will execute as usual. Flag in Mode 2 on data items locked by T1 is set to Mode 1.
All cohorts using the data items with Mode 3 flag and locked by T1 abort. Flag in Mode 3 on data items locked by T1 is set to 0. The cohorts dependent on data set of T1 will be deleted from the List.

2 T2 completes data processing before T1 receives global decision:
if T2 has locked the data item in mode 2 only, it sends WORKDONE message; otherwise, T2 does not send WORKDONE message and blocked until its deadline expires or T1 gets decision. In the first case, T2 aborts and is deleted from the List. In the second case, if T1 aborts/commits, the system will execute as the first case;

3 T2 aborts before T1 receives decision:
In this case, T2's updates are undone and T2 will be removed from the List.
At time of YES-Voting, if cohort is still dependent, YES-Voting message is deferred

The complete pseudo code of the protocol is given as below:
if (T1 receives global decision before, T2 ends execution) then
{One: if (T1's global decision is to commit) then
    {T1 commits;
    All cohorts using data items in Mode 2 or 3 locked by T1 will execute as usual;
    Flag either in Mode 2 or Mode 3 of data items locked by T1 is set in Mode 1;
    The cohorts dependent on T1 will be deleted from the List ;}
   else {T1 aborts;
       All cohorts using data items with Mode 2 flag and locked by T1 will execute as usual. Flag in Mode 2 on data items locked by T1 is set in Mode 1;
       All cohorts using data items with Mode 3 flag and locked by T1 abort. Flag in Mode 3 on data items locked by T1 is set to 0. The cohorts dependent on data set of T1 will be deleted from the List;}}
else if (T2 ends execution before T1 receives global decision)
{if(T2 has locked the data items in Mode 2 only)
T2's WORKDONE message is sent;
else{
T2's WORKDONE message is blocked;
Do {T2 wait for next event/message;
    Switch (type of event/message)
      {Case 1: if (T2 misses deadline)

{Undo computation of T2;Abort T2;Delete T2 from the List; }
  Case 2: if (T1 commits/aborts) GoTo One;}
} while (1);}}
else //T2 is aborted by higher transaction
// before T1 receives decision
{Undo the computation of T2; Abort T2; Delete T2 from the List;}

3 Optimization of Memory

It is assumed that the number of data items in the database at each site is N. The memory required for maintaining the record of data items lent by a single cohort is computed below.

Case 1: Memory Required in 2SC [1]
At least, a flag is required corresponding to every data item to show its locking status when it is locked by a cohort. So, the minimum memory required to keep the information of locking status of the data items is N/8 bytes (a flag needs at least single bit storage). Again, each site maintains a list of lenders, and also each lender maintains two lists: commit dependent cohorts and abort dependent cohorts with dirty data used by them. This can be implemented as given below.

Linked List: A dependency list is maintained which contains the id of committing cohorts (lenders) who have lend their modified data to newly arrived cohorts. Each lender in this dependency list also maintains two lists which contain id of abort and commit dependent cohorts with dirty data items utilized by them. The memory required for keeping the record of data items lend by a single cohort is computed below. Let us assume that on an average each lender has p cohorts in dependency list and q cohorts in abort dependency list.

$$M = M1 + (M2 \text{ OR } M3) * N, \text{ where}$$

M	Total Memory Required by one node of lender.
M1	Memory required for the dependency list is 14 bytes (4*3 bytes for address + 2 bytes for id).
M2	Memory required for the list of commit dependent cohorts and dirty data item is 8*p bytes.
M3	Memory required for the list of abort dependent cohorts and dirty data item is 8*q bytes.
N_d	No. of data items lent by the cohort = p + q.

Case 2: Memory Required in Proposed scheme
Minimum memory required to keep the record of Modes of every data item at a site is two bit. So, the total required memory is N/4 bytes. Here, a single list is being maintained for keeping the information of borrower and dirty data used by it. This requires 8 bytes of memory (2 bytes for borrower id + 2 bytes for dirty data + 4 bytes for address of the next node). Comparing to case 1, there is additional need of N/4 bytes memory at each site to keep the information about the Mode of every data item. With the increase in the transaction arrival rate and transaction size, there are chances of more conflicts resulting in more number of dependent cohorts on committing cohorts. If there are L cohorts lending data at any instant of time, the additional

memory required is 14*L bytes in case 1 as compared to case 2 (see in table 1). Although, it seems initially more number of bytes are needed for keeping the Mode information of data items, the proposed protocol competes with [1] at high transaction arrival rate and long transaction size.

Table 1. Study of Memory Requirement

Commit Protocol	flags at each site (in bytes)	Single lender (bytes)
2SC	N/8	$14+8*N_d$
MEFCP	N/4	$8*N_d$

It is clear from table 1 that MEFCP will require lesser memory than 2SC, whenever L > N/112.

4 Model Parameters and Simulation Results

A distributed real time database system [5], [6] consisting of N sites have been simulated. The default values of different parameters and concurrency control scheme are same as in [5]. MEFCP is compared with PROMPT and 2SC.

4.1 Impact of Transaction Arrival Rate

Fig.1 and Fig. 2 show impact of transaction arrival rates on transaction miss percentage at transaction length 3-20 (uniform distribution). Miss percentage increases with increase in transaction arrival rate. At higher arrival rate, the probability of lock conflicts for the data items and queuing delay for the use of system resources are more. The performance of the MEFCP is better with 2SC and PROMPT due to the better approach used for resources utilization, minimizing the queuing delay and sending of WORKDONE message if cohort's locked data is in Mode 2 only.

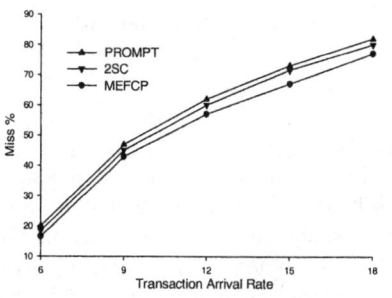
Fig. 1. Miss % with(RC+DC),Communication Delay 0ms,Normal & Heavy Load

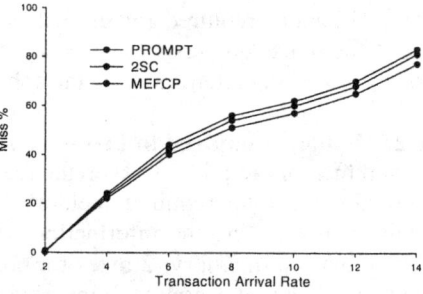
Fig. 2. Miss % with(RC+DC),Communication Delay 100ms Normal & heavy load

4.2 Impact of Transaction Size

Fig. 3 and Fig. 4 show the miss percentage for the protocols at different transaction size at communication delay 100ms & 0ms and transaction arrival rate 10. In this case, MEFCP outperforms as compared to PROMPT and 2SC at higher & low transaction size due to better buffer management and sending of WORKDONE message if locked data is in Mode 2 only.

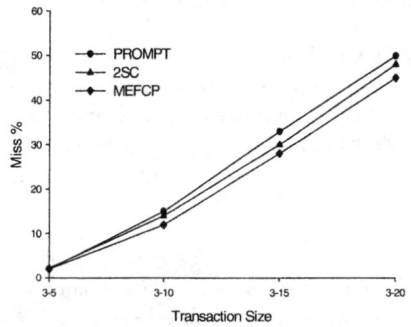
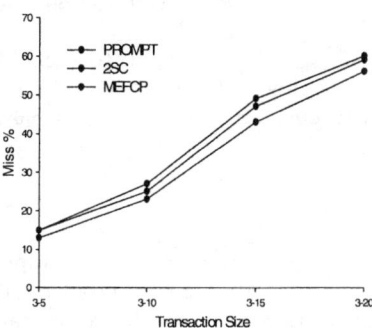

Fig. 3. Miss % with (RC+DC) at Communication Delay=0ms

Fig. 4. Miss % with (RC+DC) at Communication Delay=100ms

5 Conclusion

This paper deals with a new commit protocol with blind and update, and optimizes the storage space by only maintaining the information of borrower cohort along with data item used by it in conflicting way. This protocol outperforms as compared with PROMPT and 2SC due to not blocking the WORKDONE message of cohort locked the data item in Mode 2 only. It is well suited to data intensive application where the transaction arrival rate and transaction size are high. Further, the devices used in mobile applications have limited memory capacity. MEFCP may be useful for these applications due to its reduced memory requirements.

References

1. Qin, B., Liu, Y.: High performance distributed real time commit protocol. Journal of Systems and Software, Elsevier Science Inc. (2003) 1-8
2. Haritsa, J.R., Ramamritham, K., Gupta, R.: The PROMPT real time commit protocol. IEEE Transaction on parallel and distributed systems. 11(2). (2000) 160-181
3. Lam, K.Y., Pang, C.L., Son, S.H., Cao, J.: Resolving executing-committing conflicts in distributed real-time database systems. Computer Journal. 42 (8). (1999) 674-692
4. Garcia-Molina, H., Salem, K.: Main memory database systems: an overview. IEEE Transactions on Knowledge and Data Engineering. 4(6). (Dec. 1992) 509 – 516
5. Shanker, U., Misra, M., Sarje, A.K.: A Modified Distributed Real–Time Commit Protocol. Proceedings of the International Conference on Systemics, Cybernetics and Informatics. Hyderabad. India. (2005) 783-786
6. Lam, K.: Concurrency Control in Distributed real time database systems. PhD Thesis. City University of Hong Kong. (1994)

A Model for the Distribution Design of Distributed Databases and an Approach to Solve Large Instances

Héctor Fraire H., Guadalupe Castilla V., Arturo Hernández R.,
Claudia Gómez S., Graciela Mora O., and Arquimedes Godoy V.

Instituto Tecnológico de Cd. Madero,
Primero de Mayo S/N, Cd. Madero, Tamaulipas, 89440 México
hfraire@prodigy.net.mx, {gpe_cas, otero1250, arquigv}@yahoo.com.mx,
{ahram66, cggs71}@hotmail.com

Abstract. In this paper we approach the solution of large instances of the distribution design problem. Traditional approaches do not consider that the size of the instances can significantly affect the efficiency of the solution process. This paper shows the feasibility to solve large scale instances of the distribution design problem by compressing the instance to be solved. The goal of the compression is to obtain a reduction in the amount of resources needed to solve the original instance, without significantly reducing the quality of its solution. In order to preserve the solution quality, the compression summarizes the access pattern of the original instance using clustering techniques. In order to validate the approach we tested it on a new model of the replicated version of the distribution design problem that incorporates generalized database objects. The experimental results show that our approach permits to reduce the computational resources needed for solving large instances, using an efficient clustering algorithm. We present experimental evidence of the clustering efficiency of the algorithm.

1 Introduction

Distributed databases applications are developed using Distributed Database Management Systems (DDBMS's). Despite the advanced technology of DDBMS's, the design methodologies and tools have many limitations. Consequently, database administrators carry out the distribution design using empirical and informal approaches due to the problem complexity.

The distribution design problem consists of determining data allocation so that the communication costs are minimized. Like many other real problems, it is a combinatorial NP-hard problem. The solution of large scale instances is usually carried out solving a simplified version of the problem or using approximate methods [1], [2]. General purpose nondeterministic heuristic methods are at present the best tools for the approximate solution of this class of problems [3], [4].

2 Related Work

The distribution design problem has been dealt with by many investigators [5], [6], [7], [8], [9], [10], [11], [12], [13]. The approach proposed in [5] has been the most successful in solving large scale instances. The main limitation of these approaches is that they do not consider that the size of the instances can significantly affect the efficiency of the solution process. Conversely, in [14] the relevance of the compression is recognized, but the effect of compression on the solution quality is not considered; consequently, the compression methods proposed are inefficient and do not guarantee the scalability of the tools for automatic database design.

In order to overcome these limitations, we propose an instance compression processing. We test this approach on a new model of the replicated version of the distribution design problem that incorporates generalized database objects, and applying a method for efficient instance compression that uses clustering techniques [15], [16].

3 Description of the Distribution Design Problem

The DDB distribution design problem consists of allocating DB-objects, such that the total cost of data transmission for processing all the applications is minimized. A DB-object (or simply object) is an entity of a database that requires to be allocated at the sites of the network, which can be an attribute, a tuples set, a relation or a file.

A formal definition of the problem is the following: let us consider a set of objects $O = \{o_1, o_2, \ldots, o_{no}\}$, a set of sites $S = \{s_1, s_2, \ldots, s_{ns}\}$, where a set of operations $Q = \{q_1, q_2, \ldots, q_{nq}\}$ are executed, the objects required by the operations, an initial allocation schema, and the access frequencies of the operations in a given time. The problem is to obtain a new allocation schema that adapts to a new database usage pattern and minimizes transmission costs.

4 Mathematical Model

Traditionally it has been considered that the DDB distribution design consists of two sequential phases. Contrary to this widespread approach, it is simpler to solve the problem using our approach which combines both phases. A key element of this approach is the formulation of a mathematical model that integrates both phases [5].

The problem is modeled using binary integer linear programming. The mathematical model objective function (1) includes four terms that model the costs of processing read-only operations, read-write operations, object migration, and storage cost. A detailed description of the model can be found in [17].

$$\min Z = \sum_k \sum_j f_{kj} \sum_m \sum_i q_{km} l_{km} c_{ji} w_{jmi} \qquad (1)$$
$$+ \sum_k \sum_j f'_{kj} \sum_m \sum_i q'_{km} l'_k c_{ji} x_{mi}$$
$$+ \sum_j \sum_m \sum_i w'_{jmi} c'_{ji} d_{mi} + \sum_m \sum_i CA_i b_m x_{mi}$$

5 Instance Transformation Using Clustering

5.1 Description of the Transformation Method

When an instance of the DDB distribution design problem has repetitive operations, it is possible to transform it into an instance with fewer operations, since repetitive queries are represented by similar rows in the access matrix. Therefore, such operations can be considered as a single (clustered) operation that is issued with larger frequency. The instances reported in [12], characterized as typical on the Internet, are an example of instances with this property.

All the objects needed by a clustered operation constitute a single object of the compressed instance. The reduction such transformation can yield is directly proportional to the number of repeated operations.

The binary vector that indicates from which sites a operation is issued is called access pattern. In the access pattern matrix P_{ki}, for every k and i, $P_{ki} = 1$ if and only if $f_{ki} \neq 0$. This binary vector can be used to code a decimal number.

5.2 Adjustment of the Transformed Instance Formulation

Once the query and object clusters are created, it is necessary to adjust the access frequencies, the selectivity and the objects (cluster) size of the transformed instance. The adjustment process is carried out as follows.

Given the original instance i, matrix f_k, operation selectivities s_{km} and object sizes b_m, then the access frequency of each clustered operation c at site i is given by: $f^\star_{ci} = \sum_{k \in OperCluster(c)} f_{ki}$, $\forall c, i, k$. The size of object cluster c is given by $b^\star_c = \sum_{m \in ObjCluster(c)} b_m$, $\forall c, m$. The selectivity of query k to DB-object group c is given by: $s^\star_{kc} = (\sum_{m \in ObjCluster(c)} s_{km} \times b_m)/(\sum_{m \in ObjCluster(c)} b_m)$, $\forall c, k, m$.

5.3 Codifier Clustering Algorithm

In this method, a decimal code is assigned to each pattern, which is used to identify the cluster to which the operation belongs (Figure 1). This algorithm clusters the operations with the same decimal code. Since the algorithm complexity is $nq \times ns$, if we limit the number of sites (ns) the complexity will be linear. This is a most desirable characteristic for solving large problem instances.

6 Experimental Results

To validate our approach, a set of experiments were conducted using randomly generated instances of different sizes and characteristics, configured according

```
For Each query k of f_ki
    Code ← 0
    For Each site i of f_ki
        If f_ki > 0 then
            Code ← Code + 2^i
    Group_k ← Code
    Card_Code ← Card_Code + 1
End For
```

Fig. 1. Codifier clustering algorithm

to typical access patterns found on the Internet. To simulate several access patterns, test cases with 10, 20, 30, and 40 % access probability of the operations to the sites were generated. For each instance of a particular experiment, the clustering method was applied to compress it. Once the compression was performed, the original instance i and the compressed instance i' were solved using an exact method, and the costs of both solutions were compared. Table 1 shows the characteristics of a representative sample of test cases used in the experiments. The table includes a test case identifier (C_i), the numbers of DB-objects (O), sites (S), and operations (Q) of the included instances in the test case, the size in bytes (Size) of the test case, and the operations to sites ratio (Q/S).

Table 1. Test cases used in the experiments

Test case	Characteristics				Q/S
	Objects (O)	Sites (S)	Operations (Q)	Size in Bytes	
C_1	100	3	100	86,060	33
C_2	200	5	200	338,560	40
C_3	500	7	500	2,062,252	71
C_4	1,200	15	1,200	11,823,420	80
C_5	1,000	10	1,000	8,172,480	100

Figure 2(a) shows the reduction observed in the experiments. Notice that, for instances with access probability of 10% and 20%, the reduction of resources is at least 65%. The minimal and maximal reductions are 65% and 99%, which constitute a considerable reduction of resources. Figure 2(b) shows the impact on the solution quality. The error percentage varies from 0.10% to 3.20%, which shows that the degradation is relatively small. Therefore, the feasibility of reducing the resources required to solve large scale instances at the expense of a reasonable loss in the solution quality, is demonstrated.

In Figure 3, we show that the clustering efficiency of the codifier algorithm (♦) is better than the clustering efficiency of two hierarchical clustering algorithms (■ and △) based on the approach proposed in [18]. We can see that the efficiency of our algorithm increases faster than that of the other algorithms.

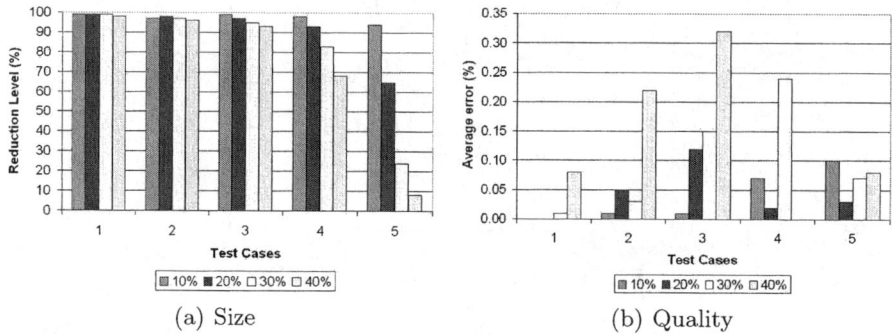

(a) Size (b) Quality

Fig. 2. Reduction level

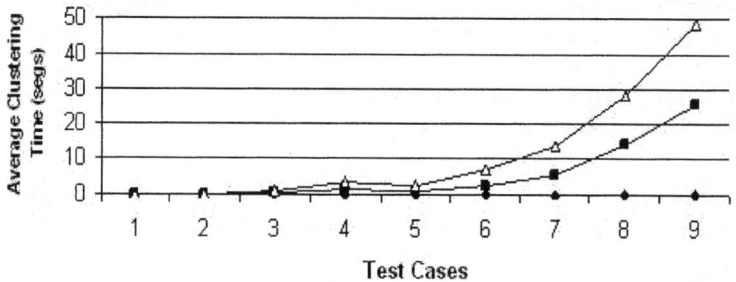

Fig. 3. Efficiency of the clustering algorithm

7 Conclusions and Future Work

This paper shows the feasibility to solve large scale instances of the distribution design problem by compressing the instance to be solved. The goal of the compression is to obtain a reduction in the amount of resources needed to solve the original instance, without significantly reducing the quality of its solution. In order to preserve the solution quality, the compression *summarizes the access pattern* of the original instance using clustering techniques.

A set of experiments, using instances configured with typical access patterns on the Internet were conducted for evaluating quantitatively the size reduction that can be achieved and its effect on the solution quality. The compression process shows high levels of reduction in the amount of resources needed, without a significant loss in the solution quality, and its efficiency increases faster than that of the other clustering algorithms. We present experimental evidence of the clustering efficiency of the proposed algorithm, which shows that, given a set of computing resources, it is now possible to solve instances larger than those previously solvable.

References

1. Garey, M., Johnson, D.: Computer and Intractability: A guide to the theory of NP-Completeness. Freeman (1979)
2. Papadimitriou, C., Steiglitz, K.: Combinatorial Optimization: Algorithms and Complexity. Dover Publications (1998)
3. Barr, R., Golden, B., Kelly, J., Steward, W., Resende, M.: Guidelines for designing and reporting on computational experiments with heuristic methods. In: Proceedings of International Conference on Metaheuristics for Optimization, Kluwer Publishing (2001) 1–17
4. Michalewicz, Z., Fogel, D.: How to Solve It: Modern Heuristics. Springer Verlag (1999)
5. Pérez, J.: Integración de la Fragmentación Vertical y Ubicación en el Diseño Adaptativo de Bases de Datos Distribuidas. PhD thesis, ITESM, Morelos, México (1999)
6. Pérez, J., Pazos, R., Vélez, L., Rodríguez, G.: Automatic generation of control parameters for the threshold accepting algorithm. Lectures Notes in Computer Science. Springer Verlag, Berlin Heidelberg New York **2313** (2002) 119–127
7. Pérez, J., Pazos, R., Frausto, J., Romero, D., Cruz, L.: Data-object réplication, distribution and mobility in network environment. Lectures Notes in Computer Science. Springer Verlag, Berlin Heidelberg New York **2890** (2003) 539–545
8. Ceri, S., Navathe, S., Wiederhold, G.: Distribution design of logical database schemes. In: IEEE Transactions on Software Engineering. Volume SE-9. (1983) 487 – 503
9. Navathe, S., Ceri, S., Wiederhold, G., Dou, J.: Vertical partitioning algorithms for database design. Volume 9. (1984) 680–710
10. Apers, P.: Data allocation in distributed database systems. Volume 13. (1988) 263–304
11. Johansson, J., March, S., Naumann, J.: The effects of parallel processing on update response time in distributed database design. In: Proceedings of the 21st International Conference On Information Systems. (2000) 187–196
12. Visinescu, C.: Incremental data distibution on internet-based distributed systems: A spring system approach. Master's thesis, University of Waterloo, Ontario, Canada (2003)
13. Baiao, F., Mattoso, M., Zaverucha, G.: A distribution design metodology for objects dbms. Distributed and Parallel Databases. Kluwer Academic Publishers **16** (2004) 45–90
14. Zilio, D., Rao, J., Lightstone, S., Lohman, G., Storm, A., Garcia-Arellano, C., Fadden, S.: Db2 design advisor: Integrated automatic physical database design. In: Proceedings of the Thirtieth International Conference on Very Large Data Bases 2004, Toronto, Canada (2004) 1087–1097
15. Halkidi, M., Batistakis, Y., Vazirgiannis, M.: On clustering validation techniques. In: Journal of Intelligent Information Systems. Volume 17., Kluwer Academic Publishers (2001) 107–145
16. Berkhin, P.: Survey of clustering data mining techniques. Technical report, Accrue Software (2002) http://www.accrue.com/products/rp_cluster_review.pdf.
17. Fraire, H.: Una Metodología para el Diseño de la Fragmentación y Ubicación en Grandes Bases de Datos Distribuidas. PhD thesis, CENIDET, Cuernavaca, Morelos, México (2005)
18. Beeferman, D., Berger, A.: Agglomerative clustering of a search engine query log. In: Proceedings of the sixth ACM SIGKDD international conference on Knowledge discovery and data mining. (2000) 407–416

Tracking of Mobile Terminals Using Subscriber Mobility Pattern with Time-Bound Self Purging Indicators and Regional Route Maps*

R.K. Ghosh[1], Saurabh Aggarwala[2], Hemant Mishra[2], Ashish Sharma[2], and Hrushikesha Mohanty[3]

[1] Department of CSE, Indian Institute of Technology, Kanpur 208016, India
rkg@cse.iitk.ac.in
[2] Department of CSE, Indian Institute of Technology, Guwahati 781039, India
{ashishs, saurbha, hemant}@iitg.ernet.in
[3] Department of CIS, University of Hyderabad, Hyderabad 500046, India
hmcs@uohyd.ernet.in

Abstract. This paper presents a new location management scheme that integrates two key ideas, namely, (i) Subscriber Movement Profile (SMP) based on spatial and temporal locality of the movement of a mobile terminal and (ii) localized updates known as Time-bound Self Purging Indicators (TSPI) generated by a probabilistic approach when a mobile terminal does not adhere to registered profile. The SMP registered by a mobile host is used to predict its cell location based on time specific movement history. The transient deviations from the registered SMP are handled efficiently by TSPIs using a Regional Route Map (RRM).

1 Introduction

Personal Communication Service (PCS) networks enable people to communicate independent of their locations. The system facilitates communication by maintaining a location database that maps a subscriber number to its current location. However, wireless network pose significant challenges due to several factors including unconstrained movements of subscribers, limited radio frequency spectrum, radio channel impairments, among others. A number of efficient location management schemes have been proposed [1], [2], [3] for tracking locations of mobile users. The focus of this work is on some of the recently proposed location management schemes [4], [5], [6] which subsume location data with various degree of imprecisions. In particular we are concerned with the location management scheme proposed in [4], [5]. Both the schemes use subscriber's mobility pattern (SMP) to predict a small set of probable locations of the callee when a call arrives. The call delivery will then be possible through a selective paging executed on the set of those predicted locations. The scheme proposed in this paper is a hybrid of above two schemes. Our motivation here is three-fold:

---

* Supported by MHRD, Govt of India sponsored project on Contex-Aware Programming and Information Dissemination over Mobile/Ad Hoc Networks.

- To reduce signaling traffic for tracking a mobile by making it responsible for predicting, maintaining and registering its own SMP.
- To efficiently handle the predictive transient deviations from registered SMP by using secondary cluster of cells determined through a regional route map.
- To reduce global signaling traffic at the cost of limited increase in local traffic to tackle unexpected deviation of a mobile from its registered SMP. The deviation of a mobile is considered as unexpected when it is not covered by secondary cluster.

The paper is organized as follows. Section 2 is concerned with the system model. It presents an overview of the database to be maintained in the mobile terminal as the network changes dynamically. The location management scheme proposed in this paper is described in Section 3; and the call setup protocol is discussed in Section 4. Experimental results have been presented in Section 5. The papers ends with concluding remarks in Section 6.

2 System Model

The coverage area of a mobile service provider can be represented by a finite set of cells $C = \{c_1, c_2, c_3, \ldots, c_n\}$. Location areas (LAs) define a partition of cells of C. An LA may consist of one to few geographically adjacent cells of C. The union of LAs form a complete covering of C. The information of a user U_i will be maintained at the base station of his/her Home Location Area (HLA). This will include the currently registered Subscriber Mobility Patterns (SMPs) of U_i and his/her call profiles. Call profiles will be used to calculate the call to mobility ratio (CMR) values. Let ρ denote the arrival rate of calls for the user U_i and κ denote the mobility rate of U_i, then $\omega = \rho/\kappa$ represents CMR for U_i.

A Subscriber Mobility Database (SMD) will be maintained by the mobile terminal itself. An SMD consists of a number of Subscriber Mobility Patterns (SMPs), one of which will be registered with the network. The registered SMP is a sequence of LAs (base stations) which the Mobile Terminal (MT) is expected to visit. If a mobile terminal deviates from its registered SMP, it will rely the local network for fuzzy determination of the location area where the local Time bound Self Purging Indicator (TSPI) has to be stored. The important SMP related data are as follows:

- SMP: $\{LA_1 \rightarrow LA_2, \ldots, LA_{n-1} \rightarrow LA_n\}$
- P_a: Prob(MT) moves to a new LA not in SMD.
- P_b: Prob(MT) moves to a LA $\in$ SMD but not in registered SMP.
- P_c: Prob(MT) moves in its registered SMP.

An LA ID which is a part of two SMPs will be stored twice in SMD. The advantage this is apparent when a new record gets added to the database. The SMP which has aged most, will be removed. Deletion of an LA ID will not affect its other duplicates which may be present in other SMPs. The overhead of maintaining duplicate LA IDs in different SMPs also retrieved in the form of amortization of cost for extra deletions.

The set of global attributes and cell related attributes stored in SMD are provided in the table below.

Global attributes	Cell attributes
N_{nd}: Number of consecutive deviations from registered SMP.	Cell Id: Cell identifier
P_{sid}: SMP Id currently in use.	SMP Id: SMP identifier
SC_{cids}: Secondary clusters' cell Ids	E_{at}: Expected arrival time
$REP_{callerSMP}$: Replicated caller SMPs	E_{et}: Expected exit time
P_{cid}: Present Cell Id	E_{st}: Expected slack limit for arrival
T_r: Total number of records	Next LA: Next location area
NR_{smp}: Registrations count each SMP	Previous LA: Previous location area
LF_{SMP}: Lifetime of each SMP	

3 Algorithm Description

There are three key concepts, namely, secondary clusters, regional route map and TSPI, which categorize the input requirements for the proposed location management scheme.

3.1 Secondary Clusters

Our protocol requires some inputs from the mobile users at the time of subscription to a new service. This information will be utilized by the HLA in selection of places that have a high probability of being frequented by the user apart from the those specified by the user profile. These alternative cells are collectively referred to as secondary clusters. One of the important attribute for secondary cluster, for example, could be the information about the subscriber's occupation. A mobile user may be an official, a travel agent, a student, or a house wife. Such an information will help to select certain cells which are not in user's profile, but may be visited in future.

3.2 Regional Route Map (RRM)

Another key input is the Regional Route Map (RRM). A RRM indicates the physical route map of roads in the neighborhood of a location area. Such a map along with the user supplied profiles will be stored at the base station of home location. It helps in a quick discovery of cells that fall in the vicinity of the registered SMP of the user. These are LAs in which the user can roam without including them either in the registered SMP or generating a TSPI. The motivation of employing RRM is driven by the fact that the users might not usually go to certain cells. But these places match the places of highest interest to the users. So, a high probability is attached to these places of being visited by a user. If an MT visits one of these places, it need not send a location update.

3.3 Time-Bound Self Purging Indicators (TSPI)

If a user has moved into a location which is not a part of its registered SMP or its secondary cluster, it first creates a Time Bound Self Purging Indicator (TSPI). TSPI is a special register entry maintained at the VLR of the expected LA as per the registered SMP of the mobile terminal. TSPI stores the new location where the MT is found at present. It may be considered as a forwarding pointer [3] stored in the VLR of the base station of the expected LA. An expiry time is also associated with the TSPI. The motivation here is to increase small amount of local traffic in place of multiple hops that may be needed to register in the HLR, thus saving valuable network bandwidth. Associating an expiry time with TSPI eliminates the need for extra signaling which MT otherwise have to do for removing the pointer [3].

The LA where the TSPI will be registered is decided by a fuzzy network controlled machine colocated with the VLR location. It chooses between the base stations of the expected LA and the HLA of the MT, based on a number of dynamic network parameters such as: (i) relative number of hops required to reach MT at its current position from HLA and from the expected LA in the registered SMP, (ii) available network bandwidth, (iii) tariff, and (iv) certain other network specific runtime parameters.

3.4 Creation of SMP

Each MT creates and maintains its SMPs. A new SMP creation procedure buildNewSMP is given below.

```
procedure buildNewSMP(LA) {
   if (Day is over) { /* New SMP creation is complete */
      reset the smpBuild flag;
      Increment the LF values of all SMPs;
   }
   else {
      if (SMPbuild == false ) {
         SMPbuild ← true; newSMP_Id ← getNewSMPId;
         /* Create a replication of registered SMP from
            beginning till the LA visited last.*/
         newSMP ← copyAllNode(currSMP,currTIME);
      }
      Create a new node of this LA;
      LastNode_of_SMP.next ← newNode;
      Prev.newNode ← LastNode_of_SMP;
      newNode.next ← NULL;
   }
}
```

When an MT moves into a LA (base station) not in its SMP and its `smpBuild` flag is `false`, then MT sets the flag to `true`. The MT then copies the nodes (LAs) of the current SMP as the preceding nodes of the new SMP till the last perfectly followed LA as per the registered SMP. The MT continues to add subsequent new nodes into the new SMP, till the time of the day is over. The span of each SMP is 24 hours. Thus after an SMP has been created a 24-hour period, it is included as a complete SMP in the SMD and the `smpBuild` flag is reset.

3.5 Maintenance of SMP

It may be argued that instead of building a new SMP, we could do by only updating the time field in an already present SMP when a time lag is noticed in visiting an LA but the order of the nodes in the SMP remains same. The reason for not following this approach is: even a small amount of transient deviation can make us loose a carefully formed otherwise frequently used SMP. So it is always better to keep track of the deviations as a freshly built SMP.

If the MT moves into a new LA that is present in its SMD but not a part of the registered SMP, then we store this LA into a separate data structure called the Transient Deviation Pattern (TDP). A TDP is allowed grow to as a sequence of maximum of three LAs to which the MT has deviated. When a TDP grows to its full size, the SMD is searched for the TDP. If the TDP is found to be a part of an existing SMP then the update algorithm returns a value 3. This indicates that the MT is following a new SMP; and, hence, needs to register the same. It also resets the current SMP building process or purges the partially built SMP. The reason for this can be explained as follows. If an MT consecutively follows three locations of another SMP in the SMD leaving its own registered SMP, then it is highly likely that MT would continue to move according to the former SMP as indicated by its stored movement pattern. If the TDP match returns a `false`, it means that the user is moving in an entirely new patten which is not present in the SMD. So this TDP is retained as a part of a new SMP which is under creation. Notice that we keep track of only the last three consecutively deviated locations, and update them by swapping the locations.

There might be a condition of SMD overflow. It occurs when no more nodes can be added to it. In this case an entire SMP deleted. There are two factors in support of this deletion, namely

(i) The partial SMP can not be registered unless it is padded with the missing LAs by arbitrary data.
(ii) Since the process of building a new SMP has been initiated, it is likely that there will be a requirement of additional space to accommodate more newly created data.

The victim SMP to be selected for deletion is decided on the basis of $\varphi = NR_i/LF_i$, which is the ratio of number of times a SMP has been registered to the number of days it has been formed. Since NR_i gives a count of the number of times SMP i has be registered it also determines the aging SMPs. The SMP which have been registered to the network the least number of time provides a low NR_i value. The ratio φ helps in determining obsolescence of an SMP. It ensures that recently formed SMPs are not selected as victims. After the deletion of an SMP, NR_i and LF_i values for all other SMPs are reduced by a factor φ. This ensures that these values do not increase to some large values for other SMPs present.

Let expected slack time E_{st} (see the cell attributes described in section 2) be defined on the basis of certain fuzzy runtime conditions represented by θ. The argument θ is decided by the central network taking into account the importance of the place. A MT need not update the network registration if it is within the slack bound. So the permissible time of stay in an LA before a TSPI creation procedure is initiated is specified by the following bounds.

$$E_{at} - (E_{st} + \theta) \leq T(stay) \leq E_{et} + (E_{st} + \theta)$$

Based on the above bounds it is straightforward to determine the when TSPI generation could be initiated. The procedure CurrentSMPfollowed provide below returns true if a mobile user is following the registered SMP or has deviated.

```
procedure CurrentSMPfollowed(LA) {
   if (LA-ID matches expected LA)
      if ($E_{at} - (E_{st} + \theta)$ $\leq t \leq$ $E_{et} + (E_{st} + \theta)$) return true;
      else return false;
}
```

On the basis of above discussion, we can now put together the process of maintenance of SMP in the form of a procedure UpdateSMP. We summarize the key points concerning update process as follows. When a mobile terminal moves to a new LA it gets the corresponding LA ID in its incoming signal. If the user has moved into a location which matches the expected LA in its currently registered SMP, then the update process exits with a value 1, indicating no action. There will not be any increase in network traffic due to the MT in this situation. Though the above scenario appear to be the best case, indeed this scenario is the usual case. Most users will likely to continue a fixed schedule for weeks at a stretch [7]. More precisely, there are only two cases under which an MT creates traffic for location update. These cases are characterized as: (i) when an MT is switched on for the first time, or (ii) when an MT moves out of a region of its registered SMP.

The updateSMP procedure will be executed by an MT whenever it moves to a different LA (base station). The values returned by this procedure represent four states of location updates. These are: (i) state 1: perfect schedule is followed, (ii) state 2: deviation is expected, (iii) state 3: deviations match with another stored SMP which should be registered, and (iv) state 4: deviations indicate a new SMP to be built. In states 3 and 4, a TSPI has to be registered in expected LA.

```
procedure UpdateSMD (currentLA) {
  expectedLA = getPredictedLA(SMP);
  if (currentLA == expectedLA)
      return 1;
  TDPcount++; TDP[TDPcount] = currentLA;
  if (currentLA ∈ SecondaryCluster) {
     if (TDPcount == 3) {
         TDPcount--; TDP[1] ← TDP[2]; TDP[2] ← TDP[3];
     }
     return 2;
  }
  else {
     CreateNewTSPI(currentLA, expectedLA, validTime);
     if (TDPcount==3) {
        if (new SMP is identified) {
           TDPcount=0; return 3;
        }
        if (SMD is full) {
           find victim SMP with least φ value
           remove victim-SMP from SMD;
           for (each remaining SMPs ∈ SMD) {
               NR_i ← [NR_i * φ]; LF_i ← [LF_i * φ]
           }
        }
        smpBuild = true;
        BuildNewSMP(LA);
        return 4;
     }
  }
}
```

Whenever a new SMP is registered in the HLA, this new SMP is also cached at m most frequently calling mobile terminals. The value of m can be decided based on network parameters, and this value may vary over the network.

4 Call Set-Up Protocol

There are two distinct cases for call setup. First one is where the callee MT's SMP is replicated in the SMD of caller MT. The second case is when the SMP of callee is not available in SMD of the caller.

4.1 Callee's SMP Replicated in Caller's SMD

In this case the call setup process is realized as follows.

- The caller looks up the current location area of the callee from latter's profile and sends a message to its local LA/MSC specifying the expected LA of the callee to which it wants to connect.
- The local LA/MSC instead of going to HLA of the callee, just pages the LA in which callee is expected to be present, so the callee is located either in its first page or it pages to the link pointed by the TSPI. So the lookup at the home location is saved.

4.2 Callee's Registered SMP Not Replicated in Caller's SMD

This case caters to the situation where caller does not frequently call up the callee. The call setup in this case is more expensive than the previous case as indicated below.

- The caller sends a request to the local LA(BS)/MSC specifying the callee's MT Id. The LA finds the home location of the callee, and sends a call connection request.
- The home location of the callee then looks up the SMP registered by the MT and fetches the current expected location. After which the paging protocol is initiated. The callee is located either in its first page or on second page to the link pointed by the TSPI if the callee has deviated from its registered SMP.

4.3 Paging Procedure

The paging procedure has to take several cases into account.

Case 1: *The callee is in an LA in registered SMP or has a TSPI.*
The first page is done at the expected LA which is obtained from the registered SMP based on the current time. If there is no answer at the first page, but a TSPI register set to true, the second page is done at the LA pointed to by the TSPI. The MT must answer at the second page otherwise the mobile is not in this state.

Case 2: *The callee is not present in its SMP or TSPI indicated LAs.*
In this case the Home LA prepares a priority list of the secondary cluster LAs present in its registered SMP. The priority list is created taking the following into account: (i) The Regional Route Map (RRM) which is a map of the paths in the city indicating places and routes. (ii) The Slack time specified in the SMP. (iii) The importance of the place based on the preferences given by the user at the time of registration. (iv) Congestion rates at those points. The paging is done in all these secondary clusters based on the priority list thus created above.

Case 3: *MT can not be located in any of the secondary clusters.*
A second level priority list will be created which will include the LA based on: (i) the proximity to the expected LA in its SMP. (ii) proximity to the expected

LA in the time frame. The paging is done again in all these LA given in the secondary level priority list thus created above till a page time-out is reached.

Case 4: *A page time-out is encountered.*
All the LAs in the current registered SMP are paged.

Case 5: *If callee still can not be located.*
It is considered to have been one of the following state: (i) has switched off without de-registering itself, (ii) has encountered a power failure, (iii) has moved to a region not covered by the network.

5 Simulation and Performance Evaluation

The performance was evaluated for a MSC with 50 cells. The topology of the network, i.e. the connectivity between the nodes and their weights is randomly generated. The simulation time is taken to be 200 units. The calls for the host were also randomly generated during this time. The deviation for the host were randomly generated. The scheme is simulated for varying call to mobility ratios (CMR) and percentage deviations ranging from 5% to 90%. The total number of moves made by host is taken to be 10. Here, we only mention some extracts of the simulations due to lack of space.

The simulation results are compared with the two previous location management schemes, viz., User Mobility Pattern Based [4] and Movement Prediction Based [5]. The various costs are averaged over a number of simulations and varying input patterns.

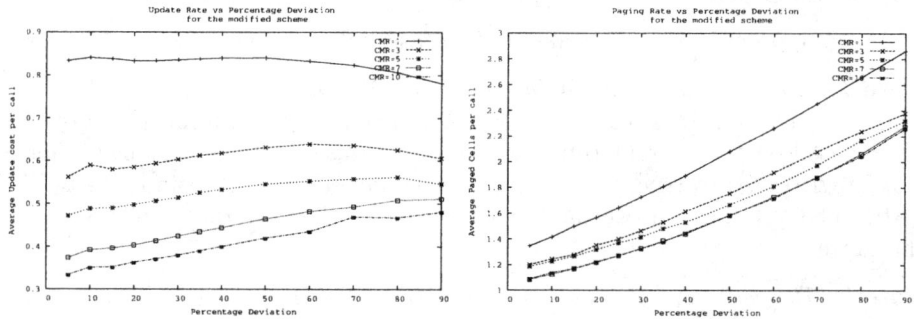

Fig. 1. Update and paging costs for the present scheme

Figures 1 and 2 respectively show the location update costs and paging costs and the network signaling costs respectively for the present scheme with respect to varying percentage deviations and CMR values. It may be observed that the number of paged cells reduces considerably due to the TSPI entry and secondary cluster search even for large deviations. Of course, the TSPI updates increase signaling cost slightly. The average paging cost is found to be 1-3 cells per call. Even for large deviations, it is found to restrict the number of paged cells to

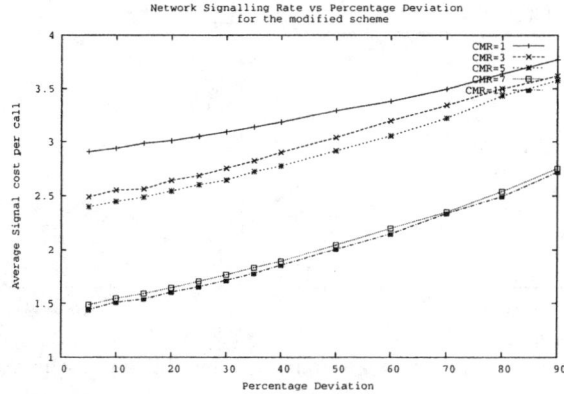

Fig. 2. Network signaling costs for the present scheme

less than 3 which will be quite high in two other schemes. This is possible since the TSPI entry allows fast tracking of the transient deviations. Although it will increase in case the host deviates appreciably without coming back to regular patterns, resulting in expiration of TSPI entry more frequently. Hence, selection of TSPI expiration timer should also be set carefully.

The increase in update rate for larger deviations reduces for higher mobility as the pattern is registered only if TDP entry is found to be full, i.e the pattern is registered only if the host possess high probability to follow it and the transient deviations are tackled through the TSPI entry.

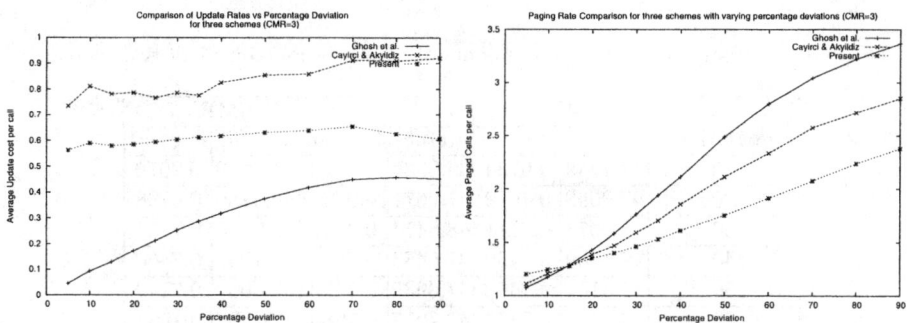

Fig. 3. Comparison of update and search costs for the three schemes for CMR=3

Figures 3, 4 and 5 provide comparison of update and paging costs for the three schemes for different CMR ratios. From the results it is clear that the present scheme leads in significant reduction in the paging costs even for larger deviations compared to two other schemes. The present scheme restricts the paging costs to 3 cells per call while the other schemes reach higher values for high mobility.

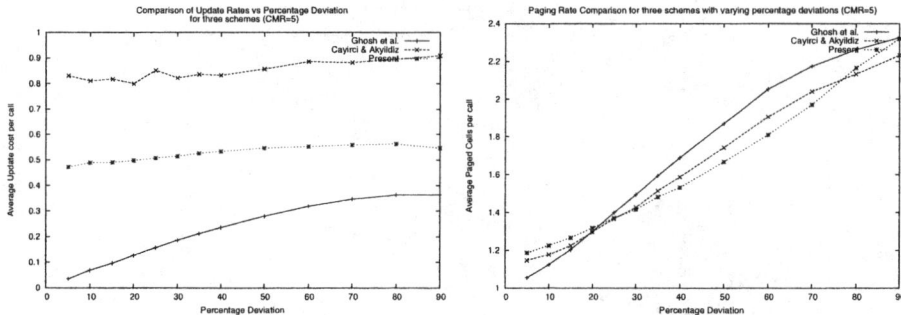

Fig. 4. Comparison of update and search costs for the three schemes for CMR=5

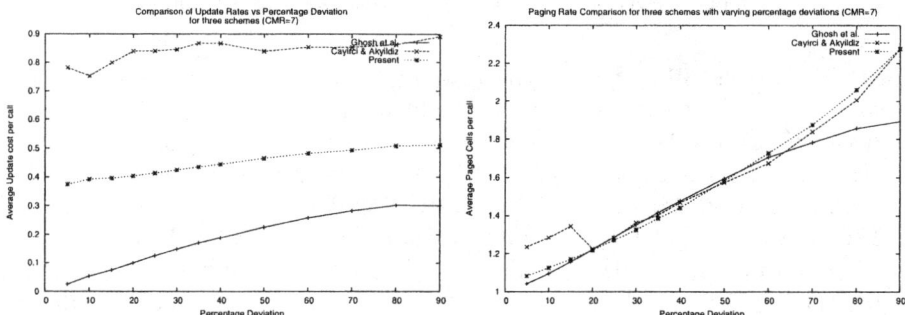

Fig. 5. Comparison of update and search costs for the three schemes for CMR=7

Table 1. Comparison of the requirements of computation times* of two schemes

| Percentage of | CMR=3 | | CMR=5 | | CMR=7 | |
deviation	Ref [5]	Present	Ref [5]	Present	Ref [5]	Present
10	48.1539	0.6734	46.8386	0.6722	51.4070	0.7078
20	47.0098	0.7010	47.2671	0.6761	46.5180	0.6798
30	49.2073	0.7111	48.8414	0.6898	46.9925	0.7131
40	49.0009	0.7409	47.5483	0.7045	52.1262	0.7460
50	47.5154	0.7452	47.3935	0.7290	47.0002	0.7149
60	48.5142	0.7874	47.7017	0.7986	47.2307	0.7567
70	48.2995	0.8181	47.6270	0.7757	47.2038	0.7694
80	50.0146	0.8291	47.4702	0.8172	47.4376	0.8028
90	49.3896	0.8503	47.6759	0.8597	47.1307	0.8541

* All times are in microseconds

The movement prediction based location management scheme proposed in [5] employs best-first search in the neighborhood of the expected locations of the MT with the help of a regional route map to find its location in case of unexpected deviations. The present scheme handles transient deviations through

TSPI entries which helps in quickly locating the mobile host instead of best-first search. The computational requirements for best-first search puts significant load at base stations. Therefore the scheme proposed in [5] can not scale up easily. Table 1 provides a comparison of the requirements of computation time of the scheme proposed in [5] and the present scheme.

6 Conclusion

In this paper we proposed a new location tracking scheme for mobile users. The scheme can be viewed as an amalgamation of the two previously proposed schemes [4], [5]. Each mobile device stores a small number of movement patterns as in [4]. One of the stored patterns is registered by the user as current pattern of movements. When the mobile user deviates from the registered SMP he/she is most likely to follow one of the other pre-declared SMPs. But there may be instances when a user makes either a predictive or an unexpected deviation. Unexpected deviations result in creation of a new movement pattern. But a predictive deviation is transient in nature and typically the user is expected resume registered pattern very soon. To minimize the extra computation load on base station and also to minimize signaling, we introduced the use TSPI. The scheme proposed in this paper relies on the MT to update the TSPI entries. Though it increases signaling cost slightly, but leads to significant reduction in paging costs. Moreover, the present scheme also tackles the expected deviations through secondary clusters found with the aid of regional route map. This is not handled by any of the two previous schemes [4], [5]. So, the proposed scheme not only designed to keep track of evolving new movement patterns but also to handle predictable but transient deviations.

References

1. Wong, V.W.S., Leung, V.C.M.: Location managment for next-generation personal communications networks. IEEE Network Magazine (2000) 18–24
2. Rose, C.: Minimizing the average cost of paging and registration: a timer-based method. ACM/Baltzer Journal of Wireless Networks **2** (1996) 109–116
3. Pitoura, E., Samaras, G.: Locating objects in mobile computing. IEEE Transaction on Knowledge and Data Engineering **13** (2001) 571–592
4. Cayirci, E., Akyldiz, I.D.: User mobility pattern scheme and paging in wireless system. IEEE Transaction on Mobile Computing **1** (2002) 236–247
5. Ghosh, R.K., Rayanchu, S.K., Mohanty, H.: Location management by movement prediction using mobility patterns and regional route maps. In: Proceedings of IWDC 2003, LNCS. Volume 2981. (2003) 153–162
6. Chackraborty, G.: Efficient location mangement by movement prediction of the mobile host. In: Proceedings of IWDC 2002,LNCS. Volume 2571. (2002) 142–153
7. Bhattacharya, A., Das, S.K.: Lezi-update: An information-theoretic framework for personal mobility tracking in pcs networks. Wireless Networks **8** (2002) 121–135

SEBAG: A New Dynamic End-to-End Connection Management Scheme for Multihomed Mobile Hosts[*]

B.S. Manoj, Rajesh Mishra, and Ramesh R. Rao

California Institute of Telecommunications and Information Technology,
Department of Electrical and Computer Engineering,
University of California at San Diego, CA 92093, USA
{bsmanoj, ramishra, rrao}@ucsd.edu

Abstract. The next generation wireless communication devices are expected to be capable of communicating with the best possible network as well as to utilize multiple networks, simultaneously. Existing solutions such as the interoperability mechanisms and the Always Best Connected (ABC) paradigm limit the access of wireless devices to only one, preferably the best possible network. Such schemes, though found to be better than the traditional single interface communication, are limited in their ability to utilize the services in the best possible way. Existing work focuses mainly on network layer or transport layer bandwidth aggregation mechanisms which either need to change the existing TCP protocol or require proxy nodes to perform the bandwidth scheduling process. We, in this paper, propose a new wireless access paradigm for multi-homed hosts based on a session layer bandwidth aggregation mechanism. The major advantages of our solution are the high end-to-end throughput, glitch free transition during both mobility and interface changes, dynamic selection of number of end-to-end paths, and above all our solution can work with existing transport and network layer protocols in today's Internet. In this paper, we provide the architectural and protocol solutions for the proposed scheme and results from extensive simulations and a Linux based implementation.

1 Introduction

The wireless spectrum is shared by several terrestrial wireless access mechanisms ranging from Wireless Personal Area Networks (WPANs), Wireless Local Area Networks (WLANs), Wireless Wide Area Networks (WWANs), and the satellite wireless communication systems. Traditionally wireless enabled communication devices were designed to use only one kind of access mechanism. However, the developments in the last two decade resulted in several new medium access mechanisms such as Bluetooth [1], IEEE 802.11b [2], GSM, GPRS, and CDMA-based

---

[*] This work is supported by NSF grants (Nos.0331707 and 0331690) and Ericsson CalIT2 Research.

data cellular systems such as 1xEVDO. In the recent past, a number of new network access mechanisms are proposed which provided a new kind of wireless access scheme which selectively uses one of the many possible multiple communication interfaces to provide a seamless communication experience for the end users. The interoperability mechanism for communication systems provided by Joseph et al. in [3] and [4], the Always Best Connected (ABC) [5], [6], [7], and [8] and the Bandwidth Aggregation (BAG) mechanisms proposed in [9] are examples for solutions for using multiple networks.

It is essential for the next generation wireless communication devices to seamlessly utilize multiple network interfaces simultaneously to achieve the following: (i) high end-to-end throughput, (ii) glitch free communication experience, (iii) connectivity in the presence of heterogeneous wireless networks, and (iv) ability to aggregate bandwidth over multiple interfaces. We, in this paper, propose a new wireless communication access paradigm called Session Layer Bandwidth Aggregation mechanism (SEBAG) which utilizes multiple network interfaces to provide end-to-end multiple paths in order to improve the end users' communication experience. Our solution is considered to be at session layer because it operates between transport and application layer. Therefore, applications that use either TCP or UDP protocol need not be aware of the use of multiple end-to-end paths. This mechanism also need not make any modifications at the transport or network layer protocols. Existing multipath transport layer solutions such as parallel TCP (pTCP) [10] require radical changes at the transport layer and hence extending it to traditional TCP is not easily possible. Application layer solutions such as XFTP proposed in [11] use multiple TCP streams but over a single interface and hence it cannot fully exploit the benefit with multi-homed hosts. Even if the XFTP is modified to operate over multiple interfaces, extending that solution to a wide variety of applications is unforeseeable. The motivation behind our work is to provide an efficient and transparent solution that operate at session level to exploit the presence of a wide variety of access networks including wired and wireless access networks. We provide a cross layer interaction based solution centered at the session layer in order to provide a very high end-to-end throughput, a solution for the transport layer head-of-line blocking problem, provisioning of seamless transition across networks, and a glitch free communication experience while switching from one network to the other. We propose the SEBAG framework to achieve the above mentioned objectives. The rest of this paper is organized as follows: Section 2 describes our solution and Section 3 presents the experimental results. The existing work in this area is described in Section 4 and Section 5 summarizes the paper.

2 Our Work

In this section, we present the Session layer Bandwidth Aggregation (SEBAG) scheme which is an always-all-connected communication access paradigm that provides end-to-end multiple paths over multiple communication interfaces. In this case, we have a thin module that operates logically between the application layer and the transport layer. This session layer module operates both at the client-side and at the server-side. The primary responsibility of these modules is to dynamically initiate and manage end-to-end transport layer connections.

At the client node, we have a module called Client-side SEBAG Aggregator Module (CSAM) and its equivalent at the server's end is Server-side SEBAG Aggregator Module (SSAM). The CSAM and SSAM are generally implemented as thin layers between application and (transport) TCP layer. The network layer solution proposed in [9] depends on a proxy node in the network that schedules the data traffic over to multiple interfaces. The most important disadvantage for this system is the single point of failure formed by the proxy node. pTCP [10] is a recently proposed transport layer solution which requires radical changes at the transport layer and is a serious limitation for large scale deployment in today's Internet. Therefore, in our solution, we intend to keep the transport layer protocol intact. Existing work in [11] shows that delinking the bandwidth aggregation process from application layer is important as it simplifies and optimizes the operation of bandwidth aggregation mechanisms. The primary responsibility of SEBAG Aggregator modules (SAMs) is to manage packet striping and aggregation based on the availability of network resources over multiple end-to-end paths. For example, when a mobile node moves into the coverage area of a new network, the CSAM identifies it and communicates to the SSAM and initiates another transport layer connection to utilize the new access network. The SSAM would now start including the new transport connection in the packet scheduling process at the server side. Similarly, when mobile hosts move out of the coverage of a particular network, it removes the transport layer connection which was setup through that network.

Figure 1 shows the schematic diagram of SEBAG scheme. When users open an application that needs file transfer, the application layer requests a transport layer connection. This request is captured by the CSAM which in turn sets up the first transport layer connection. Over this transport layer connection, CSAM transfers data and control information to SSAM module. The CSAM has several associated modules, such as the Cross Layer Interaction Module, which monitor the availability of additional active network interfaces through which end-to-end connections can be setup. CSAM communicates the information about the additional network connections to the SSAM and additional end-to-end paths are setup. This end-to-end multi-path setup is dynamic and therefore, it adds and removes additional paths as and when the mobile node comes in the presence of multiple heterogeneous networks. The communication model we used for studying this scheme is similar to a multi-homed mobile host which communicates over multiple heterogeneous wireless or wired interfaces to a server in the Internet. The number of multiple transport layer connections is determined by the number of different wireless interfaces which can connect to the wired networks. This connection establishment policy is called Always-All-Connected (AAC) where the CSAM and SSAM establish as many connections as the number of access networks available to the mobile host. Once a control channel is established between CSAM and SSAM, it proceeds to the identification of

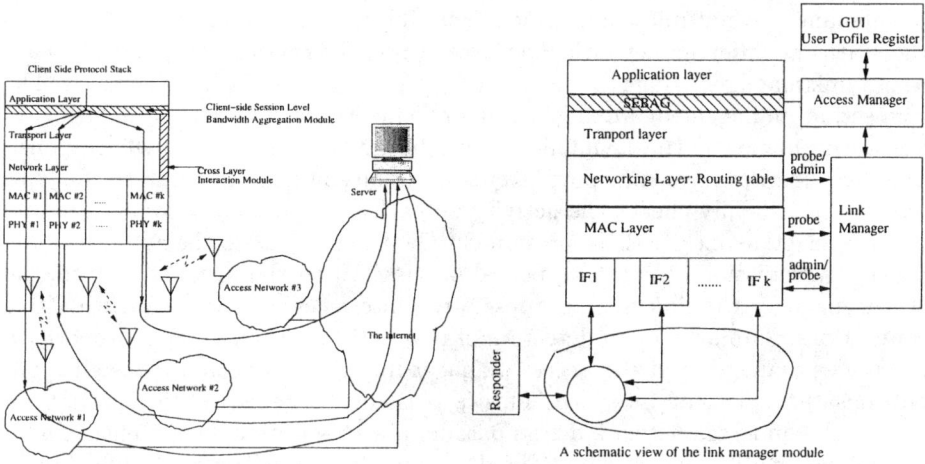

Fig. 1. The schematic diagram of SEBAG scheme cross-layer interaction module

Fig. 2. Modules in the cross layer interaction module

potential networks that can be used with the server through CSAM. The CSAM generates multiple TCP connections, each over a different interface. The CSAM also aggregates the throughput received from each TCP pipe and delivers it to the application layer in order. At the server side, the TCP connections are similar to normal connections and it need not go through the SSAM.

2.1 The Cross-Layer Interaction Module

The Cross Layer Interaction Module (CLIM) of SEBAG interacts with transport, network, and MAC layers. The major advantage of using CLIM is that we can dynamically update the properties of the end-to-end transport session depending upon the changes in the network access system. The number of TCP connections serving a single communication session is limited by the number of access networks present. Choice of access networks include 802.3, Bluetooth, 802.11b/g/a, 3rd generation cellular networks, CDMA data networks, and satellite links. The number of multipath transport connections can be decided based on policies such as cost of access, bandwidth of access networks, and power consumed by the interfaces. As an initial approach, we, in this work, studied only one scheme in which we use the number of transport layer connections (TCP connections) proportional to the number of active interfaces. Figure 2 shows Cross Layer Interaction Module within the CSAM and its interactions with different modules. Figure 2 shows the major modules in the CLIM and its interactions with other layers. The user interface provides a user with the facility to do the following: (i) control the interfaces manually, (ii) overriding the default policy of AAC operation, (iii) definition of new policies, (iv) monitoring the current end-to-end bandwidth achieved, and (v) other configuration utilities. The access manager handles the overall aggregation management and maintenance of multiple connections. The link manager monitors and updates the status of different network interfaces. Upon instructions from access manager a link manager brings the interfaces up or down as and when necessary. The connection management intelligence is implemented at the access manager. In the access manager, the DynamicMultipathControlModule holds the responsibility to add or remove transport layer connections over multiple interfaces. This module periodically checks the activity on each transport layer connection and removes inactive ones or those presumed to be inactive due to the absence of any packet transmission over a long period of time. Upon the instruction from the access manager the link manager resets the interfaces associated with inactive connections. Similarly, upon detection of new networks, link manager reports to the access manager and the access manager initiates additional transport connections through that interface. This is periodically repeated to maintain an always-all-connected network access paradigm with additional capability of end-to-end bandwidth aggregation. The DynamicMultipathControlModule briefly listed below.

DynamicMultipathControlModule()
   While(Communication Session Active)
      CheckForActivePaths()
      RemoveInActivePaths()
      CheckNewActiveInterfaces()
      SetupNewTransportConnsThroughTheNewInterfaces()
   end while

2.2 Connection Setup Process

In order to achieve high throughput for an end-to-end data transfer session, SEBAG utilizes an efficient traffic aggregation mechanism. For example, consider the end-to-end transport layer connection where each connection (for example TCP) is identified by IP addresses and TCP ports of both the sender and the receiver. When an application layer at the client side needs to open a connection, it passes the request to the Client-side SAM. Instead of setting up a single TCP connection, the CSAM sets up multiple TCP connections with the server end-point. All these connections are listed as part of a session layer connection. Now each of these connection can operate as an independent TCP connection. Each TCP connection established between CSAM and SSAM transfers data and in-band signaling packets for SEBAG. For example, the first packet sent over a new transport layer connection contains the connection identifier. Application layer protocols need not be aware that it uses multiple transport layer connections. Therefore, SEBAG is transparent to the higher layer protocols. This is achieved by handling all the interface primitives between application and transport layers. The CSAM communicates with a Cross Layer Interaction Module (CLIM) which in turn interacts with other layers. CLIM understands the number of communication interfaces, status of connectivity, and the raw data rate of each of them. By periodically updating the network access capability of the node, the CLIM aids the CSAM to change the number of transport connections dynamically.

2.3 Connection Management and Congestion Control

SEBAG manages and dynamically updates the number of transport layer connections depending on the availability of access networks at the client's end. It completely disassociates itself from the congestion control mechanism and depends on the transport layer to provide it. That is, when TCP is used at the transport layer, SEBAG depends on the TCP's congestion control. Similarly when transport layers such as UDP are used for communication, our scheme does not interfere with the use of congestion control at the transport layer. The dynamic connection management and packet scheduling at the SSAM can also lead to indirect congestion mitigation. In this work, we do not focus on the implicit congestion control caused by the SEBAG's scheduling mechanism.

2.4 Operation of Link Manager

Figure 3 shows the flow diagram representation of the algorithm that is running at the link manager. The major actions carried out by the link manager are the monitoring and conveying the status of each of the available communication interfaces and executing the commands from the access manager. Every interface at the MH has one link manager. Upon the instructions from the access manager, a link manager initiates several actions such as turning up or down, enables IP connectivity to the responding node (either server or client), and monitoring the availability of link. In addition to all the above, the link manager has the capability to attach a specific route to a particular node.

2.5 Dynamic Packet Scheduling Algorithm

When we use multiple end-to-end transport layer solutions, it is important to utilize the independent paths to achieve maximum efficiency. We studied two packet (or segment) scheduling systems here (i) Round Robin scheduling and (ii) an Expected Earliest Delivery Path First (EEDPF) scheduling. The application layer data units to be transferred will be divided into large number of semi coarse chunks (data segments) which will be scheduled for transmission over chosen paths. The round robin mechanism allocates equal number of data segments for each path. In cases where the segments are of variable sizes, then a surplus round robin scheme can be used. In our studies we used a fixed sized segments while allotting data segments to different transport layer connections. We discuss both the schemes in the following sections.

Round Robin Scheduling: As per this scheme, after setting up multiple (say k connections) transport connections through k different interfaces, the CSAM starts sending or receiving segments through each of the connections. SSAM assigns the data segments (or application layer protocol data units) in a round robin fashion so that each connection will be assigned an equal size data segment to be transferred. Note that we do not intend to duplicate the data through multiple connections. In the case of variable length of application layer protocol data units, one needs to use surplus round robin scheme to achieve an equal share of data packets across all the paths.

Expected Early Delivery Path First Scheduling: The use of above mentioned round robin mechanism may cause congestion and buffer overflow for some wireless interfaces which have low end-to-end transmission rate. Therefore, a better policy is to allot more data segments over the end-to-end path that has the highest end-to-end data rate. Since we do not have the accurate information about the end-to-end data rate of each of the paths, we propose to use a new scheme called Expected Earliest Delivery Path First (EEDPF) scheduling. In this case, we estimate the end-to-end bandwidth and assign a data segment to the connection which is expected to deliver the data segment at the earliest. When a data segment is to be scheduled, the SEBAG module finds the connection which could deliver it to the destination at the earliest. The SAM modules obtain the end-to-end throughput as well as the average end-to-end latency of each connection to determine this. The expected delivery time is estimated as the sum of the end-to-end latency and the segment transmission time. The segment transmission time is obtained as $\frac{L_p}{B_e}$ where L_p is the length of the segment to be assigned to the transport layer and B_e is the average end-to-end bandwidth obtained on that path. These information are obtained by the sender-side SAM module from the received SAM-level acknowledgment packets. Each TCP connection is assigned a number of chunks that is effectively proportional to its end-to-end data rate. The end-to-end data rate is obtained from acknowledgment packets received by SSAM from CSAM over each of these connections. This in turn leads to a scheme where a faster interface is more likely to be provided with more segments.

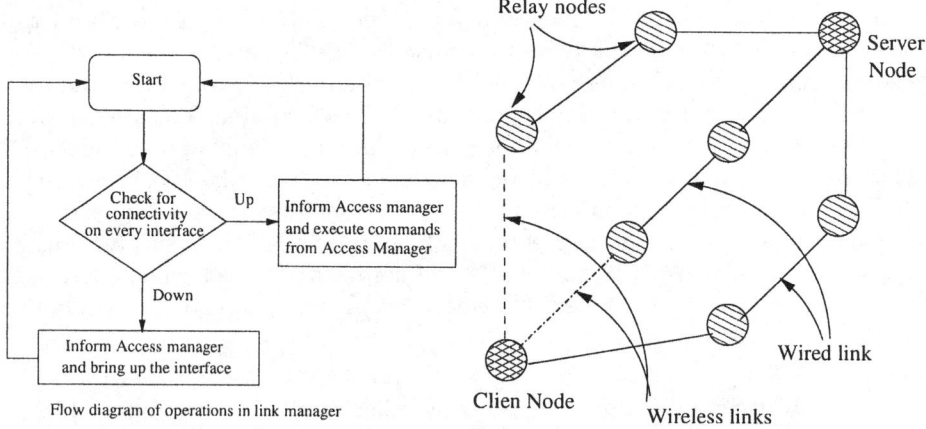

Fig. 3. The flow diagram of link manager   **Fig. 4.** The network topology used for simulation setup

3 Experiments and Results

We have carried out extensive experiments to study the performance of our system. Figure 4 shows the network topology used for simulation experiments. We have used a server and a client with three end-to-end disjoint paths between them. The client node is connected to the network using three interfaces. Two of these interfaces are wireless and the third one is a wired interface. We studied different bandwidth values for the wired and wireless interfaces. Unless otherwise specified, we used a 10Mbps 802.11b interface for the first wireless interface. For the second wireless interface, a 1Mbps wireless WAN interface is used. The default value for all the wired links including the client node's wired link is 10Mbps. The MAC protocol used for WLAN interface is IEEE 802.11b. The transmission power used for WLAN and WWAN interfaces are 7.5dBm and 13.5 dBm, respectively. The simulation engine is built around Glomosim. The number of nodes in the network is 10 with a topology shown in Figure 4. The network is exposed to background traffic that contents with the SEBAG flow under study. Figure 5 shows the throughput result obtained by the Always Best Connected (ABC) mechanism, SEBAG with round robin scheduling over two interfaces, SEBAG with round robin scheduling over three interfaces, and SEBAG with EEDPF scheduling mechanism. In this case, we have the client node with three interfaces: the WLAN interface with 10Mbps, the WWAN interface with 1Mbps, and a wired interface with 1Mbps. In this case, we find that the ABC system performs worse in comparison to other solutions and SEBAG with EEDPF scheduling scheme performs better than all others.

Figure 6 provides the throughput performance of our SEBAG system with three interfaces the WLAN interface with 1Mbps, the WWAN interface with 10Mbps and the wired interface with 1Mbps. Similar to the above result, our solution out performs existing systems.

We have also studied the throughput performance when all the interfaces are with 1Mbps in Figure 7. In this case, we noted a reduction in throughput in all the results. Even then SEBAG with round robin scheduling and EEDPF scheduling out perform the ABC scheme.

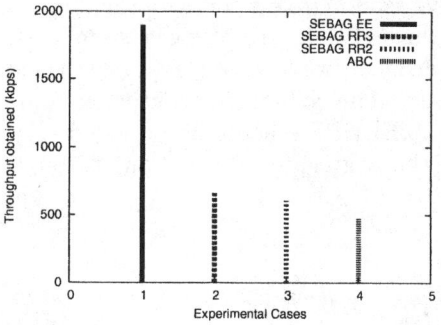

Fig. 5. Throughput performance of SEBAG

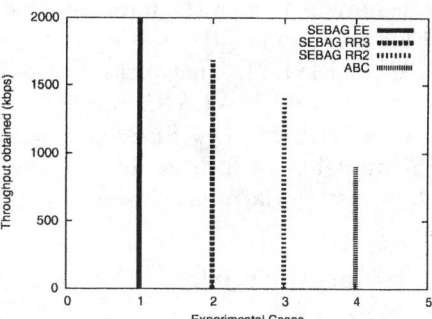

Fig. 6. Throughput performance of SEBAG

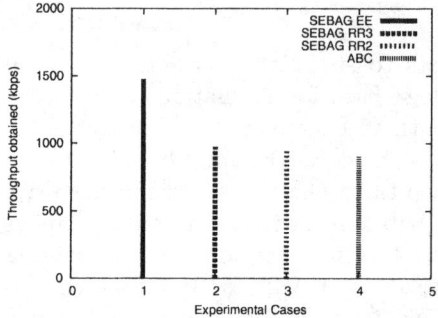

Fig. 7. Throughput performance of SEBAG

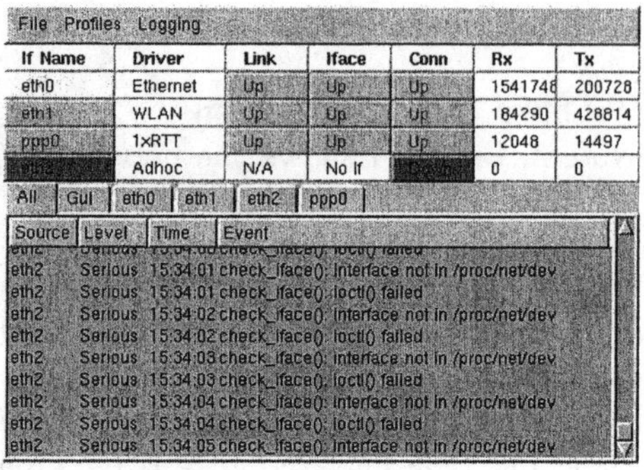

Fig. 8. A snapshot of the user and administration interface for SEBAG system

3.1 Prototype Implementation Details

In this section, we discuss the implementation details of our prototype. We have developed a prototype of the proposed SEBAG scheme at the University of California at San Diego, in association with Ericsson CalIT2 Research Center. Figure 8 shows the Multi Access Control Center which is the user interface for SEBAG. Multi Access Control Center provides information that can be used to manage different interfaces. Multi Access Control also provides the current end-to-end bandwidth obtained through each one of the currently active interfaces. Users can also activate or deactivate appropriate interfaces to control the access mechanism if they so desire. SEBAG maintains connection through all interfaces all the time and hence the throughput achieved is found to be maximum. We used a Linux based laptop computer with Pentium III processor as a client node. This client node is fitted with an Ethernet card, an IEEE 802.11b card, and a 1xEVDO card. The 1xEVDO card provides a maximum data rate of 2.4Mbps with a sector throughput of 700Kbps. WLAN interface operates at 11Mbps and the wired interface for this setup is operating at 10Mbps. The experimental prototype topology is illustrated in Figure 9. The measurements for 1xEVDO link is obtained using the data services of Verizon Wireless service provider.

Figure 10 shows the results obtained through our prototype system with WLAN and 1xEVDO networks. In this experiment, we have not used the Ethernet interface. With WLAN interface, we observed much higher performance compared to 1xEVDO. Our SEBAG scheme with EEDPF scheduling provided very high throughput which is almost equal to the aggregated throughput through each of the interfaces, as shown in Figure 10.

4 Related Work

The Always Best Connected system [5], [6], and [7] proposes to use the best network among the available ones. The decision of best network is based mainly on available bandwidth. The ABC system enables a client node to seamlessly

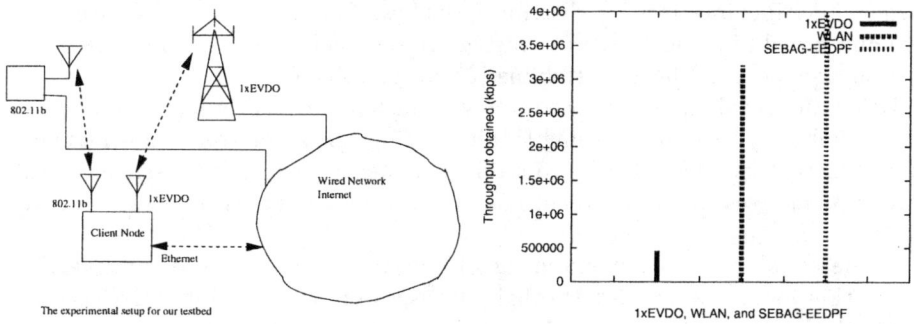

Fig. 9. The prototype testbed

Fig. 10. The network topology used for simulation setup

switch across the networks. The two major approaches used for implementation are the end-to-end approach and the proxy based approach. In the end-to-end approach, the network connection is re-initiated by a connection manager. In the end-to-end approach, during the network switching process, the connection manager would request the end-node – in most cases a server in the Internet, the remaining part of the file which was already under transmission. This is mainly achieved through application protocols such as HTTP, FTP, or RTSP. The ABC project [8] at University of California at San Diego is an example for an experimental system. The proxy-based approach uses a network layer proxy to split the end-to-end connection. In this case, the use of multiple interfaces or the choice of the best possible network service is to communicate with the proxy. The main issue of the proxy-based system is the scalability requirement of the proxy node and the head-of-line blocking problem faced at the proxy node.

The solution proposed by Joseph et al. in [3] and [4] provides an application layer mechanism which enables seamless communication across WWANs and WLANs. At the wireless LAN side, they used 802.11b based system. They also listed out issues related to the timing of the switching, billing and revenue sharing, user profiles, load balancing, and hand-off issues. They proposed three user profiles such as bandwidth conscious, cost conscious, and glitch conscious user profiles. A bandwidth conscious user would switch whenever they find a network which has better bandwidth to offer. In addition, they discussed the important issue of revenue sharing such as pricing, revenue sharing, and use of pricing system to balance the load across different networks. They also proposed a dynamic pricing system by which the system decreases the access cost for the networks with light load such that the large number of cost conscious users may shift to lightly loaded networks. They also built the interoproxy – an application layer proxy-based system that provides interoperability across WANs and LANs.

The Bandwidth Aggregation (BAG) mechanism proposed in [9] utilizes a network layer proxy to achieve bandwidth aggregation at the nodes. An IP-in-IP tunneling is used to connect the proxy with the ABC client. The proxy may identify multiple interfaces to distribute the load. The work in [9] also provides a dynamic scheduling mechanism to distribute the load across the interfaces. This system also faces the head-of-the-line blocking problem.

Hsieh and Sivakumar proposed a new transport layer protocol in [10] which has the inbuilt capability of using traffic striping across multiple transport layer micro flows. Even though they obtain good performance, their scheme needs extensive changes in the protocol and it does not interwork with the existing TCP.

Allman et al. proposed an extension of FTP in [11] that can use multiple TCP connection for handling with throughput issues on satellite links. Extension of their solution to every other application demands enormous effort.

5 Summary

The presence of heterogeneous wireless and wired access mechanisms raises several challenges in choosing and using the best possible access mechanism. We, in this work, propose a new network access mechanism by proposing to make use of multiple interfaces simultaneously. This is achieved by using a session level traffic aggregation mechanism which provides multiple end-to-end paths. Such a mechanism not only provides very high throughput compared to existing schemes, also provides high flexibility in using any available network interfaces. This system can also dynamically choose a specific set of interfaces that are available. We propose a framework with a cross layer interaction module, which enables the interaction between session layer and lower layers. We studied the performance of our system using simulations and a prototype implementation and found that it provides very high throughput and flexibility.

References

1. www.bluetooth.com
2. www.wi-fi.org
3. D. A. Joseph, B. S. Manoj, and C. Siva Ram Murthy, "The Interoperability of Wi-Fi Hotspots and Packet Cellular Networks and the Impact of User Behavior", *Proceedings of IEEE PIMRC 2004*, vol. 1, pp. 88-92, September 2004.
4. D. A. Joseph, B. S. Manoj, and C. Siva Ram Murthy, "Interoperability of Wi-Fi Hotspots and Cellular Networks", *Proceedings of ACM WMASH 2004*, pp. 127-136, September-October 2004.
5. E. Gustafsson and A. Jonsson, "Always Best Connected", *IEEE Wireless Communications*, vol. 10, no. 1, pp. 49-55, February 2003.
6. V. Gazis, N. Houssos, N. Alonistioti, and L. Merakos, "On The Complexity of Always Best Connected in 4G Mobile Networks", *Proceedings of IEEE VTC 2003*, vol. 4, pp. 2312-2316, October 2003.
7. G. Godor, A. Eriksson, and A. Tuoriniemi, "Providing Quality of Service in Always Best Connected Networks", *IEEE Communications Magazine*, pp. 154-163, July 2003.
8. http://www.calit2.net/briefingPapers/ABC.html

9. K. Chebrolu and R. Rao, "Communication Using Multiple Wireless Interfaces", *Proceedings of IEEE WCNC 2002*, vol. 1, pp. 327-331, March 2002.
10. H. Y. Hsieh and R. Sivakumar, "A Transport Layer Approach for Achieving Aggregate Bandwidths on Multi-homed Mobile Hosts", *Proceedings of ACM MOBICOM 2002*, pp. 83-94, September 2002.
11. M. Allman, H. Kruse, and S. Ostermann, "An Application Level Solution to TCP's Satellite Inefficiencies", *Proceedings of WOSBIS 1996*, November 1996.
12. T. Hacker, B. Athey, and B. Noble, "The End-to-end Performance Effects of parallel TCP Sockets on a Lossy Wide-area Network", *Proceedings of IPDPS 2002*, April 2002.
13. T. Nandagopal, K. W. Lee, J. R. Li, and V. Bharghavan, "Scalable Service Differentiation Using Purely End-to-end Mechanisms: Features and Limitations", *Proceedings of IWQoS 2000*, June 2000.
14. D. A. Maltz and P. Bhagawat, "MSOCKS: An Architecture for Transport Layer Mobility", *Proceedings of IEEE INFOCOM 1998*, 1998.
15. C. Siva Ram Murthy and B. S. Manoj, *Ad hoc Wireless Networks: Architectures and Protocols*, Prentice Hall PTR, New Jersey, USA, 2004.

Efficient Mobility Management for Cache Invalidation in Wireless Mobile Environment

Narottam Chand, R.C. Joshi, and Manoj Misra

Department of Electronics & Computer Engineering,
Indian Institute of Technology, Roorkee – 247 667, India
{narotdec, joshifcc, manojfec}@iitr.ernet.in

Abstract. Caching has been widely used to improve system performance in wireless mobile environment. Mobility management is a key component in allowing a client to maintain normal communication with the server while on the move, independent of its location. A cache invalidation strategy ensures that any cached item by a client has same value as on the origin server. Mobility of clients makes the cache maintenance task more complex. This paper extends our previous caching strategy in a wireless environment to handle inter-cell clients' mobility. Experiments are performed to evaluate the proposed strategy and compare the results with other existing strategies.

1 Introduction

Today, the most popular use of wireless networks is for personal communication applications such as mobile phone calls, and short messaging service (SMS). As the price of mobile computing devices drops and the cost of accessing wireless networks decreases, users have begun looking for a wide variety of services including on-demand information access, the ability to surfing the World Wide Web (WWW) while on the move, access to remote data independent of location and time, etc. Caching at mobile client can relieve bandwidth constraints imposed on wireless and mobile computing. This not only reduces the uplink and downlink bandwidth consumption but also the average query latency.

Barbara and Imielinski [4] provide a caching solution where the server periodically broadcasts *invalidation report (IR)* in which the changed data items are indicated. Early work in mobile caching focused on solving the problem of client disconnection [5], [6], while recent work [1], [2], [3] addresses the issues of long query latency with *updated invalidation report (UIR)*. Similar to IR strategy, the UIR based strategy also suffers from the disadvantage of long query latency due to cache miss.

To overcome the limitations of IR and UIR strategies, we have developed *update report (UR)* based synchronous stateful caching strategy [7], [8]. The design idea of UR strategy includes reducing the query latency, minimizing the client disconnection overhead, better utilization of wireless channel and conserving the client energy. The track of cached items for each client is maintained at the home mobile support station (MSS) in the form of *cache state information (CSI)*. Use of CSI reduces the size of IR by filtering out non-cached data and handles arbitrarily long disconnection. Our previous work [7], [8] doesn't support inter-cell mobility of clients. In this paper, we extend the UR based caching strategy to maintain consistent cache for dynamic data accessed from remote database server while a client is on the move.

The rest of the paper is organized as follows. Section 2 gives a brief description of UR caching strategy. Section 3 presents the caching framework to handle client mobility. Section 4 describes the simulation model and studies the performance through a number of experiments. Concluding remarks are given in the last Section.

2 UR Based Caching Strategy

The model consists of two distinct sets of entities: *Mobile Hosts (MHs)* and *Fixed Hosts (FHs)*. Some of the FHs called *Mobile Support Stations (MSSs)*, are augmented with a wireless interface in order to communicate with the MHs, which are located within a cell. MSSs are also known as *Base Stations (BSs)*. Server database D is a collection of N data items. Each data item d_i is of same size S_{data} (in bits) and has two timestamps: t_i is the most recent timestamp when d_i got updated at the server and t_i^r, called *latest request time*, represents the most recent time when d_i was last requested by any client. Clients only issue simple requests to read the most recent copy of data items. Frequently accessed data items are cached on the client side. Each client has same cache capacity of C items. To ensure cache consistency, the server broadcasts UR every L seconds and it also broadcasts (m-1) RRs between two URs as shown in Fig. 1. To answer a query, the client listens to the IR/UIR part of UR/RR report and uses it to decide cache validity. If there is a valid cached copy of the requested item, the client returns the item immediately; otherwise, it sends an uplink query request.

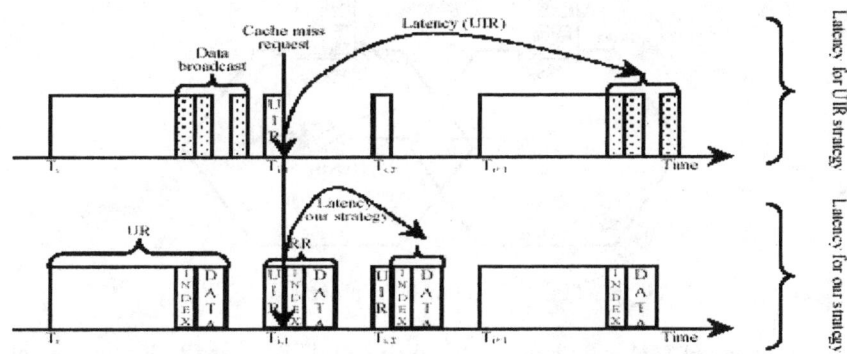

Fig. 1. Reducing the query latency.

For each client MH_x, the home MSS maintains cache state information CSI_x (i.e. list of cached items). To save energy a client may power off and only turns on during the report broadcast. Our strategy reduces the size of IR by filtering out non-cached items, thus enhancing the overall performance. Reduced size of IR/UIR reports improves the downlink channel utilization. Further, it enhances the uplink channel utilization by adopting *delayed uplink (DU)* technique [8].

3 Handling Client Mobility

A client wants its connection to be retained even if it moves to another cell. Handoff is the process which guarantees that the client will keep its connection without any interruption. The cell residency time (the time for which a mobile client stays in a particular cell) has an impact over the client caching strategy. The cell residency time of a client depends on the mobility behavior of the user carrying the mobile host. For example, when users are driving their cars, they can move between cells faster than while they are walking. When MH_i moves from old cell to a new cell, it will be registered in the new cell and a copy of CSI_i from the home MSS will be replicated at the new MSS. If the old MSS is not the home MSS, it (old MSS) deregisters MH_i by deleting CSI_i from its local disk and transfers pending request (if any) of MH_i to the new MSS. MH_i resumes its operation in the new cell and more requests can be directed to the new MSS after the pending data items have been received. While away from the home cell, the changes which occur in the contents of CSI_i at current MSS will also be propagated back at the home MSS so that both the copies are consistent.

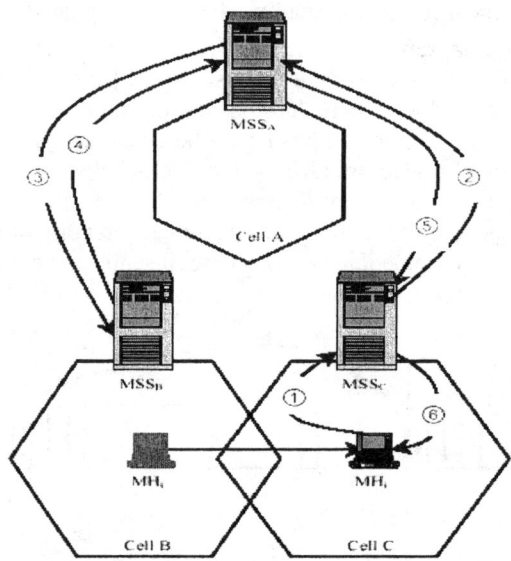

Fig. 2. Client mobility process.

Consider a client MH_i who belongs to home cell A, is presently located in cell B. As shown in Fig. 2, when MH_i moves from old cell B to new cell C, six messages namely 1 - ADMIT, 2 – ENTRY, 3 – DEREGISTER, 4 – DELETED, 5 – REGISTER, and 6 – PERMIT are exchanged among home MSS, old MSS and new MSS.

The following sequence of events takes place:
1. On entrance to new cell, the MH_i sends ADMIT message to the new BS MSS_C. The message contains the id of host and address of its home BS. When MSS_C receives the ADMIT message, it sends ENTRY message to the MSS_A, saying that MH_i has entered into its cell.

2. When MSS$_A$ receives the ENTRY message, it sends DEREGISTER message to the old BS (MSS$_B$). The DEREGISTER contains id of MH$_i$ and when received at MSS$_B$, it forwards pending request (if any) of the MH$_i$ to MSS$_A$ as part of message DELETED and deregisters MH$_i$ by deleting related information including CSI$_i$.
3. MSS$_A$ sends a message REGISTER to MSS$_C$. The message contains CSI$_i$ and list of all the pending items of MH$_i$.
4. On receiving REGISTER from MSS$_A$, the MSS$_C$ registers MH$_i$ in the new cell. The received CSI$_i$ is used as state information about MH$_i$ and list of pending items will be processed during the next UR/RR broadcast whichever arrives earlier.
5. MSS$_C$ sends a PERMIT message to MH$_i$ confirming the registration. MH$_i$ on receiving PERMIT resumes its operation so that more requests can be directed to the new MSS after the pending request have been processed.

The above scheme can be easily integrated with Mobile IP (MIP). The CSI of a mobile host can be maintained by the home agent. Further, the CSI can be replicated with the foreign agent when the mobile host moves to a foreign network.

4 Performance Evaluation

Table 1 shows system parameters and corresponding values. Clients generate queries following exponential distribution. The mean inter-arrival time of queries generated by all clients is T_q. The inter-arrival time of updates at the server is distributed exponentially with a mean of T_u. A client has a probability p_d to enter the disconnection mode only when the outstanding query has been served. A client follows exponential distributed disconnection with mean time T_d. The client's roaming process is also assumed to follow exponential distribution with the mean residency time T_r in a cell.

Table 1. Simulation parameters.

Parameter	Value	Parameter	Value
Server database size (N)	1000 items	Uplink channel bandwidth (B_{up})	100 Kbps
Number of clients (M)	100 clients	Downlink channel bandwidth (B_{down})	2 Mbps
Item size (S_{data})	4096 bits	Mean update arrival time (T_u)	10 sec
Item id size S_{id}	32 bits	Percentage of updates on hot data (p_u)	40
Update timestamp size (T_{data})	32 bits	Client cache size (C)	50 items
RR/UIR broadcasts (m-1)	4	Mean query generate time (T_q)	100 sec
Broadcast interval (L)	20 sec	Mean disconnection time (T_d)	0-400 sec
Broadcast window (w)	10 intervals	Hot data access percentage (p_a)	80
Hot data subset (N_H)	1-50	Client disconnection probability (p_d)	0.10
Cold data subset (N_C)	Remaining D	Mean residency time in a cell (T_r)	1-4096 sec

To evaluate the performance of proposed strategy, we consider four performance parameters: cache hit ratio, query latency, throughput (number of requests served per UR interval) and number of uplink requests. For performance comparison with proposed strategy, UIR and IR strategies are also implemented.

Fig. 3 shows the cache hit ratio as a function of the cell residency time (T_r). It can be seen that the cache hit ratio for all the strategies improves with an increase in T_r. Due to frequent movement of a client at very low T_r (e.g., less than 8 seconds), the cache hit ratio is very low because the client hardly has time to download the requested items. In the IR, when a client moves to a new cell, all the data items are purged from the cache, thus the cache hit ratio is almost zero for T_r less than 8 seconds.

Our strategy outperforms the UIR strategy because for a cache miss request, the data broadcast is made every 2 (i.e., L/(2*m)) seconds. As T_r grows longer than 2 seconds, the cache hit ratio of our strategy starts improving, whereas in case of UIR strategy, the hit ratio improves only for $T_r > 10$ (i.e., L/2) seconds.

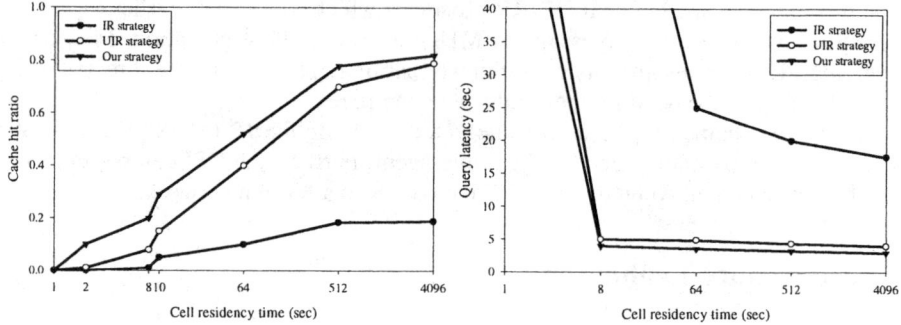

Fig. 3. Cache hit ratio **Fig. 4.** Query latency.

From Fig. 4, it is clear that our strategy has the lowest query latency. When the T_r is very low (e.g. less than 8 seconds), a client frequently moves among the cells and a request due to cache miss is hardly processed by the server. Thus under low cell residency time the query latency is very high.

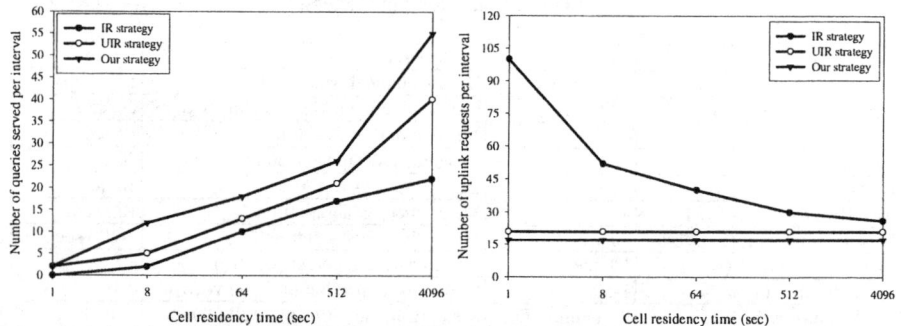

Fig. 5. Throughput **Fig. 6.** Number of uplink requests

Fig. 5 shows that the throughput increases with an increase in T_r. Since our strategy has higher hit ratio than the IR and UIR at all T_r, it can serve more queries locally, and clients send lower number of requests to the server. Due to *delayed uplink (DU)* [8], our strategy further reduces the number of uplink requests. As a result our strategy has always the highest throughput. As shown in Fig. 6, our strategy has the lowest number of uplink requests.

5 Conclusions

Due to various constraints of mobile environment and frequent mobility of clients from one cell to another, the task of cache maintenance becomes more complex. This paper presents a mobility handling strategy so that a client can maintain consistent data while on the move. Simulation experiments show that our strategy performs better than IR and UIR schemes.

References

1. Cao, G.: On Improving the Performance of Cache Invalidation in Mobile Environments. ACM/Kluwer Mobile Network and Applications, 7(4) (2002) 291-303.
2. Cao, G.: Proactive Power-Aware Cache Management for Mobile Computing Systems. IEEE Transactions on Computers, Vol. 51, No. 6 (2002) 608-621.
3. Cao, G.: A Scalable Low-Latency Cache Invalidation Strategy for Mobile Environments. IEEE Transactions on Knowledge and Data Engineering, Vol. 15, No. 5 (2003) 1251-1265.
4. Barbara, D., Imielinski, T.: Sleepers and Workaholics: Caching Strategies in Mobile Environments. Proceedings of the ACM SIGMOD Conference on Management of Data, (1994) 1-12.
5. Jing, J., Elmagarmid, A., Helal, A., Alonso, R.: Bit-Sequences: An Adaptive Cache Invalidation Method in Mobile Client/Server Environments. Mobile Networks and Applications, (1997) 115-127.
6. Tan, K.L., Cai, J., Ooi, B.C.: An Evaluation of Cache Invalidation Strategies in Wireless Environments. IEEE Transactions on Parallel and Distributed Systems, Vol. 12, No. 8 (2001).
7. Chand, N., Joshi, R.C., Misra, M.: Broadcast Based Cache Invalidation and Prefetching in Mobile Environment. International Conference on High Performance Computing, Springer LNCS, 3296 (2004) 410-419.
8. Chand, N., Joshi, R.C., Misra, M.: Energy Efficient Cache Invalidation in a Disconnected Wireless Mobile Environment. International Journal of Ad Hoc and Ubiquitous Computing, (2005).

Analysis of Hierarchical Multicast Protocol in IP Micro Mobility Networks

Seung Jei Yang[1] and Sung Han Park[2]

[1] Mobile Communications Company, LG Electronics Inc.,
459-9 Kasan-dong, Kumchon-gu, Seoul, 153-023, Korea
sjyangub@lge.com
[2] Department of Computer Science and Engineering,
Hanyang University, Ansan, Kyunggi-Do, 425-791, Korea
shpark@cse.hanyang.ac.kr

Abstract. In this paper, we analyze the performance of hierarchical multicast protocol in IP micro mobility networks. The most important parameters to provide multicast service in IP micro mobility networks are the connection recovery time and the reconstruction cost of multicast tree according to the host mobility. So we derive the average connection recovery time and total tree reconstruction cost due to the handoff of mobile hosts by considering the hierarchical structure of IP micro mobility networks. We also verify the analytical results by simulation.

1 Introduction

Mobile IP [1] enables host mobility by allowing global IP mobility by transparently maintaining IP connections regardless of changes in the location of the mobile host. However, in the cellular based mobile access networks providing micro mobility of hosts, Mobile IP has many constraints when the network is handling frequent handoffs [2]. A new address has to be obtained for every handoff and the address should be registered to Home Agent (HA) which is located at a far distance. Therefore, Mobile IP increases handoff delay time and increases the burden on global internet. Moreover, severe decrease in quality occurs during the handoff period. To solve these problems, researches to support micro mobility such as Cellular IP [3], HAWAII [4] have been proposed. These researches aim at supplementing Mobile IP rather than replacing it by another and enabling Mobile IP to handle micro mobility without mutual interaction with global Internet. To do this, in IP micro mobility networks, access network is being configured as tree type for easier routing. By proposing domain based method, problems of triangle routings delivered via home agent (HA) have been resolved even if node delivering datagram is located near to the mobile host. However, these researches are only focusing on the transmission of unicast datagram and the study of multicast service is required. In this paper, we analyze the performance of hierarchical multicast protocol in IP micro mobility networks. The most important parameter is handoff of mobile hosts on multicast service. To deal with the dynamic group membership and the dynamic member location due to handoff of mobile hosts, the multicast protocol in mobile networks reconstructs a multicast tree every time when a host moves to a new network. So the analysis of the connection recovery time and the reconstruction cost of multicast tree with respect to the host mobility is required.

2 Analytical Model

To process the micro mobility without frequent mutual interaction with global internet backbone network and for easier routing, IP micro mobility networks have the network domain based hierarchical structure. That is, routers are configured with tree type structure. Therefore, in this paper, modeling of multicast agents is performed for tree type network structure for compatibility with IP micro mobility networks. Fig. 1 shows the structure of network for the performance analysis of the hierarchical multicast protocol in this paper.

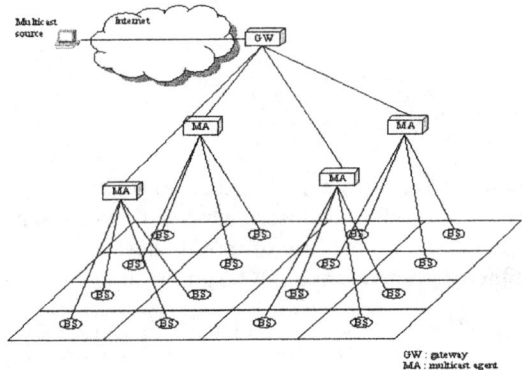

Fig. 1. The structure of network for performance analysis

Network structure for the performance analysis has a hierarchical structure that forms a tree which binds 4 base stations by multicast agents. In this paper, the number of cells increases by multiples of 4 by extending the hierarchy of the multicast tree. In performance analysis, we derive the average connection recovery time and total tree reconstruction cost by considering handoff of mobile hosts. We define assumptions on the characteristics for the analysis of hierarchical multicast as follows.
- There exist calls that require multicast service only.
- The arrival rates for the new call and the handoff call follow a Poisson distribution.
- Service time for a call follows an Exponential distribution.

Because of limited number of wireless channels, a base station can provide a service to a limited number of mobile hosts at a time and since each mobile host is getting service independently from other mobile services, performance analysis is performed through M/M/∞//M queuing model [5]. In this paper, we represent the probability that there are k mobile hosts in a cell as p_k.

3 Performance Analysis

3.1 Average Connection Recovery Time (T)

Average connection recovery time is the one that calculates the average of the delay time for the cell to be subscribed to a multicast group in the case that the mobile host moves to the cell not subscribed to the multicast group. Average connection recovery time is given by Eq. (1).

$$T = E[H_{CB}] \times (2TD_{LINK}+TD_{GROUP}) \times P_J + TD_{GROUP} \qquad (1)$$

where $E[H_{CB}]$ is average number of hops required for a cell to join a multicast group when a mobile host moves to the cell, which is not subscribed to the multicast group. As shown in Fig. 1, the tree that binds four areas together is defined with hierarchical structure. This is done by expanding binary tree to two dimensions. Therefore, when tree is defined by K levels, the value for $E[H_{CB}]$ can be obtained by dividing average distance between two adjacent cells, that are randomly selected, by 2. $E[H_{CB}]$ can be given by Eq. (2).

$$E[H_{CB}] = \frac{\sum_{j=1}^{K} 2^j(K-j+1)}{2^K-1} \times \frac{1}{2} \qquad (2)$$

P_J is the probability of the cell to be subscribed to multicast group in the case that the mobile host moves to the cell not subscribed to multicast group. P_J is derived from the average of $P_J(t)$ that is the probability of mobile host to move to the cell which is not subscribed to multicast group at time t. $P_J(t)$ and P_J are given by Eq. (3).

$$P_J(t) = \frac{\lambda_H e^{-\lambda_H t}}{(1+\frac{\lambda}{\mu})^M}, \quad P_J = \int_t^\infty \frac{\lambda_H e^{-\lambda_H t}}{(1+\frac{\lambda}{\mu})^M} dt = \frac{1}{(1+\frac{\lambda}{\mu})^M} = p_0 \qquad (3)$$

From the Eq. (3), the probability for the cell to be subscribed to multicast group in the case that the mobile host moves to the cell which is not subscribed to multicast group, P_J is equal to p_0.

Consequently, when the derived equations are applied to Eq. (1), the average connection recovery time, T is obtained as follows.

$$T = (\frac{\sum_{j=1}^{k} 2^{j-1}(k-j+1)}{2^k-1}) \times (2TD_{LINK}+TD_{GROUP}) \times \frac{1}{(1+\frac{\lambda}{\mu})^M} + TD_{GROUP} \qquad (4)$$

3.2 Total Tree Reconstruction Cost (C)

Total tree reconstruction cost is calculated in accordance with tree paths that is newly connected and released due to the handoff of mobile host. In this paper, the cost has been applied by converting it into delay time.

$$C = N \times P_J \times (E[H_{CB}] \times (2TD_{LINK}+TD_{GROUP})+ TD_{GROUP})$$
$$+ N \times P_L \times (E[H_{RB}] \times (2TD_{LINK}+TD_{GROUP})+ TD_{GROUP}) \quad (5)$$

where $E[H_{RB}]$ is average number of hops required for cell to be released from multicast group due to the handoff of mobile host. As shown in Fig. 1, when tree is defined by K levels, the value for $E[H_{RB}]$ can be obtained by dividing average distance between two adjacent cells, that are randomly selected, by 2.

$$E[H_{RB}] = \frac{\sum_{j=1}^{K} 2^j (K-j+1)}{2^K - 1} \times \frac{1}{2} \quad (6)$$

$N \times P_J$ is the average value of cells that are newly connected to the multicast tree due to the handoff of mobile host. That is, $N \times P_J$ is to be $N \times 1 / (1+\frac{\lambda}{\mu})^M$. P_L is the probability of removal of previous cell from multicast group due to the handoff of mobile host. P_L is derived from the average of $P_L(t)$ as the probability for handoff or service closure after mobile host's receiving of service being completed at time t, in the case that there is only one mobile host in the cell. $P_L(t)$ and P_L are given by Eq. (7).

$$P_L(t) = \left[\frac{\mu e^{-\mu} \frac{M\lambda}{\mu}}{(1+\frac{\lambda}{\mu})^M} \right], \quad P_L = \int_0^\infty t\, p_L(t)\, dt = \frac{M\lambda}{\mu\left(1+\frac{\lambda}{\mu}\right)^M} = p_I \quad (7)$$

From the Eq. (7), the probability of removal of previous cell from multicast group due to the handoff of mobile host, P_L is equal to p_I.

Therefore, when the derived equations are applied to Eq. (5), total tree reconstruction cost, C is obtained as follows.

$$C = \frac{N(\mu + M\lambda)}{\mu(1+\frac{\lambda}{\mu})^M} \times \left(\frac{\sum_{j=1}^{k} 2^{j-1}(k-j+1)}{2^k - 1} \times (2TD_{LINK}+TD_{GROUP})+TD_{GROUP} \right) \quad (8)$$

3.3 Performance Results

Table 1 shows parameters for performance analysis. The parameter values refer to those in existing studies [6]. In this paper, we assume that λ_N, λ_H and μ are 2.75E^{-04} calls/sec, 5.91E^{-02} calls/sec and 4.88E^{-02} calls/sec, respectively. We also assume that TD_{SG}, TD_{LINK} and TD_{GROUP} are 50ms, 3.5ms and 10ms, respectively.

Table 1. Simulation Parameters

Parameters	Description
N	Number of cells
M	Average number of multicast mobile hosts in a cell
B	Number of base stations managed by a multicast agent
λ_N	Arrival rate of new multicast call
λ_H	Arrival rate of handoff multicast call
$1/\mu$	Average service time of a call
TD_{LINK}	Packet delivery time between links
TD_{GROUP}	Delay time which the router joins multicast group
TD_{SG}	Average transmission delay time between multicast source and gateway
H_{GB}	Hop counts between gateway and base station
H_{CB}	Hop counts for the cell to subscribe a multicast group when multicast mobile host moves to new cell

Fig. 2 shows the graph that compares the mathematical analysis results and simulation results on average connection recovery time, that is, the average of delay time for the cell to be subscribed to multicast group when the mobile host moves to the cell not subscribed to multicast group. The performance of the average connection recovery time is determined according to the value of M, that is, average number of multicast mobile hosts in a cell. If the value of M is small, the probability that a mobile host moves to a cell that is not subscribed to the multicast group is very high. Hence it results in relatively high connection recovery time compared with that for other values of M. If the value of M is over 4, most cells are subscribed to the multicast group and so the average connection recovery times have constant value regardless of the number of cells.

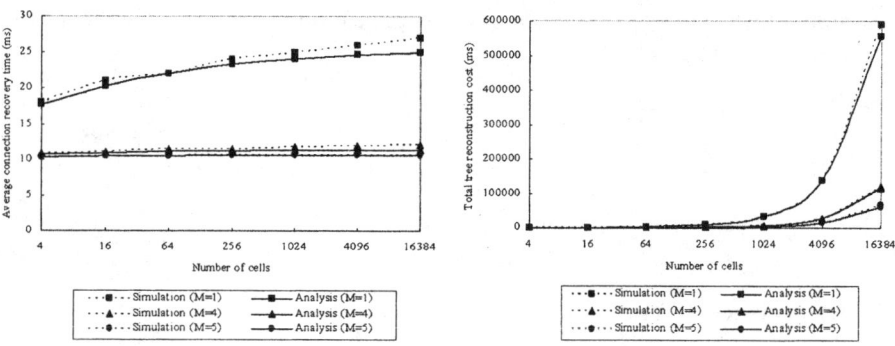

Fig. 2. Average connection recovery time   **Fig. 3.** Total tree reconstruction cost

Fig. 3 is the graph that compares the mathematical analysis results and simulation results on total tree reconstruction cost due to host mobility. Total tree reconstruction cost is calculated by total delay time of multicast links that is newly connected and released due to the handoff of mobile host. The performance is determined according to the average value of cells that are newly connected to the multicast tree due to the handoff of mobile host and the average value of cells that are released from multicast group due to the handoff of mobile host. In the case that the value of M is 1, the probability that mobile host moves to a cell that is not subscribed to multicast group is very high because around half of the total number of cells are not members of the multicast group. So it shows a relatively high total tree reconstruction cost because the number of cells that is subscribed to and released from the multicast group is high compared with that for other value of M. It can be found that as the value of M increases, the cost decreases. Fig. 2 and Fig. 3 also show that the simulation results match closely with the analysis results and perfect match was not done since infinite simulation time is physically not achievable.

4 Conclusions

In this paper, we derive the average connection recovery time and total tree reconstruction cost to analyze the performance of a hierarchical multicast protocol in IP micro mobility networks. Modeling of multicast agents is performed for it having tree type structure for compatibility with IP micro mobility networks. The performance of the average connection recovery time is determined with respect to the average number of multicast mobile hosts in a cell. The performance of total tree reconstruction cost is also determined with respect to the average value of cells that are newly connected to the multicast tree due to the handoff of mobile hosts and the average value of cells that are released from multicast group due to the handoff of mobile hosts. The analytical and simulation results are useful to design and implement IP micro mobility networks to provide efficient mobile multicast service.

References

1. Perkins, C.: IP Mobility Support. RFC2002. Mobile IP networking group
2. Silva, P., Sirisena, H.: A Mobility Management Protocol for IP-Based Cellular Networks. IEEE Wireless Communications (2002) 31-37
3. Campbell, A., Gomex, J., Valko, A: Cellular IP. Internet draft (2000). draft-ietf-mobileip-cellularip-00.txt, work in progress
4. Ramjee, R., et al.: IP Micro Mobility Support Using HAWAII. Internet draft (2000). draft-ietf-mobileip-hawaii-01.txt, work in progress
5. Kleinrock, L.: Queueing Systems Volume1. John Wiley (1975) 89-114
6. Orlik, P., Rappaport, S.: A Model for Teletraffic Performance and Channel Holding Time Characterization in Wireless Cellular Communication with General Session and Dwell Time Distributions. IEEE Journal on Selected Areas in Communications (1995) 868-879

Efficient Passive Clustering and Gateway Selection in MANETs

T. Shivaprakash[1], C. Aravinda[1], A.P. Deepak[1], S. Kamal[1], H.L. Mahantesh[1], K.R. Venugopal[1], and L.M. Patnaik[2]

[1] University Visvesvaraya College of Engineering, Bangalore 560 001, India
[2] Indian Institute of Science, Bangalore 560 012, India

Abstract. Passive clustering does not employ control packets to collect topological information in ad hoc networks. In our proposal, we avoid making frequent changes in cluster architecture due to repeated election and re-election of cluster heads and gateways. Our primary objective has been to make Passive Clustering more practical by employing optimal number of gateways and reduce the number of rebroadcast packets.

1 Introduction

Mobile Ad hoc Network (MANET) is an infrastructure-less network which consists of a collection of wireless mobile hosts to form a temporary network without the aid of any base station. Since bandwidth is limited in an ad hoc network, it is important to construct a virtual backbone consisting of only a subset of nodes that have the privilege to forward packets. Such a virtual backbone called *spine* plays an important role in routing, broadcasting and connectivity management in wireless ad hoc networks. An effort should be made to keep this backbone thin and connected [1].

A cluster is a set of nodes which can be treated as a single entity during packet transmission. Each node in a cluster assumes a role depending on its position and other topological information. The most important role in a cluster is played by the *Clusterhead*. A node which belongs to more than one cluster becomes a *Gateway*. A gateway is responsible for routing packets across two clusters as they are reachable from both the clusters in a single hop. Passive Clustering mechanism does not use any explicit control messages to maintain clusters. Instead, it piggybacks the control information on the out-going data packets and has the advantage of reducing the control overhead. The active clustering algorithm was proposed by Lin and Gerla [2] based on Least Id principle. An innovative mechanism for cluster formation called Passive (On Demand) clustering is provided in [3]. This method does not use any explicit control messages. The existing reactive protocols such as DSR [4], AODV [5] have high control overhead and rebroadcast messages.

This paper addresses the issue of scalability with respect to increase in the number of control packets using Passive clustering. A new Gateway Selection Heuristic which eliminates redundant gateways during Passive Clustering has been proposed.

2 Problem Definition

Given a wireless network $G_w(V, E, n)$ of a finite set of nodes, $V = \{v_1, v_2, ..., v_n\}$ and a finite set of links $E = \{(v_i, v_j) \mid v_i, v_j \in V \wedge v_i \neq v_j\}$, a link is said to exist between two nodes v_i and v_j if they are within the transmission range of each other. The objectives are to (i) reduce the number of rebroadcasts by reducing the number of redundant gateways between the overlapping clusters and (ii) reduce the quantity of control information loaded on the data packets.

2.1 Topological Problems Associated with Passive Clustering

Problem 1: *An ordinary node may move into other clusters and generate a spurious gateway.*

When a node moves from one cluster to another cluster, it starts receiving packets from the new cluster head. It updates the cluster table with the information about the new cluster head, while retaining the information about the previous cluster head. In this situation, it enters into a gateway ready state and further, it may become a gateway.

This is highly unacceptable, because (i) after the movement, it may not be in the common region of both clusters (ii) it may cause the real gateway candidate to become *ordinary*, resulting in the loss of connectivity between two clusters. (iii) it will have privilege to rebroadcast, which it should not have, resulting in an increase in the number of rebroadcasts and hence an increase in the traffic.

Problem 2: *A gateway may move away from the intersection area into a single cluster without relinquishing the status of the Gateway.*

Ideally, such a gateway must become an ordinary node, since it now belongs to one cluster only. Instead, it continues to assume that it belongs to two clusters and hence it will stay in *gateway* state, rebroadcasting all the incoming packets.

Problem 3: *Spurious generation of multiple gateways.*

In a dense wireless network, there will be a number of nodes in the intersection region of any two clusters. All of them compete for the *Gateway* status and the one with the *least id* wins. However, if all the candidates do not hear from same cluster heads, then all of them become gateways. This creates redundant gateways and causes a broadcast storm [6] in the wireless network.

Problem 4: *Formation of redundant clusters.*

During the initial setup, all the nodes that receive packets from the ordinary nodes, become cluster heads. This results in dense and overlapped clusters.

Problem 5: *Problems associated with the cluster head moving out of a cluster.*

If an ordinary node does not receive packets from its cluster head for a long time, it assumes that the cluster head is still present but it has no packets to send. The ordinary node knowing nothing about its cluster head's absence continues to send packets to the cluster head to route them to the destination resulting in the loss of packets and redundant broadcasts by the source.

3 Algorithm: Efficient Passive Clustering (EPC)

In the cluster architecture, a node can be in any of the following states: *initial, ordinary_node, gw_ready, gateway, dist_gw, cluster_head*. The algorithm is as follows:

1. All nodes are in the *initial* state and they are assigned a unique ID.
2. A node that first wants to transmit packets becomes the source node. It sends a packet to all its neighbors and declares itself as a Cluster Head.
3. If the *initial* node hears from a *cluster_head*, it becomes an *ordinary_node*.
4. If a node (other than *initial* and *cluster_head*) hears from a non-Cluster Head,
   (a) It checks whether the sender node was a Cluster Head before. This check is carried out by scanning its cluster table in search of the sending node's ID. (Cluster Table maintains a list of Cluster Heads reachable from the node).
   (b) If the sender node was a Cluster Head before, then its entry is cleared from the cluster table of the receiving node. Packets from this node are not forwarded henceforth.
   (c) If cluster set of the node becomes null, the node changes its state to *cluster_head*.
5. Contention between the Cluster Heads is resolved by the Least ID method. This is because the Cluster Head does not monitor the cluster. The purpose of this step is to have only one Cluster Head per cluster.
6. An *ordinary_node* receiving packets from more than one *cluster_head* enters into *gw_ready* (gateway ready) state.
7. A *gw_ready* node becomes a *gateway* based on the *Intelligent Gateway Selection Heuristic*.
8. A *gateway* on receiving packets from other *gateway* or *gw_ready* nodes, may change its state based on the *Intelligent Gateway Selection Heuristic*.
9. If an *ordinary_node* hears from another *ordinary_node* or *dist_gw* of another cluster, and if there are no gateways in the intersection area, it becomes a Distributed Gateway (*dist_gw*).
10. If a *dist_gw* hears from *gateway* or *gw_ready* of the same cluster-pair, it becomes *ordinary_node*.
11. No node remains in the intermediate state for a long time.
12. If the node times out its state is set to *initial*.

3.1 Intelligent Gateway Selection

The number of rebroadcast packets is directly proportional to the number of gateways. Redundant gateways increase the number of rebroadcasts. Hence, we give a heuristic that selects a optimum number of gateways. The Intelligent Gateway Selection Heuristic takes into account the history of competitions that a node underwent using Competition_count(C_c), while deciding its status [7]. The Competition Count (C_c) of a node is the number of times a node competes for the *gateway* status. It is set to zero, each time a node acquires either *initial* or Cluster Head status. The Redundancy Factor (R_f) of the network is the maximum number of common clusters that any two neighboring gateways can connect. Every node has a data structure called a Cluster Set, which is the set of all cluster heads from which it can receive packets.

Case 1: *Only one node in the intersection area:* When the node receives packets from two cluster heads, it enters into the *gw_ready* state and it becomes a gateway.

Case 2: *Two or more nodes in the region of intersection of clusters:* When a node receives packets from the other Gateway or *gw_ready*, it compares its cluster set with that of the sending node. If both the sets are same, then the one with the least ID becomes the gateway.

Case 3: *The cluster-set of one node in the intersection area is a subset of the cluster-set of another node:* Suppose there are two nodes in the intersection area of clusters such that, the cluster-set of one node is a subset of the cluster-set of another node. Then the node with the superset will be selected as the gateway. Every gateway performs this comparison by intercepting the packets from its neighboring gateways.

Case 4: *Two nodes such that (cluster-set(node1)∼cluster-set(node2))≠0:* In this case both the nodes have a tendency to declare themselves as gateways when they receive packets from each other. But this may not be optimal, since there may be a difference of just one cluster head between the cluster-sets. This leads to creation of redundant gateways. The receiving node computes the number of clusters that are common to both the sending node's and receiving node's cluster-sets. If this value is less than or equal to the Redundancy Factor(R_f), then both nodes are designated as Gateways. Otherwise, the node with the least Competition_count(C_c) is designated as the Gateway. The heuristic intelligently selects the best gateway in the intersection area of two or more clusters.

4 Performance Analysis

Passive clustering is simulated in the NS-2 (version 2.26) simulation environment. Simulation results reveal that there is a reduction in the control overhead and the number of rebroadcasts by the application of the EPC algorithm. The number of gateways and the number of cluster heads are also reduced. The IEEE 802.11 DCF and two-ray propagation model is employed for simulation. The broadcast range for each node is 250 meters and the area of experiment is 2x2 sq. km. Mobility is measured in meters per minute. Both the simple passive clustering and improved passive clustering algorithm are implemented on AODV [5].

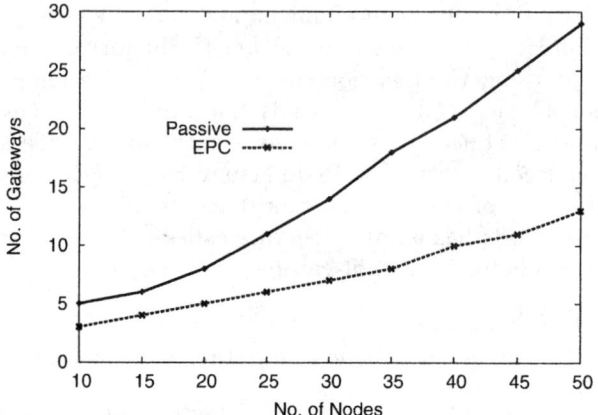

Fig. 1. No. of Gateways vs. No. of Nodes

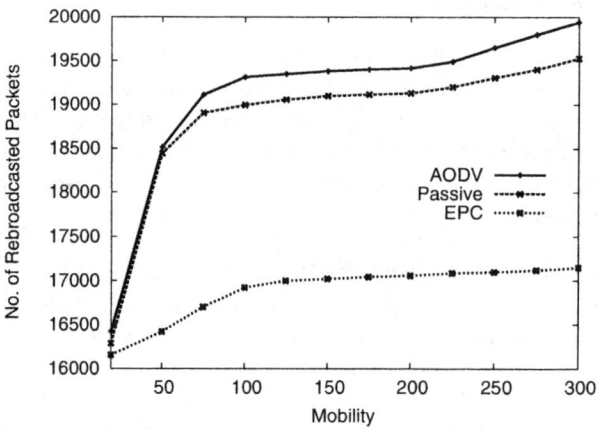

Fig. 2. No. of Rebroadcasted Packets vs. Mobility

By employing the efficient gateway selection heuristic, with the Redundancy Factor set to one, minimal number of gateways are chosen. Not more than one gateway is chosen between two clusters. The gateways form a *thinner* backbone while maintaining the connectivity among all the clusters within the designated area. Also, inclusion of more nodes will not increase the number of clusters and the number of gateways will remain fairly constant. Hence, the gateway curve of our algorithm is linear compared to that of the simple passive clustering as shown in Fig. 1.

The Number of Rebroadcasted Packets (NRP) is the total number of packets that are broadcast and rebroadcast from all the nodes, irrespective of their states. This is a very important parameter because an increase in NRP results in broadcast storm. The number of rebroadcasts is directly proportional to the total number of cluster heads, gateways and distributed gateways in the ad hoc network. This is because in passive clustering, only the cluster heads, gateways and distributed gateways of a cluster have the privilege to forward the packets they receive. As depicted in Fig. 2, the number of rebroadcasts is the lowest for EPC. With the application of the gateway selection heuristic and other improvements over passive clustering, the number of rebroadcasts is reduced considerably. The curve corresponding to our EPC algorithm is more stable (flatter) than others. The number of rebroadcasts is the highest for AODV since every node forwards the incoming packets. The number of rebroadcast messages in passive clustering is lower than AODV, but much higher than EPC.

5 Conclusion

The simulation results show that the EPC clustering algorithm is inexpensive, efficient and stable even under mobile conditions. The number of clusters is found to be optimal in dense wireless networks. This paper has proved that Passive Clustering becomes practically possible by implementing the intelligent gateway selection heuristic and on-demand timeout mechanism. Frequent changes in cluster architecture are avoided by precluding repeated re-election of cluster heads. This improves the network performance. Future work can be carried out by employing distributed gateways to route packets.

References

1. Bhargavan, V., Das, B.: Routing in Ad Hoc Networks Using Minimum Connected Dominating Sets. International Conference on Communications'97, Montreal, Canada, June. (1997)
2. Lin, C. R., Gerla, M.: Adaptive clustering for mobile wireless networks. IEEE Journal on Selected Areas in Communication. (1996) 1265–1275
3. Yi, Y., Gerla, M., Kwon, T. J.: Efficient flooding in Ad Hoc Networks using On-Demand (Passive) Cluster Formation. Proc. of MobiHoc. (2002)
4. Johnson, D. B., Maltz, D. A.: Dynamic source routing in Ad Hoc Networks. Mobile Computing. (2000) 153–181
5. Perkins, C., Royer, E., Das, S.: Ad hoc on Demand Distance Vector (AODV) Routing. Internet draft, IETF, October. (1999)
6. Tseng, Y. C., Ni, S. Y., Chen, Y. S., Sheu, J. P.: The broadcast storm problem in a Mobile Ad Hoc Network. Proc. of the Mobicom. (1999)
7. Shivaprakash, T., Venugopal, K.R., Patnaik, L.M.: Selection of Gateways in MANETs. Technical Report, Department of Computer Science Engineering, University Visvesvaraya College of Engineering, Bangalore University, Bangalore, December 2003.

Mobile Agent Based Message Communication in Large Ad Hoc Networks Through Co-operative Routing Using Inter-agent Negotiation at Rendezvous Points

Parama Bhaumik[1] and Somprakash Bandyopadhyay[2]

[1] Dept. of Information Technology,
Jadavpur University, Kolkata, India
parama@it.jusl.ac.in
[2] MIS group, Indian Institute of Management Calcutta, India
somprakash@iimcal.ac.in

Abstract. The wide availability of mobile devices together with the technical possibility to form ad-hoc networks paves the way for building highly dynamic communicating communities of mobile users. A challenge is how to deliver messages in such networks incurring least routing overhead. Cooperative routing is a mobile-agent assisted team approach, which utilizes a set of fixed cluster head nodes to provide proper coordination and cooperation for exchanges and sharing of messages in the team. Our routing strategy aims at reducing routing overheads, message traffic and unnecessary random node visits in the network for delivering data. The main benefit provided by cooperative routing is considerable network traffic reduction at high load. We highlight the main components of the system and discuss the agent life cycle in detail together with the parameters and strategies governing the migration of agents, their merging and termination.

1 Introduction

A mobile ad-hoc network is a multihop fully autonomous network that can be set up anywhere any time. An interesting application in such an environment is decentralized rapid message delivery services while incurring the least routing overheads. For this environment to operate properly several of the well-established protocols at the different telecommunications layers are revisited. Most of these protocols are based on flat architecture where each node maintains complete routing information and thus the overhead increases considerably with the network size and traffic [1], [2], [6].

In this paper we propose a mobile agent-based cooperative routing protocol for delivering short messages in a large clustered network whose performance increases with the increasing traffic in the network due to high degree of cooperation among the agents. In our previous works [3],[4],[5] the mobile agents were used to deliver messages where they acted as a messenger that will migrate from a source to a destination individually to deliver the message. Thus when there are a number of sources to send messages to a common destination simultaneously; a group of parallel redundant traffic vested with similar responsibility will be generated. This traffic will eventually consume the bandwidth and other crucial resources of the ad-hoc wireless network.

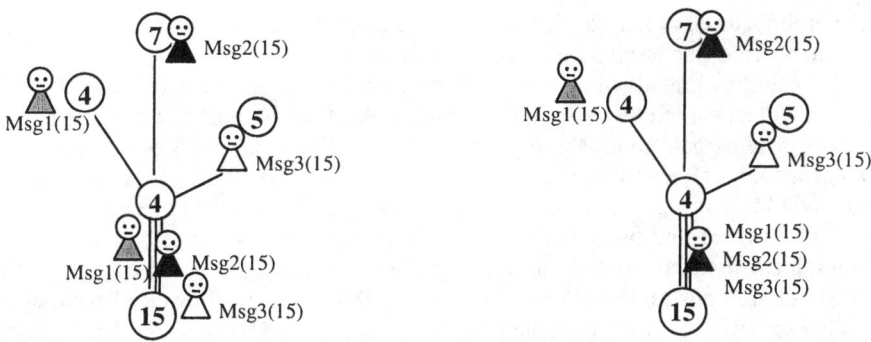

Fig. 1. Delivery of messages without and with agent cooperation at cluster head node 15

The novelty of this paper is the introduction of a cooperating agent team that will meet for the purpose of cooperative task- oriented behaviors like sharing of each others responsibilities, exchanging network information and merge all the individual agents carrying messages for the same destination into a single agent. The agents, which are relieved of their responsibilities in the process, will be terminated if possible and thus reducing the traffic load heavily. This mode of cooperation can be made clear from the Fig. 1. All these cooperation essentially works directly through inter-agent communication. The entire algorithm works on the fact that agents "need to be on the same place at the same time" i.e., they must know the existence of each other.

Now in our work to extend the chance of meeting of agents, we have made arrangement of meeting points at the fixed cluster heads distributed randomly within the network. Agents navigating through the network for delivering messages must visit these cluster heads whenever they are entering a new cluster domain. This compulsory agent visit increases the degree of spatial coordination (agents must be on the same place) at the cluster heads. The temporal coordination (agents must meet at the same time) has been enhanced with the introduction of a short waiting delay offered to each mobile agent by the cluster heads. This waiting time will further increase the chance of meeting with other agents and can highly reduce the agent-chasing problem. The meeting place hosted at the cluster heads can be called as the *Rendezvous points* within the network and the detainment period of the agents can be called *Rendezvous periods*.

The rest of the paper is structured as follows. We discuss design view of the proposed framework for message communication in Section 2. Agents and the message delivery using cooperative routing protocol have been described in Section 3. Simulation results are presented in Section 4. The paper is concluded in Section 5.

2 Proposed Framework for Message Communication

A hierarchical partitioning of networks into clusters offers several advantages like improvement of routing and mobility management, increment of system capacity, reduction in signaling and control overhead that makes the network more scalable.

Such architecture is relatively stable due to the localized nature of route computation and can be used in a large mobile ad-hoc wireless environment.

Motivated by the advantages of clustering we have proposed a framework consisting of a collection of clusters of mobile nodes. We have assumed of a wireless ad-hoc network where each node is equipped with GPS (Geographical Positioning System) for extraction of geographical co-ordinates, routing and direction of movement of each node [4]. To support the idea of stable adaptive clusters we have assumed cluster heads to be fixed and have distributed some fixed nodes as our cluster heads over a geographical area covered by ad-hoc wireless networking infrastructure (randomly placed autonomous nodes). These stationary nodes are then allowed to form clusters within a specific geographical boundary from their own geographical coordinate position. The philosophy of forming the clusters is a fixed node with say coordinate (X=40, Y=30) forms a cluster with boundary X=X+15 to X=X-15 and Y= Y+20 to Y=Y-20. This way each cluster head forms almost rectangular partitioned overlapping clusters. The nodes whose GPS lies within the boundary, receive the membership binding request from the cluster head and send their node identity number along with their GPS to the head as acknowledgement. Once the clusters have been formed all packet transfer take place through geographical routing i.e. knowing the GPS of the destination the route decision to the next hop must minimize the geographical path length between source and destination [4],[5].

When a member node becomes mobile it informs the cluster head about its migration and on traveling to a new region boundary it will send request packet to the current cluster head for membership. Thus our cluster heads are vested with the responsibility of keeping neighborhood integrity record with periodic refreshment.

The functionalities of the cluster head are specifically designed for providing maximum cooperation with the mobile agents and also to act as the mail server within the cluster. The responsibility vested on to a cluster head can be classified under the following headings: i) Membership List Formation and Modification, ii) Accepting and delivering the messages to its members, iii) Handling mobile nodes by acting as temporary mail servers, and iv) Detainment of mobile agents for some time.

3 Agents and the Co-operative Routing Protocol

Information carried by mobile agents for extending cooperation: Here the mobile agents are allowed to carry the information of already visited clusters along with them. The idea behind this is to capture and share the partial network information present with roaming agents. The integration of all such partial information at a common *Rendezvous point* helps cooperative tasks like taking the decision for next destination, suitable exchange of messages between agents, getting up-to-date knowledge of the network and reducing unnecessary redundant visit of nodes without the need for exchanging huge chunks of routing table data.

Navigation policies followed by mobile agents: The movement from node to node is made in a way to minimize the distance between the agent's current location (current location of the node where the agent is residing) and the cluster head location using the GPS technology [4], [5]. Though the order of cluster visits take place in a random manner still the redundancy in the path visit has been avoided by maintaining the path visit list.

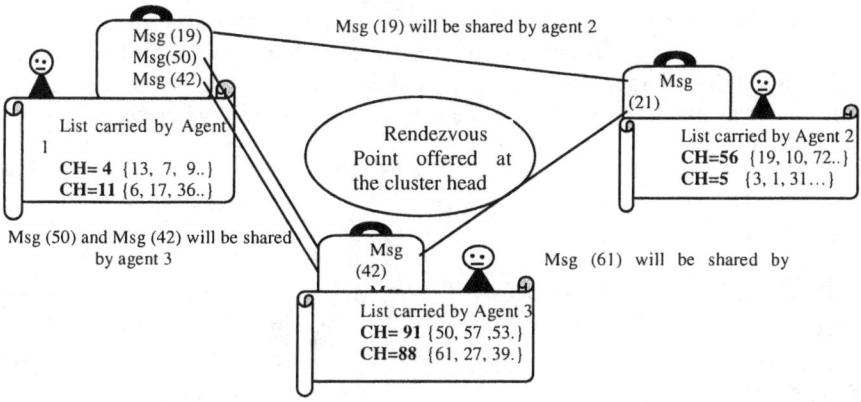

Fig. 2. Cooperative message exchanges at rendezvous points using the information list carried by each agent

Creation and termination of mobile agents: When a mobile node wants to send some message, it immediately creates an agent. Each such agent attaches with itself a bag to carry the message and puts the message in its bag. These agents will be terminated automatically when there is, no more messages to deliver in their bags.

Mobile agents at Rendezvous points: When an agent visits any cluster head the cluster head cooperates with the agent by allowing it to consult the node membership list maintained by it .The cluster head will further take the responsibilities of delivering the messages if the mobile agent has any message for this local cluster. If a mobile agent does not have any message to deliver to the current cluster head (Rendezvous point) then the cluster head will host the mobile agent for a pre specified period of time (Rendezvous period). The main idea behind the detainment of any agent is to give it a fair chance to meet with other mobile agents currently present in the system.

Inter-agent cooperation: The sharing of network information carried by them willable the agents to have free consultation on the already visited clusters along with their members. The agent starts passing on the undelivered message to another if the destination node of one's message lies on the back home journey path of another. This cooperative view is clear from Fig. 2.

4 Performance

In this section, we evaluate the performance of the cooperative routing scheme for short message delivery within an ad-hoc wireless network setup. The results confirm that the cooperative routing algorithm using Rendezvous periods at Rendezvous points are very efficient in delivering messages under high traffic. The network used for the simulation consists of 1000 nodes in a 500m X 500m simulation area. Nodes are allowed to move randomly at speeds between 0 and 30m/sec. The performance of the entire routing protocol has been evaluated using the following two criteria: i) total number of agent traffic in the network ii) the time period selected as Rendezvous period for the network.

Following figures give the impact of Rendezvous periods on network traffic, required number of node visits and the impact is clear when the load in the network is considerably high. A larger Rendezvous period ensures the increased chance of meeting with other agents. It is clear from the graphs of Fig. 3 that when there is a single agent in the system there is no impact at all.

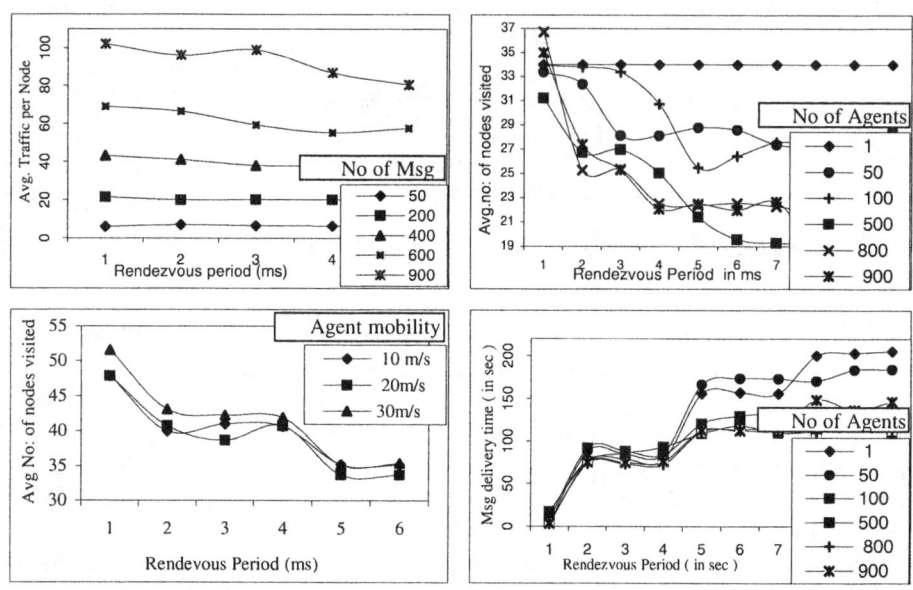

Fig. 3. a) Average traffic reduction in the network with increasing Rendezvous period. b) Average number of node visits with Rendezvous period. c) The impact of mobility on number of node visits with 500 agents in the network. d) Average message delivery time with increasing Rendezvous Period.

The graph of Fig. 3d gives the end-to-end delay for message delivery. At the beginning, the increasing nature of the curves show a definite delay of message delivery due to the waiting delay introduced at each cluster head. After the *Rendezvous period* of 5 secs the end-to end delay remains same. With further increase in this delay period the effect of cooperation among the agents cannot be realized any more.

The impact of Rendezvous points on mobility is clear from Fig. 3c where the number of node visits for delivering messages with mobility 30m/sec is higher than the node visits for delivering messages with low mobility like 10m/sec. The average number of nodes required to deliver a message also gets increased if the member node is not currently available. In that case after waiting for a pre specified time for the node to come back the cluster head handovers the undelivered message to any one of the agents available at its site. Thus the message begins a fresh journey with the newly attached agent. This case has been encountered when the mobility of the nodes are kept high i.e., 30m/sec.

5 Conclusion

In this paper we have developed an agent-based message transfer system ensuring minimal consumption of network resources, which accomplish the task of delivering messages through groups of cooperative agents. The paper has tried to formalize cooperative autonomous mobile agents through capturing the dynamic character of agent group. In essence, cooperative routing may be employed in any large ad-hoc wireless network with little overhead. Our future work will include addressing the impact of number of Rendezvous points i.e., the number of cluster heads on the scheme and as well as completion of the protocol for mobile Rendezvous points.

References

1. Roy Choudhury, R., Bandyopadhyay, S. and Paul, K.: A Distributed Mechanism for Topology Discovery in Ad Hoc Wireless Networks using Mobile Agents. In Proc. of MOBIHOC 2000 in conjunction with IEEE/ACM Mobicom 2000, Massachusetts,USA (2000).
2. Bandyopadhyay, S. and Paul, K.: Evaluating the performance of mobile agent based message communication among mobile hosts in large Ad Hoc wireless network. In Proc. of IEEE/ACM MobiCom'99, Seattle, Washington, USA (1999).
3. Bandyopadhyay, S. and Paul, K.: Using Mobile Agents For Off-Line Communication Among Mobile Hosts In A Large, Highly-Mobile Dynamic Networks. In Proc. of IEEE/ICPWC Jaipur, India, (1999).
4. Roy Choudhury, R., Bandyopadhyay, S. and Paul, K.: A Mobile Agent Based Mechanism to Discover Geographical Positions of Nodes in Ad Hoc Wireless Networks. In Proc. of APCC 2000, Seoul, Korea, (2000).
5. Roy Choudhury, R., Bandyopadhyay, S. and Paul, K.: Topology Discovery in Ad Hoc Wireless Networks Using Mobile Agents. In Proc. of IWMATA, Paris, France, (2000).
6. Marwaha S., Khong Tham C. and Srinivashan, D.: Mobile Agent based Routing Protocol for Mobile Ad Hoc Networks. GLOBECOM.2002., Taipei, Taiwan, (2002).

Network Mobility Management Using Predictive Binding Update[*]

Hee-Dong Park[1], Yong-Ha Kwon[2], Kang-Won Lee[2], Young-Soo Choi[2], Sung-Hyup Lee[2], and You-Ze Cho[2]

[1] Department of Computer Engineering,
Pohang College, Pohang, 791-711, Korea
hdpark@pohang.ac.kr
[2] School of Electrical Engineering & Computer Science,
Kyungpook National University, Daegu, 702-701, Korea
{skymiso, kw0314, yschoi, tenetshlee, yzcho}@ee.knu.ac.kr

Abstract. This paper proposes an efficient network mobility management scheme for mobile networks such as trains and buses moving on the predetermined path. In this scheme, each mobile router maintains a database about the list of access routers, their network prefixes, and cell radii on the moving path. The mobile router therefore knows in advance network prefixes and its care-of addresses of all subnets without beacon signals. Using this database and current location information, the mobile router can prepare network layer hand-over with predictive binding update before link layer handover occurs, thereby the service disruption time due to handover will be reduced to the link layer handover latency.

1 Introduction

The IETF working group on network mobility (NEMO) is currently standardizing basic support for moving networks [1]. The NEMO basic protocol will be built on Mobile IPv6 with minimal extensions [2]. Therefore, the handover mechanism of a mobile router (MR) is essentially the same as that of a mobile node (MN) with Mobile IP. The handover is classified into two components, L2 (or link-layer) handover and L3 (or network-layer) handover. Usually, the L3 handover is not dependent on the L2 handover, although it must precede the L3 handover. The L2 handover involves channel scanning, authentication, and MR-AR (Access Router) association, which lead to about 100 to 300 msec latency. And the L3 handover involves movement detection, new care-of address (CoA) configuration, and binding updates, which lead to about 2 to 3 seconds latency. This handover latency causes service disruption and packet loss, which lead to performance degradation. Fig. 1 shows the L3 handover procedure in the NEMO basic support [3].

---

[*] This work was supported in part by the KOSEF (contract no.: R01-2003-000-10155-0), the ITRC of the Ministry of Information and Communication (MIC), Korea, and the BK21 project.

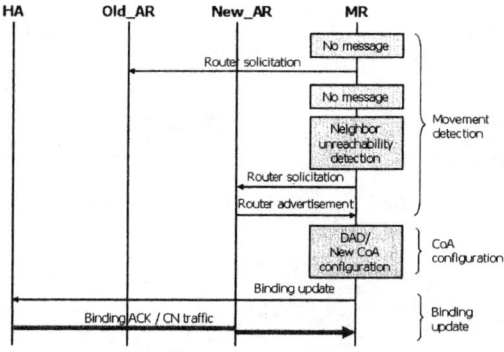

Fig. 1. The L3 handover procedure in the NEMO basic support

There are a number of proposals to reduce the latency and the number of packets lost due to the handover. Among these proposals, we focus on proactive fast handover schemes performing the L3 handover before the L2 handover in advance [4]-[9]. Generally, these proactive handover schemes provide better performance, because the L3 handover is performed before the L2 handover. But, these schemes can cause unnecessary handover preparations and forward too many duplicated packets. Furthermore, these schemes require the L2 trigger. This means that the L3 handover is not fully independent of the link layer, plus erroneous and imprecise anticipation of the L3 handover may be taken. This paper proposes a fast handover scheme with movement prediction to minimize the handover latency and packet loss. This scheme uses the peculiar mobility characteristics of public vehicles such as trains and buses. Their moving pattern has a tendency to be predictable, because the moving path and direction are very routine. This enables an MR to predict a next handover and to prepare the L3 handover before the L2 handover occurs. Therefore, the service disruption and packet loss during a handover can be reduced significantly.

2 The Proposed Handover Scheme

In this scheme, each MR maintains a mobility database about the list of ARs, their network prefixes, and cell radii on the moving path. The MR therefore knows in advance the network prefixes and CoAs of all subnets and can configure a next CoA without beacon signals. With this database and the current location information, the MR can prepare L3 handover by pre-registration to HA before L2 handover occurs. This makes total handover latency to be close to that of L2 handover. In order to predict a handover execution time, the MR should sense or estimate the Predictive handover decision point shown in Fig. 2. For estimation, the MR uses the information in the mobility database and recognizes its current geographic location as well as the moving speed. The MR can pinpoint its exact location with the aid of GPS (global positioning system) or sensors laid

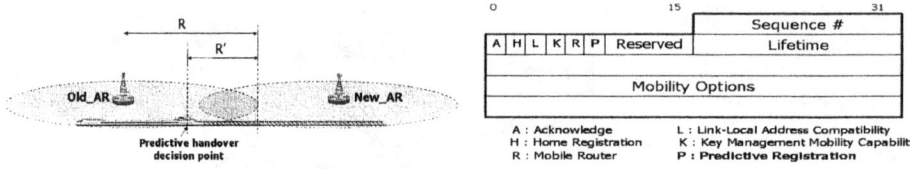

Fig. 2. Predictive handover decision point   **Fig. 3.** Predictive BU message format

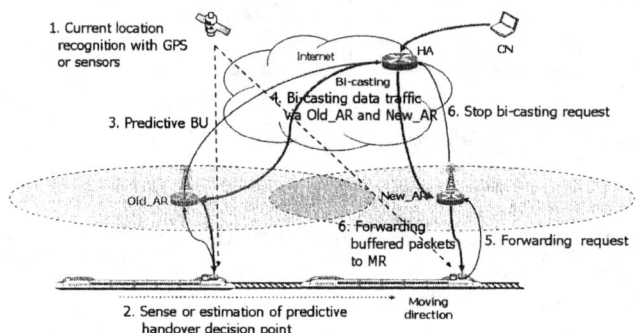

Fig. 4. Handover procedures of the proposed scheme

on the side of railroad. When the MR reaches the predictive handover decision point, prior to entering the new AR's coverage area, it sends a Predictive BU message (shown in Fig. 3) to its HA. After receiving the Predictive BU message, the HA bi-casts data packets to the MR through the new AR and the old AR simultaneously, until the completion of the handover.

Unlike the proactive handover schemes described in Section 1, this scheme does not utilize the L2 trigger mechanism. This allows a clean separation between layer 2 and layer 3 of the protocol stack. Fig. 4 and Fig. 5 show the handover procedures and message diagram of the proposed scheme, respectively. The handover procedures are described as the following:

① The MR on a train should always recognize its current location with the aid of the GPS or sensors laid on the side of railroad.

② Based on the mobility database, the MR estimates the predictive handover decision point and time.

③ When the MR reaches the handover decision point, it sends a Predictive BU message to its HA through the old AR before the L2 handover. The message contains the next CoA. In case that the predictive handover decision point is located in the overlapping area between neighboring ARs, the MR may confirm the reachability to the next AR at the same time.

④ After receiving the Predictive BU message, the HA bi-casts data packets to the MR through the new AR and the old AR simultaneously, until the completion of the handover. The new AR can buffer the packets to minimize packet loss.

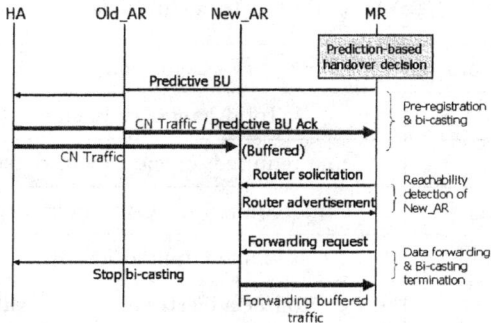

Fig. 5. Message diagram in the proposed handover scheme

⑤ As soon as the MR detects reachability to the next AR on the new link, it sends a Forwarding request message to the new AR.
⑥ When the new AR receives the Forwarding request message, it forwards the buffered data packets to the MR, and sends a Stop bi-casting message to the HA in order to prevent the HA sending packets through the old AR.

With the Predictive BU, the proposed scheme performs the L3 handover before the L2 handover. Therefore, the total handover latency is close to that of the L2 handover. If the MR cannot receive a Predictive binding ACK message from the HA, it considers that the pre-registration proves to be a failure, then performs general L3 handover after the L2 handover.

3 Performance Evaluation

This section compares the performance of the proposed handover scheme with that of the NEMO basic solution. Two critical performance issues are service disruption time and packet loss during handovers. Table 1 shows the parameters for performance evaluation.

Service disruption time during a handover can be defined as the time between the reception of the last packet through the old AR until the first packet is received through the new AR. In this paper, we regard the service disruption time as the total handover latency, T_{HO}. The total handover latency of the NEMO basic solution can be expressed as a sum of its components and with the signaling delay time shown in Fig. 1. This is given by:

$$T_{HO} = T_{MD} + T_{CoA-conf} + T_{BU}$$
$$= 2\tau + 2RTT_{MR-AR} + RTT_{AR-HA} \qquad (1)$$

where the delays for encapsulation, decapsulation, and the new CoA creation are not taken into consideration.

The total handover latency of the proposed scheme, however, will be close to T_{L2}, the L2 handover latency, because the MR performs the L3 handover before

Table 1. Parameter definitions

Parameters	Definition
T_{HO}	Total handover latency
T_{MD}	Time required for movement detection
$T_{CoA-Conf}$	Time required for CoA configuration
T_{BU}	Time required for BU
τ	Router advertisement interval
RTT_{MR-AR}	Round-trip time between MR and AR
RTT_{AR-HA}	Round-trip time between AR and HA

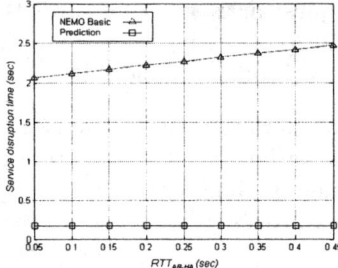

Fig. 6. Service disruption time

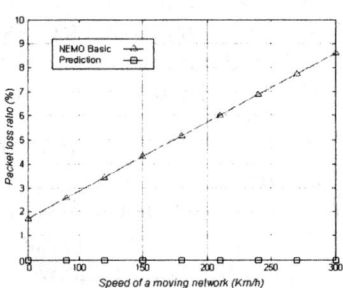

Fig. 7. Packet loss ratio

the L2 handover in advance, with keeping the reachability to the old AR. This makes the L3 handover latency to be minimized in the new AR's coverage area. Since packet loss does not occur during the time when the CN traffic travels from the HA to an MR after the completion of the BU, the packet loss period during a handover can be expressed as T_{HO} - 0.5 RTT_{MR-HA}. Therefore, from (1) the packet loss period is given by:

$$T_{loss} = 2\,\tau + 1.5 RTT_{MR-AR} + 0.5 RTT_{AR-HA} \qquad (2)$$

On the other hand, the packet loss time of the proposed scheme will be around T_{L2}. Nevertheless, there is no packet loss during a handover, due to the mechanism of the HA's bicasting and the new AR's buffering.

Packet loss ratio (ρ_{loss}) is defined as the ratio of the number of lost packets during a handover to the total numbers of transmission packets in a cell. This can be also expressed as:

$$\rho_{loss} = \frac{T_{loss}}{T_{cell}} \times 100 \quad (\%) \qquad (3)$$

where T_{cell} is the time it takes an MR to pass through a cell.

Fig. 6 and 7 compare the service disruption time and packet loss ratio between the proposed scheme and the NEMO basic support, respectively. We assume that TL2 is 200 msec, the router advertisement interval is 1 second, the radius of AR cell coverage is 1 Km, and RTT_{MR-AR} is 10 msec. RTT_{AR-HA} is assumed to be 100 msec in Fig. 7. As shown, the service disruption time of the NEMO basic solution is about 2 to 2.5 seconds, while the service disruption time of the proposed scheme is close to 200 msec. On the other hand, the packet loss ratio of the NEMO basic solution is proportional to the speed of a moving network, while the packet loss ratio of the proposed scheme will be constantly zero when the new AR buffers data sent from HA.

4 Conclusion

This paper proposed a fast handover scheme with movement prediction for public transportation such as trains and buses. This scheme uses the peculiar mobility characteristics of them. Their moving pattern has a tendency to be predictable. Therefore, without the L2 trigger, the MR can forecast its next handover and perform L3 handover before L2 handover in the old AR's coverage area. This makes the total handover latency and packet loss to be reduced significantly. The analytical results shows that the proposed scheme can provide excellent performance as compared with the existing IP layer handover scheme.

References

1. V. Devarapalli: Nemo basic support protocol. *IETF RFC* 3963, Jan. (2005).
2. D. Johnson et al.: Mobility Support in IPv6. *IETF RFC* 3775, June. (2004).
3. E. K. Paik and Y. H. Choi: Prediction-Based Fast Handoff for Mobile WLANs. In *Proc. of ICT*, vol. 1, pp. 748-753, Feb. (2003).
4. E. Shim et al.: Low Latency Handoff for Wireless IP QoS with Neighborcasting. In *Proc. of ICC* 2002, Apr. (2002).
5. R. Koodli, Ed.: Fast Handovers for Mobile IPv6. *IETF RFC* 4068, July (2005).
6. R. Hsieh et al.: S-MIP: A Seamless Handoff Architecture for Mobile IP. In *Proc. of INFOCOM* 2003, Mar. (2003).
7. F. Feng and D. Reeves: Explicit Proactive Handoff with Motion Prediction for Mobile IP. In *Proc. of WCNC* 2004, vol. 2, pp. 855-860, Mar. (2004).
8. N. Van den Wijngaert and C. Blondia: A Location Augmented Low Latency Handoff Scheme for Mobile IP. In *Proc. of ICMU*04 Jan. (2004).
9. E. Hernandez and A. Helal: Predictive Mobile IP for Rapid Mobility. In *Proc. of IEEE WLN* 2004, Nov. (2004).

Planning in a Distributed System

Rajdeep Niyogi and Sundar Balasubramaniam

Computer Science and Information Systems Group,
Birla Institute of Technology and Science, Pilani 333 031, India
{rajdeep, sundarb}@bits-pilani.ac.in

Abstract. We identify a type of distributed system where the notion of space in planning is important. We give a formal modeling of the distributed system where planning is done. We show that several interesting planning goals in the model can be specified in a spatio-temporal logic STL. We develop an efficient planning procedure for these goals in the distributed setting.

1 Introduction and Motivation

In planning, the task of an agent is to devise a sequence of actions (plan), from an initial state, that achieves some desired property (goal). Our goals are similar to the temporally extended goals [1]. However, planning does not involve reasoning about spatial properties. But if we want to do planning in a distributed system [2], then the meaning of space is quite intuitive: a node in a network, or a subnet in a network is a location. A location means space and the location has to be given explicitly. Now for the locations there are different spatial properties. Let us consider a scenario where we wish to do planning.

Example: The task of a recruiting agent is to select some students, with good experience in planning, who pass the qualifying test. The students are attached to laboratories (AI, Computer graphics(CG), VLSI), that are attached to departments, which are in turn attached to Institutes. Here the locations are Institutes, departments, and laboratories. Refer to figure 1.

Similar hierarchical structures exist in networking, the file system in UNIX, and bibliographical databases to mention a few.

Containment Relation: A containment relation is asymmetric and transitive. In the example, a department is contained in an institute, a laboratory is contained in a department.

We consider settings where the level of containment is always bounded and it is small. In general, the level of containment may be unbounded but finite. (The level of containment is 2 for the above example.)

For the above example, a plan should transfer the agent to a location—where a student with good experience in planning exists. In this paper we show how to find such a plan. The activities like **conducting a test** or **executing a code** can be easily planned for as in conventional planning. We now give some motivations.

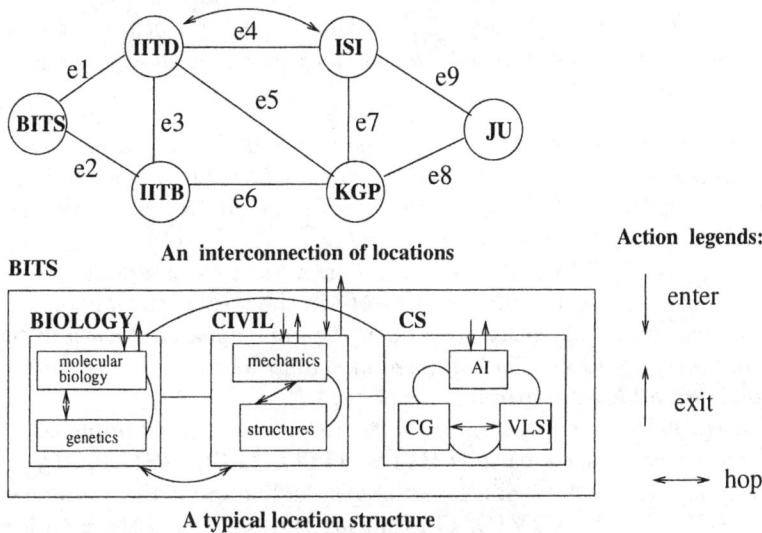

Fig. 1. A planning scenario with nested locations

A mobile agent is a program that can move between machines in a distributed environment performing tasks [3]. We look at mobile agents from a planning perspective. We show that in some situations it will be helpful to have a plan for the agent. Mobile ambients [4] are a generalization of the mobile agents. An ambient is a location that can move. In order to specify the properties of the ambients, a modal logic of space and time, called ambient calculus, has been developed in [5].

Our aim is to find plans for a mobile ambient. As an initial step we consider a simplistic situation where the locations are static and the agent is not communicating. We are interested in reasoning about temporal properties that can be satisfied at location(s) in the space. This is unlike conventional planning where we are interested only in temporal properties. For the planning goals, we give a spatio-temporal logic specification in section 4. The logic has been inspired by the tree-logics developed in [6, 7] for the purpose of specifying query languages for semistructured data, like XML.

We assume planning is done on a spatial model that is given explicitly. We discuss the model in section 2. We use generic planning operators(actions) that are specified as in conventional planning. Our planning domain, given in section 3, takes as input the spatial model, an initial location, a set of actions, and a goal formula. We give the planning procedure in section 5.

2 The Spatial Model

Definition 1 (Location Graph). *A location graph (LG) is a graph whose vertices are locations and edges are links between locations.*

Each location may contain other locations. Each location has a name.

Definition 2 (Location Tree). *The location tree of a location named Loc is a directed tree with root Loc, whose nodes are locations and an edge $u \to v$ denotes that v is contained in u.*

We formally denote a location by a 4-tuple $\langle Loc, LT, LG, prop \rangle$ where, Loc is the name of a location, LT is the location tree of Loc, LG is a location graph of Loc where the nodes in LG are the children of Loc in LT, and $prop$ represents atomic propositions(AP) that are true at Loc.

We refer to the children of a location l in a location tree as the immediate sublocations of l, and to the descendants of l in the tree as the sublocations of l.

If n is a location in a location tree, then the immediate sublocations of n may be interconnected by links. These immediate sublocations and their interconnections constitute a location graph.

We illustrate the above meanings with respect to the example in figure 1. The set consisting of six locations **BITS, IITD, IITB, ISI, KGP, JU**, and the set of links $e1, \ldots, e9$ form a location graph. The immediate sublocations of **BITS** are **BIOLOGY, CIVIL, CS** that are the vertices of the location graph of **BITS**. The immediate sublocations of **CS** are **VLSI, CG, AI** which are also the sublocations of **BITS**.

3 Planning Domain

Planning is performed in the spatial model given in the previous section. We define the planning domain as $\mathcal{D} = \langle M, l_0, A, G \rangle$ where, M is the spatial model (described in section 2), l_0 is the initial location (a location in M), A is a set of actions that the agent can perform on M, (actions transfer the agent from one location to another), G is a goal formula expressed in a spatio-temporal logic, (given in section 4).

A plan $\mathcal{P}$ is a finite sequence of actions $a_1, a_2, \ldots, a_m$. We say that the plan path corresponding to the location l_0 and the plan $\mathcal{P}$ is the sequence of locations $l_0, l_1, \ldots, l_m$. The *planning problem* is, given the planning domain $\mathcal{D}$, and a goal formula Φ, to find a plan for which the plan path satisfies Φ. We present a method of solving the planning problem in section 5.

We consider STRIPS like formalism for specifying actions. We use three generic actions, *hop*, *enter*, and *exit* that are defined formally as:

Definition 3 (hop). *The precondition of $hop_{i,j}$: there is a link between two locations i, j in a location graph, and the agent is at location i. The effect: the agent is at location j.*

Definition 4 (enter). *The precondition of $enter_{i,j}$: the agent is at location i, location j is an immediate sublocation of i. The effect: the agent is at location j.*

Definition 5 (exit). *The precondition of $exit_{i,j}$: the agent is at location i, location i is an immediate sublocation of j. The effect: the agent is at location j.*

In figure 1 we show the actions applicable to different locations.

4 A Spatio-Temporal Logic (STL)

4.1 Syntax and Semantics of STL

$c[p]$ refers to a particular location c where an atomic proposition p is true. We use a spatial operator called *inside* (denoted by I) that allows reasoning about a sublocation of any location. We use a temporal operator *along* (denoted by H) that allows reasoning about locations with time.

Syntax of STL: $\psi \longrightarrow c[\phi] \mid H\,\phi \qquad \phi \longrightarrow \psi \mid p \mid \neg\phi \mid \phi_1 \wedge \phi_2 \mid I\,\phi$

Semantics of STL: The satisfaction of a goal formula ψ with respect to a location l in a location graph LG is defined inductively as:

Case I: $\psi = c[\phi]$
I(a): $\phi = p \qquad (LG, l) \models c[p] \qquad$ iff $l = c$ and p holds in c, where $p \in AP$
I(b): $\phi = \neg\phi_1 \qquad (LG, l) \models c[\neg\phi_1]$ iff $(LG, l) \not\models c[\phi_1]$.
I(c): $\phi = \phi_1 \wedge \phi_2$
$\qquad (LG, l) \models c[\phi_1 \wedge \phi_2] \quad$ iff $(LG, l) \models c[\phi_1]$ and $(LG, l) \models c[\phi_2]$.
I(d): $\phi = I\,\phi_1 \qquad (LG, l) \models c[I\,\phi_1] \quad$ iff $l = c$ and there exists a sublocation n of c such that $(LG', n) \models n[\phi_1]$ where LG' is the location graph of m; m is the parent of n in the location tree of c.
I(e): $\phi = H\,\phi_1$
$\qquad (LG, l) \models c[H\,\phi_1] \qquad$ iff $l = c$ and $(LG, l) \models H\,\phi_1$
I(f): $\phi = c_1[\phi_1]$
$\qquad (LG, l) \models c[c_1[\phi_1]] \qquad$ iff $l = c$, and there exists an immediate sublocation n of c such that $(LG', n) \models c_1[\phi_1]$, where LG' is the location graph of c.
Case II: $\psi = H\,\phi$
$\qquad (LG, l) \models H\,\phi \qquad$ iff there exists a path $l_0, \ldots, l_k, l = l_0$ in LG such that $(LG, l_k) \models l_k[\phi]$

4.2 Examples of Planning Goals

Notation: phrase that is in bold represents a particular location, and the phrase in italics represents a proposition.

1. visit an Institute where the **CS Department** *has a supercomputer*.
This is expressed as: $\qquad H$ CS$[p]$.

2. there is a researcher *working on voting theory* and there is another researcher *working on learning theory* in the **Statistical Institute** (stat).
This is expressed as: $\qquad$ stat$[(I\,p) \wedge (I\,q)]$.

3. visit a department in an Institute that has been *accredited by NAAC*.
This is expressed as: $\qquad H\,I\,p$.

However, there are situations that cannot be expressed in our logic; for instance, locations cannot be quantified.

5 A Planning Procedure

The planning procedure **FindPlan** takes a goal formula as input and returns a satisfying plan, if any, as output.

procedure FindPlan(goal formula $\psi = H \phi$) {
   case 1. [initial location (l_0) is labeled by ϕ]
       (i) **LabelPhi**($node, \phi$);
       (ii) **if** $\phi \in$ label set of *node* **then return PlanForPhi**(l_0, ϕ);
   case 2. [initial location (l_0) is not labeled by ϕ]
       (i) create a queue Q;
       (ii) insert l_0 into Q, mark l_0;
       (iii) **while** (Q is not empty) {
            remove a node s from Q;
            for all adjacent nodes s' of s {
                if s' is not marked {
                    mark s'; insert s' into Q;
                    LabelPhi(s', ϕ);
                    if $\phi \in$ label set of s' **then** {
                        $P_\phi = $ **PlanForPhi**(s', ϕ);
                        plan is obtained by prefixing P_ϕ by *hop* actions;
                        return plan;}}}}
   case 3: [no locations are labeled by ϕ]
       return no satisfying plan exists for the goal;}

The labeling algorithm **LabelPhi** for locations is inductively defined on ϕ.

procedure PlanForPhi(*node*, ϕ)
[Plans for different forms of subformulas are defined inductively as follows.]
   1. let ϕ be of the form $\phi = node[c_1[c_2[\ldots[c_k[\phi']]]]]$
      $plan = enter_{node,c_1}, enter_{c_1,c_2}, \ldots, enter_{c_{k-1},c_k}$;
      if $\phi' = p$ **then return** $plan$;
      else if ϕ' is not in the form of ϕ **then**
         return $plan + P_{\phi'}$; where $P_{\phi'}$ is the plan for ϕ'.
   2. let ϕ be of the form $\phi = I\ \phi'$
      let $P_{\phi'}$ be the plan for ϕ'; $i = node$; initialize $plan$ (a list) to empty;
      traverse the tree from i and stop at k whose label set contains $I\ \phi'$ or ϕ';
      while the child j of i is not k {
         $plan = plan + enter_{i,j}; i = j;$ }
      $plan = plan + enter_{i,k}$;
      return $plan + P_{\phi'}$;
   3. let ϕ be of the form $\phi = (I\ \phi_1) \wedge (I\ \phi_2)\ldots \wedge (I\ \phi_k)$
      let $plan_1, \ldots, plan_k$ denote plans (that are found separately) $I\ \phi_1, \ldots, I\ \phi_k$ respectively; store the plans in a list α in the increasing order of length;
      initialize $plan$ to empty;
      for $i = 1$ to $length(\alpha) - 1$ {
         let T_i contain the corresponding *exit* actions for $plan_i$;
         $plan = plan + plan_i + T_i;$ }
      return $plan + plan_i$;

5.1 Properties of the Planning Procedure

Theorem 1 (Decidability). *The logic STL is decidable.*

Proof: By induction on the procedure that update the label set. □

Theorem 2 (Model checking). *Given a location l, and a formula ψ, the problem of finding whether $l \models \psi$ can be done in time $O(|\psi| \cdot (|LG| + |LT|))$, where $|\psi|$ is the length of the formula ψ, $|LG|$ is the maximum size of a location graph, and $|LT|$ is the maximum size of a location tree in the spatial model.*

Proof: By induction on the length of ψ. □

Theorem 3 (Complexity of Planning). *Given a location l, and a formula ψ, the problem of finding a plan for ψ at l can be done in time $O(|\psi| \cdot (|LG| + |LT|))$.*

Proof: The complexity of procedure **FindPlan** is the sum of the time taken by the labeling procedure and the plan finding procedure. In **PlanForPhi** for the first two cases the time taken is at most $|LT_l| \times |\psi|$. In case 3 we need in addition, time to sort k plans that takes $O(k \lg k)$. Now $k < |\psi|$. Thus for case 3, time needed is at most $|LT_l| \times |\psi|$. Hence, the overall complexity is that needed for labeling. □

6 Conclusions

In this paper we give a formal modeling of a simplified distributed system (without communication). We show that several interesting planning goals in the setting can be specified using STL and an efficient planning procedure can be found for such goals. To the best of our knowledge the introduction of the notion of space in planning has not been done before. As part of our future research we would like to see whether an efficient planning procedure can be obtained for mobile ambients as well.

References

1. Bacchus, F., Kabanza, F.: Planning for Temporally Extended Goals. Annals of Mathematics and Artificial Intelligence. 22 (1998) 5–27
2. DesJardins, E M., Durfee, H E., Ortiz, L Charles Jr., Wolverton J Michael.: A Survey of Research in Distributed, Continual Planning. AI Magazine, Winter (1999) 13–22
3. Kotz, D., Gray, S R.: Mobile Agents and the Future of the Internet. ACM Operating Systems Review, 33 (1999) 7–13
4. Cardelli, L., Gordon, A.: Mobile Ambients. Proc. of FoSSaCS, LNCS (1998) 140–155
5. Cardelli, L., Gordon, A.: Anytime, Anywhere: Mobile Logics for Mobile Ambients. Proc. of Principles of Programming Languages, ACM press, (2000) 365–377
6. Cardelli, L., Ghelli, G.: A Query Language Based on Ambient Logic. Proc of European Symposium on Programming, LNCS 2028 (2001) 1–22
7. Zilio Dal, S., Lugiez, D.: XML Schema, Tree logic and Sheaves Automata. Proc of 14th International Conference on Rewriting Techniques and Applications, LNCS 2706 (2003) 246–263

Using Inertia and Referrals to Facilitate Satisficing Distributions*

Teddy Candale, Ikpeme Erete, and Sandip Sen

University of Tulsa
{teddy-candale, ikpeme-erete, sandip-sen}@utulsa.edu

Abstract. We study the problem of distributed, self-interested agents searching for high-quality service providers where the performance of a service provider depends on its work load. Agents use referrals from peers to locate satisfactory providers. While stable environments may facilitate fast convergence to satisfying states, greedy and myopic behaviors by distributed agents can lead to poor and variable performances for the entire community. We present mechanisms for resource discovery that involve learning, over interactions, both the performance levels of different service providers as well as the quality of referrals provided by other agents. We study parameters controlling system performance to better comprehend the reasons behind the observed performances of the proposed coordination schemes.

1 Introduction

We study the problem of autonomous agents choosing between several service providers to obtain desired services. We assume a completely distributed environment without central authority or knowledge. Our research goal is to develop mechanisms by which such agent communities can stabilize on states where all agents are satisfied with the service provider they are currently using.

Locating high-quality services is a challenging problem when sharing resources with a large population. Number of service providers are typically limited and their performances depend both on their intrinsic capabilities and workload. Myopic, self-interested behavior can lead to poor performances for the individual and can result in system-wide instability. There is thus a need for non-myopic mechanisms to promote performance and stability of such decentralized systems.

While ideal rational agents may aspire for optimal satisfaction levels, dynamic, partially known, and open environments can render the realization of this ideal improbable. Such an agent is unable to accurately assess the impact of its own decisions, including choice of service providers and making referrals, on the system. As such, it is unrealistic to expect strategies that will always optimize performance. Rather, we posit that agents should concentrate on finding service providers that provide a quality of service which exceeds an acceptable

---

* This work has been supported in part by an NSF award IIS-0209208.

performance threshold. This formulation is consistent with Simon and others view of bounded rationality of decision makers within the context of complex organizations [1, 2, 3, 4].

Referrals from other agents can help agents find more satisfying service providers. But such referrals may cost the referring agent since the load on the referred provider may increase, with corresponding performance deterioration. This is particularly true with referral chains, i.e., if an agent can refer providers it located through referrals from other agents. While referral systems have been widely studied both in theory and in practical applications, the negative side-effects of referrals have not received adequate treatment. We seek to analyze the benefits and disadvantages of referrals in domains where the cost of referrals is uncertain. The goal is to identify situations where an agent should or should not use referrals. Our goal is to develop strategies by which a system of autonomous agents can quickly reach stable configurations where all agents are satisfied with the choice of their current service providers.

2 Framework

Environment: We present an environment where agents share a set of service providers to perform daily tasks. Let $\mathcal{E} = <\mathcal{A}, \mathcal{R}, perf, L, S, \Gamma >$ where: $\mathcal{A} = \{a_k\}_{k=1..K}$ is the set of agents, $\mathcal{R} = \{r_n\}_{n=1..N}$ is the set of providers, $f : \mathcal{R} \times \mathbb{R}_+ \rightarrow [0, 1]$ provides the intrinsic performance of a provider given a load, $L : \mathcal{A} \rightarrow \mathbb{R}_+$ is the load function for the agents, $S : \mathcal{A} \times [0, 1] \rightarrow [0, 1]$ is the satisfaction function of agents, $\Gamma = \{\gamma_1, \ldots, \gamma_K\}$ is the set of satisfaction thresholds, representing aspiration levels of agents. Each day d, agent a_k has a load $L(a_k)$ to perform. a_k assigns its load to a selected provider to handle it in its behalf. At the outset, a_k knows the set of providers that can process its task without the knowledge of their intrinsic capabilities represented by their performance function, $f(r_n, \cdot)$, for provider r_n. a_k is also unaware of the current load on the providers. If $\mathcal{A}_n^d$ is the set of agents using the provider r_n at day d then the provider's performance after processing all these orders is $perf = f(r_n, \sum_{a \in \mathcal{A}_n^d} L(a))$. $perf$ is the service quality received at the end of the day d by every agent in $\mathcal{A}_n^d$. $a_k \in \mathcal{A}_n^d$ will evaluate the performance of r_n by the value $s = S(a_k, perf)$. a_k will be satisfied if $s \geq \gamma_k$.

Our aim is to design interaction protocols and behaviors that allow all agents to find satisfying providers. The concept of *distribution* represents how agents distribute themselves over the providers. We call $D = \{\mathcal{A}_n\}_{n=1..N}$ a distribution where $\mathcal{A}_n$ is the set of agents which use provider r_n. A Γ-acceptable distribution is a distribution where every agent is satisfied, i.e, each agent receives a satisfaction above its own satisfaction threshold. A Γ-acceptable distribution is expected to be a stable distribution since no agent will have the incentive to change their choice of provider. Consequently, it is an equilibrium concept and our goal is to enable agents to reach such distributions.

Inertia: Oscillations in our environment will happen if at a distribution close to a Γ-acceptable distribution the system has the tendency to evolve to a worse distribution and vice versa. We assume that the total load applied by all agents in the system is approximately equal to the total capacity of all service providers to produce satisfactory performance for all agents if they are properly distributed. Intuitively, a distribution where almost everyone is satisfied contains very few under-used or over-used providers and the rest are occupied by the right number of agents. Those under-used providers $\mathcal{R}_u$ are very attractive. Consequently, agents will be inclined to move to them, which leads the system to a distribution where providers in $\mathcal{R}_u$ will be overcrowded. This key, problematic effect can be mitigated by increasing the inertia in the system, where inertia is an inverse function of the number of agents moving at any given time.

An agent may decide to switch resources relying on its own information or on a referrer or to explore to discover either unknown resources or to be able to adapt to changes in the environment. Inertia can be controlled by the following methods:

Exploration: Fast convergence requires learning about provider and referral qualities: more informed decisions will expedite system convergence to satisfactory distributions. Consequently, some systematic exploration of providers is necessary. However, such exploration decreases inertia and can impact convergence rate. An environment where agents explore too much will produce system instability where agents will not have accurate estimations of provider performances since loads vary significantly. In this context, referral systems can be useful since agents may substitute their exploration with others' experiences.

Decision Process: When designing our agents, we chose a "move when you think you can do better"-principle. Consequently, agents never move when they are satisfied. If unsatisfied, agents pick with probability α a resource randomly to ensure exploration. With probability $1 - \alpha$ they try to locate a resource. Henceforth, we refer to the processing in this step as the *decision process*. We present five different decision processes: with and without use of referrals and with more or less inclination to move. We first present decision processes without the use of referral.

NRLI (*No Referral Low Inertia*): This decision process consists in picking a resource for which the agent expects to get at least a minimum level of satisfaction, γ_k^-. Let $es_{k,n}$ be the expected satisfaction agent a_k believes it will get by using resource r_n. Let $\mathcal{R}_{k,\gamma_k^-} = \{r_n \mid \gamma_k^- \leq es_{k,n}\}$ be the set of resources expected to provide satisfaction more than γ_k^-. A resource r_{n_k} is chosen in $\mathcal{R}_{k,\gamma_k^-}$ with likelihood $es_{k,n}$. In the case $\mathcal{R}_{k,\gamma_k^-} = \emptyset$, a_k does not move.

NRHI (*No Referral High Inertia*): NRHI is a variant of NRLI. Agent a_k using NRHI will not move to a provider expected to provide lesser satisfaction than the provider, $r_{n_k^c}$, it is currently using. a_k does not move if $es_{k,n_k} < es_{k,n_k^c}$.

RTLI (*Referral Truthful Low Inertia*): RTLI is also a variant of NRLI. If $\mathcal{R}_{k,\gamma_k^-} = \emptyset$, a_k asks another agent a_{k_h} for referral. a_{k_h} provides both the name of a resource $r_{n_{k_h}}$ and an estimation of the satisfaction it will get ($es_{k_h,n_{k_h}}$).

a_k is trustful in the sense it does not try to correct the value $es_{k_h,n_{k_h}}$. a_k will use the referral if $\gamma_k^- < es_{k_h,n_{k_h}}$. Besides, when approached for help, an agent using RTLI is truthful in the sense that it reports its actual estimate[1]. It refers a resource it would have chosen itself. In other words, it provides the outcome of NRLI.

RTHI (*Referral Truthful High Inertia*): RTHI is a mixture of NRHI and RTLI. When looking for a resource using its own information, an agent uses NRHI and when looking for a referral the agent uses RTLI. When answering a request, it provides the outcome of NRHI.

BRLI (*Balance Referrer Low Inertia*): BRLI is a variant of BRLI. An agent using RFLI will answer a request only from agents with which it has a negative or null balance of exchange. A balance of exchange is the difference between the sum of what it has given and what it has received. More formally, let $bal_{k,k'}$ be the balance maintained by a_k with agent $a_{k'}$. a_k increases $bal_{k,k'}$ by es_{k,n_k} when it provides r_{n_k} as a referral to $a_{k'}$. a_k decreases $bal_{k,k'}$ by $s_{k,n_{k'}}$ where $s_{k,n_{k'}}$ is the satisfaction obtained by a_k if it uses $r_{n_{k'}}$, 0 otherwise.

3 Experimental Results

In the previous section, we propose two methods to control the inertia: the coefficient of exploration and the use of decision processes. We will evaluate the two controlling methods while also providing comparisons between referral methods and those without referral.

Experiments comprise a large number, $K = 200$, of identical agents. We use sufficient resources to exactly satisfy the agents present in the environment. In other words, if C_n is the capacity of resource r_n then $\sum_{n=1}^{N} C_n = K \cdot L$ where L is the load imposed by each agent. Hence, we are always sure of the existence of a Γ-acceptable distribution.

We ran experiments to see the influence of the coefficient of exploration α on the speed of convergence. In other words, we measured the number of iterations needed to reach a Γ-acceptable distributionwhen agents use protocols defined in Section 2 given the value of α. Figure 1 presents the result. One environment comprises a high number of resources ($N = 100$) and one comprises a lower number of resources ($N = 40$). We highlight the following observations:

HI performances are much better than those of LI for most values of α. This shows that the speed of convergence is improved greatly if agents decide to move less often. By not moving when it thinks no other resource can satisfy it better than its current resource, an agent avoids conflict of interest since many agents are likely to choose the same resource. Besides, an agent can benefit from the departure of others by staying in its current resource. However, when $N = 100$ and $\alpha \leq 0.02$ both HI and LI have poor results but for different reasons. Detailed analysis of the system given the inertia show us HI have poor performance due to too high inertia; the performance of the system improves very slowly, while

---

[1] In other work, we consider the motivations and the effect for untruthful referrals.

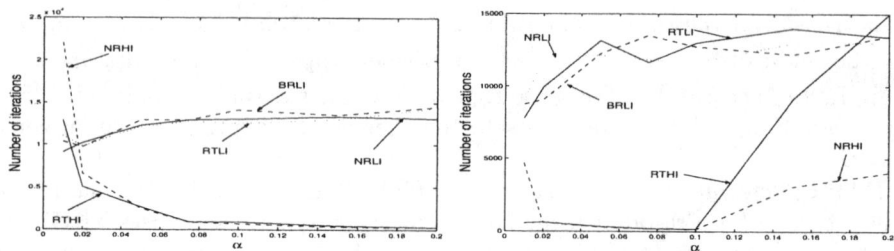

Fig. 1. Number of iterations to reach convergence given α for $N = 100$ (left) and $N = 40$ (right) (200 agents)

LI have poor performances due too low inertia; the system oscillates. In spite of the fact that HI≤LI[2] when $N = 100$ and $\alpha \leq 0.02$, HI is preferable to LI because they work better in more environments. Improving HI performance can be achieved more easily by tuning the parameter α.

For high inertia referral decision processes works better, i.e., NRHI ≤ RTHI for $\alpha \leq 0.1$. For $\alpha < 0.1$, the inertia is higher with much less exploration, thus preventing substantial improvement of the entropy. The situation is improved by using a referral system. The use of others' information accelerates the resource discovery process. We observe the opposite phenomenon when $\alpha \geq 0.1$, RTHI≤NRHI. With higher values of α, agents are more inclined to explore the environment and hence move more often. This is amplified by the referral system. The use of other's information makes RTHI agents switch resources when NRHI will not.

Performances of LI schemes are equivalent for $N = 100$ and 40. There exists a range of α values in which HI schemes has desirable performances. Detailed studies showed us that for very small values of α the system evolves very slowly with HI since very few agents moves leading to slow convergence. When the values of α are too high, too many agents move simultaneously leading to instable system, i.e., the system oscillate between good states and undesirable states. HI has better scale-up performances when $N = 100$ compare to when $N = 40$. In fact, when the number of resources decreases, assuming the number of agents fixed, more agents have the inclination to move leading to a diminution in the inertia. The range of α values for which HI have desired performances is smaller with lower number of resources.

4 Related Work

Sen & Sajja have studied the use of referrals to locate service providers when an agent first enters a new community with no prior knowledge of the quality of service providers or the reliability of the referrers [5]. In their work, peers have a short term cost of processing the referral request, which can be negligible in most domains. In our setting, referrals have a long term cost as the asking agents

---

[2] HI≤LI denotes that LI converges faster than HI.

may use the referred provider in the future and also refer it to others and hence possibly reduce the performance of that provider.

Coordination is a key issue in multiagent systems. Sen et al. [6] show that information can negatively impact agent coordination over resources. They allow agents to move to providers only in the neighborhood of the one they are currently using, thereby achieving perfect coordination faster. They conclude that too much information available to agents lead to oscillating provider loads. This leads to variable provider performances and low convergence speed. Rustogi & Singh [7] study the influence of inertia for system convergence in the same domain. They proved that high inertia speeds up convergence when knowledge increases but low inertia perform better with little knowledge.

5 Conclusion and Future Work

We have investigated different decision processes to locate satisfactory service providers. These decision processes give agents differing inertia of switching resources given their current and expected satisfactions from different resources and can include referrals from other agents. The main conclusion of our experiments is that decision processes with higher inertia of movement (HI procedures) produce faster convergence and better scale-up than those with lower inertia. Even faster convergence with the HI schemes can be produced by using referrals or by tuning the exploration coefficient α. Desirable performances are more difficult to obtain when using LI decision processes regardless of the use of referral systems.

We are currently exploring the effect of non-identical agents and resources. Planned future work includes use of deceptive referral agents and minimizing such disruptive behavior.

References

1. Goodrich, M., Stimpson, J.: Learning to cooperate in a social dilemma: A satisficing approach to bargaining. In: International Conference on Machine Learning. (2003) 728–735
2. March, J.G., Simon, H.A.: Models of Man, New York (1958)
3. Simon, H.A.: Models of Man, New York (1957)
4. SJ Russell, D.S.: Provably bounded-optimal agents. Journal of Articial Intelligence Research (1995)
5. Sen, S., Sajja, N.: Robustness of reputation-based trust: boolean case. In: AAMAS '02: Proceedings of the first international joint conference on Autonomous agents and multiagent systems, New York, NY, USA, ACM Press (2002) 288–293
6. Sen, S., Arora, N., Roychowdhury, S.: Using limited information to enhance group stability. Int. J. Hum.-Comput. Stud. **48** (1998) 69–82
7. Rustogi, S.K., Singh, M.P.: Be patient and tolerate imprecision: How autonomous agents can coordinate effectively. In: IJCAI '99: Proceedings of the Sixteenth International Joint Conference on Artificial Intelligence, San Francisco, CA, USA, Morgan Kaufmann Publishers Inc. (1999) 512–519

Privacy Preserving Decentralized Method for Computing a Pareto-Optimal Solution*

Satish K. Sehgal** and Asim K. Pal

Indian Institute of Management Calcutta, Kolkata, India
satish.sehgal@gm.com, asim@iimcal.ac.in

Abstract. Distributed methodologies to find pareto-optimal frontier with concern to privacy, of objectives and constraints, of parties is of interest in scenarios like negotiations. Adaptation of lagrangian method to solve distributed weighting method for both strictly concave and not strictly concave (e.g. linear) value functions is proposed for a maximization problem.

1 Background and Motivation

The methods currently available for the multi-objective optimization even in the distributed scenario do not consider the disclosure aspect and thus cannot be directly adopted to the scenario of negotiation where participating parties decide to find and then negotiate on Pareto-optimal frontier. *Secure Multi-party Computations* (SMC) attempts to preserve the privacy of information as the central aspect of the computation in a distributed situation. However, SMC algorithms are often very computation and communication intensive.

One of the traditional methodology to find a pareto-optimal point is *weighting method* [1]. In the distributed version of this method the scalarized objective is decomposed by introducing a decision variable for each participant or decision making agent (DMA) and then applying the dual decomposition method [2]. The decomposition results in a separable problem which is solved iteratively with each DMA solving its own optimization problem whereas the mediator agent (MA), if used, updates the parameters of the optimization problems. When DMA's optimal solutions converge, the common optimum is guaranteed to be Pareto-optimal (po). Disclosure to be avoided are:

1. In any iteration: the derivative at a point.
2. Across iterations: disclosure due to repeated information exchange, e.g. while transitioning to better points for all parties in each iteration a disclosure of only the points (and not even the slopes) enables the DMAs (and MA) to construct strategically equivalent functions of other DMAs (due to concavity of the problem).

---

* The work is partly supported by the AICTE project ISISAMB.
** Presently working at India Science Lab, GM R&D, Bangalore, India.

2 Problem and Assumptions

An *option*[1] in n dimensional decision space U is represented as $\mathsf{x} = (x_1, x_2, \ldots, x_n)^\mathrm{T}$, e.g. $\mathsf{U} = \Re^n$. There are m parties involved. Each DMA^i has a convex *decision space* $\mathsf{X}^i \subset \mathsf{U}$; $i = 1, 2, \ldots, m$. The *feasible space* $\mathsf{X} = \bigcap_i \mathsf{X}^i$ is common to all DMAs. Each DMA^i has a concave objective function, $v^i : \mathsf{X}^i \to Y^i \subset \Re$ representing its preference structure. A decision vector $\mathsf{x}^* \in \mathsf{X}$ is *Pareto-optimal (po) for a maximization problem* if and only if there is no other $\mathsf{x} \in \mathsf{X}$ s.t. $v^i(\mathsf{x}) \geq v^i(\mathsf{x}^*); \forall i = 1, \ldots, m$, where the inequality is strict for at least one i.

In the paper we present algorithms to *find a single po point*. The frontier can be obtained by varying the weights in the weighting method. The algorithms consider the privacy of DMAs as a prime concern. The assumptions are:

1. There is one DMA per side and coalition formation is prohibited.
2. The DMAs do not like to disclose either their decision space or the value function over that space.
3. The parties (including mediator) are *semi-honest*. A *semi-honest* agent follows the protocol properly but it keeps record of all its computations [3].
4. The communication channel is secure.

3 Distributed Weighting Method (Heiskanen, 1999)

Weighting method combines multiple objectives into a single objective:

$$\max_{\substack{\mathsf{x} \in \mathsf{X}^j \\ j=1,\ldots,m}} \sum_{i=1}^{m} w^i \, v^i(\mathsf{x}) \tag{1}$$

where $w^i \geq 0$ for all DMA^i; $i = 1, \ldots, m$, and $\sum_{i=1}^{m} w^i = 1$. The value functions in the scenario considered here are distributed among (and are private to) the DMAs and each DMA has its own set of constraints. The union of these constraints form the feasible region. As DMAs would solve the problems individually (i.e. locally), it is required to make the problem distributed. For distribution, an n-dimensional variable $\mathsf{x}^i = (x_1^i, x_2^i, \ldots, x_n^i)^\mathrm{T}$ for each DMA^i is introduced, leading to an alternative formulation

$$\max_{\substack{\mathsf{x}^j \in \mathsf{X}^j \\ j=1,\ldots,m}} \sum_{i=1}^{m} w^i \, v^i(\mathsf{x}^i) \tag{2}$$

$$\text{subject to} \quad \mathsf{x}^i - \mathsf{x}^{i+1} = 0, \; i = 1, \ldots, m-1$$

where, v^i is *strictly concave*.

Using Lagrange's multiplier vector λ^i, we get the dual

$$\min_{\lambda^1, \ldots, \lambda^{m-1}} \left[\max_{\substack{\mathsf{x}^j \in \mathsf{X}^j \\ j=1,\ldots,m}} \left\{ \sum_{i=1}^{m} w^i v^i(\mathsf{x}^i) + \sum_{i=1}^{m-1} (\lambda^i)^\mathrm{T}(\mathsf{x}^i - \mathsf{x}^{i+1}) \right\} \right] \tag{3}$$

---

[1] We use option, alternative, point, solution and decision interchangeably.

Decomposing the problem we get every DMA^i's concave optimization problem which it can solve independently. Let, $V^i(z) = w^i v^i(z) + (\hat{\lambda}^i)^T z$ where

$$\hat{\lambda}^i = \begin{cases} \lambda^i & \text{if } i=1 \\ \lambda^i - \lambda^{i-1} & \text{if } i=2,\ldots,m-1 \\ -\lambda^{i-1} & \text{if } i=m \end{cases} \quad (4)$$

Thus, for $i = 1,\ldots,m$,

$$DMA^i\text{'s problem}: \quad x^{i*} = \arg\max_{z \in X^i}(V^i(z)) \quad (5)$$

$$\text{Common problem}: \quad \min_{\hat{\lambda}^1,\ldots,\hat{\lambda}^{m-1}} \sum_{i=1}^{m} V^i(x^{i*}) \quad (6)$$

Since the value function v^i is private to DMA^i, the above problem (2)-(3) cannot be solved in traditional centralized ways. In the iterative methodology [2] a mediator agent (MA) was utilized. MA sends the DMAs initialized values of λ's, DMAs solve their problems and find the optimal x^{i*} and send those back to MA, who then modifies the λ's based on these values. The iterations continue till convergence. However, this exchange of x^i's causes disclosure to MA (see below). We have proposed modifications to this method for reducing disclosure. MA can compute the min of Equation (6) for a given $w = (w^1,\ldots,w^m)$. MA will announce when the convergence is achieved. Then all the DMA's x's are supposed to have converged. To check convergence, the DMAs can compute $\bar{x}^* = \frac{1}{m}\sum_{i=1}^{m} x^{i*}$ using secure summation protocol [4] and calculate $\delta^i = \bar{x}^* - \bar{x}^{i*}$. They accept $\bar{x}^*$ as solution if $\delta^i < \delta$ ($\delta > 0$ is a predefined threshold) for all i. The last condition can be checked through a secure computation of 'and' of m logical truths.

3.1 Information Disclosure

Each DMA^i solves Eqn (5) at iteration k and sends the solution $x^{i*}(k)$ to MA. The solution to the problem is such that for[2] $j = 1,\ldots,n$: $\left(\frac{\partial V^i}{\partial x^i_j}\right)_{x^{i*}(k)} = 0$. Knowing the slopes at various points (one for each iteration) a strategically equivalent function in the vicinity of the x^i's can be constructed, because the derivatives of concave functions are monotonically decreasing functions. From the point of view of the in-progress negotiation this disclosure about the solution points may not be of much use though, where only the last, i.e. the po point is required for negotiation. However, this disclosure can affect future negotiations.

4 Secure Distributed Weighting Algorithm (with Mediator)

The basic procedure remains same as [2], however modifications have been proposed in the next level of details. The λ^i's are updated by the DMAs themselves (initially λ^i's are

---

[2] The partial derivative $\frac{\partial V^i}{\partial x^i_j}$ is being computed at $x^{i*}(k)$.

set at random). The process continues till a solution is found with all x^i's converging. In the following algorithm in Step 4, the condition checking for all DMAs are to be held through secure and distributed 'and'ing of logical truths. Similarly, for summations in Step 4.

The inputs are $v^i(x)$ where $x \in X^i$, X^i is a convex set (This includes the constraints of DMA^i.), $i = 1, \ldots, m$, $w^i > 0$ for DMA^i such that $\sum_{i=1}^{m} w^i = 1$. The output is the po point corresponding to weights $w = (w^1, \ldots, w^m)$ for the DMAs. In the algorithm one semi-honest mediator is used.

1. DMAs together perform (This can be done by any DMA who informs the rest of the DMAs.)
   (a) Initialize α (< 1), β (> 1), ε (> 0), $\mu^i(0)$, $x^i(0)$, $i = 1, \ldots, m$, $\lambda^i(0)$, $i = 1, \ldots, m-1$ and r an n dimensional random vector.
   (b) Compute $\hat{\lambda}^i(0)$, $i = 1, \ldots, m$.
2. MA sets $k = 1$ (k represents iteration number), $flag \leftarrow 0$ and $toss \leftarrow 0$.
3. Each DMA^i performs
   (a) Solves its problem (Equation (5)) for the current values of $\hat{\lambda}^i(k)$.
   (b) Sends MA $y^{i*}(k) = x^{i*}(k) + r$ and $\overline{\lambda}^i(k) = \lambda^i(k) + r^i(k)$, where $r^i(k)$ is a random vector chosen by DMA^i.
4. MA performs
   (a) **If** $toss = 1$ goto Step 4 (d) iv.
   (b) Finds $\overline{y}(k) = \frac{1}{m} \sum_{i=1}^{m} y^{i*}(k)$.
   (c) **If** $\varepsilon^i = \frac{1}{m} \sum_{i=1}^{m} \frac{1}{\sqrt{n}} \|y^{i*}(k) - \overline{y}(k)\| < \varepsilon$, $\forall i = 1, \ldots, m$ **then** sends to all DMA's: $flag \leftarrow 0, \overline{y}(k)$.
   (d) **else** for each DMA^i do
      i. Finds $s^i(k) = y^{i*}(k) - y^{(i+1)*}(k)$.
      ii. **If** $k = 1$ **then** sets $\mu^i(k+1) = \mu^i(k)$
         Else If $k > 1$ and for all DMA^js $D^j = -\frac{(s^j(k))^T s^j(k-1)}{\|s^j(k-1)\|} \leq 0, \forall j$
         then sets $\mu^i(k+1) = \beta \mu^i(k)$ **else** sets $\mu^i(k+1) = \alpha \mu^i(k)$
      iii. Updates multipliers $\overline{\lambda}^i(k+1) = \overline{\lambda}^i(k) + \mu^i(k) \otimes s^i(k)$ (Here $\otimes$ is defined as $(a_1, a_2, \ldots, a_n)^T \otimes (b_1, b_2, \ldots, b_n)^T = (a_1 b_1, a_2 b_2, \ldots, a_n b_n)^T$.)
      iv. Tosses a coin. **If** Head **then** sets $k = k+1$, $toss \leftarrow 0$ and sends $\overline{\lambda}^i(k)$ to DMA^i (nothing is sent to DMA^m) **else** sets $toss \leftarrow 1$ and sends a random vector from the domain of λ's. (Note, a random vector will affect one entire round. Hence it has to be chosen with a probability bounded on both side. An upper bound would ensure convergence. Dummy λ's should ideally be in the vicinity of the actual λ of that iteration.)
5. Each DMA^i performs: **If** $flag = 1$ **then** finds $\overline{x} = \overline{y} - r$
   else (i) computes λ^i from $\overline{\lambda}^i$ and sends to DMA^{i+1} (except DMA^m does not send this). (ii) computes $\hat{\lambda}^i$ using Equation (4) and go to Step 3.

As evident from the algorithm above, here we have adopted *Jacobi* type (parallel) iterations rather than the *Gauss-Seidel* type (sequential) [5], where we would have updated λ^i only after first $(i-1)$ DMA's λ's are received. Although the Gauss-Seidel method sometimes converges faster, it is not completely parallelizable. This would increase the time taken to reach the solution. However, Gauss-Seidel method might allow us to incorporate preferential treatment in the system where there are more iterations with a subset of DMAs. We can also have a system where updating is first carried out with one half of the DMAs, and later with the rest. This scheme has the benefits of both algorithms, is faster and incorporates the new information in the same iteration. However this needs to be explored further.

Information disclosure: The disclosure, if any, is from the communication steps of the algorithm. Each DMA^i other than DMA^1 learns about $\lambda^{i-1}(k)$, for all $k \geq 1$. Since it does not learn $\hat{\lambda}^{i-1}(k)$ and the corresponding $\mathbf{x}^{(i-1)*}$ there is no disclosure of the sort discussed in Section 3.1. MA learns nothing as the only information it receives is $\mathbf{y}^i$ and $\overline{\lambda}^i$ for each DMA^i which appears random due to r and r^i. At the cost of some additional communication, DMAs can select a new r for each iteration to avoid any interpretation due to large number of iterations of the algorithm. By deduction from the $\overline{\lambda}^i$ DMA^i can ascertain about $\mathbf{x}^{i+1}$ of DMA^{i+1}. But since in some iterations MA sends random vectors DMAs may not be able to deduce any relationship between the vectors. However, construction of these vectors could be a challenging task for the MA.

5 Secure Distributed Weighting Algorithm - Augmented Lagrangian (Without Mediator)

In the previous section the value functions were assumed to be strictly concave. If the *value functions are not strictly concave* e.g. linear [6], we cannot use the dual formulation of the problem using the Lagrangian multipliers. In such circumstances we can adopt Augmented Lagrangian Method where the value function is made quadratic. The discussion below is from [5] and [2]. We modify the algorithm to make it secure. However the basic features are not altered and thus convergence of the algorithm is not jeopardized.

In this algorithm we do not use a mediator. As DMAs would solve the problems individually, the maximization problem remains same as that in Section 3, except we replace $\mathbf{x}^i = \mathbf{x}$ in place of $\mathbf{x}^i = \mathbf{x}^{i+1}$ in (2). A quadratic term is appended to make the function strictly concave. The optimization problem becomes:

$$\min_{\lambda^1,\ldots,\lambda^{m-1}} \left\{ \max_{\mathbf{z} \in X^i} \sum_{i=1}^{m} \left\{ w^i v^i(\mathbf{z}) + (\lambda^i)^T(\mathbf{x} - \mathbf{z}) - \frac{u}{2}\|(\mathbf{x} - \mathbf{z})\|^2 \right\} \right\} \quad (7)$$

where u is a positive scalar constant and $\|.\|$ is second order norm.

Large values of u make the quadratic term pointed and the method makes slow progress; while for small values of u, the quadratic term is blunt and the method makes fast progress toward the optimal solution [5]. Decomposing the above problem as before the Augmented Lagrangian is solved with iteration [5]:

$$\mathbf{x}^i = \arg\max_{\mathbf{z} \in \mathbf{X}^i}\left\{w^i v^i(\mathbf{z}) - (\lambda^i(k))^T(\mathbf{x}(k-1) - \mathbf{z}) - \frac{u}{2}\|(\mathbf{x}(k-1) - \mathbf{z})\|^2\right\} \quad (8)$$

$$\text{where} \quad \mathbf{x}(k) = \frac{1}{m}\sum_{i=1}^{m}\mathbf{x}^{i*}(k) + \frac{1}{mu}\sum_{i=1}^{m}\lambda^i(k) \quad (9)$$

The coding of the algorithm will be similar to that in algorithm in the previous section. Only, the common computation has to be performed by all DMAs together using appropriate SMC techniques such as secure summation, secure 'and'ing of logical truths, secure consensus mechanism[3] etc. This computation is possible without a mediator because the condition $\mathbf{x}^i = \mathbf{x}^{i+1}$ (in the algorithm of the previous section) has been replaced by $\mathbf{x}^i = \mathbf{x}$. For the same reason there needs to be appropriate changes in the definitions of $\varepsilon^i = \frac{1}{m}\sum_{i=1}^{m}\frac{1}{\sqrt{n}}\|\mathbf{x}^{i*}(k) - \mathbf{x}(k)\|$, $s^i(k) = \mathbf{x}^{i*}(k) - \mathbf{x}(k)$ and $\lambda^i(k+1) = \lambda^i(k) + u\,\mu^i(k) \otimes s^i(k)$. There is no need for random vectors $\mathbf{r}$ and $\mathbf{r}^i$'s.

6 Conclusions

In this paper we secured the distributed weighting method to find a po point maintaining the privacy of the participating parties. The algorithm is modification of the existing algorithm to find a po point based on the concept of secure multi-party computation. The convergence, thus, is already established. The issue of computational efficiency is yet to be studied.

References

1. Miettinen, K.M.: Nonlinear Multiobjective Optimization. Kluwer Academic Publishers (1999)
2. Heiskanen, P.: Decentralized method for computing pareto solutions in multiparty negotiations. European Journal of Operational Research **117** (1999) 578–590
3. Goldreich, O.: Secure multi-party computation. (working draft) Version 1.1, http://www.wisdom.weizmann.ac.il/~oded/pp.html, accessed on 10 Feb. 2004 (1998)
4. Huang, M.A., Teng, S.: Secure and verifiable schemes for election and general distributed computing problems. In: Seventh Annual ACM Symposium on Principles of Distributed Computing, ACM Press (1988) 182–196
5. Bertsekas, D.P., Tsitsiklis, J.N.: Parallel and Distributed Computation. 2 edn. Prentice–Hall International Editions (1989)
6. Sehgal, S.K., Pal, A.K.: A hybrid method to find pareto-optimal extreme points for linear optimization in multi-party negotiations. In: Thirteenth Workshop on Information Technology and Systems (WITS), Seattle, Washington, USA (2003)
7. Sehgal, S.K., Pal, A.K.: Finding pareto-optimal set of distributed vectors for negotiation with minimum disclosure. In: Sixth International Workshop on Distributed Computing (IWDC), Kolkata, India (2004)

---

[3] These mechanisms can be based on *Preference hiding scheme* proposed in [7].

Author Index

Afroz, Nahid 183
Aftosmis, Michael 293
Aggarwala, Saurabh 512
Aravinda, C. 548

Bae, Hae-Young 87
Baek, Yunju 105
Bakshi, Amol 451
Balasubramaniam, Sundar 566
Baldoni, Roberto 226
Bandyopadhyay, Somprakash 275, 554
Bandyopadhyay, Subir 183
Bansal, Tarun 201
Basu, Ananya 362
Bhadoria, Ranjana 147
Bhaumik, Parama 554
Biegel, Bryan 293
Biswas, Rupak 293
Brooks, Walter 293

Candale, Teddy 572
Cao, Chongying 416
Castilla, V. Guadalupe 506
Chae, Oksam 93
Chakrabarti, Dibyendu 329
Chand, Narottam 536
Chatterjee, Mainak 99
Chen, Y. 189
Choi, Young-Soo 439, 560
Cho, You-Ze 439, 560
Ciotti, Robert 293
Cohen, Reuven 13
Cui, Wei 99

Das, Abhijit 404
Das, Aritra 349
Das, Ashok Kumar 404
Das, Gautam K. 57
Das, Manik Lal 398
Das, Rajib K. 475
Das, Sandip 57
Das, Shukti 147
Deepak, A.P. 548
Dhar, Subhankar 306

Dhurandher, Sanjay Kumar 281
Dou, Wenhua 318

Engelhardt, Kai 25, 32
Erete, Ikpeme 572

Fraigniaud, Pierre 13
Fraire, H. Héctor 506
Freeman, Kenneth 293

Ghanshani, Pankaj 201
Ghosh, Rahul 349
Ghosh, R.K. 512
Godoy, V. Arquimedes 506
Gómez, S. Claudia 506
Gopalan, N.P. 153
Guha, Ratan 99
Gulati, V.P. 398

Halder, Pranab 362
Henze, Christopher 293
Hernández, R. Arturo 506
Hinke, Thomas 293
Hossain, M. Julius 93
Huang, Chun 99
Huang, Xinli 213
Hwang, Soyoung 105

Ilcinkas, David 13
Islam, Rabiul 183
Iwai, Hisato 275

Jaekel, A. 189
Jalal, Anand S. 487
Jang, Yong-Il 87
Jang, Youn-Seon 338
Jariyakul, Nattaphol 368
Jayanti, Prasad 45
Jiang, Jie 318
Jiang, Q. 117
Jin, Hai 238
Jin, Haoqiang 293
Jin, Yongmoon 355
Joshi, R.C. 536
Jung, Ji Won 380

Author Index

Kamal, S. 548
Kang, Mikyung 355
Kim, Hanil 355
Kim, Hye-Soo 338
Kim, Jae-Hong 87
Kim, Jae-Won 338
Kim, Junghwan 428
Kim, Jung-Hyun 87
Kim, Woo-Hun 410
Ko, Myeong-Cheol 428
Konorski, Jerzy 262
Korman, Amos 13
Ko, Sung-Jea 338
Ko, Young-Bae 287
Kundu, Sukhamay 463
Kwon, Yong-Ha 560

Lee, Donggeun 171
Lee, Jeong-Ook 428
Lee, Junghoon 355
Lee, Kang-Won 560
Lee, Soon-Jo 87
Lee, Sung-Hyup 439, 560
Li, Yin 213

Ma, Fanyuan 213
Mahantesh, H.L. 548
Maitra, Subhamoy 329
Mandal, Partha Sarathi 141
Mandal, Purna Ch. 362
Manivannan, D. 117
Manoj, B.S. 63, 524
Meenakshi, M. 159
Mian, Adnan Noor 226
Mishra, Hemant 512
Mishra, Rajesh 524
Misra, Manoj 147, 500, 536
Mohanty, Hrushikesha 512
Mohapatra, Surjyakanta 404
Mora, O. Graciela 506
Moses, Yoram 25, 32
Mukhopadhyaya, Krishnendu 141
Murthy, C. Siva Ram 63

Nagarajan, K. 153
Nandi, R. 349
Nandi, Subrata 362
Nandy, Subhas C. 57
Niyogi, Rajdeep 566

Obana, Sadao 275

Pal, Asim K. 578
Park, E.K. 171
Park, Gyungleen 355
Park, Hee-Dong 439, 560
Park, Soon-Young 87
Park, Sung Han 542
Patnaik, L.M. 111, 548
Peleg, David 1, 13
Persson, K.E. 117
Petrovic, Srdjan 45
Prasad, T. Siva 195
Prasanna, Viktor K. 451
Prathombutr, Passakon 171

Ramani, A.K. 487, 493
Rao, Ramesh R. 524
Rhee, Kyung Hyune 380
Rho, Jungkyu 428
Rieck, Michael Q. 306
Roy, Bimal 329
Roy, Siuli 275

Saminadan, V. 159
Sanyal, S.K. 349
Sarje, A.K. 147, 500
Saxena, Ashutosh 398
Scipioni, Sirio 226
Sehgal, Satish K. 578
Sekhar, Archana 63
Sen, Sandip 572
Shanker, Udai 500
Sharma, Ashish 512
Shivaprakash, T. 548
Shrestha, Deepesh Man 287
Singhal, M. 117
Singh, G.V. 281
Sinha, Bhabani P. 57, 183
Sinha, Koushik 250
Song, Zhen 318
Sreenath, Niladhuri 195
Srimani, Pradip K. 75, 129
Sur, Chul 380

Tak, Sungwoo 171
Tanwani, S. 487
Thigpen, William 293
Tokekar, Sanjiv 493
Tokekar, Vrinda 493
Tucci-Piergiovanni, Sara 226

Ueda, Tetsuro   275

Vavilapalli, Srihari   404
Venkateswaran, P.   349
Venugopal, K.R.   548

Xu, Jie   238
Xu, Zhenyu   75, 129

Yang, J.   117
Yang, Jing   416

Yang, Jong-Phil   380
Yang, Seung Jei   542
Yang, Soomi   392
Yoo, Kee-Young   410
Yun, Jae-Woong   338

Zhang, Guoqing   416
Zhang, Hao   238
Zhang, Heying   318
Znati, Taieb   368
Zou, Bin   238

Lecture Notes in Computer Science

For information about Vols. 1–3742

please contact your bookseller or Springer

Vol. 3850: R. Freund, G. Păun, G. Rozenberg, A. Salomaa (Eds.), Membrane Computing. IX, 371 pages. 2006.

Vol. 3838: A. Middeldorp, V. van Oostrom, F. van Raamsdonk, R. de Vrijer (Eds.), Processes, Terms and Cycles: Steps on the Road to Infinity. XVIII, 639 pages. 2005.

Vol. 3837: K. Cho, P. Jacquet (Eds.), Technologies for Advanced Heterogeneous Networks. IX, 307 pages. 2005.

Vol. 3835: G. Sutcliffe, A. Voronkov (Eds.), Logic for Programming, Artificial Intelligence, and Reasoning. XIV, 744 pages. 2005. (Sublibrary LNAI).

Vol. 3833: K.-J. Li, C. Vangenot (Eds.), Web and Wireless Geographical Information Systems. XI, 309 pages. 2005.

Vol. 3829: P. Pettersson, W. Yi (Eds.), Formal Modeling and Analysis of Timed Systems. IX, 305 pages. 2005.

Vol. 3828: X. Deng, Y. Ye (Eds.), Internet and Network Economics. XVII, 1106 pages. 2005.

Vol. 3827: X. Deng, D. Du (Eds.), Algorithms and Computation. XX, 1190 pages. 2005.

Vol. 3826: B. Benatallah, F. Casati, P. Traverso (Eds.), Service-Oriented Computing - ICSOC 2005. XVIII, 597 pages. 2005.

Vol. 3824: L.T. Yang, M. Amamiya, Z. Liu, M. Guo, F.J. Rammig (Eds.), Embedded and Ubiquitous Computing – EUC 2005. XXIII, 1204 pages. 2005.

Vol. 3823: T. Enokido, L. Yan, B. Xiao, D. Kim, Y. Dai, L.T. Yang (Eds.), Embedded and Ubiquitous Computing – EUC 2005 Workshops. XXXII, 1317 pages. 2005.

Vol. 3822: D. Feng, D. Lin, M. Yung (Eds.), Information Security and Cryptology. XII, 420 pages. 2005.

Vol. 3821: R. Ramanujam, S. Sen (Eds.), FSTTCS 2005: Foundations of Software Technology and Theoretical Computer Science. XIV, 566 pages. 2005.

Vol. 3820: L.T. Yang, X. Zhou, W. Zhao, Z. Wu, Y. Zhu, M. Lin (Eds.), Embedded Software and Systems. XXVIII, 779 pages. 2005.

Vol. 3819: P. Van Hentenryck (Ed.), Practical Aspects of Declarative Languages. X, 231 pages. 2006.

Vol. 3818: S. Grumbach, L. Sui, V. Vianu (Eds.), Advances in Computer Science – ASIAN 2005. XIII, 294 pages. 2005.

Vol. 3815: E.A. Fox, E.J. Neuhold, P. Premsmit, V. Wuwongse (Eds.), Digital Libraries: Implementing Strategies and Sharing Experiences. XVII, 529 pages. 2005.

Vol. 3814: M. Maybury, O. Stock, W. Wahlster (Eds.), Intelligent Technologies for Interactive Entertainment. XV, 342 pages. 2005. (Sublibrary LNAI).

Vol. 3810: Y.G. Desmedt, H. Wang, Y. Mu, Y. Li (Eds.), Cryptology and Network Security. XI, 349 pages. 2005.

Vol. 3809: S. Zhang, R. Jarvis (Eds.), AI 2005: Advances in Artificial Intelligence. XXVII, 1344 pages. 2005. (Sublibrary LNAI).

Vol. 3808: C. Bento, A. Cardoso, G. Dias (Eds.), Progress in Artificial Intelligence. XVIII, 704 pages. 2005. (Sublibrary LNAI).

Vol. 3807: M. Dean, Y. Guo, W. Jun, R. Kaschek, S. Krishnaswamy, Z. Pan, Q.Z. Sheng (Eds.), Web Information Systems Engineering – WISE 2005 Workshops. XV, 275 pages. 2005.

Vol. 3806: A.H. H. Ngu, M. Kitsuregawa, E.J. Neuhold, J.-Y. Chung, Q.Z. Sheng (Eds.), Web Information Systems Engineering – WISE 2005. XXI, 771 pages. 2005.

Vol. 3805: G. Subsol (Ed.), Virtual Storytelling. XII, 289 pages. 2005.

Vol. 3804: G. Bebis, R. Boyle, D. Koracin, B. Parvin (Eds.), Advances in Visual Computing. XX, 755 pages. 2005.

Vol. 3803: S. Jajodia, C. Mazumdar (Eds.), Information Systems Security. XI, 342 pages. 2005.

Vol. 3802: Y. Hao, J. Liu, Y.-P. Wang, Y.-m. Cheung, H. Yin, L. Jiao, J. Ma, Y.-C. Jiao (Eds.), Computational Intelligence and Security, Part II. XLII, 1166 pages. 2005. (Sublibrary LNAI).

Vol. 3801: Y. Hao, J. Liu, Y.-P. Wang, Y.-m. Cheung, H. Yin, L. Jiao, J. Ma, Y.-C. Jiao (Eds.), Computational Intelligence and Security, Part I. XLI, 1122 pages. 2005. (Sublibrary LNAI).

Vol. 3799: M. A. Rodríguez, I.F. Cruz, S. Levashkin, M.J. Egenhofer (Eds.), GeoSpatial Semantics. X, 259 pages. 2005.

Vol. 3798: A. Dearle, S. Eisenbach (Eds.), Component Deployment. X, 197 pages. 2005.

Vol. 3797: S. Maitra, C. E. V. Madhavan, R. Venkatesan (Eds.), Progress in Cryptology - INDOCRYPT 2005. XIV, 417 pages. 2005.

Vol. 3796: N.P. Smart (Ed.), Cryptography and Coding. XI, 461 pages. 2005.

Vol. 3795: H. Zhuge, G.C. Fox (Eds.), Grid and Cooperative Computing - GCC 2005. XXI, 1203 pages. 2005.

Vol. 3794: X. Jia, J. Wu, Y. He (Eds.), Mobile Ad-hoc and Sensor Networks. XX, 1136 pages. 2005.

Vol. 3793: T. Conte, N. Navarro, W.-m.W. Hwu, M. Valero, T. Ungerer (Eds.), High Performance Embedded Architectures and Compilers. XIII, 317 pages. 2005.

Vol. 3792: I. Richardson, P. Abrahamsson, R. Messnarz (Eds.), Software Process Improvement. VIII, 215 pages. 2005.

Vol. 3791: A. Adi, S. Stoutenburg, S. Tabet (Eds.), Rules and Rule Markup Languages for the Semantic Web. X, 225 pages. 2005.

Vol. 3790: G. Alonso (Ed.), Middleware 2005. XIII, 443 pages. 2005.

Vol. 3789: A. Gelbukh, Á. de Albornoz, H. Terashima-Marín (Eds.), MICAI 2005: Advances in Artificial Intelligence. XXVI, 1198 pages. 2005. (Sublibrary LNAI).

Vol. 3788: B. Roy (Ed.), Advances in Cryptology - ASIACRYPT 2005. XIV, 703 pages. 2005.

Vol. 3785: K.-K. Lau, R. Banach (Eds.), Formal Methods and Software Engineering. XIV, 496 pages. 2005.

Vol. 3784: J. Tao, T. Tan, R.W. Picard (Eds.), Affective Computing and Intelligent Interaction. XIX, 1008 pages. 2005.

Vol. 3783: S. Qing, W. Mao, J. Lopez, G. Wang (Eds.), Information and Communications Security. XIV, 492 pages. 2005.

Vol. 3781: S.Z. Li, Z. Sun, T. Tan, S. Pankanti, G. Chollet, D. Zhang (Eds.), Advances in Biometric Person Authentication. XI, 250 pages. 2005.

Vol. 3780: K. Yi (Ed.), Programming Languages and Systems. XI, 435 pages. 2005.

Vol. 3779: H. Jin, D. Reed, W. Jiang (Eds.), Network and Parallel Computing. XV, 513 pages. 2005.

Vol. 3778: C. Atkinson, C. Bunse, H.-G. Gross, C. Peper (Eds.), Component-Based Software Development for Embedded Systems. VIII, 345 pages. 2005.

Vol. 3777: O.B. Lupanov, O.M. Kasim-Zade, A.V. Chaskin, K. Steinhöfel (Eds.), Stochastic Algorithms: Foundations and Applications. VIII, 239 pages. 2005.

Vol. 3776: S.K. Pal, S. Bandyopadhyay, S. Biswas (Eds.), Pattern Recognition and Machine Intelligence. XXIV, 808 pages. 2005.

Vol. 3775: J. Schönwälder, J. Serrat (Eds.), Ambient Networks. XIII, 281 pages. 2005.

Vol. 3774: G. Bierman, C. Koch (Eds.), Database Programming Languages. X, 295 pages. 2005.

Vol. 3773: A. Sanfeliu, M.L. Cortés (Eds.), Progress in Pattern Recognition, Image Analysis and Applications. XX, 1094 pages. 2005.

Vol. 3772: M. Consens, G. Navarro (Eds.), String Processing and Information Retrieval. XIV, 406 pages. 2005.

Vol. 3771: J.M.T. Romijn, G.P. Smith, J. van de Pol (Eds.), Integrated Formal Methods. XI, 407 pages. 2005.

Vol. 3770: J. Akoka, S.W. Liddle, I.-Y. Song, M. Bertolotto, I. Comyn-Wattiau, W.-J. van den Heuvel, M. Kolp, J. Trujillo, C. Kop, H.C. Mayr (Eds.), Perspectives in Conceptual Modeling. XXII, 476 pages. 2005.

Vol. 3769: D.A. Bader, M. Parashar, V. Sridhar, V.K. Prasanna (Eds.), High Performance Computing – HiPC 2005. XXVIII, 550 pages. 2005.

Vol. 3768: Y.-S. Ho, H.J. Kim (Eds.), Advances in Multimedia Information Processing - PCM 2005, Part II. XXVIII, 1088 pages. 2005.

Vol. 3767: Y.-S. Ho, H.J. Kim (Eds.), Advances in Multimedia Information Processing - PCM 2005, Part I. XXVIII, 1022 pages. 2005.

Vol. 3766: N. Sebe, M.S. Lew, T.S. Huang (Eds.), Computer Vision in Human-Computer Interaction. X, 231 pages. 2005.

Vol. 3765: Y. Liu, T. Jiang, C. Zhang (Eds.), Computer Vision for Biomedical Image Applications. X, 563 pages. 2005.

Vol. 3764: S. Tixeuil, T. Herman (Eds.), Self-Stabilizing Systems. VIII, 229 pages. 2005.

Vol. 3762: R. Meersman, Z. Tari, P. Herrero (Eds.), On the Move to Meaningful Internet Systems 2005: OTM 2005 Workshops. XXXI, 1228 pages. 2005.

Vol. 3761: R. Meersman, Z. Tari (Eds.), On the Move to Meaningful Internet Systems 2005: CoopIS, DOA, and ODBASE, Part II. XXVII, 653 pages. 2005.

Vol. 3760: R. Meersman, Z. Tari (Eds.), On the Move to Meaningful Internet Systems 2005: CoopIS, DOA, and ODBASE, Part I. XXVII, 921 pages. 2005.

Vol. 3759: G. Chen, Y. Pan, M. Guo, J. Lu (Eds.), Parallel and Distributed Processing and Applications - ISPA 2005 Workshops. XIII, 669 pages. 2005.

Vol. 3758: Y. Pan, D.-x. Chen, M. Guo, J. Cao, J.J. Dongarra (Eds.), Parallel and Distributed Processing and Applications. XXIII, 1162 pages. 2005.

Vol. 3757: A. Rangarajan, B. Vemuri, A.L. Yuille (Eds.), Energy Minimization Methods in Computer Vision and Pattern Recognition. XII, 666 pages. 2005.

Vol. 3756: J. Cao, W. Nejdl, M. Xu (Eds.), Advanced Parallel Processing Technologies. XIV, 526 pages. 2005.

Vol. 3754: J. Dalmau Royo, G. Hasegawa (Eds.), Management of Multimedia Networks and Services. XII, 384 pages. 2005.

Vol. 3753: O.F. Olsen, L.M.J. Florack, A. Kuijper (Eds.), Deep Structure, Singularities, and Computer Vision. X, 259 pages. 2005.

Vol. 3752: N. Paragios, O. Faugeras, T. Chan, C. Schnörr (Eds.), Variational, Geometric, and Level Set Methods in Computer Vision. XI, 369 pages. 2005.

Vol. 3751: T. Magedanz, E.R.M. Madeira, P. Dini (Eds.), Operations and Management in IP-Based Networks. X, 213 pages. 2005.

Vol. 3750: J.S. Duncan, G. Gerig (Eds.), Medical Image Computing and Computer-Assisted Intervention – MICCAI 2005, Part II. XL, 1018 pages. 2005.

Vol. 3749: J.S. Duncan, G. Gerig (Eds.), Medical Image Computing and Computer-Assisted Intervention – MICCAI 2005, Part I. XXXIX, 942 pages. 2005.

Vol. 3748: A. Hartman, D. Kreische (Eds.), Model Driven Architecture – Foundations and Applications. IX, 349 pages. 2005.

Vol. 3747: C.A. Maziero, J.G. Silva, A.M.S. Andrade, F.M.d. Assis Silva (Eds.), Dependable Computing. XV, 267 pages. 2005.

Vol. 3746: P. Bozanis, E.N. Houstis (Eds.), Advances in Informatics. XIX, 879 pages. 2005.

Vol. 3745: J.L. Oliveira, V. Maojo, F. Martín-Sánchez, A.S. Pereira (Eds.), Biological and Medical Data Analysis. XII, 422 pages. 2005. (Sublibrary LNBI).

Vol. 3744: T. Magedanz, A. Karmouch, S. Pierre, I.S. Venieris (Eds.), Mobility Aware Technologies and Applications. XIV, 418 pages. 2005.